REMOTE
EXPERIENCES

DAVID DE VLEESCHAUWER

REMOTE EXPERIENCES

EXTRAORDINARY TRAVEL ADVENTURES FROM NORTH TO SOUTH

Words by Debbie Pappyn,
Essays by Sophy Roberts, Pico Iyer and Stanley Stewart

TASCHEN

About Remote Experiences

by David De Vleeschauwer

This book is a record of my travels along the world's longitude from the Arctic to the Antarctic. Twelve remote destinations, twelve remote experiences. It is testament to the beauty of the natural world, and the staggering range of landscapes, seascapes, and people that lie in the spaces in between.

I've devoted the greater part of my life to seeking out remote places: regions outside my everyday experience as a European citizen, far beyond the tourist trails that criss-cross our world. Nothing gives me a greater thrill than swapping skyscrapers for steppe, streetlights for snow glare. Those moments where it's me, my camera, and an otherworld.

The images and essays that follow are an exploration of what 'remote' really means in our modern reality. Because it's no longer as simple as getting away from it all. Newspaper headlines and the social-media circus follow us to nearly every corner of the globe. Even without the white noise, there is still the evidence of your own eyes: natural disasters, deforestation, rising sea levels, war. That African safari park, teeming with game, is intensively managed. A visit to Antarctica comes with so much red tape that every movement is catalogued. Even the Arctic — about as 'remote' as you can get — is a hotbed of geopolitical maneuverings. The region's ownership is deeply contested among world powers greedy for a slice of its vast oil and gas reserves, including Russia, the United States, Canada, Denmark, Finland, Iceland, Norway, and Sweden. There is no such thing as pristine. Yet there is still beauty, on a dizzying scale. Places where the balance between people and planet speaks of potential, not destruction. In Upper Mustang, I see hope in the way a vibrant people thrive in the rain shadow of the Himalayas. In Abruzzo, the southern Italian sheep migration, a generations-deep tradition, is a reminder of the quiet we can still find in a life lived slowly. In Papua New Guinea there is joy in the fierce pride and strong belief systems nurtured over thousands of years. In Botswana's Okavango Delta I see the passion with which humans are still motivated to protect what is good in our world. It is on the fringes of our world that I find optimism for our planet's future.

Yet the ecosystems in which I find myself are fragile. In travelling to these precious remnants of wild — and sharing them with you — am I contributing to their decline? In looking for the 'remote', am I exoticising someone else's reality for my own fleeting moment of escape? There's no easy answer. But I believe that it is through experiencing pockets of remoteness, and seeing the evidence that balance can still exist in our world, that we are motivated to protect it.

So I continue to seek the remote, travelling with a manifesto stamped into my brain: to travel thoughtfully, gently, deeply, and with respect for the place I'm visiting. To be aware, above all, that 'remote' is a privilege that can only exist as long as we protect it.

Into the Remote

by Debbie Pappyn

As the current world turns faster and faster, the need to find places and spaces where time stands still or life moves slowly resonates even more to the modern traveler. Places with few people but with big skies, where streets are scarce, or even non-existent, and the joy of going off-road, off the map is overwhelming. Places where the origin of us is palpable. Places often untouched by any modern society because they are too ancient or too wild. Sites where the sense of place is so strong that almost no mobile-phone signal can get through. The only status update that can be made is a mental one: I am here and that is enough.

A rock next to a stupa somewhere in a remote and tiny village in the lost Kingdom of Lo with views over the Tibetan Himalayan range turns out to be the best chair ever to sit, calm down, and listen to the monk's chanting in the distance. The sensation of a bitingly cold wind blowing over a snowy plain in the Mongolian Altai Mountains is almost healing. It blows away any sense of direction and leaves you feeling almost lost. A blissful moment, because nowadays we are too much found. The sound of three-metre- (10 ft.) thick ice cracking under the weight of an enormous nuclear icebreaker slamming its way up to the geographical North Pole. A reverberating sound flowing over a sea of endless white ice and a feeling of nothingness, of lightness of being in a place where there is not a lot but it is probably the grandest thing you ever saw.

Travelling to remote places is not about ticking off lists or being brave and adventurous. It is allowing yourself to be in a place that will allow you to be yourself again. To connect again with things that might be forgotten or overseen in our modern world. Unpolluted skies, clean horizons, plains and forests, or frozen oceans where we, humans, are not essential, just visitors, passers-by. Where solitude is not a bad thing.

Just like the Portuguese writer José Saramago wrote in his book *Journey to Portugal*, an account of his solo exploration in a vintage car taking Saramago to the most remote villages and off-the-map roads of his country: "The valley is an immense green meadow, with trees crisscrossing and marking it in all directions: black, gold, tawny, amid a profound silence: complete, rare, anxious, but essential to the solitude, to this unforgettable moment."

Remote places like this lonely valley allow a traveller to admire and enjoy again the overlooked, the forgotten, the humble things, from colours in nature to shadows and cloud formations, the sound of silence and yes, without a doubt, that essential feeling of solitude and contentment.

Remote is a Point of View

by Sophy Roberts

Sophy Roberts is the author of the critically acclaimed travel book *The Lost Pianos of Siberia*. She is also a journalist who writes frequently about remote travel for *The Wall Street Journal*, *Financial Times* and other titles worldwide.

Before visiting the Altai Mountains — one of Russia's most inaccessible wilderness areas, stitching the boundary where Russia, China, and Mongolia meet — I'd been obsessed about it for a while. It started withan image I'd seen of space junk from Soviet rockets in a cloud of white butterflies; the picture thrummed with vibrations in the air, a rushing whirr against the Altai's summer flush of grass. The landscape felt like a kind of paradise, a throwback to an untamed planet — raw and unpopulated, the rocket the poisoned apple in a field of blooming life. I started to gather more research, including images from an extraordinary archaeological dig on the Ukok Plateau: a Scythian mummy buried some 2,500 years ago, her body like a sleeping child, preserved in an underground tomb. She was buried with her horses, a bag of cosmetics, a yellow silk top with maroon piping, and flakes of gold.

I find images a powerful call to the remote — which is why I love David's photographs. But when I get to these places, I am also soon reminded that remote is a question of perspective — a lesson I was to learn when I eventually got to the Altai Mountains to find a piano. I was in the middle of a three-year search for an instrument worthy of a musical friend, a search that became a book. My journey had taken me to the furthest reaches of the Russian Far North, and to Kamchatka, but something about the Altai felt further away than anywhere I'd been before. Even the local dialect contributes: the Ukok Plateau — a jewel of the region— translates as 'The End of Everything'.

I travelled down to the neck of mountains on the Altai's southern edge, to a village called Ust-Koksa, where I met a retired Aeroflot aeroplane navigator, Leonid Kaloshin. It turned out that Kaloshin was searching for instruments too: he regularly bought pianos and transported them to the Altai for local children to play. We discussed Russian literature, Ernest Hemingway, and Lord Byron in a room lined with tens of thousands of books.

It was the last thing I expected to find in a Siberian backcountry, 3,200 kilometres (2,000 mi.) from Moscow. At the end of our conversation, Kaloshin turned to me and asked if I would look out for a piano for him too. "But this place is so remote," I said. "The world is very remote," he replied. "We are at the centre." It was an important moment of realisation for me: 'remote' is a point of view.

The concept of remoteness is one that is fading: the truth is, there are very few unoccupied places on earth. In recent years, with the modern realities of overpopulation, cheap tourism, and overconsumption, places once considered remote — Machu Picchu, the Masai Mara, the jungles of Costa Rica — have become easily accessible and overrun. Increasingly, the 'remote' is found in landscapes that nature has reclaimed after humans have left, due to war, climate change, natural and manmade disasters. But to me, 'remote' can still be a romance. It means a place with no fences — I'm drawn to steppe, taiga, and desert, places where the only visible divide is between sky and earth.

I think of Chad, a place I last visited in 2020, travelling to the Ennedi Massif, a 40,000-square-kilometre (15,500 sq. mi.) plateau about the size of Switzerland. Part of the Sahel's 'Arc of Instability', this region is marked as a chequerboard of red by Western governments, with war, climate, and geopolitics combining to create a 'no-go' zone for Westerners. It's magnificent — and best appreciated from the skies—with thousands of narrow spires teetering upwards from the sunbaked desert, and honeyed sandstone cliffs scooped into vast ridges.The scorching heat of the desert can reach 50°C (120°F). It's harsh, hostile, and, from my perspective, nearly unsurvivable in its remoteness. At a first glance, the only trace of humankind is the burned-out wreckage of Libyan tanks — remnants of a brief battle in 1987 between Colonel Gaddafi's Libyan forces and the Chadian army.

But look to the nearest oasis, and there's a group of locals with their herds of camels, goats, and sheep, custodians of an area their ancestors have lived in for generations. What feels impossibly remote to me is in fact the centre of their world. Further into the desert, there are ancient ironworks, with 3,500-year-old stoves still leaning into the sand-drifts. The sandstone cliffs have paintings dating back to 4000 BCE. Once I knew how to look, there was evidence everywhere I went that this was the cultural, geographical, and psychological heart of someone else's reality.

It's a pattern that repeats wherever I travel, whether confronted with Papua New Guinea's Crocodile Men living in total harmony with a harsh environment that seems engineered to kill, or Central Africa's indigenous forest people scratching out a living in the 'wilderness' as exoticised by the tourism industry, or even in the far-flung Kuril Islands in the Pacific, made remote by territorial dispute as much as geography. These places are as much a part of someone's ordinary existence as the West Dorset valley where I live is to mine. So when I chase the 'remote', I'm conversely seeking the centre of things. It's when you visit a maverick's lodge in a 'lost-land,' or cross 'uncharted' territory — however skewed my Western preconceptions might be — that you find yourself looking at a new perspective.

'Remote' is a point of view, but by using it as a standard by which to seek the planet's last wildernesses, lost corners, or troubled Paradises, it acts as a trigger for empathy and understanding.

The Truest Meaning of Remote

by Pico Iyer

Pico Iyer is the author of 15 books, translated into 23 languages, among them, most recently, *The Art of Stillness*, *Autumn Light*, and *A Beginner's Guide to Japan*.

In January, in the Year of the Virus, I found myself drifting around the Antarctic Peninsula. Day after day dawned silent, with an array of whites and turquoises and rich sapphires stretched across the wide horizon. At 10:30 every night, I walked around the deck in the silver light of midsummer, sometimes watching fin whales porpoising along the prow of our ship. For more than a week I saw barely a trace of human habitation: only black-browed albatrosses and leopard seals and colonies of 20,000 penguins among whom I walked as if before the birth of human civilisation.

For more than 30 years at that point I'd been travelling across Tibet and Iceland and Namibia; indeed, exactly 30 years before, I'd been wandering amidst two million penguins near Punta Tombo, in the vast, wind-whipped open spaces of Patagonia. But Antarctica was a reminder of how much empty space there remains in the world, and how deeply such emptiness can fill us up. As I returned to my two-room apartment on the crowded streets of suburban Japan, I couldn't put away memories of those broad skies in every direction, which expanded my sense of time as much as of space.

Upon disembarking from the Silver Cloud in Tierra del Fuego, I heard that 77 fresh passengers from China were about to board, and I read of a virus that had closed down the city of Wuhan. Little did I know that the word 'remote' was about to acquire a human rather than a natural meaning, as people across the world began to work from home, to talk on screens, and to practice 'social distancing'. Suddenly, for almost everyone, the next town down the road came to seem as remote as Antarctica. Abruptly, the word 'remote' became associated with working 'remotely' and not with the quiet or solitary places that open our hearts and imaginations.

The year of 'sheltering in place' surely intensified a longing in many of us to be far from noise and fear: walking across the Arctic edges of Northern Europe, or on the untroubled steppes of Mongolia, where things look very much as they might have done during the plague of the 14th century. In reaching lazily for the 'remote' to turn on the TV to catch scenes of beauty and silence, we realised afresh how much we were missing the unspoiled landscapes that really expand us. Why had I devoted my life to seeking out remote spots on Nuka Hiva, across the Himalayas, around the highlands of Ethiopia, I wondered? Partly, I'm sure, to remind myself how narrow my perspectives remain. "There are more things in heaven and earth," as Hamlet says to Horatio, "than are dreamt of in your philosophy."

The remote places I've witnessed take me to remote places within myself, those states of surrender and wonder, of suspended disbelief, that too often I forget. In Antarctica I found myself reaching for words like 'sublimity' and 'awe', words that belong to previous centuries, when explorers could look like visionaries. I recalled the days I'd spent in Camp Denali, a cluster of huts in the midst of Alaska's Denali National Park, seven hours from the main highway. There was no electricity in the cabins, and no running water. To relieve myself, I had to walk through the emptiness, braving bears.

But what that meant was that a 6,000-metre- (20,000 ft.) high mountain seemed close enough to touch. The silence opened doors inside me. The freedom from mobile phones and television made every day feel like a rich and deep adventure. Newly awakened to what was around me, I was jolted out of my habitual way of seeing, of being, and found myself stepping into surprise again and again.

Such moments of possibility, the other lives that hide out inside us that are unlocked by faraway places, come back to me powerfully as I page through the treasures in this book. Some of these places have worked their spell on me already; others are landscapes of the imagination. Yet even places not obviously delightful, such as North Korea, have expanded my sense of what is possible, left me speechless and humbled, reminded me of how provincial I remain.

The Year of the Virus gave me proof every day that there are wonders down every street. I started taking walks every morning and afternoon, in every direction from my flat, and found cherry blossoms, bamboo forests, what sounded like nightingales. The excitement we associate with the remote is never far away. But that time of remaining near home also recalled to me the wonders across the planet. No generation before mine has been able to sail to the North Pole, or fly to Papua New Guinea; to do so is to be reminded of how urgent it is to protect the pristine places around us (often by not flying or sailing too much) and how indispensable are the open places within.

Many of us may be 'working remotely' still, but it's the deeper and more ancient primal sense of 'remote' that will give us not just our livings, but our realest lives.

Remote Mongolia

by Stanley Stewart

The winner of numerous awards for his books and his travel journalism, Stanley Stewart has undertaken lengthy journeys across five continents. He is a Fellow of the Royal Society of Literature and of the Royal Geographical Society.

In the wide spaces of Outer Mongolia, the landscapes seem to have been sculpted by winds. For thousands of miles there are no roads, no fences, no boundary lines. Scattered here and there, marooned in seas of grass, in a country the size of western Europe, are only a handful of towns where clustered buildings sag into various states of dilapidation. The people here are nomads for whom migration and movement are a way of life. In this place no one cares much for towns, and the permanence they pretend to offer.

I was crossing Mongolia by horse, riding over a thousand miles from Bayan-Ölgii in the west to Dadal in the east, from the Altai to the wide grasslands of the eastern plains. The journey belonged to the romances of childhood, an odyssey of innocence. I was fortunate to find it lived up to expectations. I woke every morning to birdsong and the sound of the horses grazing just beyond the tent flaps and the sense that the world was newborn. Each day arrived unencumbered by expectations, and each day brought wonder and awe, emotions almost lost in our sophisticated worlds.

As with so many remote places, it is easy to think in Mongolia that you have dropped through some tear in the canvas of the universe, fallen down some unexpected rabbit hole in the continuum of time, to find yourself in another world, even in another age. I had crossed Asia in the footsteps of a 13th-century French friar travelling to the Mongol court. His account of the interior of a yurt, known in Mongolia as a *ger*, could stand as a perfect description of Mongol tents today. Out here, among the flocks and the encampments and the horses, little has changed in 800 years.

It was in the western Altai that I met the Kazakh eagle hunters, known as berkutchi, documented so movingly here in David's photographs. The tradition of falconry has been passed down to them from their Mongol neighbours; Genghis Khan's father was said to have been a keen falconer while his grandson Kublai Khan was reported by Marco Polo to have kept 5,000 birds. The Kazakhs use the golden eagles of the region, vast birds with a wingspan of almost two-and-a-half metres (8 ft.).

"We love them as we love our children," one of the hunters said, stroking the breast feathers of a hooded bird. In a high pass, he unleashed the eagle into vast reaches of sky until it was no more than a tiny circling speck. The hunter never took his eyes off him. He said, "The birds make me feel humble."

I had been thinking about humility and the way that it was the gift of remote places. Travel of any kind lifts us out of our lives. As our horizons change, so do our perspectives. But travel to the remotest places of the Earth, to the places that require us to leave so much of our own world behind, is liberating. In Mongolia, I felt as if I had shed several layers of old skin, had somehow shucked away the unnecessary detritus of my life. In so many vast remote spaces — in the Danakil, in the passes of the High Andes, crossing the deserts of Namibia, camping in the depths of Antarctica — my own life had been recalibrated, pared down to its essences. I felt humbled by the beauty, the scale, the emptiness of such places. Childish wonder rushed into places within me where unnecessary concerns had mushroomed, where useless personal baggage was weighing me down.

From the complexities of the Altai Mountains, I rode on into landscapes of elemental simplicities, a country of languorous horizontals. Treeless, smooth as felt, and thick with wild flowers, long vistas of grass were draped between distant hills where horses galloped. Adrift among tides of sheep, the round domes of white gers echoed the gentle curves of the country. In this place, I felt the parameters of my life changing. Enclosed by new horizons, I had adopted the priorities of the nomads — a good camp, water, grazing for the horses, the promise of clear weather.

I envied the way nomads of this remote place were so at ease with impermanence. Most of us need to mark our passage through the world, to put down roots, to establish some token of security in an impermanent world. But away across these wide steppes, the nomads are content to vanish like early mist. When the pastures grow thin, they strike camp, load their camels, and head into the hills in search of fresh grazing. All that marks their passage are the round circles of bleached grass where their tents have stood. With time, in a matter of weeks, these too will be gone. And the landscapes will look exactly as they did before they passed this way. That is the humility of nomads. They understand the world is so much bigger than they are, and that only the world is permanent. And that there is a wonderful freedom in that understanding.

TABLE OF CONTENTS

1

Navigating to the Top of the World

Sail across the frozen ocean to the geographical North Pole aboard the world's largest and strongest nuclear-powered icebreaker, and take a polar plunge.

18–47

2

Into the Realm of Arctic Blue

Venturing into the icy heart of Arctic Europe when the sun stays just below the horizon and the remote islands of Svalbard are blissfully bathing in an enigmatic blue light.

48–75

3

A Road Adventure to the Northernmost Tip of Europe

The Baltic Sea Circle Rally is 8,000 km (5,000 mi.) of solitude in the wintry depths of Scandinavia and the Baltics. Icy roads, tranquil fjords, snowy forests; from Hamburg to the North Cape — and back.

76–105

4

Golden Eagle Hunting in the Far West of Mongolia

In a forgotten corner of the Altai Mountains, join the berkutchi at the Golden Eagle Festival, to celebrate a unique tradition that dates back to the days of Genghis Khan.

106–139

5

Unexpected Discoveries in a Sealed-Off Nation

Buzzing with nationalist zeal, North Korea is fearsome and fascinating in equal measure. Come for the Arirang Mass Games, stay for the communist architecture and a glimpse of rural life.

140–185

6

A Slow-Paced Ramble into the Heart of Abruzzo

Drink, sleep, and eat like an Italian shepherd during the transumanza, a slow-paced, seasonal migration of sheep and goats into the heart of Abruzzo, where old traditions thrive.

186–207

7

Immersive Travel through the Lost Kingdom of Lo

Follow an ancient Himalayan salt trade route into the dusty Kingdom of Lo, a spectacle of pristine and picturesque landscapes hidden behind 4,000-metre (13,000 ft.) mountain passes.

208–243

8

A Wild Exploration of the Last Terra Incognita

Between active volcanoes and dreamy Pacific shores, experience an incredible diversity of tribal cultures, a natural world still largely unknown to science, and a thriving city covered in ash.

244–275

9

St Helena, an Atlantic Island Refuge

Take an eccentric and unpredictable flight over the fierce Atlantic Ocean to the world's most remote inhabited island. A rocky gem for lovers of solitude and quirkiness.

276–297

10

Old-School and Classic Safari in the Okavango Delta

In Botswana, join the legendary wildlife guide Ralph Bousfield on his deep explorations of the world's greatest inland river delta, and sleep alfresco under starry skies.

298–343

11

Into the Saddle and Out of This World

For a breathtaking perspective on emerald lakes and knife-edge mountain passes, mount an Argentine Criollo horse and follow a real gaucho guide into the Patagonian wilderness.

344–379

12

A Private Escape to Antarctica

Break the ice along the edge of the White Continent aboard the only private icebreaker superyacht. An over-the-top exclusive experience at sea.

380–419

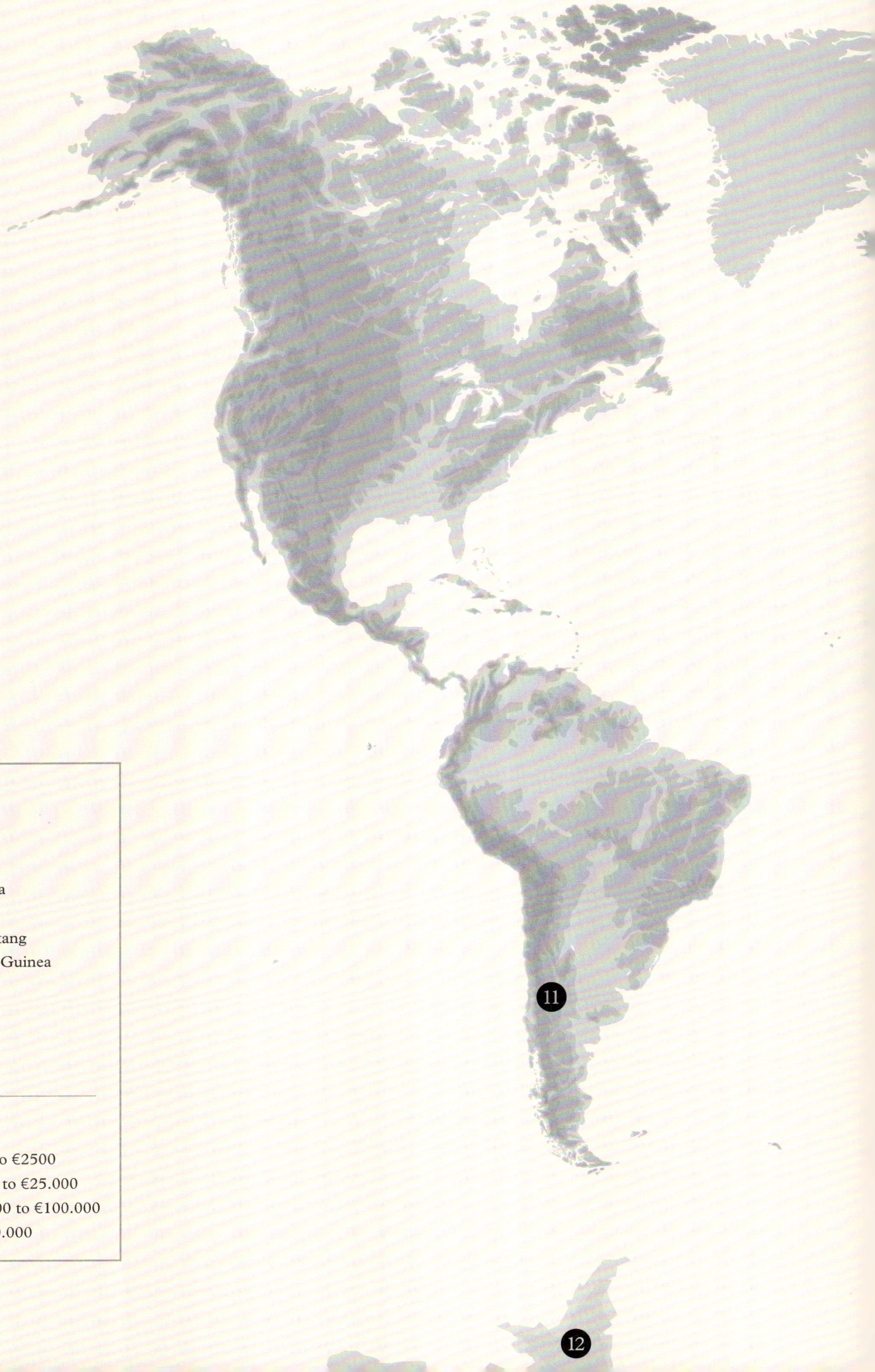

1	North Pole
2	Svalbard
3	North Cape
4	Mongolia
5	North Korea
6	Italy
7	Upper Mustang
8	Papua New Guinea
9	St Helena
10	Botswana
11	Argentina
12	Antarctica

€	from €250
€€	from €250 to €2500
€€€	from €2500 to €25.000
€€€€	from €25.000 to €100.000
€€€€€	from €1.000.000

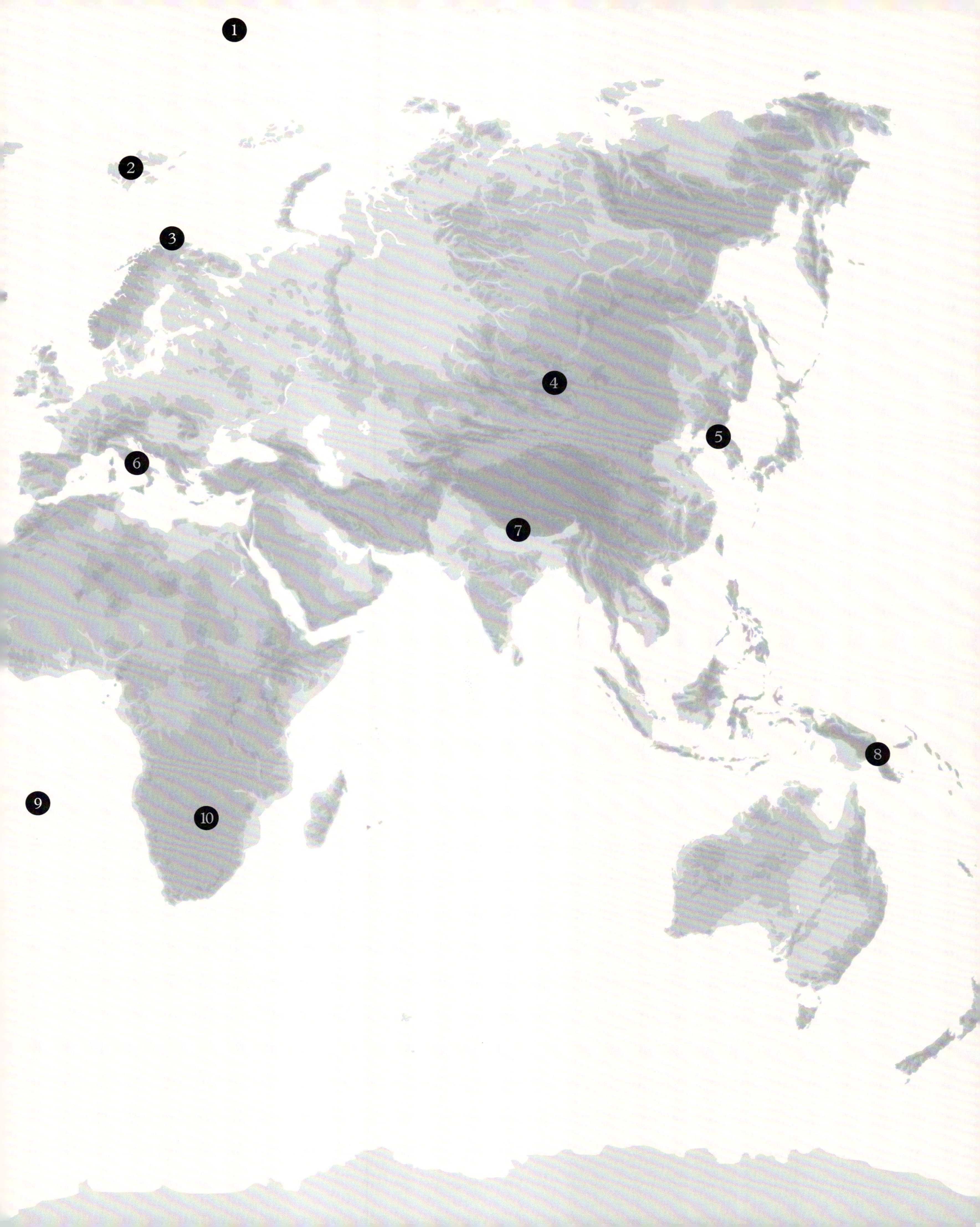
1
2
3
4
5
6
7
8
9
10

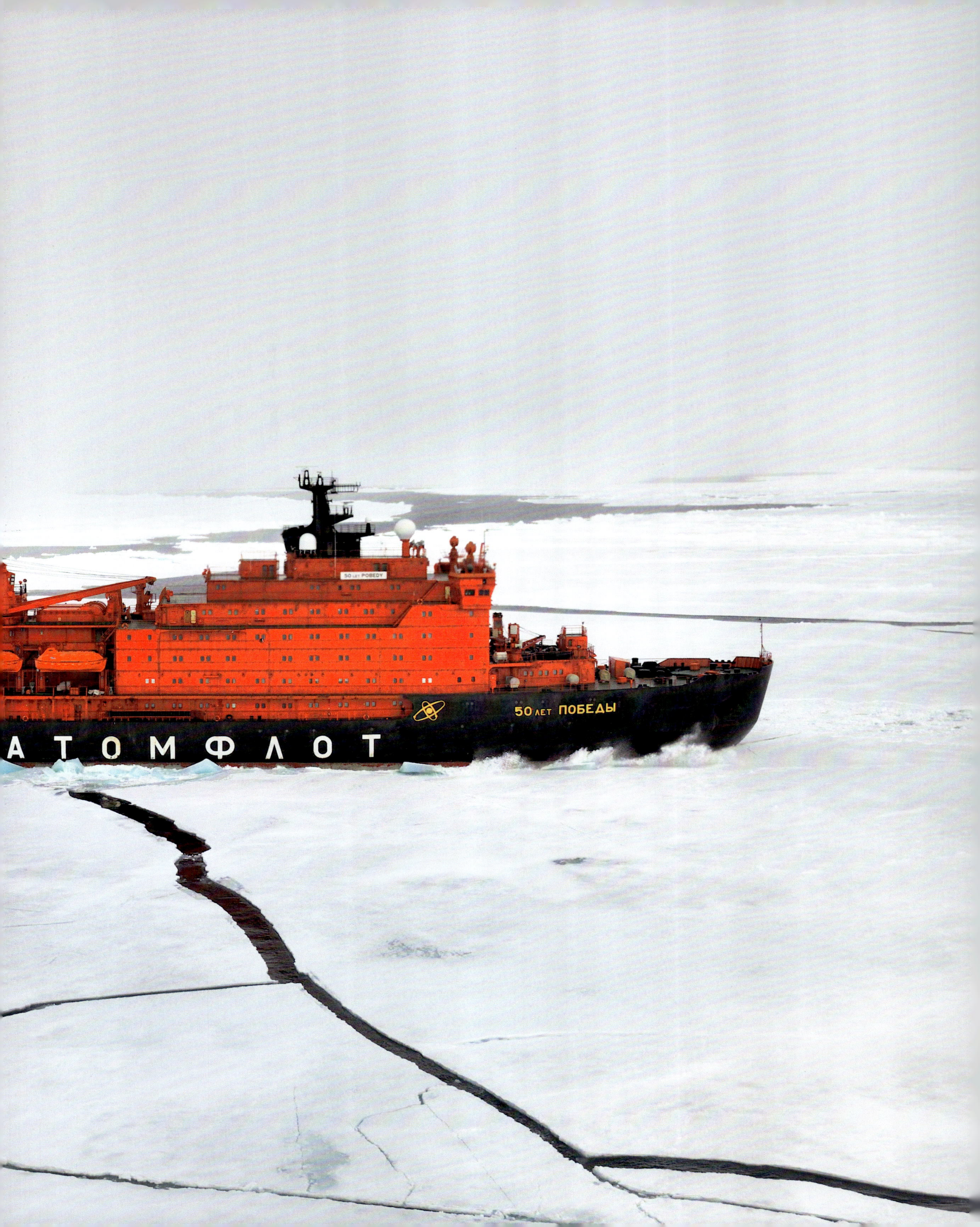
50 ЛЕТ ПОБЕДЫ
АТОМФЛОТ

CHAPTER I

NAVIGATING TO THE TOP OF THE WORLD

A NAUTICAL ODYSSEY TO 90 DEGREES NORTH

Geographical North Pole

90°0'0.00" N
133°40'41.64" E

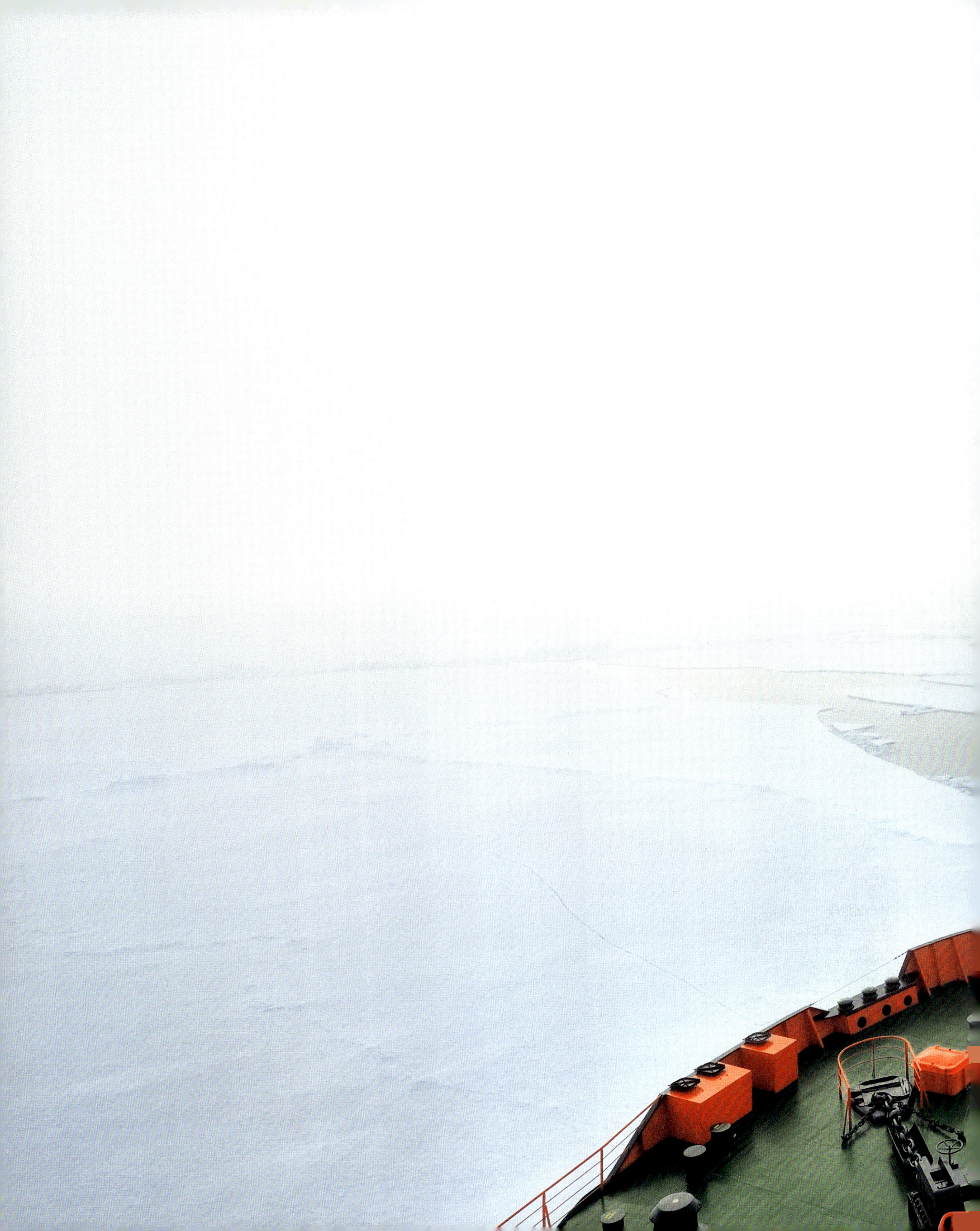

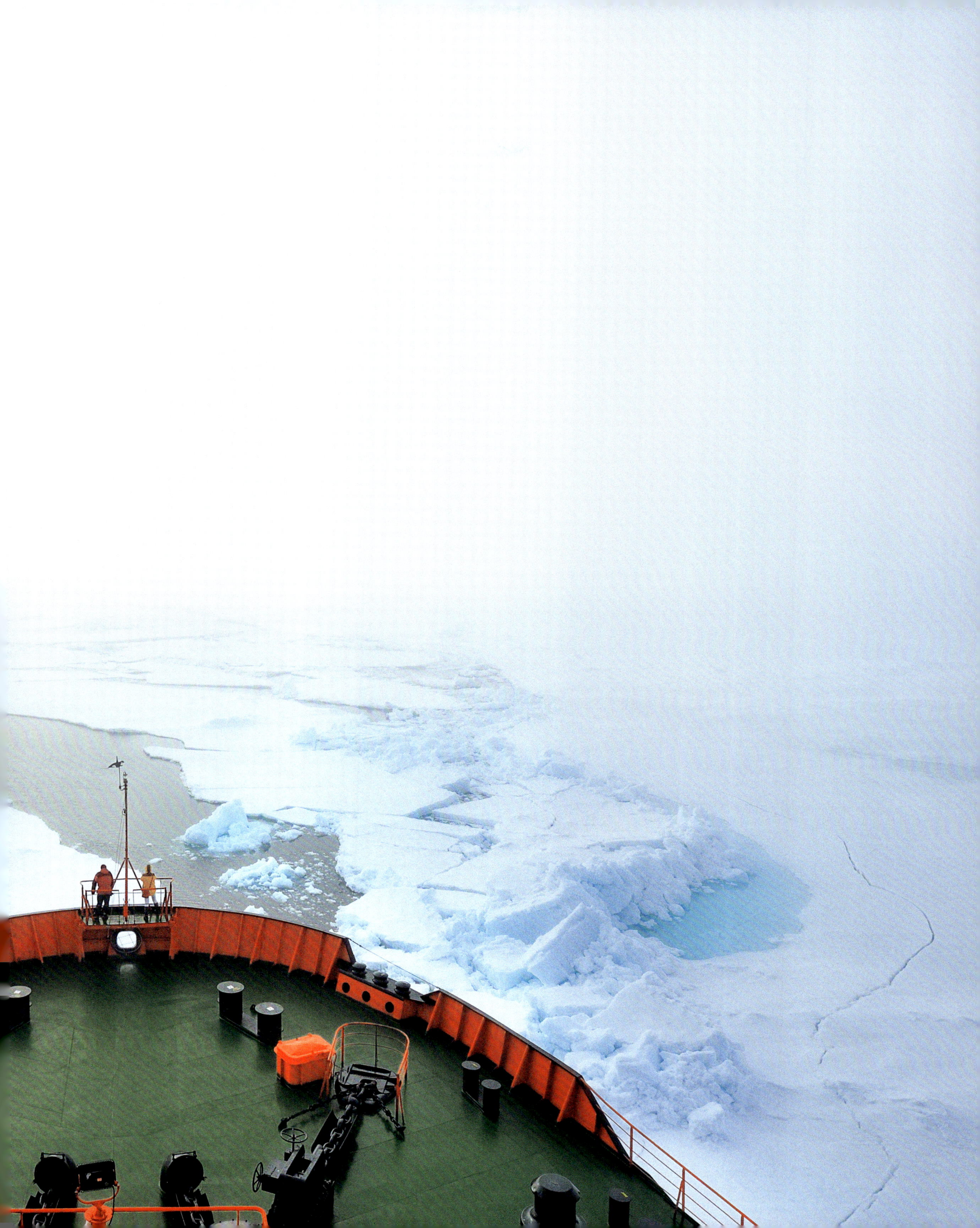

Navigating to the Top of the World

A nautical odyssey to 90 degrees north

Region:	**Geographical North Pole**
Best time to travel:	**June and July**
Closest airport:	**Murmansk, Russia**
Price tag:	**€€€€**
Style of travel:	**A singular adventure on a frozen ocean**

Introduction

Owned by no one, claimed by many, the enigmatic North Pole — the ice-covered region at the very top of the world — is nothing more than a constantly shifting sheet of ice in the middle of the Arctic Ocean. The world's largest and strongest nuclear-powered icebreaker, *50 Let Pobedy*, sails every summer to 90° north with one hundred passengers aboard, all eager to set foot on the geographical top of the world. For most of them, this moment is much more than ticking off a list. It is all about the voyage. When the ship cracks an ice sheet three-metres- (10 ft.) thick on its way there, it sounds raw and without reservation. The name of the ship means '50 Years of Victory' and refers to the 50th anniversary of the Soviet Union's triumph in World War II. To commemorate its launch, the ship carried the Olympic flame to the North Pole in 2013 during the run-up to the Sochi Winter Olympics. Powered by a pair of 171-megawatt nuclear reactors and two 27.6-megawatt steam turbo-generators, the almost 160-metres- (525 ft.) long icebreaker can reach 21.4 knots — almost 40 kilometres per hour (25 mph) — and sail nonstop for almost six years without returning to land for refuelling. With nuclear reactors on board, fuel stops become almost a thing of the past. The behemoth's primary mission is to carry super-sized cargo ships through the frozen Northeast Passage in winter. During summer, when the nautical shortcut is ice-free, the ship transforms into an expedition ship for exploration and adventure, departing from its home port of Murmansk. This quaint Arctic town is the headquarters of the Russian nuclear navy. From here the *50 Years of Victory* sails in the direction of the mythical Franz Josef Land, where only Russian ships are allowed to drop anchor. There is a throwback quality to those who travel this far north, an eccentricity reminiscent of early explorers who were not deterred by extremes or afraid of challenging situations. After leaving Murmansk, it takes two days to reach the pack ice. This is the realm of the polar bear, where human visitors face a new set of rules. From here onwards, the frozen silence, the supernaturally white panoramas, and the murky gloom of the frigid ocean below seem endless. As *50 Years of Victory* slices through the blank landscape, the thunderous groan of the ship's steady progress reverberates through the ship's red-painted reinforced hull, while passengers wrapped in thick layers of warm clothes stand on the bow, watching the ice split, snap and crack, followed by the blue of the Arctic Ocean surging through, as if gasping for breath.

The Views

This is not an ordinary sea voyage aboard a fancy cruise ship. The route to the northernmost point on the planet, where the Earth's axis of rotation meets its surface, takes 11 laborious days, moving not much faster than 20 kilometres per hour (12.5 mph) when cracking through the ice. The presence of polar bears gives this voyage an extra dimension: in this frozen world, man is not king. Travellers take this trip not only to set foot atop the world but to be immersed in the raw beauty of the High Arctic. Gazing at vast expanses of sea ice — in some places three-metres- (10 ft.) thick — is addictive and even soothing. The sun never sets in the Arctic summer and yet the light transforms constantly, reflecting off the pale-white ice sheet. Sometimes the weather is cold, moody, and dull, with a thick fog. Other days there is a velvety sheen to the light, flushed with pink or lavender tones. The frozen ocean is often delicately shrouded in a flowing palette of shades of white. It doesn't matter whether you are out on deck or gazing through a port-hole in the warm belly of the ship to this ever-changing panorama. After a couple of days sailing, when the

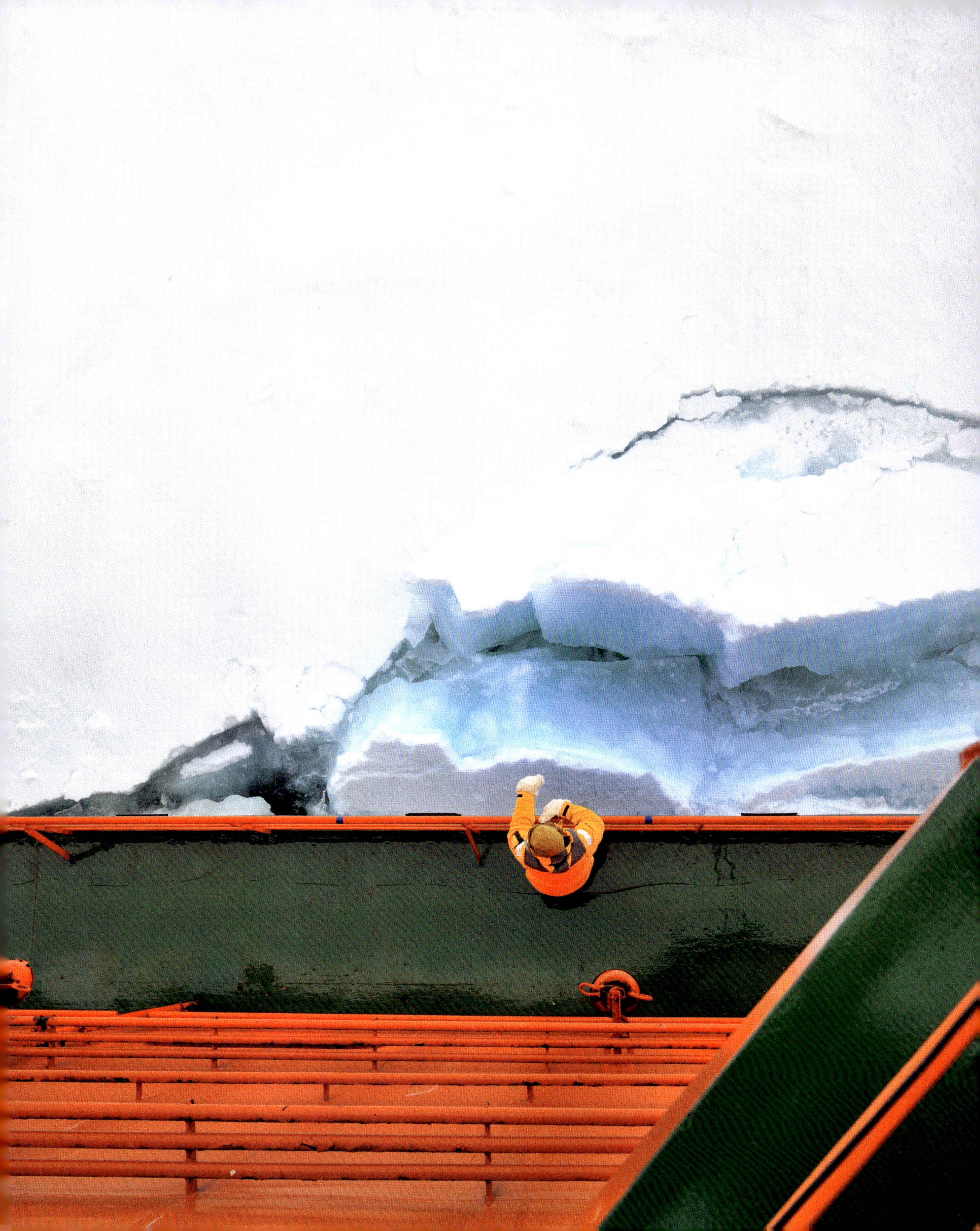

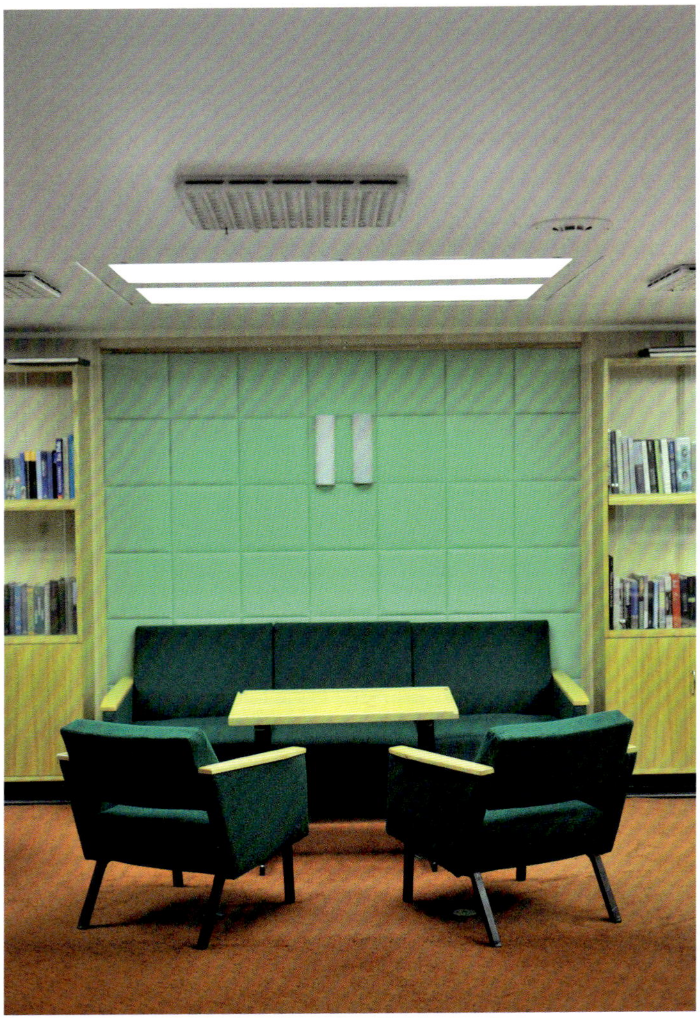
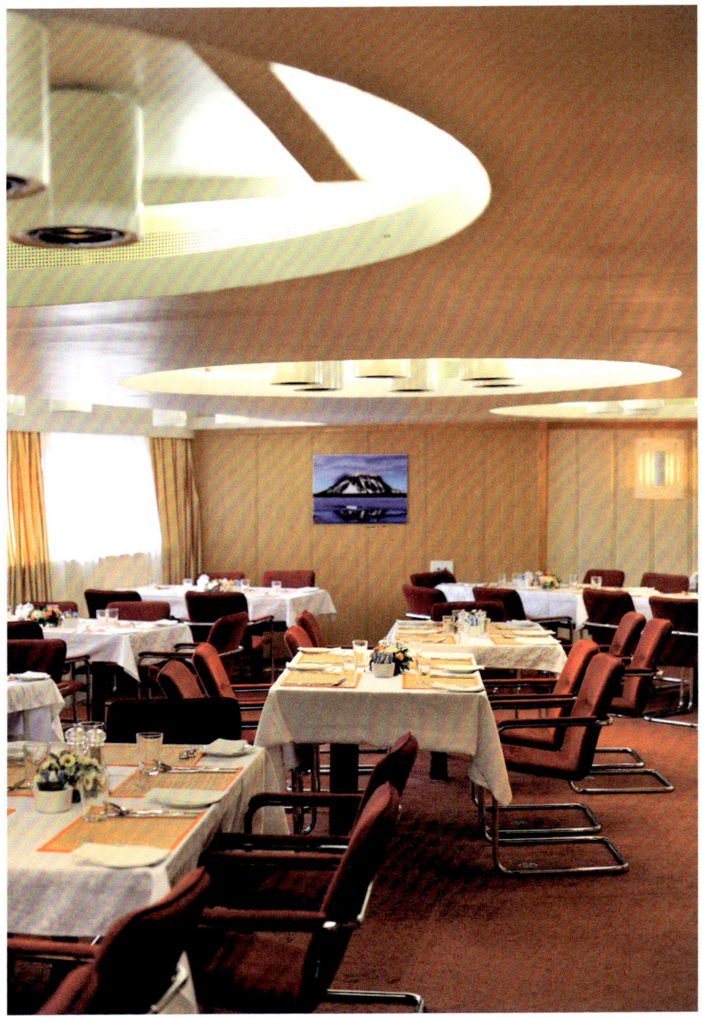
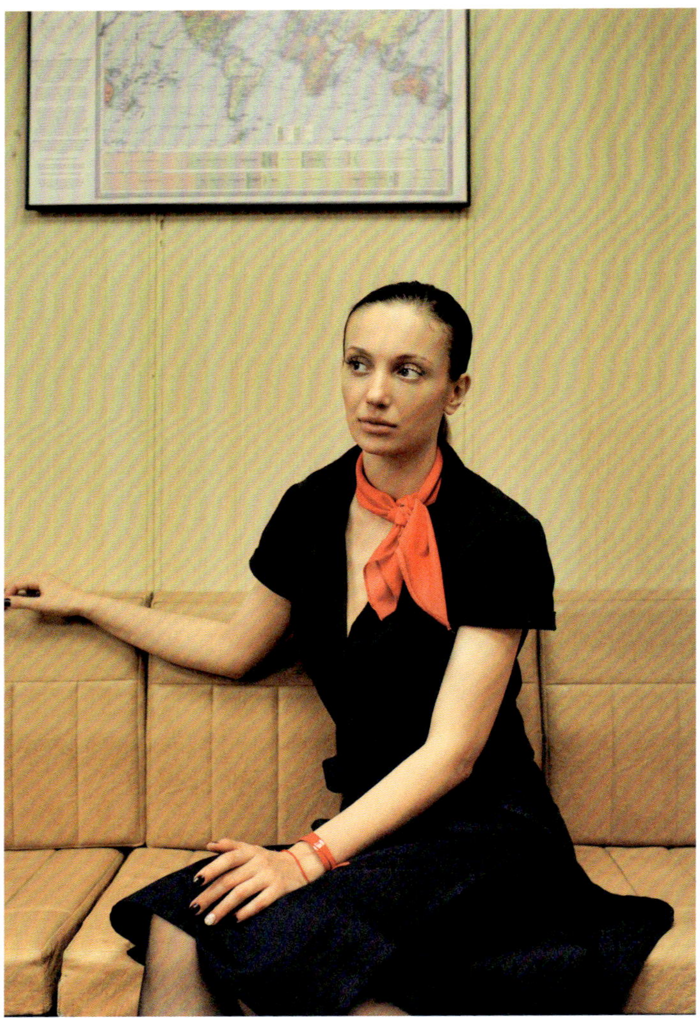
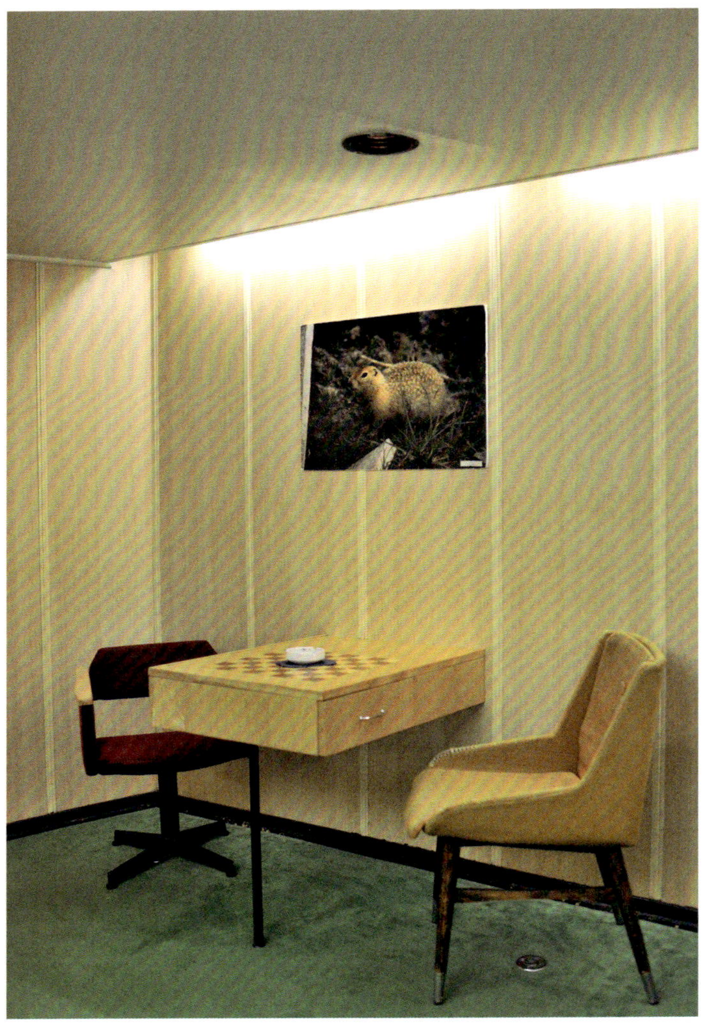

Opposite The authentic décor and warm ambience inside the *50 Years* are in stark contrast with its robust exterior and its roaring and heavy-duty engines in the belly of the ship. The salon where lectures are held is stocked with books and magazines about polar explorers and Arctic regions. In the afternoon, coffee and freshly baked cake are always served. At the Victory Bar, strong drinks mix with magical vistas over the bow of the ship.

North Pole is finally reached, excitement ripples through the ship. Some passengers describe it as a feeling of recalibration; a new beginning. On top of the world, it is time for a toast with ice-cold vodka. All passengers gather in a circle around the geographic North Pole and raise their glasses in cheer. This is a conqueror's moment, almost triumphal. And then there is the infamous Polar Plunge. This might be the most iconic swim anywhere in the world. This plunge, which is nothing more than a dip in the Arctic Ocean, is a once-in-a-lifetime thrill. Daredevils warm up with mulled wine and more vodka, while the ship's crew sets up a barbecued lunch on the ice. In the few hours spent at the Pole, the sun is out, the wind picks up, the clouds arrive, it snows, there is fog, and a planned hot-air balloon ride is cancelled. There are four seasons in one hour on top of the world. The North Pole is elusive; this place does not want to be caught. This is a very brief anchorage; remaining at the exact latitude 90° north lasts a minute or so. By the time the ship leaves, it has already drifted two nautical miles, almost four kilometres. The frozen ocean is in constant motion. On its way back south, the ship tries to retrace its tracks in the thick ice sheet. At Hooker Island, sturdy Zodiacs take passengers along towering glaciers near Rubini Rock to witness spectacular bird cliffs where more than 70,000 kittiwakes, ivory gulls, little auks, and guillemots nest. Helicopter rides allow mind-blowing bird's-eye views of the packed sea ice and the remote territory of Franz Josef Land. Passengers are invited to join these aerial jaunts to grasp the true vastness of this frozen desert. When the helicopter lifts, the massive, floating steel colossus that is the ship transforms into a tiny red speck in an ocean of bright, shimmering white, its tracks in the ice still visible like an artery filled with dark and cold Arctic Ocean. The helicopter swoops around the ship, letting the passengers grasp how endless and timeless — always bright or always dark — this landscape is. Before this voyage, nobody would be able to realise that this panorama stretches all the way up there to the elusive North Pole. Now, after setting foot on that pole at 90° north, these travellers finally understand the vastness and magnitude of this thick blanket of white ice that stretches around the extreme and surreal top of our world.

Creature Comforts

Inside *50 Years of Victory*, there is an unexpectedly warm atmosphere and cosy ambience. The vintage décor features victorious memorabilia, along with portraits of the fierce-looking captain. It almost feels like the digital age never reached the ship and analogue is the only language spoken. A Finnish sauna and a compact swimming pool are hidden in the belly of the ship, the latter filled with warm seawater heated by the steam of the turbines on board. The best spot is the bar with unlimited views over the bow of the ship and the Arctic Ocean. The cabins are simple but comfortable with wooden furniture, small en-suite bathrooms and portholes to connect with the Arctic scenery. In stark contrast with the comfortable setting inside, the outer decks are exposed to the real polar climate: always fresh and crisp, sometimes bitingly cold, slippery when freezing, dusted in white when snowing, but always beautiful, even on foggy days, when the ship is surrounded by nothing but an eerie silence.

Go Now

This voyage is all about travelling to an exceptionally rare and raw place where there is no concept of time and where every direction the traveller looks is south. In 30 years' time it's likely that the geographical North Pole will be entirely ice-free during the summer months and this unique feeling of otherworldliness could be lost forever.

How to Go

Every year in July and August, polar expedition cruise companies organise a handful of sailings to the geographical North Pole with the *50 Years of Victory*. Count on 14 days for the whole voyage. All-inclusive rates start at 30.000 Euro per person for the expedition. As an environment-friendly alternative, the French cruise company Ponant sends off its brand-new Le *Commandant Charcot* explorer ship to sail from Spitsbergen to the geographical North Pole. This hybrid-electric exploration cruise vessel became the first of its kind to reach the geographic North Pole in September 2021. The *Charcot* takes its guests during the months of July and August to 90° north. Starting from 33.700 Euro per person (in a double cabin) for a 15-day cruise.

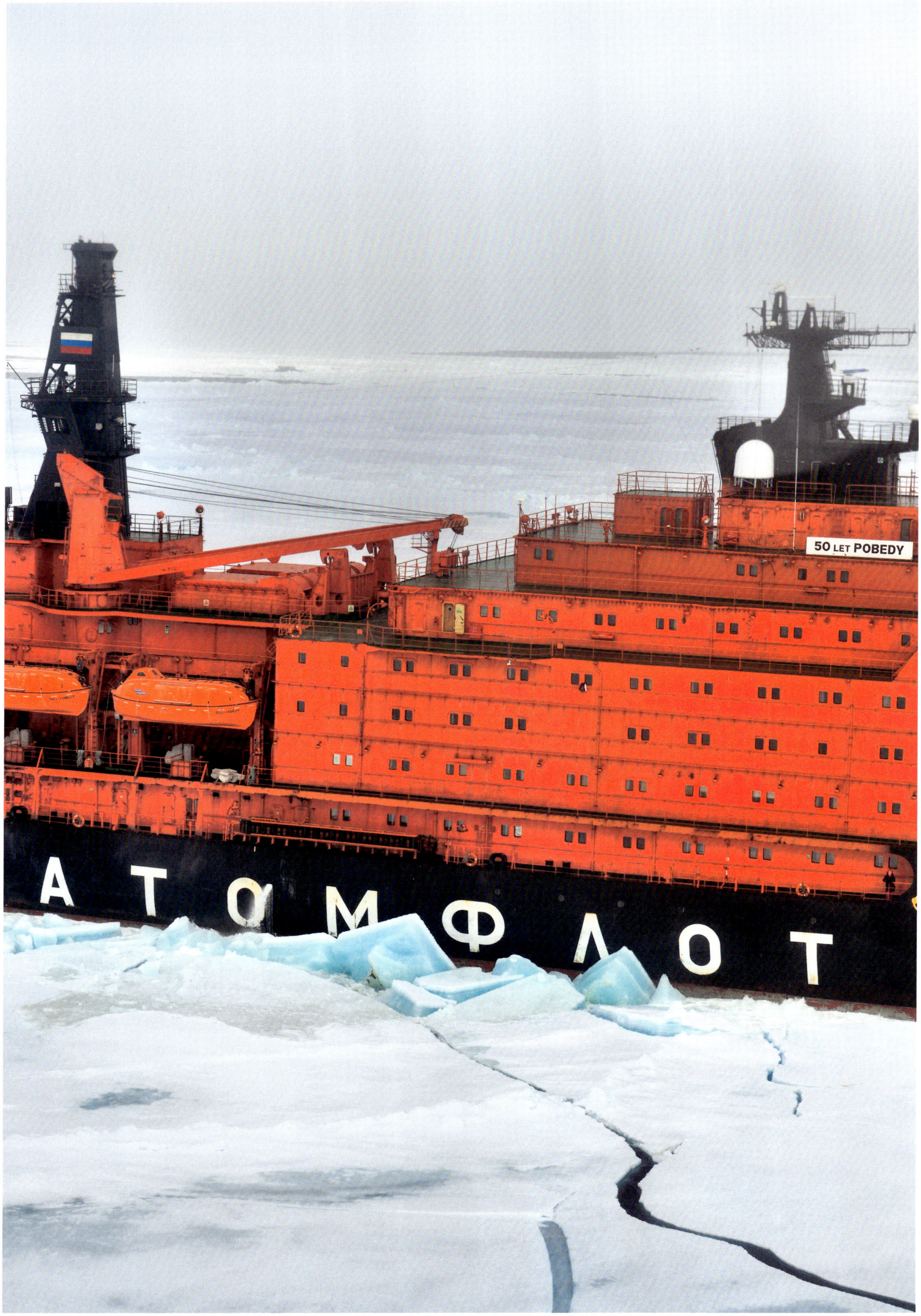
50 LET POBEDY
АТОМФЛОТ

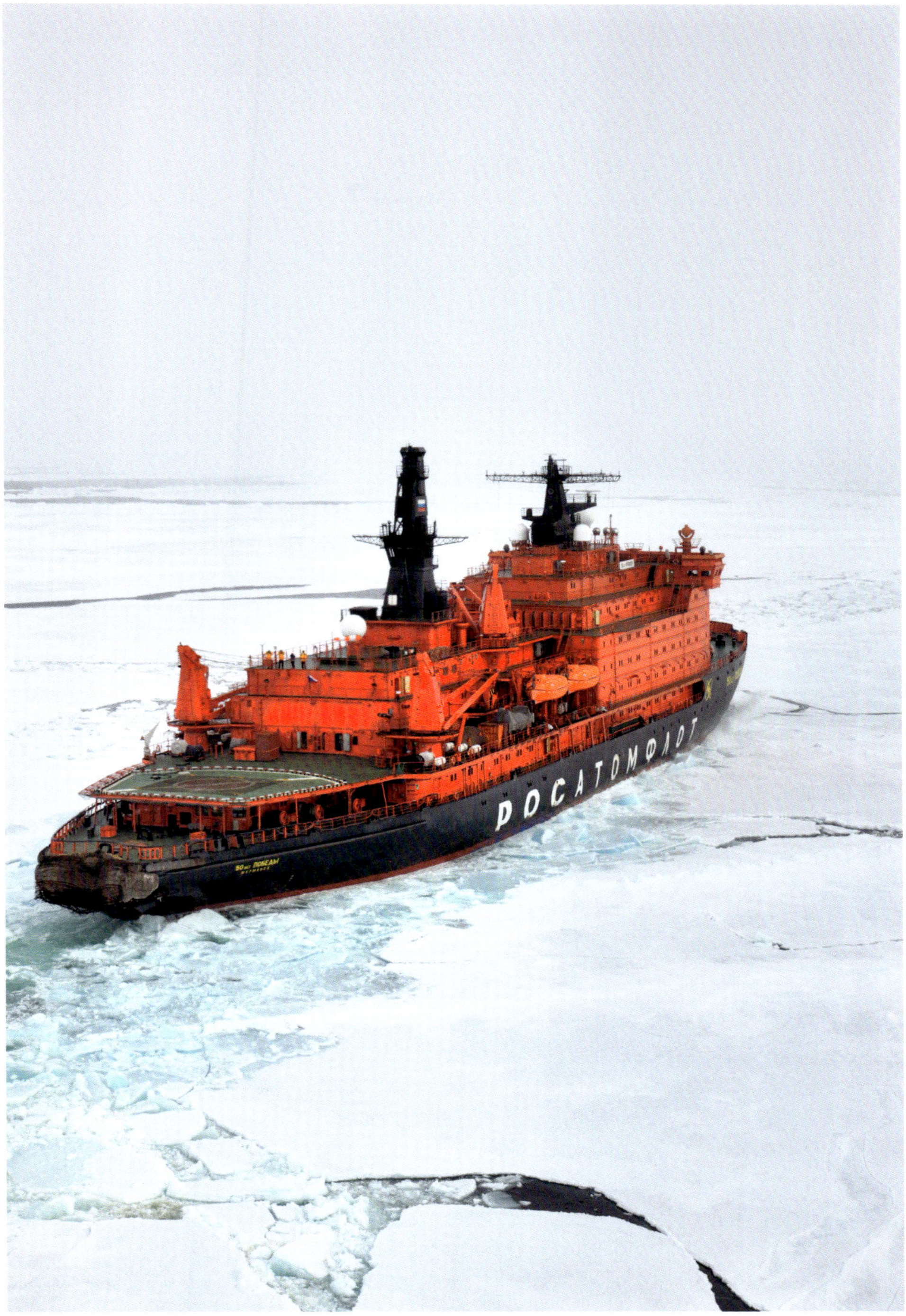

This spread *50 Years of Victory*, built in Saint Petersburg and owned by the Russian Federation, was constructed with a double hull, which makes her the strongest icebreaker ever constructed. Her unique, spoon-shaped bow prevents her from getting stuck in the ice, while her stern is designed to hook into other cargo ships, then tug them through the thick winter ice of the Northeast Passage. This Northern Sea Route is the shortest way between the West and the East. No queues or pirates along the way. It only takes 20 days to go from west to east instead of the 40 days via the Suez Canal. The Arktika-class icebreaker *50 Years of Victory* is 160 metres (525 ft.) long and 30 metres (98 ft.) wide and is capable of breaking through ice up to five metres (16 ft.) thick.

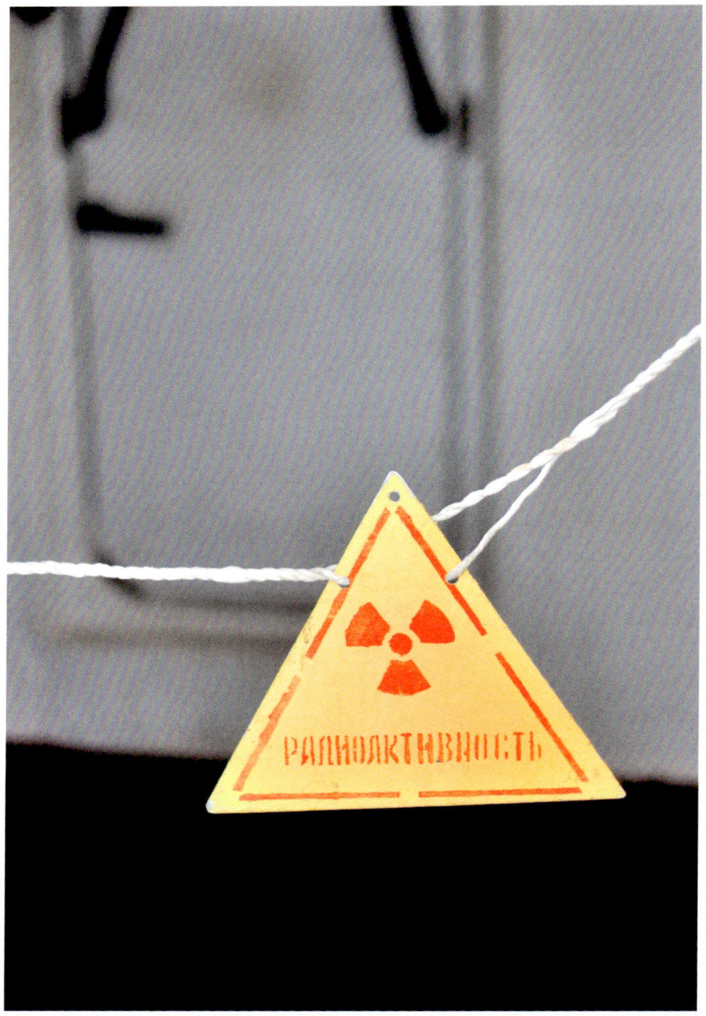
РАДИОАКТИВНОСТЬ

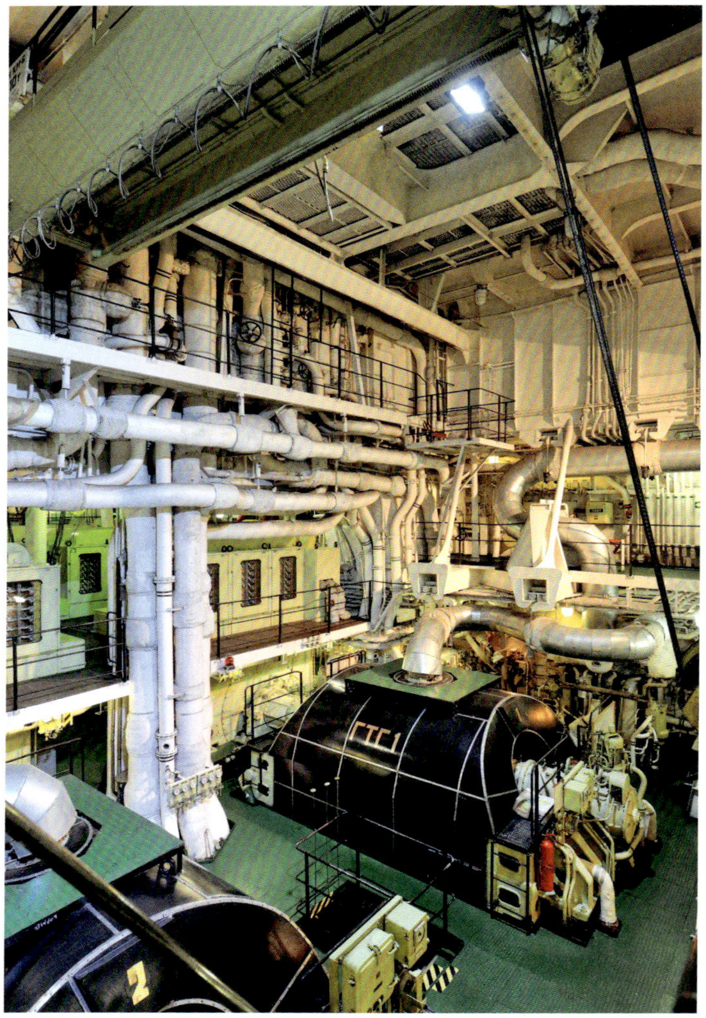

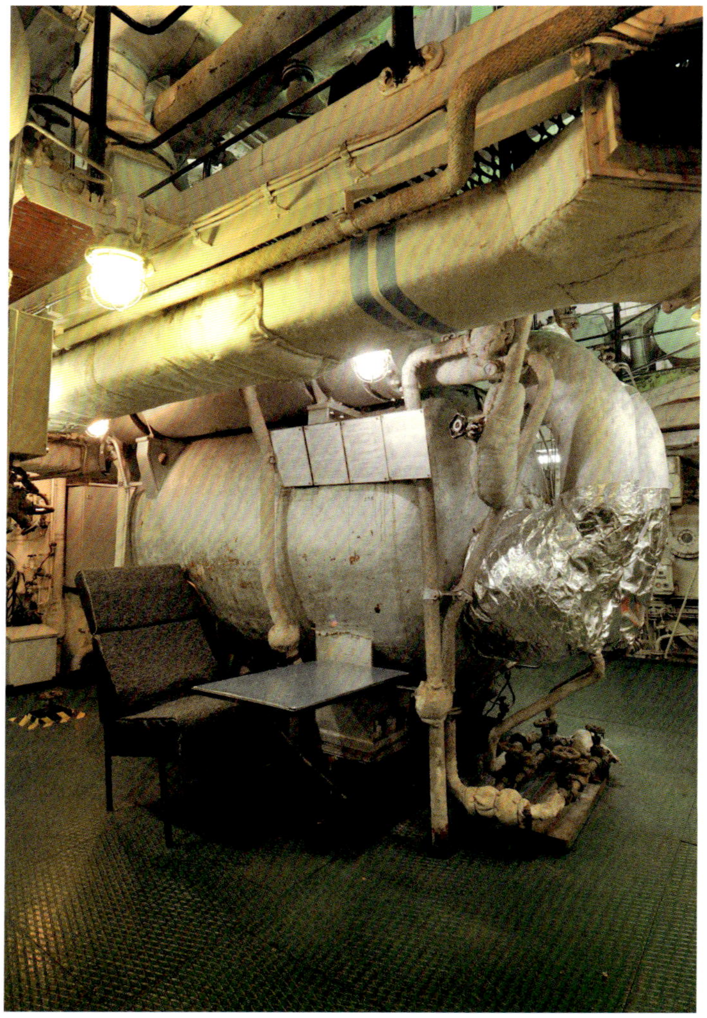

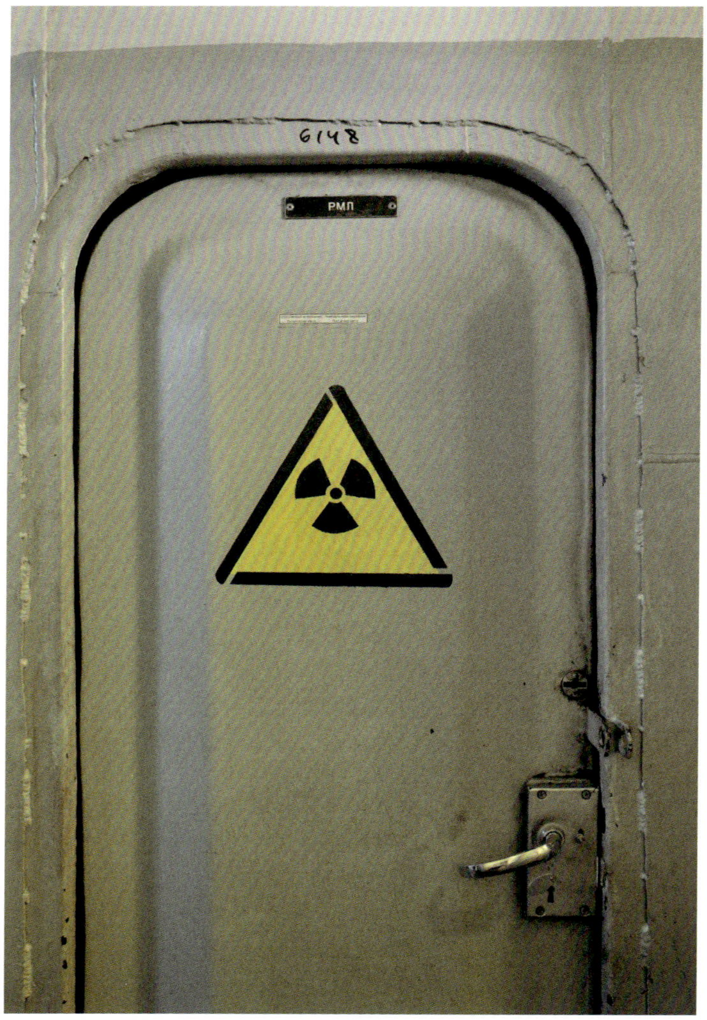
6148
РМП

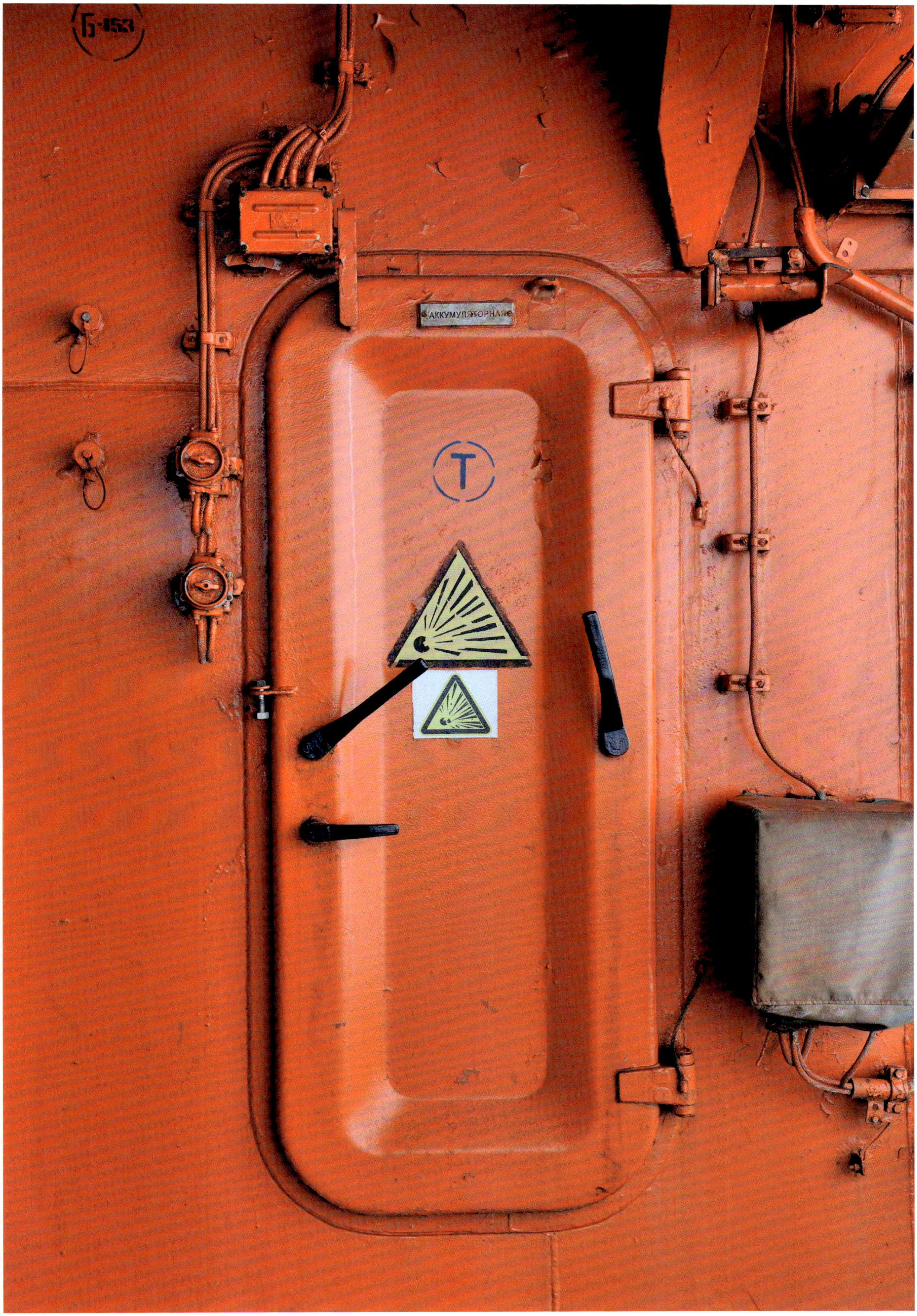

Б-153
АККУМУЛЯТОРНАЯ
Т

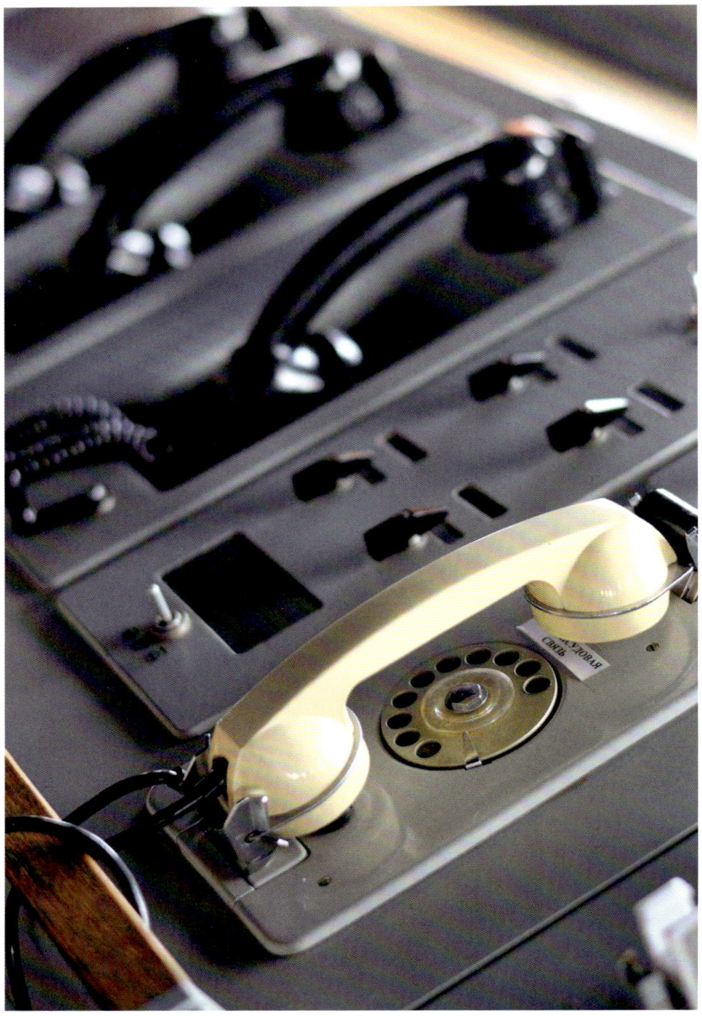

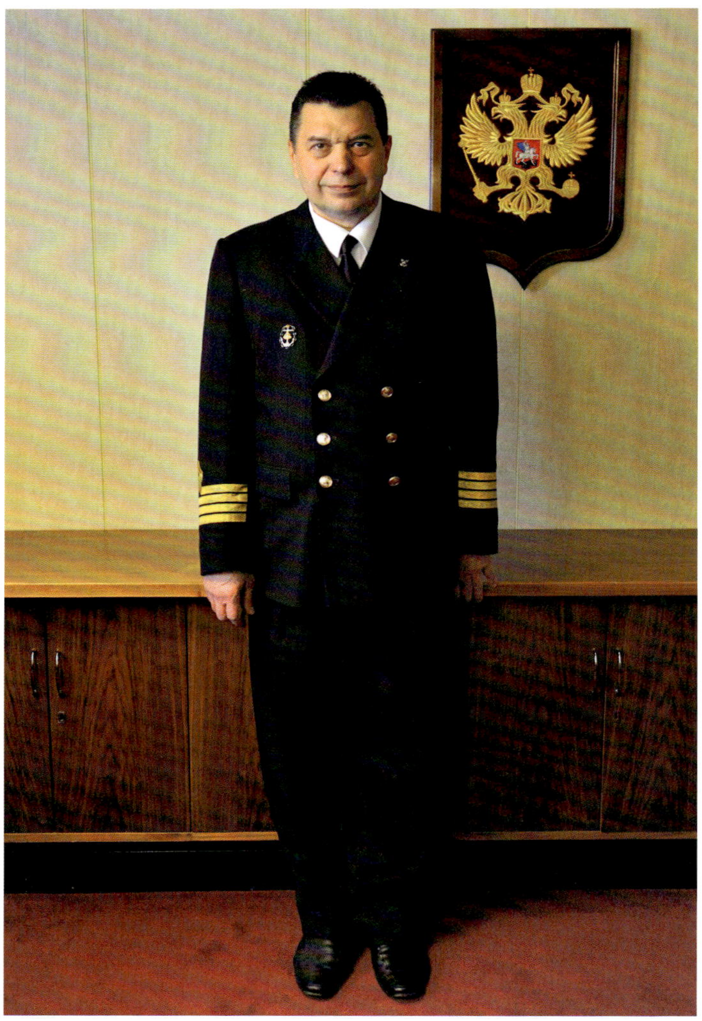

Previous spread The ship's propulsion system includes two nuclear reactors, one to provide power while the other remains in standby mode. With an output of 75,000 horsepower, the ship is the strongest icebreaker in the world and able to reach 21.4 knots or 40 kilometres per hour (25 mph) in open water. In 1977, the *Arktika*, the sister ship of the *50 Years of Victory*, was the first Soviet icebreaker to reach the geographical North Pole. The first tourist ever to set foot on top of the world followed in 1991. In 2017, on the 40th anniversary of the first sailing, another icebreaking ship sailed the exact route as the *Arktika*, only this time it took just 79 hours to reach the pole from Murmansk — less than half the time of the first voyage — and setting a new world record.

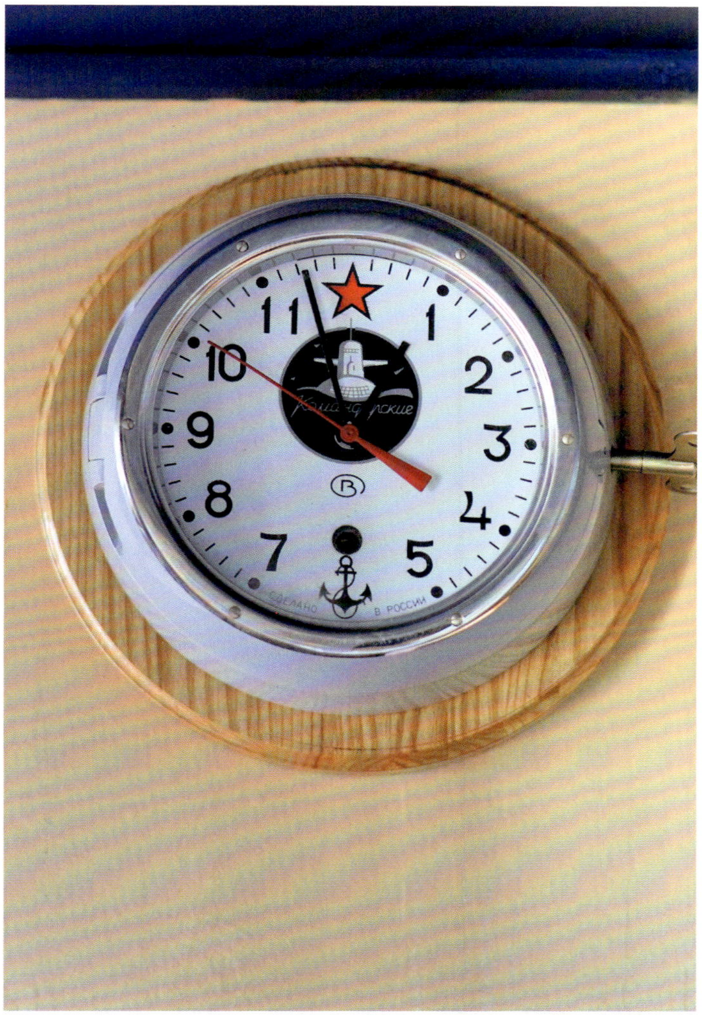
СДЕЛАНО В РОССИИ

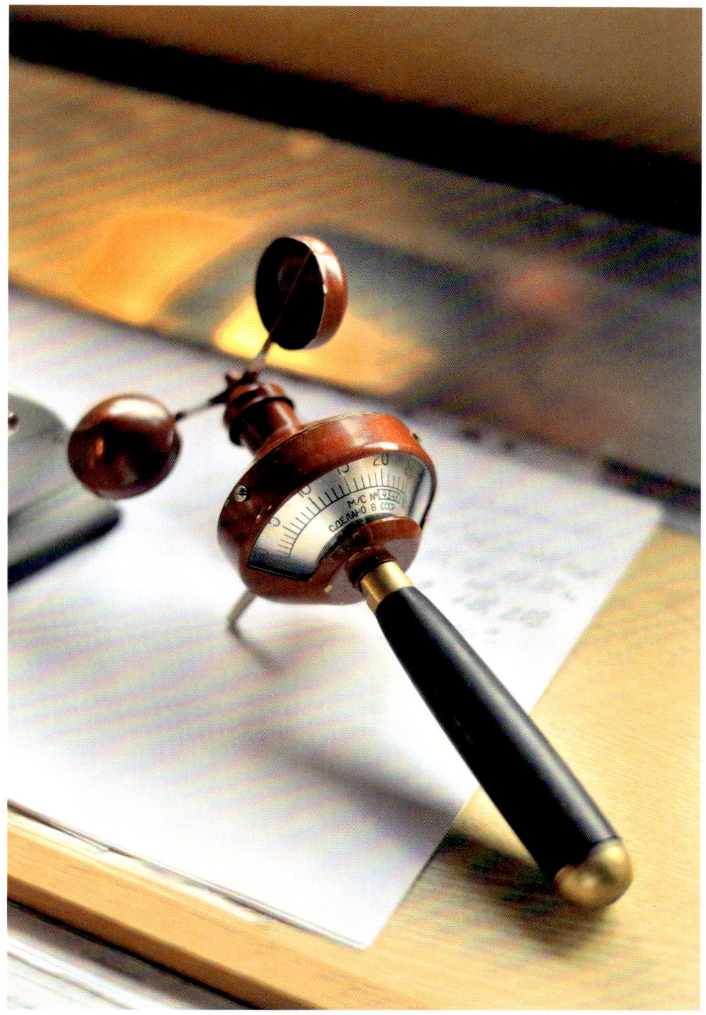

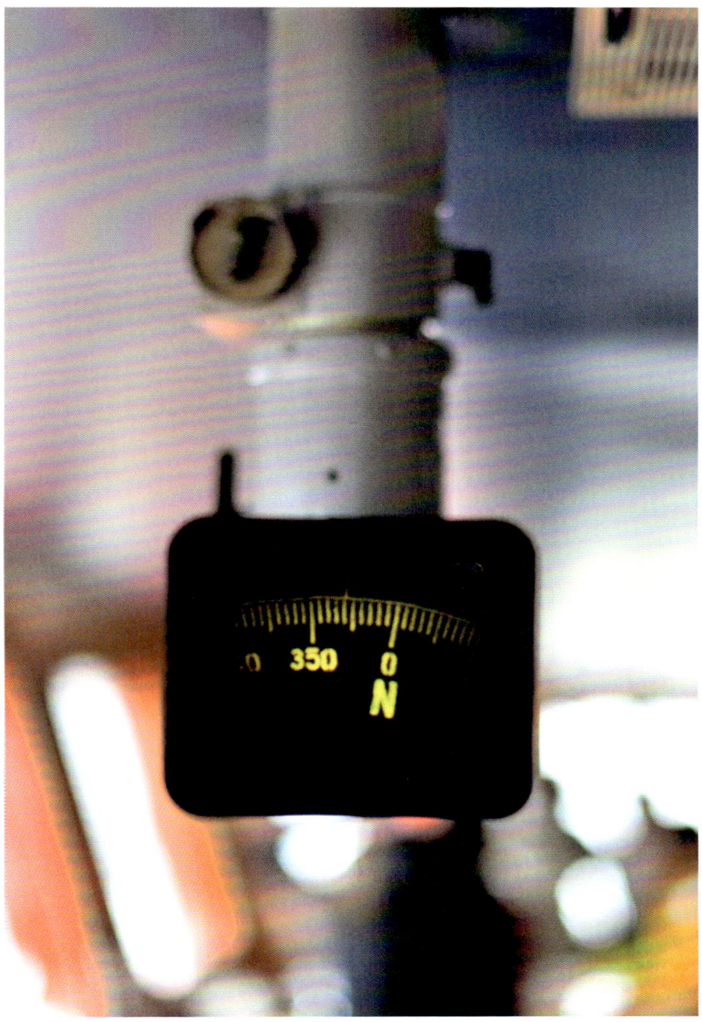
350
0
N

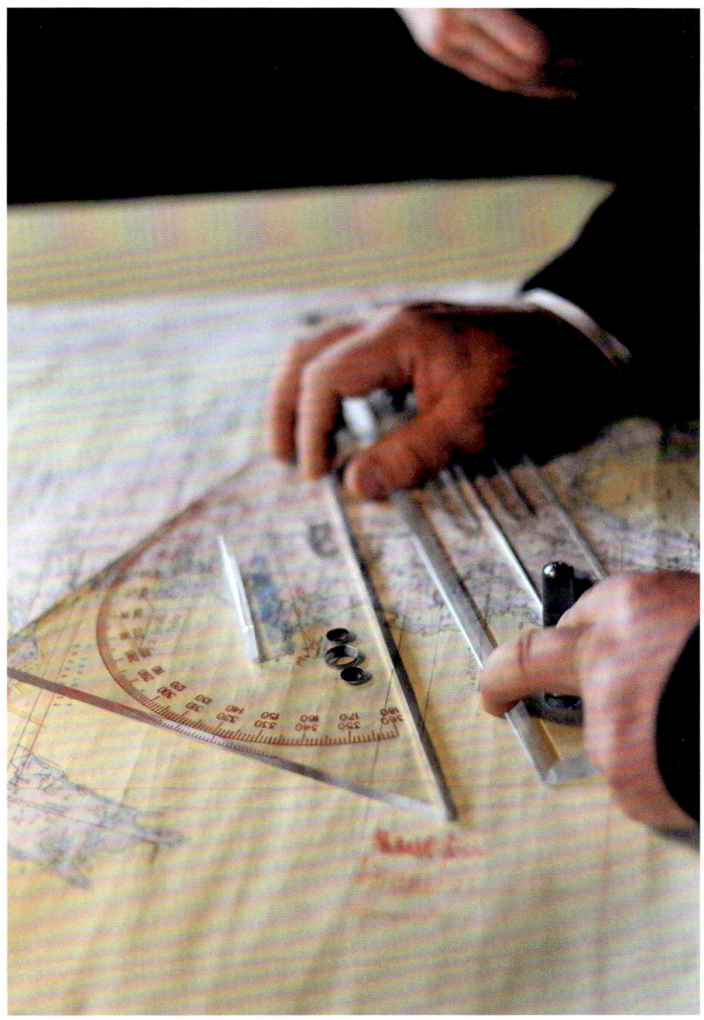

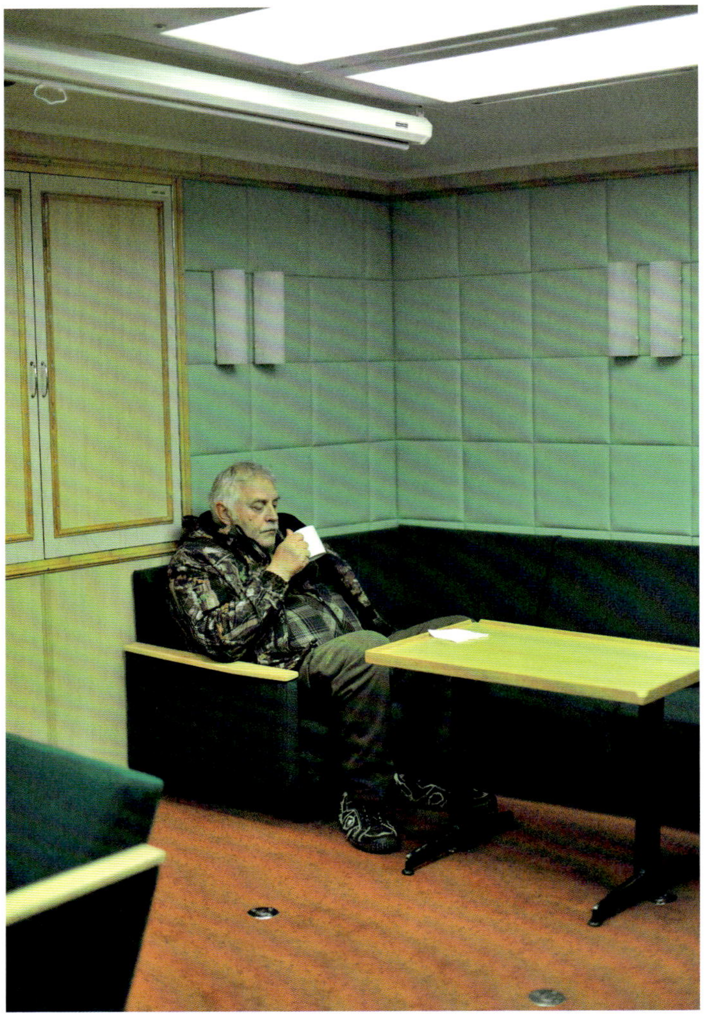

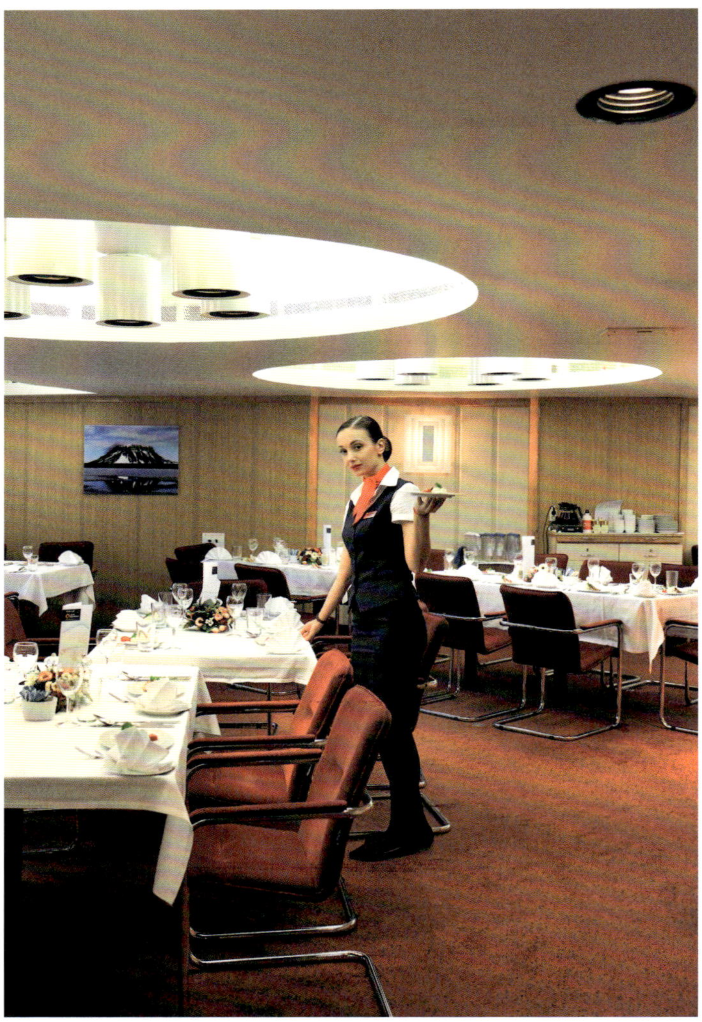

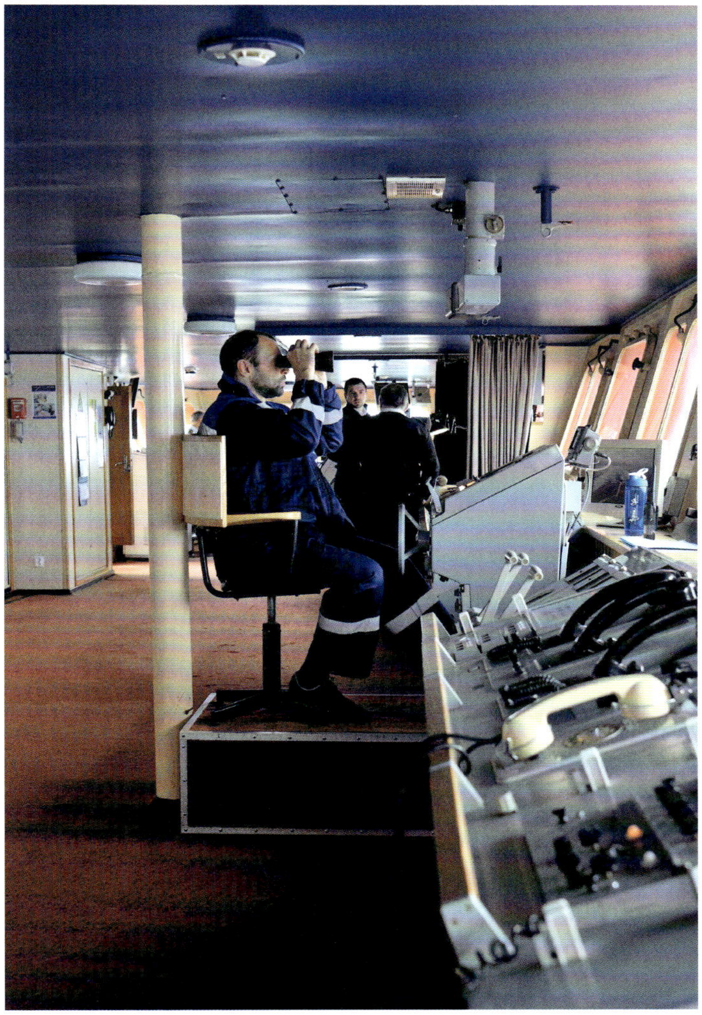
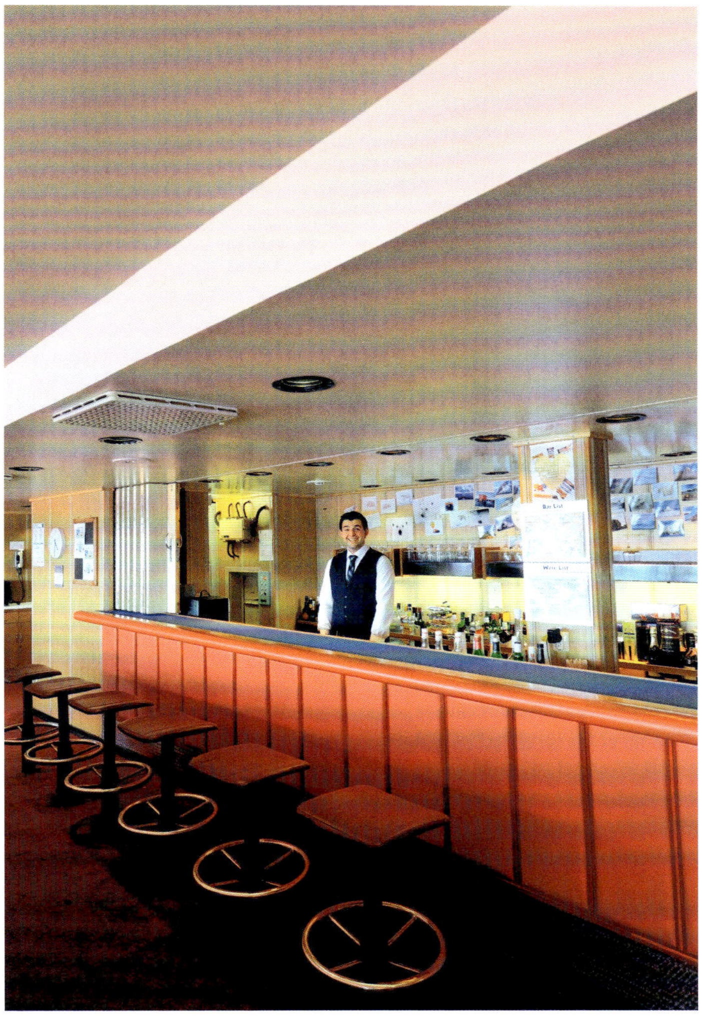
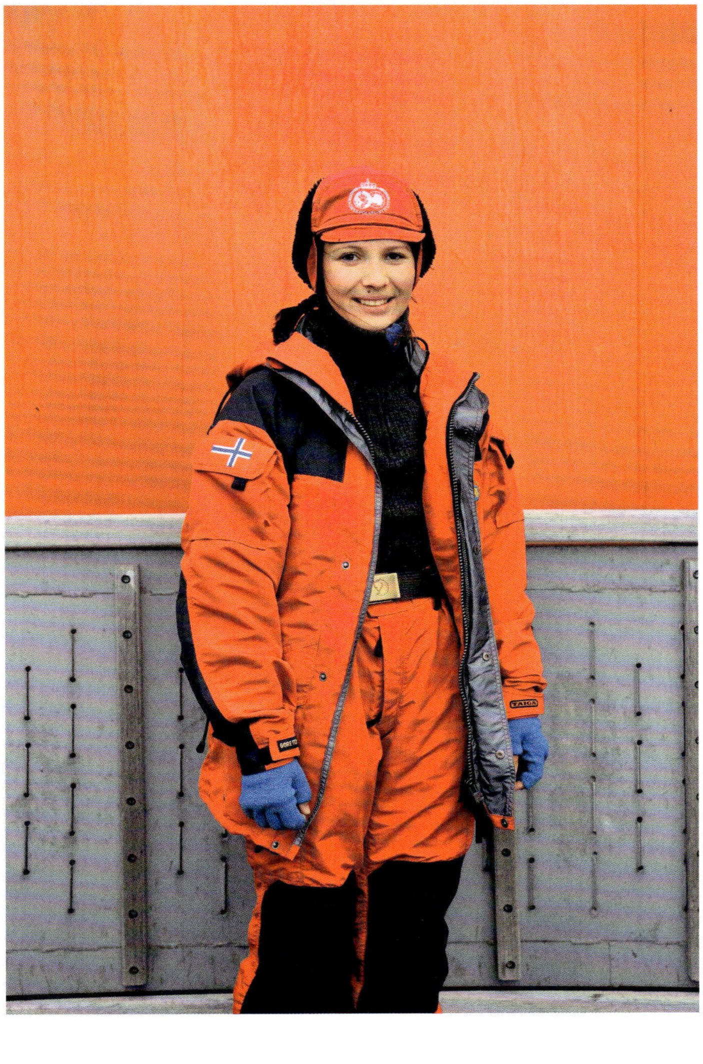
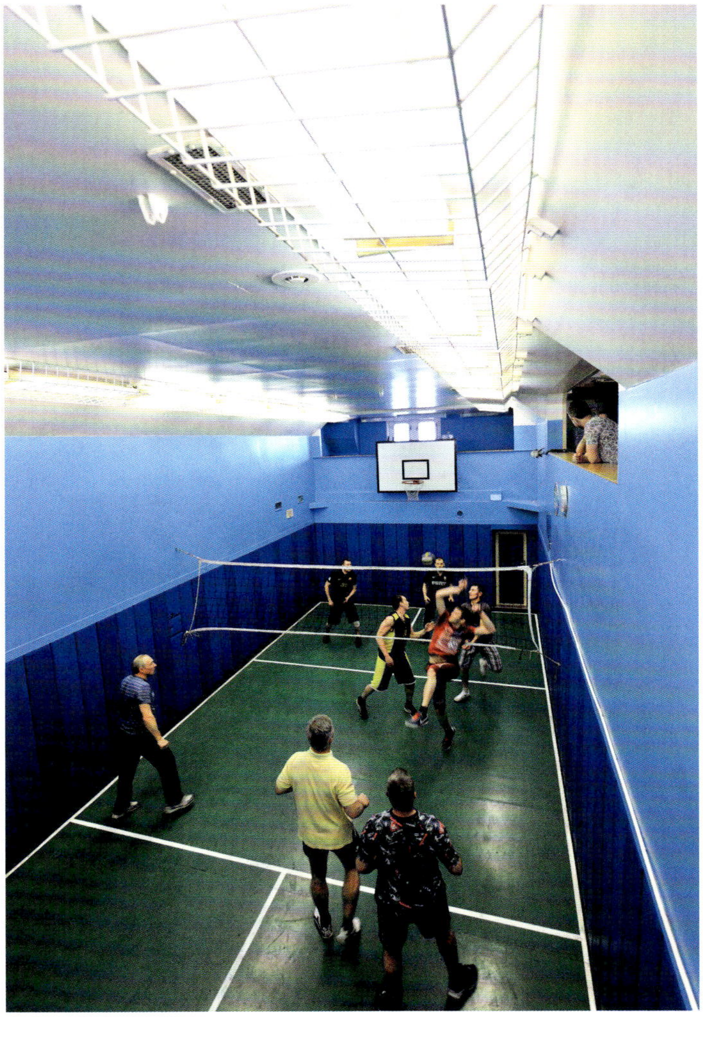

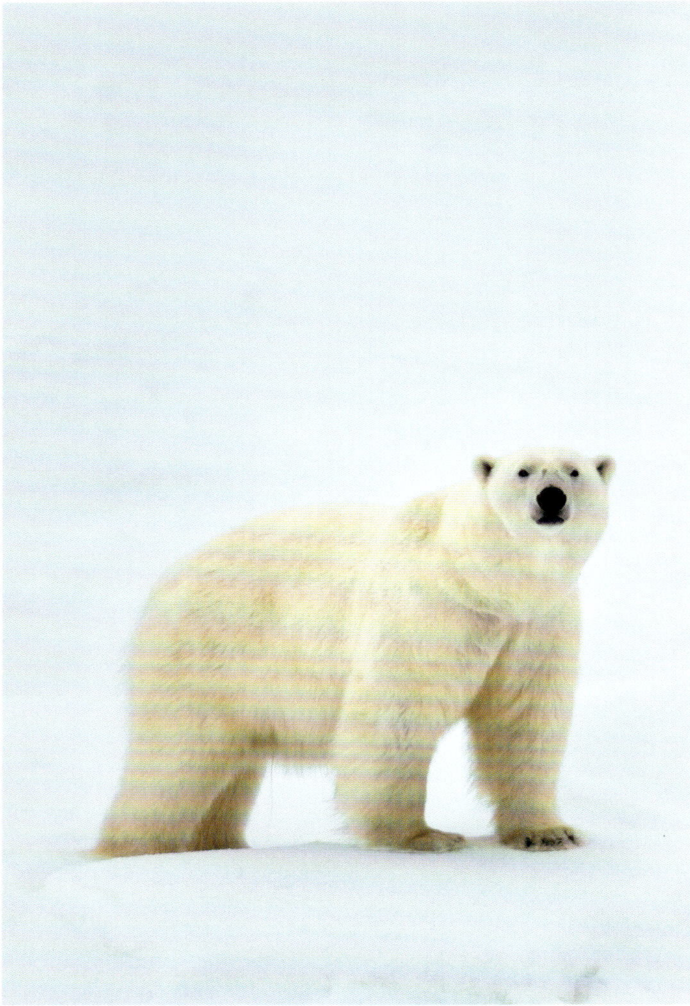

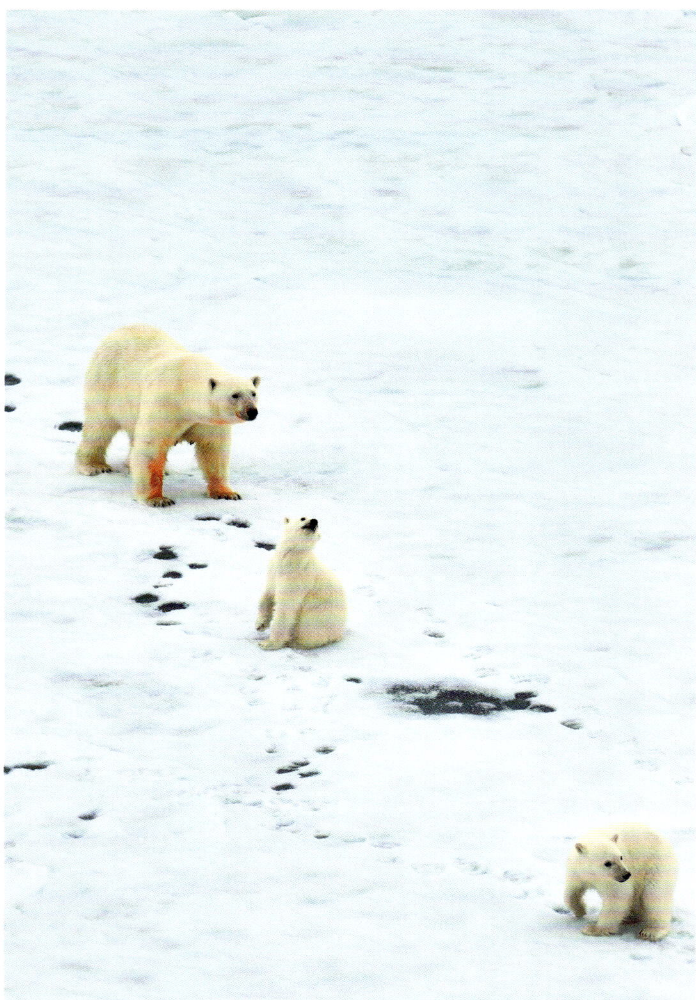

This spread All subspecies of polar bear roam the circumpolar region, and there is a good chance of spotting one prowling the edge of an ice floe or gliding through the icy waters. In summer, the bears follow the retreat of the sea ice and move eastward to hunt around the Russian archipelago of Franz Josef Land. Cape Tegetthoff and the Cambridge Strait offer some of the best opportunities on earth to see these giant creatures, who can stand more than three metres (10 ft.) tall on their hind legs. Sometimes the bears approach the ship when it is stationary.

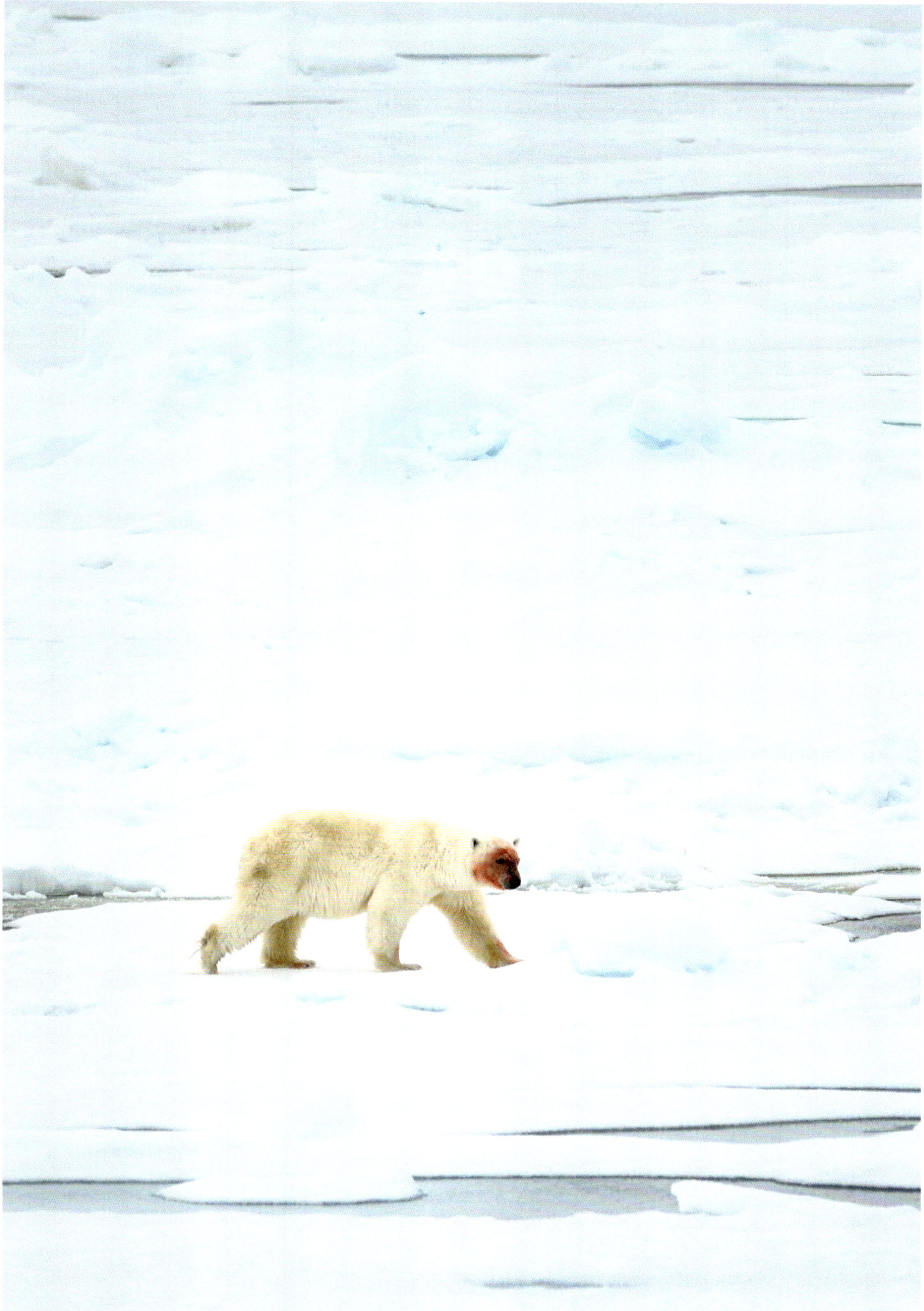

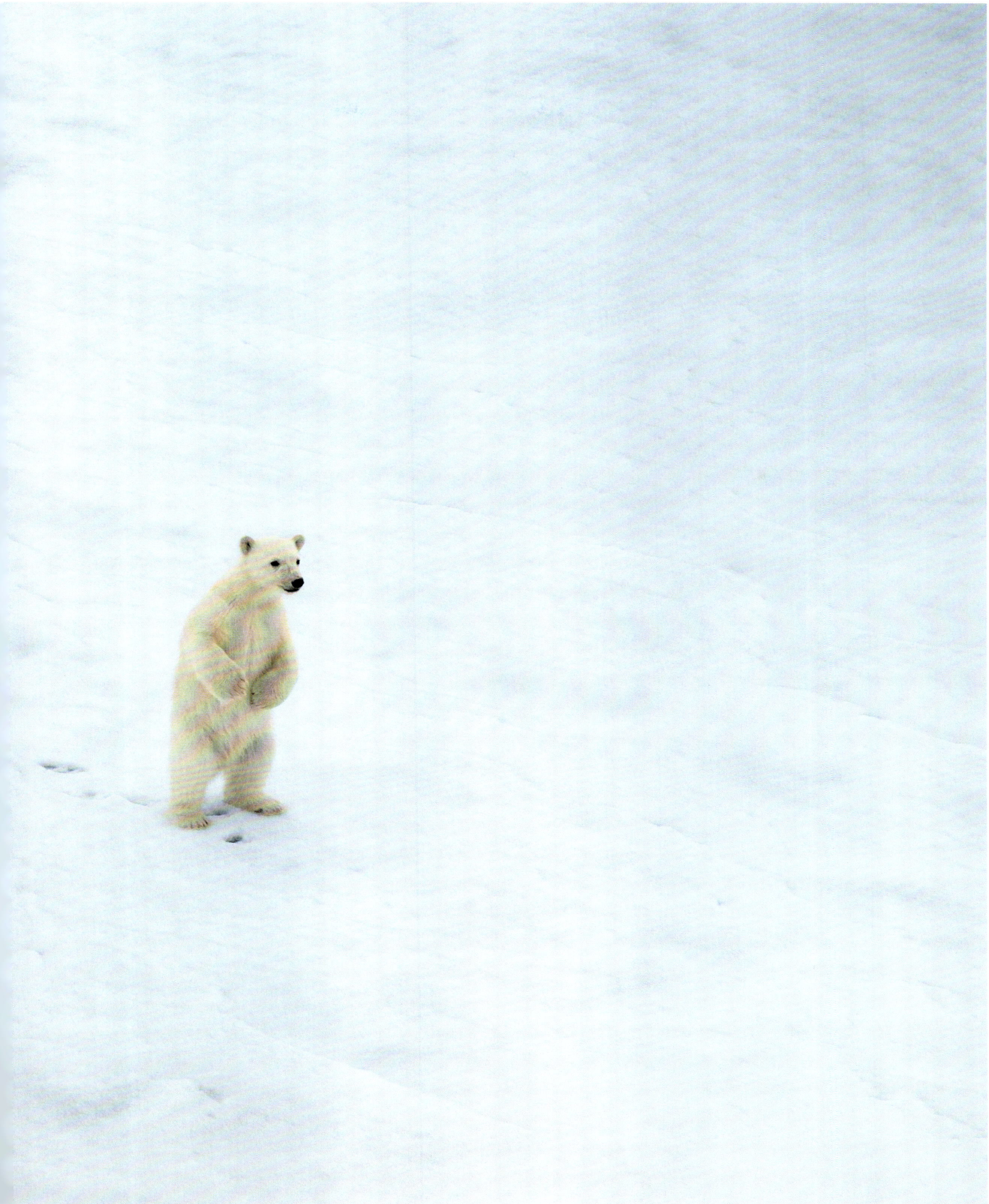

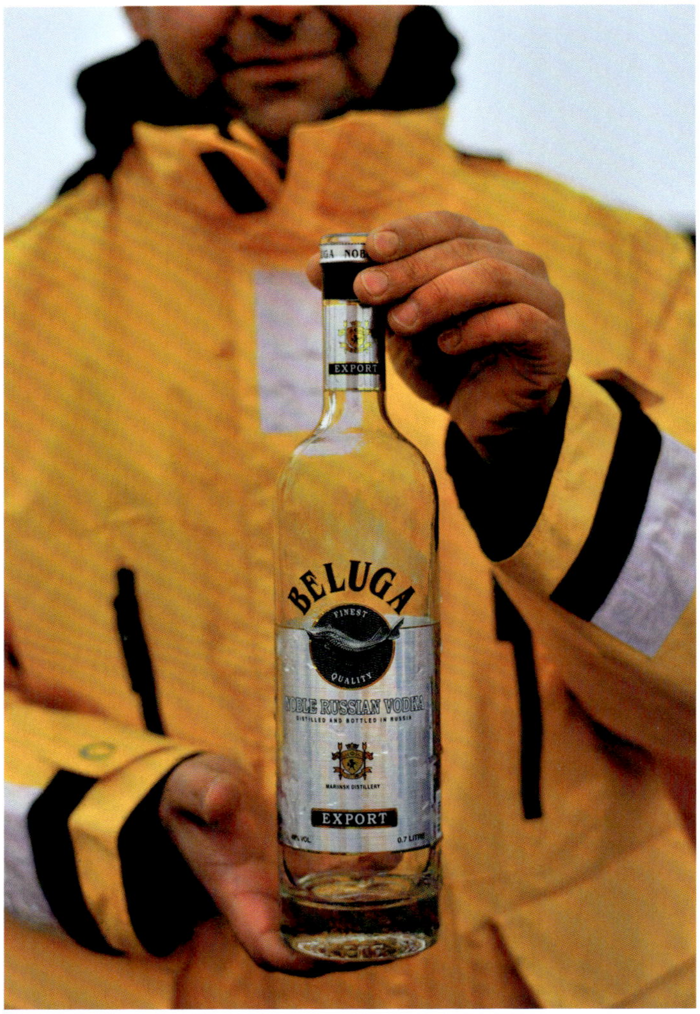

This spread With no land mass, the North Pole is made up of huge ice floes on the surface of the Arctic Ocean. Beneath the ice, the water is more than four kilometres (2.5 mi.) deep, creating an environment of extreme quiet and serenity. Only around 1,000 people travel to the North Pole every year. Unlike Antarctica, nobody lives permanently at the North Pole, although the Soviet Union launched a temporary research station there in the 1930s when a plane dropped four researchers onto the ice. One year later, their research station had drifted into the Greenland Sea, almost 2,500 kilometres (1,550 mi.) away. The *50 Years of Victory* also drifts during the few hours that it anchors at the geographical North Pole. Staying anchored at the exact latitude 90° north lasts only a minute. By the time the ship leaves, it has already drifted two nautical miles — almost four kilometres.

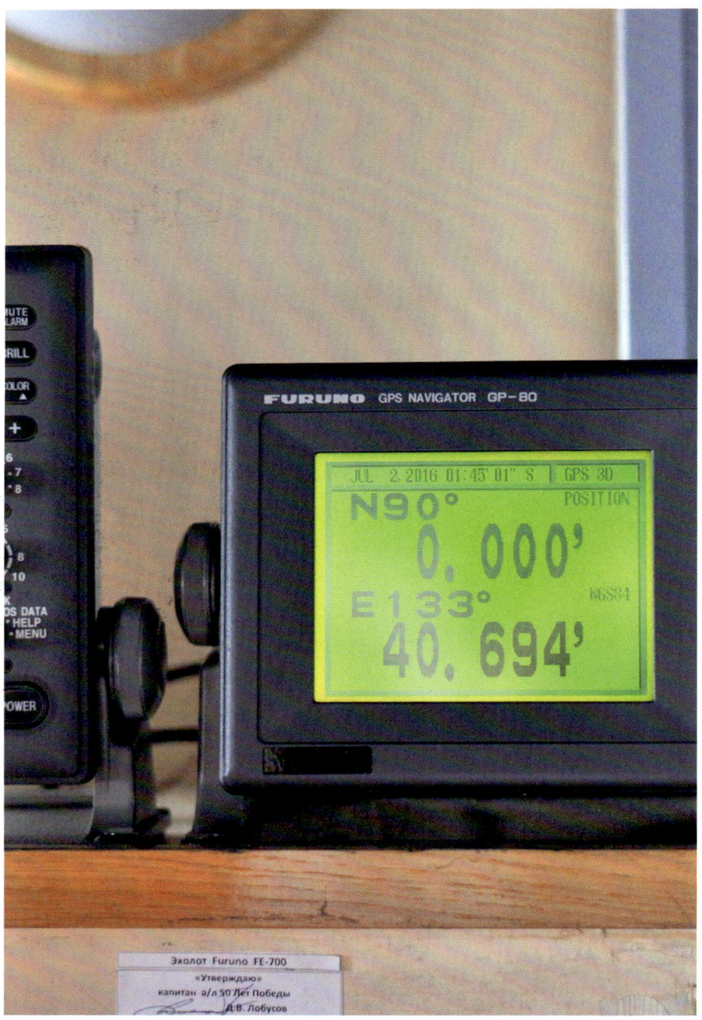

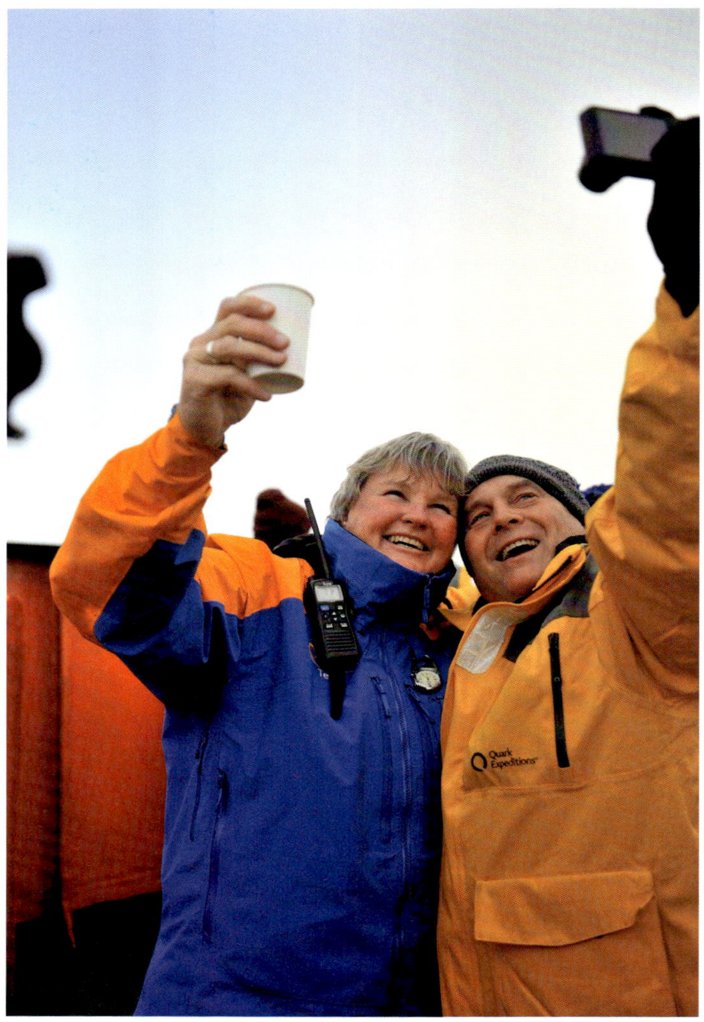

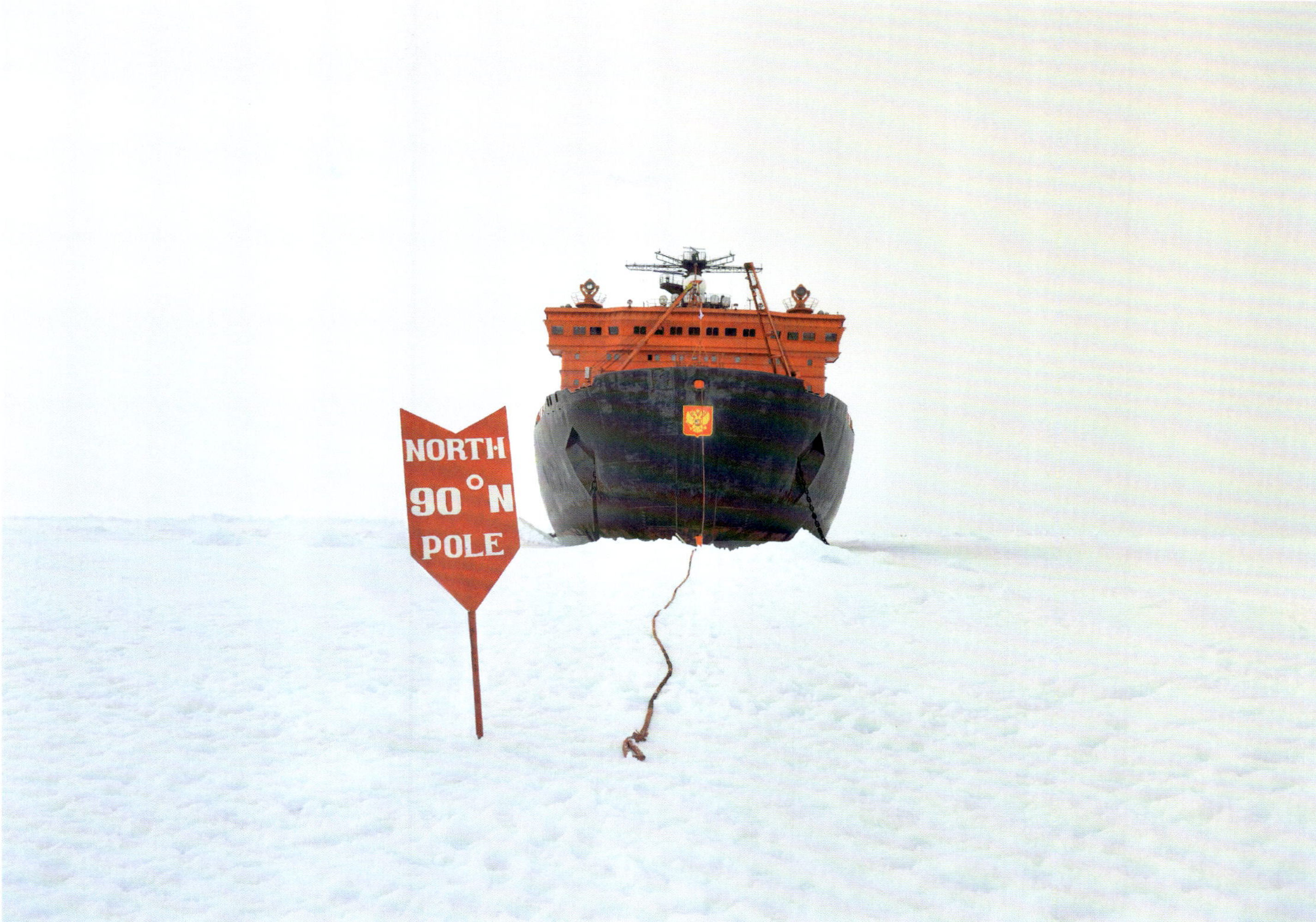
NORTH
90°N
POLE

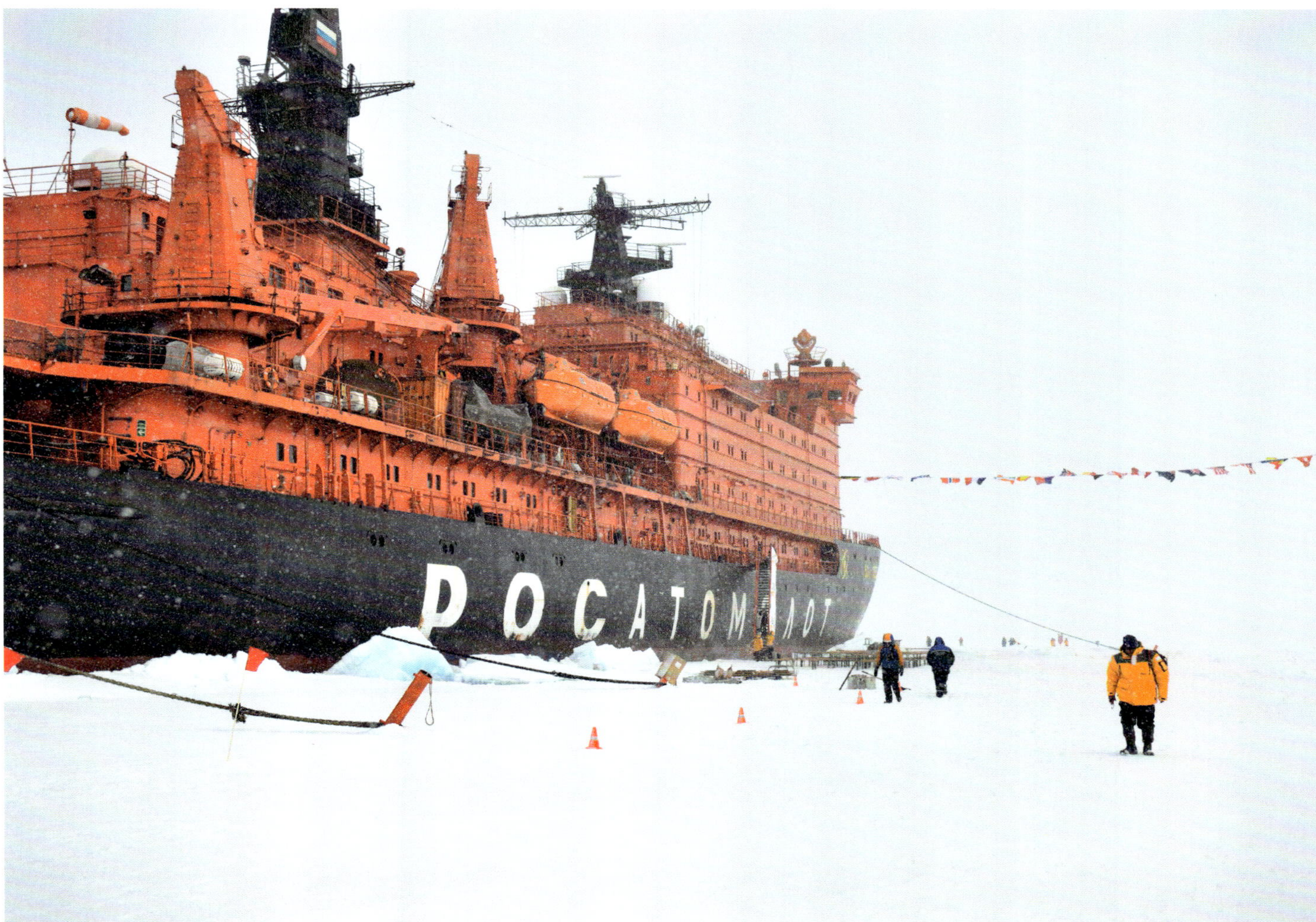
РОСАТОМ

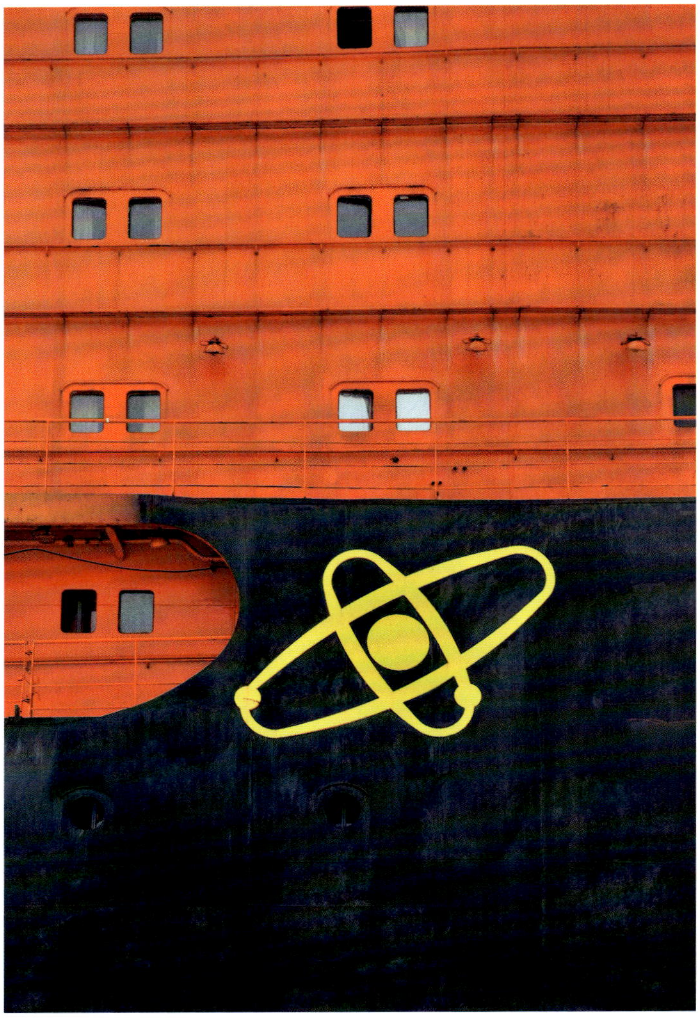

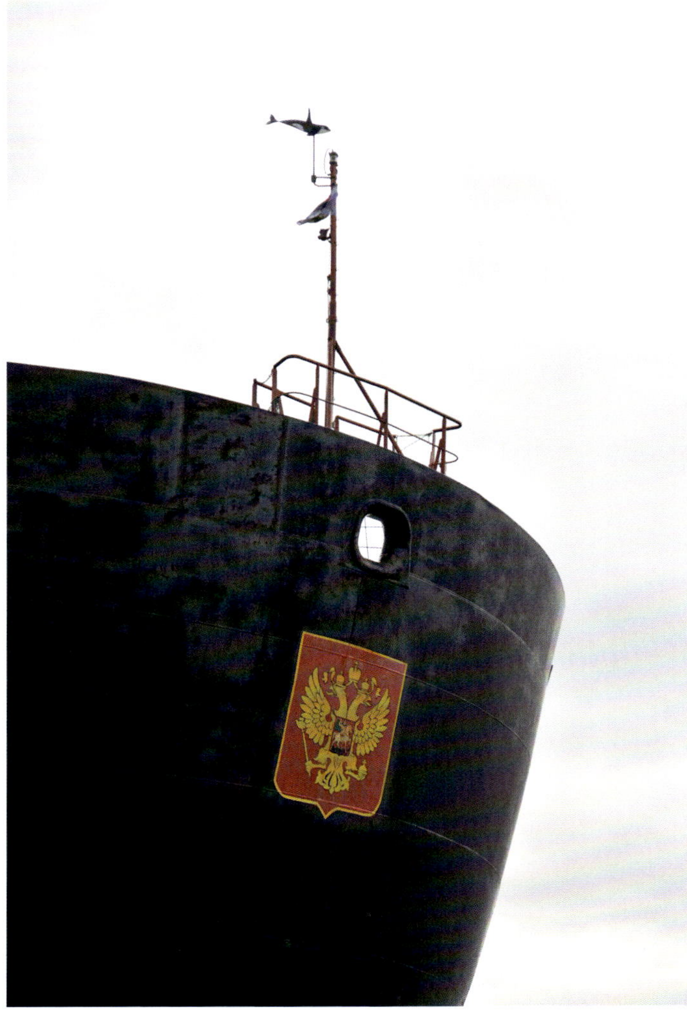

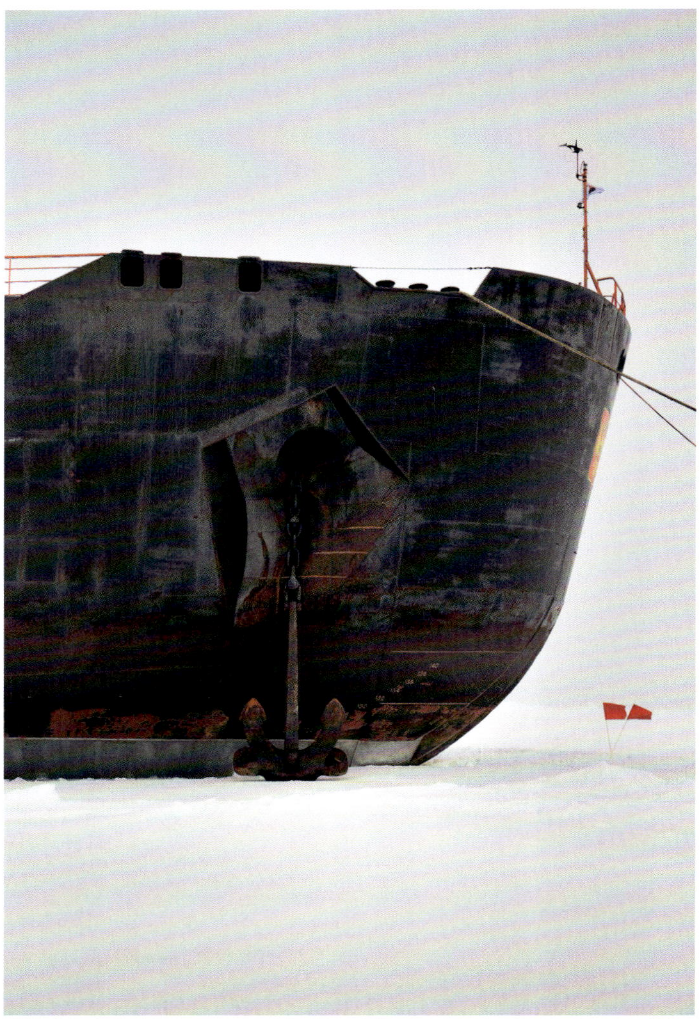

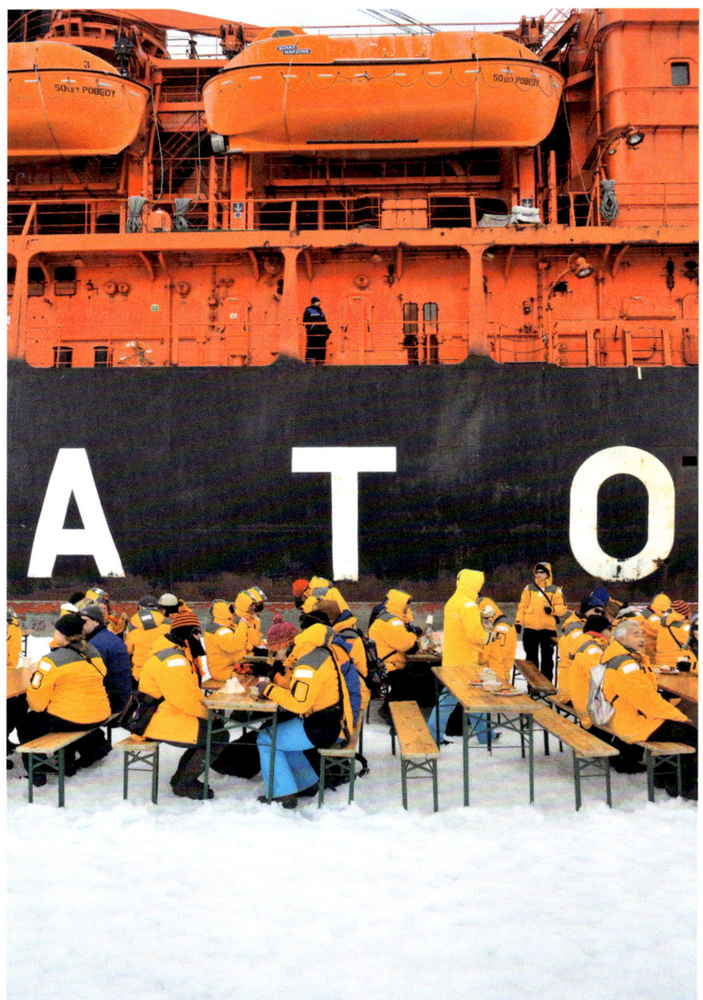
A T O

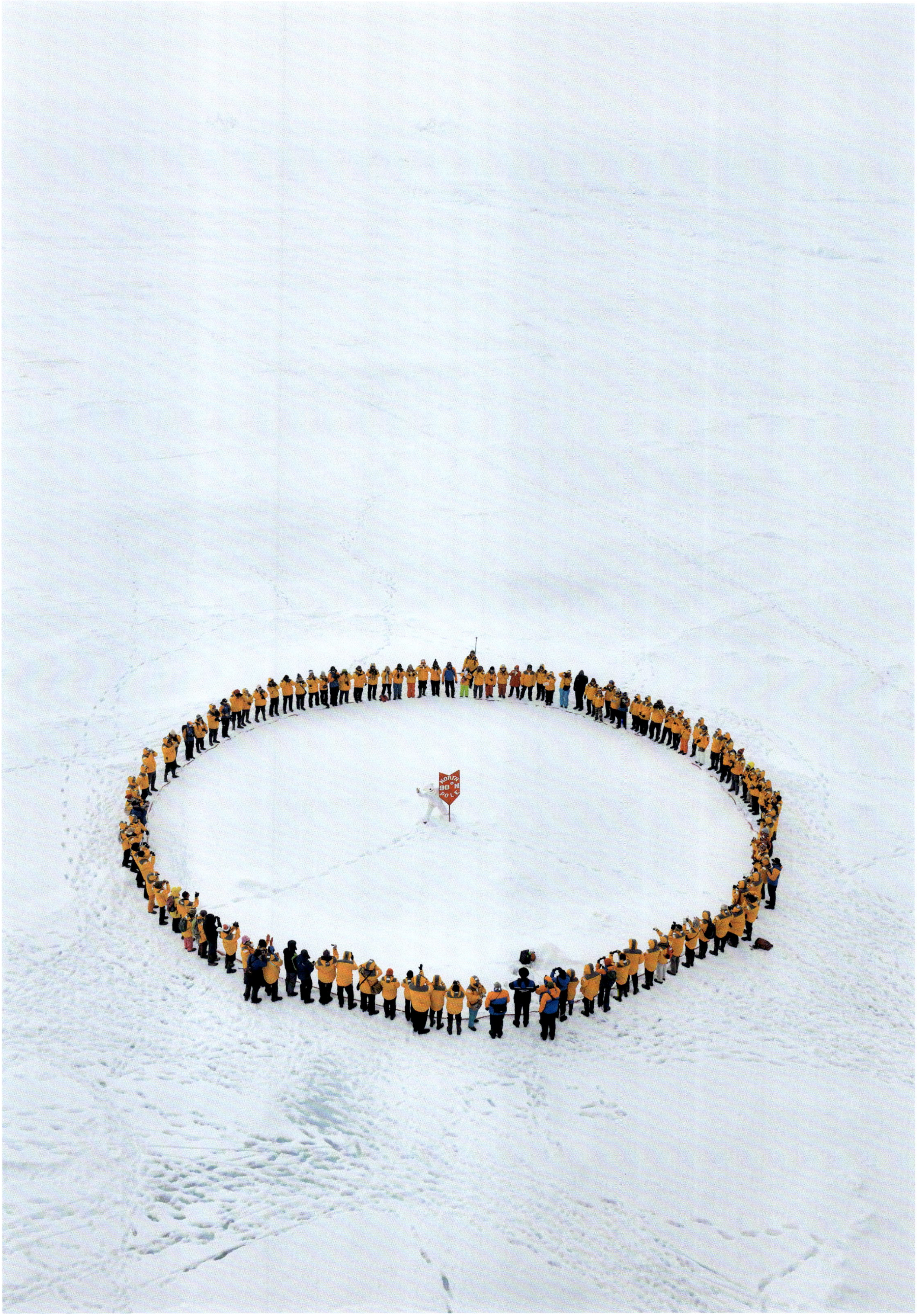
NORTH
90°N
POLE

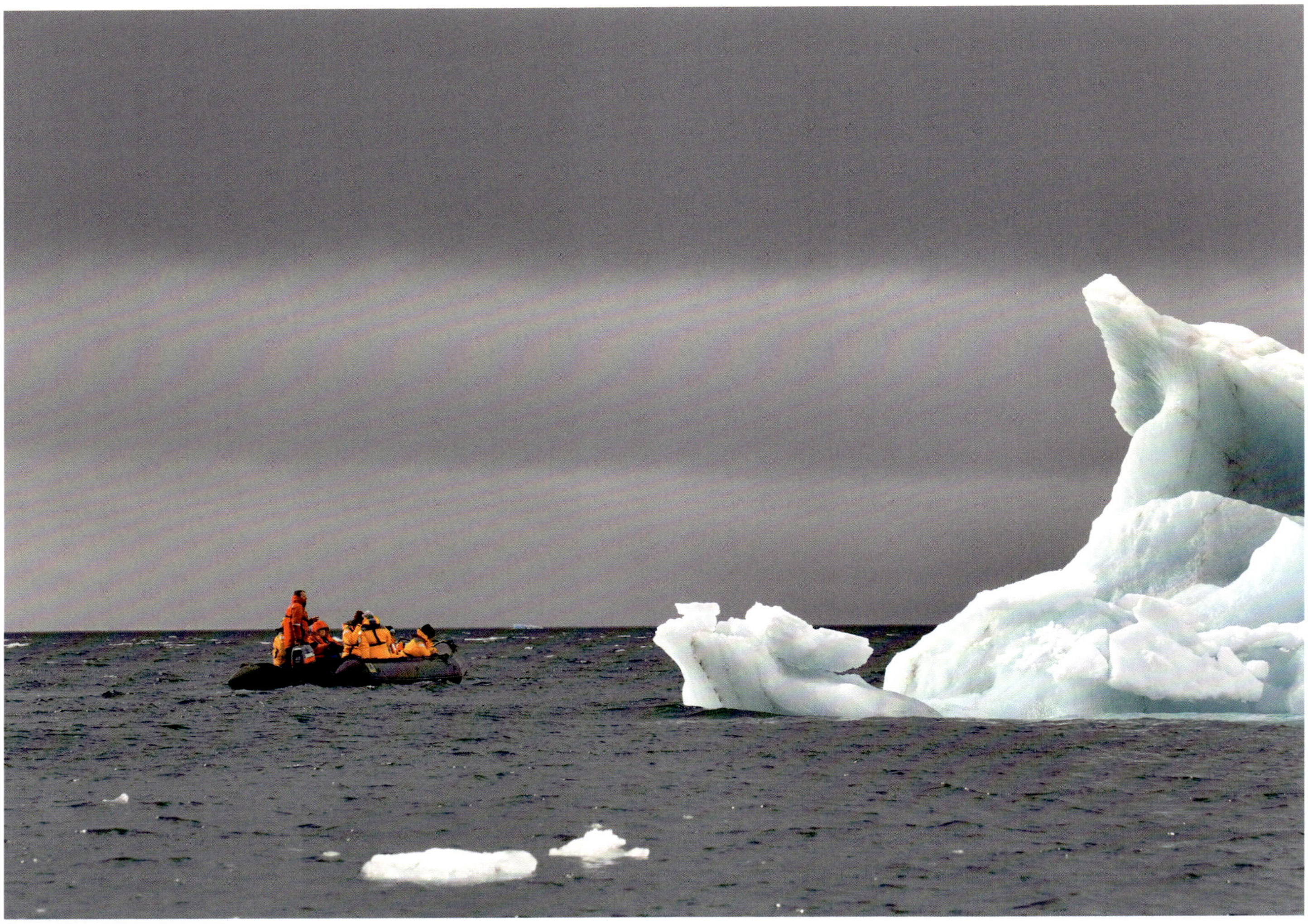

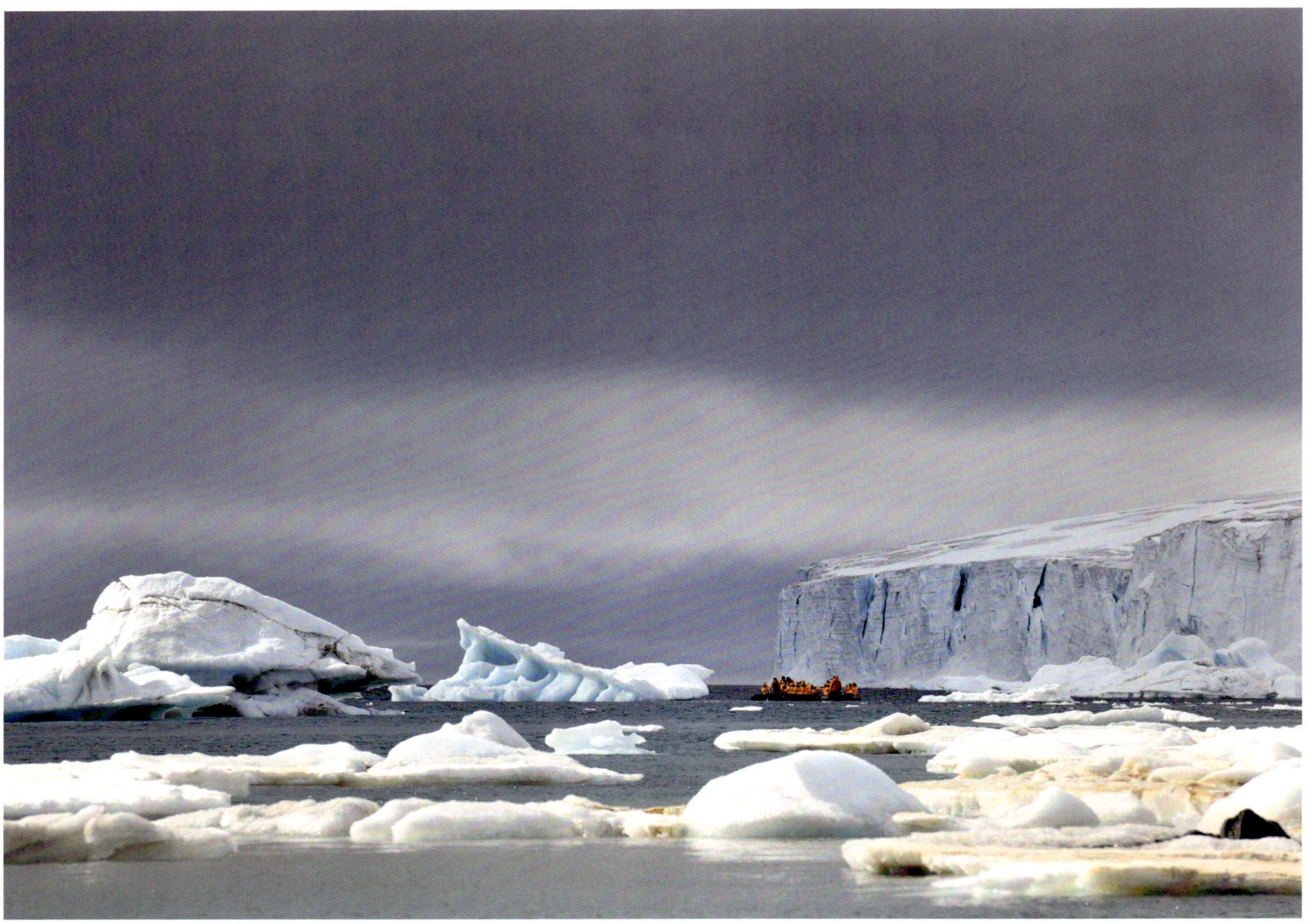

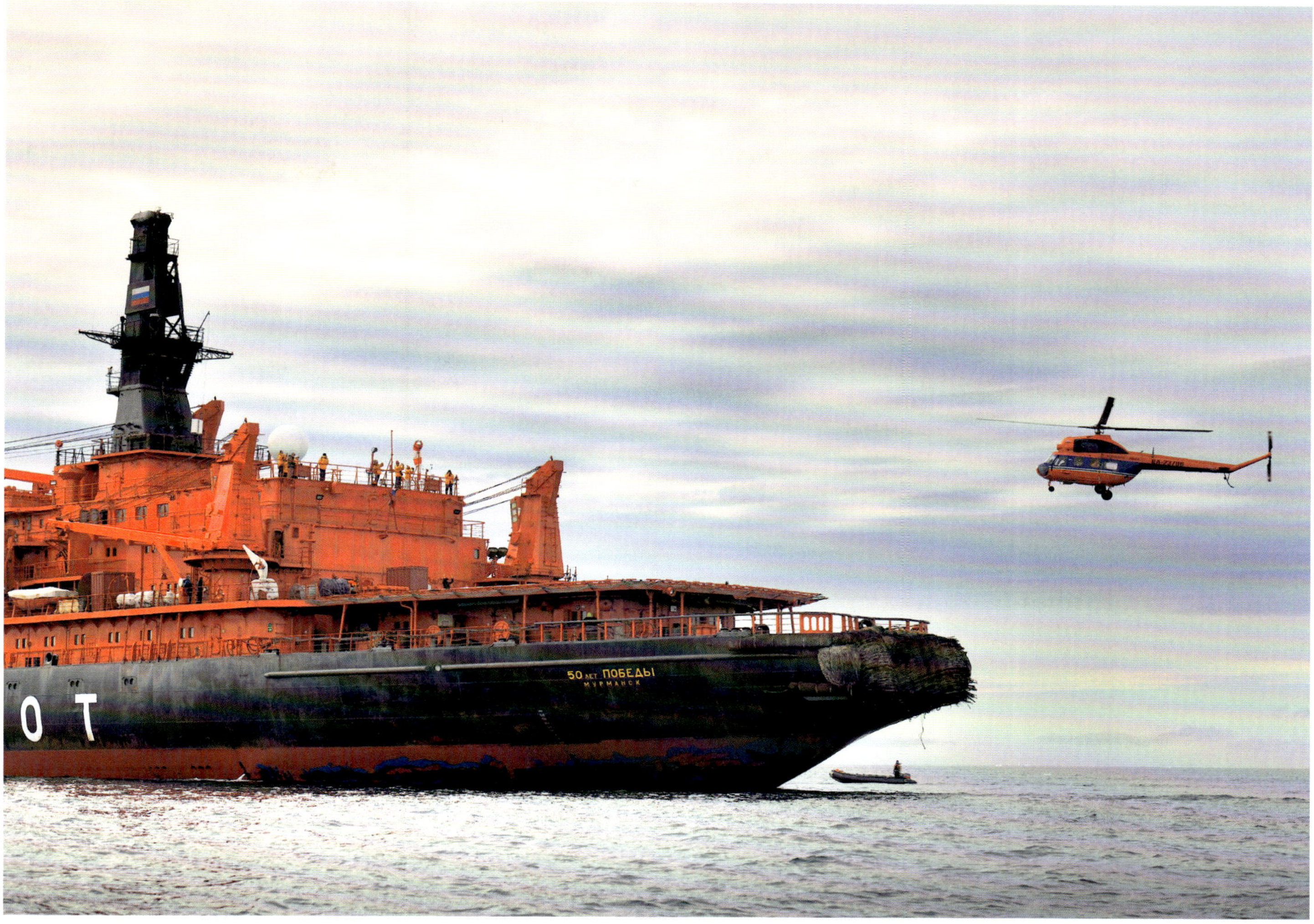

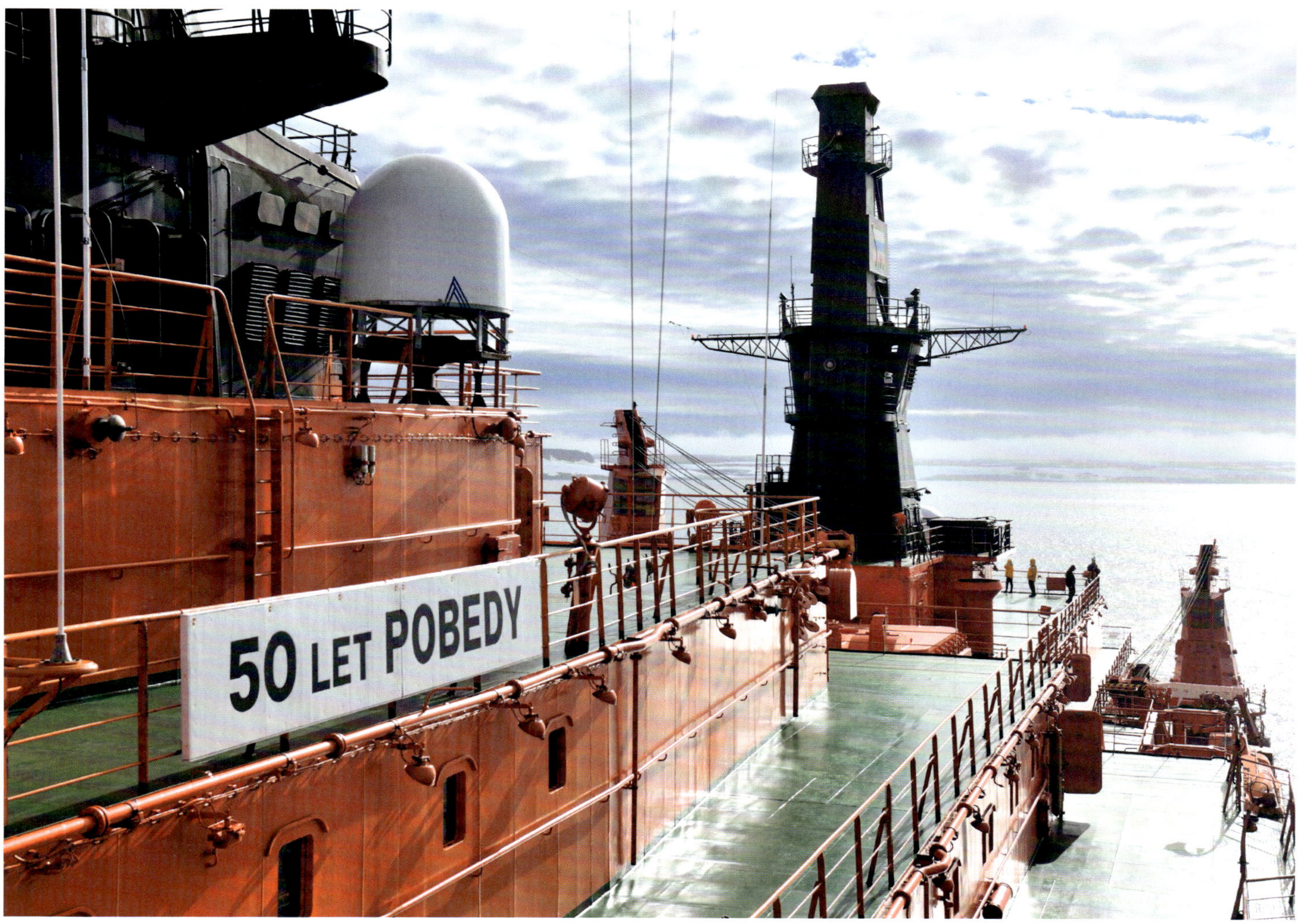
50 LET POBEDY

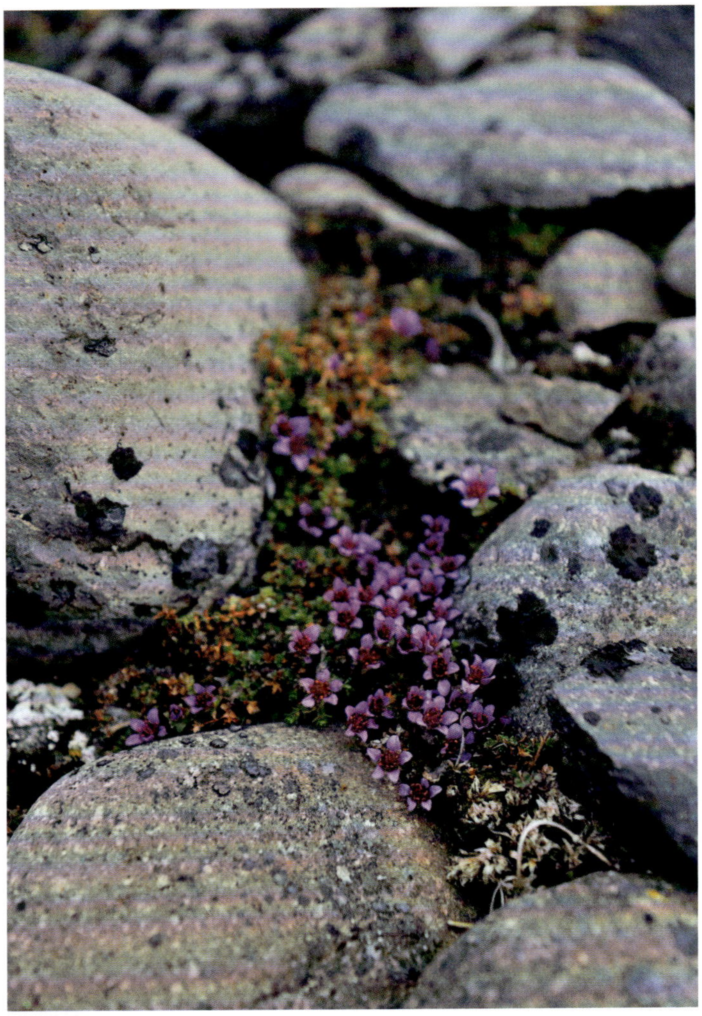

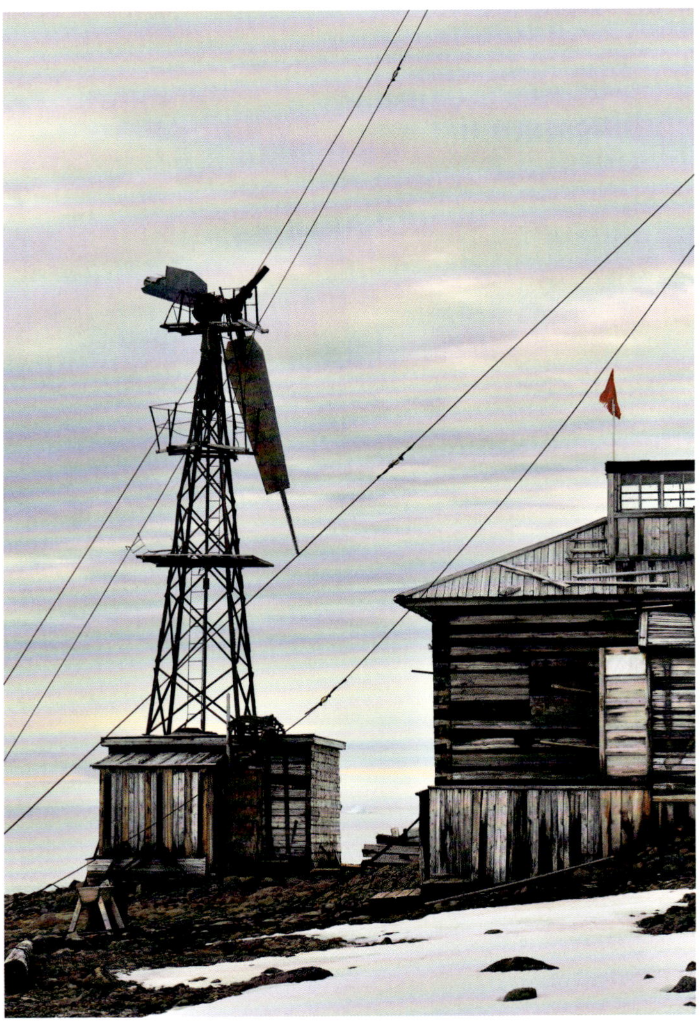

This spread Located just 900 kilometres (560 mi.) from the North Pole, Franz Josef Land is a collection of 192 islands and is one of the most inaccessible places on earth. This makes it an extraordinary place to spot Arctic birds and wildlife. Opportunities for humans to visit are limited, making it an environmentally important nature reserve within the Russian Arctic National Park administration. As a Russian vessel, the *50 Years of Victory* enjoys unique permission to anchor in Franz Josef Land.

Previous spread Travelling in Zodiacs and by helicopter, passengers are able to explore cliff colonies of Brünnich's guillemots, black guillemots, little auks, fulmars, and kittiwakes, along with observing Atlantic walruses as they haul themselves onto the sea ice while bowhead whales feed in the icy waters below.

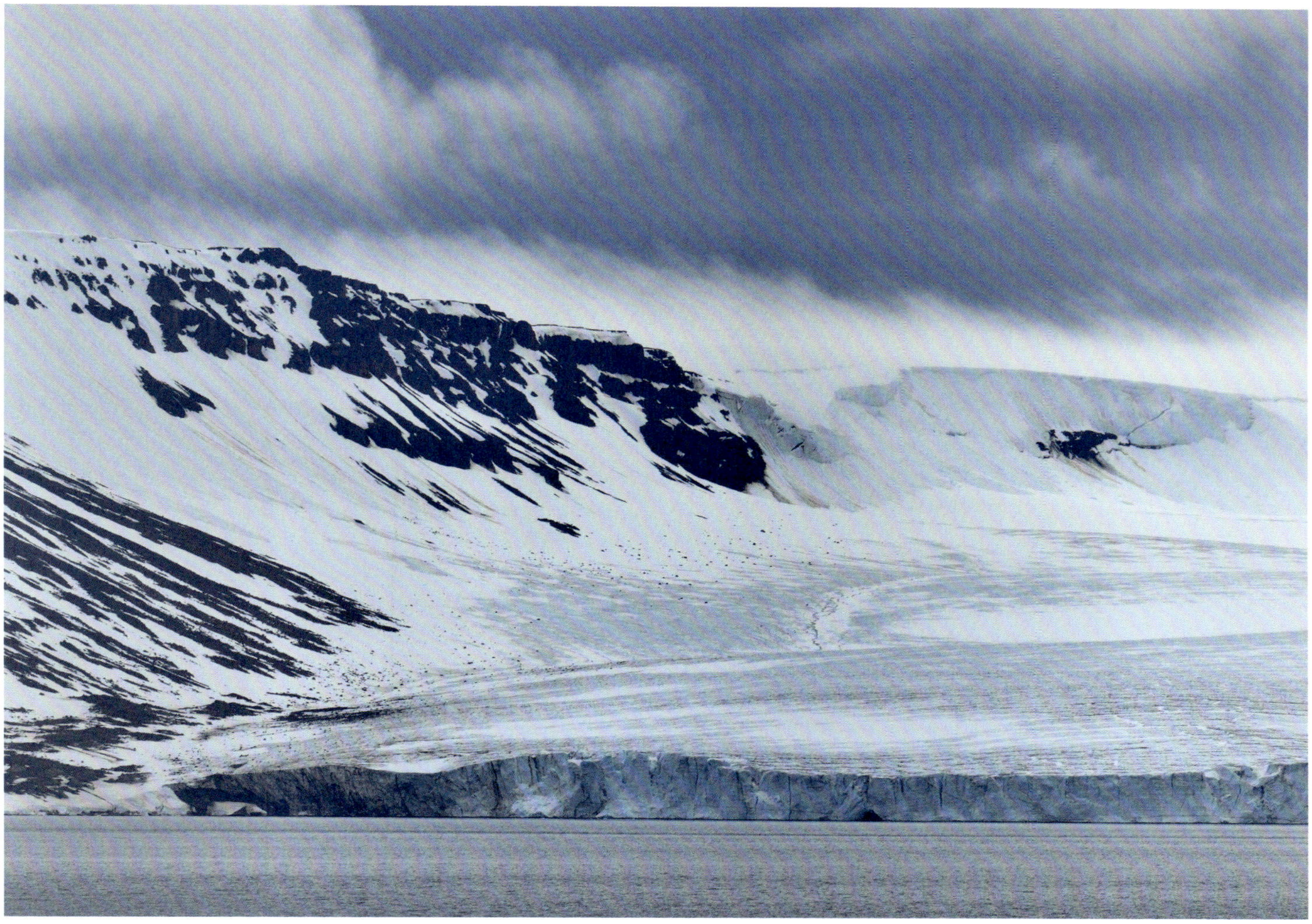

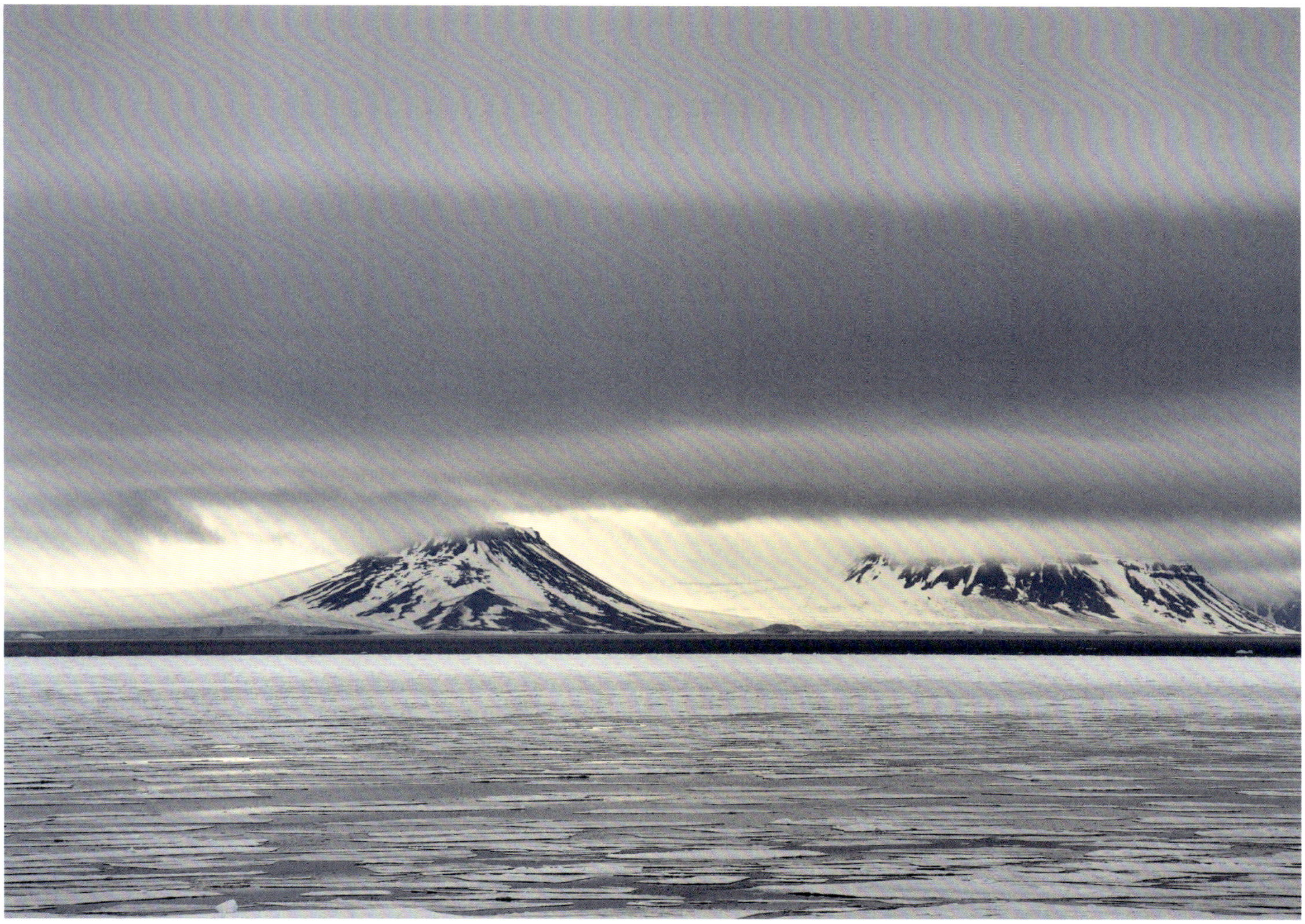

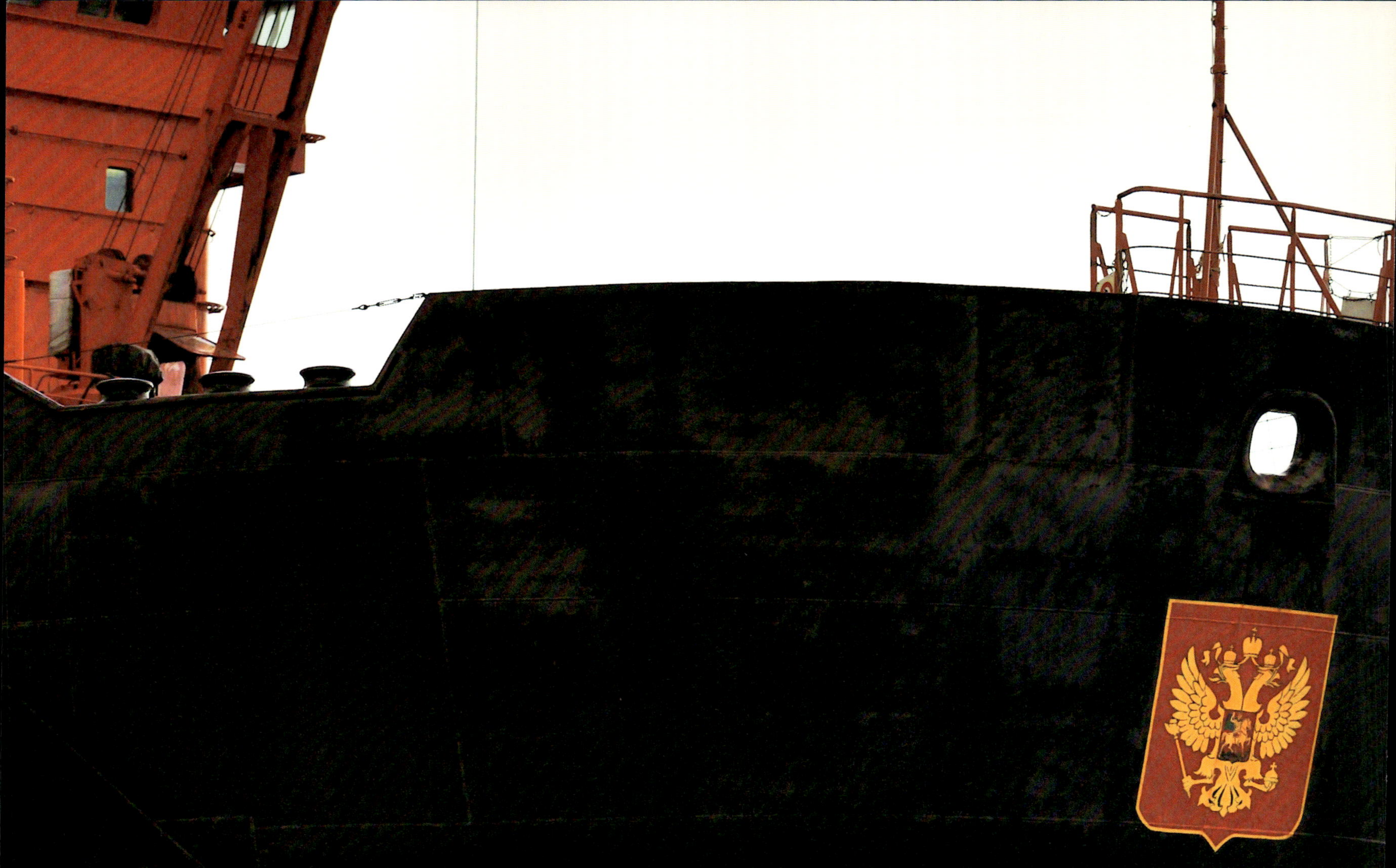

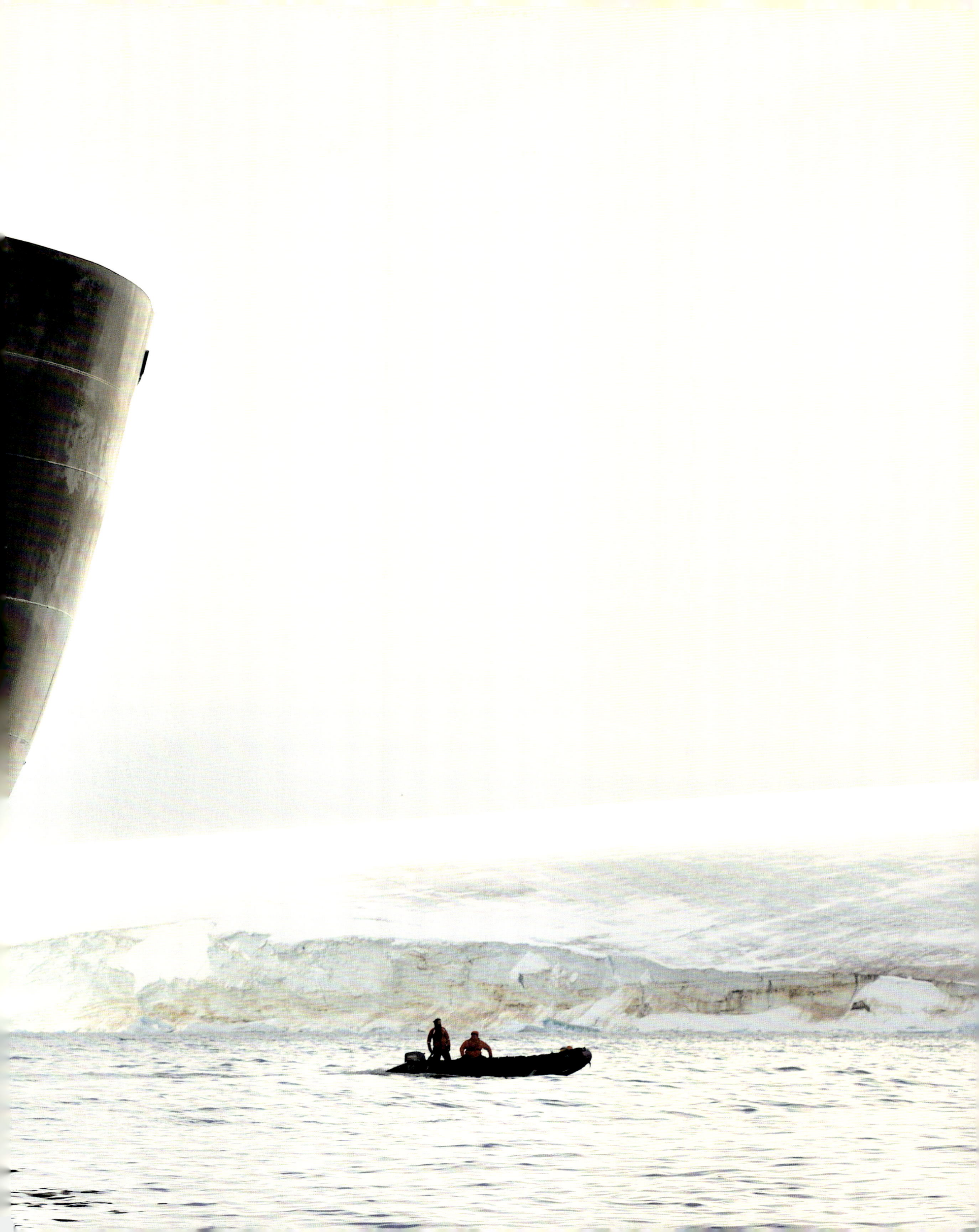

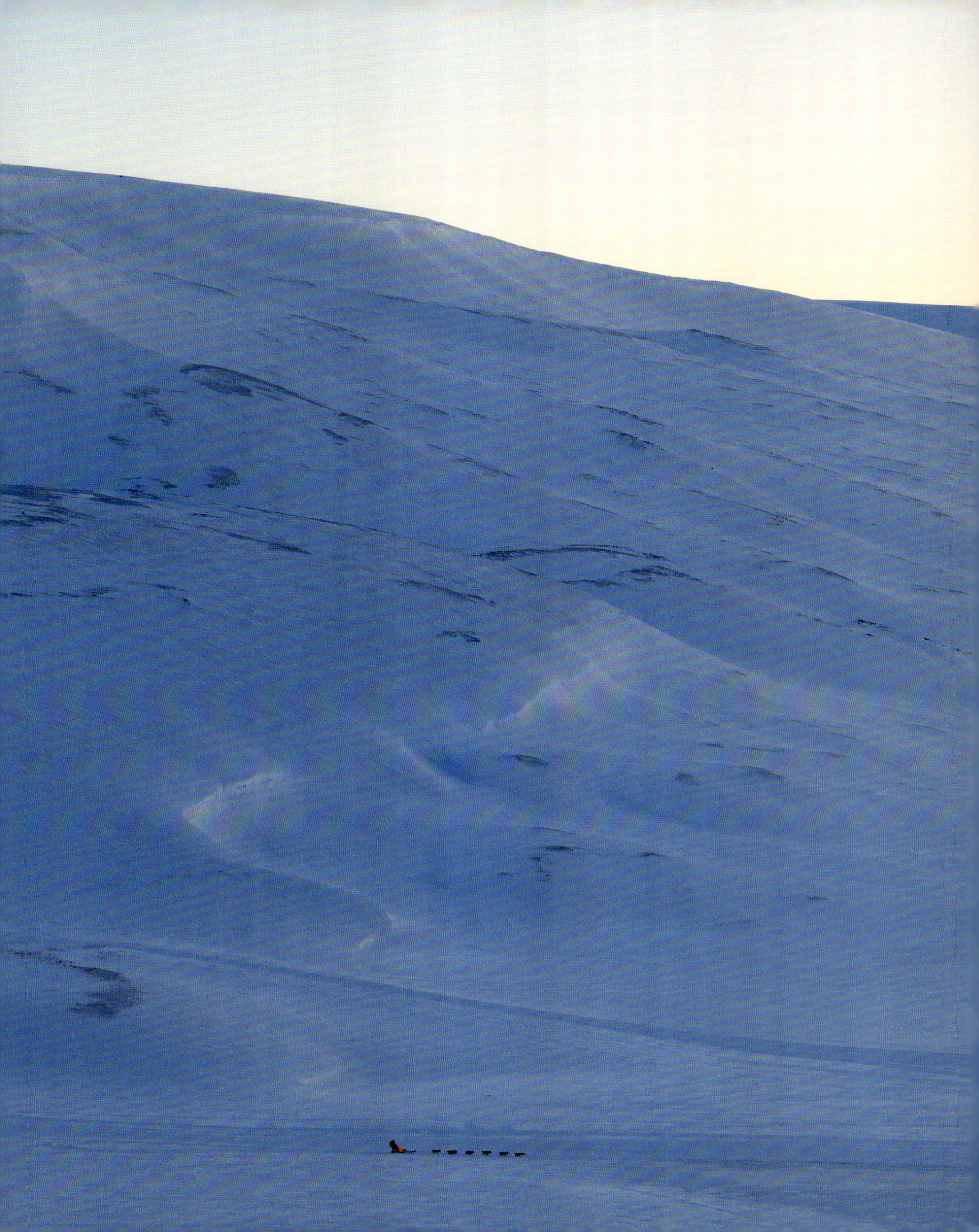

CHAPTER 2

INTO THE REALM OF ARCTIC BLUE

POLAR ISLAND VIBES

Svalbard

78°43'11.47" N
20°20'57.60" E

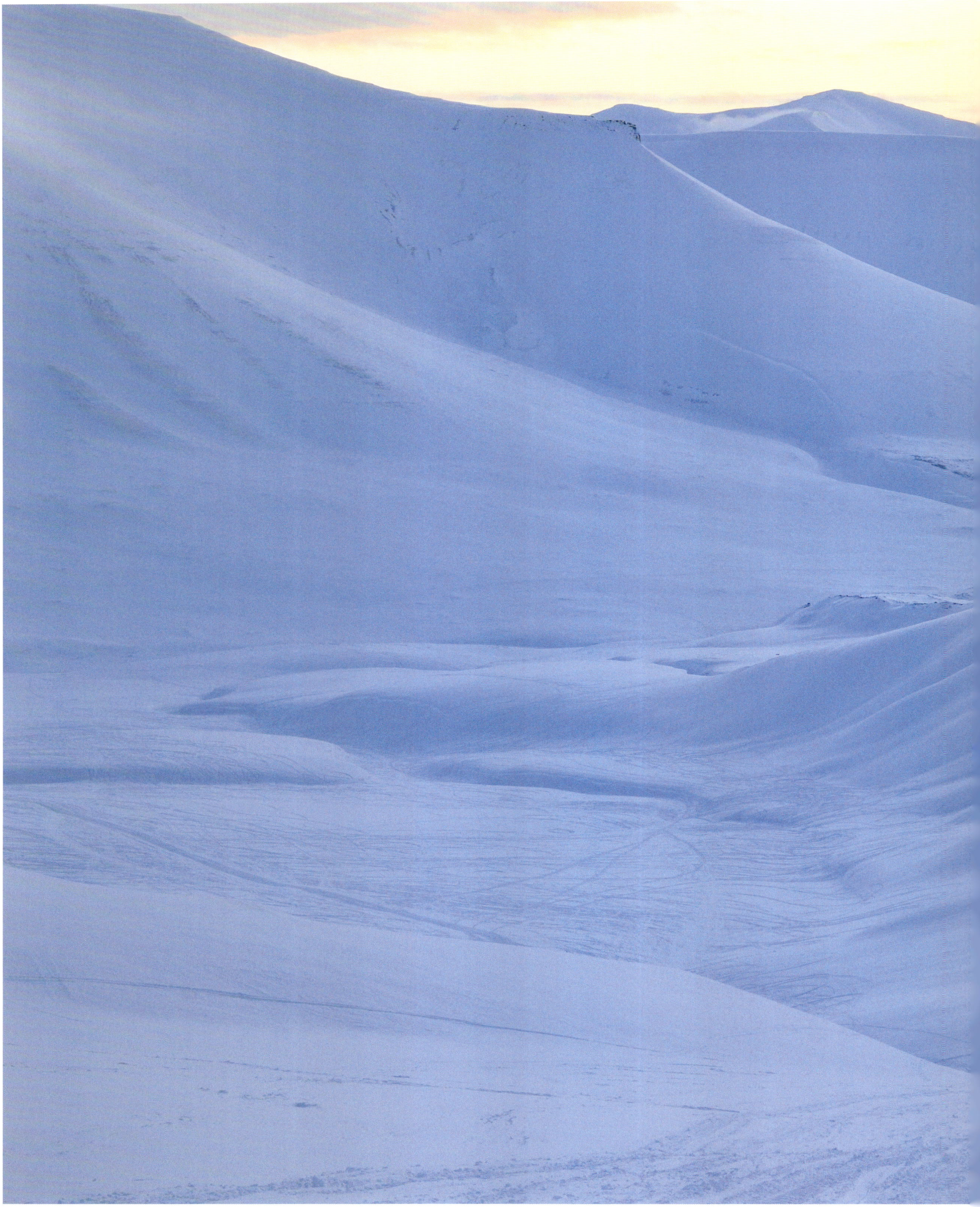

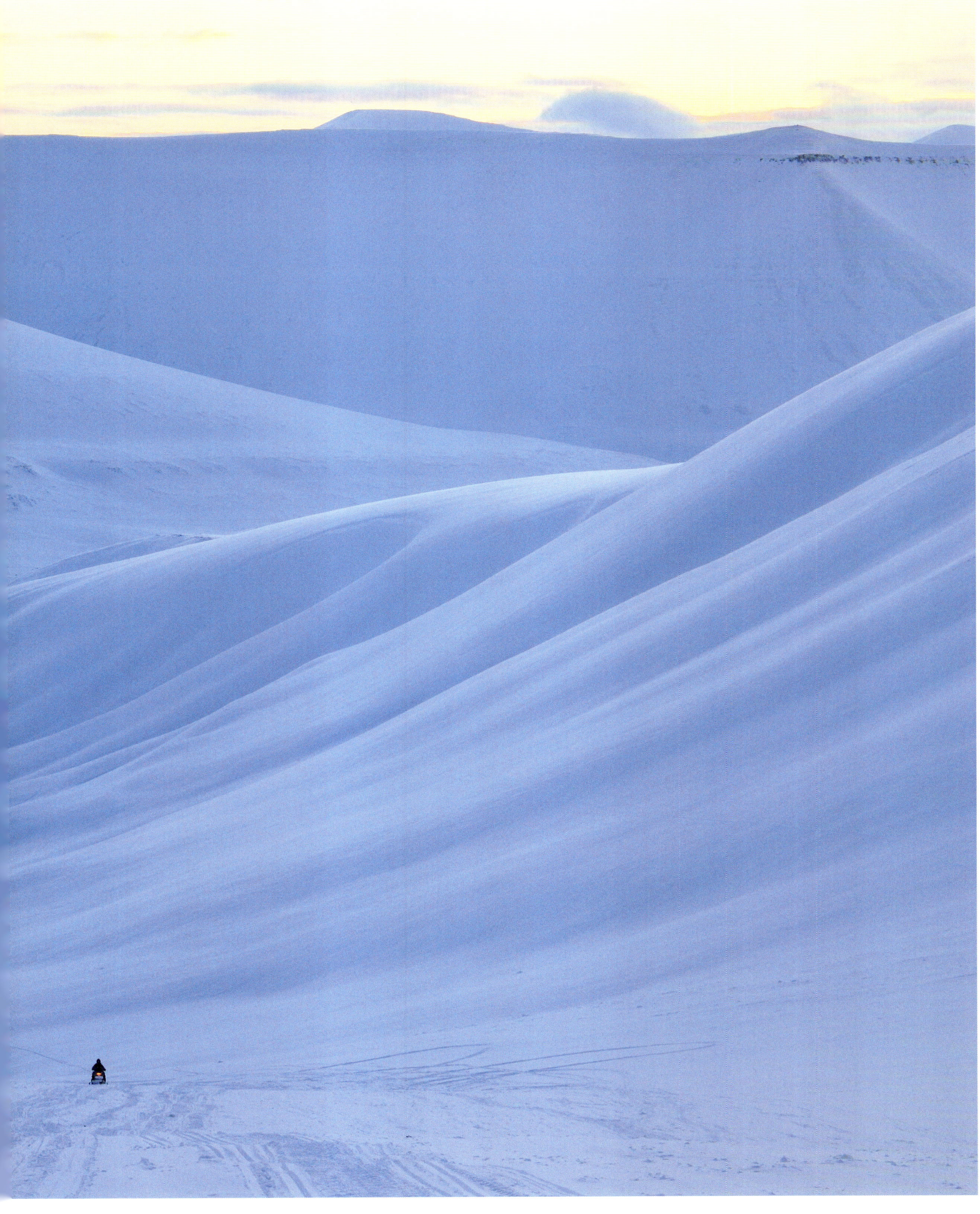

Into the Realm of Arctic Blue

Polar island vibes

Country:	**Norway**
Best time to travel:	**February and March**
Closest airport:	**Longyearbyen, Spitsbergen**
Price tag:	**€€€**
Style of travel:	**An adrenaline-filled quest into the icy heart of Arctic Europe**

Introduction

Somewhere lost in the cold Arctic Ocean, the islands of Svalbard — old Norse for 'Cold Coast' — are situated roughly midway between continental Norway and the North Pole. A few centuries ago, only tough whalers and intrepid fishermen ventured this far north. Even a decade ago, flights on Scandinavian Airlines from Tromsø to Longyearbyen, one of the northernmost towns in the world and the capital of Svalbard, were mostly filled with scientists and workers. The nearly perpetual winter darkness and hungry, free-roaming polar bears kept the rest of the world away. It was the Dutch explorer Willem Barentsz who discovered Spitsbergen in 1596, though at the time he had no clue it was an archipelago and much more than a single island. In 1920, Norway gained sovereignty over Spitsbergen thanks to the Svalbard Treaty granting 40 signatory nations the right to fish, hunt, and mine. Svalbard is an odd duck in Europe: it's tax-free, falls outside the EEA (European Economic Area), the Schengen Zone, and even the Nordic Passport Union. It belongs to Norway but there are Russian settlements (sometimes abandoned); there is a renowned international satellite station, an Arctic university and, to top it off, the largest seed vault in the world. People never leave town unless armed with a gun to prevent fatal polar bear encounters. Temperatures plunge to -46°C (-50°F) and the average snowfall is almost six metres (20 ft.). During winter, the islands are bathed in a diffuse, bluish Arctic light. The highpoint each year is when the sun finally returns after three months of total darkness and its first rays hit the steps of the old church. It's a day of celebrations — indoor and outdoor — an ode to new beginnings, even while temperatures still hover well below -20°C (-4°F). Spitsbergen is a place of extremes, where nature dominates the human lifestyle. Longyearbyen, located on Spitsbergen, Svalbard's main island, was founded as a Norwegian coalmining outpost around the beginning of the 20th century. Long before that, the area was a lawless land frequented by whalers and hunters seeking the pelts of bears and Arctic foxes. Those days are over, but rumour has it that there might be gold on the islands, and the Arctic waters surrounding them are being explored for natural gas. During summer, Longyearbyen resembles a dusty village from the Wild West, but for the rest of the year, it's a winter wonderland often bathed in a soft, pinkish Arctic light. Especially along its main street, which is car-free and lined with colourful wooden A-frame houses, quirky shops, and cosy restaurants. Some 2,200 people live here permanently, many of them employed by the University Centre in Svalbard, which has some 700 students from 50 nations. Others work at the futuristic-looking Svalbard Satellite Station, a joint venture between the Norwegian Space Agency and NASA, and one of the most important satellite stations in the world. Another peculiar detail is the presence of the Global Seed Vault, which lies hidden inside a mountain with views over Longyearbyen's tiny airport. This strikingly designed and massively protected natural freezer was funded by the Bill & Melinda Gates Foundation and has a capacity to store around 4.5 million seed varieties. Even when Spitsbergen is buried in snow, adventurous travellers still find their way to quaint Longyearbyen and then continue further into the white void of Svalbard on snowmobiles and dog sledges. It resembles a fairy-tale portal to a frozen backcountry dotted with abandoned radio stations — one of which is now a guesthouse with sauna — and sailing ships frozen stuck in the Arctic Ocean. There are old coalmining towns, where only polar bears and diehard adventurers survive.

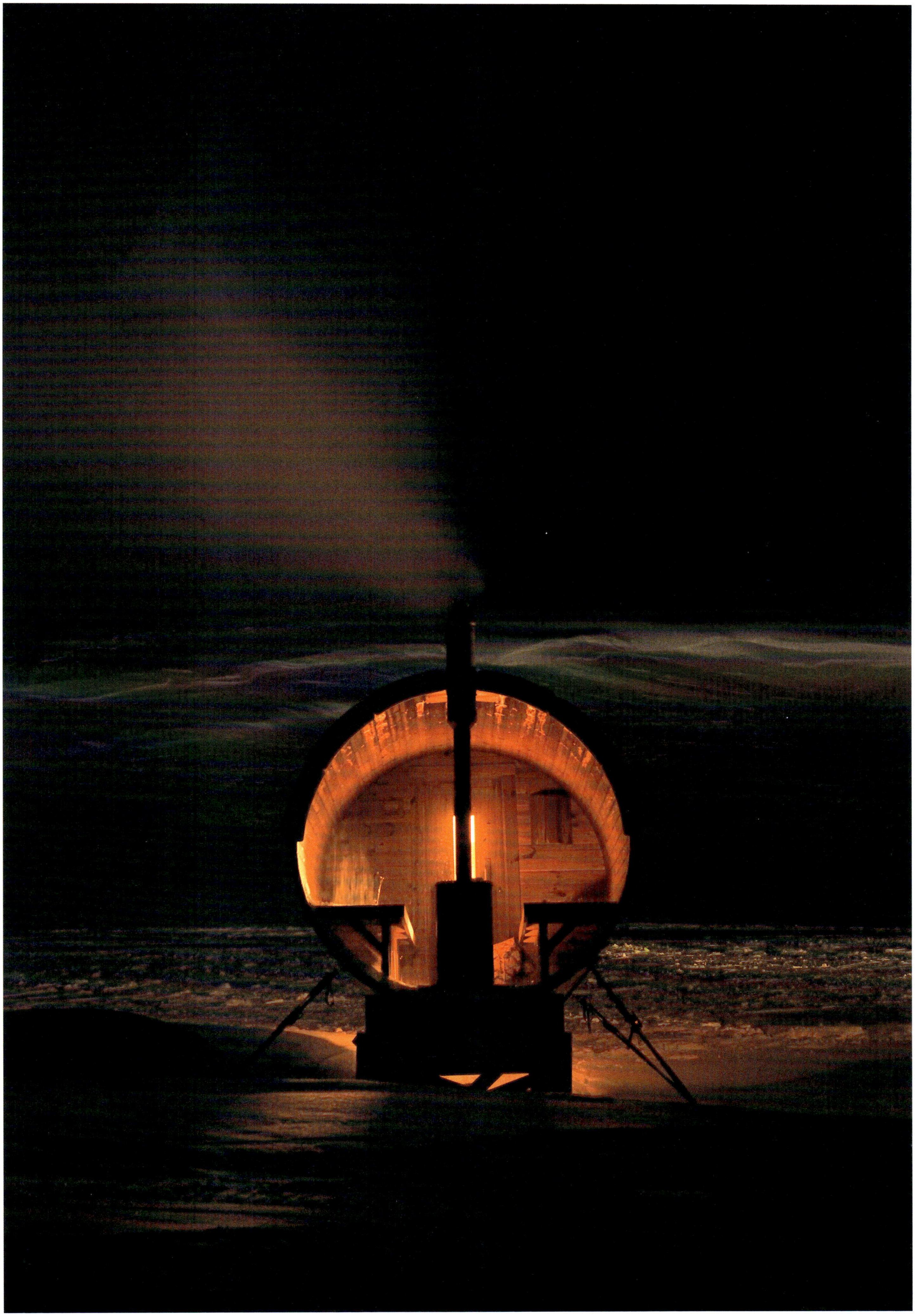

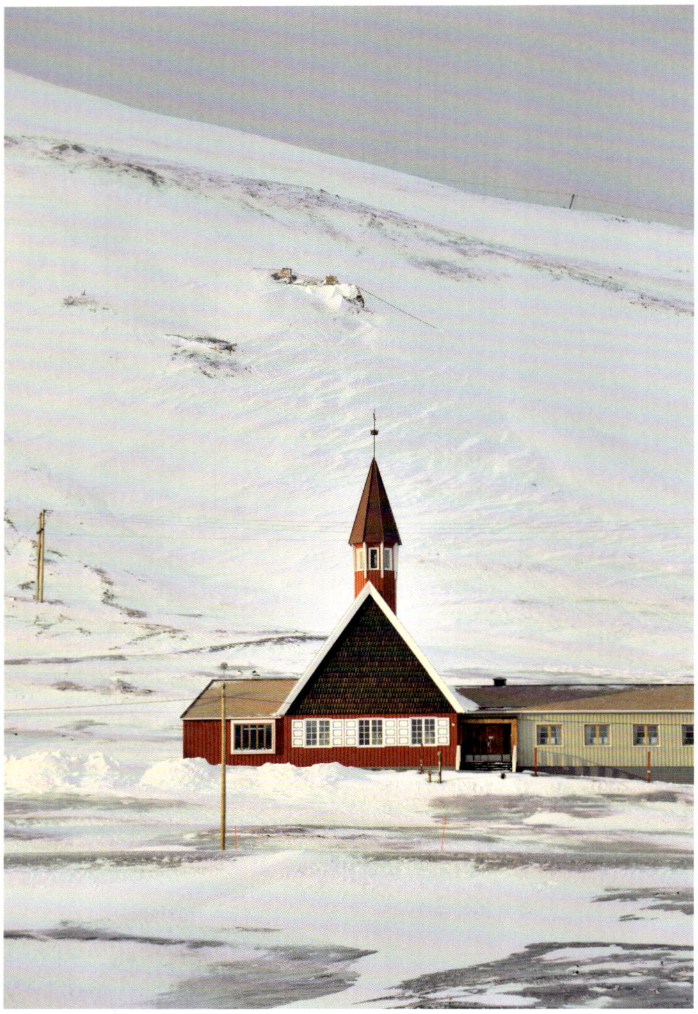

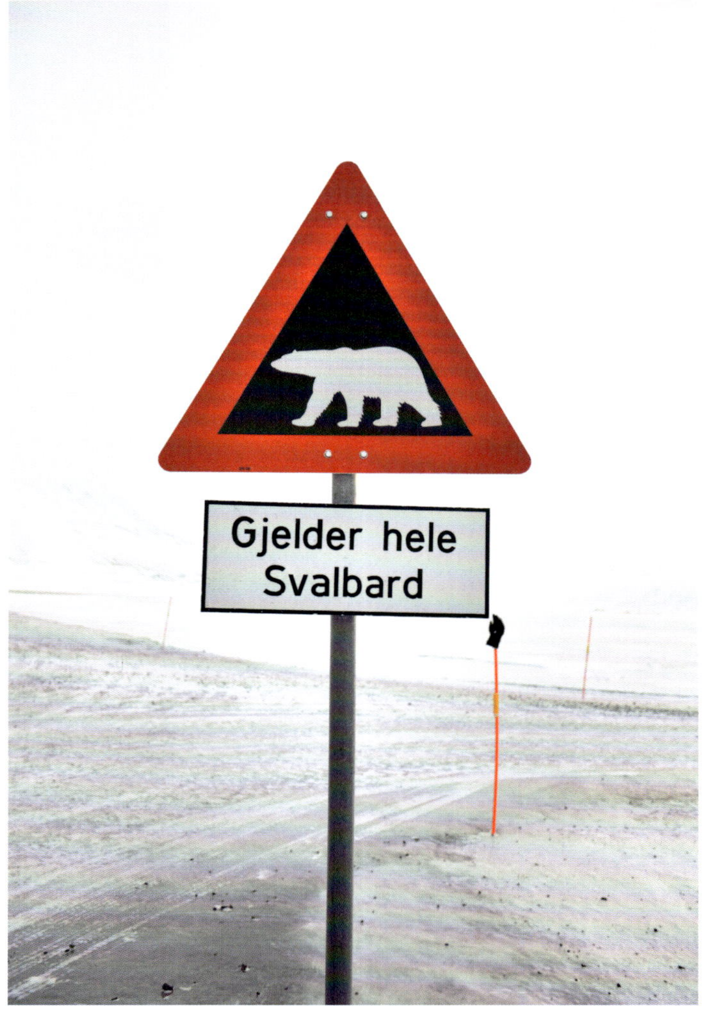
Gjelder hele
Svalbard

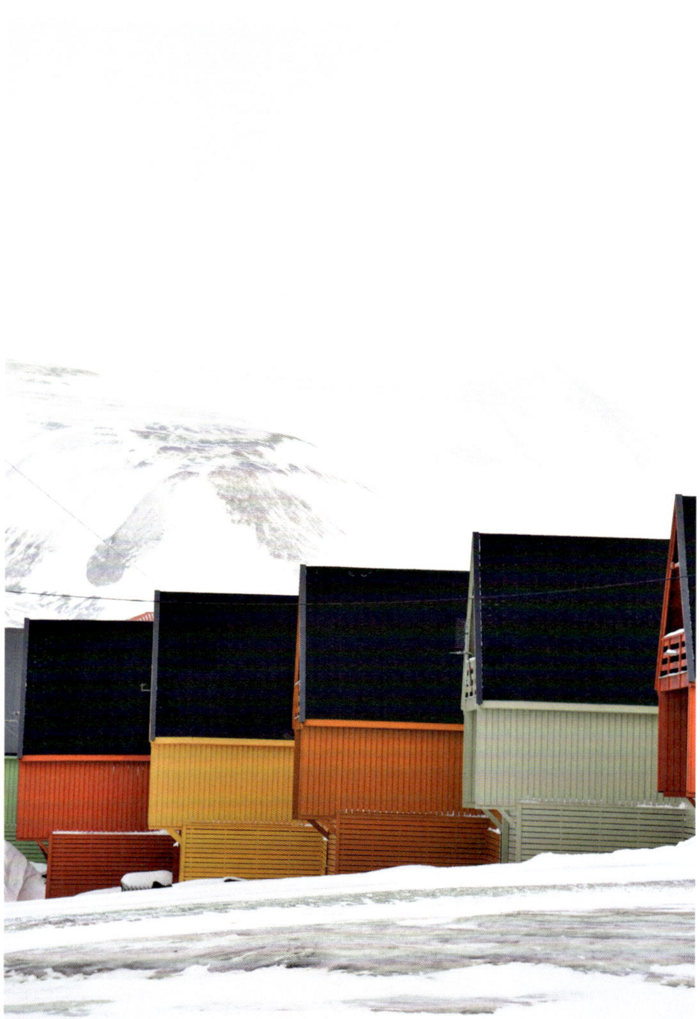

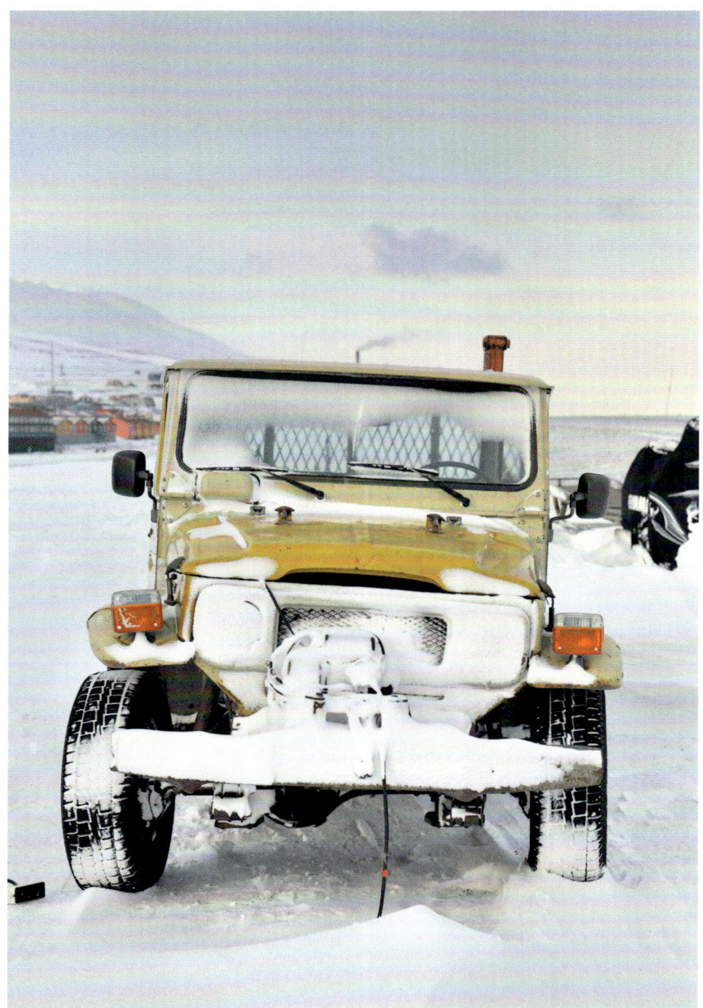

Opposite Charming and quaint, Longyearbyen is the northernmost town on the planet. With 2,200 residents of 50 nationalities, the town is a miniature human mosaic on the edge of the Arctic wilderness. When the wind blows from a certain direction, there are polar bears too, carried by drift ice in the direction of town. Because of these wandering visitors, residents living on the outskirts of Longyearbyen never venture outdoors unarmed. This must be one of the few places on earth with a 'no guns allowed inside' sign at the supermarket entrance.

The Views

When people do venture outside town, they always carry a loaded gun and pack nerves of steel. In this climate the constitution of a Viking comes in handy. Crossing the frozen plains of Spitsbergen, surrounded by a white emptiness that makes hills and valleys almost invisible, is a pure adrenaline rush. Leaving Longyearbyen behind on a sledge pulled by strong Alaskan huskies and diving deep into the wild hinterland of Spitsbergen feels like travelling to another planet. Some trips last a couple of hours and others take several days, with camping in the wild or overnight stays in remote small hotels like the Isfjord Radio Station. Sometimes the destination is Nordenskiold Glacier or the deserted Russian town of Pyramiden, nestled deep within Billefjord, which is frozen for almost half the year and only accessible by snowmobile. The town was once the largest settlement on Svalbard but was abruptly abandoned in 1998 after the coal was almost entirely depleted. On the empty square of Pyramiden, the world's northernmost Lenin statue looks out over frozen Svalbard. The vibe is 'back to the USSR', the only thing missing is the inhabitants. Another Russian settlement is Barentsburg, roughly 55 kilometres (34 mi.) from the capital, a three- to four-hour snowmobile ride, depending on snow, wind, and the presence of bears. The drive is sensational. Snowy hairpin bends, the crackling surface of the frozen Arctic Ocean, and steep, icy hills traversed at breakneck speeds, and then, just as the cold starts to penetrate the layer upon layer of thermal clothing, Barentsburg comes into sight. Soviet murals still adorn dreary apartment blocks and the old coalmine still belches pitch-black smoke, soiling the snow around town. If anyone fancies a deeper dive into Barentsburg's very Russian way of life, they can stay overnight. If not, it's a cold and dark drive back to Longyearbyen. The northern lights might flank the sky and flirt with the stars. 'The lights are a religious experience,' people say here.

Creature Comforts

Despite the rise of tourism, Svalbard's offerings remain limited, which is probably not a bad thing.
The conventional Radisson Blu Polar Hotel offers a hearty breakfast and cosy bar. With any luck, the northern lights can be enjoyed from the restaurant while puffin carpaccio is served. The newest and best place in town is Funken Lodge, located in an old red building above town where coalminers once stayed. In early 2018, the lodge was completely renovated and now offers comfy rooms and a Scandi-style restaurant and bar. Locals eat out at Huset, a sophisticated restaurant, or enjoy a pint at the bar of the Basecamp Hotel—also a good place to bed down — on the main street. Further afield, there are quaint places to sleep, such as the Basecamp Isfjord Radio Station — which, until a few years ago, was an important telecommunications link between the Norwegian mainland and Svalbard's inhabitants — or the Juva Cabin, which has a stunning tube-shaped sauna next door.

Go Now

Most visitors arrive with the sole intention of briefly experiencing the extreme Arctic landscape, only to conclude that they would rather have stayed longer, because despite its painful temperatures, an evening stroll through Longyearbyen's streets, with their colourful, warmly lit houses, is the kind of Scandinavian, or even better, Arctic charm any traveller can easily get used to.

How to Go

Easy, short SAS flights connect Longyearbyen to Oslo and Tromsø. Spitsbergen is open all year round, with a tourist season that kicks off when the sun returns mid-March. However, a better time to visit, especially for photography, is the 'blue period' just before the tourist season, when hotels are cheaper, the days are marked by a beautiful bluish aurora, and the northern lights are on display almost every night.

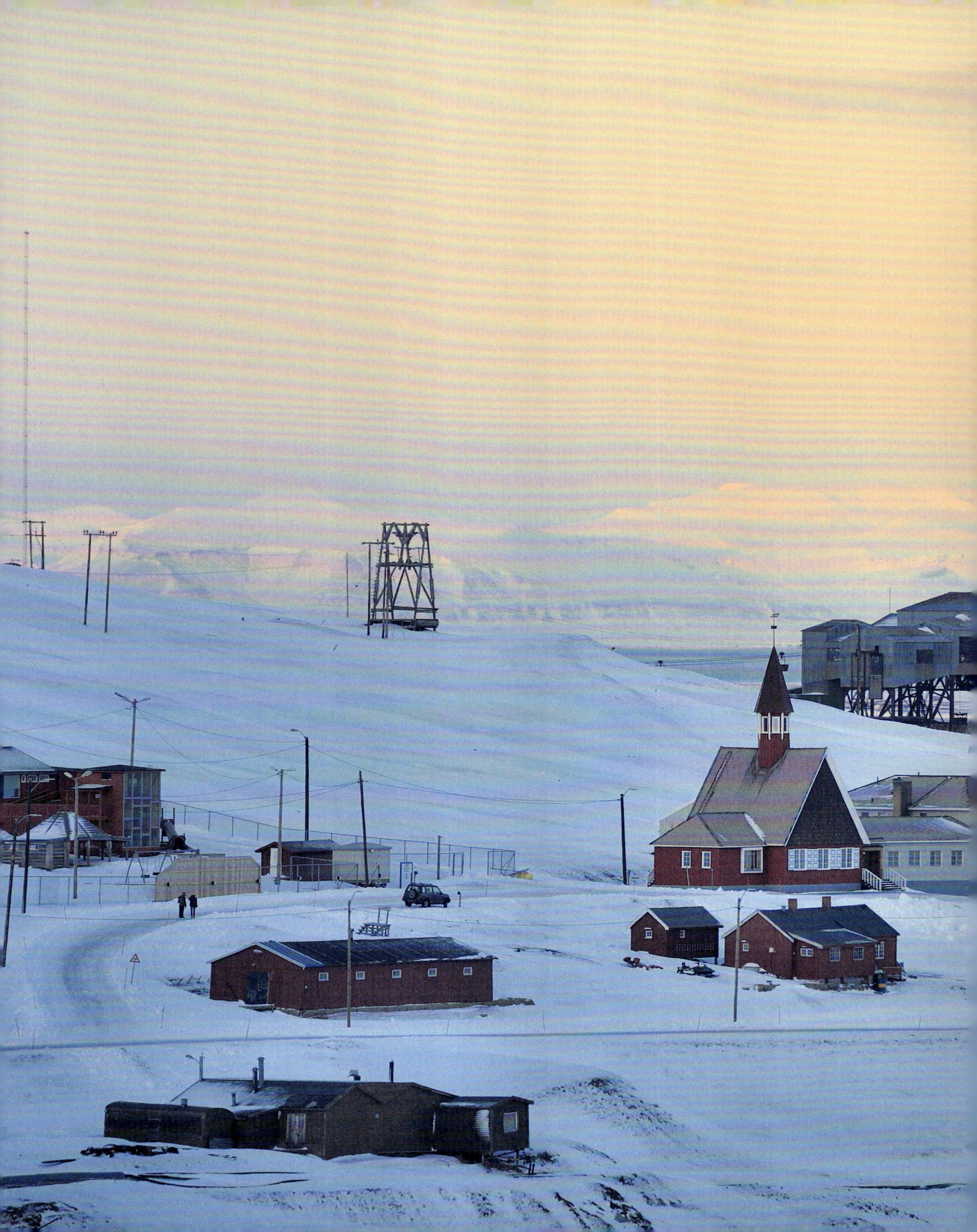

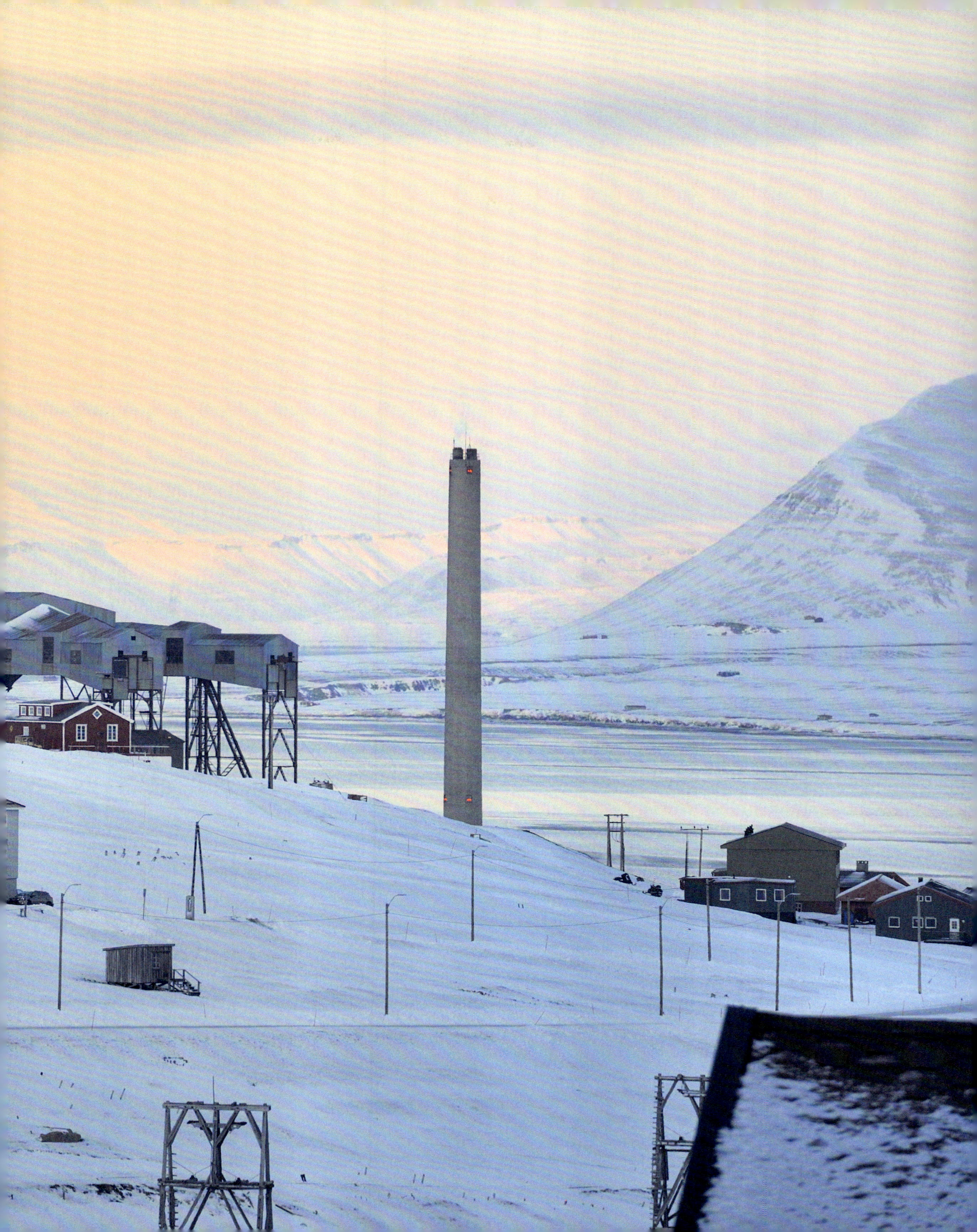

Previous spread Although the oil-rich Norwegian archipelago of Svalbard has been targeted for oil production, for now at least, Spitsbergen — the only permanently inhabited island in the chain — remains largely deserted, a tranquil paradise hidden inside Europe's freezer. While the Arctic tourism boom is being felt in neighbours such as Lapland and Iceland, Svalbard still offers an off-the-radar location for outdoor action, snowmobiling, dog sledding, and watching the northern lights, all made the more pleasant in February and early March when the islands are bathed in the glow of that quintessential bluish Arctic light.

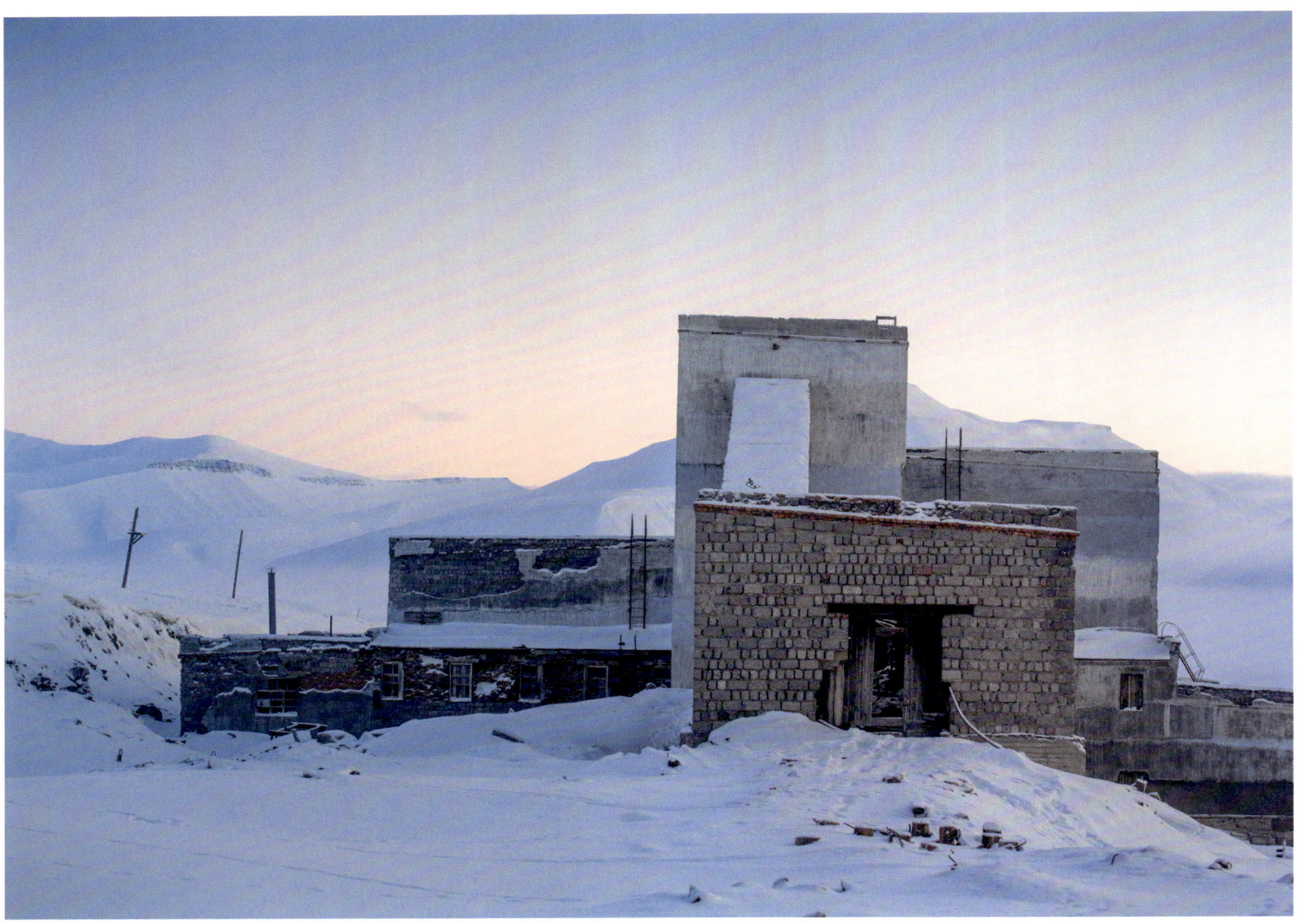

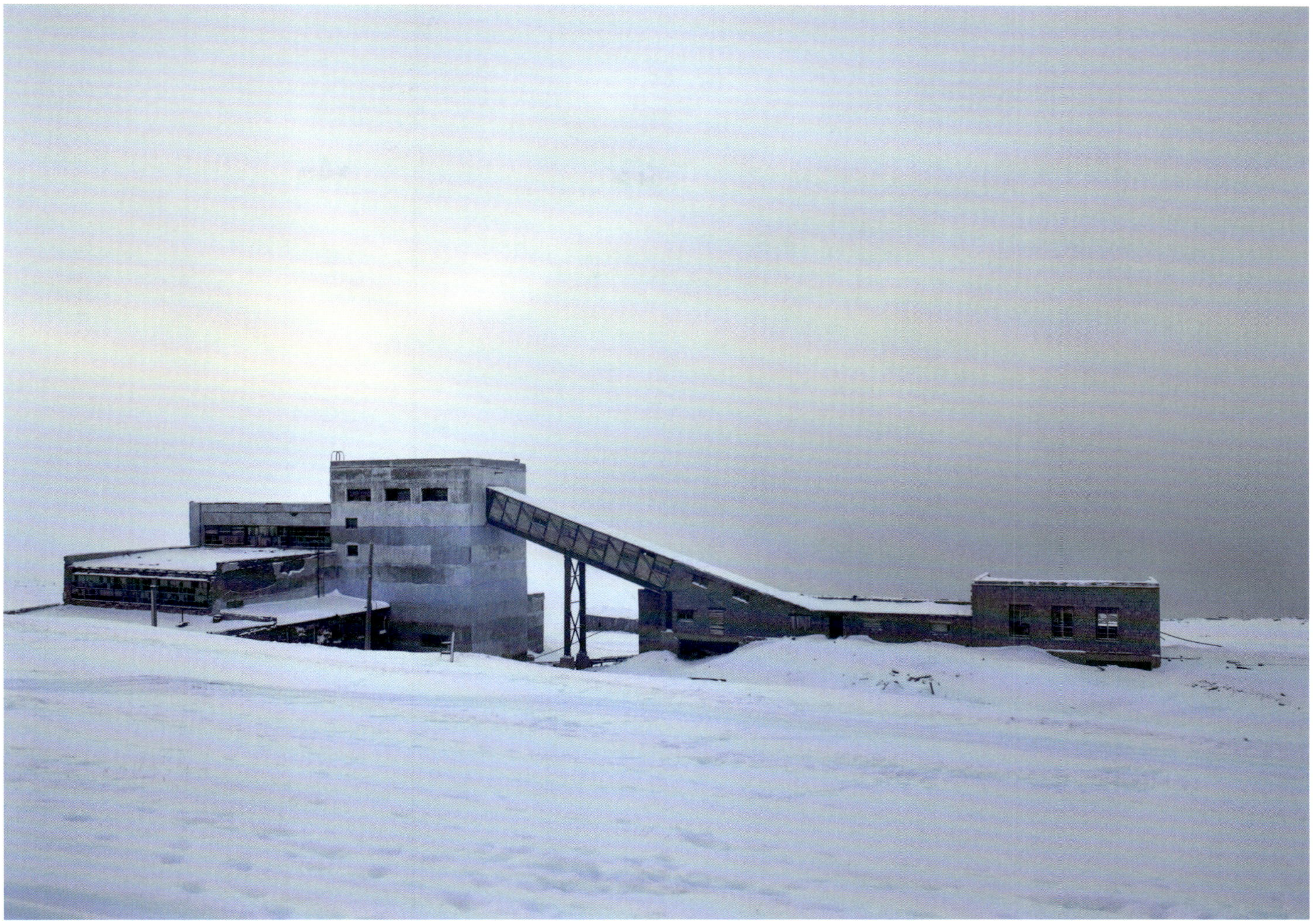

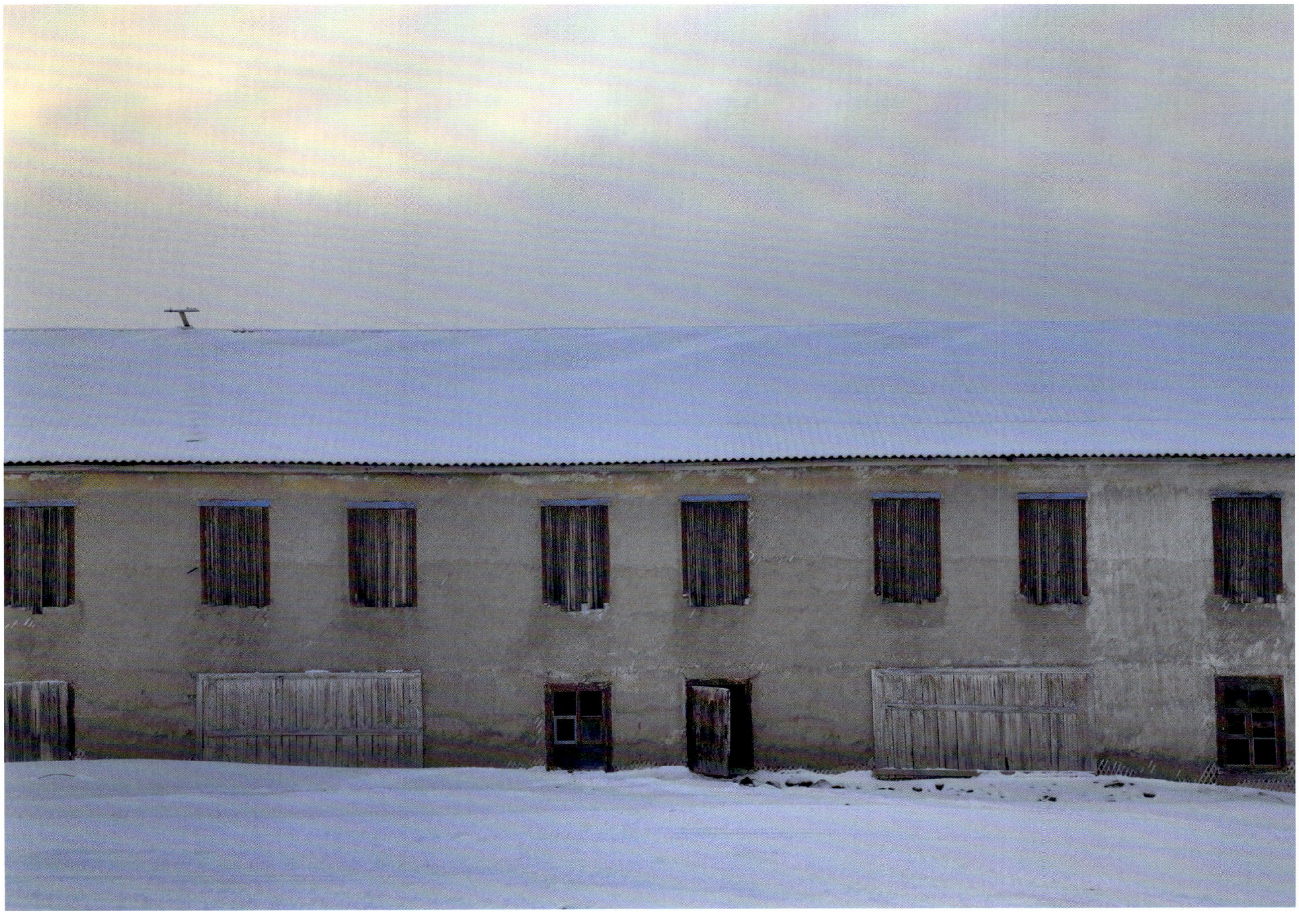

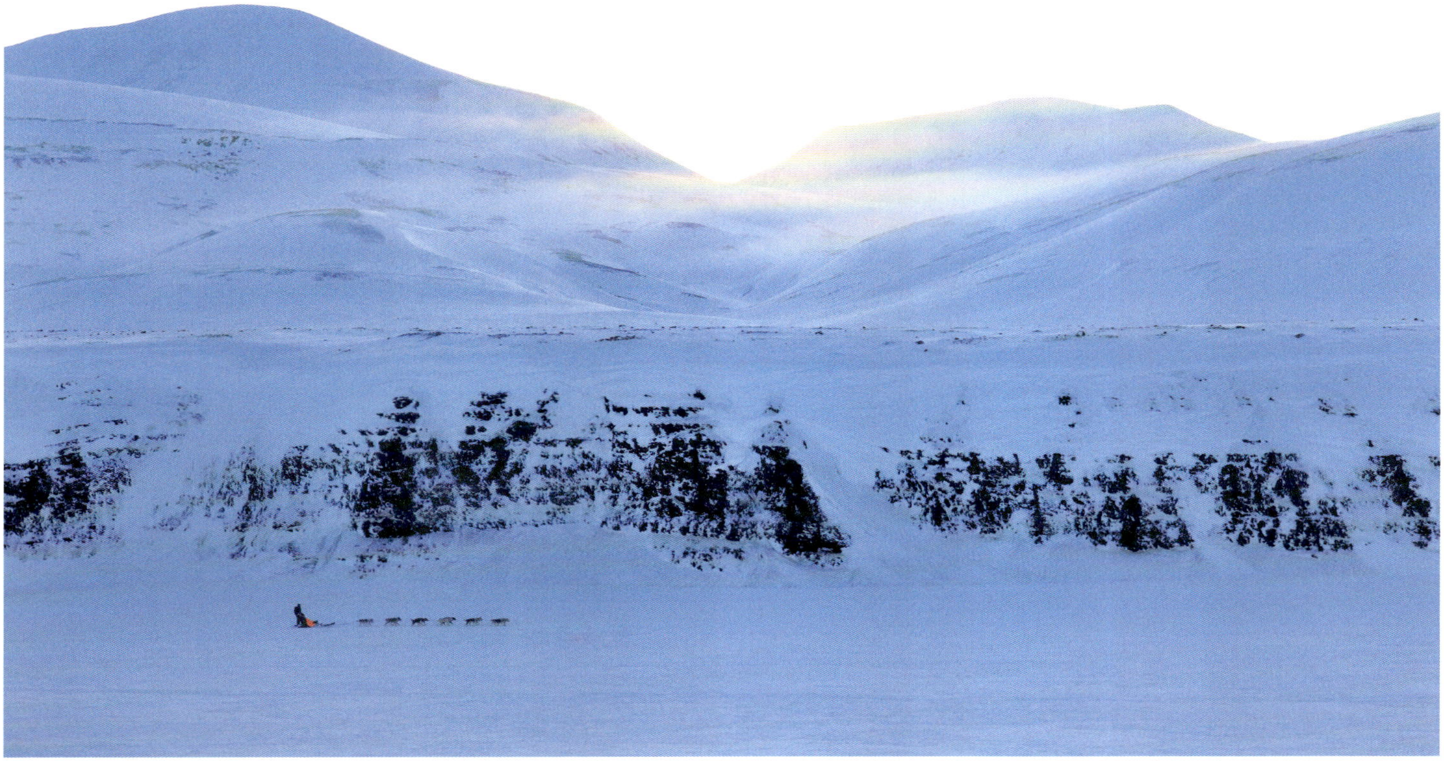

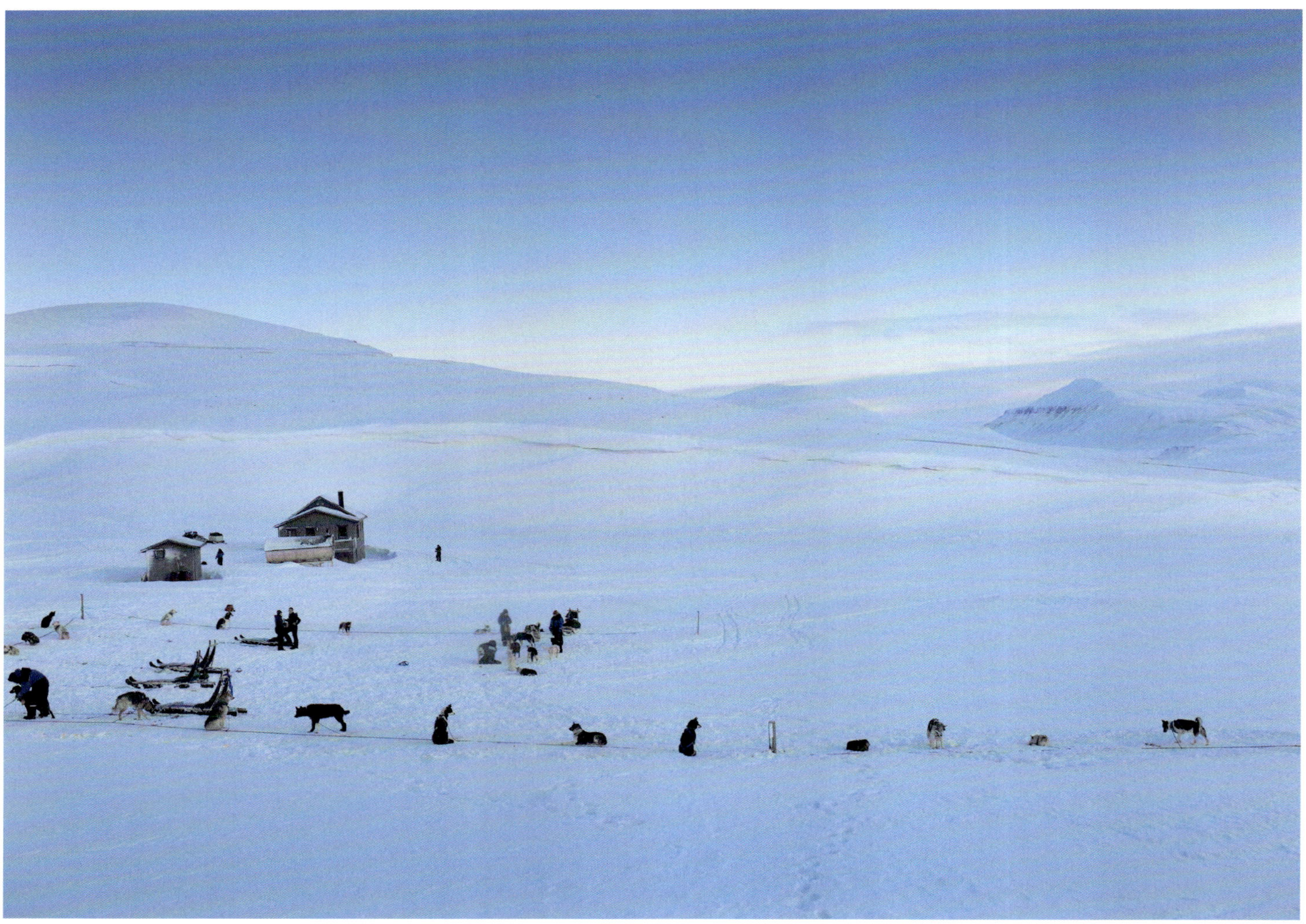

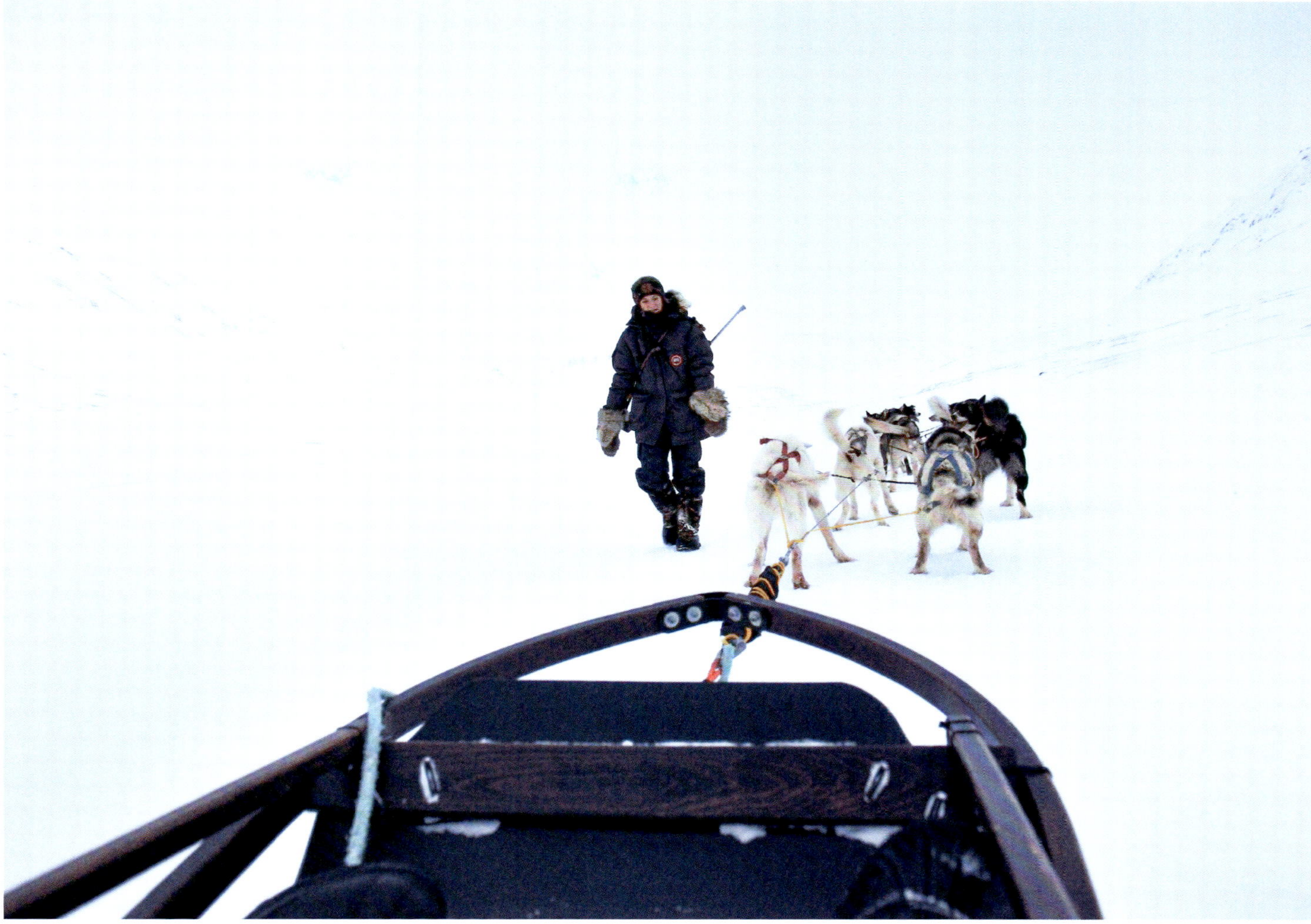

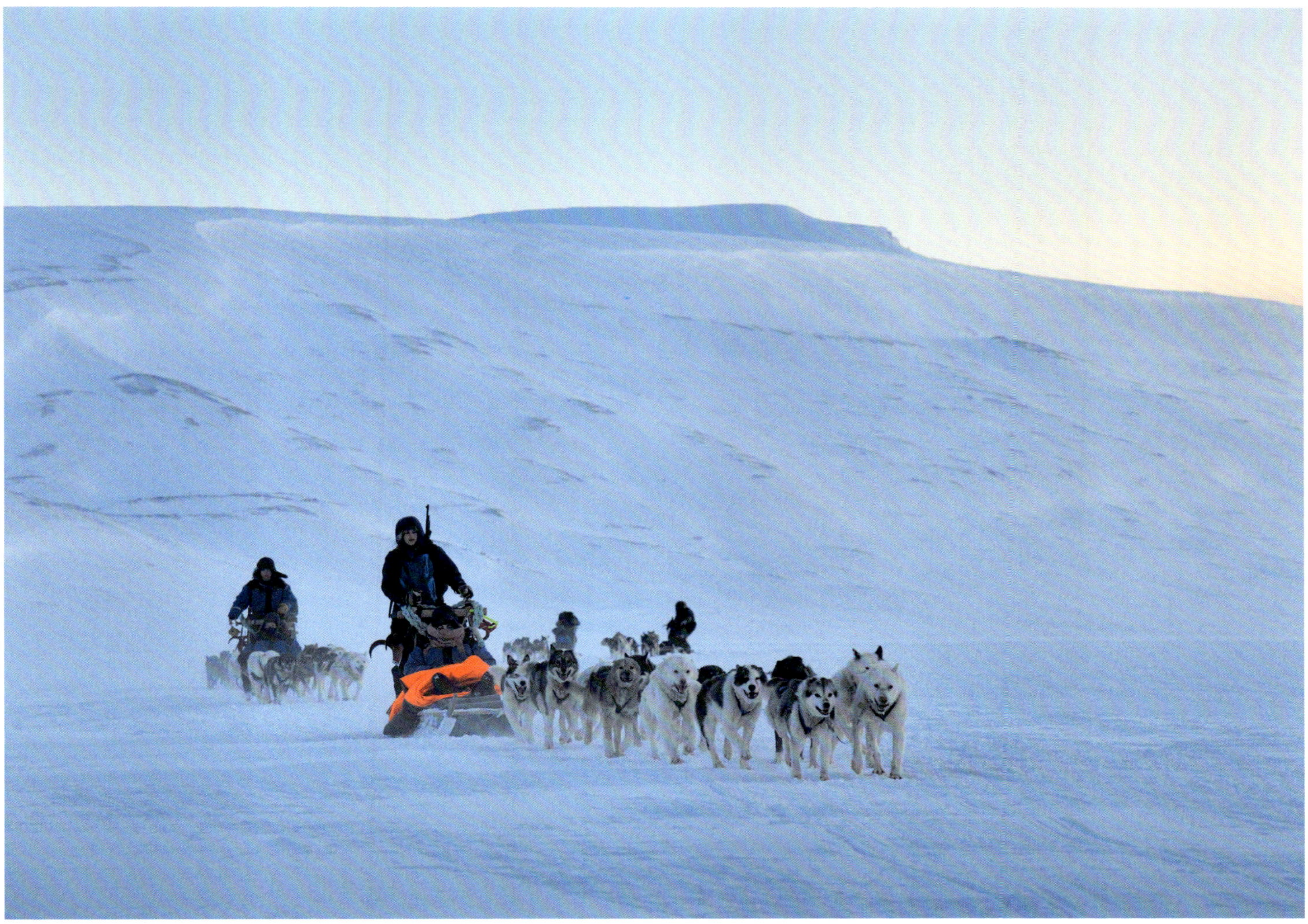

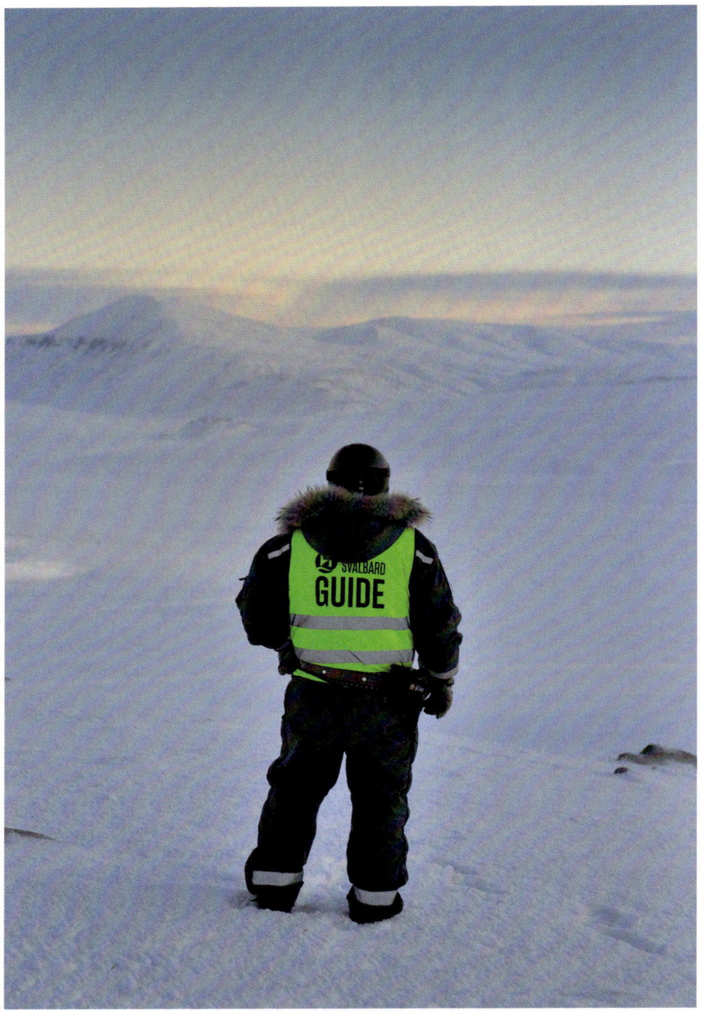
SVALBARD
GUIDE

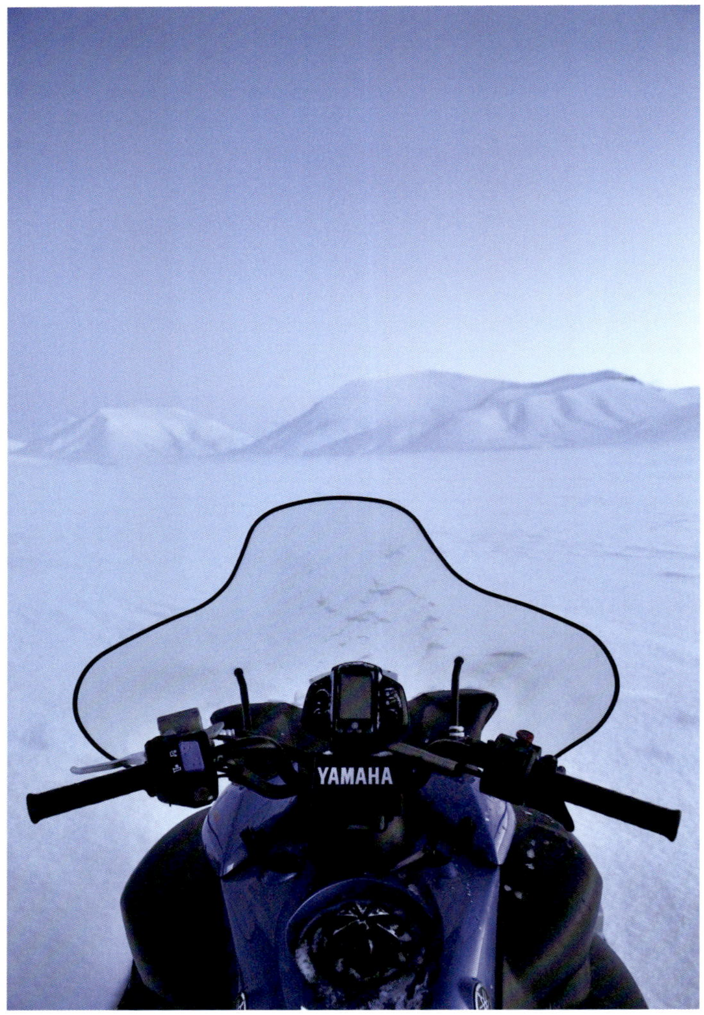
YAMAHA

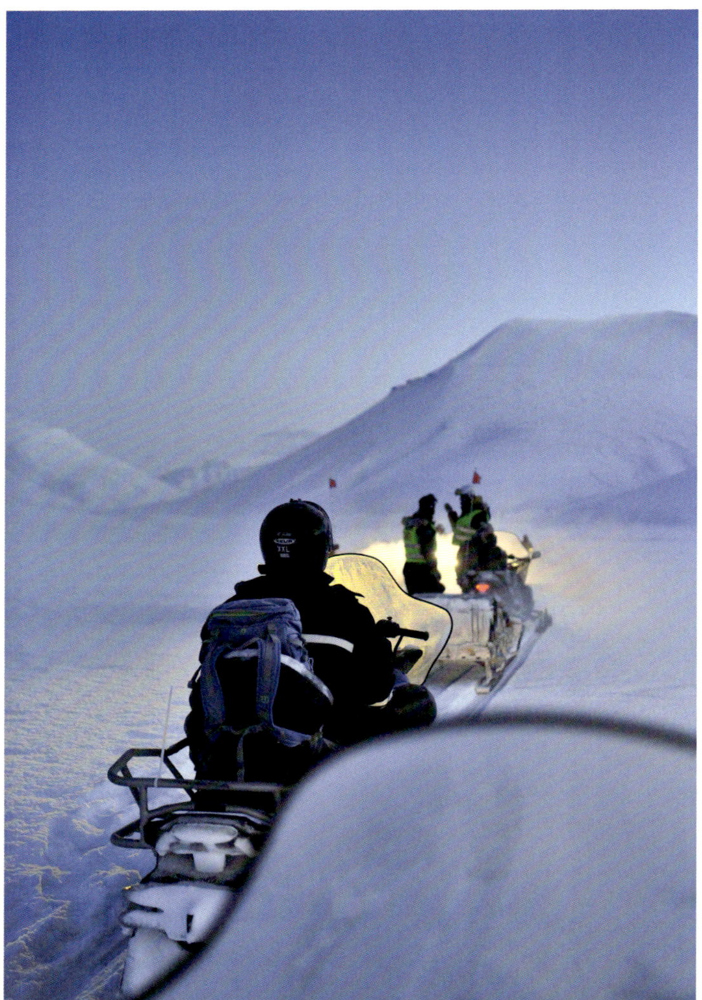

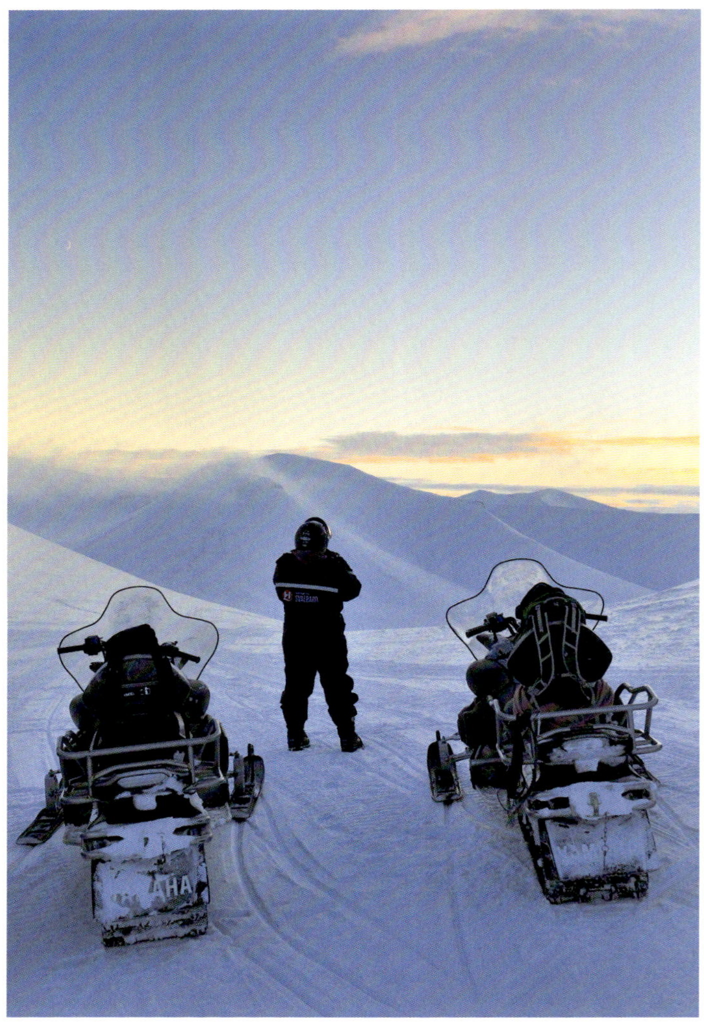

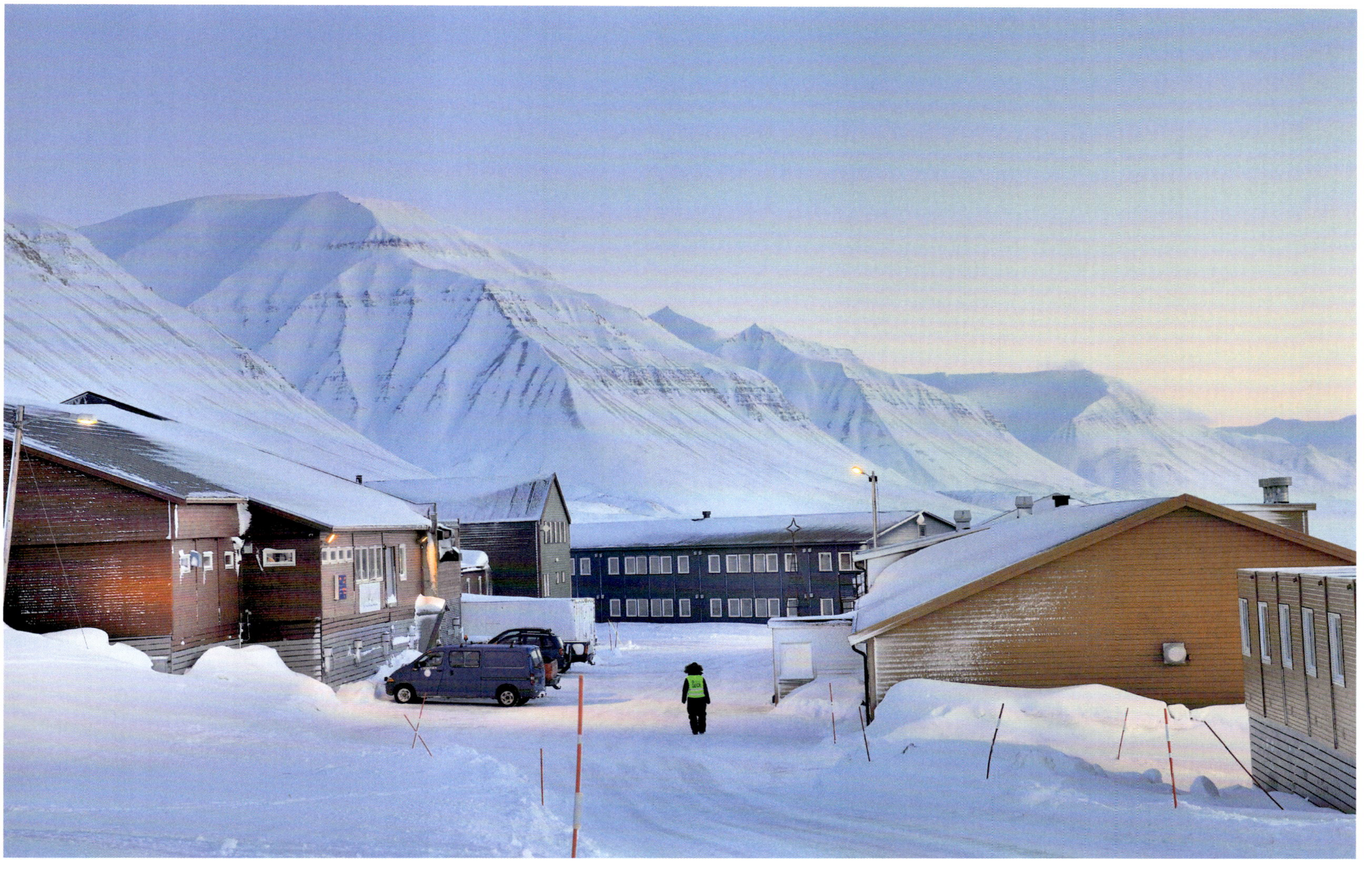

Above Sveagruva — Swedish for mine — is one of the old mining settlements, lying at the head of the Van Mijenfjord, about 60 km (37 mi.) southeast of Longyearbyen. In the past there was a small airport and port but now the only way to get to Svea is on snowmobile through a majestic landscape of glaciers, vast valleys, and frozen tundra. The whole town and the coal-mining business are being dismantled and cleaned up by the Norwegian government so only a handful of 'historic' buildings will remain.

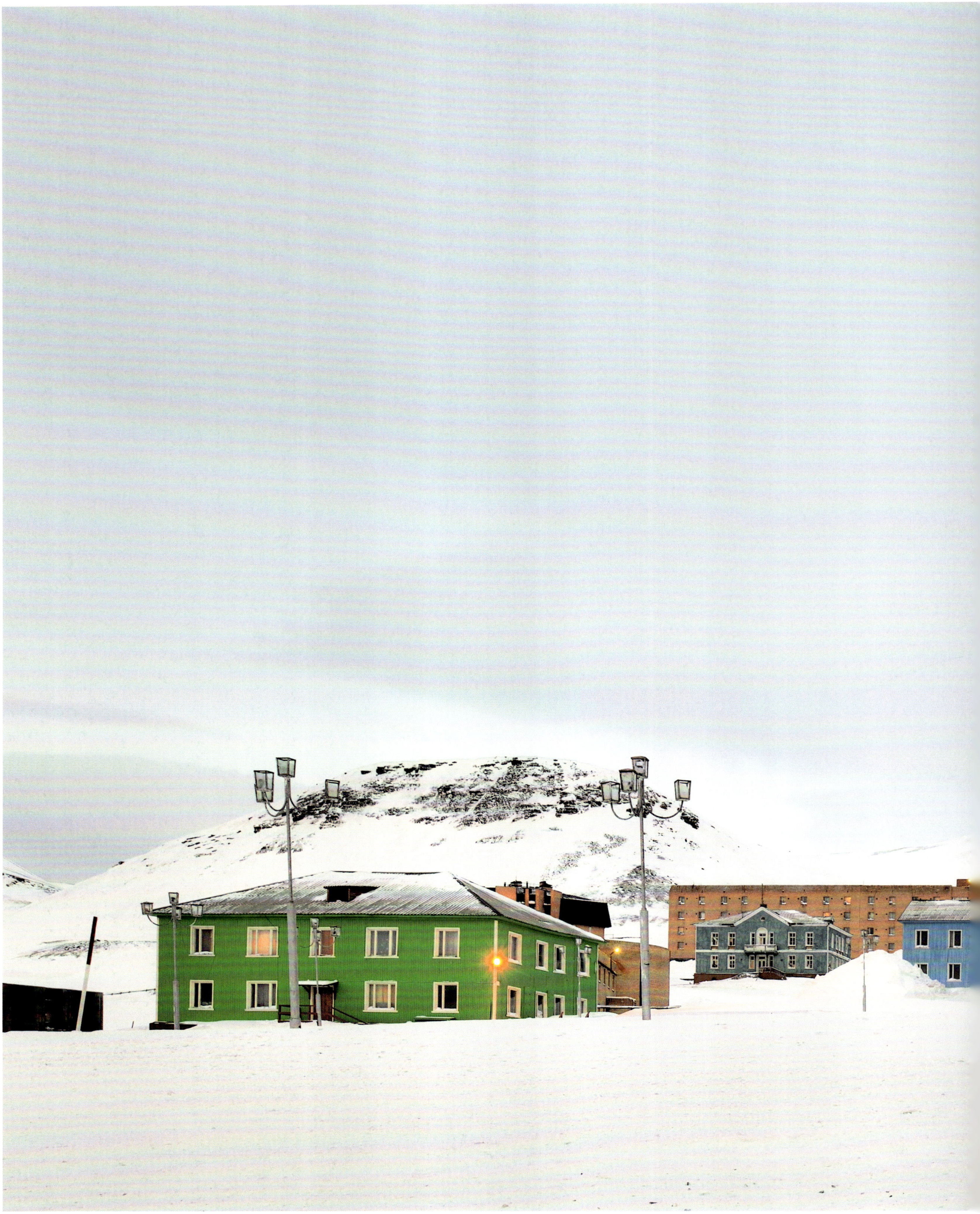

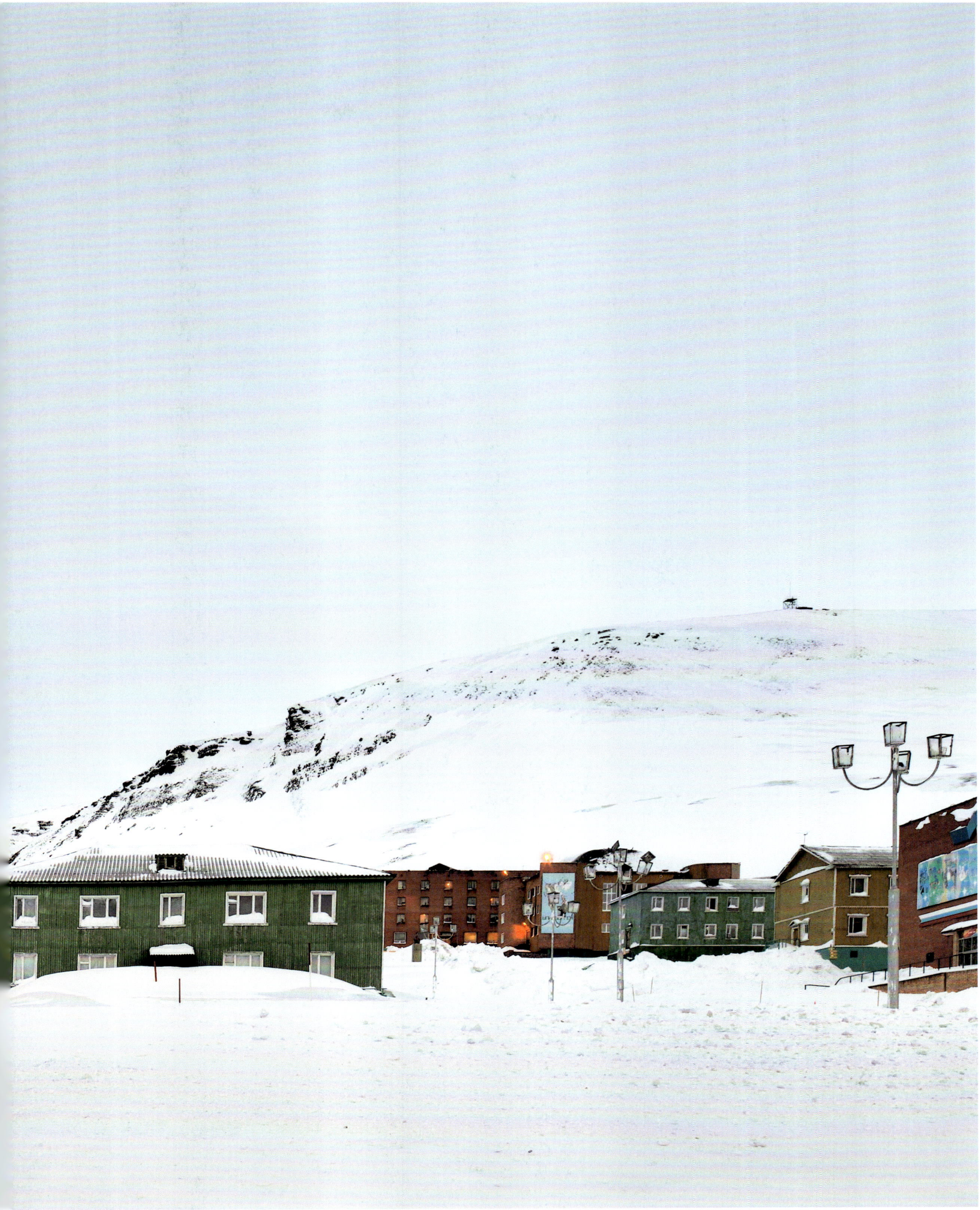

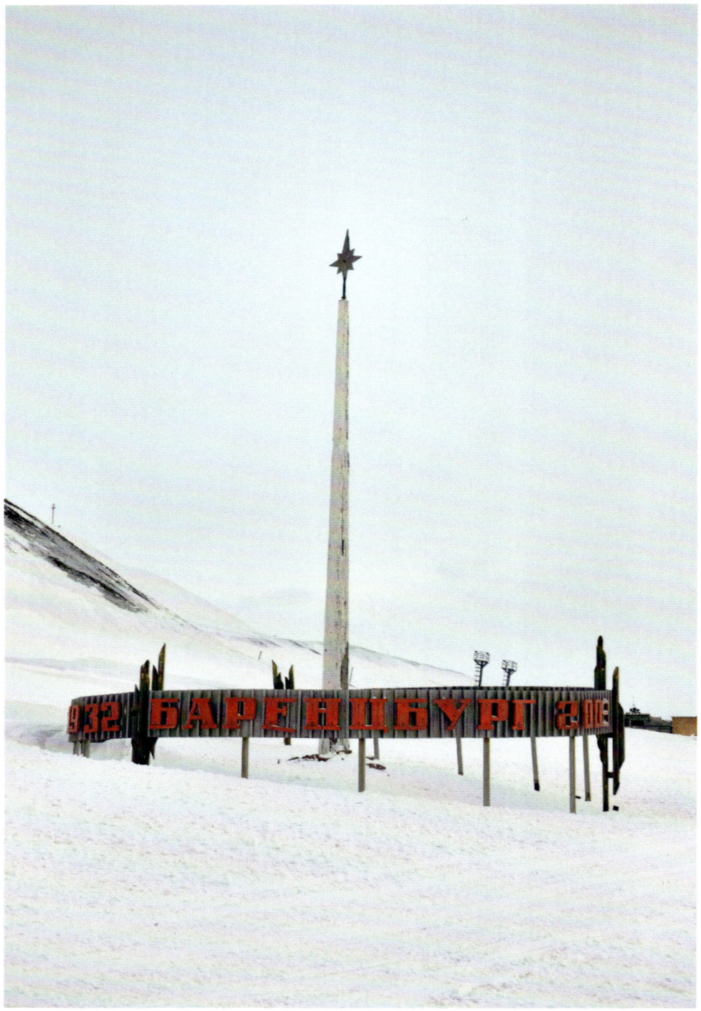
БАРЕНЦБУРГ
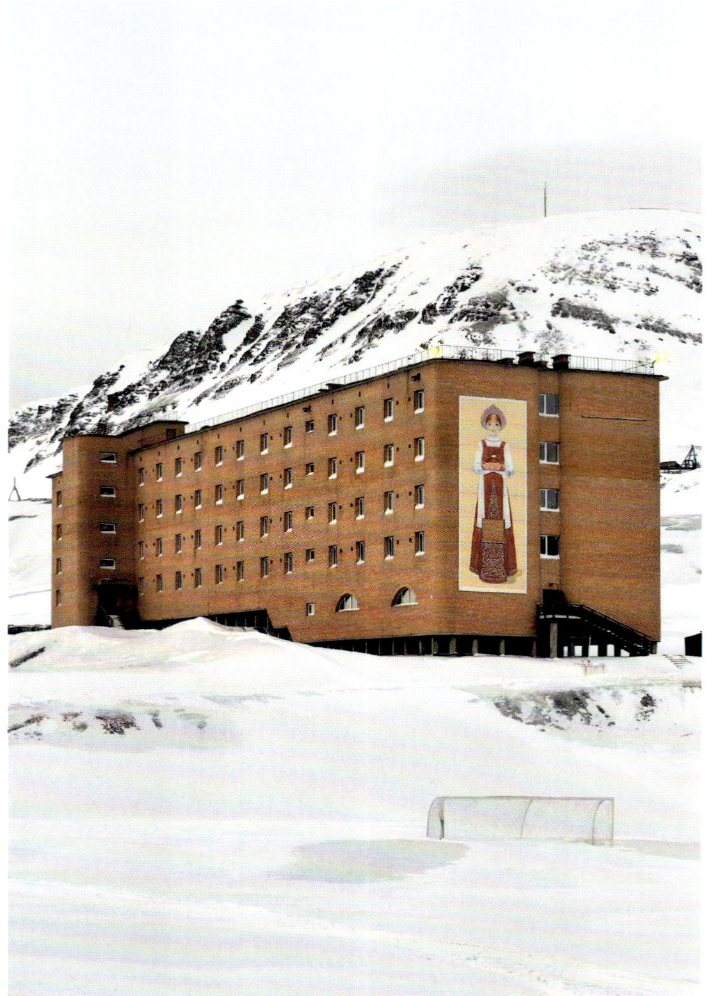
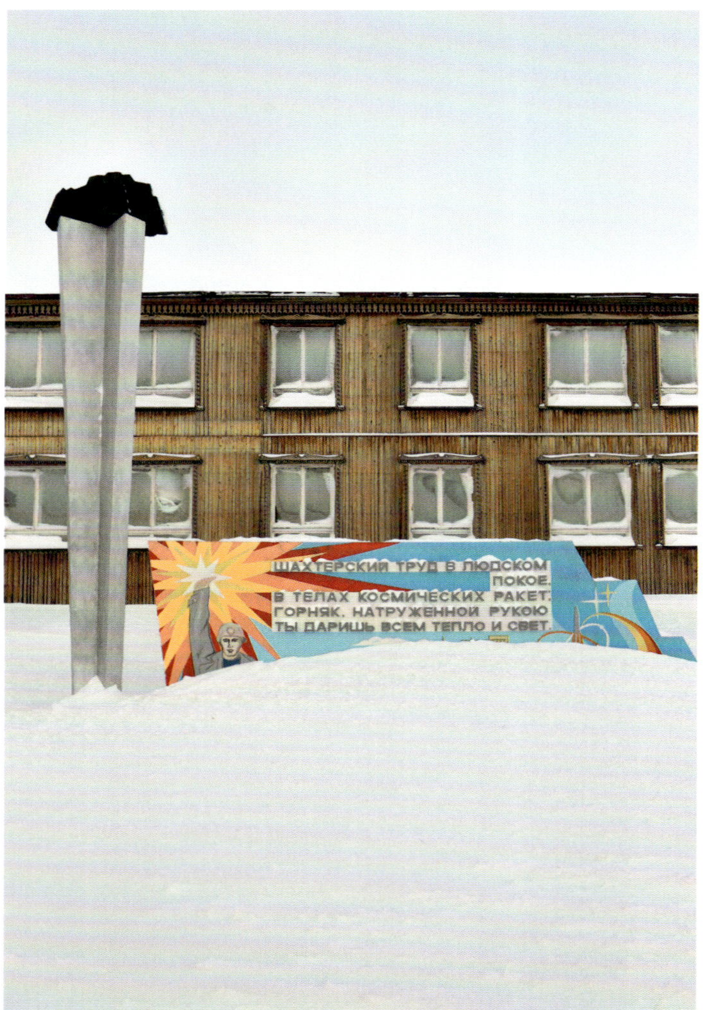
ШАХТЕРСКИЙ ТРУД В ЛЮДСКОМ
ПОКОЕ.
В ТЕЛАХ КОСМИЧЕСКИХ РАКЕТ.
ГОРНЯК. НАТРУЖЕННОЙ РУКОЮ
ТЫ ДАРИШЬ ВСЕМ ТЕПЛО И СВЕТ.
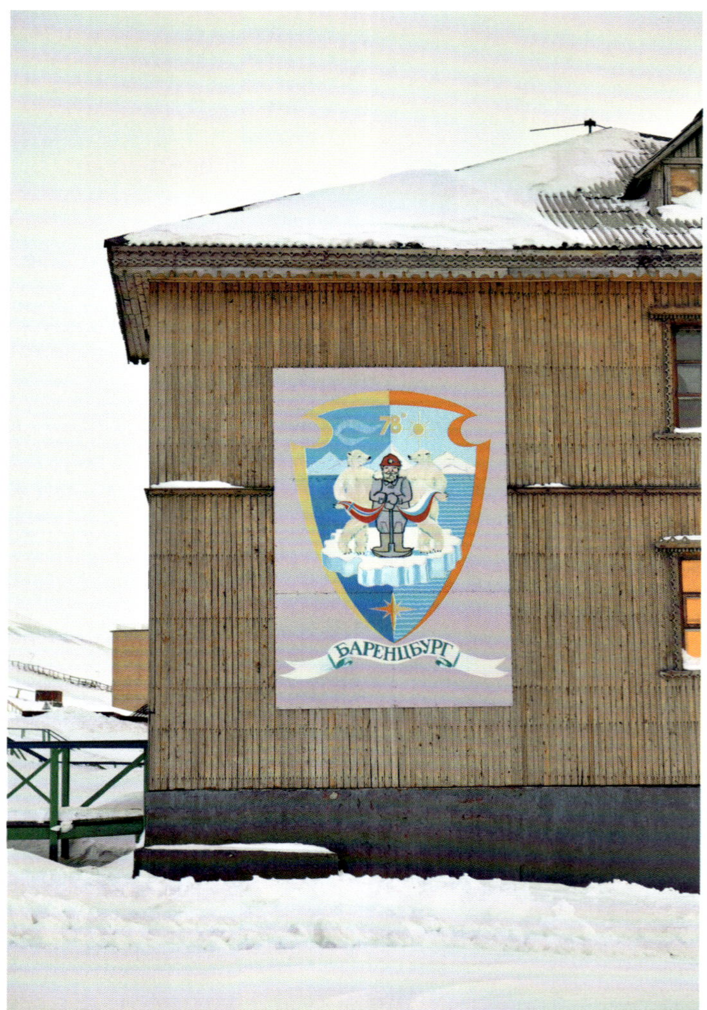
БАРЕНЦБУРГ

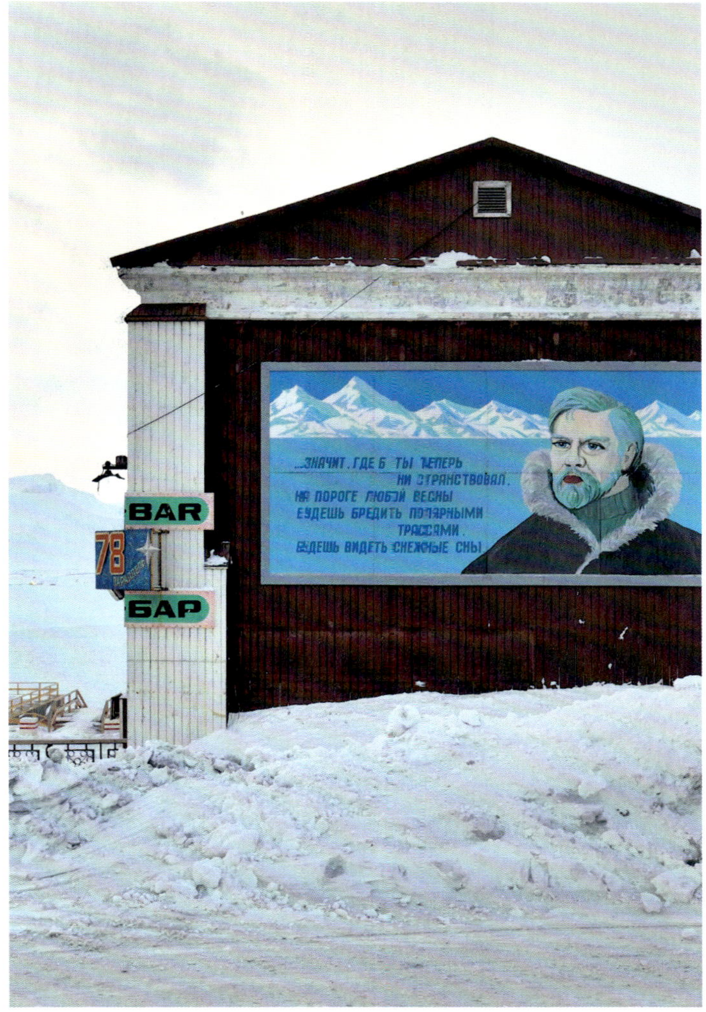

This spread Svalbard was once internationally recognised as terra nullius, or no-man's land. The 1920 Svalbard Treaty placed the archipelago under Norwegian sovereignty while giving interested parties such as Russia the right to exploit its economic resources, on condition that the island would never be militarised. One of these Russian settlements is Barentsburg, with around 500 permanent Russian residents. This town is the second-largest settlement on Svalbard and is seen as Russia's Trojan horse in Europe. Purchased from the Netherlands to extract coal, Barentsburg is primitive and ghostly during its long winters, but as the Arctic Ocean becomes increasingly ice-free, the access to Europe that it offers Russia might bring considerable benefits. Despite its eerie looks, the village has a proper hotel with a restaurant and bar where travellers can bed down after a day-long snowmobile trip out of Longyearbyen. Warming up with shots of vodka and endless cups of borscht is compulsory.

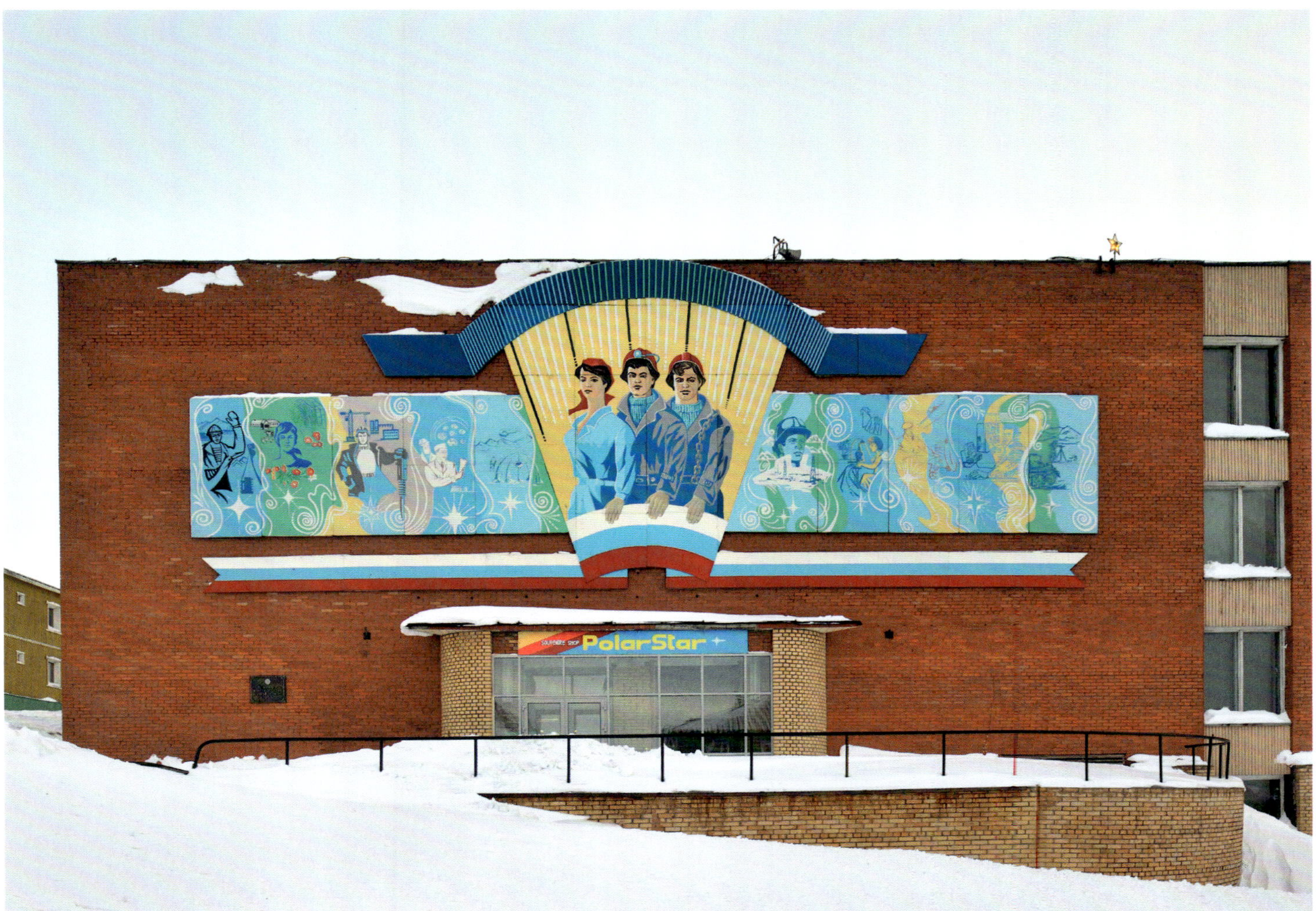

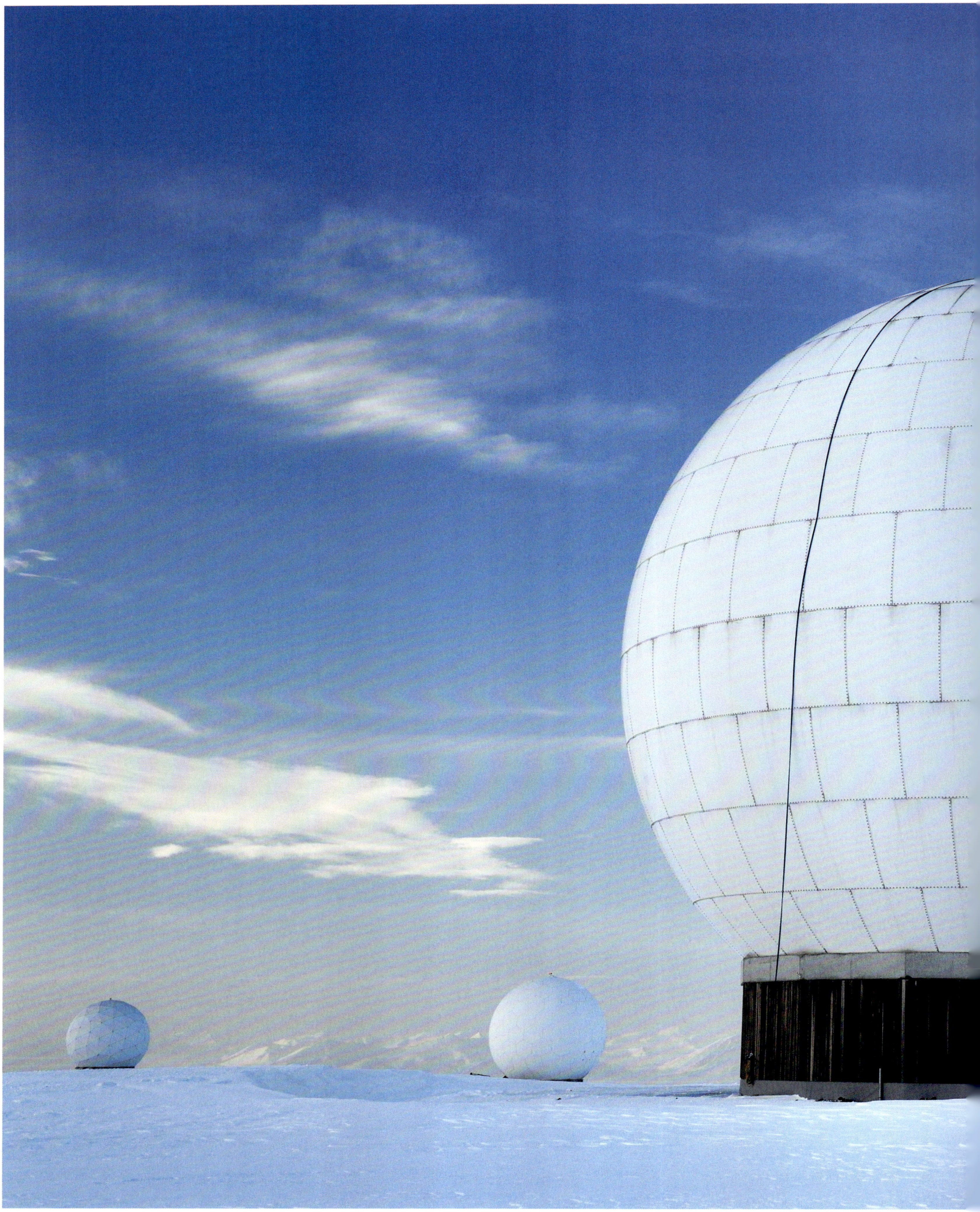

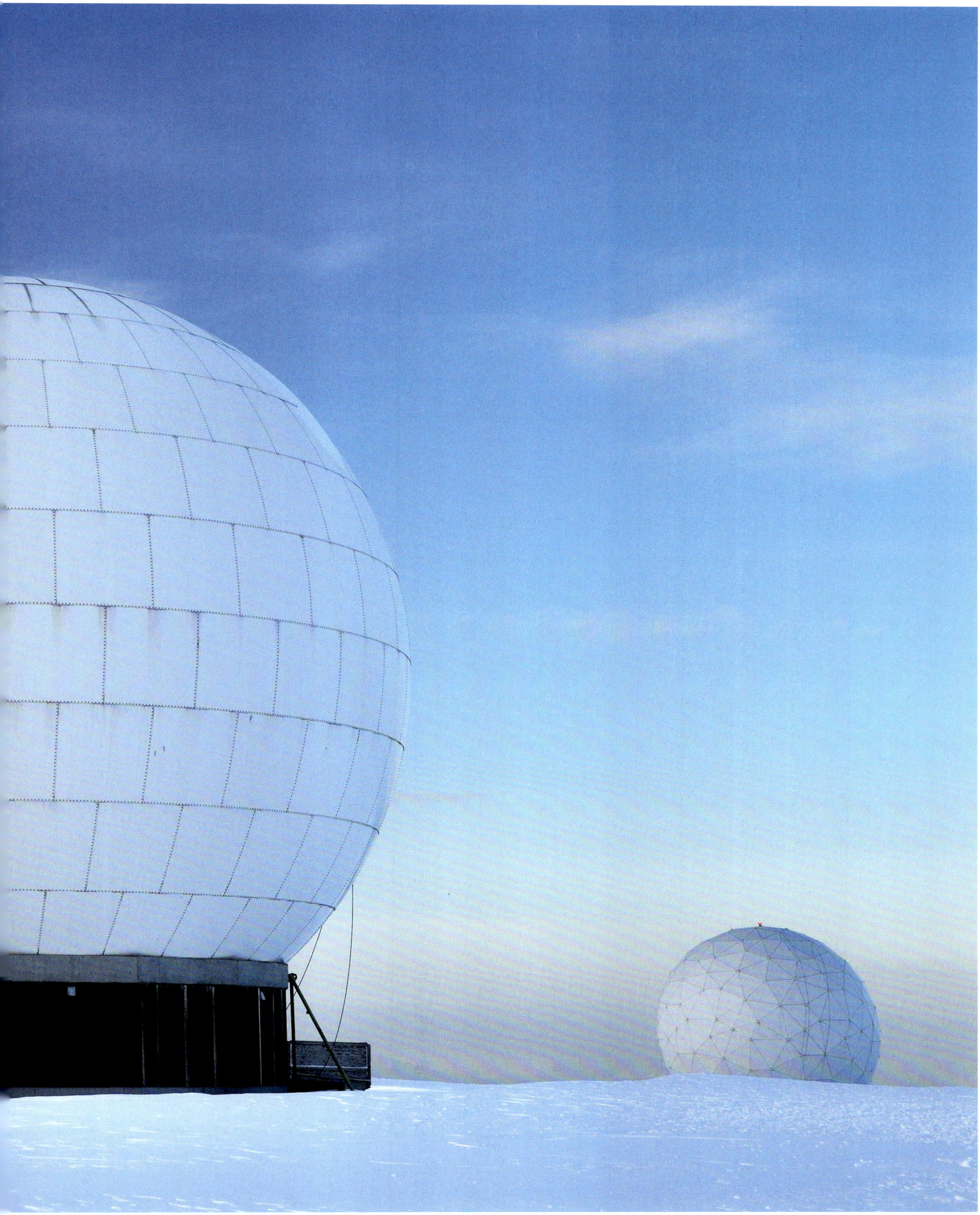

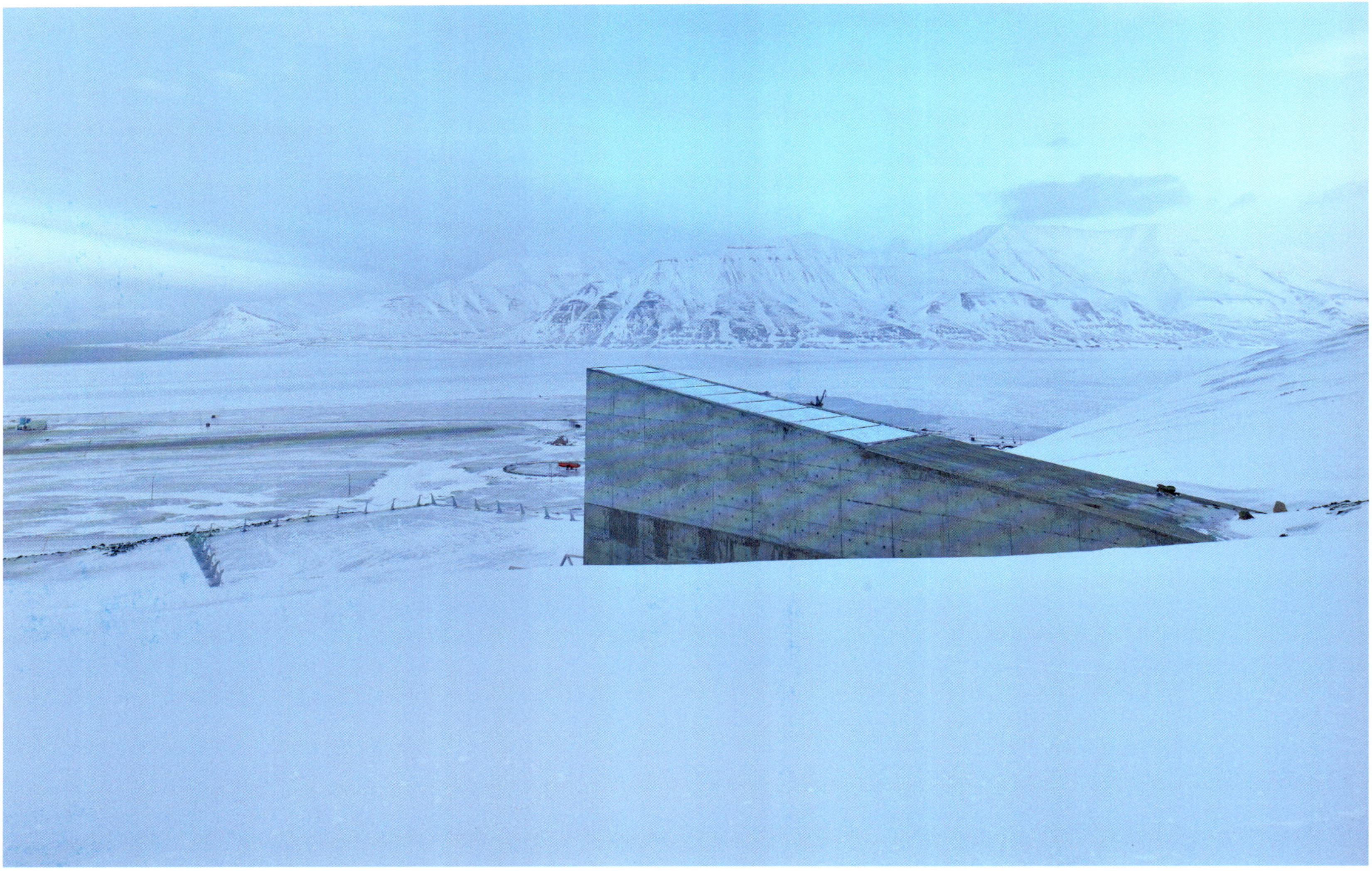

Above This strikingly designed and massively protected natural freezer was funded by the Bill & Melinda Gates Foundation and contains almost one million seeds. In the event of a catastrophe, such as war or a natural disaster, the seeds within can be reintroduced to nature. The vault also functions as a backup for 1,750 seed banks across the globe. The total number of seed samples deposited here was around 1.05 million in early 2018, but ICARDA, the International Centre for Agricultural Research in Dry Areas, previously located in Aleppo, has withdrawn nearly 90,000 samples to create new gene bank collections in Lebanon and Morocco.

Above Svalbard Satellite Station (SvalSat) is a joint venture between the Norwegian Space Agency and NASA, located on a snowy plateau on Platåberget, a mountain on Spitsbergen. Operated by Kongsberg Satellite Services, it gathers data for everything from nautical navigation to solar research — even studying the northern lights. Satellite coverage at this latitude and this high up north is unique in the world. SvalSat is the only commercial ground station in the world able to provide all-orbit support (14 of 14 orbits) to polar-orbiting satellites. The station also follows the famous Galileo and Copernicus satellites from ESA, the European Space Agency.

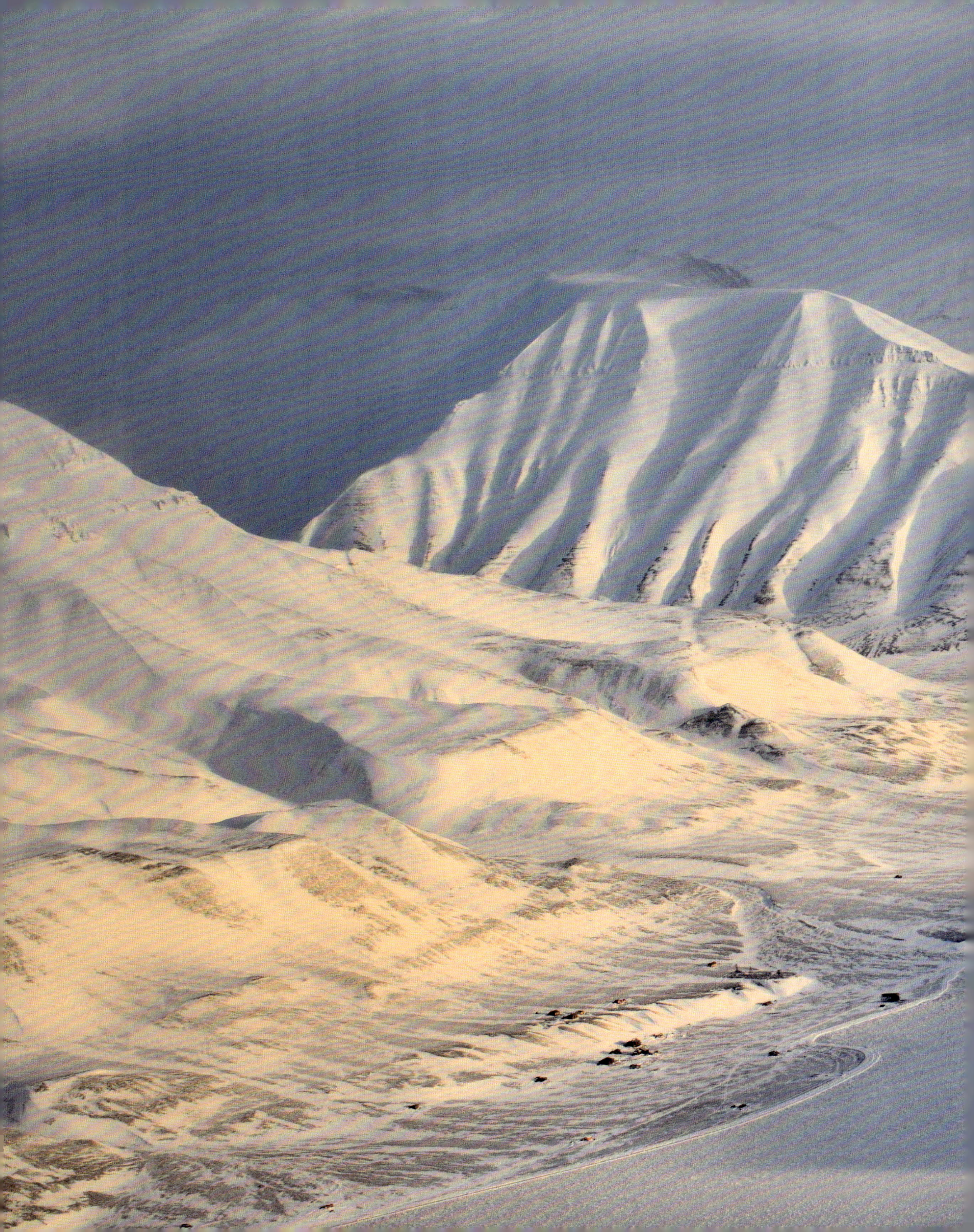

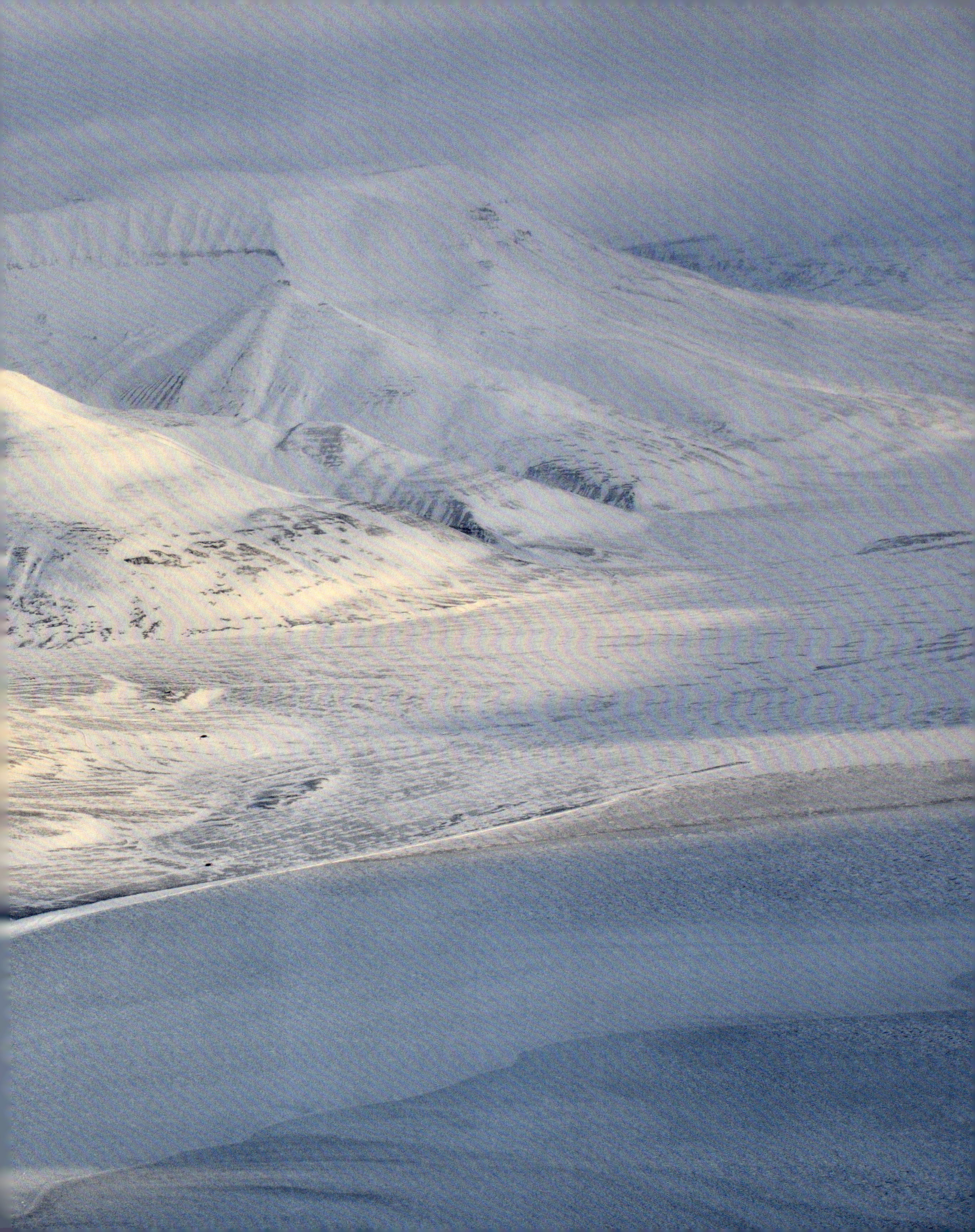

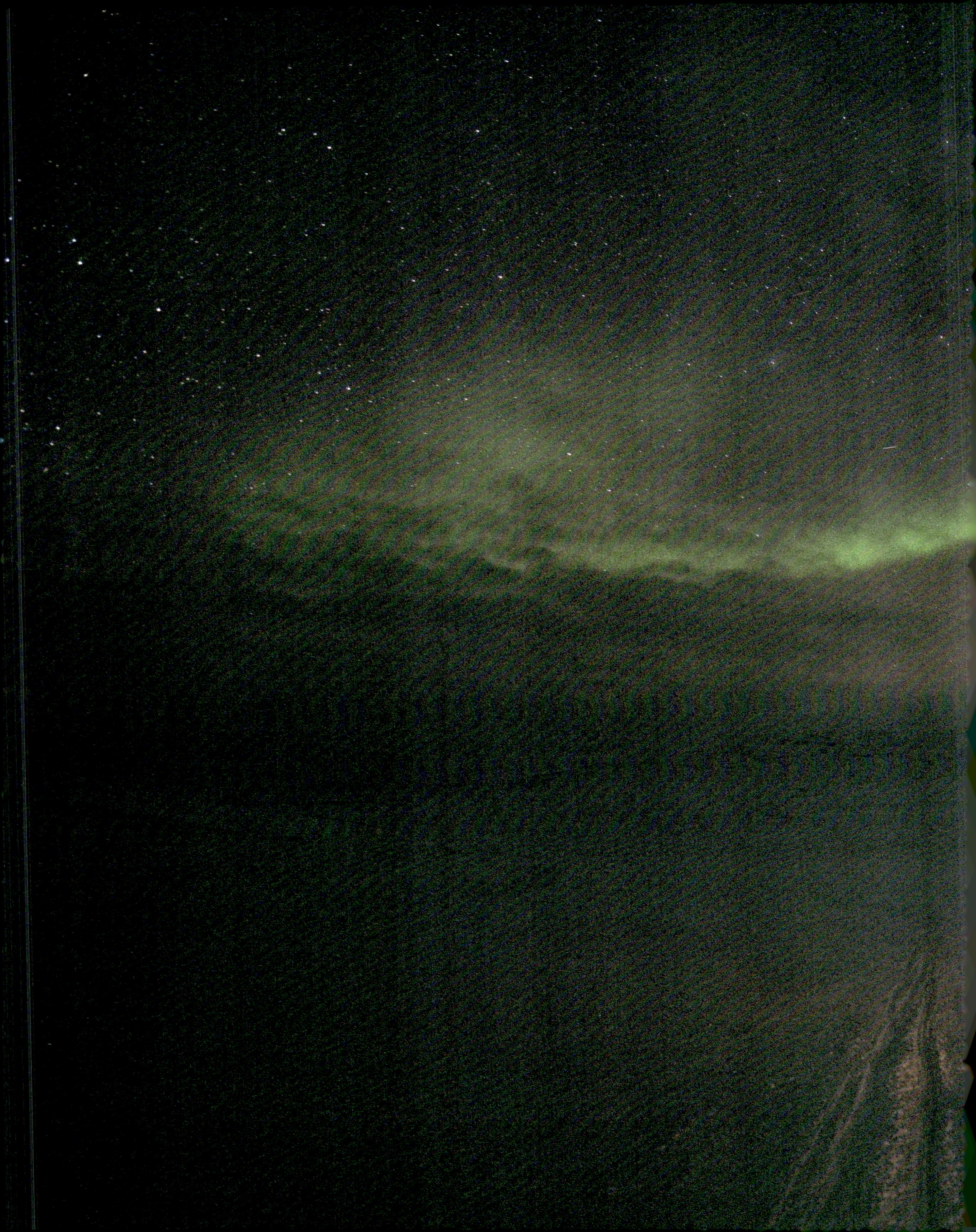

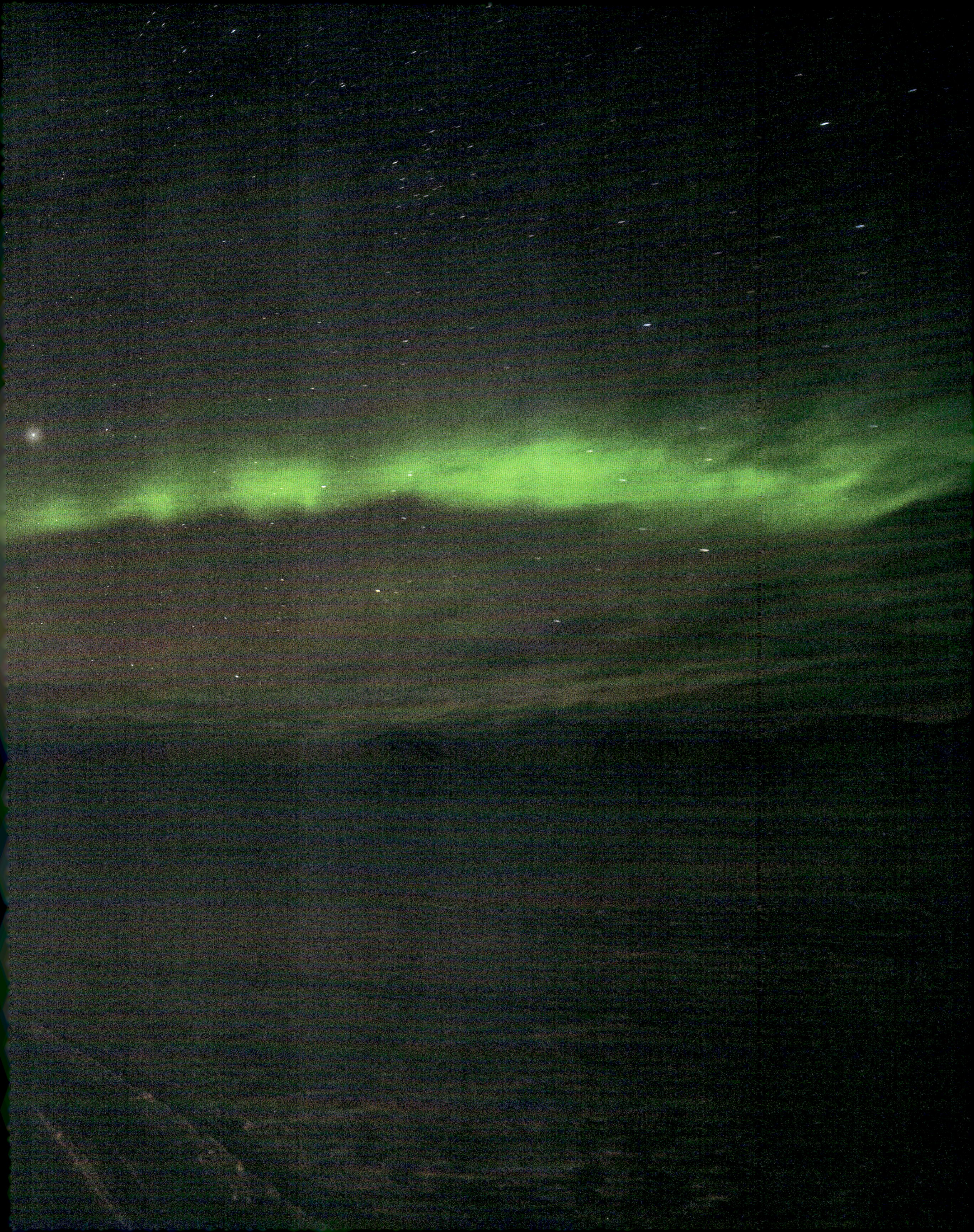

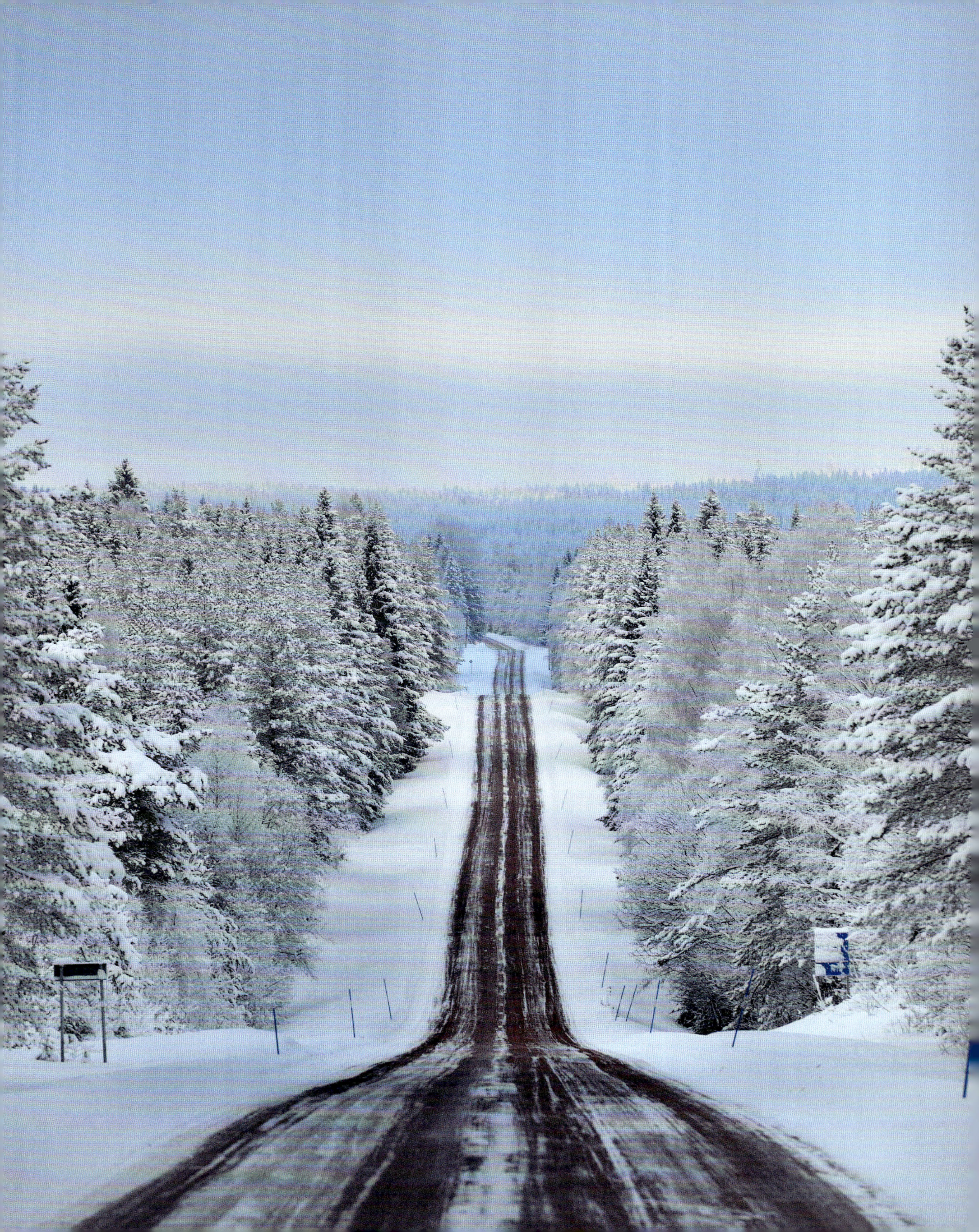

CHAPTER 3

A ROAD ADVENTURE TO THE NORTHERNMOST TIP OF EUROPE

THE BALTIC SEA CIRCLE RALLY, A WINTRY ROAD TRIP

Germany, Denmark, Sweden, Norway, Finland, Estonia, Latvia, Lithuania, Poland

71°10'12.60" N
25°47'23.99" E

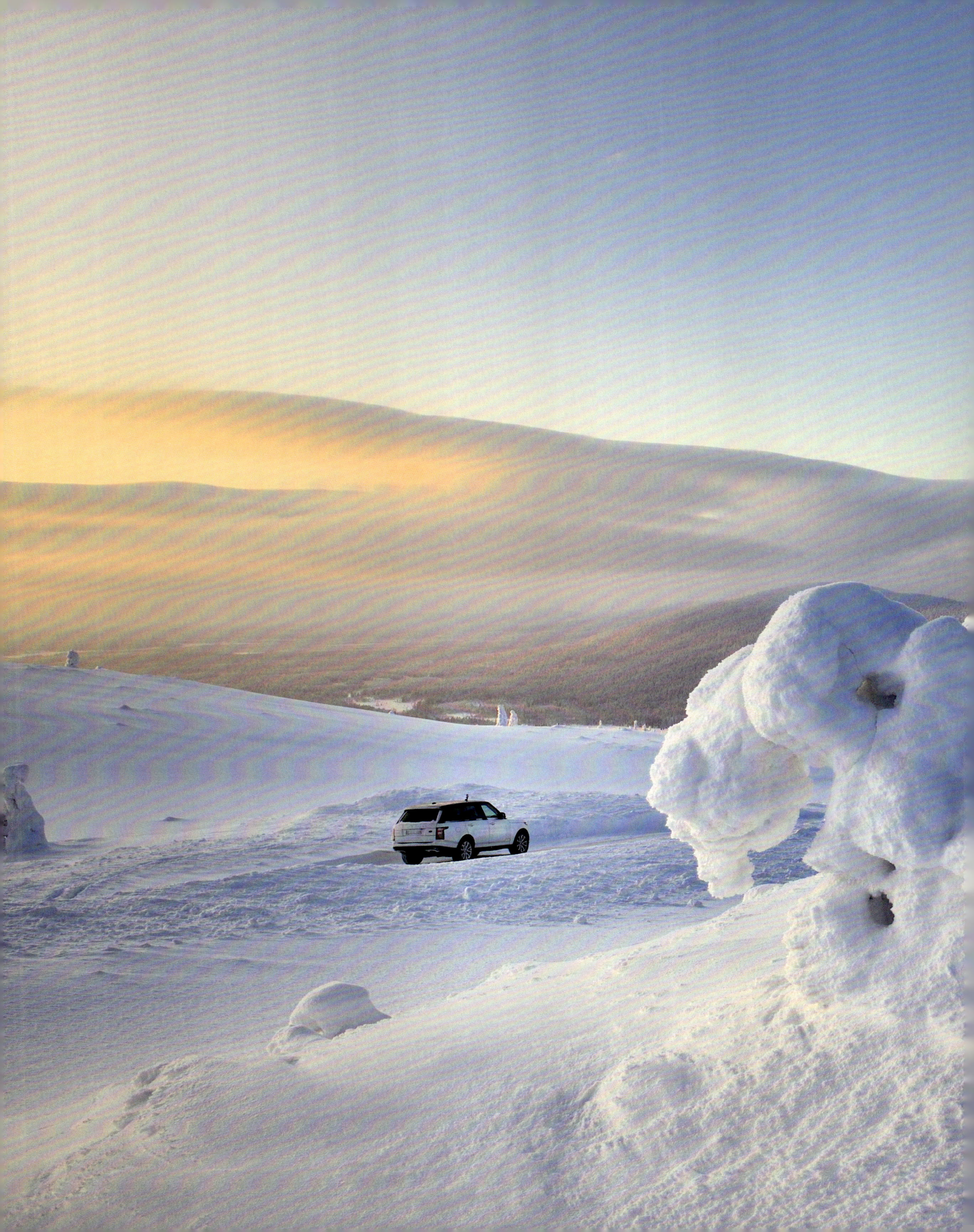

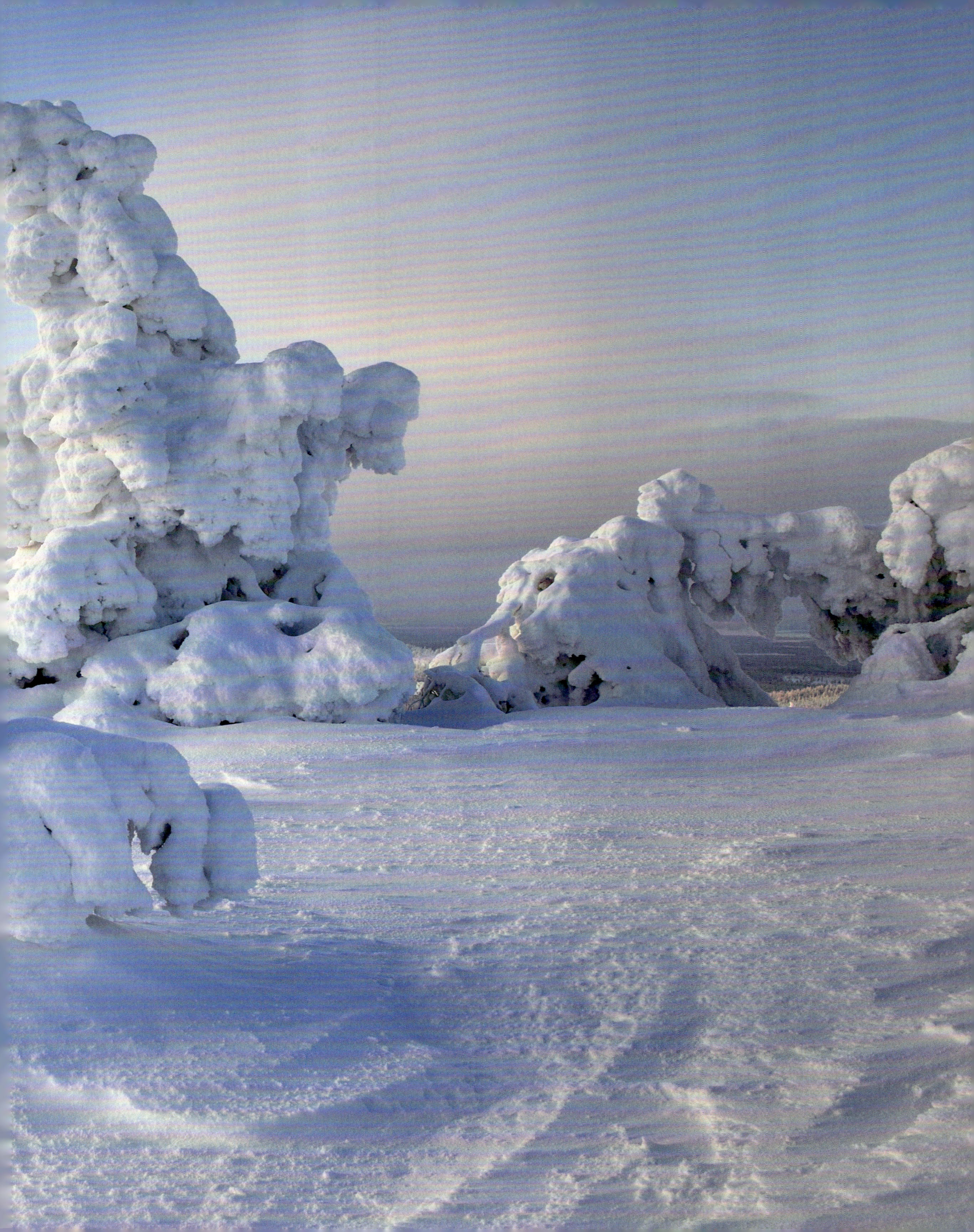

A Road Adventure to the Northernmost Tip of Europe

The Baltic Sea Circle Rally, a wintry road trip

Country:	**Norway + 8 other countries**
Best time to travel:	**Middle of winter**
Closest airport:	**Hamburg, Germany**
Price tag:	**€€**
Style of travel:	**A winter odyssey along Scandinavian back roads**

Introduction

The adrenaline-fuelled Baltic Sea Circle Rally crosses nine European countries to the most northern point of mainland Europe along blissfully empty roads. The rally takes place in the quiet depths of winter and without the use of GPS or highways. This adventure is not about posh cars, flaunting wealth, or arriving somewhere first. It's about finding joy in extremes, handling challenging roads and extreme weather, and having plenty of fun while doing so. The rally was launched in 2011 by the Superlative Adventure Club (SAC) and originally included only five teams. Today it includes more than 100. With only a few rules, none of which is enforced, it is like a gentleman's agreement: cars must fall into specific age ranges, no digital maps are allowed, and highways are to be avoided. The 8,000-kilometre (5,000 mi.) drive covers a vast amount of terrain: nine countries in all, starting in Hamburg, Germany, and leading all the way to the North Cape through Denmark, Sweden, and Norway, before looping back to Germany via Finland, the Baltics, and Poland. In warmer months, many of the locations en route are clogged with cars, filled with tourists and mobile homes, but in winter the route takes on a fairy-tale quality. The teams, which straddle various ages, backgrounds, and nationalities, get to enjoy silent, serene landscapes virtually alone — a splendid interlude interspersed with meeting up, if desired, with like-minded travellers or teams along the route. Thanks to the challenging, isolated terrain, it is comforting to know that dozens of other rally teams are nearby, heading in the same direction, also following the compass north and pursuing the same goal: the iconic North Cape, the northernmost point in Europe.

The Views

Baltic Sea Circle Rally cars and participants line up on the waterfront in Hamburg, ready to hit the frozen tarmac. Roadmaps with tips and suggested itineraries are handed out while GPS trackers are installed for the purpose of following the teams live online. After a final check, everyone heads north. Some rally teams are made up of groups of friends, others include couples, and there are even solo drivers who complete the 8,000-kilometre (5,000 mi.) adventure alone. The first night is spent in Scandinavia and sets the mood for the rest of the trip. In the Swedish countryside drivers can look forward to a cosy cabin, a piping-hot sauna, and perhaps a homemade sea-buckthorn schnapps. Some teams camp and cook their own food, others stay in luxury hotels and enjoy four-course meals. The charm of the rally is its accessibility: no matter how big the budget is, no matter whether the team drives a BMW or an old, shabby minibus, anyone can join this adventure. Although all vehicles drive in the same direction with the same goal — reaching the North Cape in Norway — no two journeys are the same. Every team, and team member, experiences the drive in a singular way, depending on the roads they choose, the stops they make, the locals they meet, and the food and drinks they try. Paradoxically, despite spending long hours in a car each day, the sense of freedom the teams experience is immense. After the first night spent in Scandinavia the route heads north through Sweden where a couple of nights later the first SAC party and meet-up is planned. Most of the teams aim for Norway's rugged Lofoten archipelago where red-painted fisherman's cabins dot the shores of tiny islands, then continue along snowy, empty fjords. During summer, this region is popular with tourists in cars and campers, but now, it's empty and silent. Further north, days shorten, and drivers get accustomed to zipping through

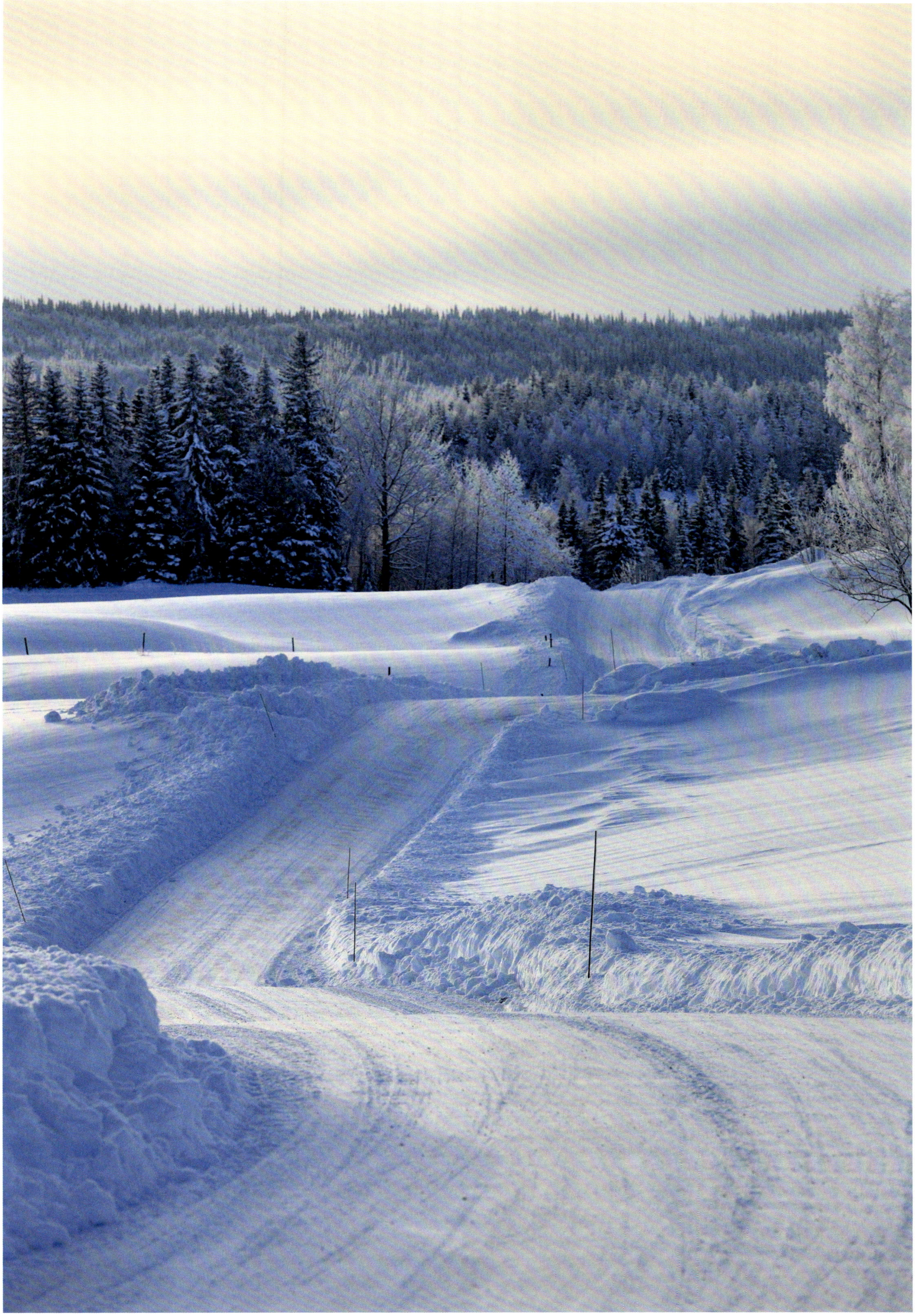

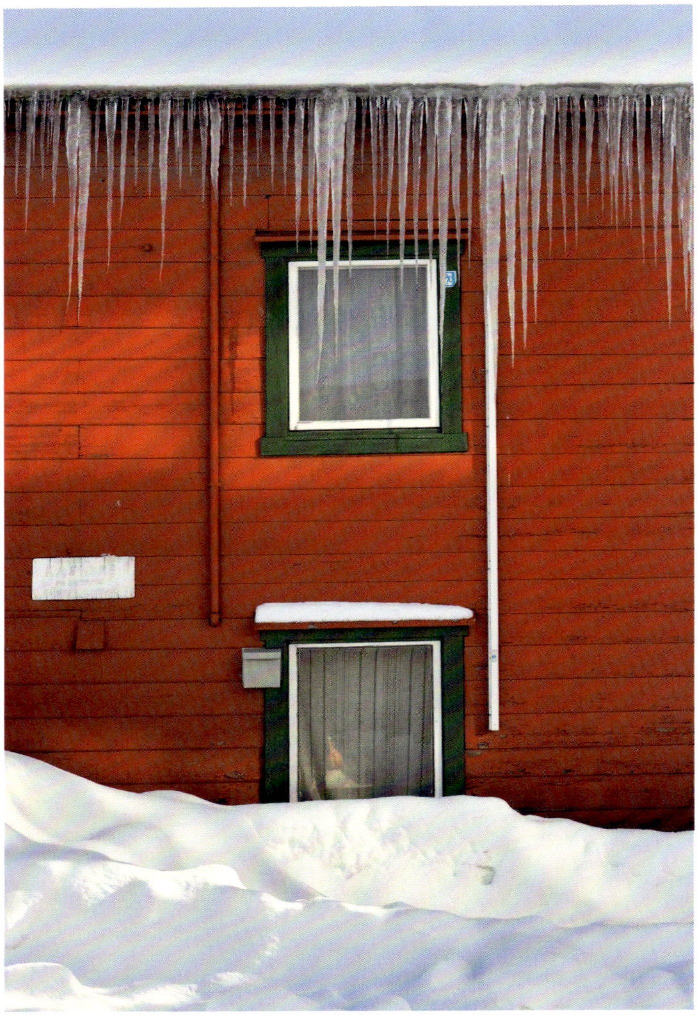

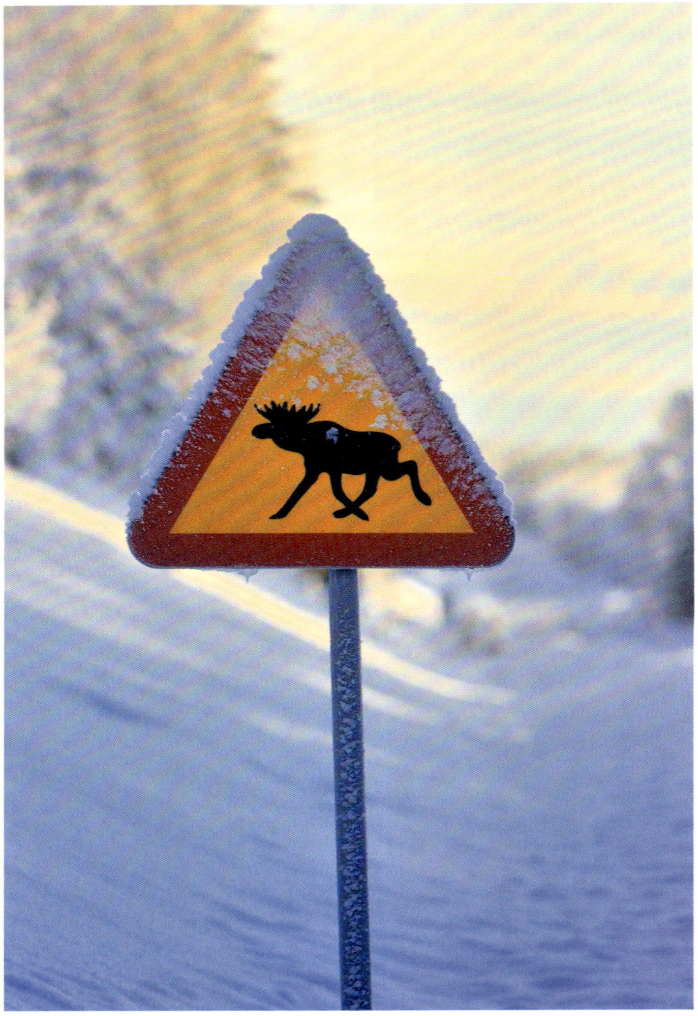

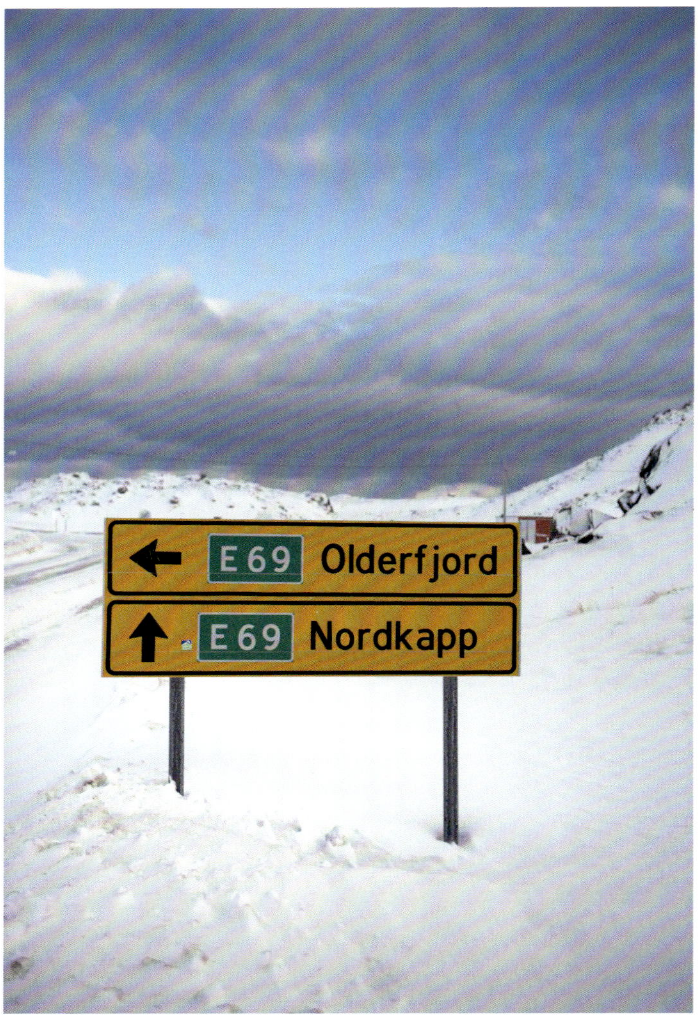
E 69 Olderfjord
E 69 Nordkapp

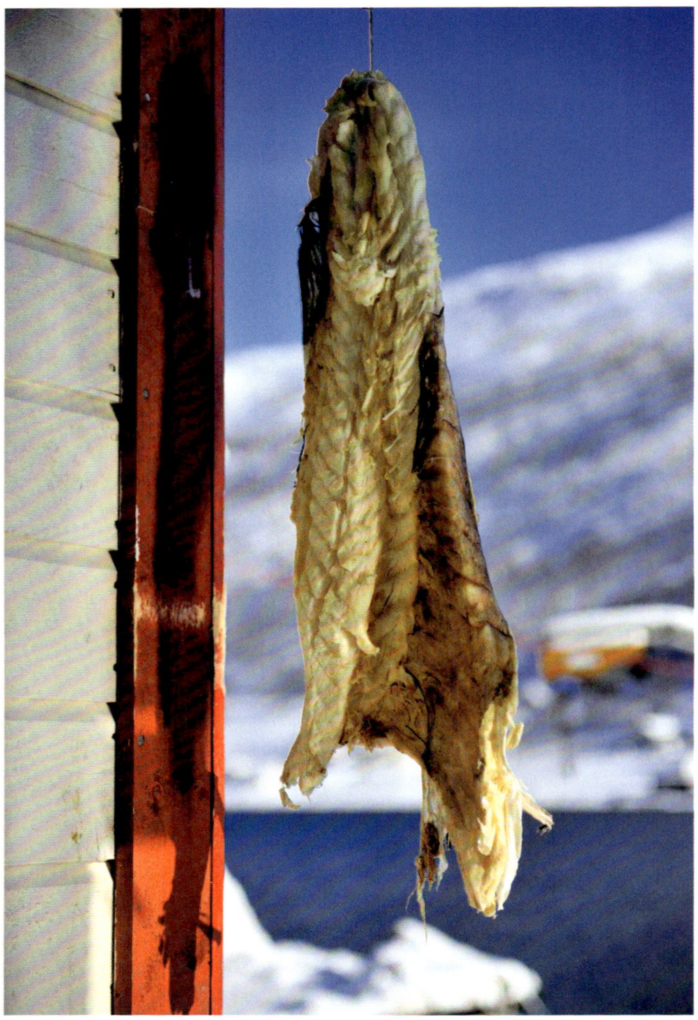

Opposite Red-painted fisherman's cabins called rorbuer dot the shores of serene fjords in Norway's High North. The advantage of crossing Scandinavia in the quiet depths of winter is that the roads are empty and the pristine landscapes can be experienced at their most authentic. Days are cold and short, but bright and beautiful when the sun is out. Long nights are often enlivened by magic displays of the northern lights.

the inky darkness. Black ice, whizzing Arctic winds, and whiteouts make the rally more challenging. Somehow the teams arrive simultaneously at a rendezvous point at the top of mainland Europe, where they form a caravan to drive safely to the legendary North Cape, the northernmost point of Europe. The mood is cheerful as virtually everyone reaches this goal together. Toasts are made with plenty of hugs and high-fives, with barely a thought for the icy Arctic wind. The following days are spent traversing Finnish Lapland. For hours at a time the roads wind through empty pine forests, the short days brightened by a distinctive, pinkish Arctic light. By now, all teams are speeding south, including one in a vintage Mercedes taxi from Berlin with more than 1.5 million kilometres (932,000 mi.) on the odometer. The old Benz still moves effortlessly through the harsh landscape. Estonia offers stunning blue skies and a frozen Baltic Sea where lonely ice roads connect remote islands to the mainland. In Kolja, near Tallinn, the capital of Estonia, a huge, wooden teepee functions as a second SAC party venue, with saunas, hot tubs, and plenty of Estonian food and beer. Like-minded souls meet again to share memories and adventures. On a monochrome morning, Lithuania's Hill of Crosses, a Christian pilgrimage site containing more than 100,000 crucifixes, appears. From here it's only a few hours to Poland, where the finish line is almost in sight. On the last evening, the BSC crowd enjoys a lively farewell party in a cosy café on the Hamburg docks. The happy event is tinged with nostalgia for the excitement and anticipation of the rally. The Baltic Sea Circle Rally is not just about the thrill of victory or being the fastest to the finish line. Neither is it about who raises the most money for the almost 100 different charities. Instead, the voyage is all about the experience between start and finish and crossing Northern Europe's marvellous, ever-shifting winter landscapes, while enjoying an exhilarating sense of freedom.

Creature Comforts

During the rally, only two nights of accommodation are pre-booked and included in the package price of 1,090 Euro per car/team. For the remaining nights, participants are free to choose where they bed down. Since this is low season in Scandinavia, finding last-minute accommodation is not an issue. Some teams set up tents and camp, others go for cheap hostels. But it might also be a small wooden chalet in the thick forests of Finland, a cosy B&B in a Swedish town, or an inviting wilderness hotel.

Go Now

Driving through Scandinavia in the middle of winter is surely the most authentic way to experience the region. Navigating the frozen shores and lonely roads, drivers will truly capture its majestic, icy essence.

How to Go

The only requirement is to be at the starting line on the docks of Hamburg. The event happens every year at the end of February and it takes 16 days. There are four categories of cars: Old Schooler (30+ years), Youngtimer (20+ years), Freshman (5+ years), and Electric (which can be brand new).

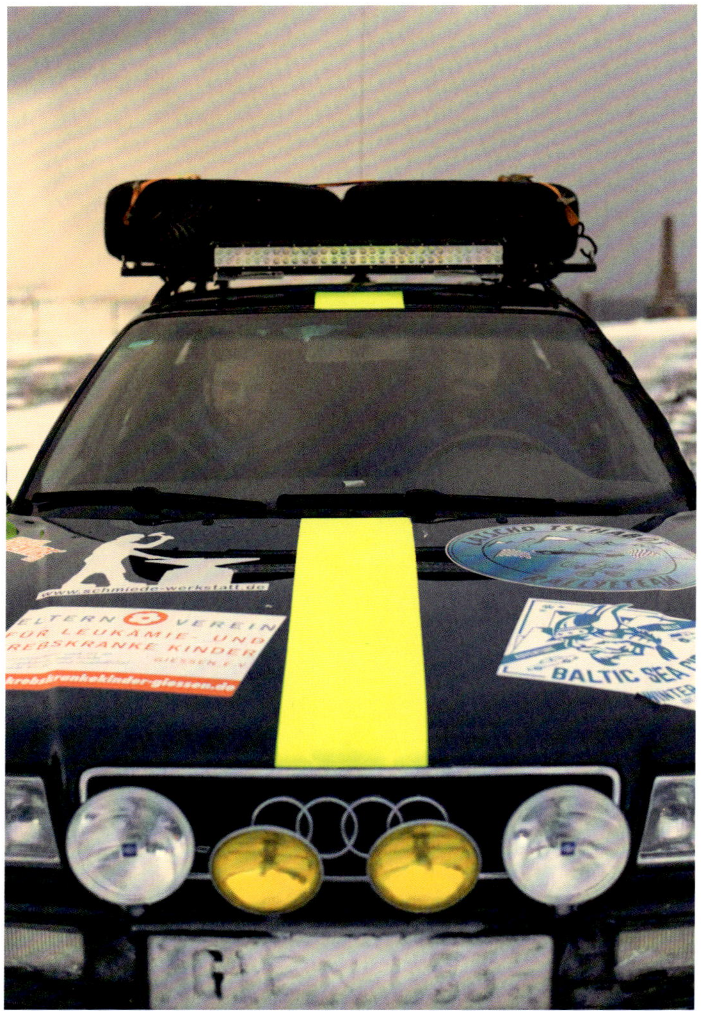
www.schmiede-werkstatt.de

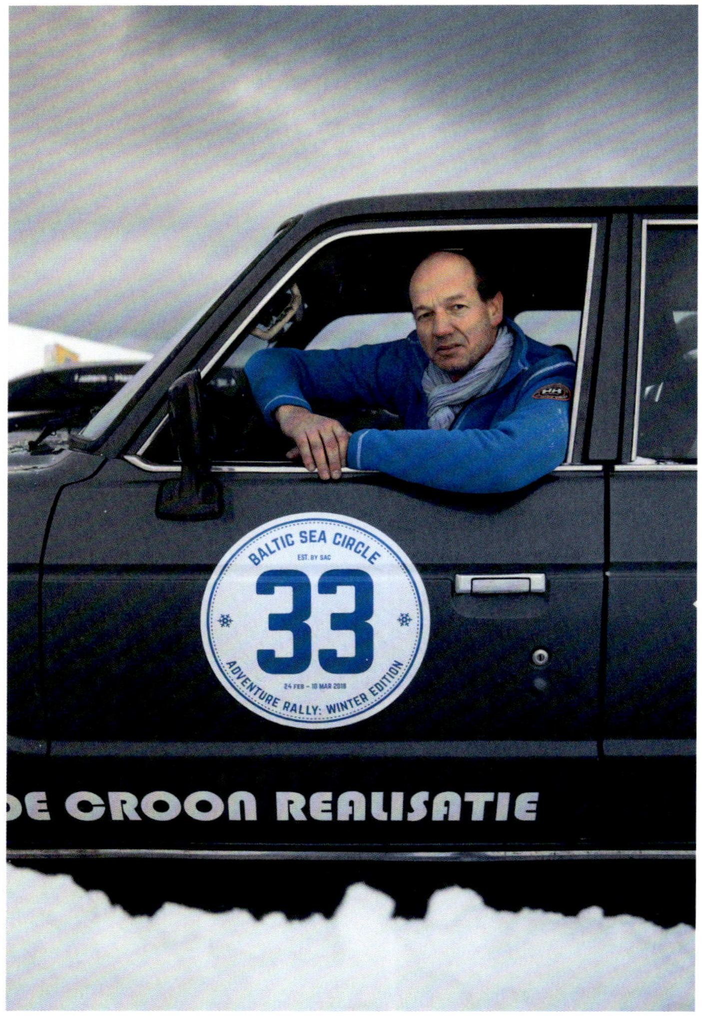
BALTIC SEA CIRCLE
EST. BY SAC
33
ADVENTURE RALLY: WINTER EDITION
DE CROON REALISATIE

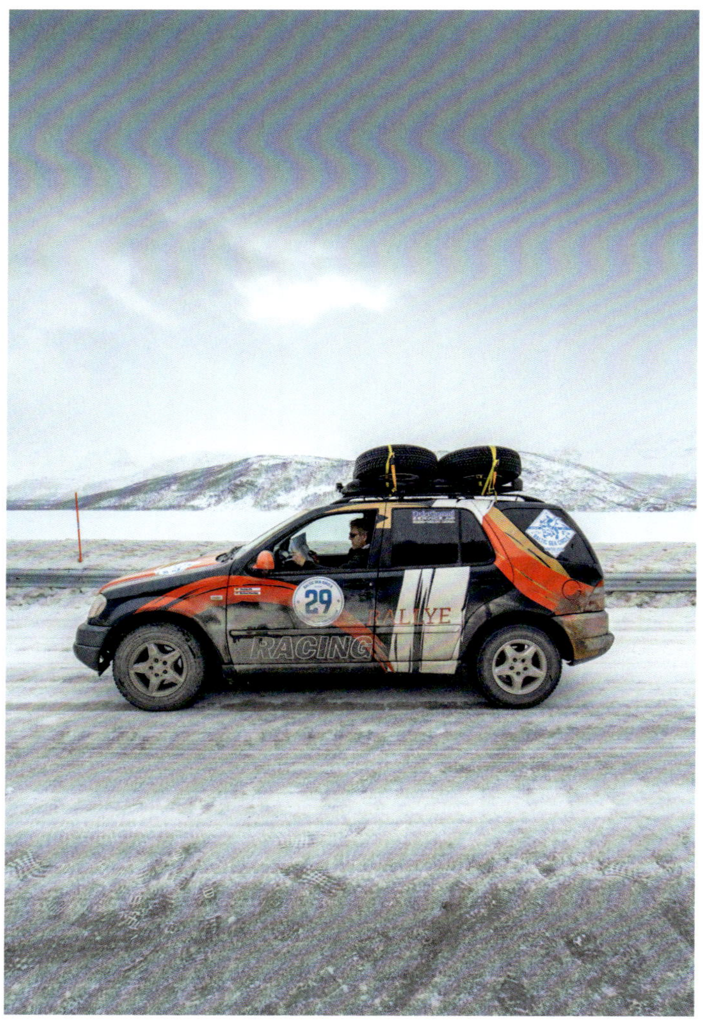
29
RALLYE
RACING

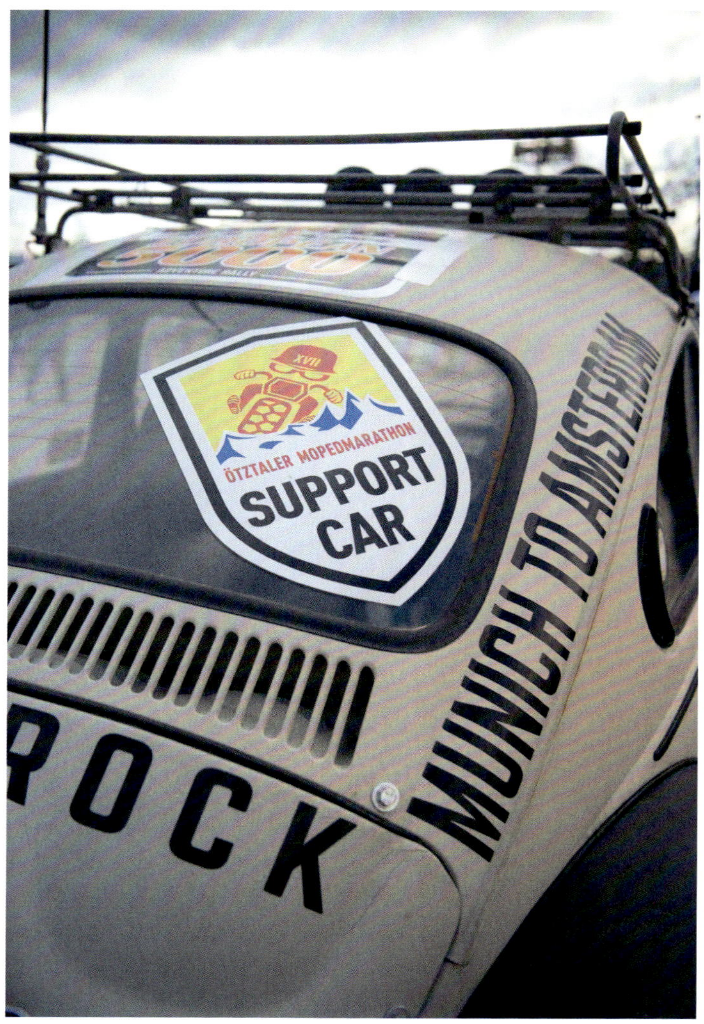
XVII
ÖTZTALER MOPEDMARATHON
SUPPORT
CAR
MUNICH TO AMSTERDAM
ROCK

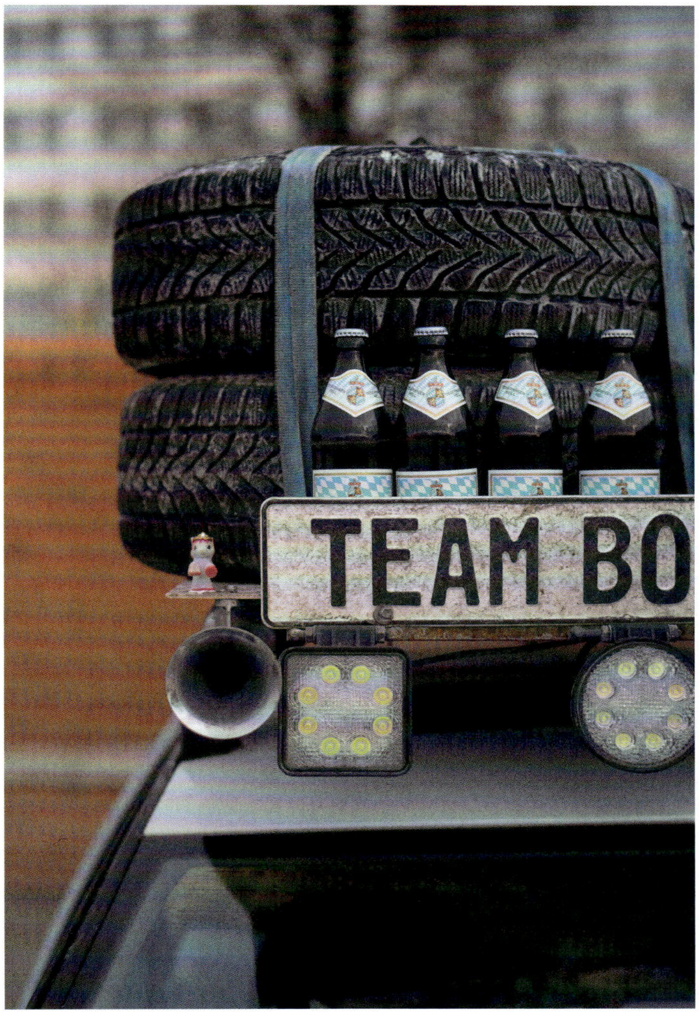
TEAM BO

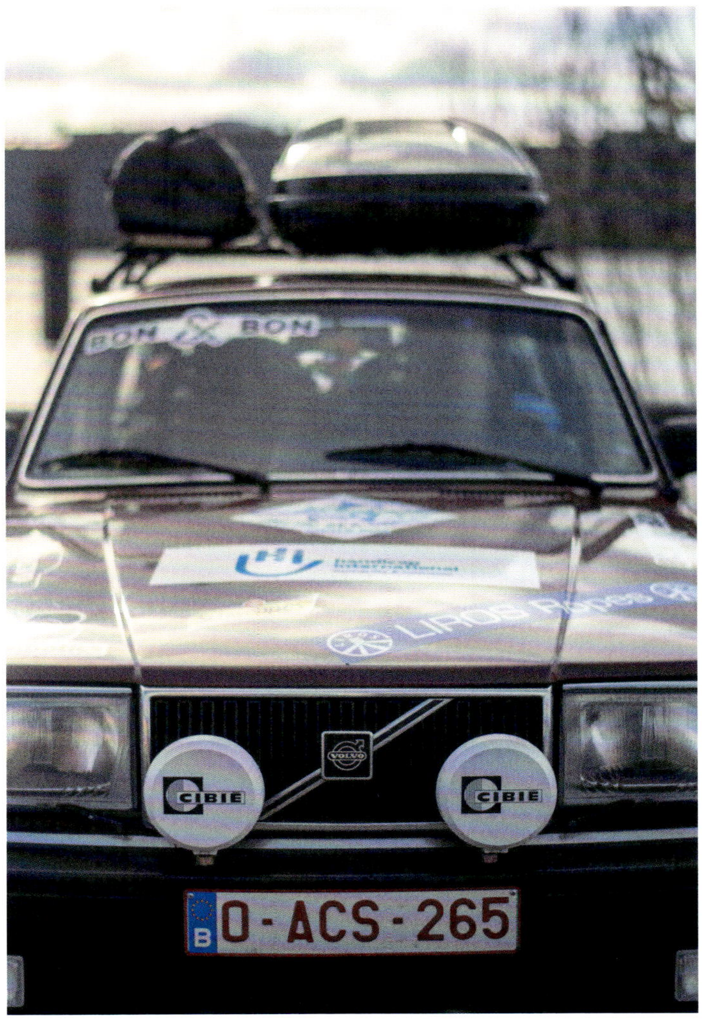
BON & BON
CIBIE
CIBIE
0-ACS-265

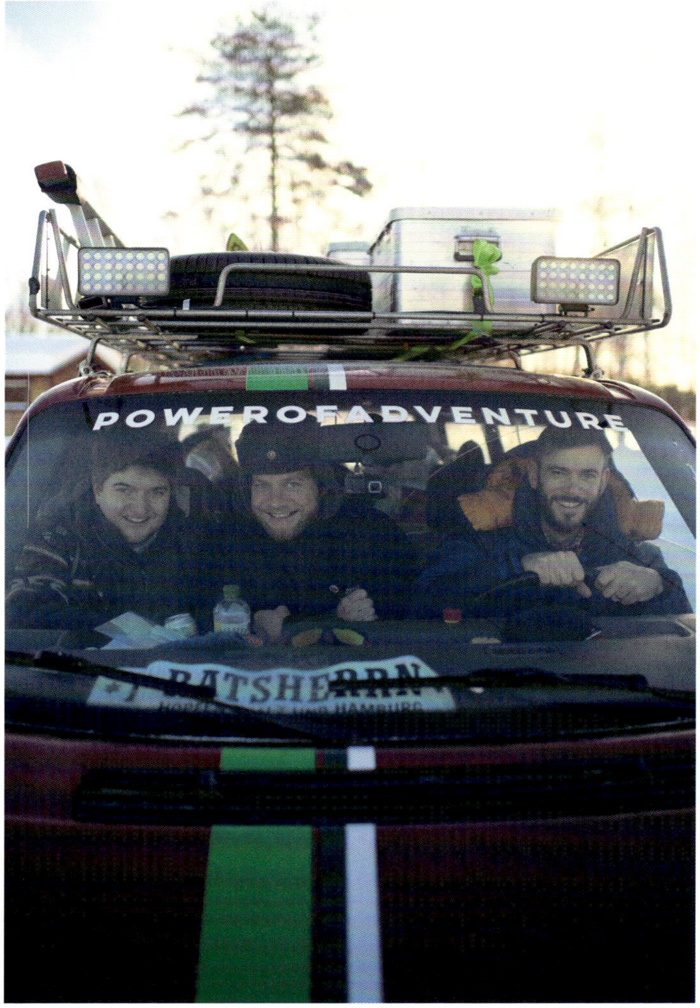
POWEROFADVENTURE

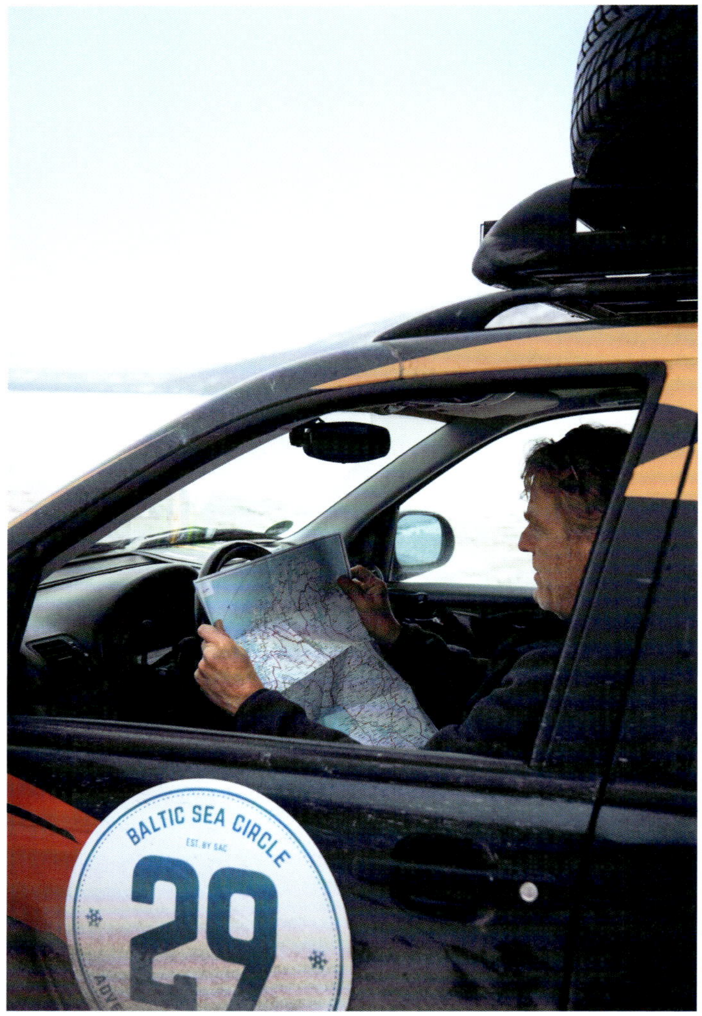
BALTIC SEA CIRCLE
EST. BY SAC
29

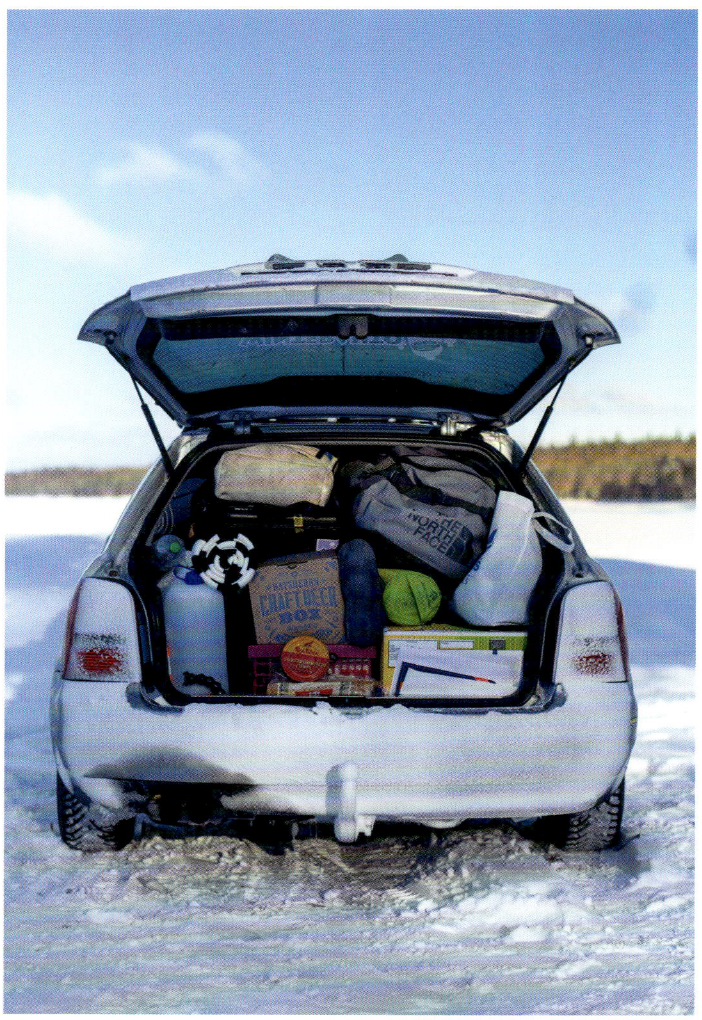
RATSHERRN
CRAFT BEER
BOX

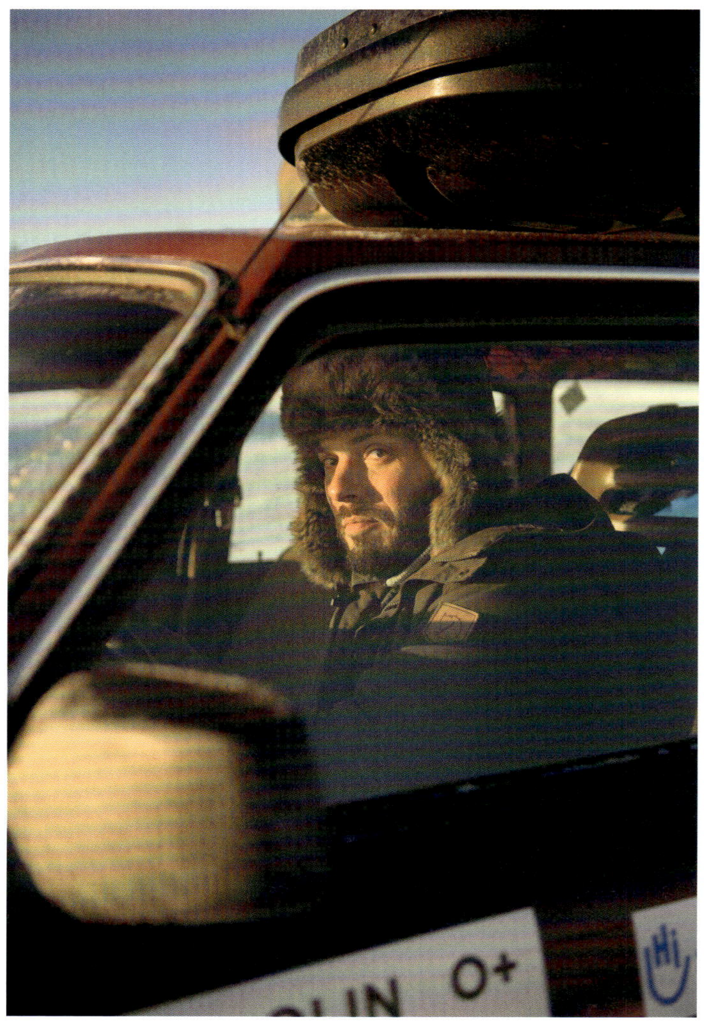

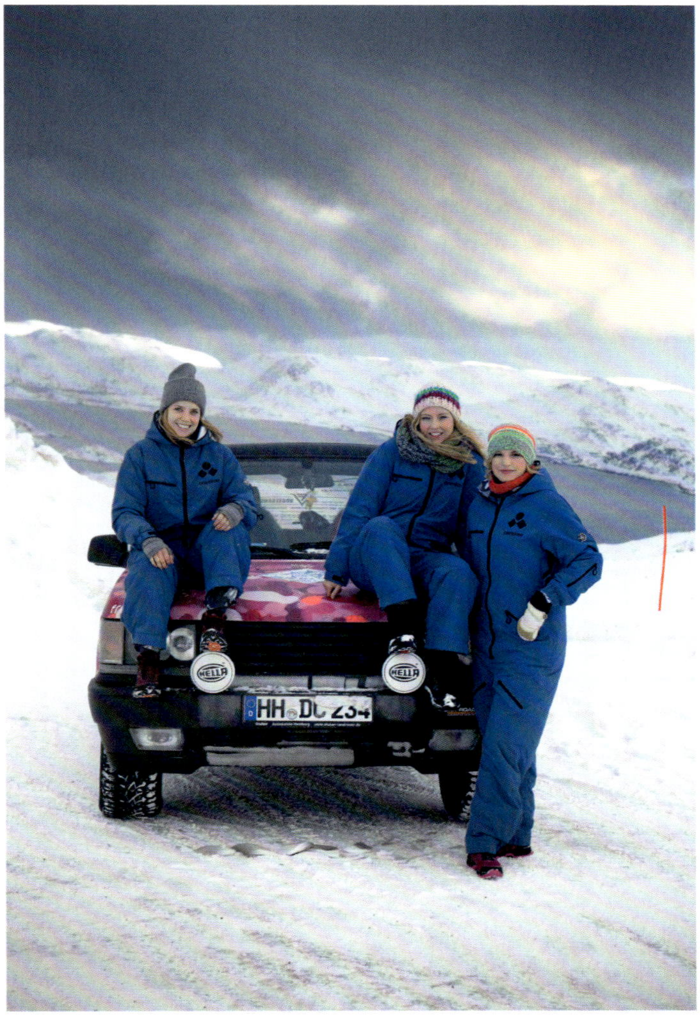

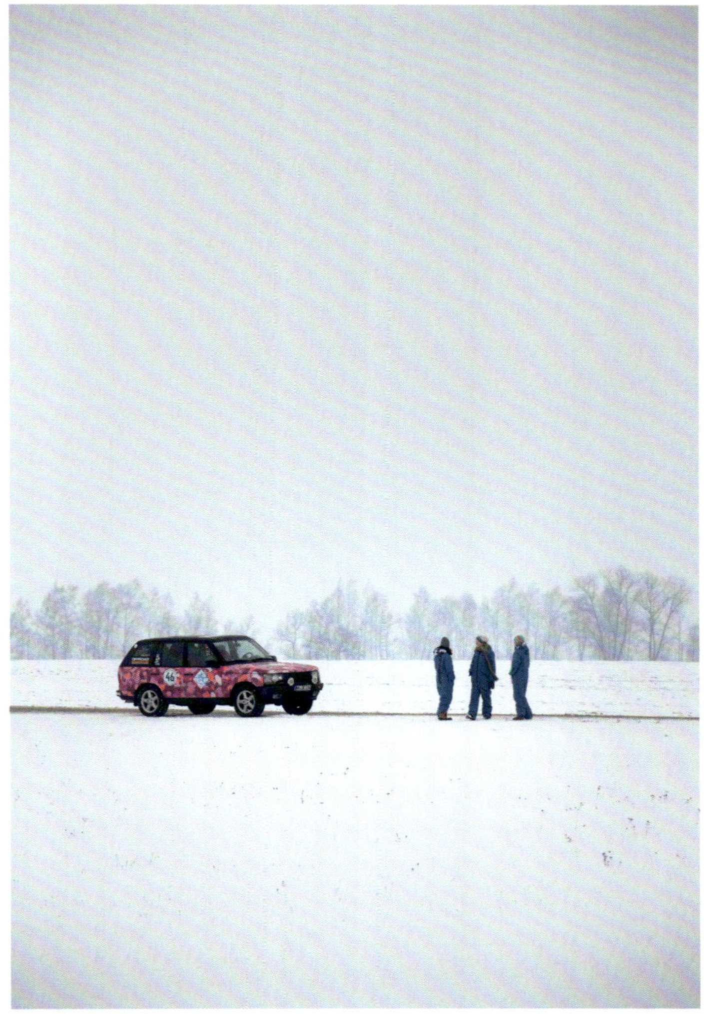

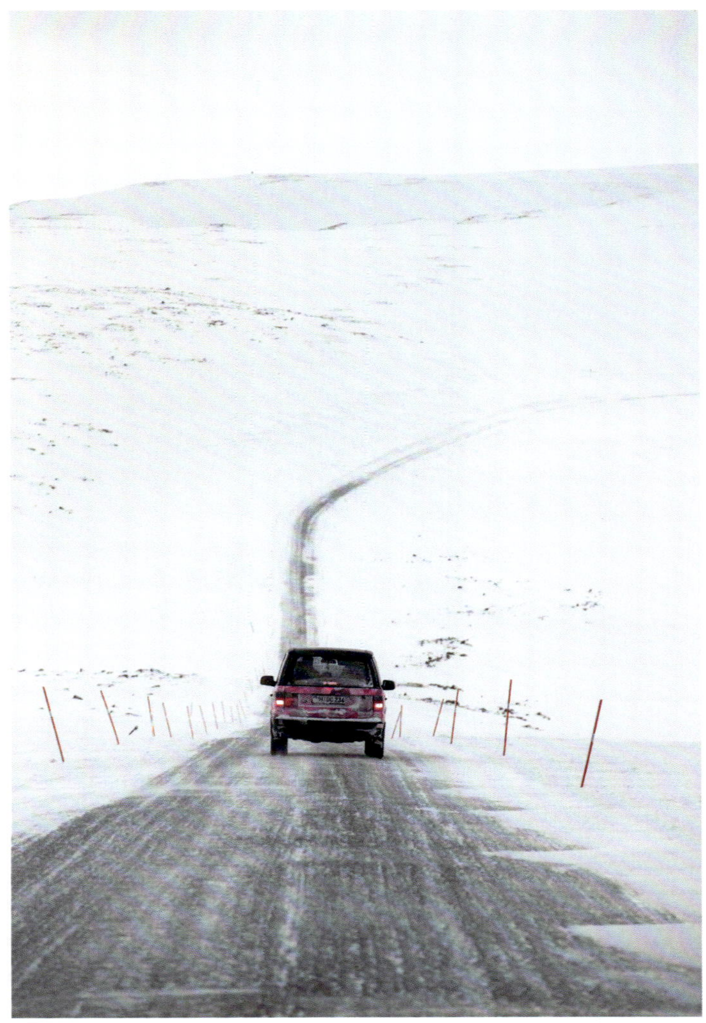

This page Team DriftChicks, three women from Hamburg with families waiting at home, drive a pimped-up pinkish Range Rover. When black ice sends them into a snowy ditch, two other teams pull them out and the DriftChicks quickly hit the road again. For them, it's the perfect opportunity to forget about their busy, corporate lives and reconnect with each other.

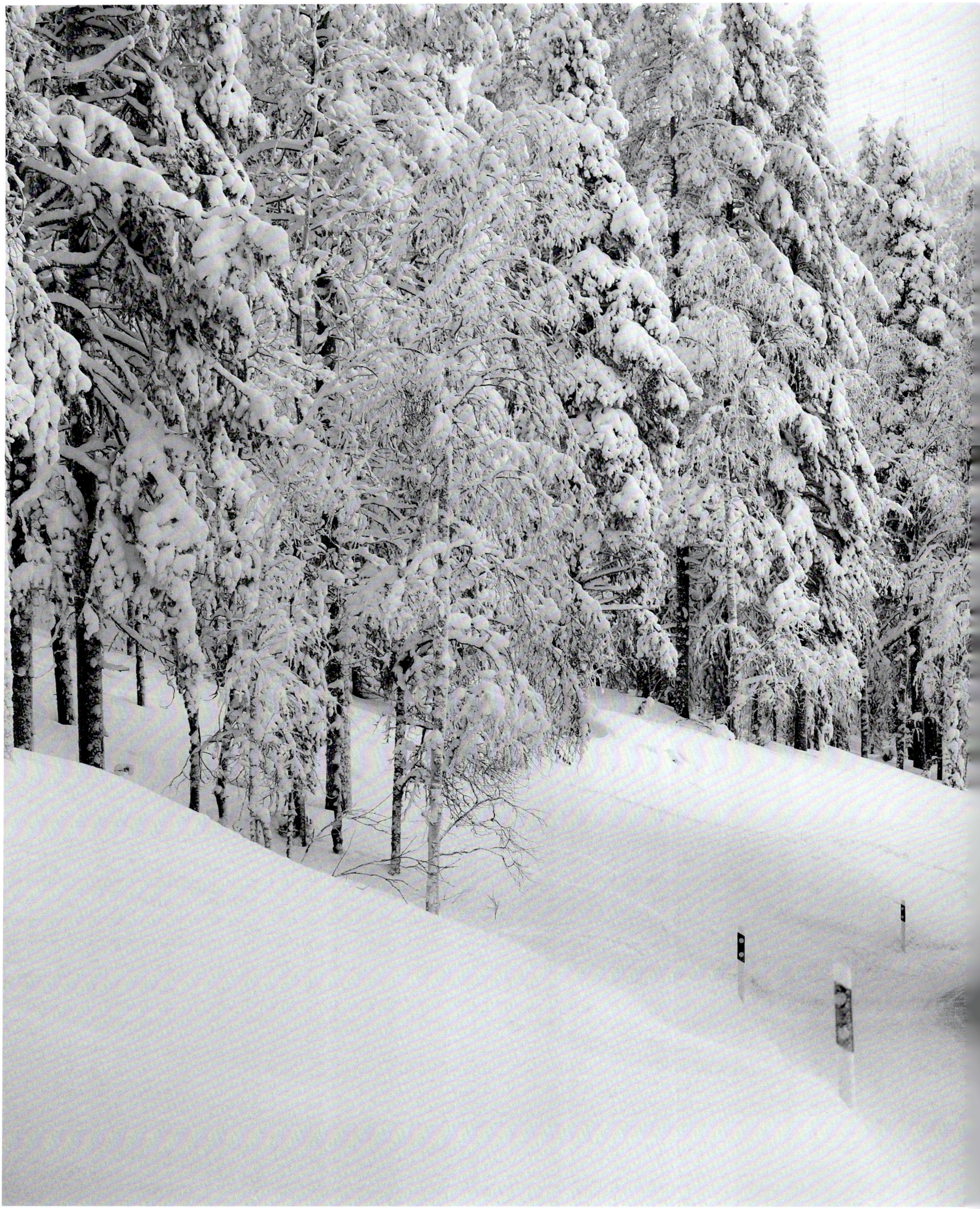

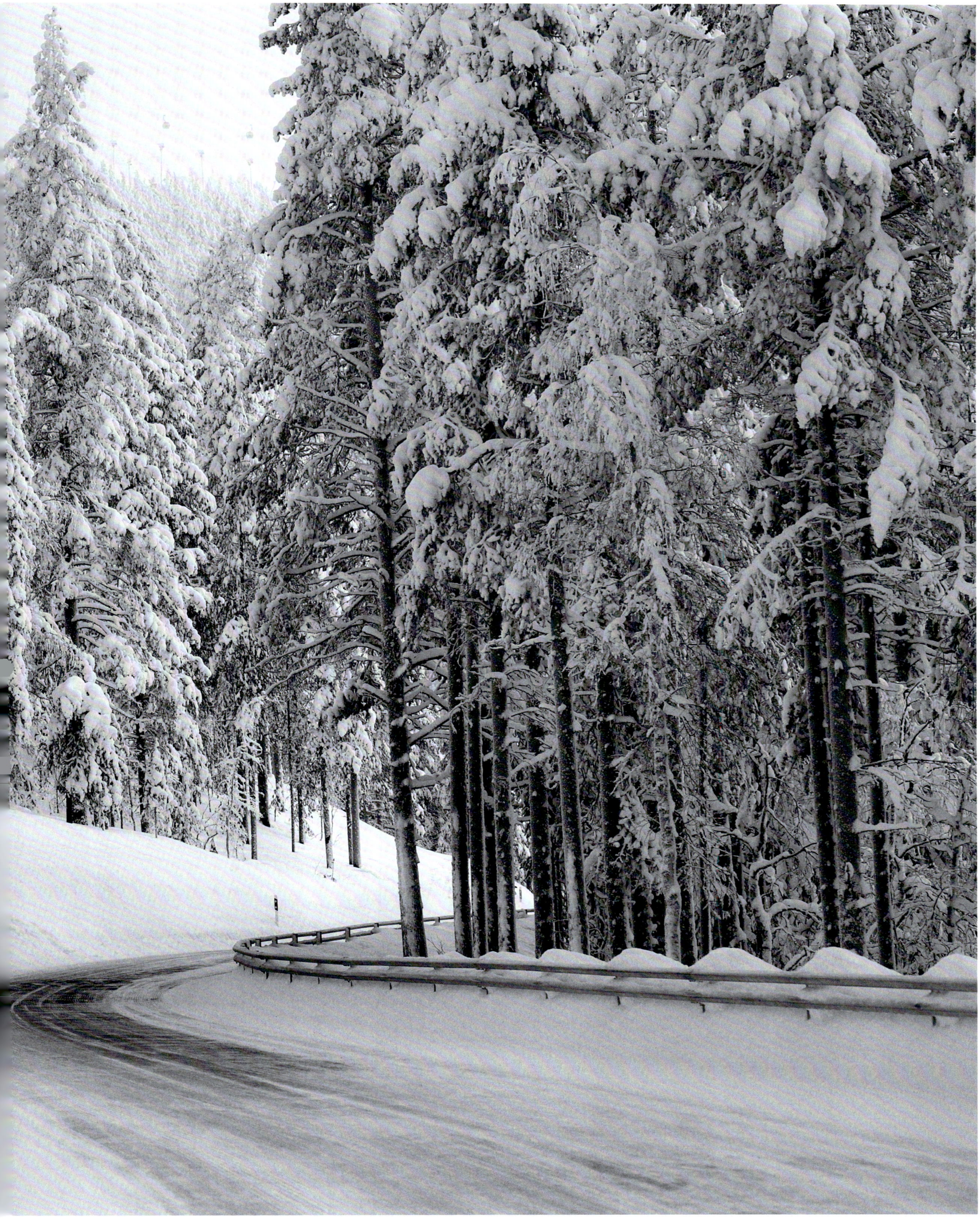

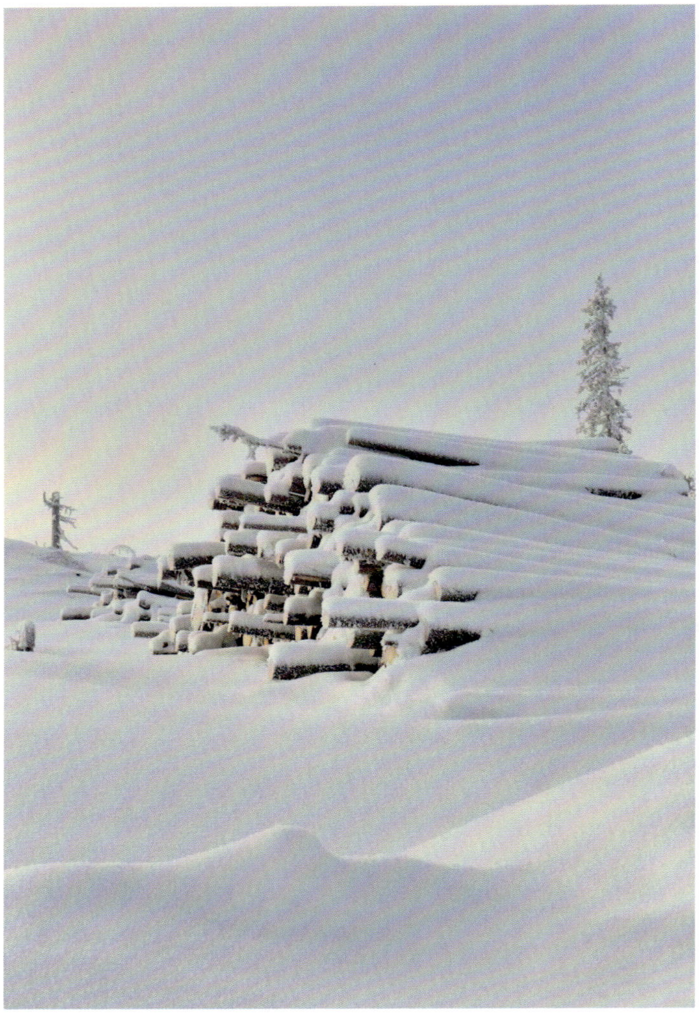
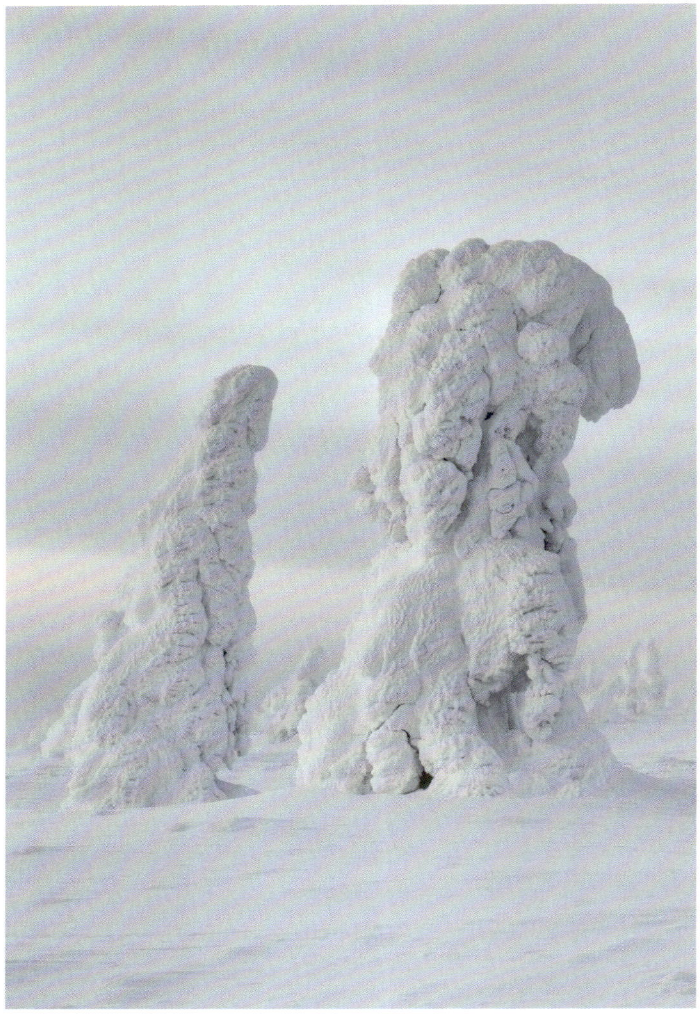
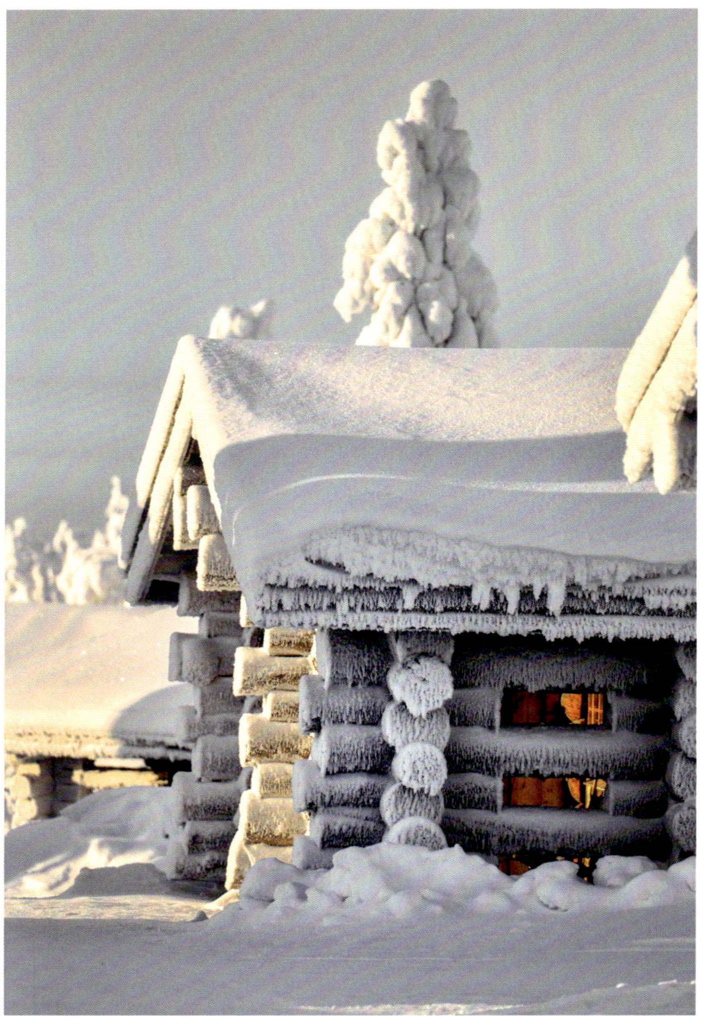
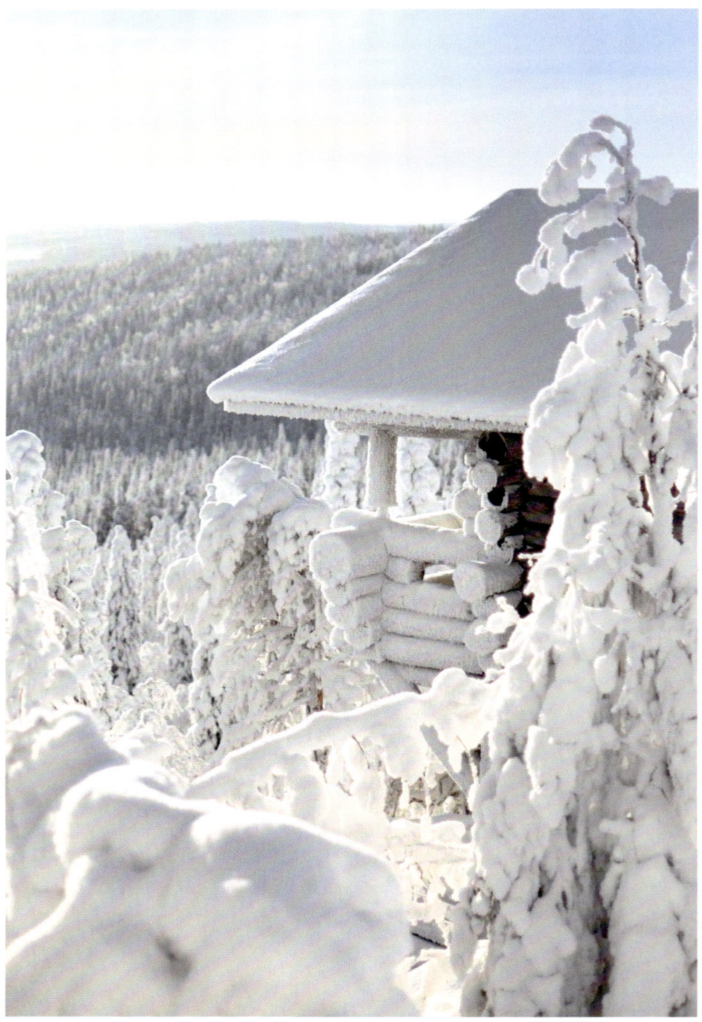

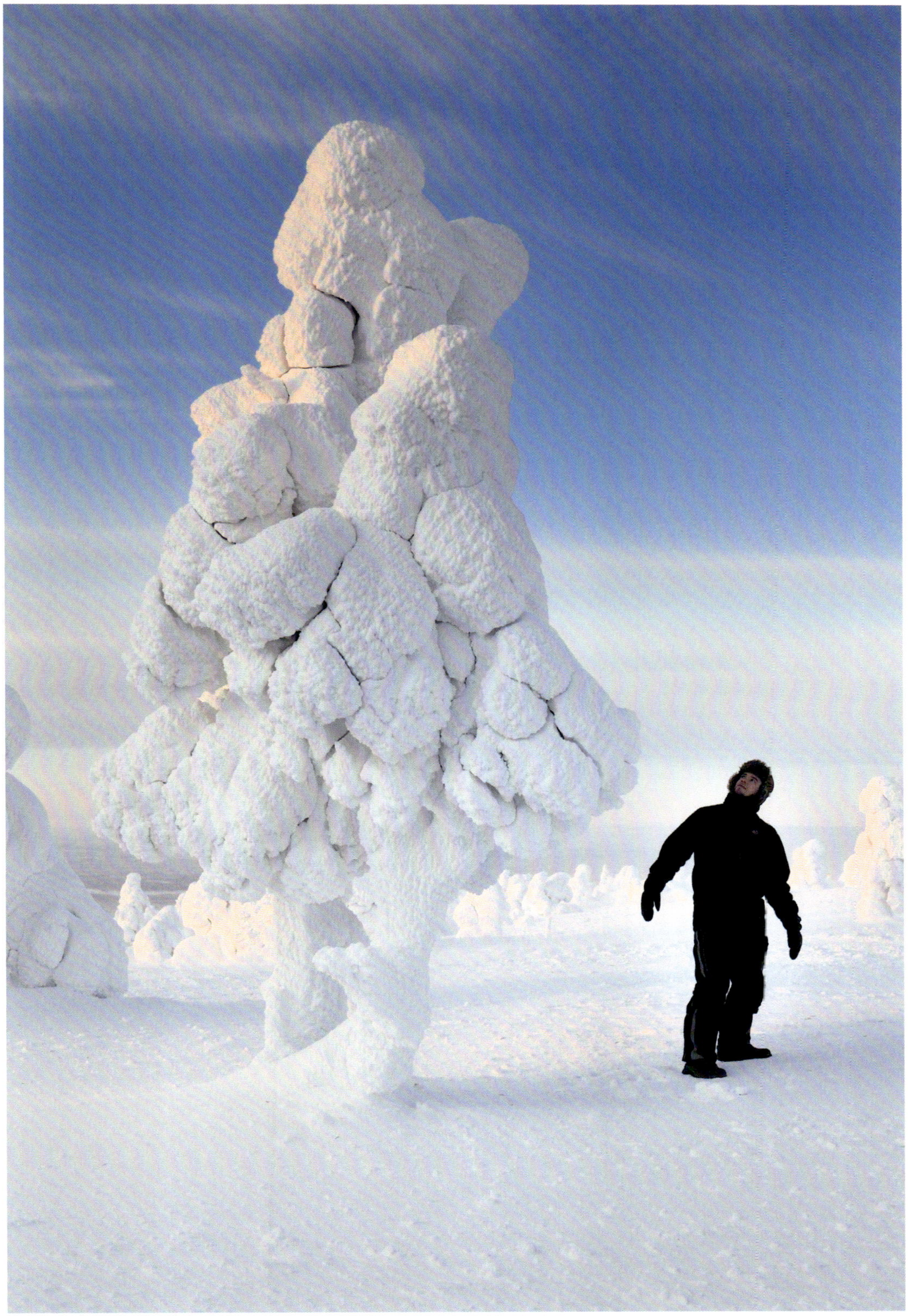

This spread Seventy per cent of Finnish Lapland is covered in forest. Driving through this otherworldly expanse of snow-laden trees is a magical experience. Some teams take the opportunity to test their skills on seasonal ice 'roads' on frozen rivers and lakes, and try some ice fishing on the way. Next to robust log cabins, reindeer soup is prepared on wood fires.

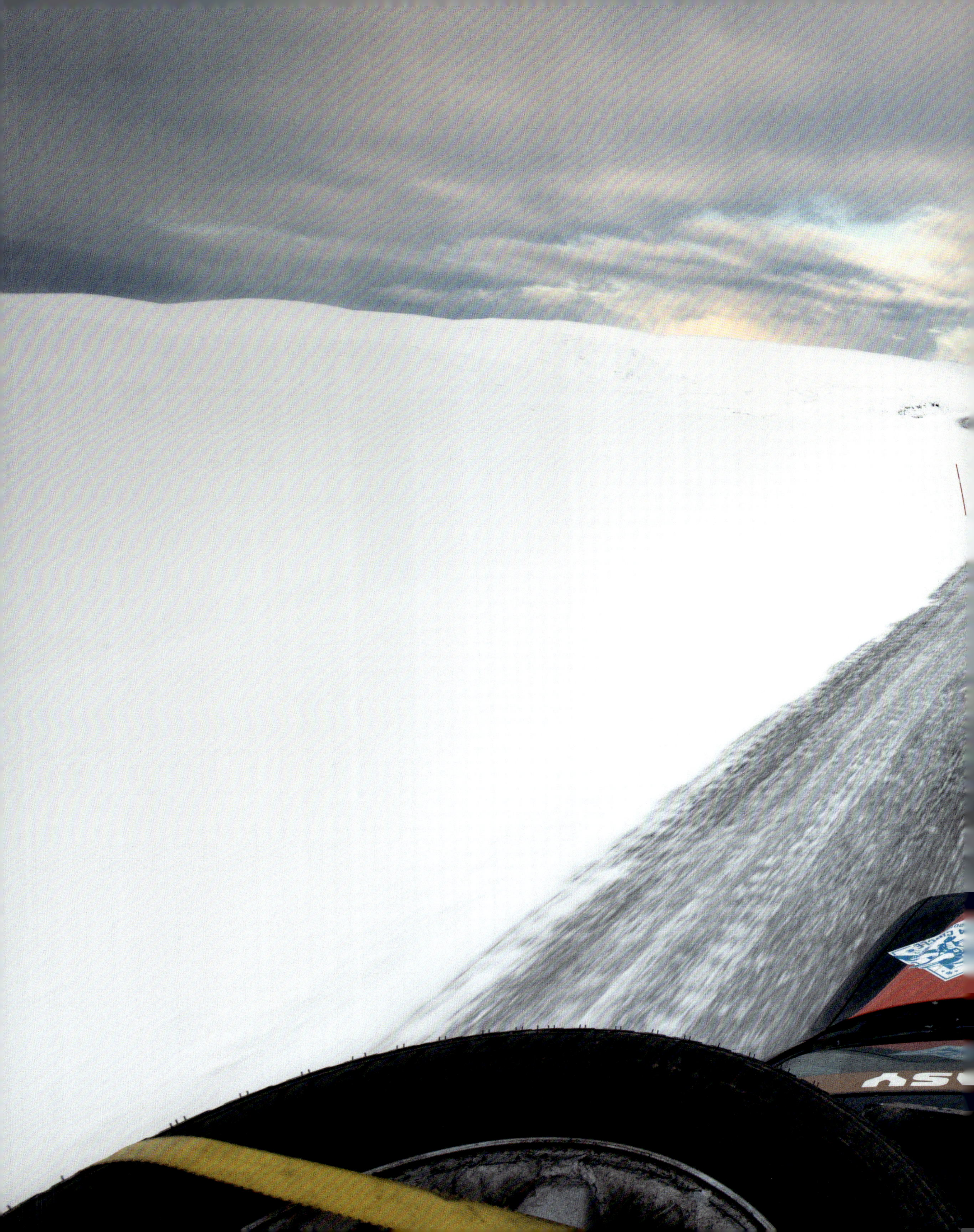

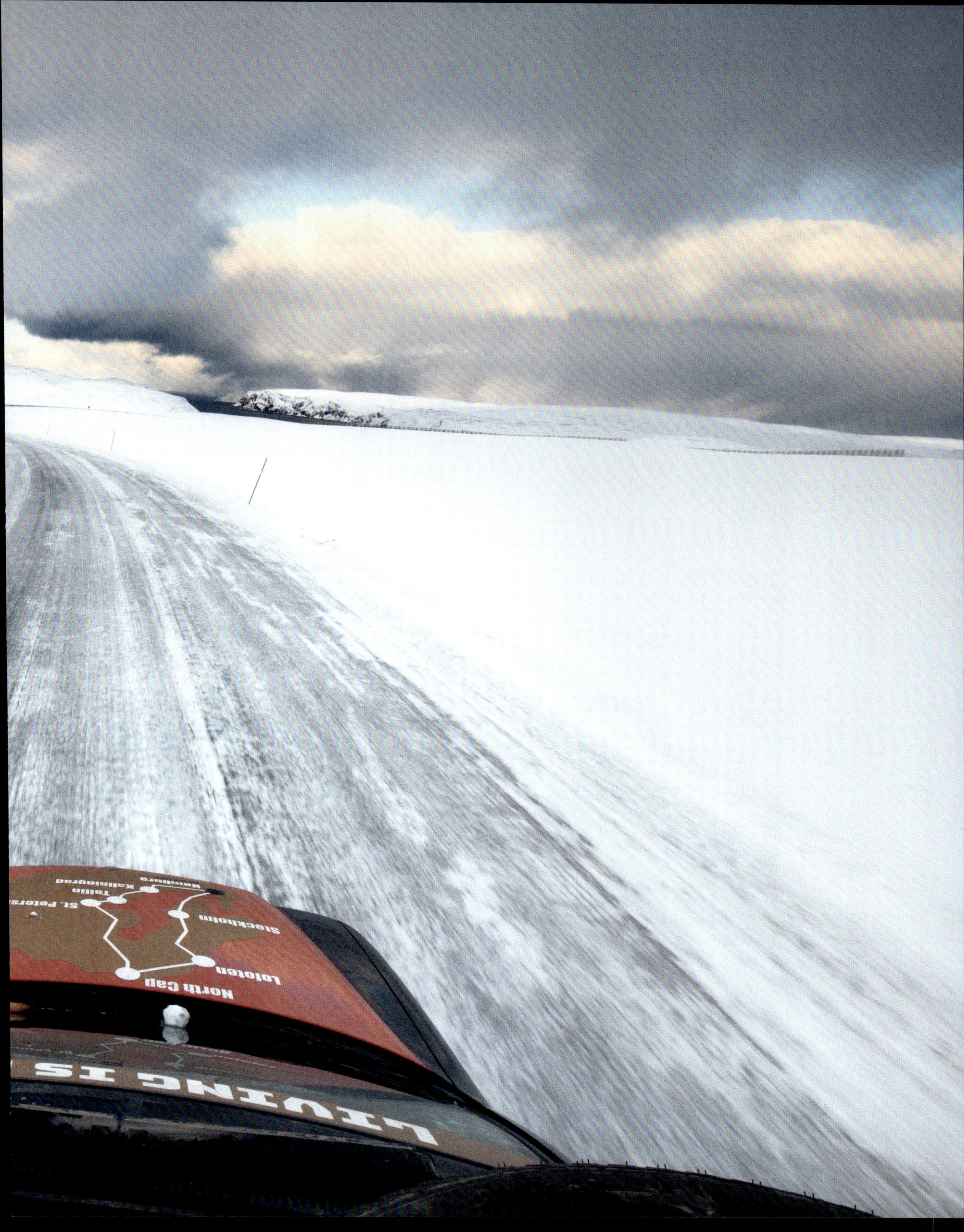
North Cap
Lofoten
Stockholm
Tallin
LIVING IS

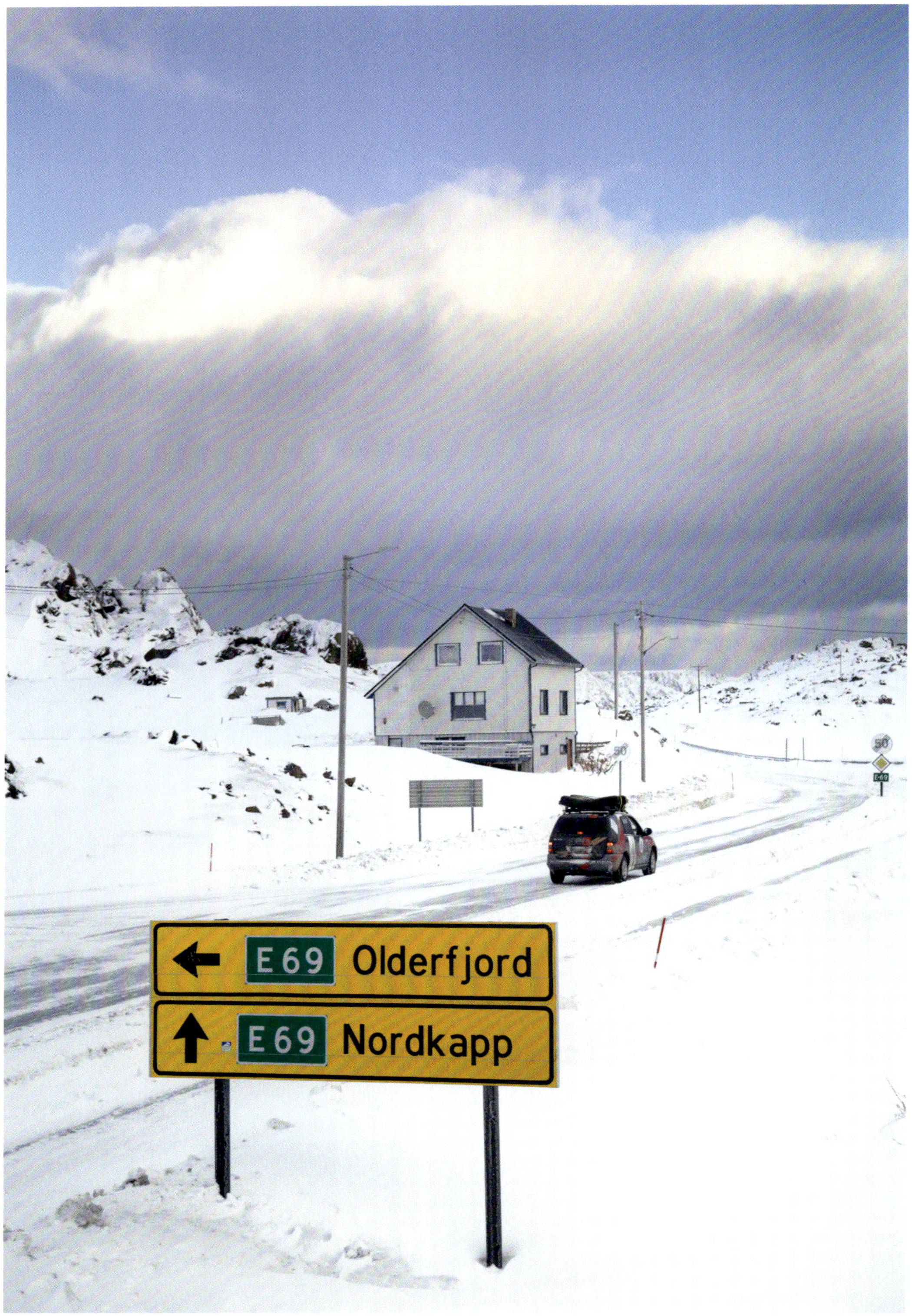

Above In winter, when Arctic winds deposit dangerous layers of ice on the road, driving in a convoy behind a powerful snowplough is the only way to safely reach the North Cape. The meeting point is a junction 13 kilometres (8 mi.) from the North Cape plateau. The northern terminus of the E69, created in 1975, is the northernmost point in continental Europe that's accessible by car. From here, the view stretches all the way to the North Cape: an impressive, 307-metre- (1,007 ft.) high cliff with a large plateau on top, and, beyond, the tempestuous Barents Sea.

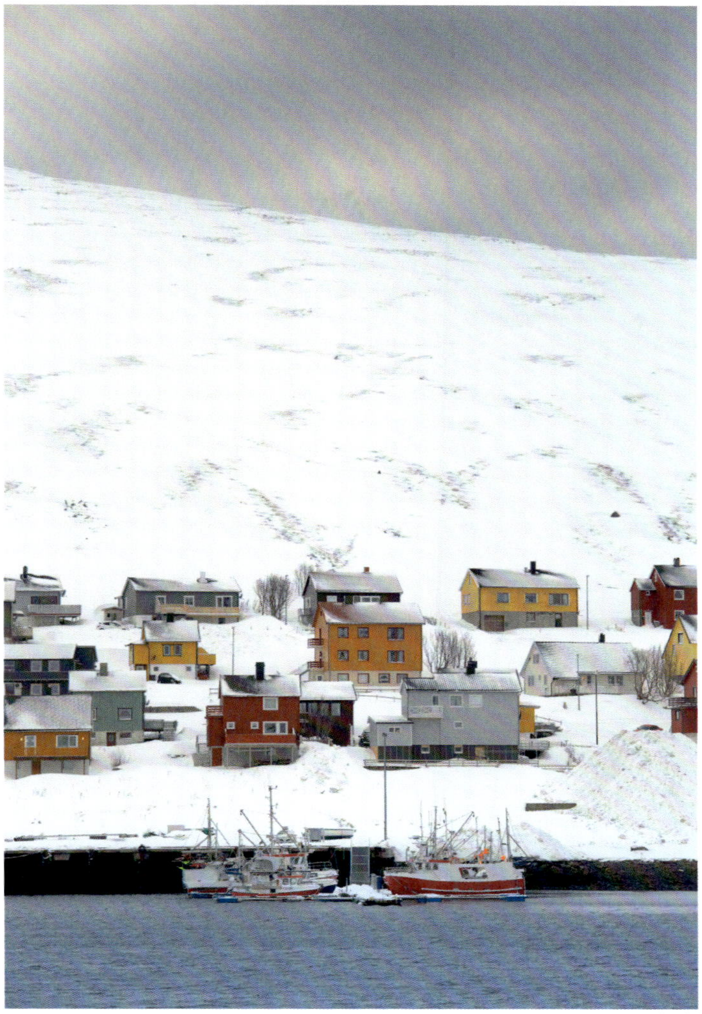

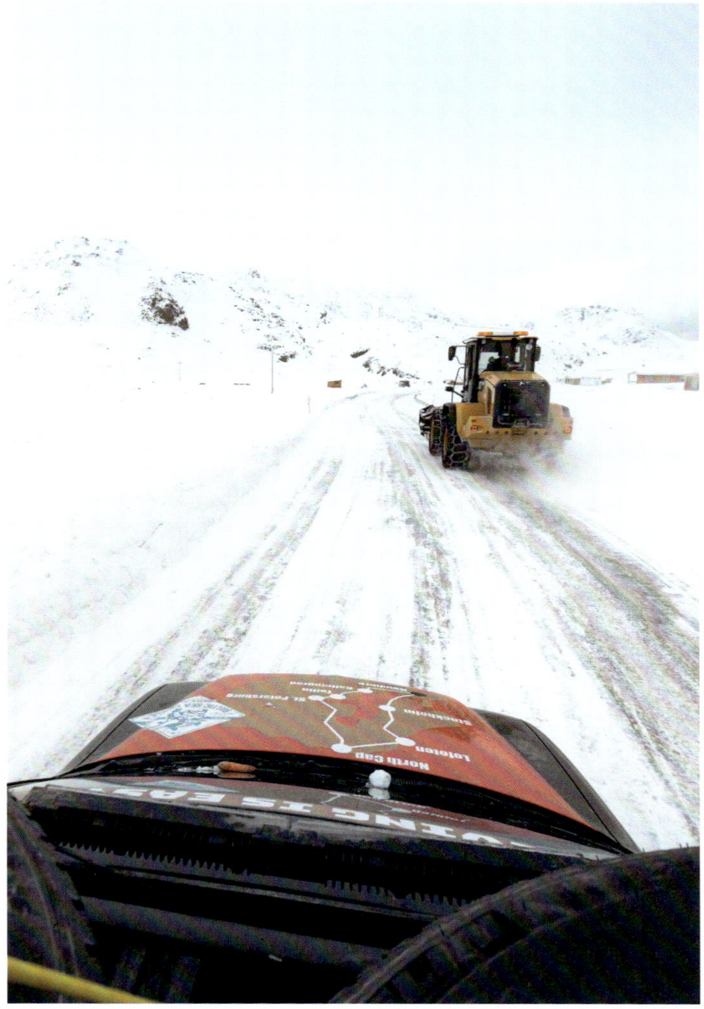

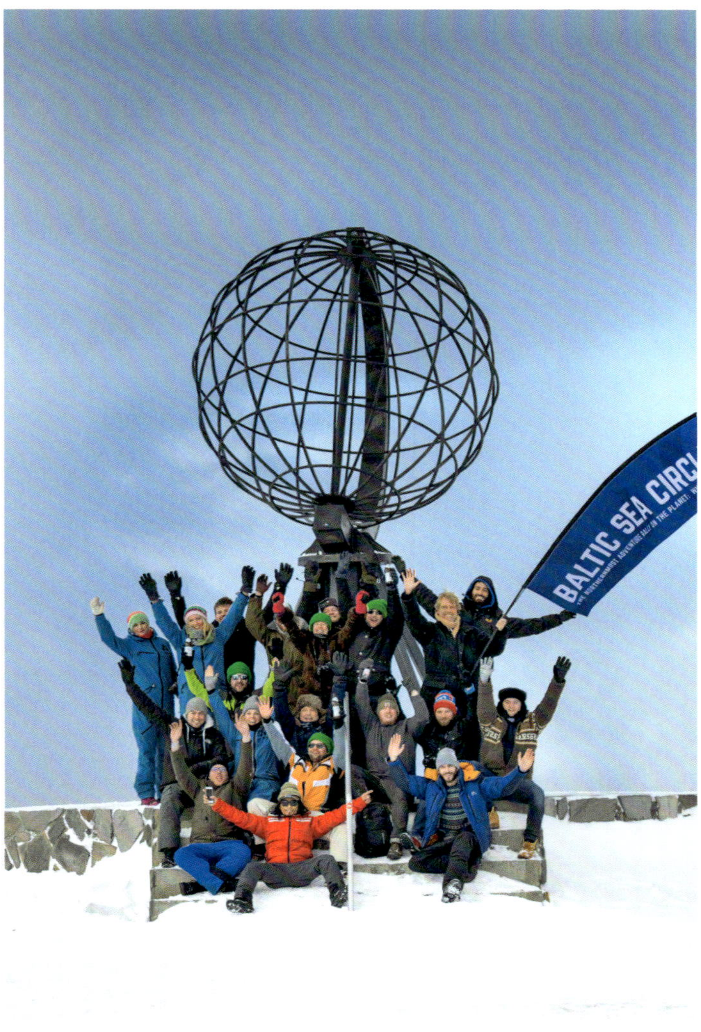
BALTIC SEA

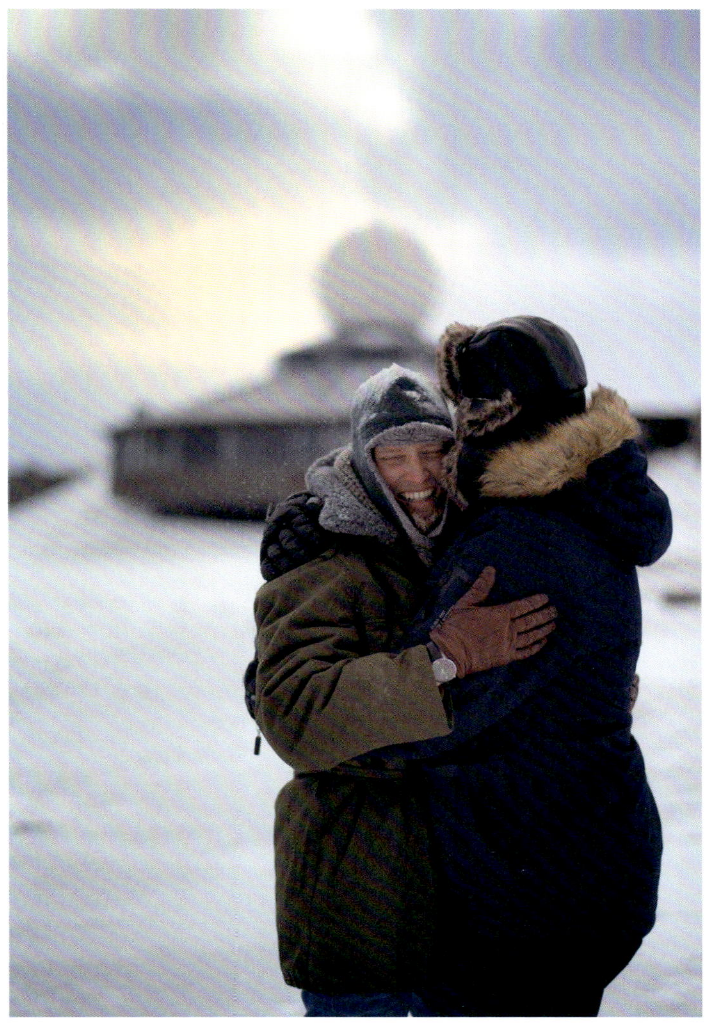

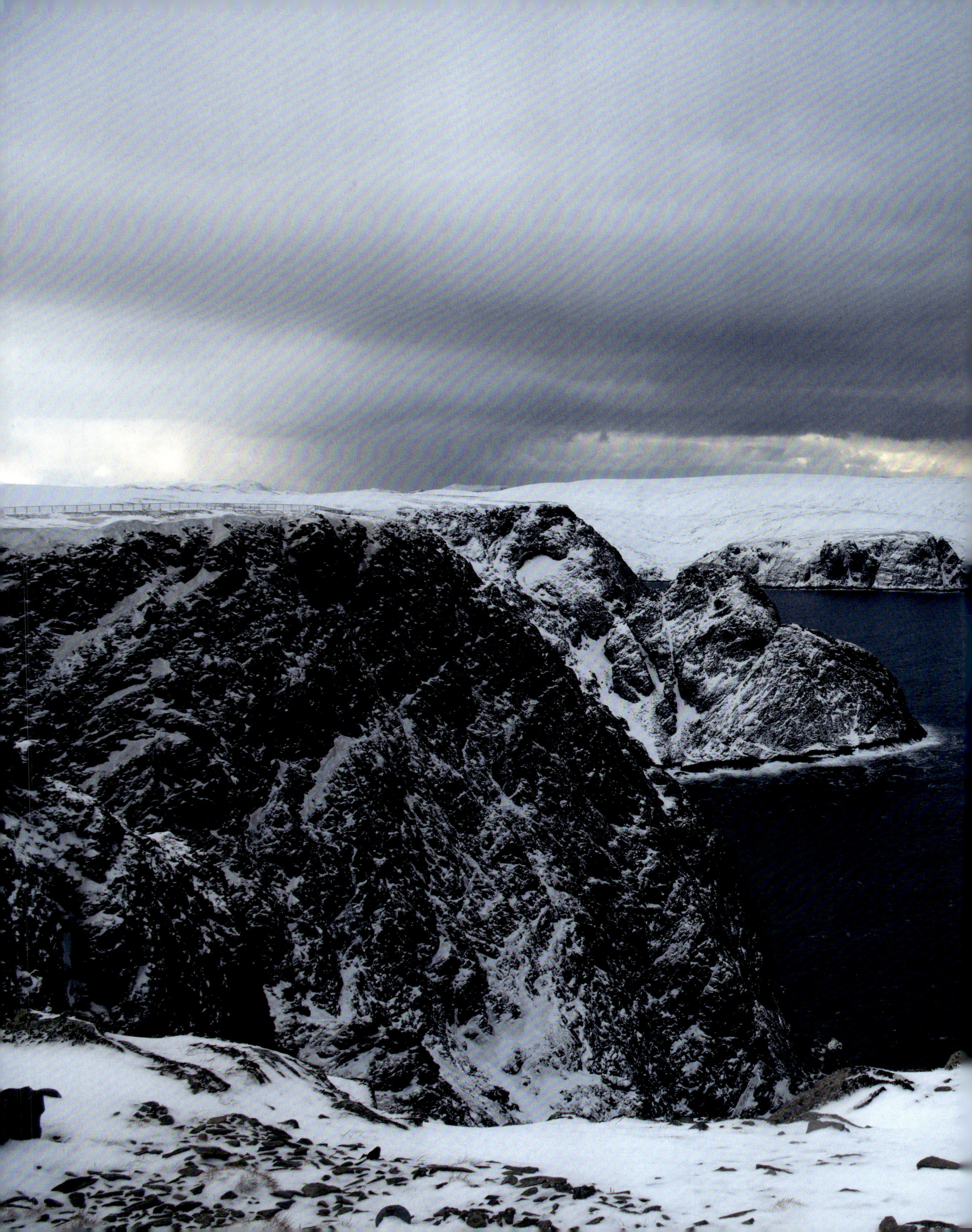

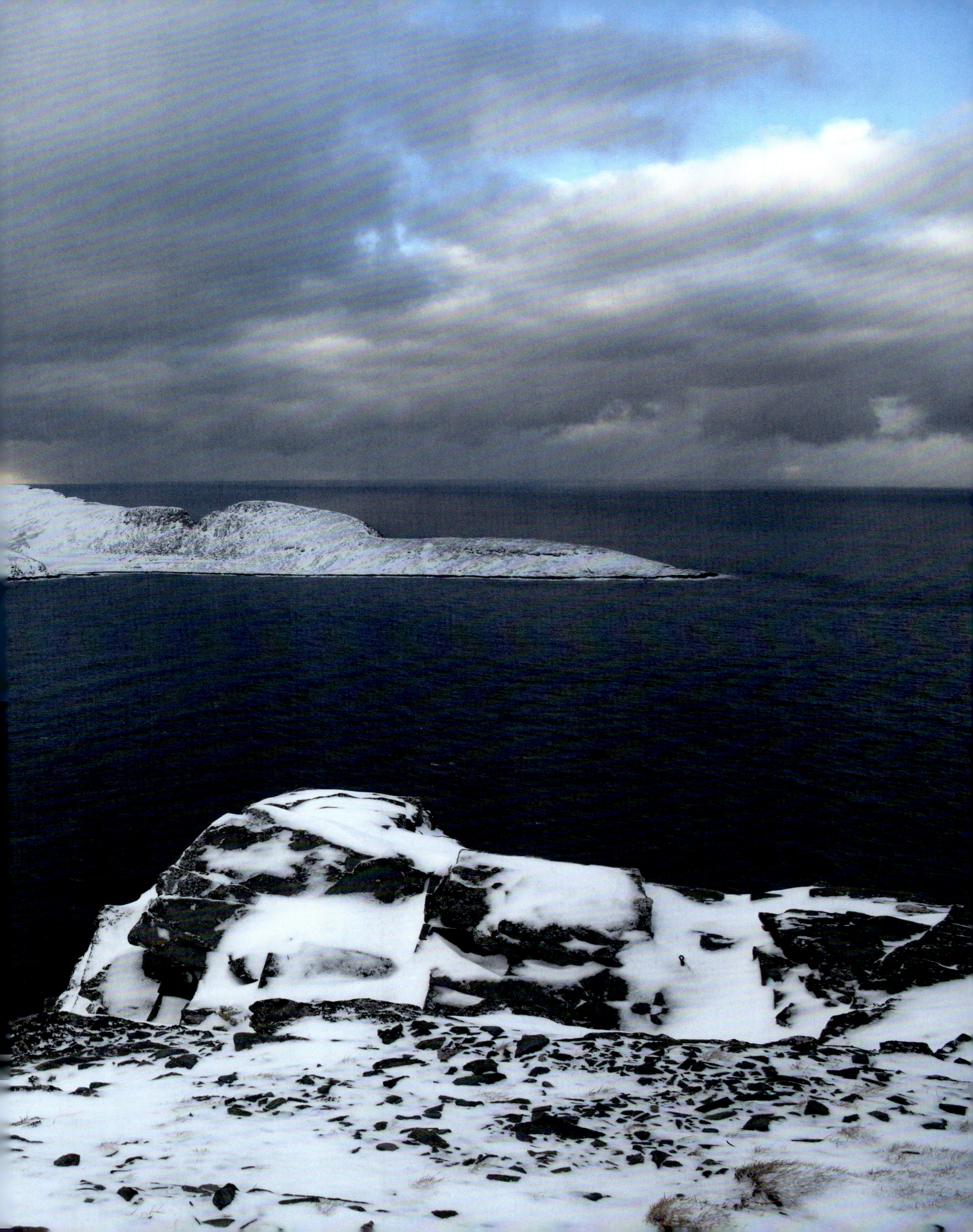

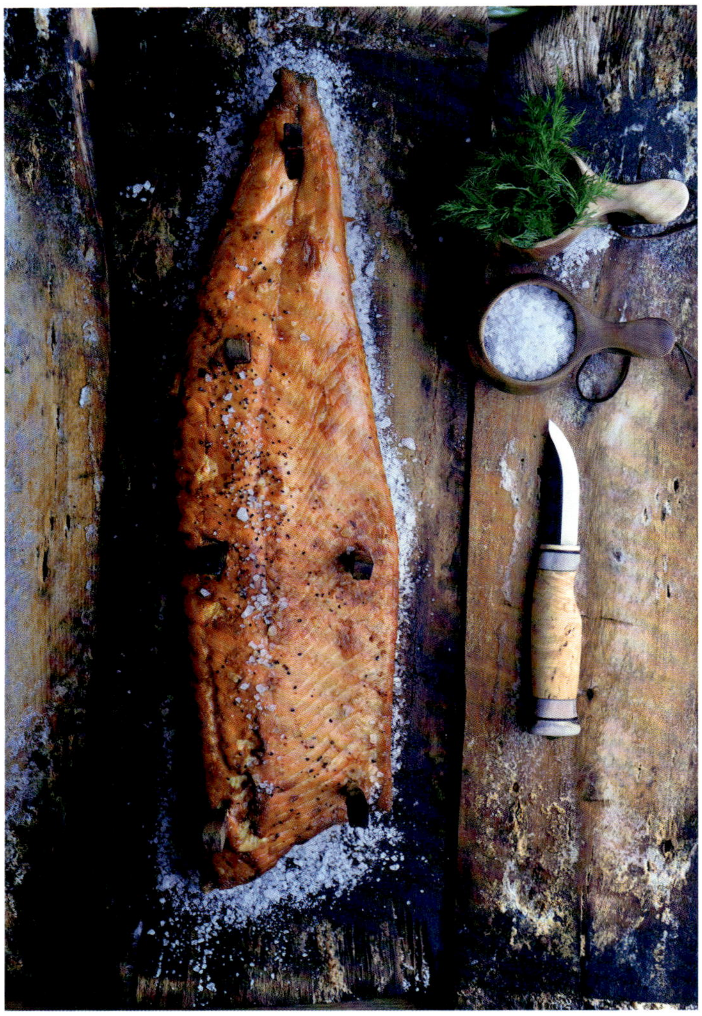

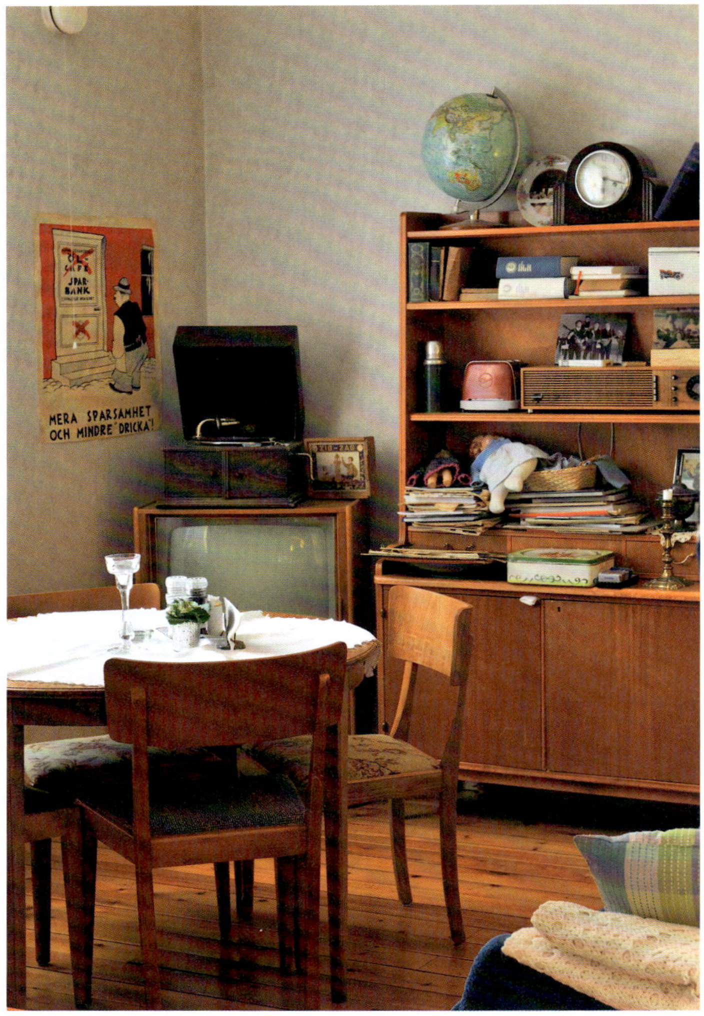

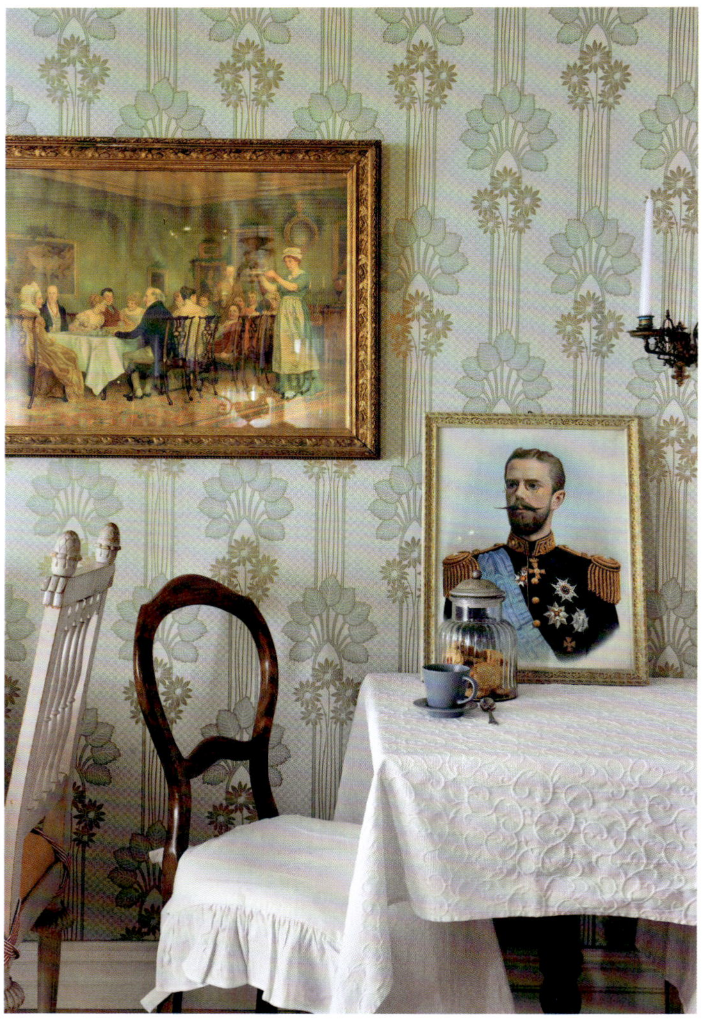

Previous spread The impressive North Cape cliff has long been a popular and important navigation mark for marine traffic in the north. The almost 600-million-year-old North Cape is at 307 metres (1,007 ft.) above sea level. A quarter of a million tourists visit the North Cape each summer, making it one of Norway's top travel destinations.

This spread Each team overnights and dines according to their own preferences. Some camp, others choose simple rental cabins or exceptional lodgings such as the iconic Ice Hotel and the Tree Hotel in Swedish Lapland. At the latter, owners Kent and Britta also run the intimate Brittas Pensionat, the perfect lodging for this intense road trip. The food is homemade, the rooms are warm and cosy, and there is always a piping-hot sauna.

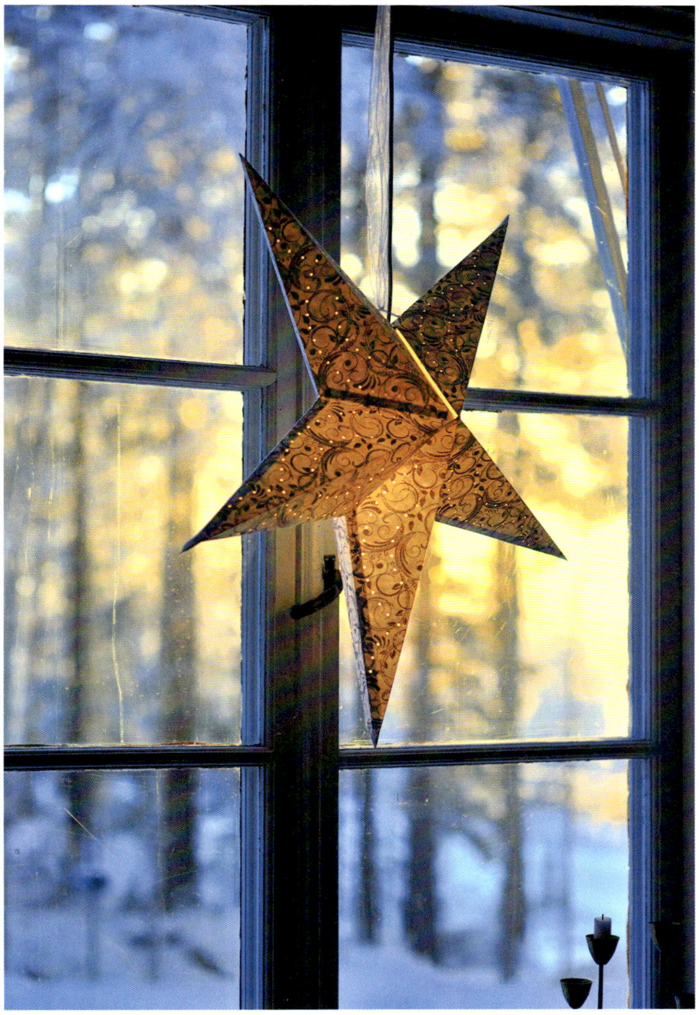

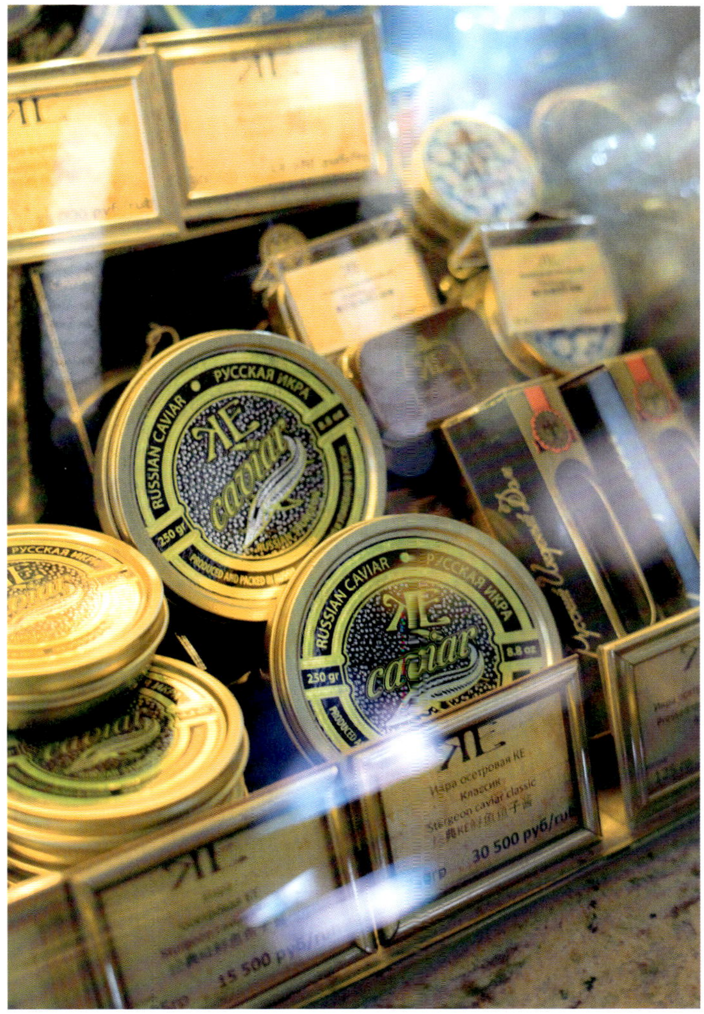
RUSSIAN CAVIAR
РУССКАЯ ИКРА
caviar
250 gr
8.8 oz
Икра осетровая KE
Классик
Sturgeon caviar classic
30 500 руб/
15 500 руб/

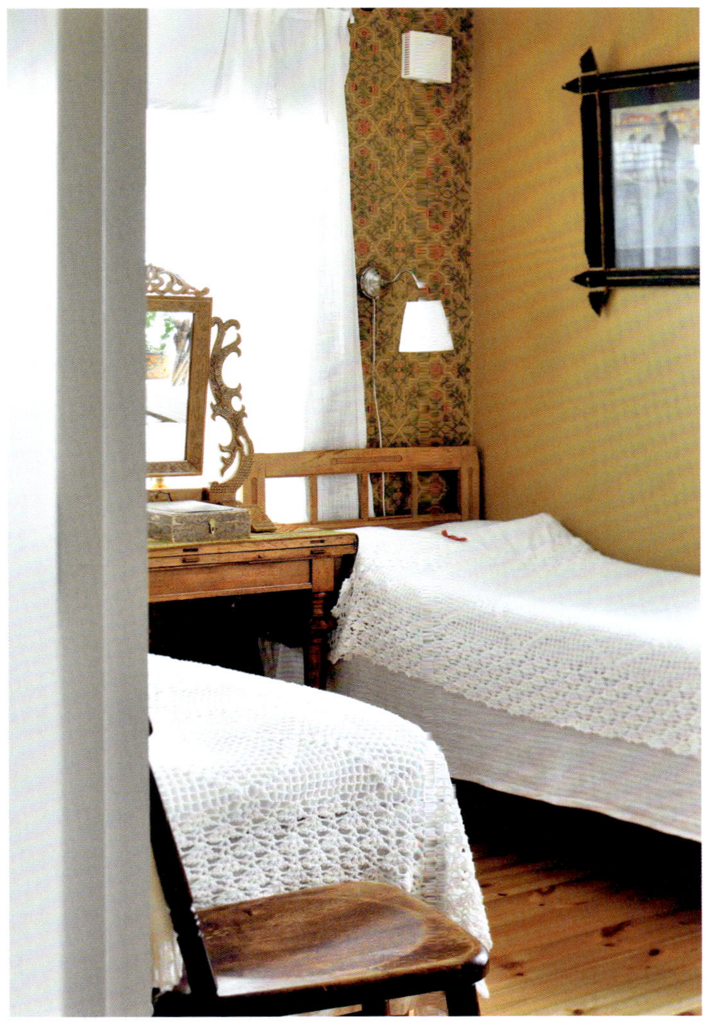

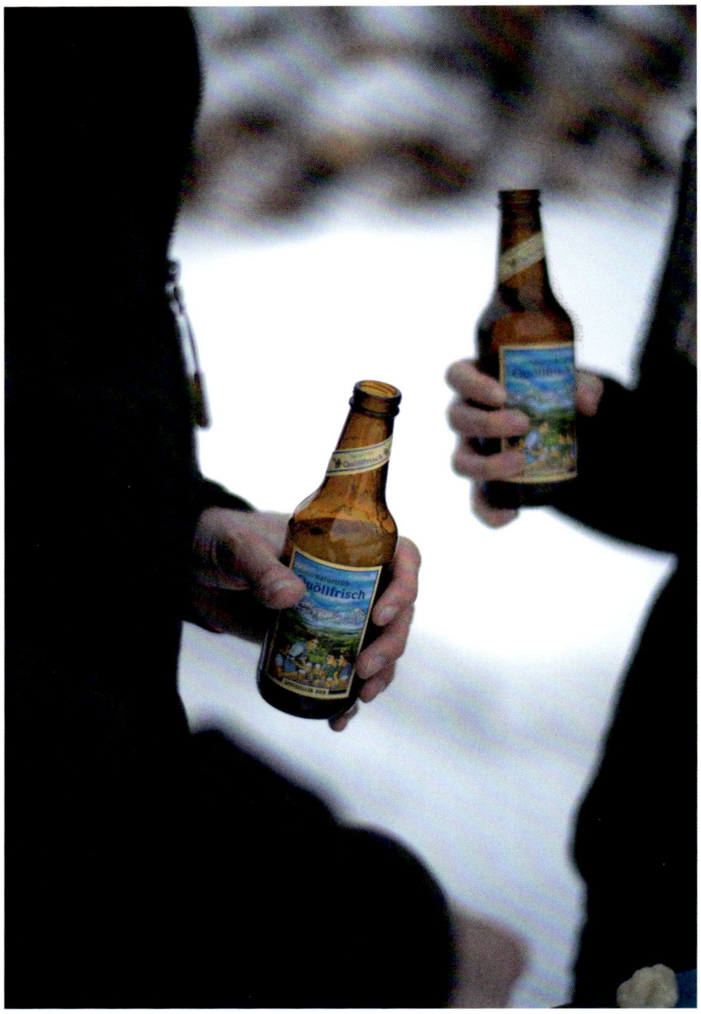
Quöllfrisch

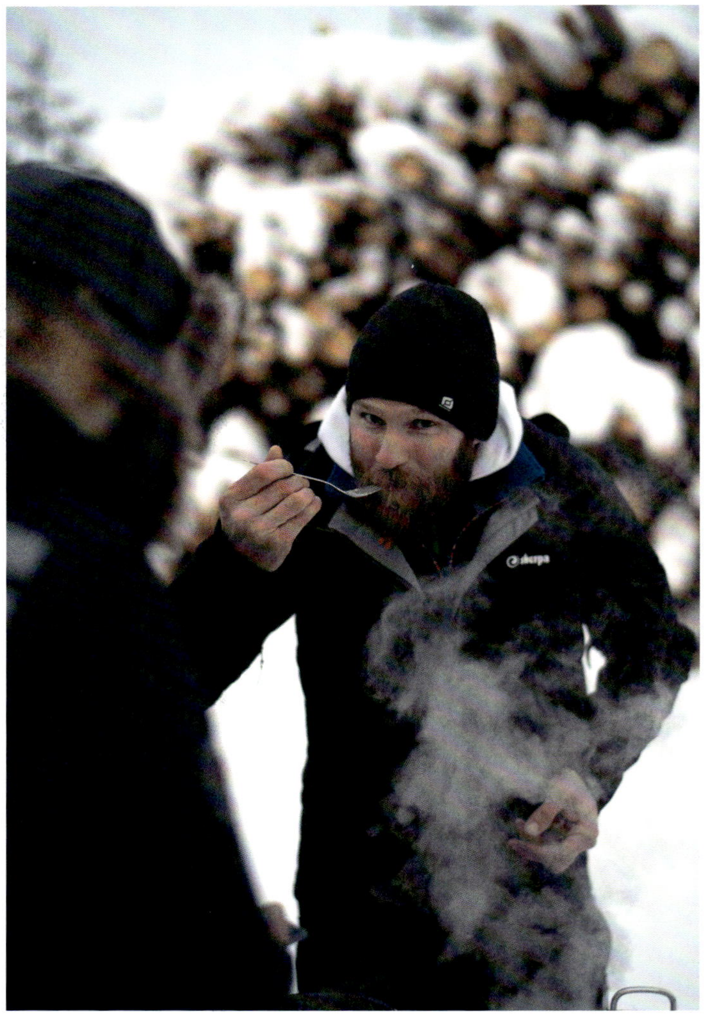
sherpa

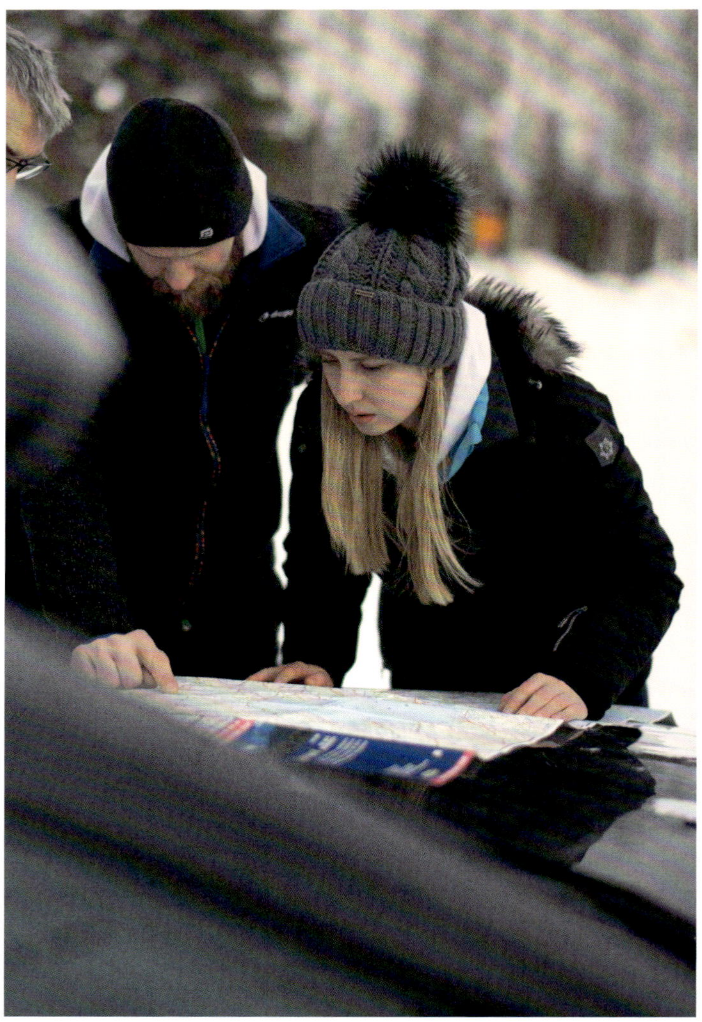

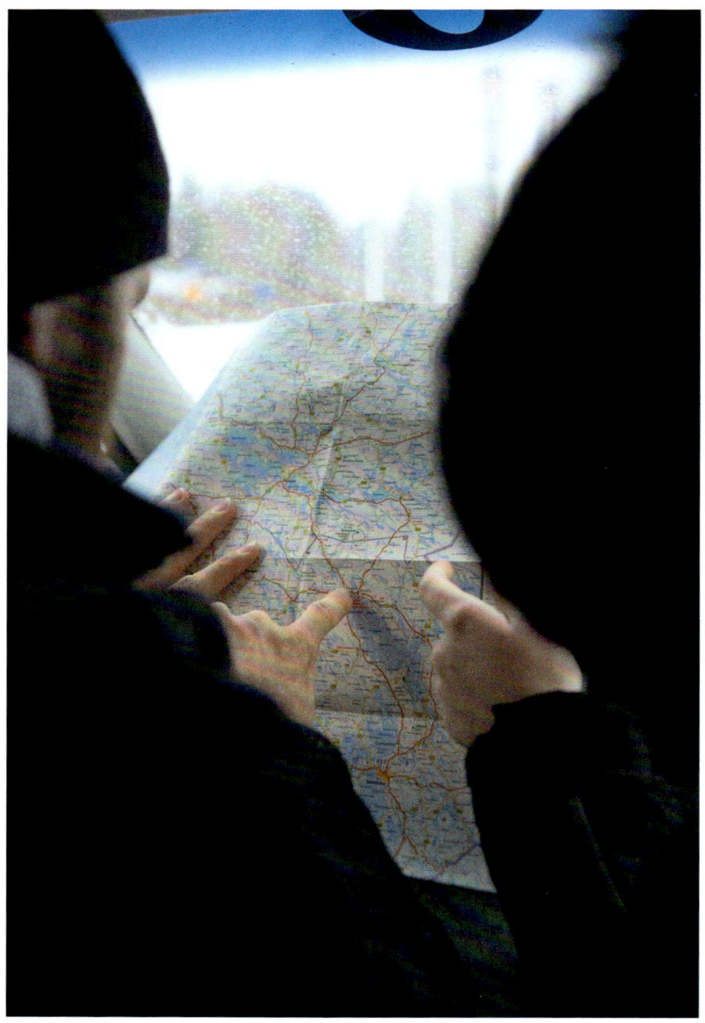

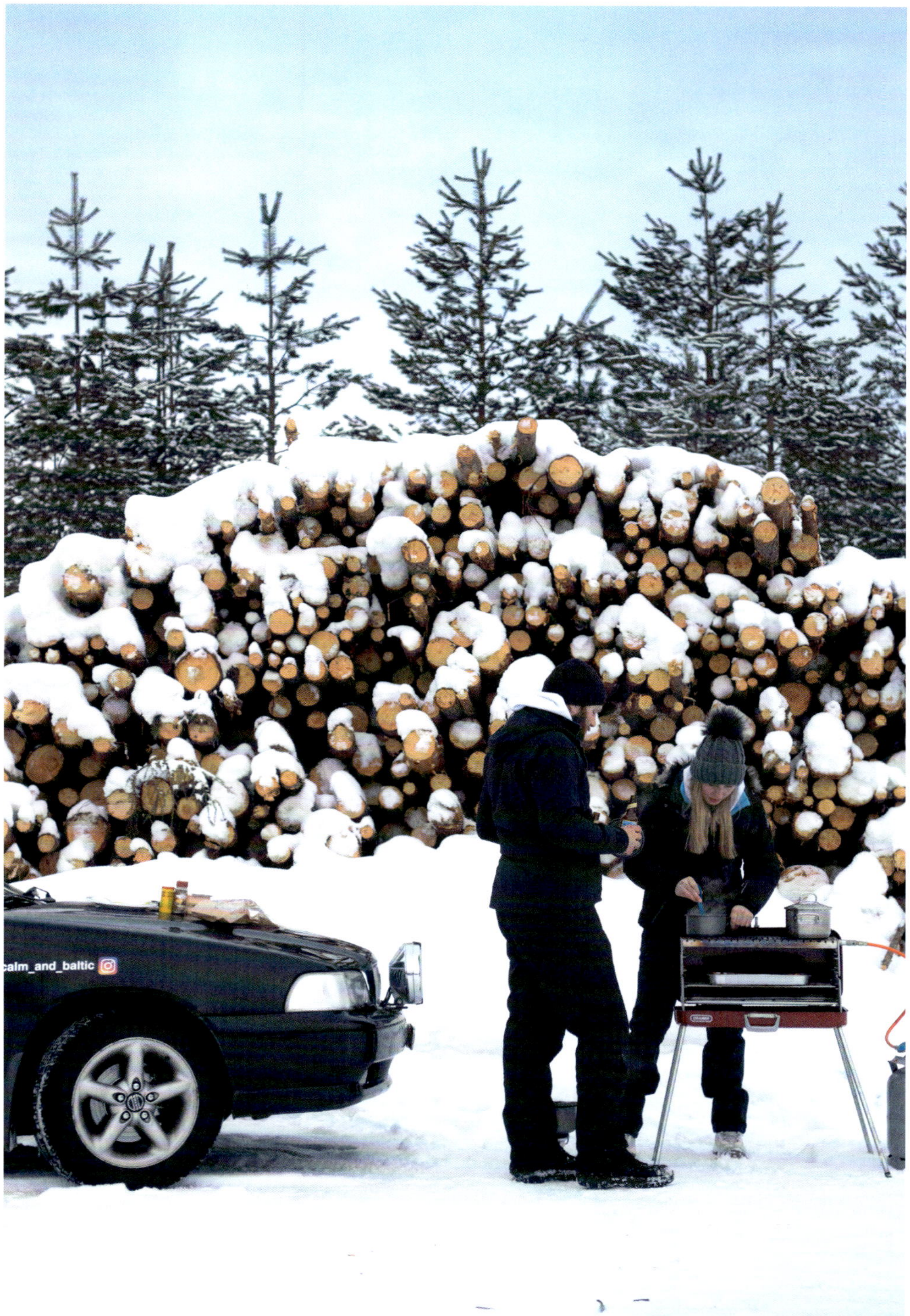

This spread Picnicking is an essential part of the rally, either with homemade treats or local delicacies bought en route. The Swiss members of team Keep Calm and Baltic prepare a roadside fondue at -20°C (-4°F) — a perfect match for the scenery of Finnish Lapland. Passing teams are invited to join in the feast and crack open a bottle of Quöllfrisch — beer from the Swiss Appenzell region.

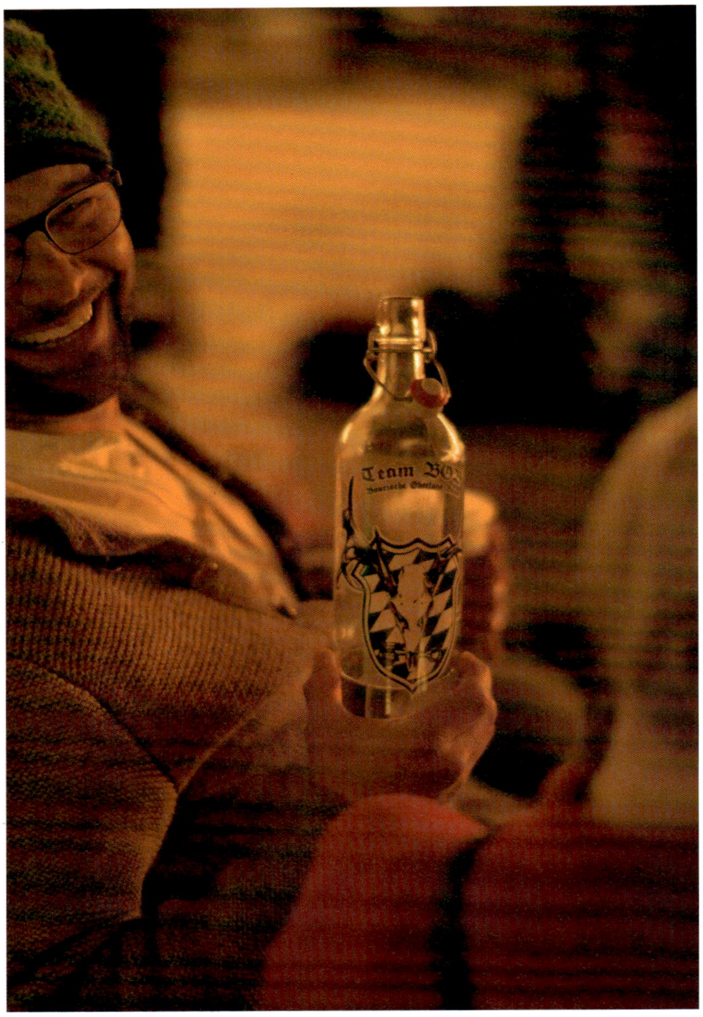

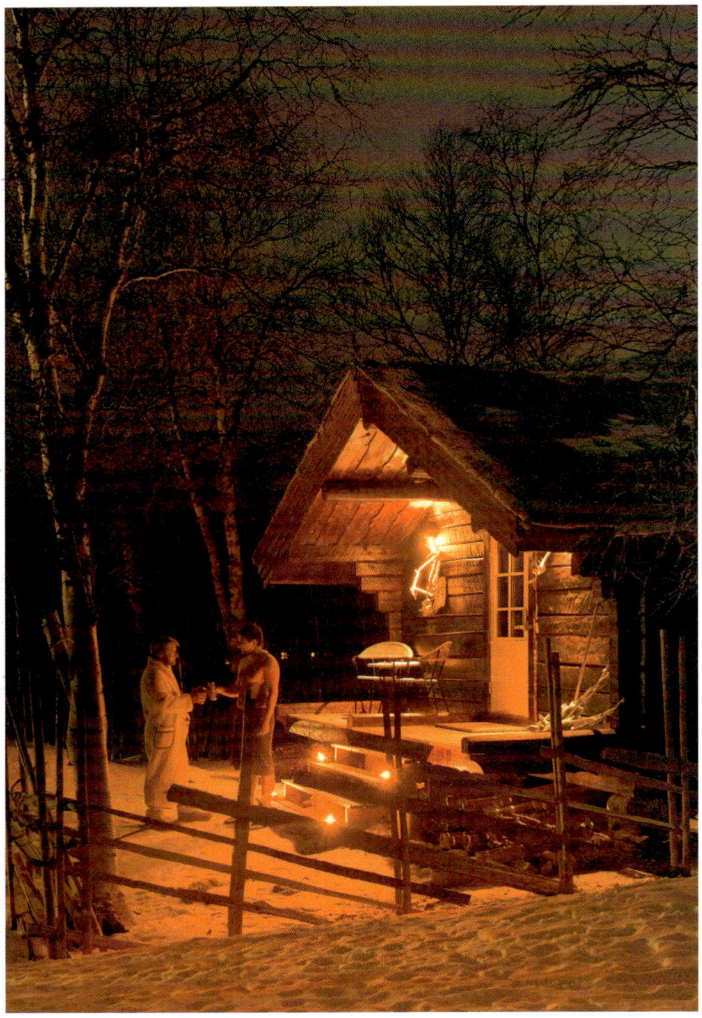

This spread On two evenings, all the rally teams are invited to eat, drink, and spend the night together. One of those evenings is at Kolja, an Arctic party venue inside a massive teepee near Tallinn — a dreamy winter wonderland of saunas, hot tubs, and plenty of Estonian food and beer.

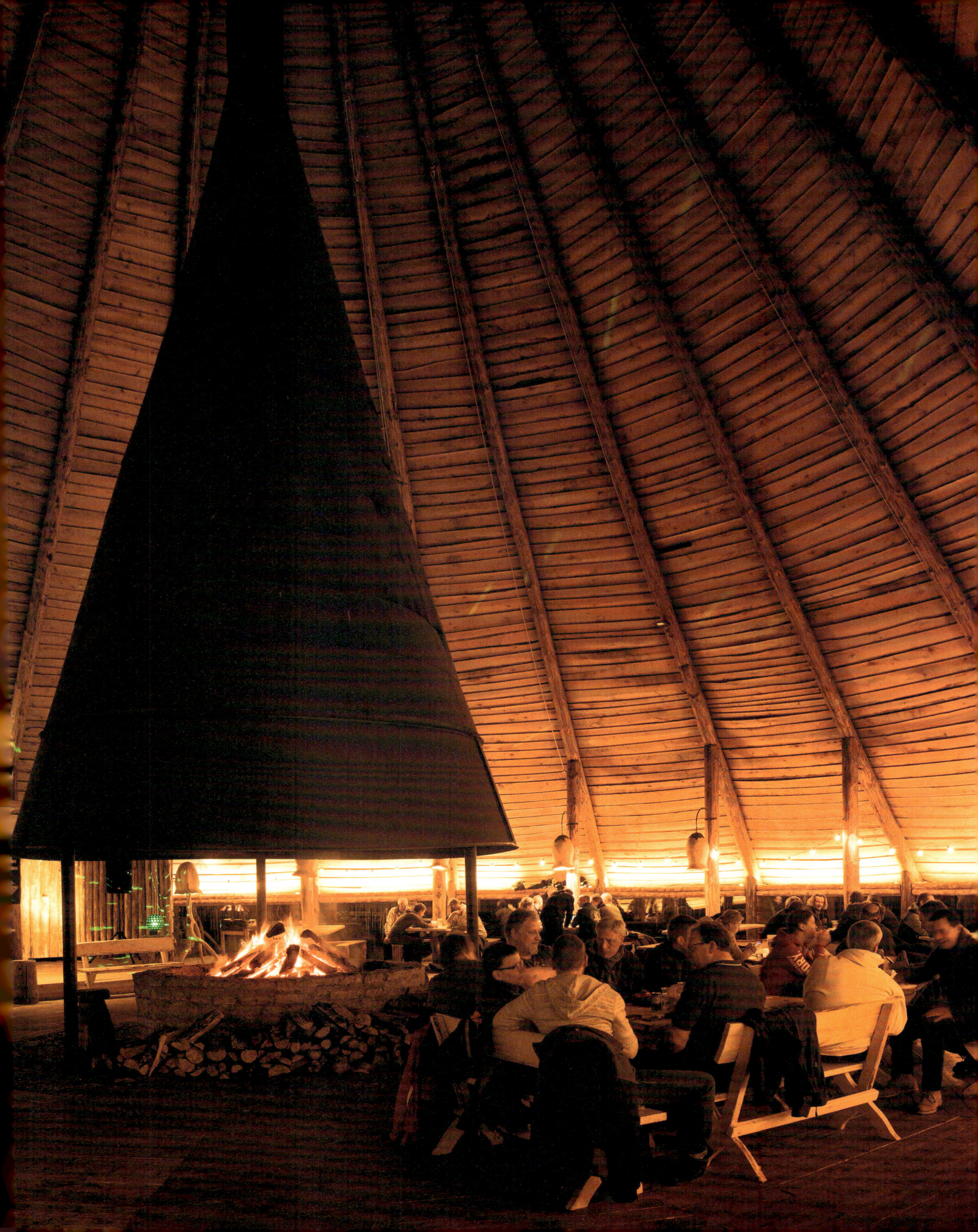

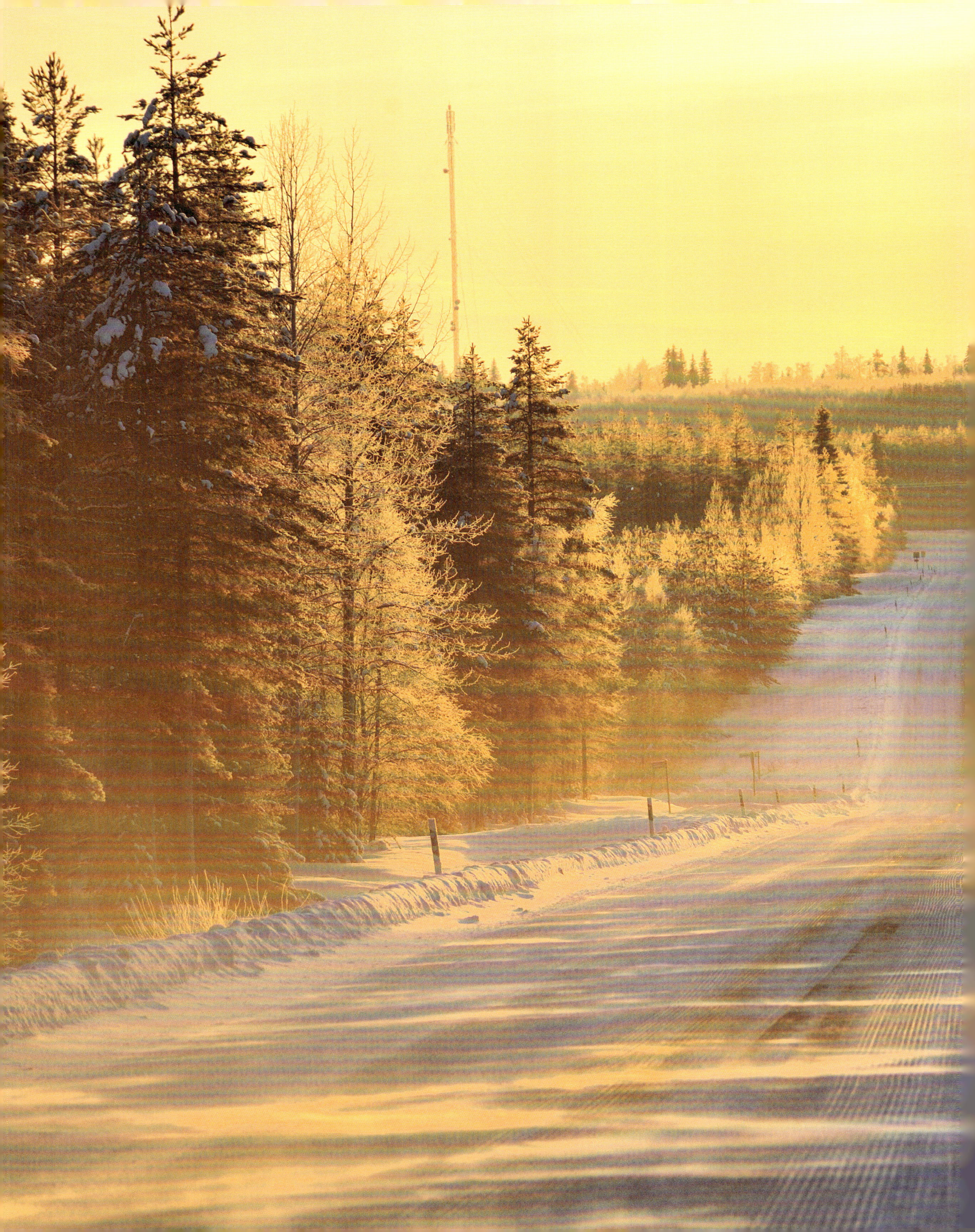

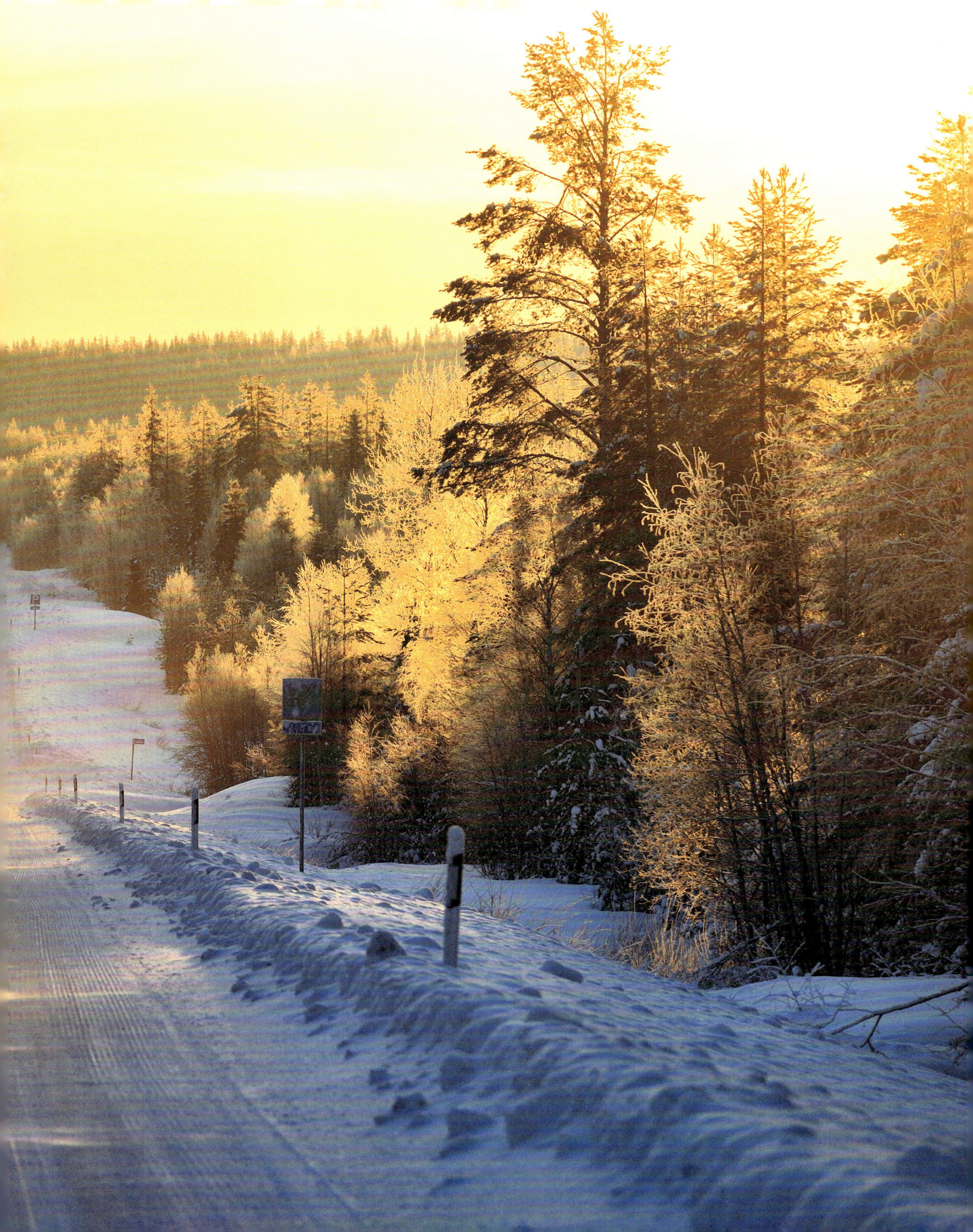

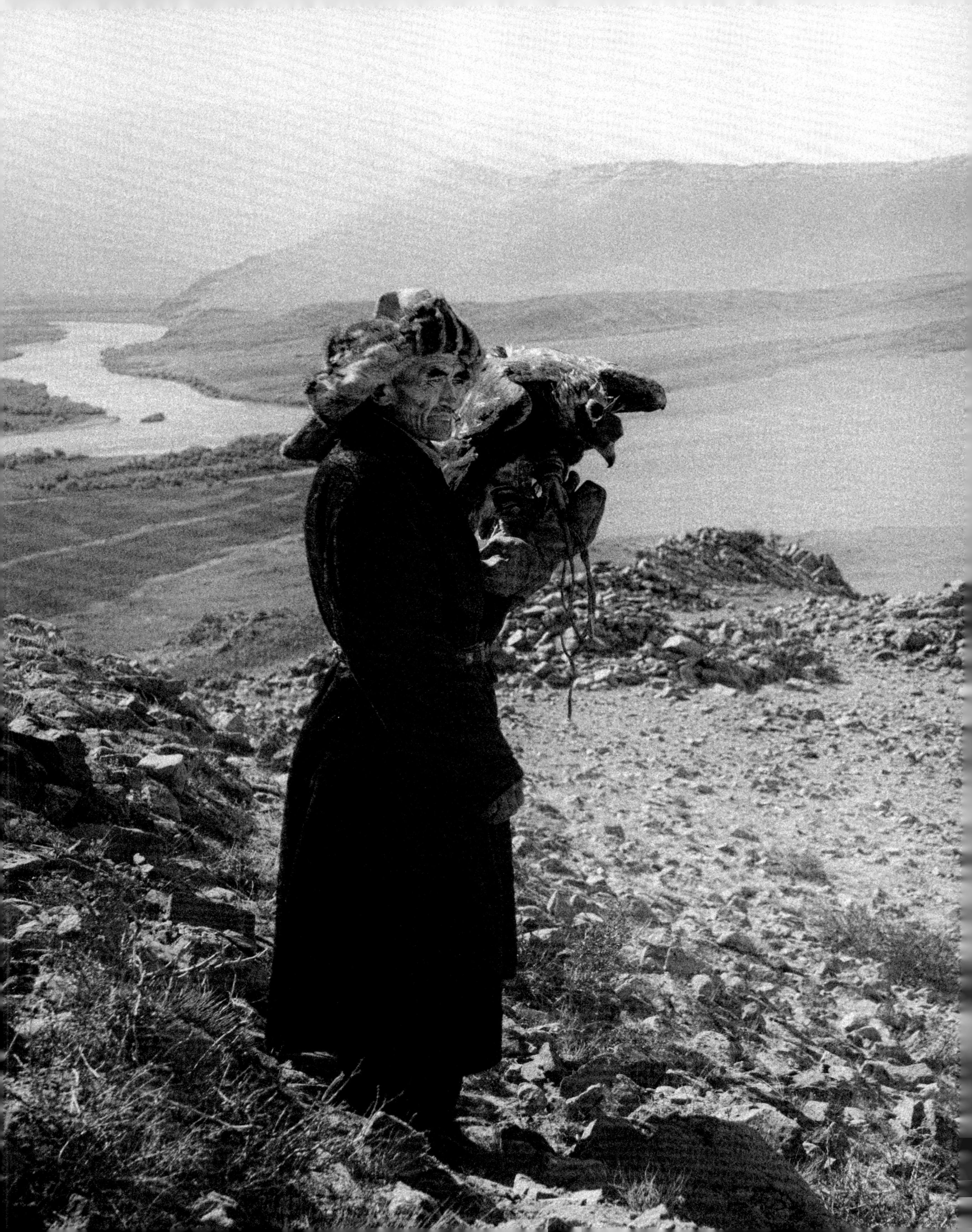

CHAPTER 4

GOLDEN EAGLE HUNTING IN THE FAR WEST OF MONGOLIA

NOMADIC TRADITIONS

Mongolia

48°52'24.61" N
89°37'21.56" E

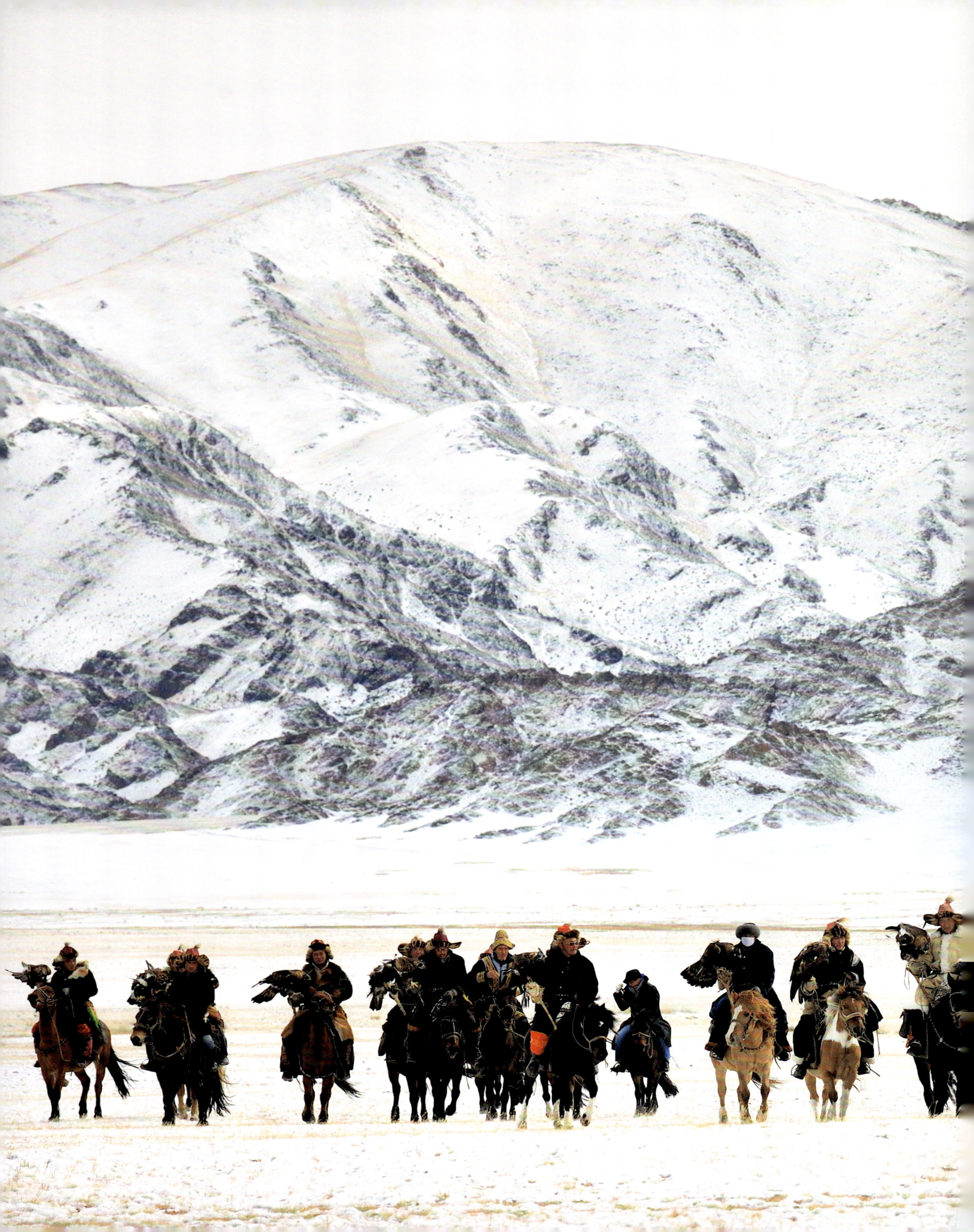

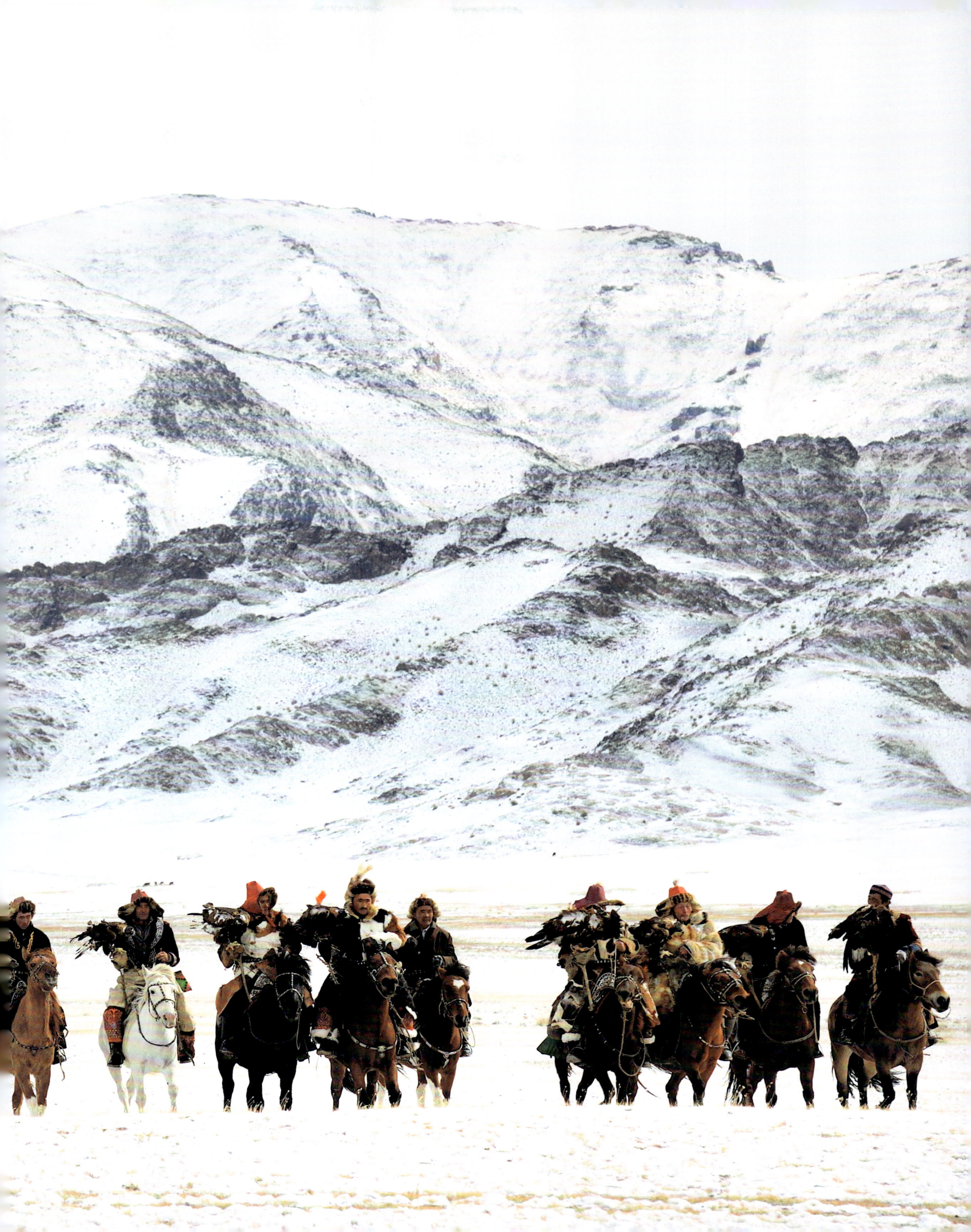

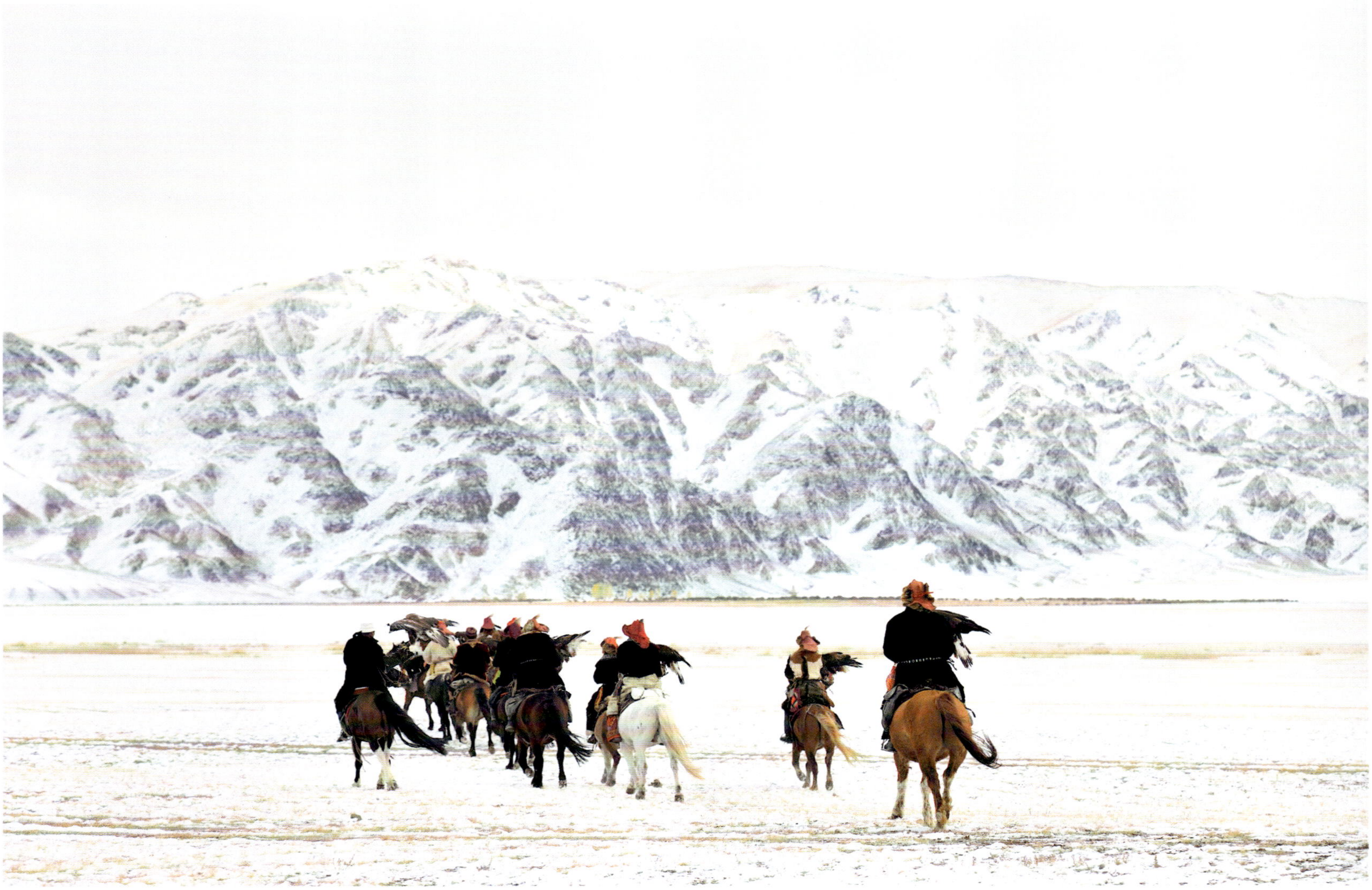

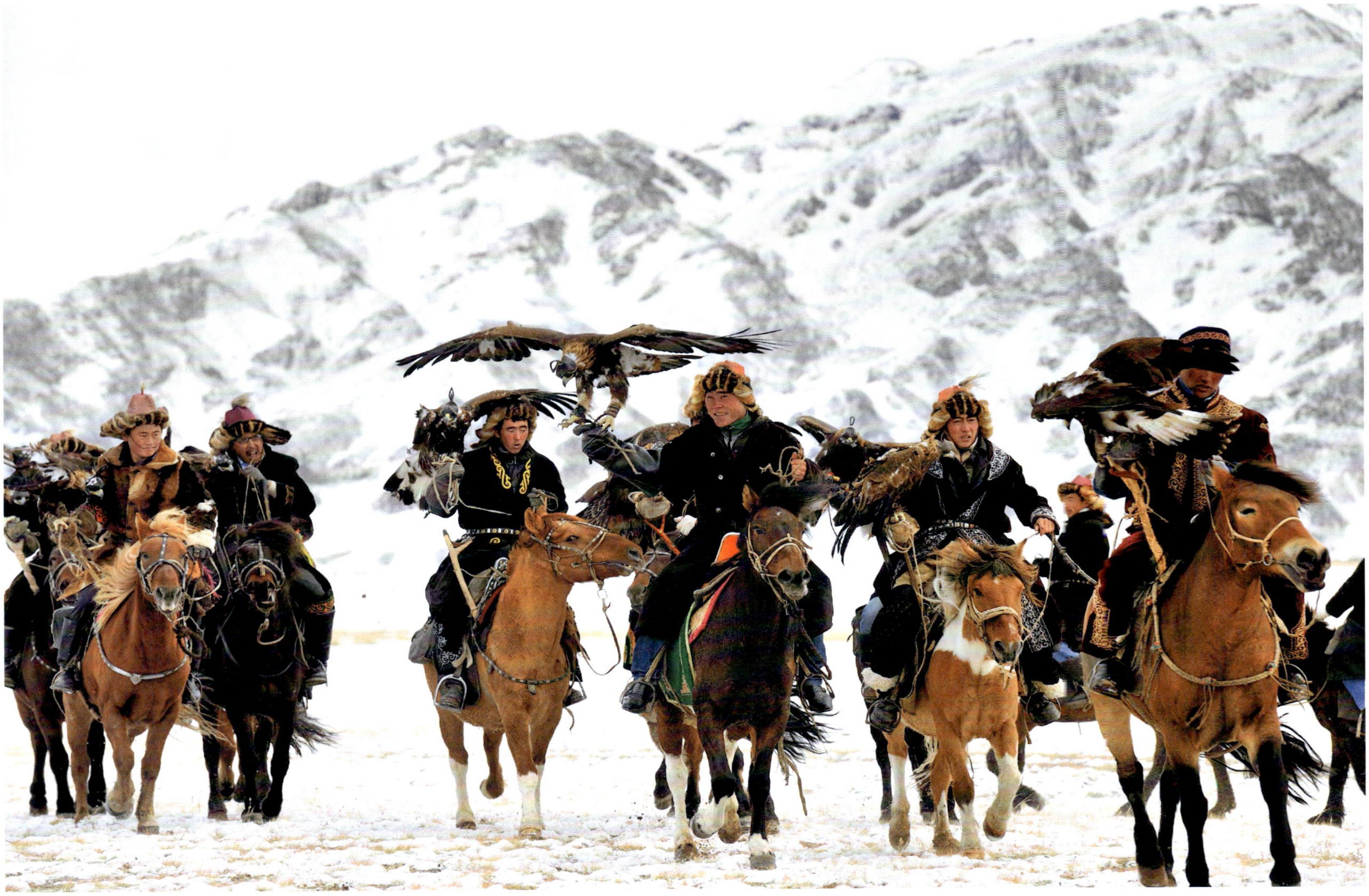

Golden Eagle Hunting in the Far West of Mongolia

Nomadic traditions

Country:	**Mongolia**
Best time to travel:	**September**
Closest airport:	**Bayan-Ölgii**
Price tag:	**€€**
Style of travel:	**A small-scale, off-the-map festival experience**

Introduction

Every year in September, Bayan-Ölgii, a remote province in the westernmost corner of Mongolia, comes alive with the Kazakh Eagle Festival. Just 75 kilometres (47 miles) from the Russian border, nomadic eagle hunters known as berkutchi gather to celebrate their age-old traditions in the windswept and dusty provincial capital, Ölgii. In the distance, the barren, snow-specked Altai Mountains serve as a backdrop to one of the most fascinating celebrations of cultural heritage in Asia. The art of eagle hunting dates to a time when the ancestors of Genghis Khan lived in a kingdom by the Aral Sea. Forced to flee the rising Russian Empire, these Kazakhs took refuge in the remote Altai Mountains of what is now Western Mongolia. Cut off from their homeland when the Soviet Union and China drew borders on either side of them, they continued to live their semi-nomadic lifestyles, carrying on traditions banned by Stalin and Chairman Mao. The remote Altai Mountains became one of the only places where the berkutchi could continue their sacred tradition of hunting with golden eagles. Each year, visitors come to Bayan-Ölgii to witness this extraordinary display. Often distinguishable by their weathered faces and colourful fur attire, the hunters live in simple houses or well-insulated *gers* (a Mongolian type of yurt) scattered across the valleys. Here, golden eagle chicks are hand-raised, sheltered from fierce snowstorms and cared for like family members. The bond between a berkutchi and his bird is strong and sacred. They will only hunt in winter, when the fur of their prey is thick and soft. In the run-up to the festival, they train intensively.

The Views

At the time of the Kazakh Eagle Festival, a fierce winter with temperatures below -40°C (-40°F) is just around the corner. In mid-September, the nights are already freezing. An arctic wind blows a flurry of dust and light snow across the endless plains. Few roads lead into this vast and remote area where lofty peaks keep modernity at bay, but this doesn't stop Kazakhs from all over the region travelling to Ölgii to participate in or witness the event. The kick-off, when 50 or so eagle hunters come riding in with the craggy, snowcapped Altai range behind them, looks like a scene from a western. Cold winds blow across the valley but the hunters are undeterred. Their fur jackets and hats are made from the foxes, rabbits, marmots, and even wolves hunted by their beloved eagles. The more extravagant their attire, the more respect they command. It is said that only eagles caught in the wild can take on wolves. Some, old eagle hunters say, can even overpower a snow leopard. The daily hunting routine is sacred. Each morning, after a few cups of steaming-hot milky tea, a hunter puts on a thick leather glove to carry his eagle on his arm. Then, resting this arm on a wooden crutch, he mounts his horse and rides off into the dawn. Patient and determined, horse, man, and bird roam the mountains and endless steppes, hoping for a glimpse of the elusive but beautiful corsac fox that is endemic to the barren subarctic deserts of Central Asia. During the festival, humans, eagles, and horses demonstrate expert skill as they work in unison to compete in traditional Kazakh games. In one of the competitions, hunters have their birds fly to the top of a hill and back while they wait in the valley below. In another, the goal is to pick up a coin from the ground while riding a horse. One game offers comic relief when horsemen who steal a kiss are consequently chased by furious horsewomen cracking whips. In yet another, a horseman attempts to place the carcass of a goat or calf in a goal. Come the evening, traditional Kazakh music is played and small shows are held in downtown Ölgii.

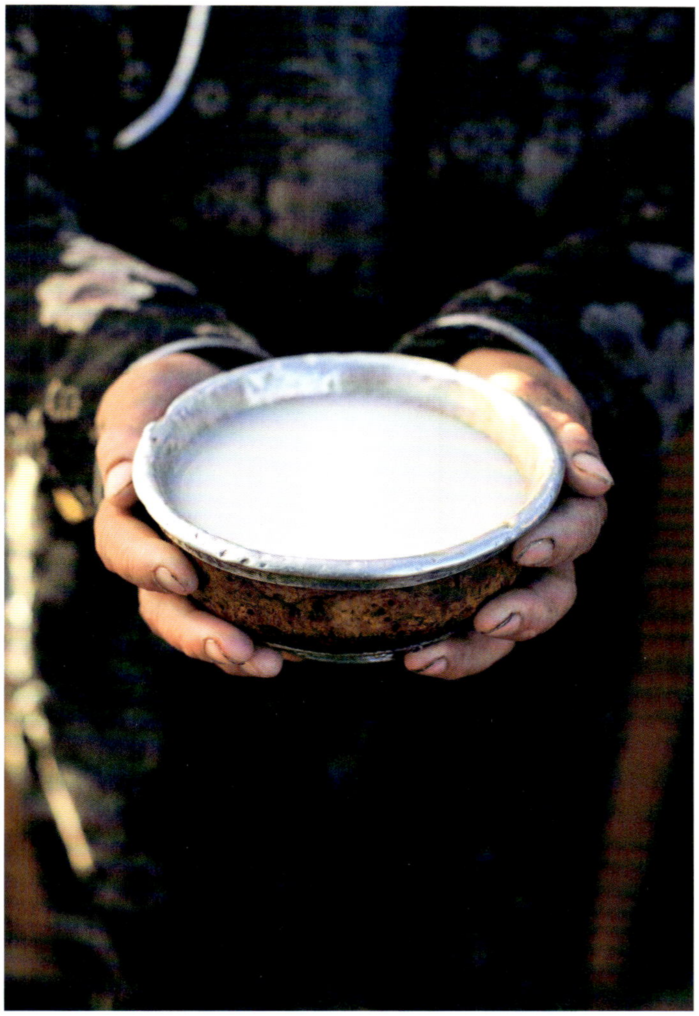
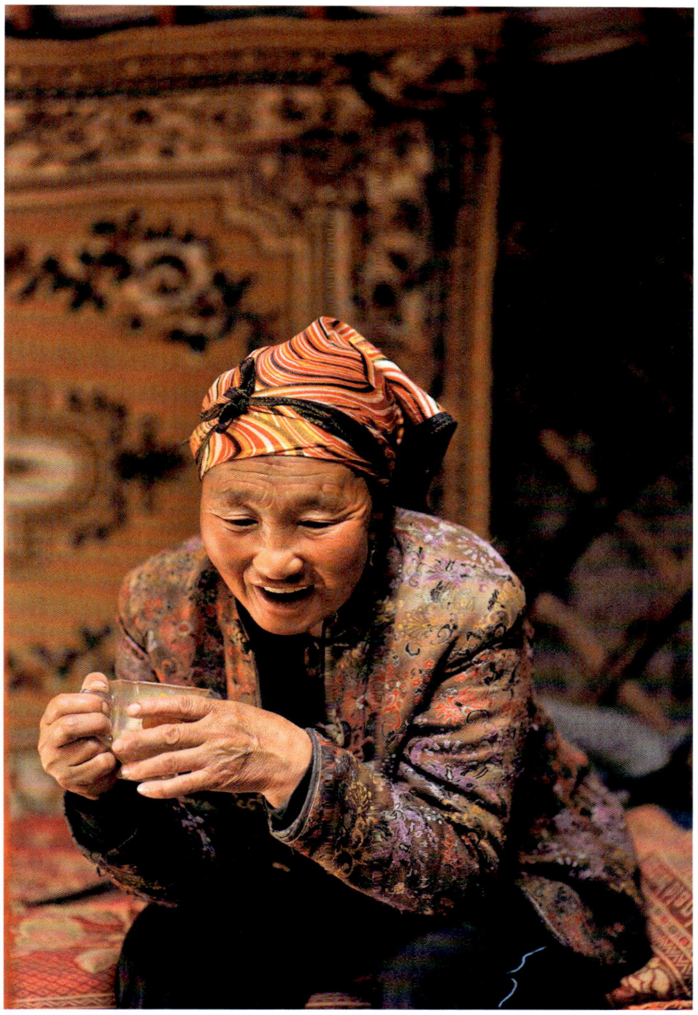
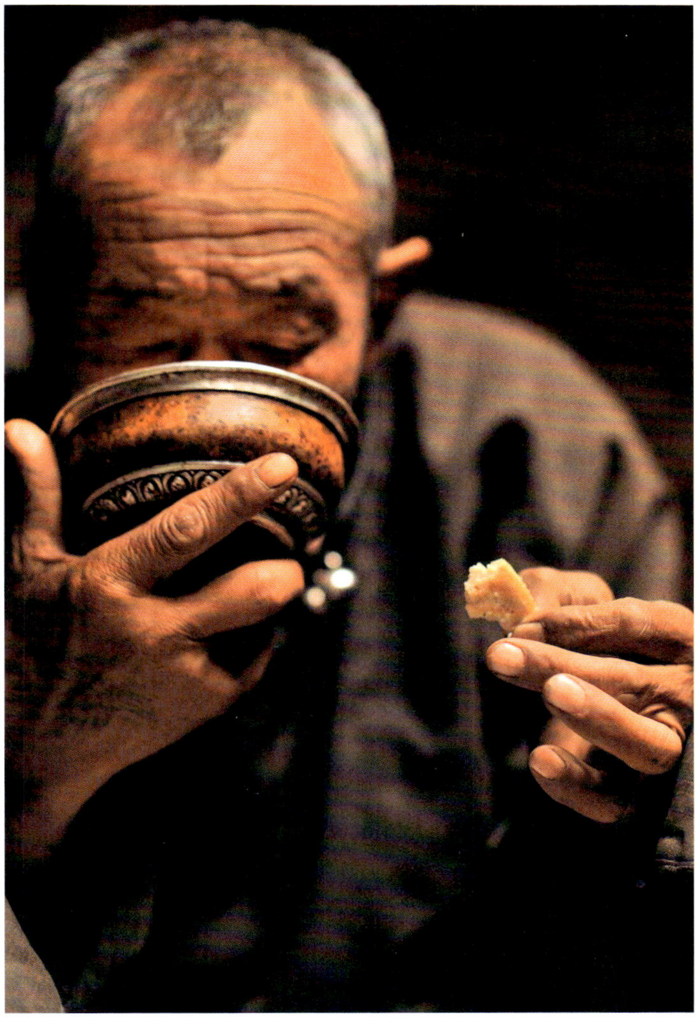
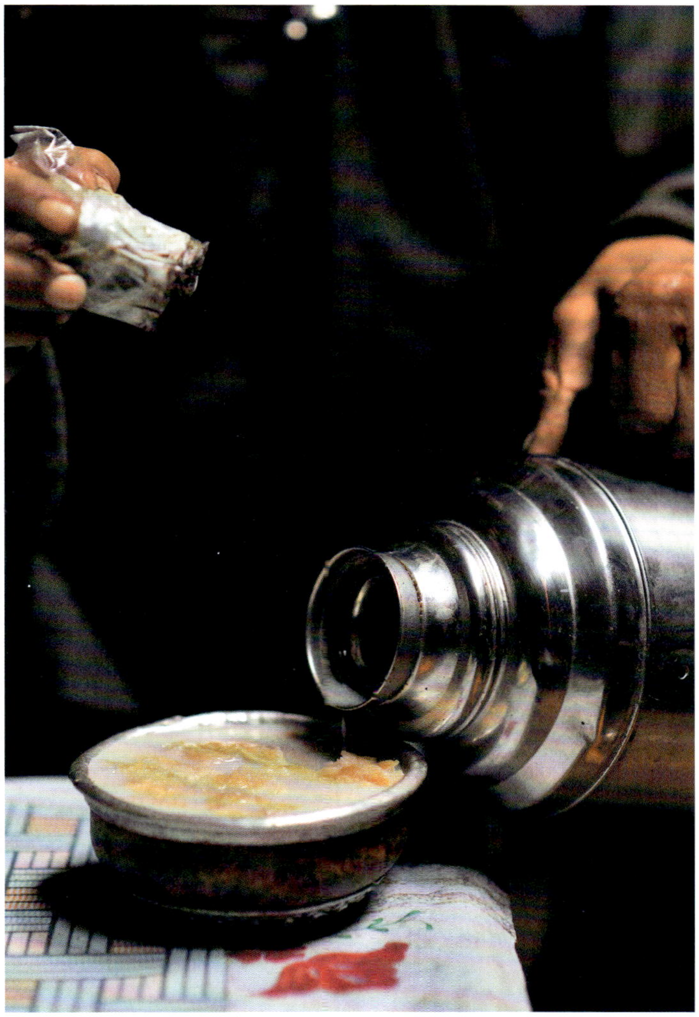

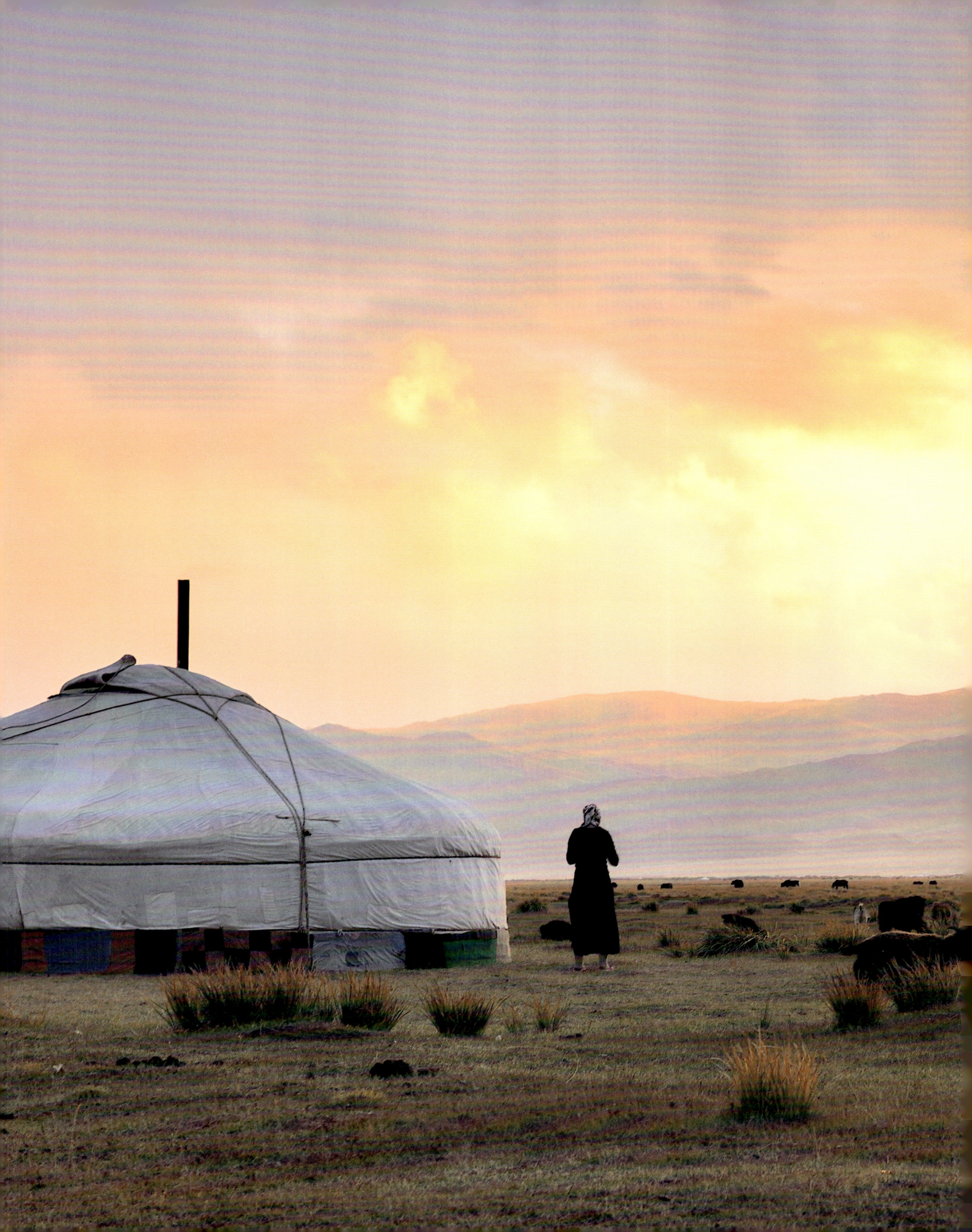

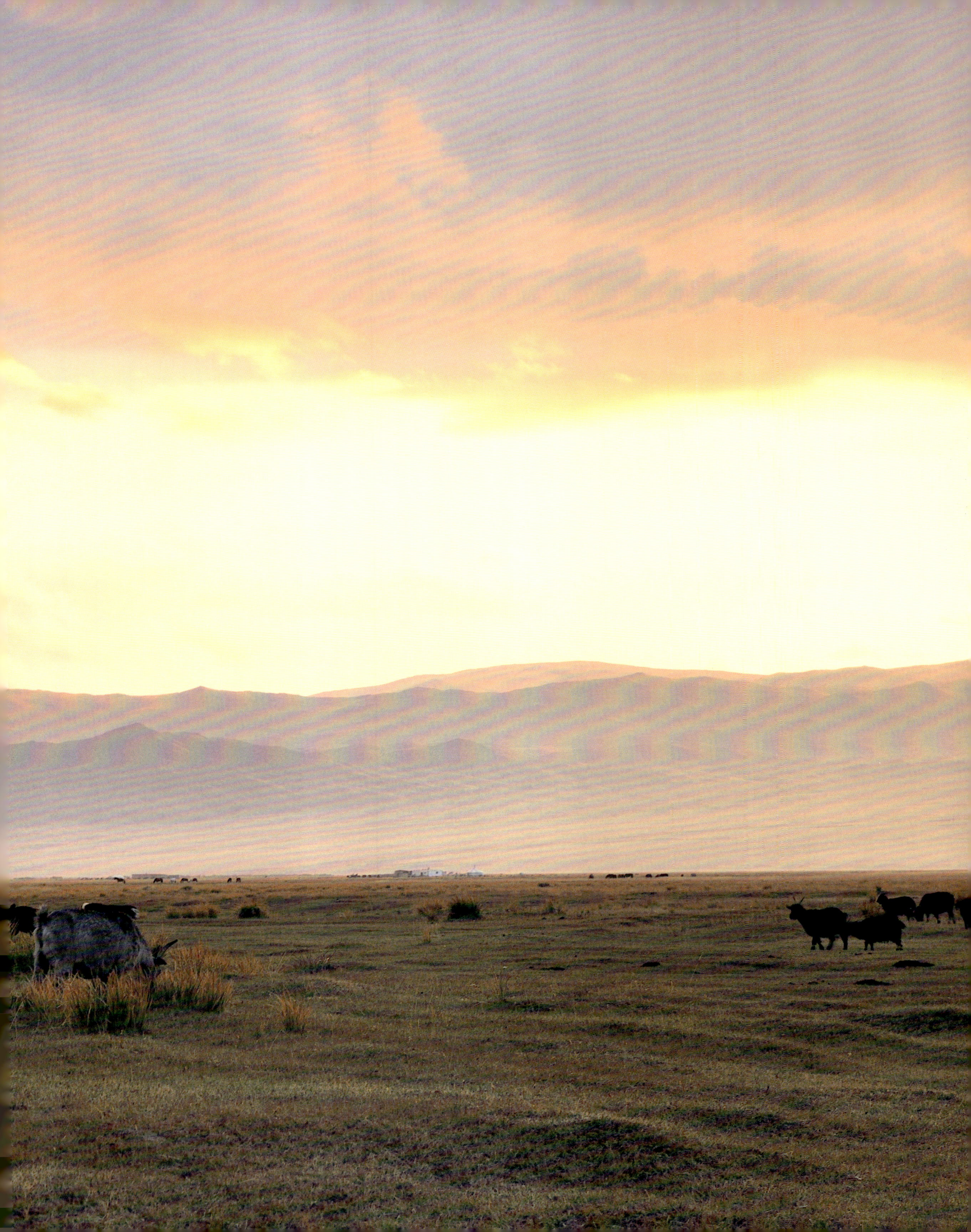

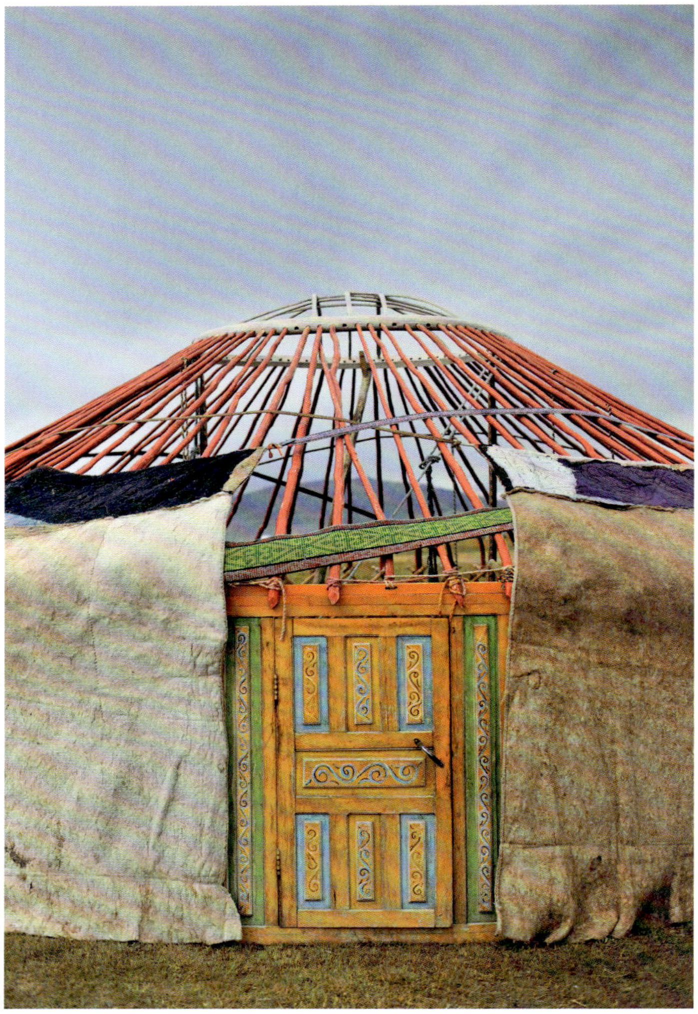

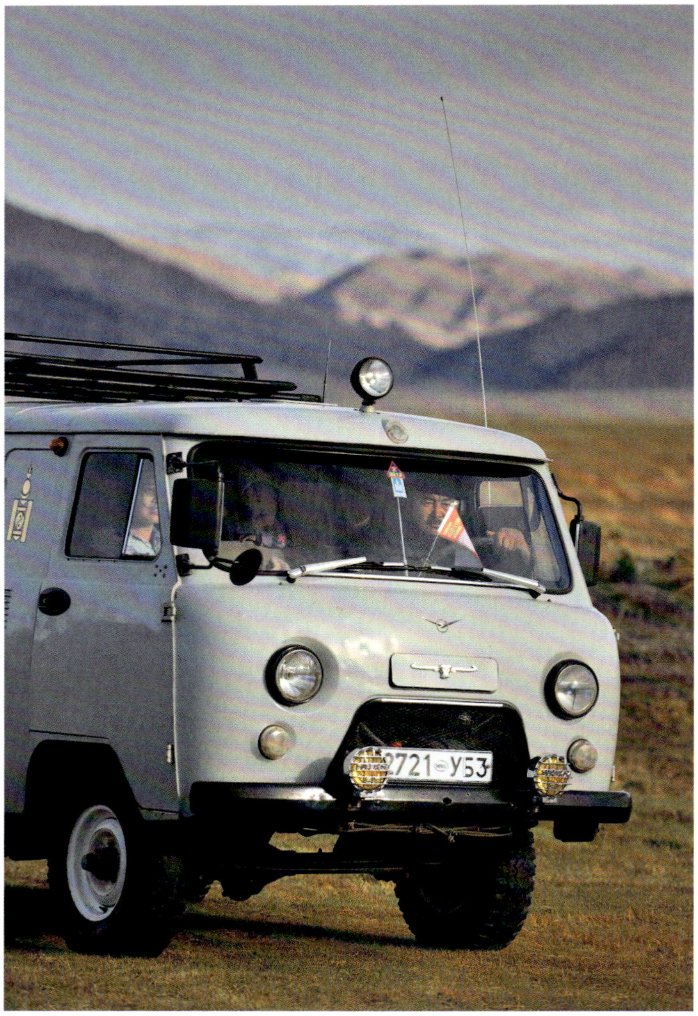

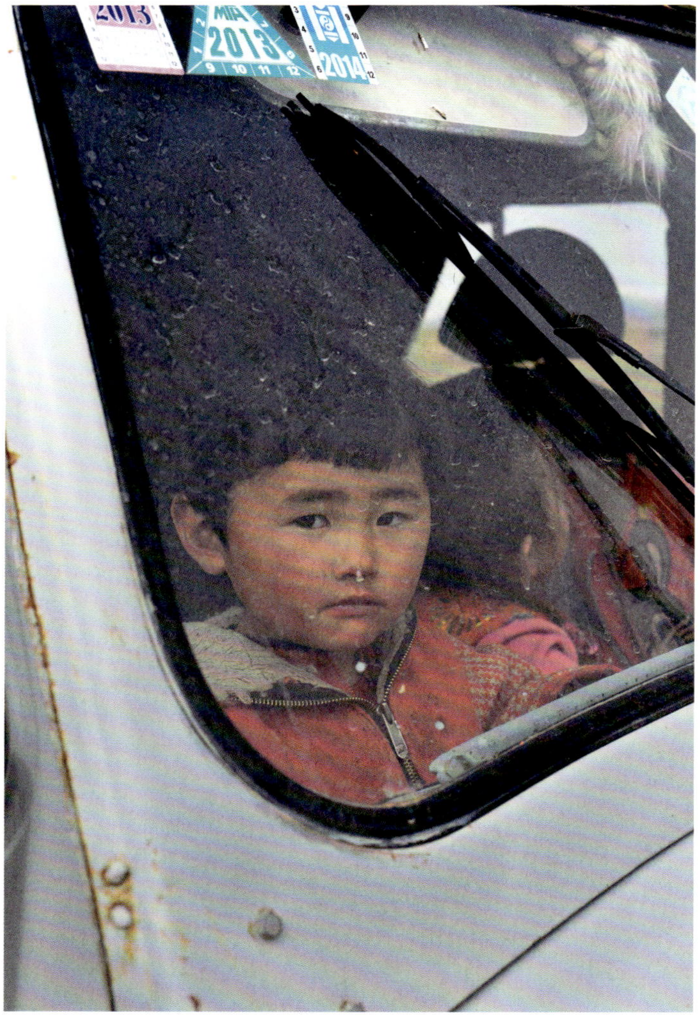
2013
2014

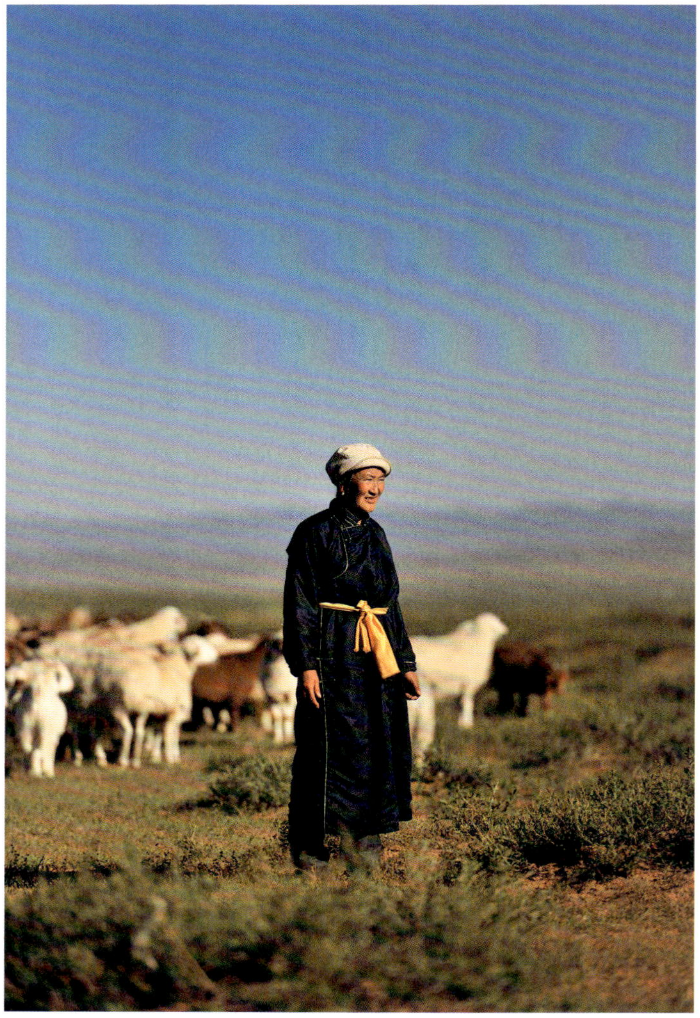

Opposite Because of the harsh climate and short wet season, many Mongolians lead pastoral and nomadic lives centred on animal husbandry. Goats, sheep, yaks, camels, and horses provide meat, wool, dairy produce, and transportation. Men manage the herds and make saddles, harnesses, and weapons for hunting. Women take care of children and make silk deels – traditional ankle-length costumes similar to caftans – and milk cows, goats, and even mares. Fermented mare's milk, airag, is Mongolia's national drink.

Back in their gers, the berkutchi let their birds rest, often in their own small quarters next to the living room. Life is tough yet simple on these snowy plains of the Altai. According to a tradition that has been passed down from father to son for 6,000 years, a berkutchi needs only three things: a fast horse, a faithful eagle, and a loyal dog to guard his ger. Every spring the nomads of Mongolia make the migration from the restraint of rather sad-looking winter villages to lush valleys where meadow pasture is plentiful and life is more free and unrestricted. Here the nomads feel more themselves, while living a simpler and natural life in their yurts, in the companionship of their horses and their feeding herds. These valleys and pastures, these humble yurts are more than a seasonal place to live. They are a way of life, a testimony to freedom and a means of continuing their relationship with nature. To preserve the culture and art of eagle hunting, some of the daughters learn to become berkutchi. In the past, the thought of female eagle hunters would have been unthinkable, but now girls are being taught the art of hunting to join or even replace their brothers who sometimes leave to work in bigger cities.

Creature Comforts

It's possible to bed down in a simple ger with an eagle hunter and his family. Call it couch surfing, or a homestay in the far west of Mongolia. All that is required is a warm sleeping bag, an open mind, and a willingness to leave modern comforts behind. In return, travellers gain a fascinating insight into daily life in one of the most dramatic corners of the world.

Go Now

The Sagsai district of Bayan-Ölgii, which is home to more than 100 berkutchi and their trained eagles, is a two-and-a-half-hour flight from Ulaanbaatar. In stark contrast with the better-known Naadam Festival held each July in the nation's capital, the small and remote Kazakh Eagle Festival flies under the radar. The few foreigners who do venture here are often photographers or filmmakers. Hence it is mostly locals who climb on their horses and rickety motorcycles, or squeeze their families into old cars to attend the festival. Despite being centuries old, the nomadic way of life and traditions of the Altai-Kazakhs are steadily disappearing. Many young nomads now head for the city, a smartphone in the pocket of their fur hunting coat. The skills and knowledge of their forebears are no longer being passed down as readily. Still, berkutchi blood runs thick here in the far west of Mongolia, and proud locals will continue to honour their cultural heritage for some time yet.

How to Go

The Kazakh Eagle Festival takes place every year during the month of September. A three-day festival experience will cost around 350 Euro per person, excluding domestic flights. Pack several layers of warm clothes, a dust-proof bag, and a decent pair of binoculars.

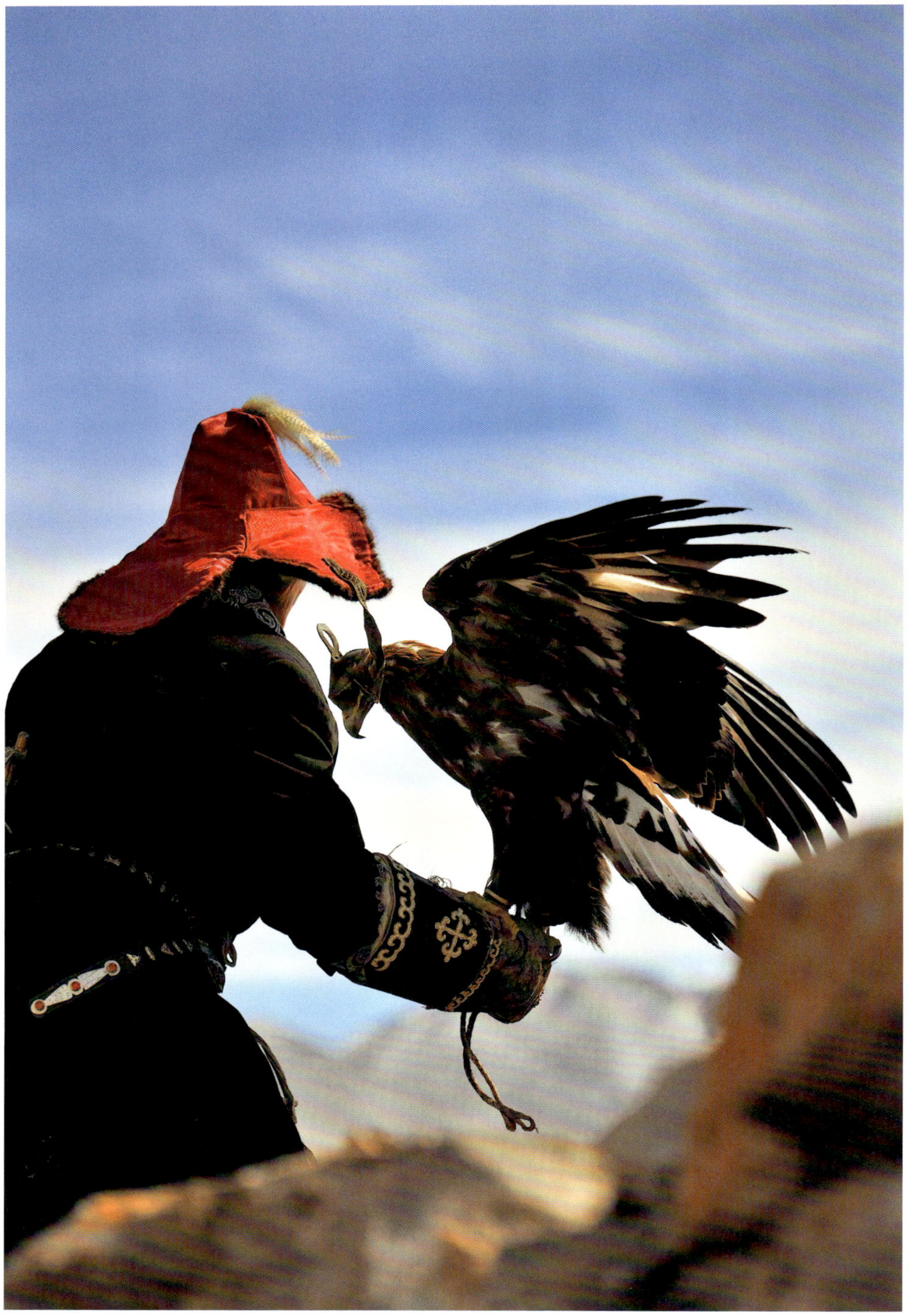

This spread According to one Kazakh proverb, a fast horse and a soaring eagle are the wings of a nomad. Another saying, "Our ancestors had three comrades: swift foot, tazi, and bürkit" refers to horses, sighthounds, and golden eagles. With these three hunting companions, the unforgiving steppes of Mongolia became a land of plenty. The nomadic hunters share a rare sympathy with their animals, raising them and training them, spending a lot of time together in nature.

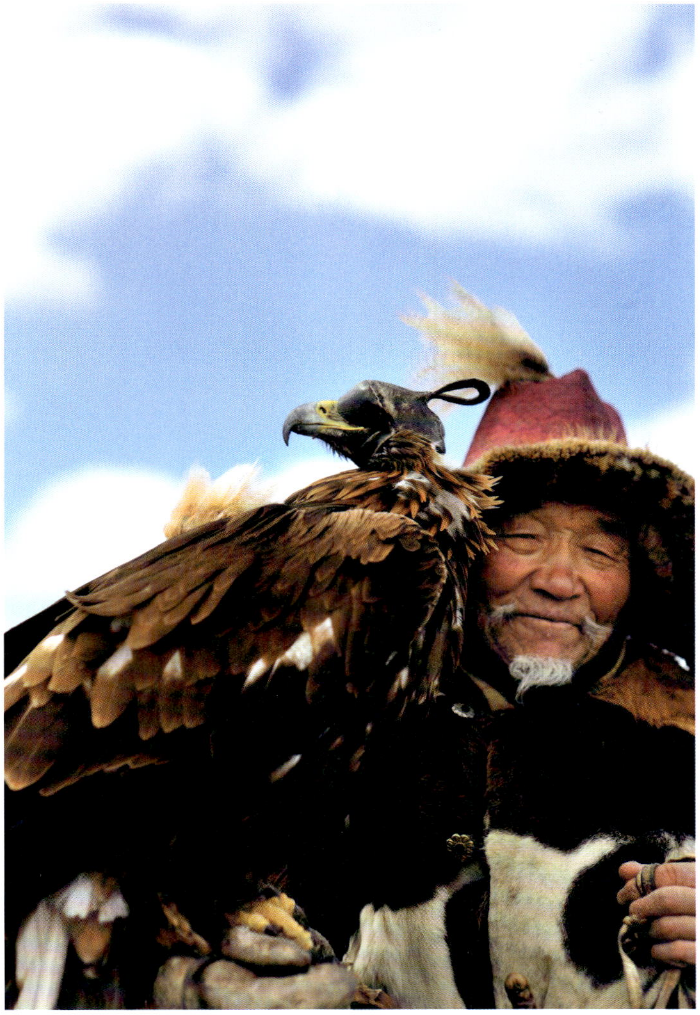
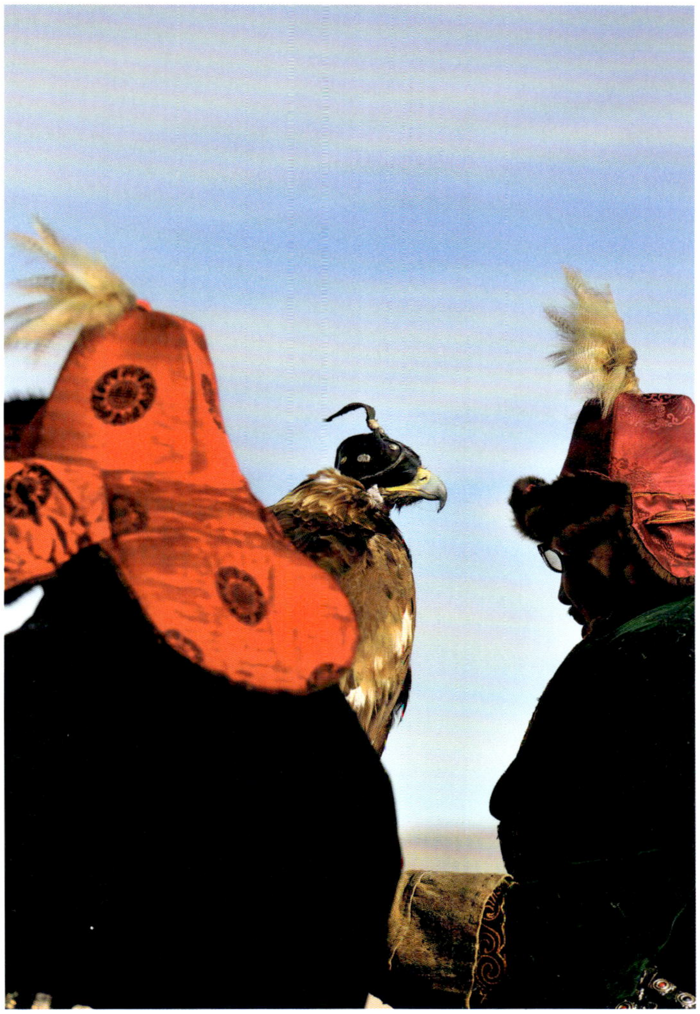
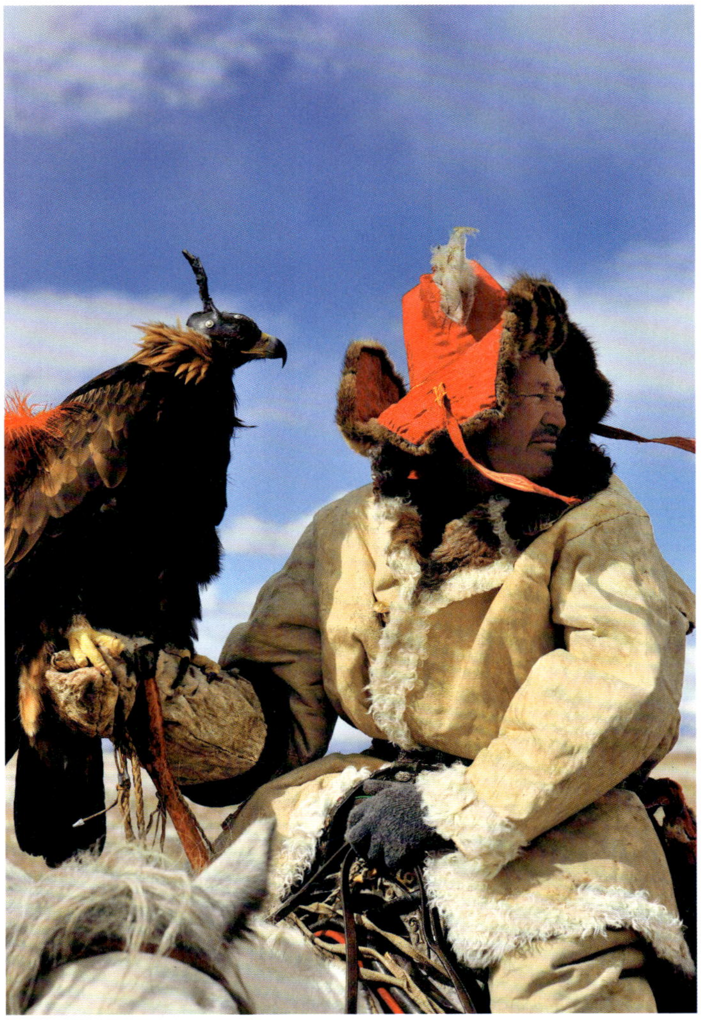
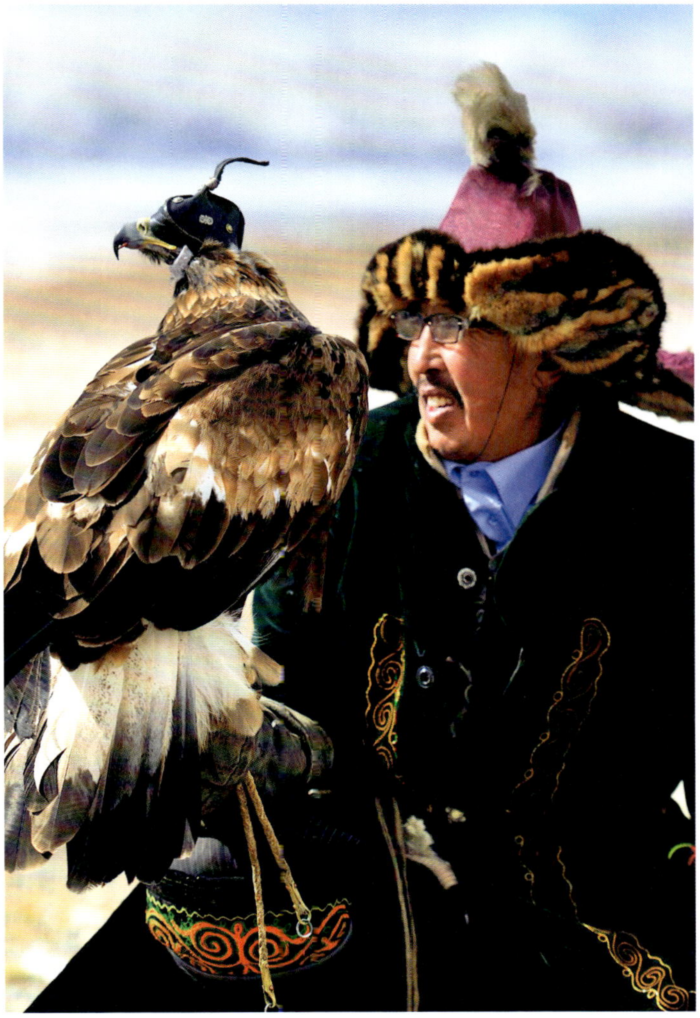

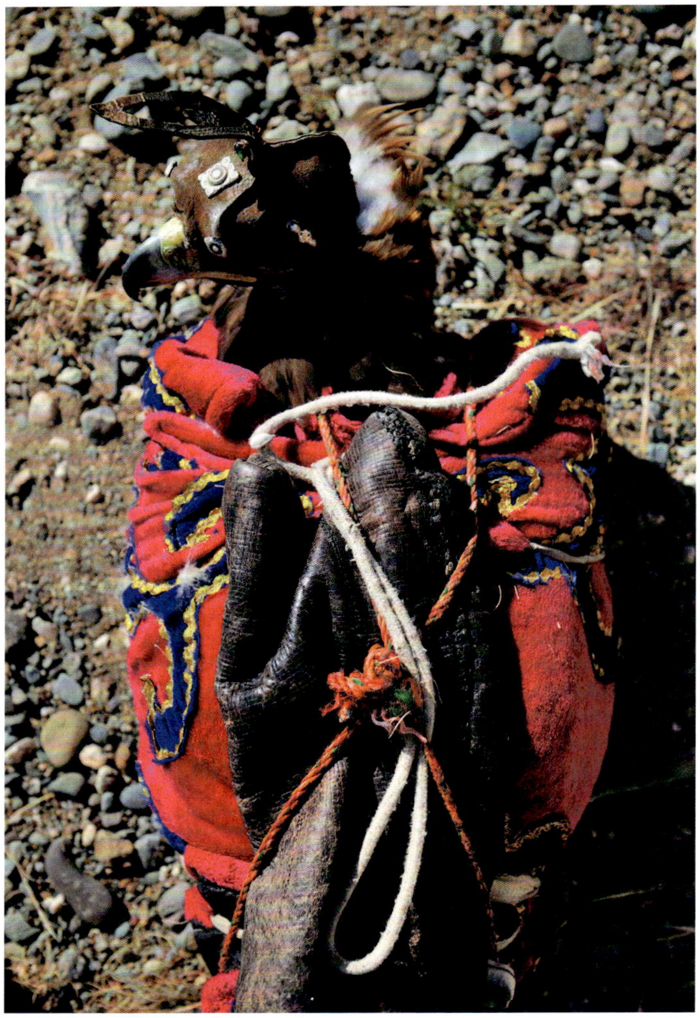
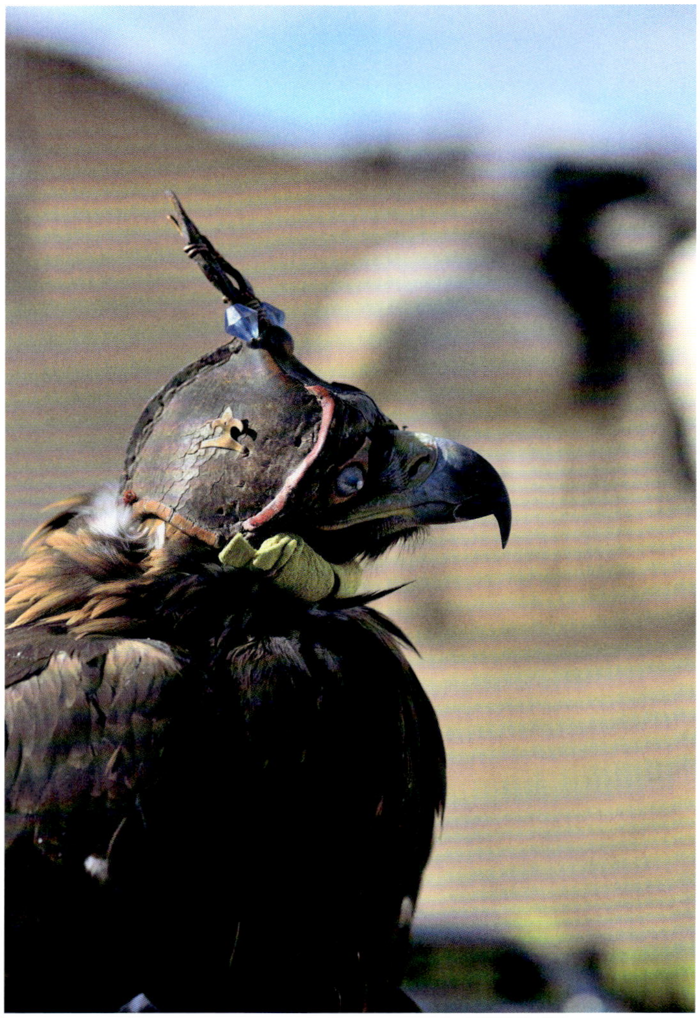
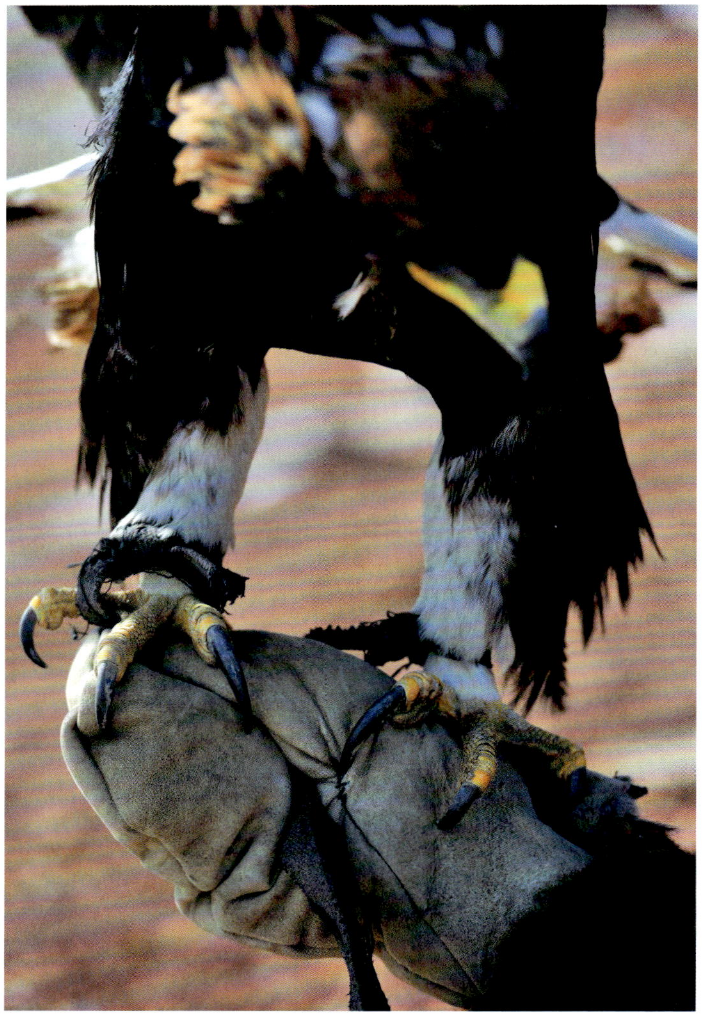
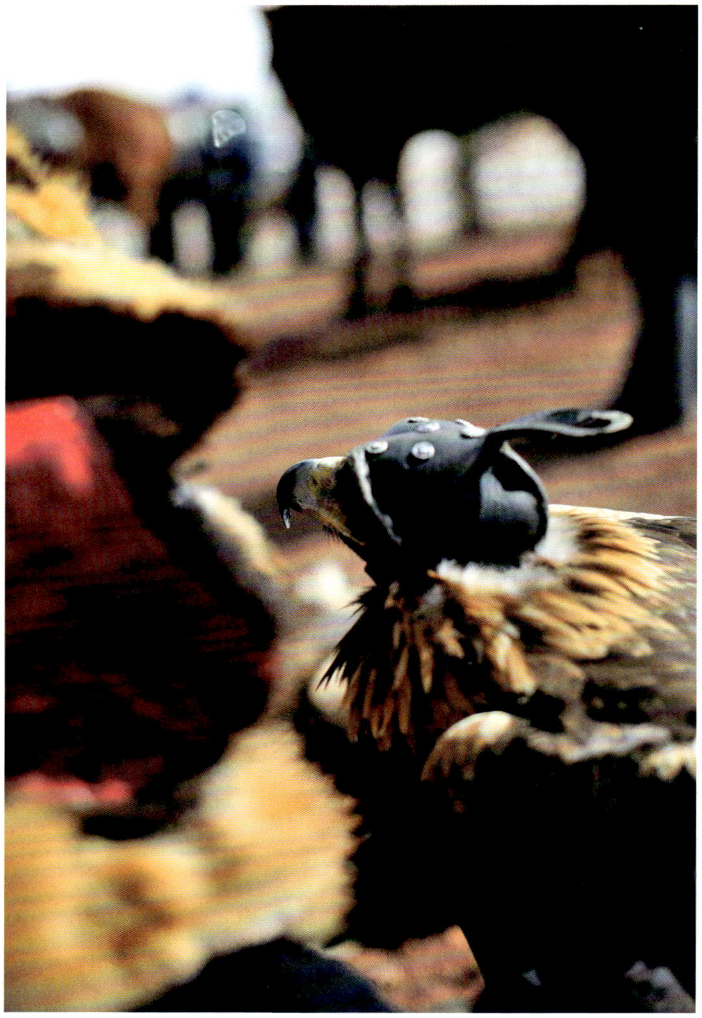

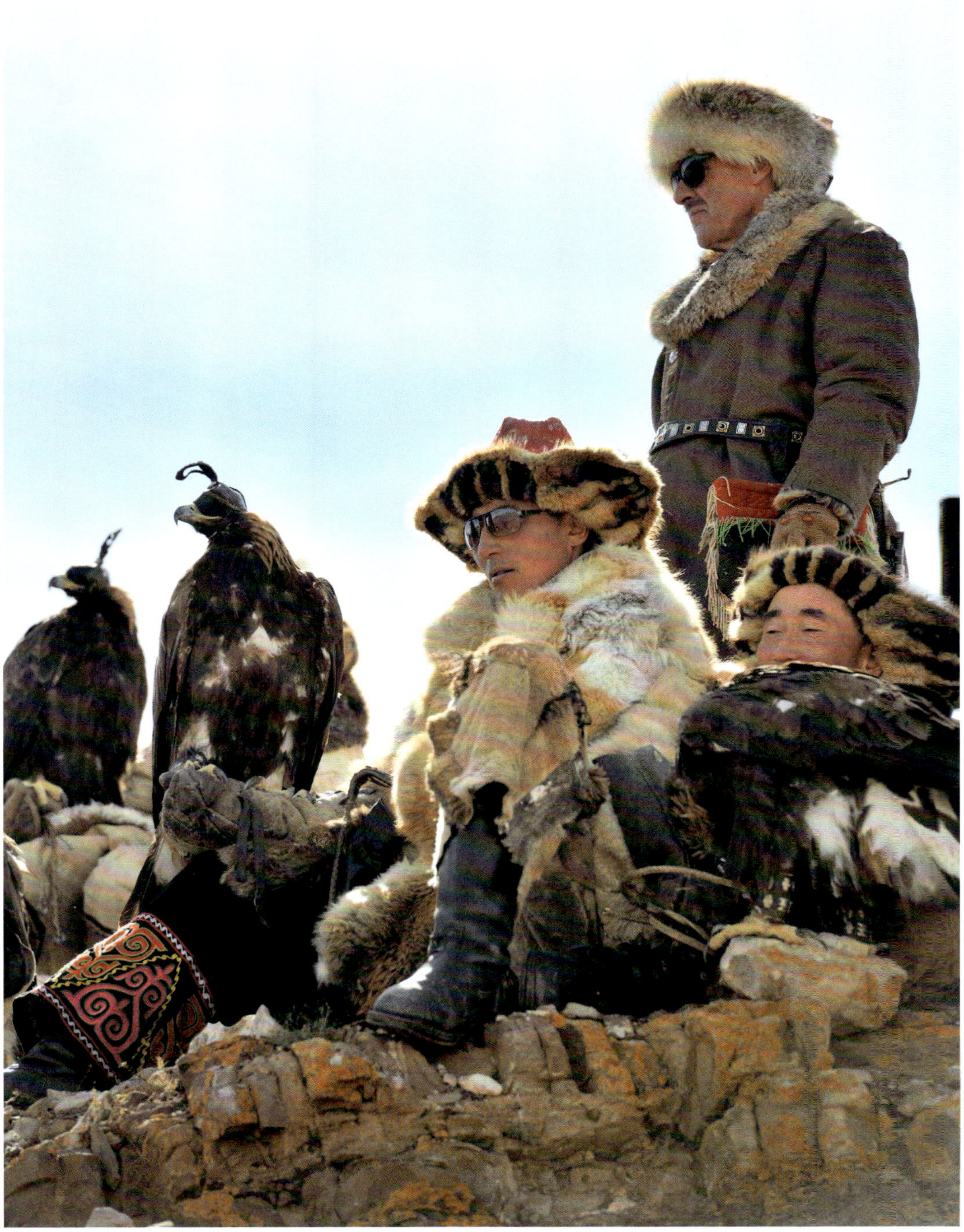

This spread For millennia, the ancient art of eagle hunting has been handed down through the generations across the northern steppes from the Caucasus to China. Today, it is kept alive by Kyrgyz and Kazakh nomadic groups in Mongolia, Kyrgyzstan, Kazakhstan, and China's Xianjiang province. They are the descendants of nomads who roamed the steppes of Inner Asia. In 2016 the spotlight was put onto the eagle hunters thanks to the acclaimed documentary *The Eagle Huntress*, which gave an insight into the life of a female berkutchi. Eagle hunting is traditionally passed down among male relatives, but there are no religious or cultural prohibitions against a girl becoming an eagle hunter. Children, including girls, often help to care for the eagles, go along on hunts, and attend eagle festivals.

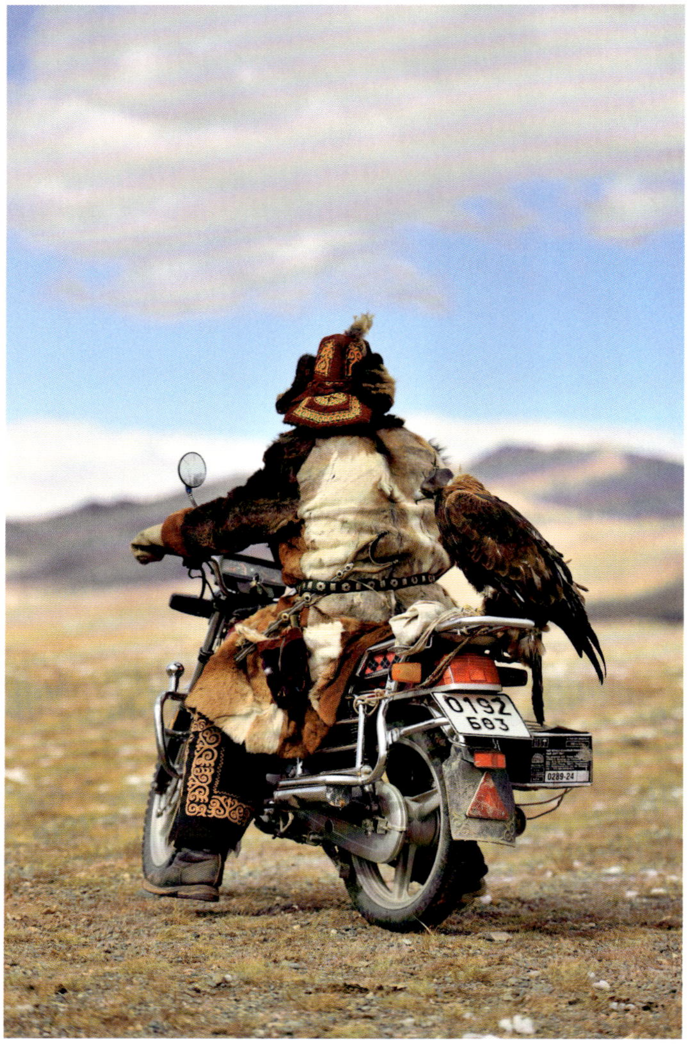

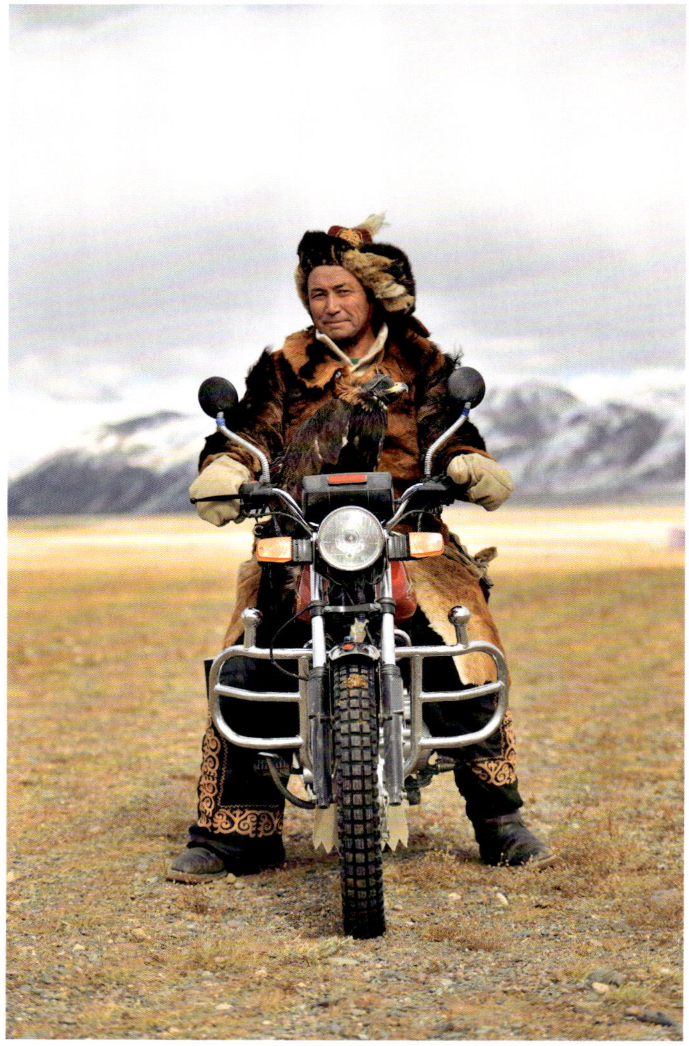

This spread It's impossible to go off-road in the vastness of the Altai Mountains. With practically no roads, everything is off the map. In the past, the eagle hunters only moved around on horseback, but nowadays they also use motorcycles to cross large distances between villages and nomadic settlements. When they take along their eagle on the motorbike, the hunters pack them carefully into a leather pouch, while wrapping the skin around the animal so the bird is protected and snug. Safe and protected, the eagles rarely protest.

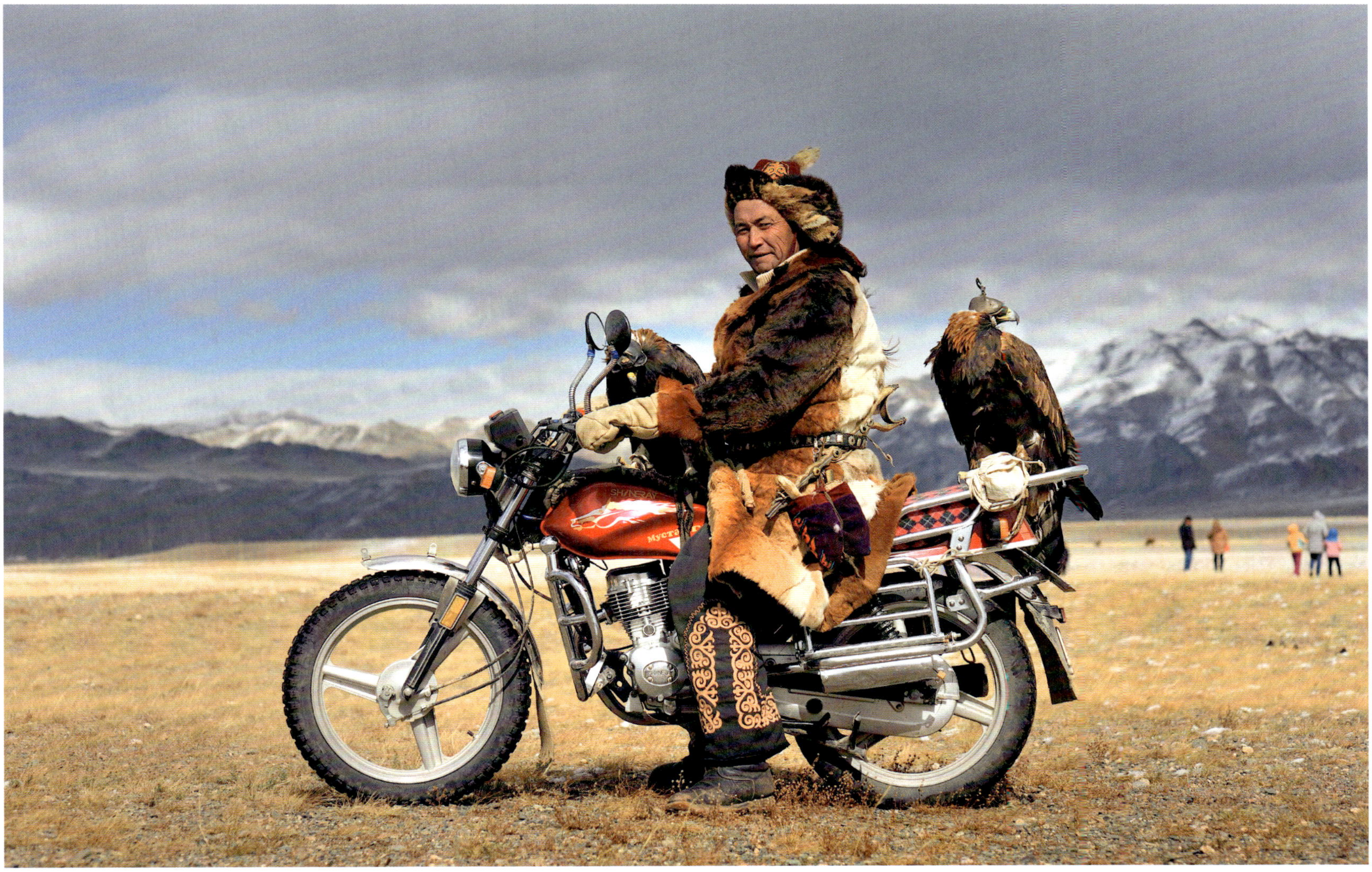

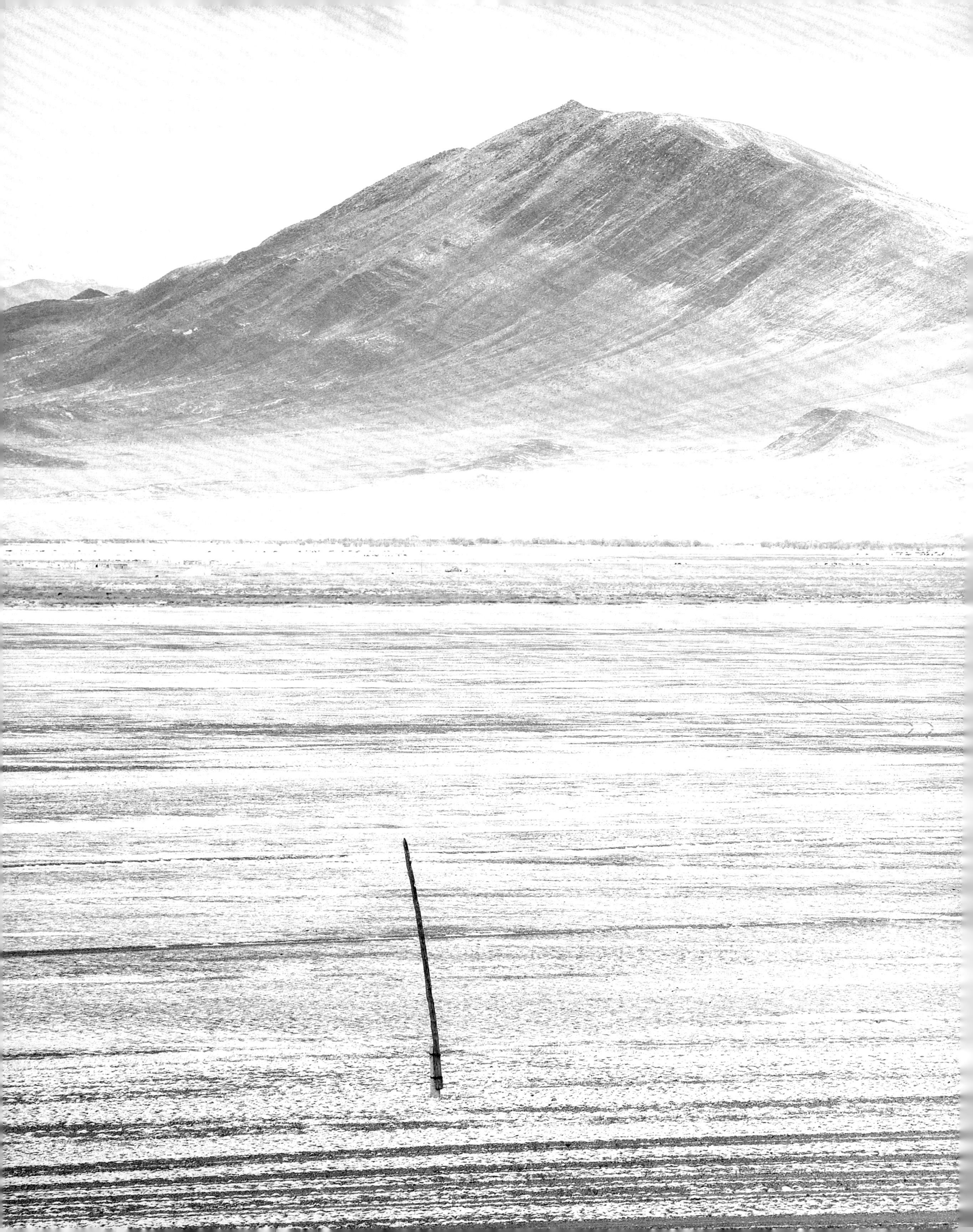

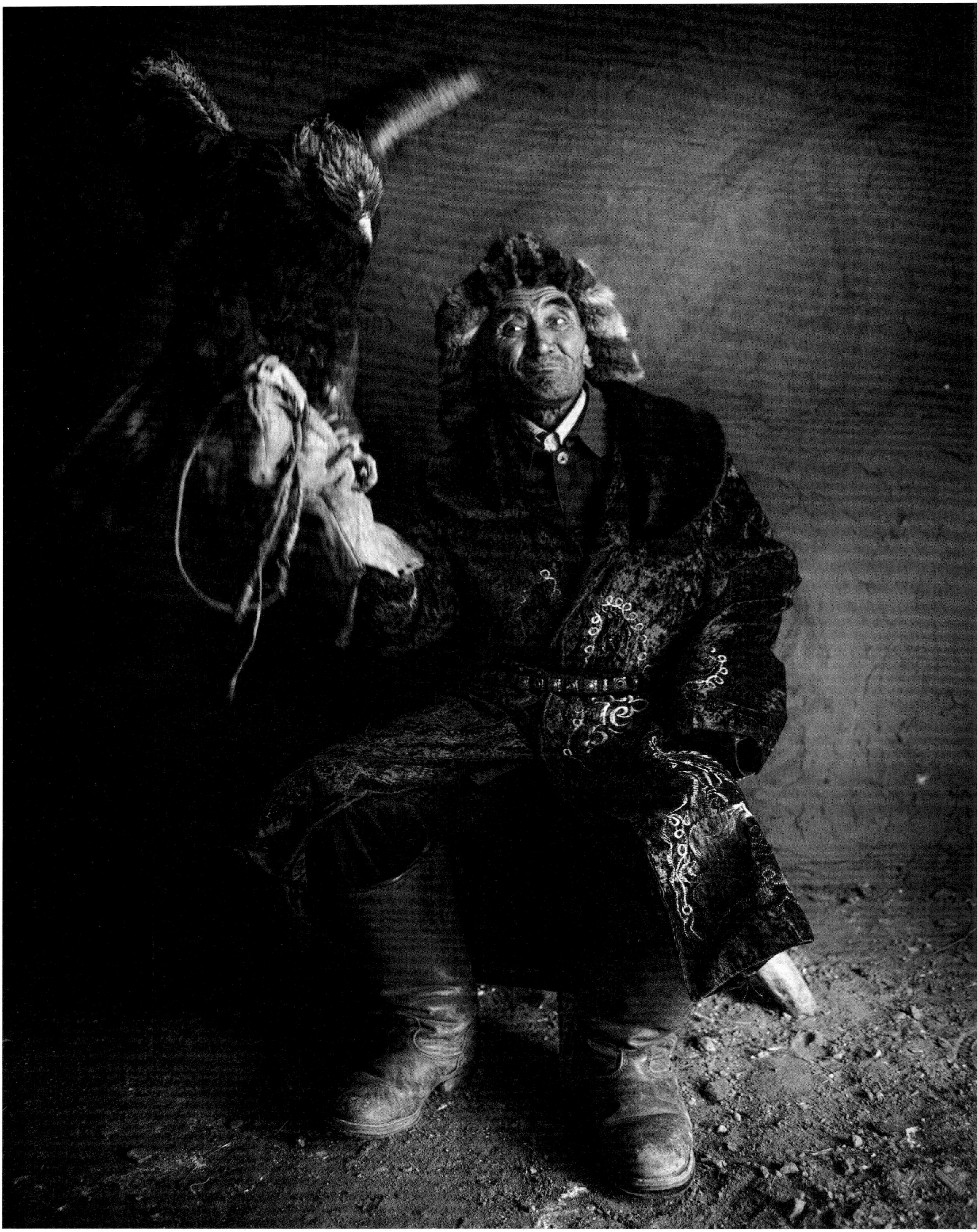

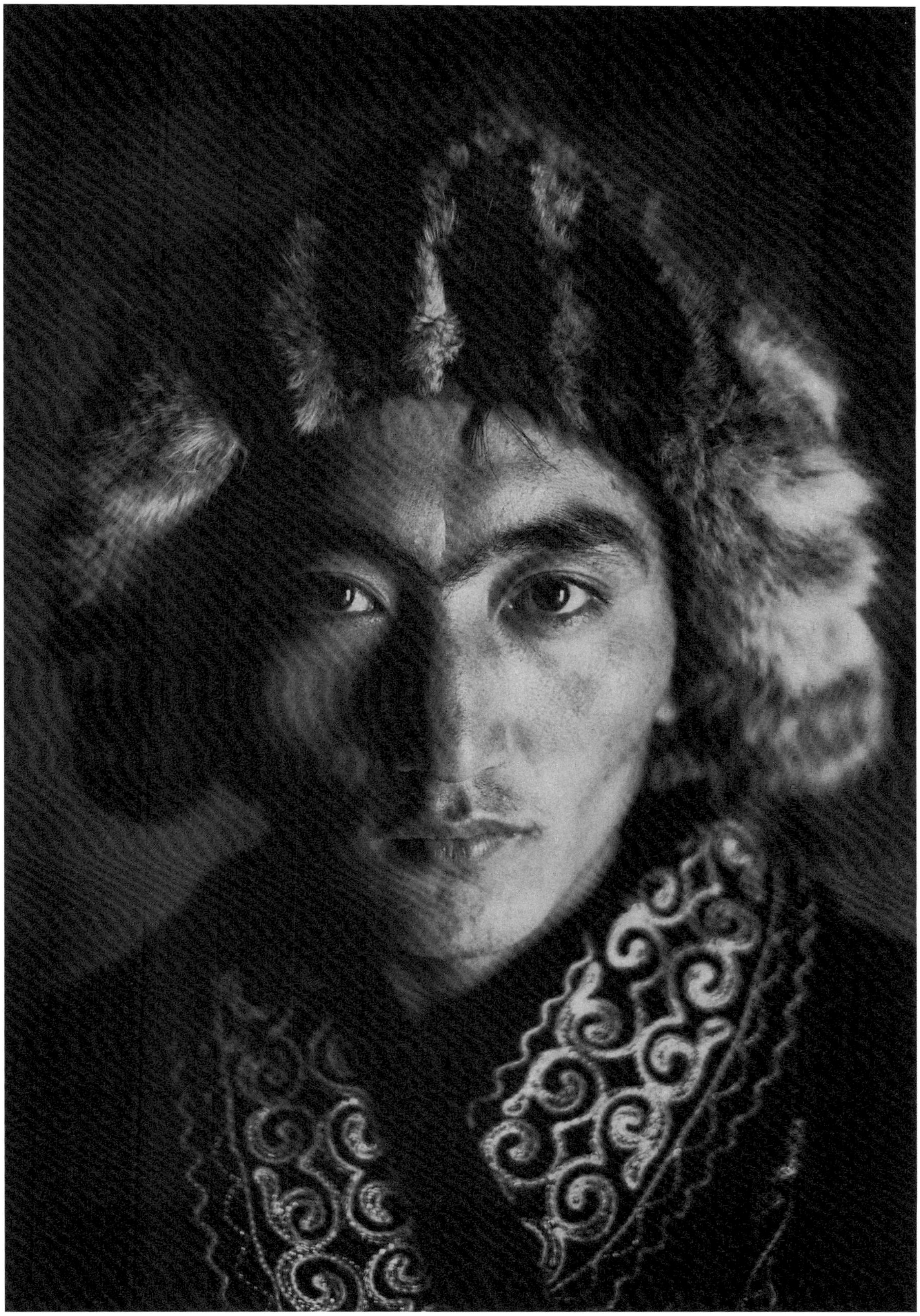

This spread Eagle hunter Mana Raiymhan and Mana's son Asylbek in the comfort and warmth of their small nomadic house. The eagle hunters are active through winter, when temperatures plunge well below -40°C (-40°F) and the pelts of rabbits, foxes, and wolves are at their most luxuriant. Golden eagles have been beloved hunting companions across the northern steppes from the Caucasus to China for millennia. The eagle hunters who train them treat them like a member of their family. After hunting with them for about 10 years, the birds are released back into the wild.

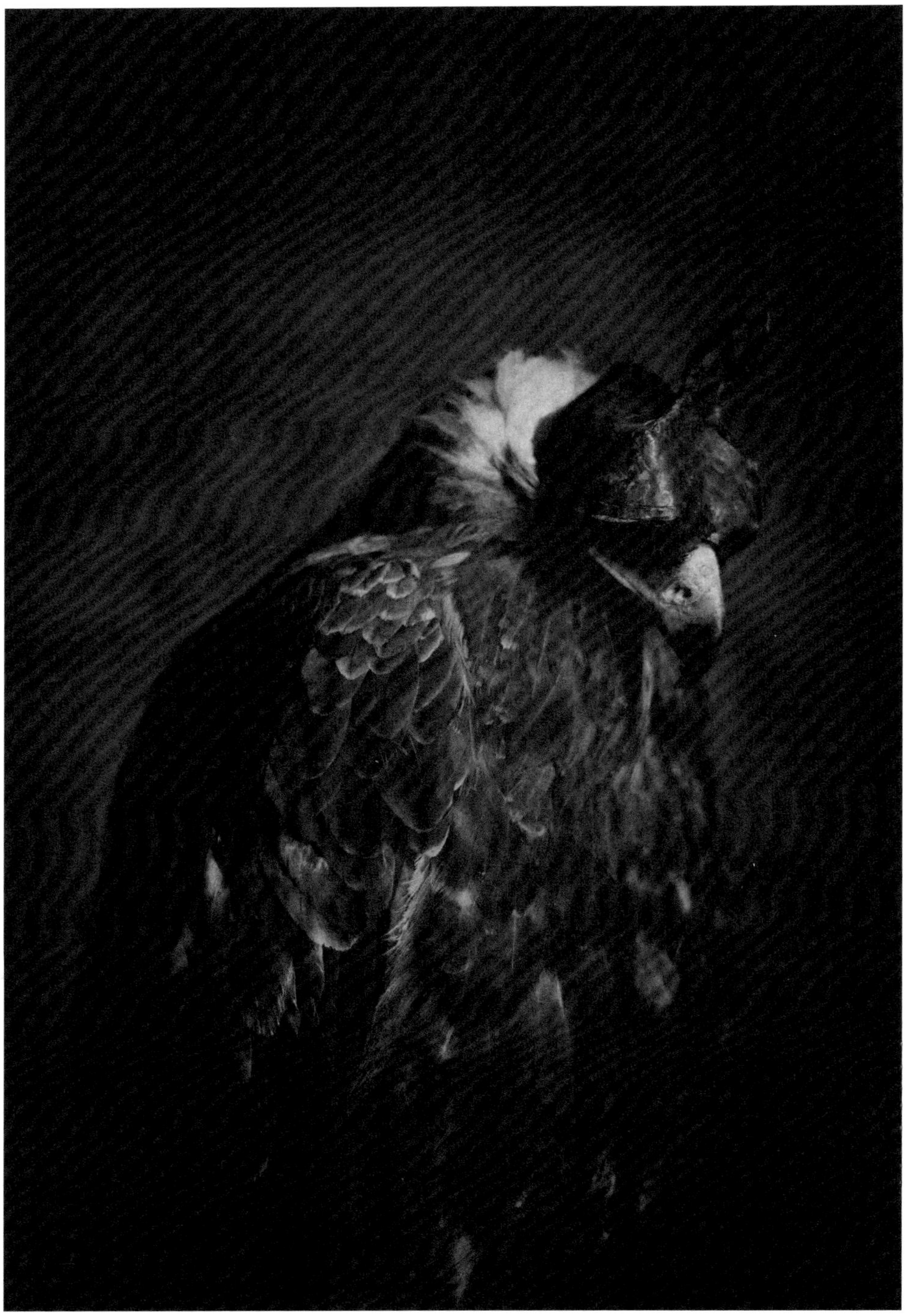

This spread Eagles can weigh up to six kilos (13 lb.) and are perfectly adapted to hunt small animals — rabbits, hares, wild goats, deer, and foxes — in winter. Their sharp claws have an extremely strong grasp; their very acute eyesight allows them to spot prey in snow-covered grasslands and mountains. Sometimes they will take on lynxes, Pallas's cats, and even wolves, especially when hunting together. Eagles are strong predators especially adapted to winter hunting when the fur of the animals is thicker and warmer. They hunt in these snow-covered grasslands and mountain bluffs during freezing, sub-zero temperatures. Female eagles are bigger, stronger, and more powerful than males.

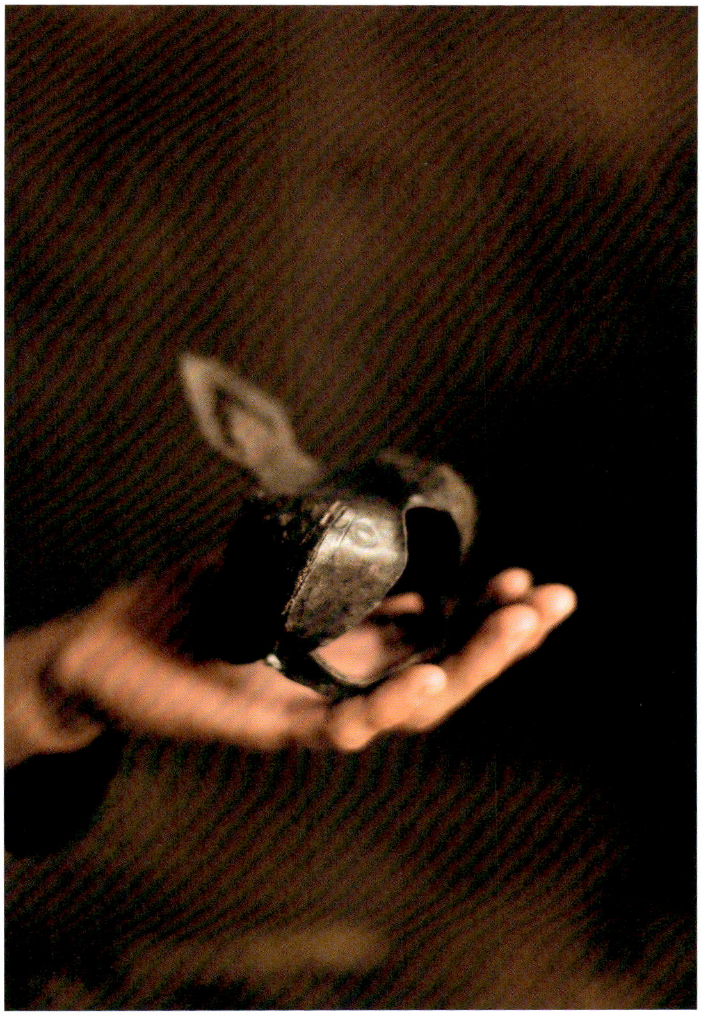
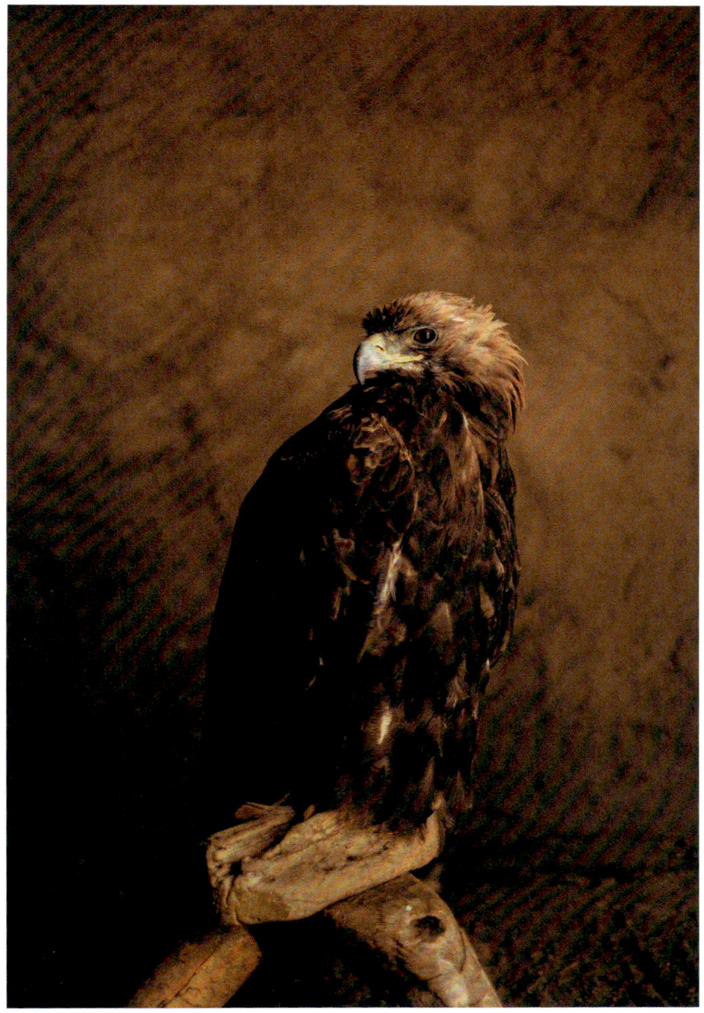
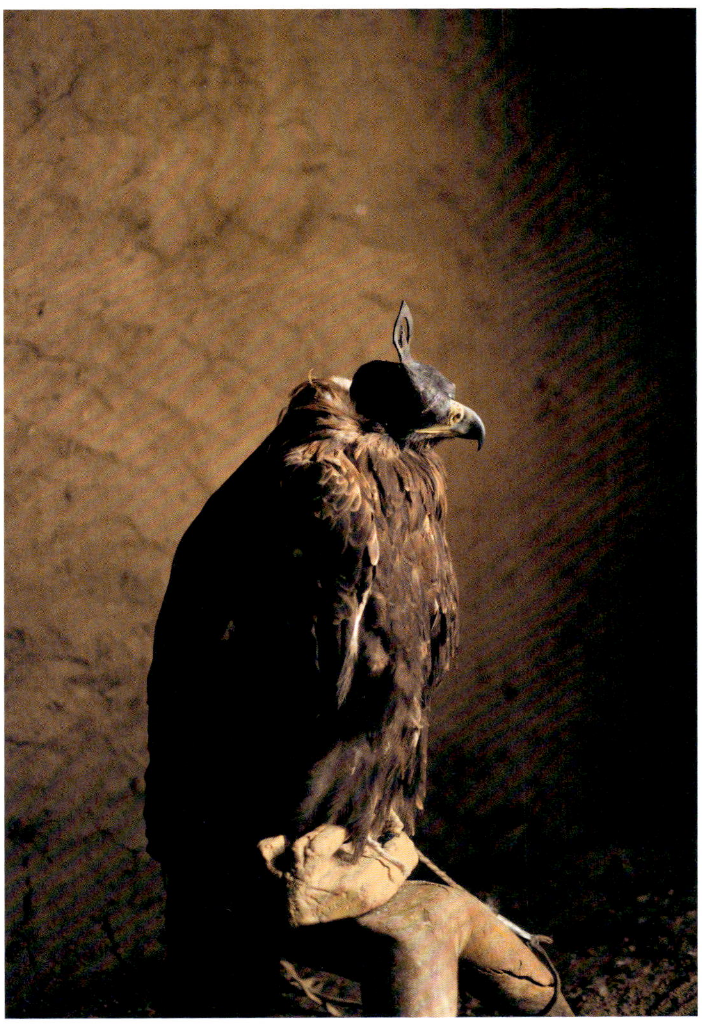
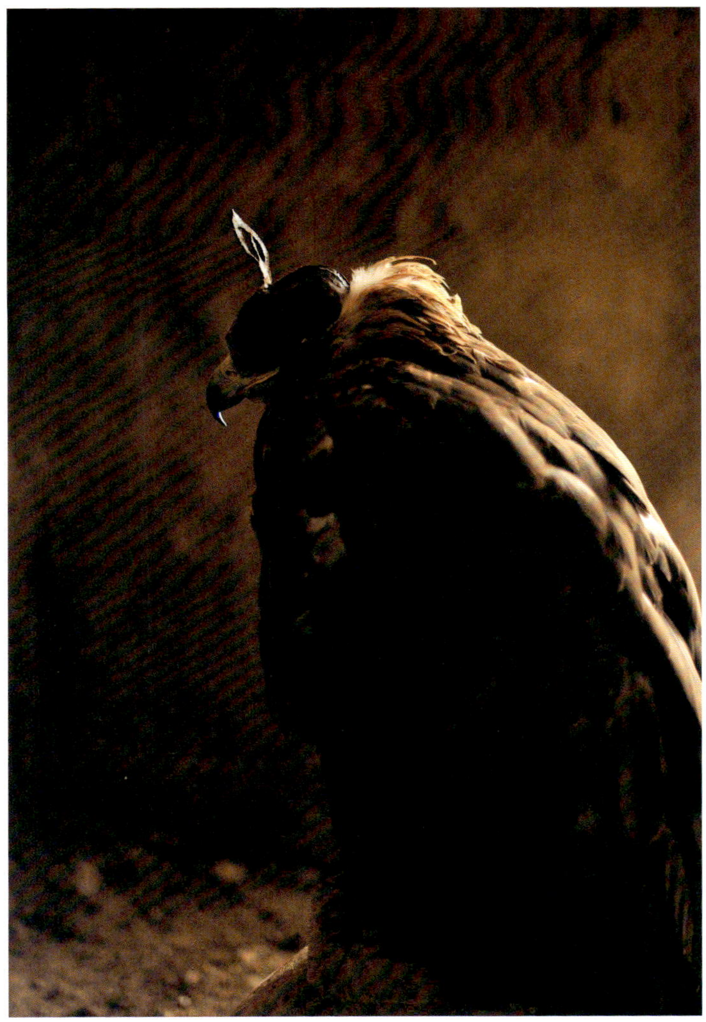

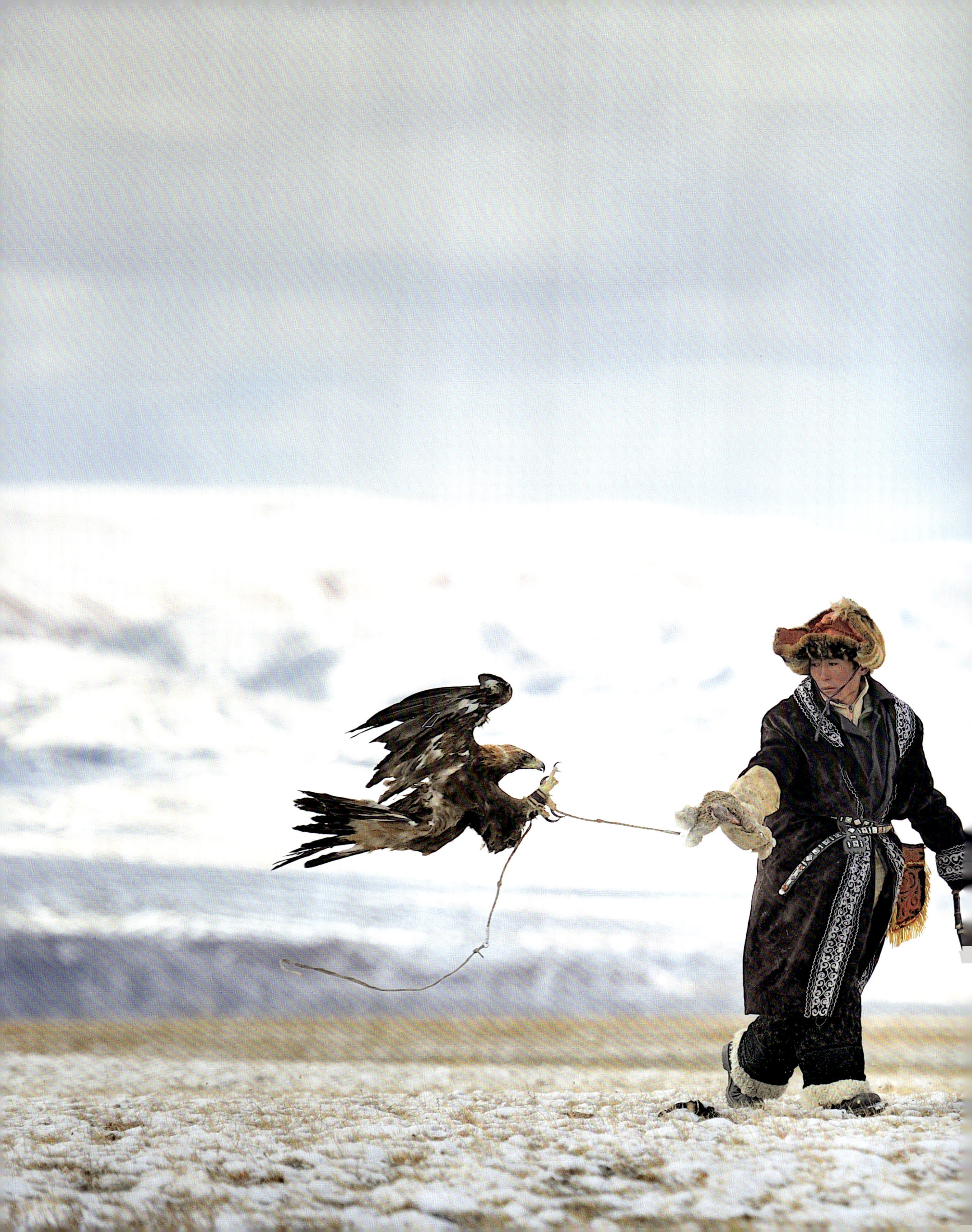

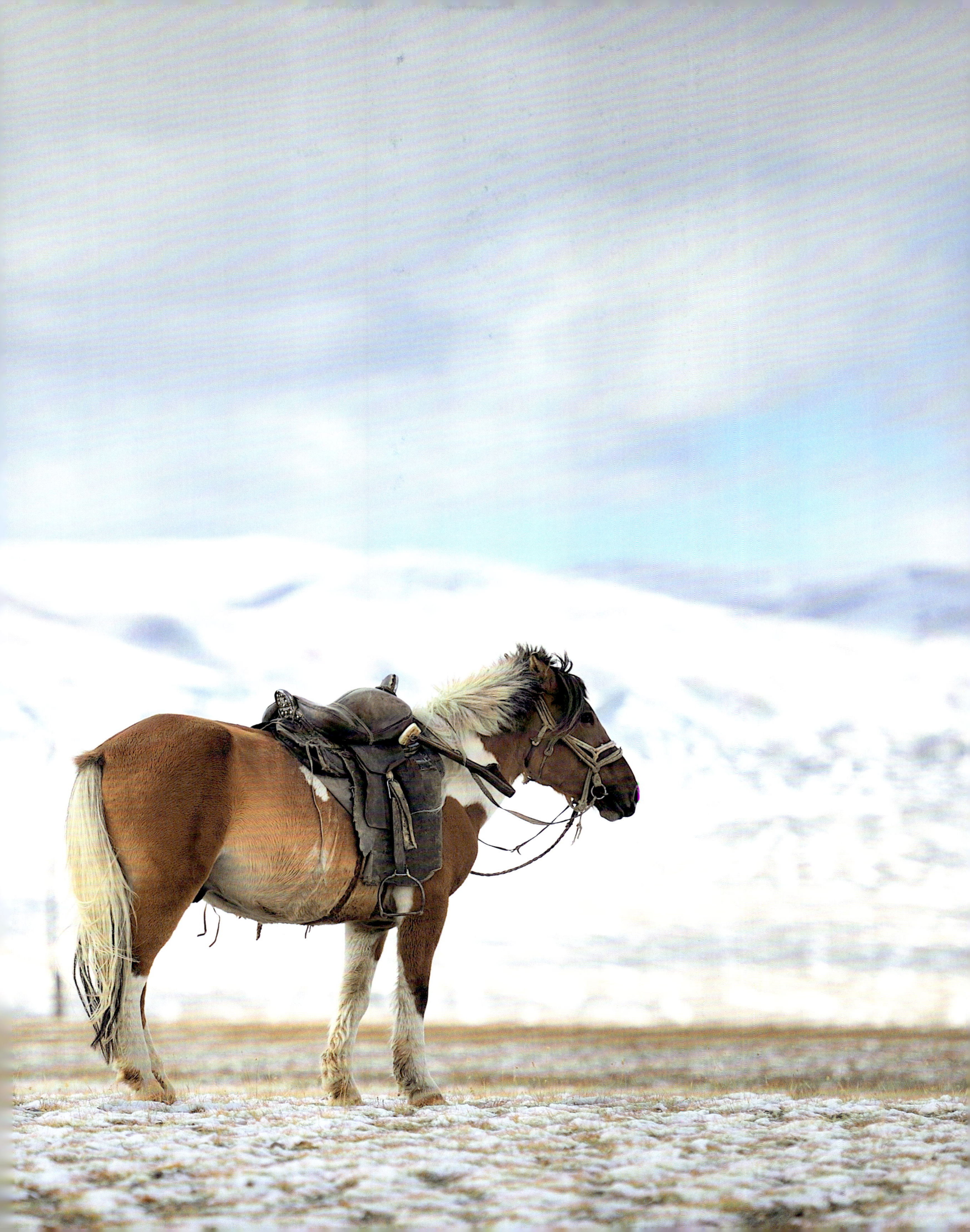

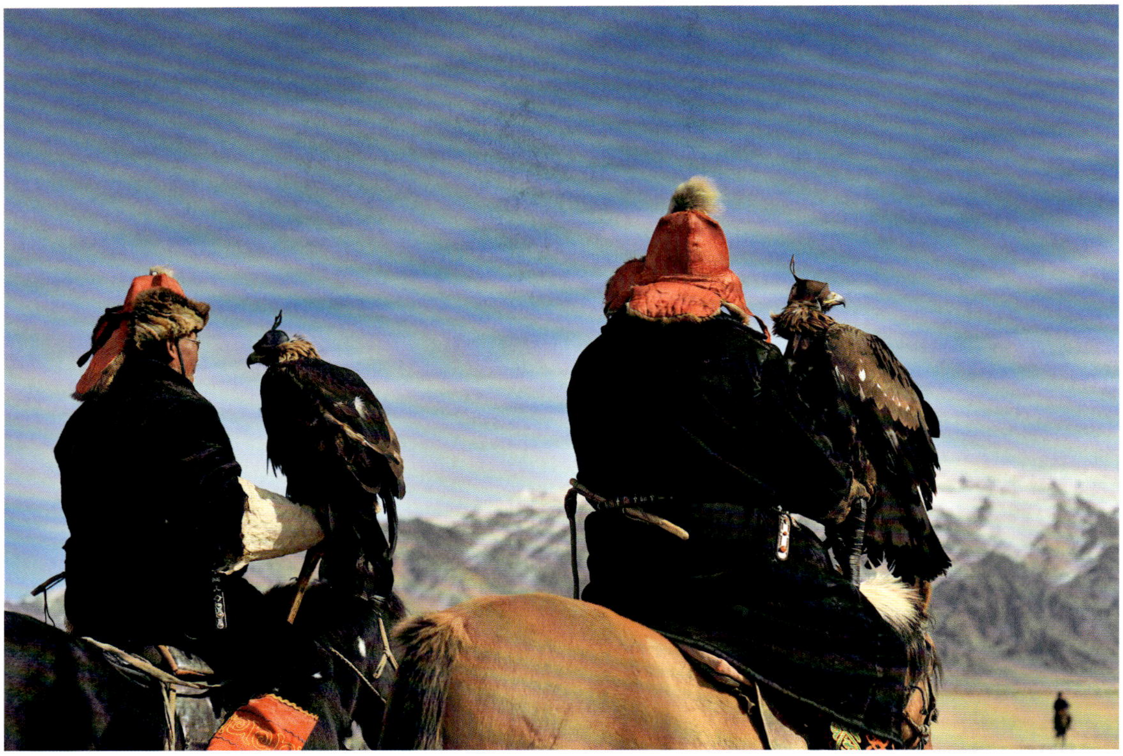

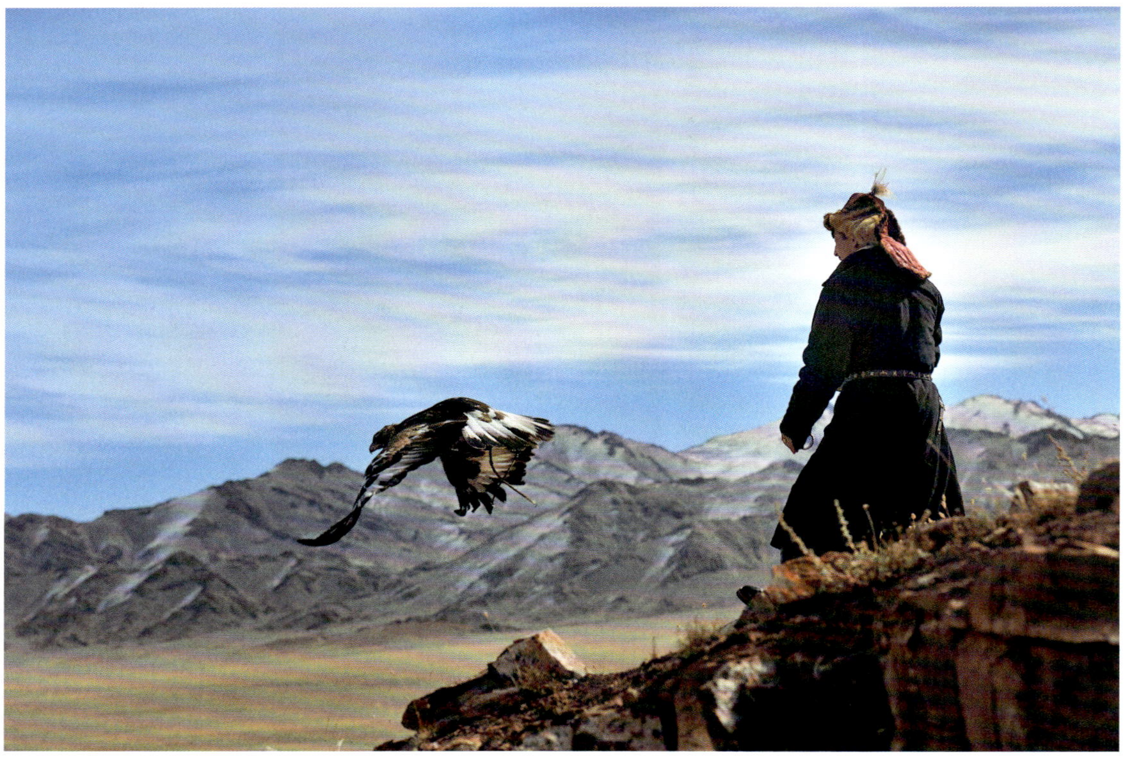

This spread Mongolians cherish their horses, take great pride in their riding skills, and love horse racing. They are one of the few horse-riding cultures left in the world. The legendary Genghis Khan established an empire that extended from Europe to Korea and from Siberia to the Himalayas. Notorious in Europe as 'Hell's Horsemen', Mongols could ride up to 120 kilometres (75 mi.) a day, across deserts and mountains considered — until the arrival of these mounted legions — to be inaccessible. Even in the 21st century, Mongolia remains a horse-based culture and retains its pastoral traditions with pride. Every year on the third weekend of September, eagle hunters gather near the village of Sagsai for the Altai Kazakh Eagle Festival, to celebrate their eagle-hunting traditions and to show off their horse-riding skills.

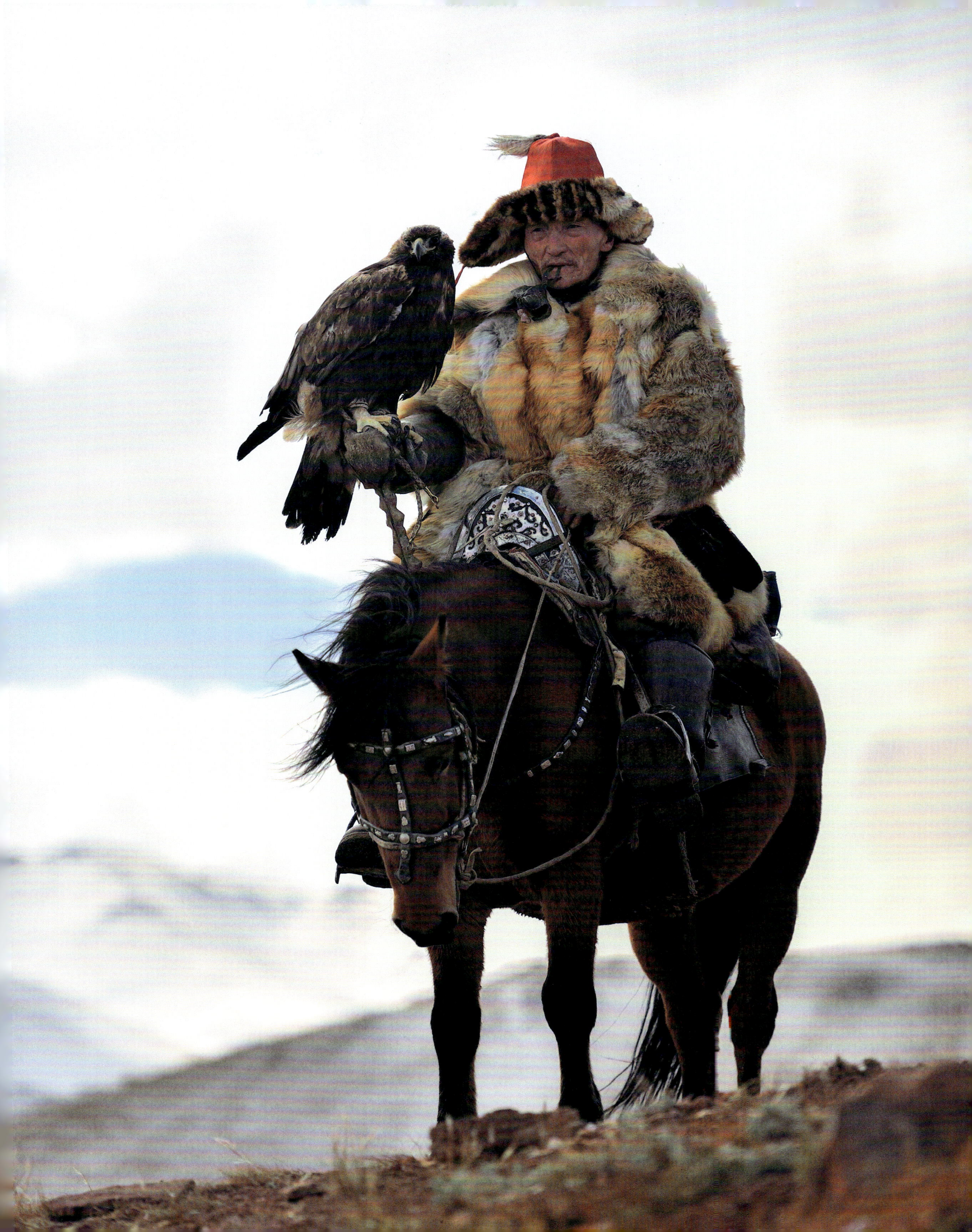

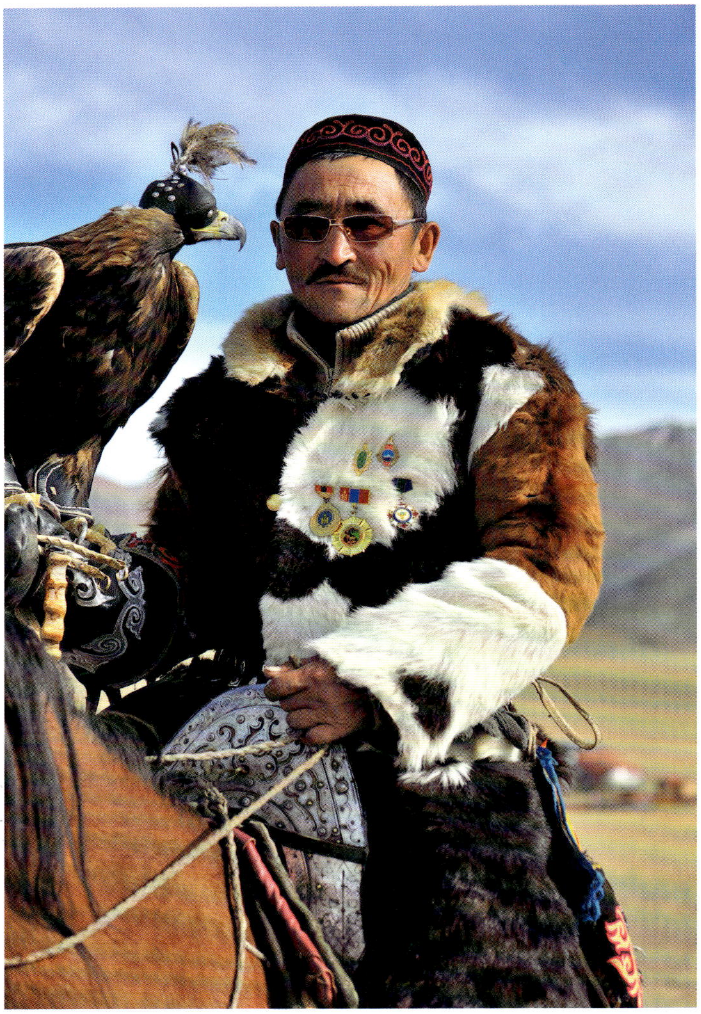
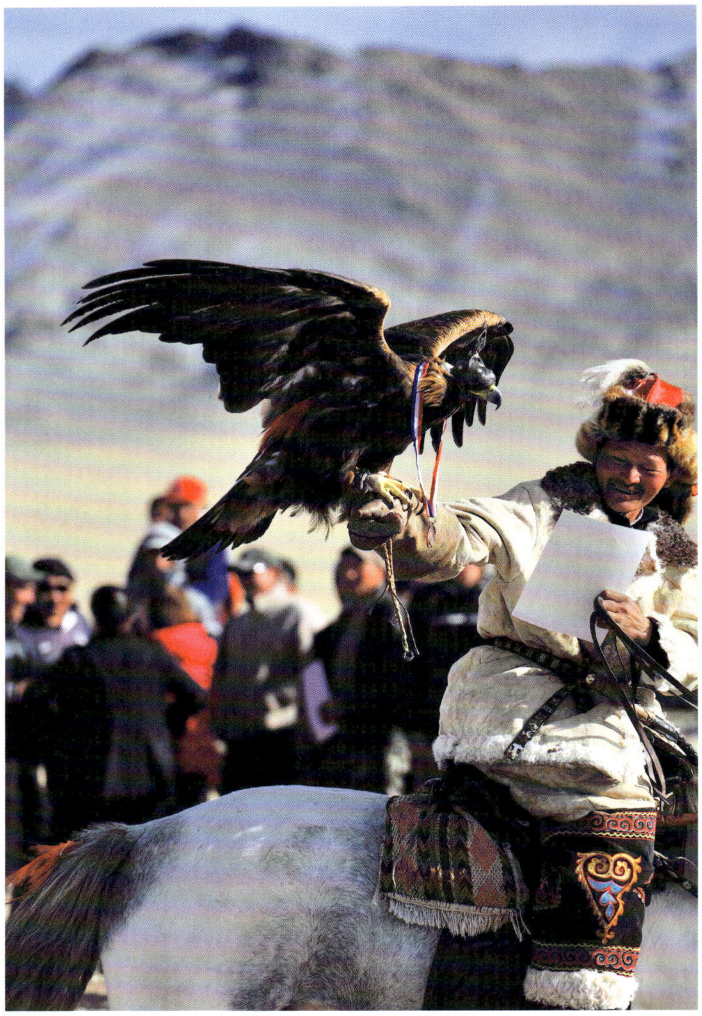
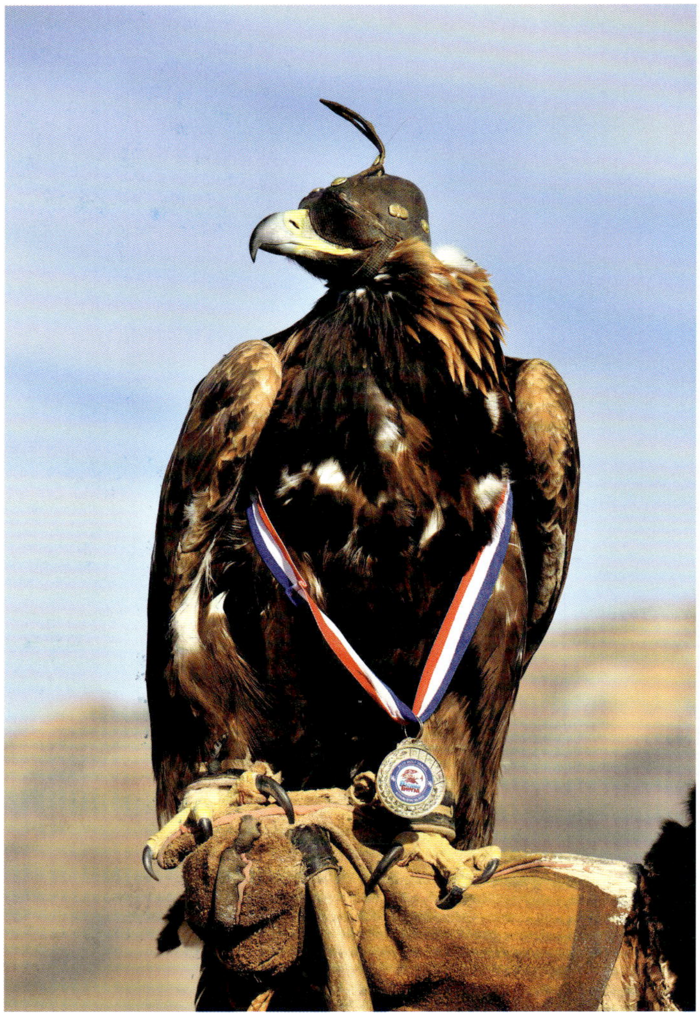
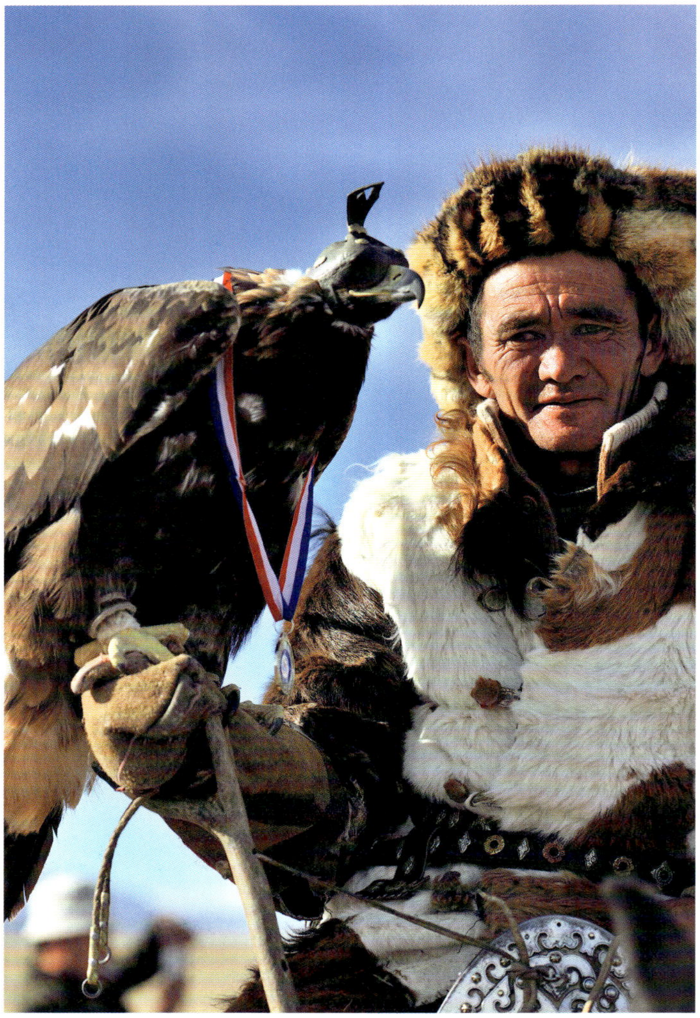

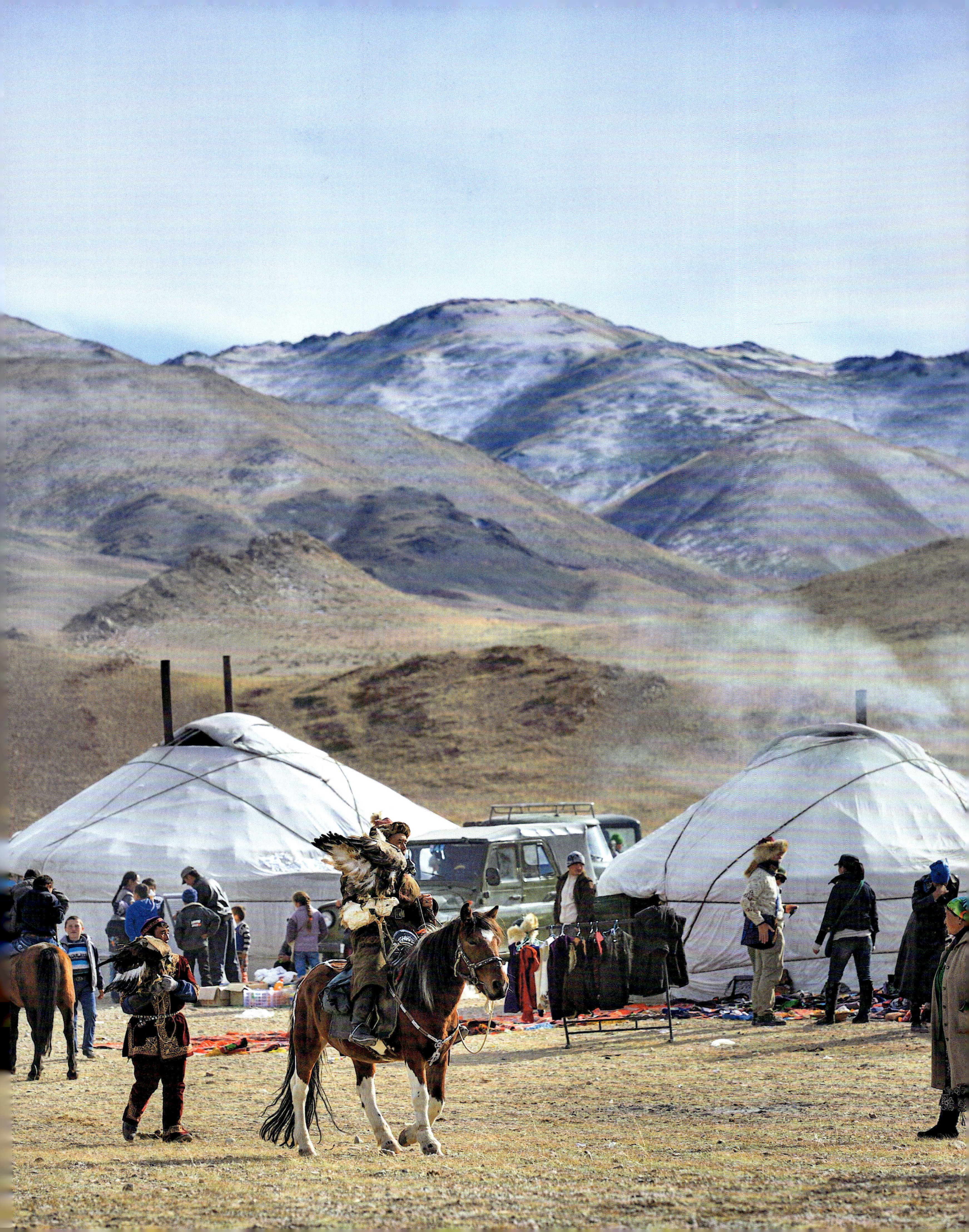

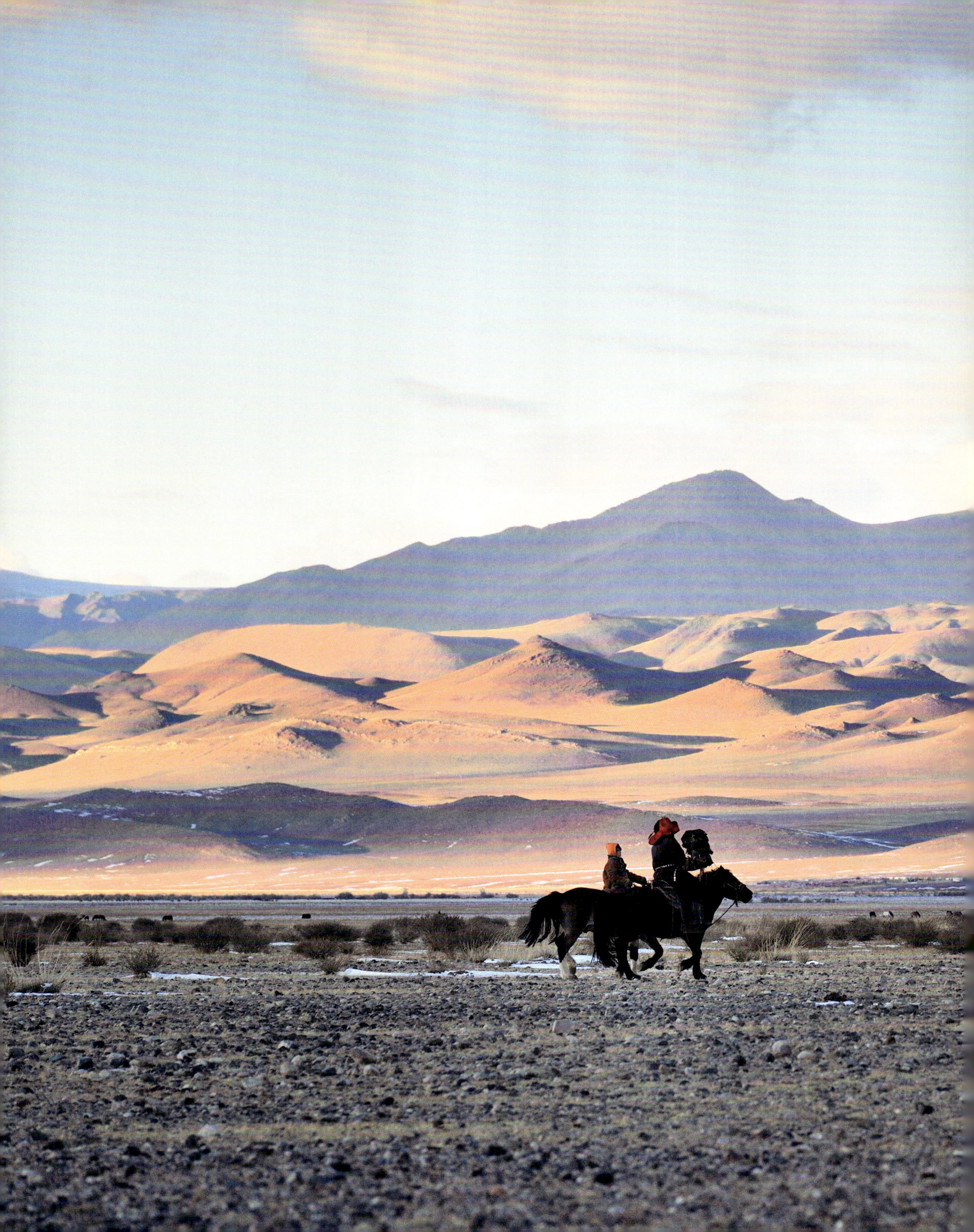

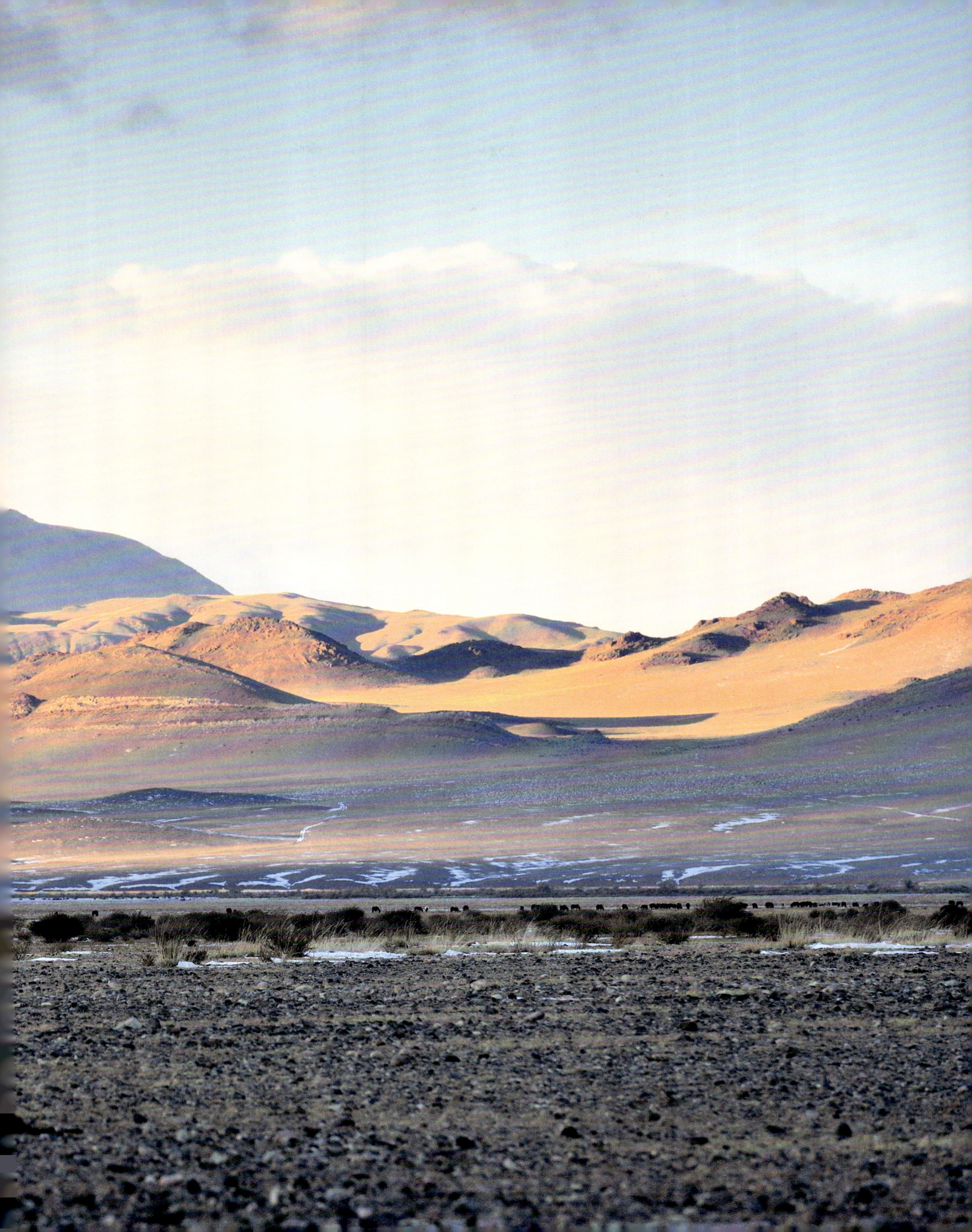

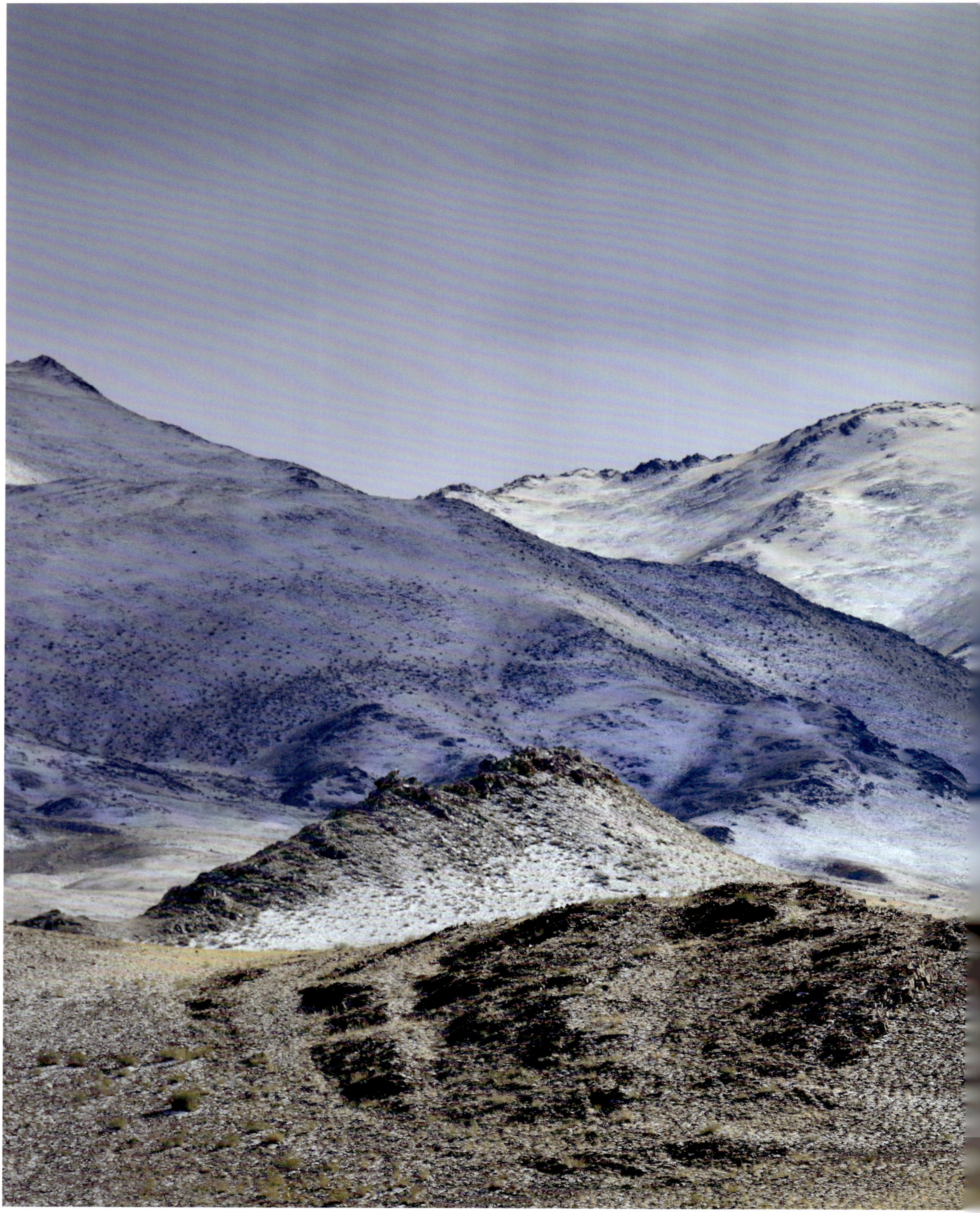

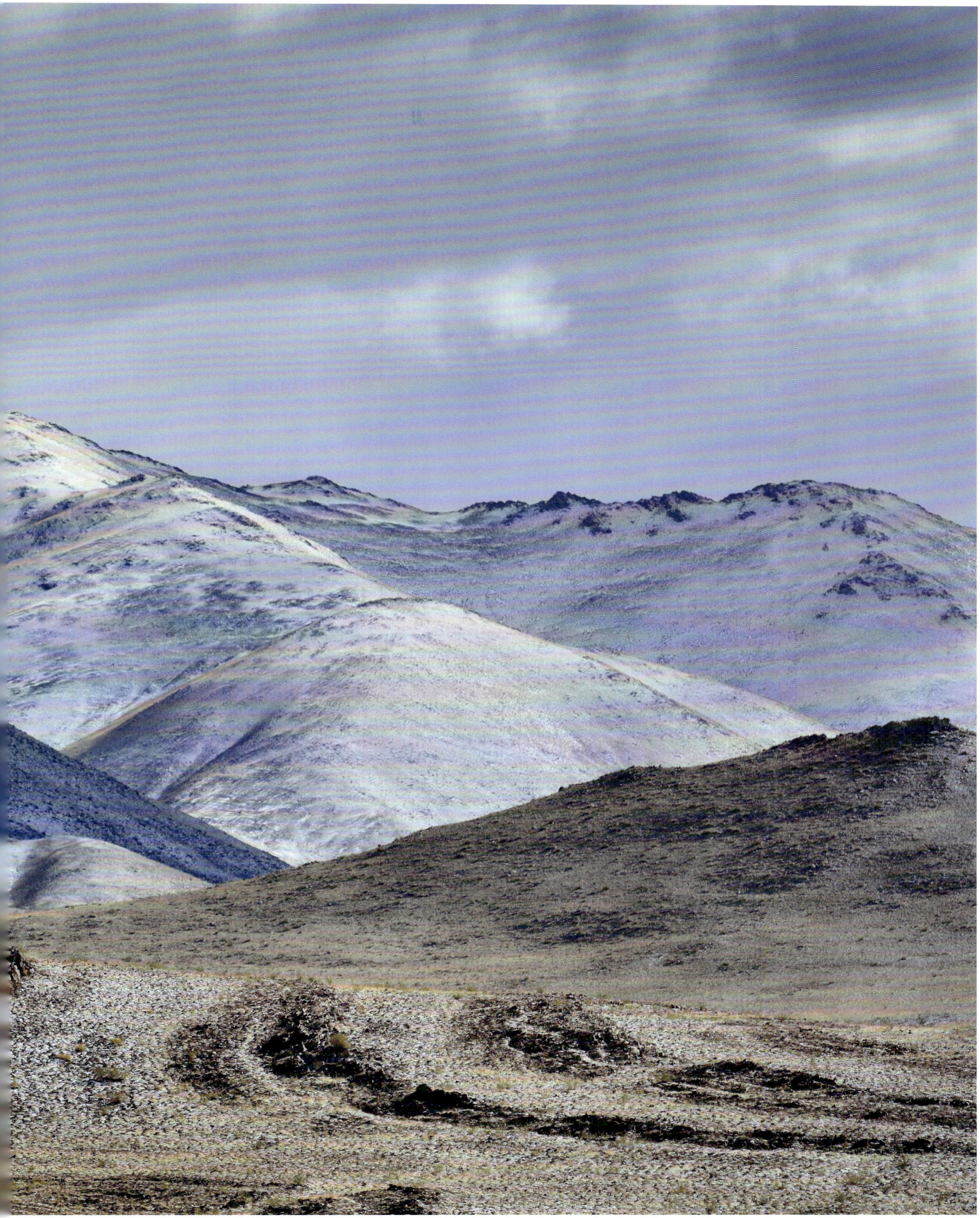

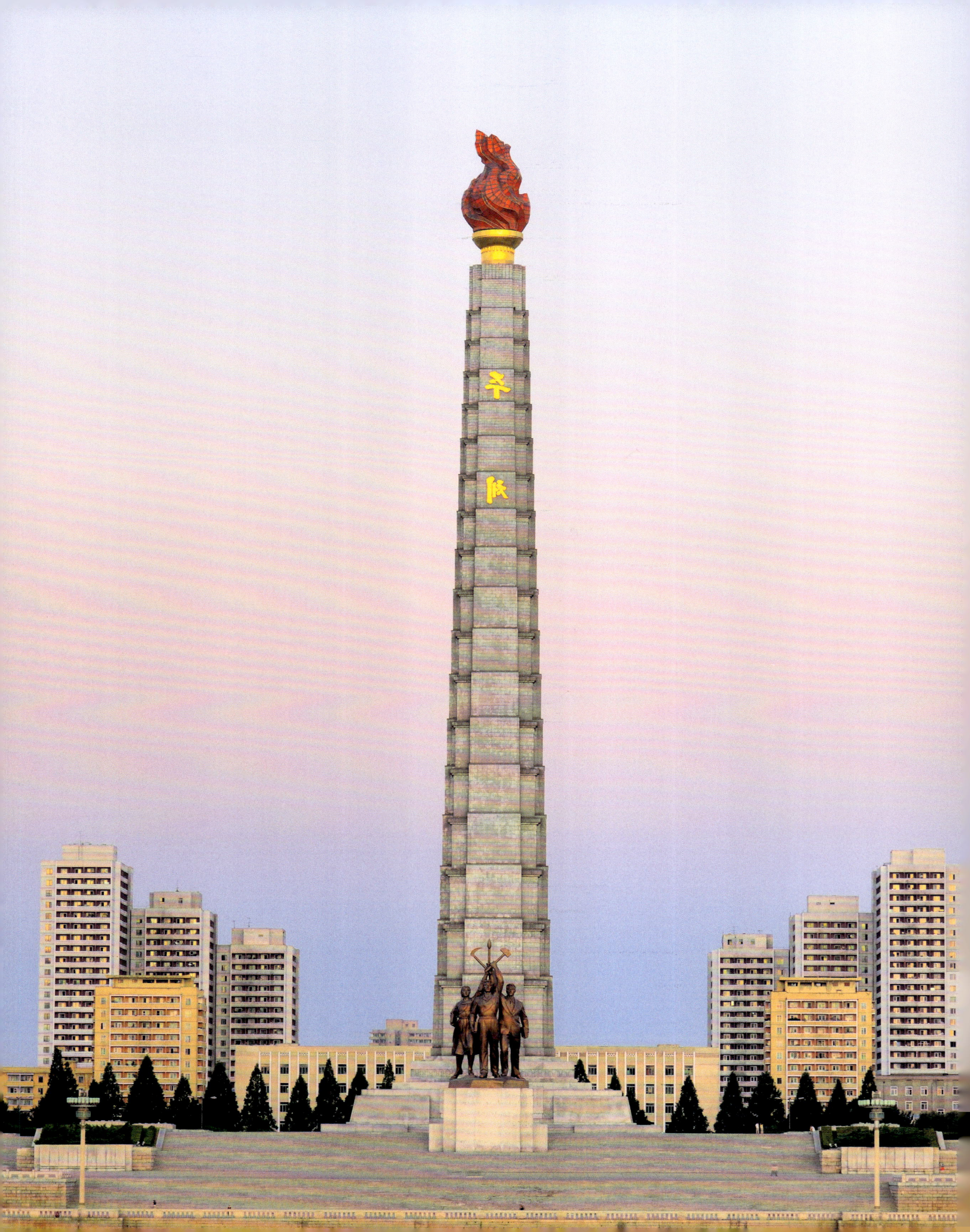
주체

CHAPTER 5

UNEXPECTED DISCOVERIES IN A SEALED-OFF NATION

OPEN MINDS AND CLOSED COUNTRIES

North Korea

39°01'03.91" N
125°45'47.75" E

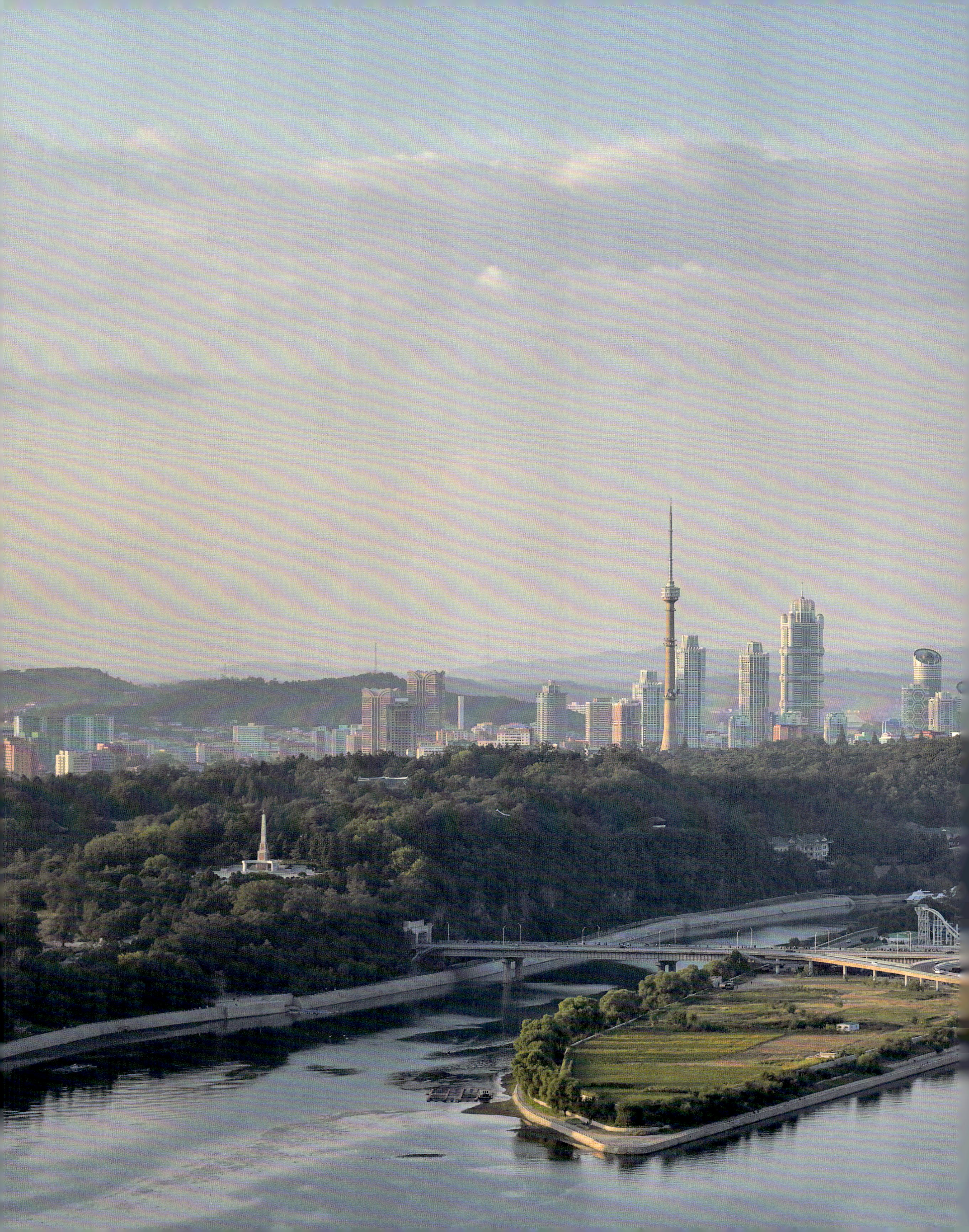

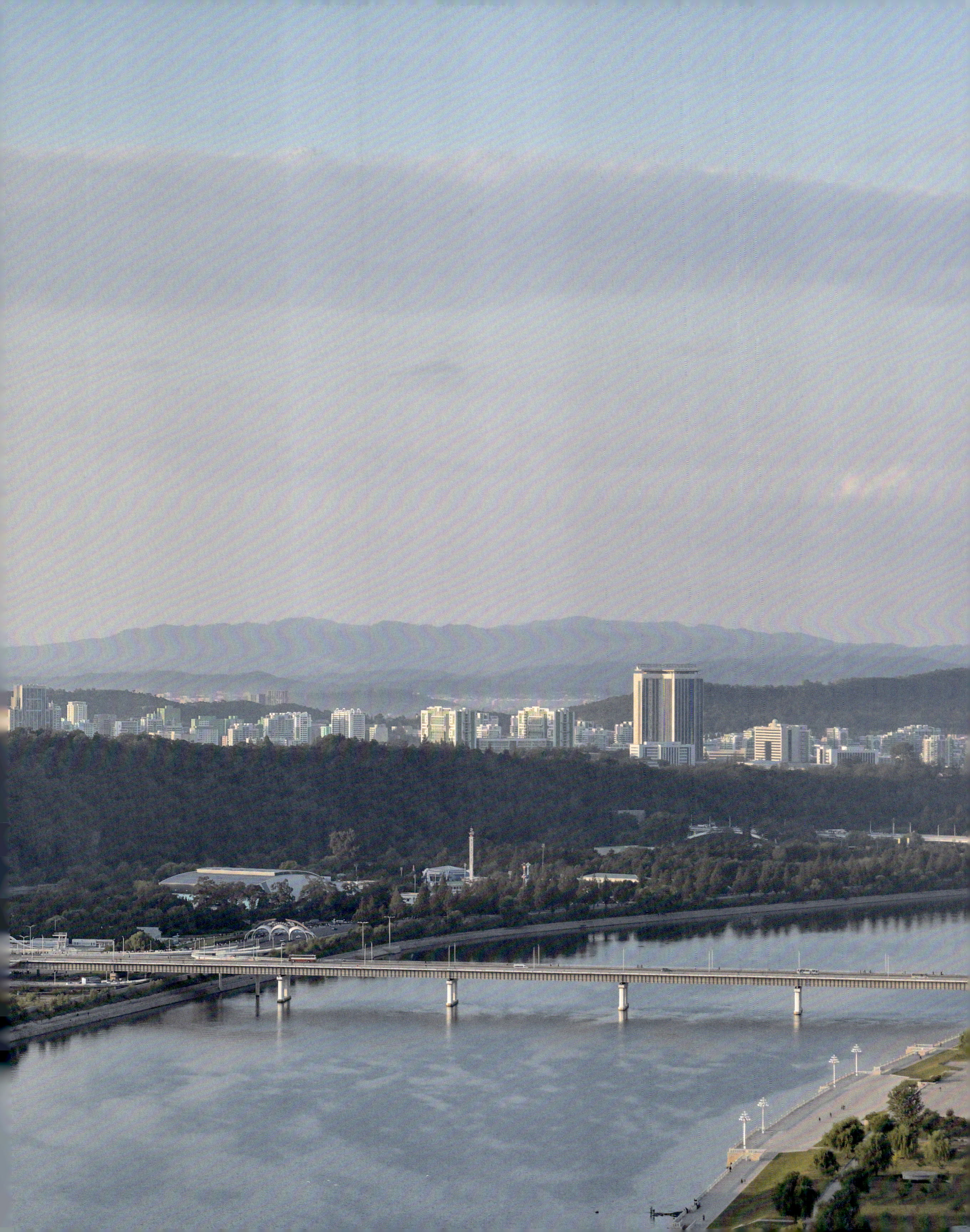

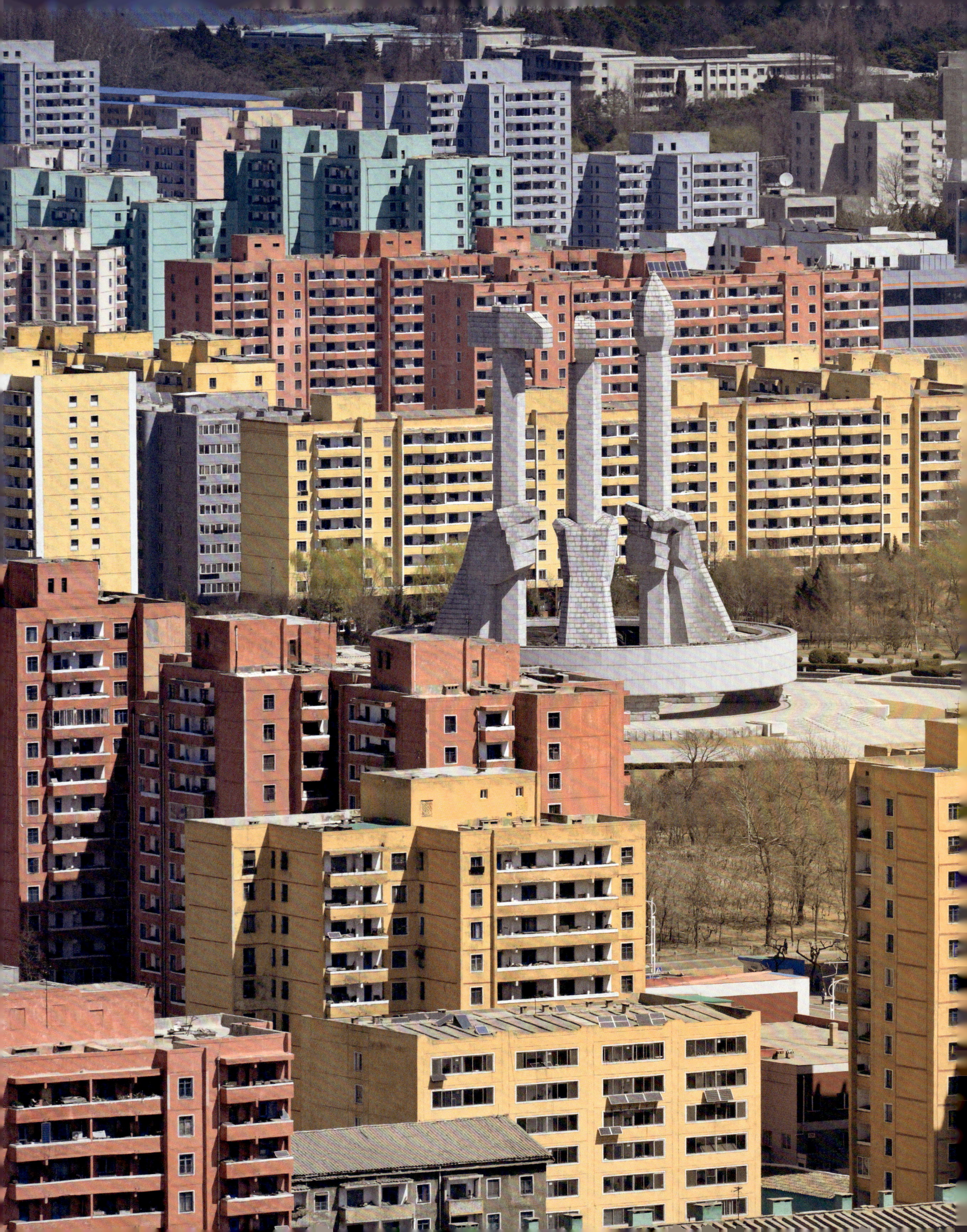

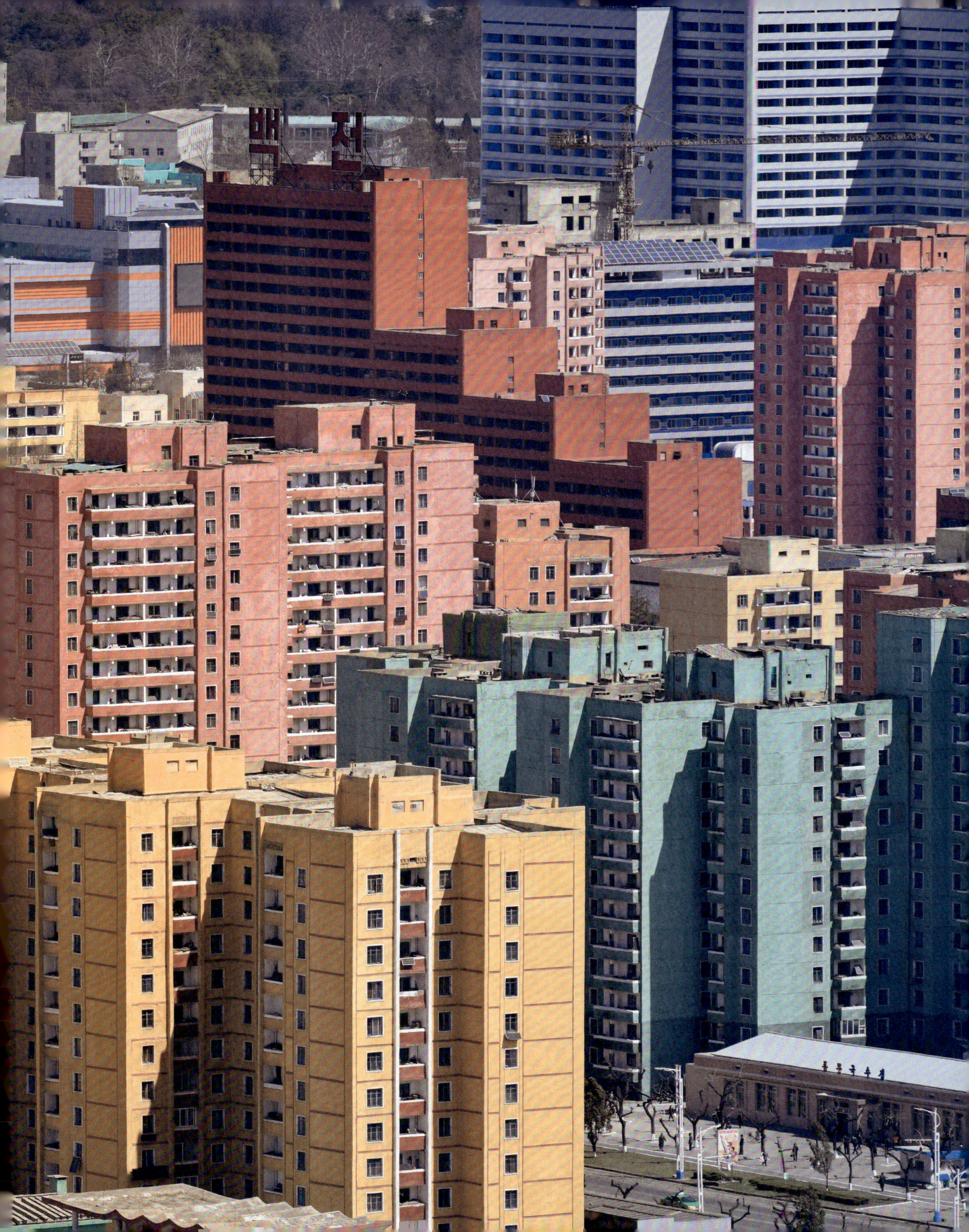
백
전

Unexpected Discoveries in a Sealed-Off Nation

Open minds and closed countries

Country:	**North Korea**
Best time to travel:	**Spring and autumn**
Closest airport:	**Pyongyang**
Price tag:	**€€**
Style of travel:	**Restricted but remarkably easy**

Introduction

Any curious traveller can step with an open mind into one of the most closed and unknown countries in the world and experience in a very singular way one of the last Communist dynasties. It's surprisingly easy to book a ticket to Pyongyang, the capital of the Democratic People's Republic — even though many assume it is totally off limits. Fewer than 6,000 Western tourists visit the country every year. Behind the headlines and the stereotypes lies a country full of astonishing communist architecture; a place where even foreigners can join the Pyongyang Marathon or experience the rare spectacle of the Arirang Mass Games, and visit forest, snowcapped mountain ranges, and long, sandy beaches. Via well-organised tours and cultural exchange, tourism can help bridge the divide between North Koreans and foreigners, fostering a better understanding between the DPRK and the rest of the world. The only thing left to do is to enter this sealed-off nation with an open mind.

The Views

The business-class cabin of North Korea's national airline, Air Koryo, is filled with passengers in suits, almost all wearing the red pins with the Leader shown on them. White-gloved flight attendants wearing red lipstick hand out copies of North Korea's only English-language newspaper, *The Pyongyang Times*. From Beijing, it's only a 90-minute flight to Pyongyang. On rare evening flights, the border is clearly visible from the air. Beyond the Yalu River that separates the two countries, the territory of North Korea looms dark, contrasting with the city lights of China. Despite its unappealing image, Pyongyang is a city in progress. New buildings are constantly being added to the skyline. In recent years, a fast-growing middle class has moved into the high-rises of a brand-new area, Changjon Street, nicknamed 'Pyonghattan' which was rumoured to have been built in a year. In Pyonghattan residents go to the gym, have fancy weddings, enjoy fine dining, and aspire to more Westernised lifestyles. Among them, women like Kim Jong Un's wife dress in the latest fashions and accessorise with groomed pet dogs — unthinkable a few years ago. More than 8 per cent own a mobile phone. In the past, the six-lane motorways running through the capital were famously empty. Now the roads are busier, with taxis and traffic lights. Pyonghattan's residents jump into those taxis to dine at fancy restaurants and bars, or perhaps to attend the Arirang Mass Games. This event is the highlight of any visit to North Korea and takes place at the Rungrado May Day Stadium — the largest stadium in the country (with a claimed 150,000 seats). Inside, the spectators sit on one side of the arena, while on the other 20,000 participants hold books with coloured pages that become part of a synchronised succession of visuals in a stadium-wide backdrop: enormous mosaics, pictures, slogans, and depictions of the North Korean flag. A giant screen, powered by human pixels. This is one of the DPRK's main attractions and an unforgettable spectacle for travellers. The Mass Games are a synchronised socialist-realist spectacular, a massive show, featuring more than 100,000 participants in a 90-minute display of gymnastics and dance, accompanied by music and all wrapped in a highly politicised package underpinned by propaganda. The near-annual event runs every week during September and October. Almost 80,000 dancers and gymnasts train throughout the year to perform for their compatriots and especially for their leader. As victorious music blasts across the arena, thousands of dancers and gymnasts arrive, and a surreal display of mass choreography unfolds. The décor changes constantly;

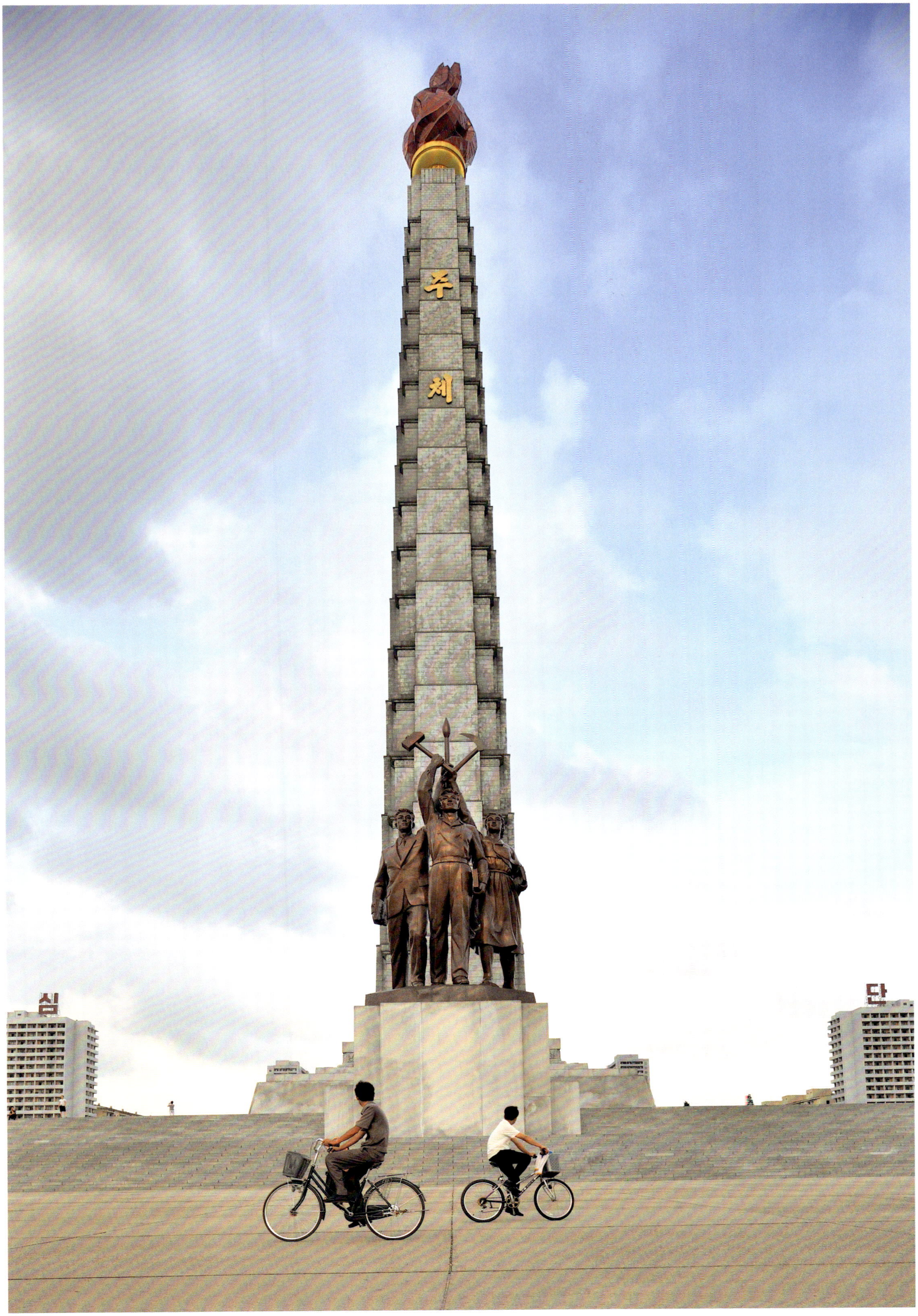
주
체
심
단

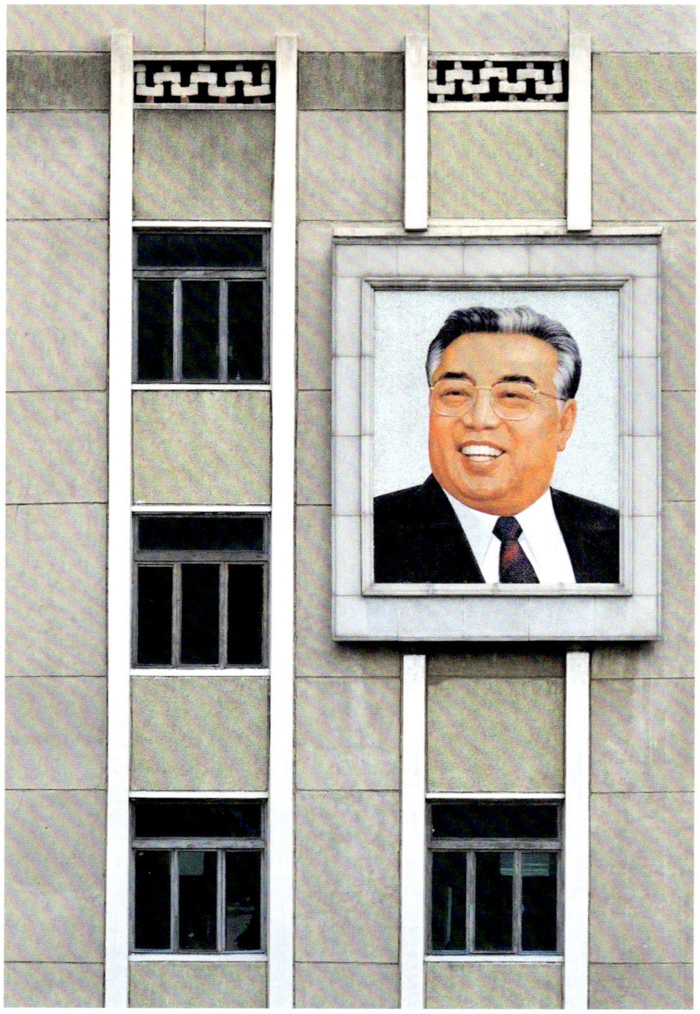

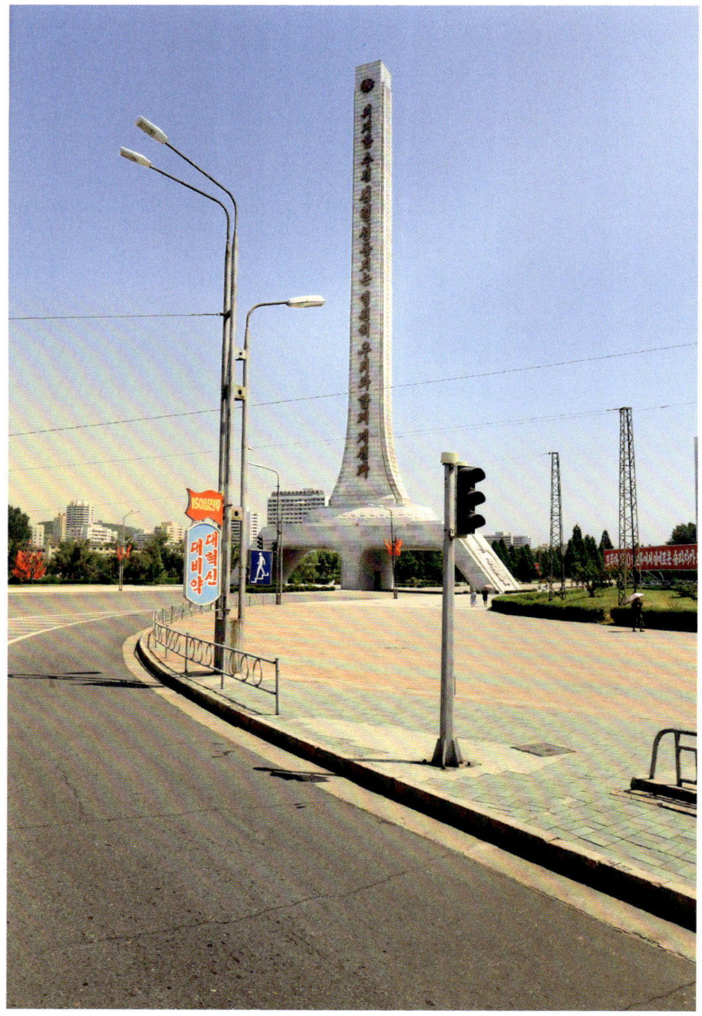
대혁신
대비약

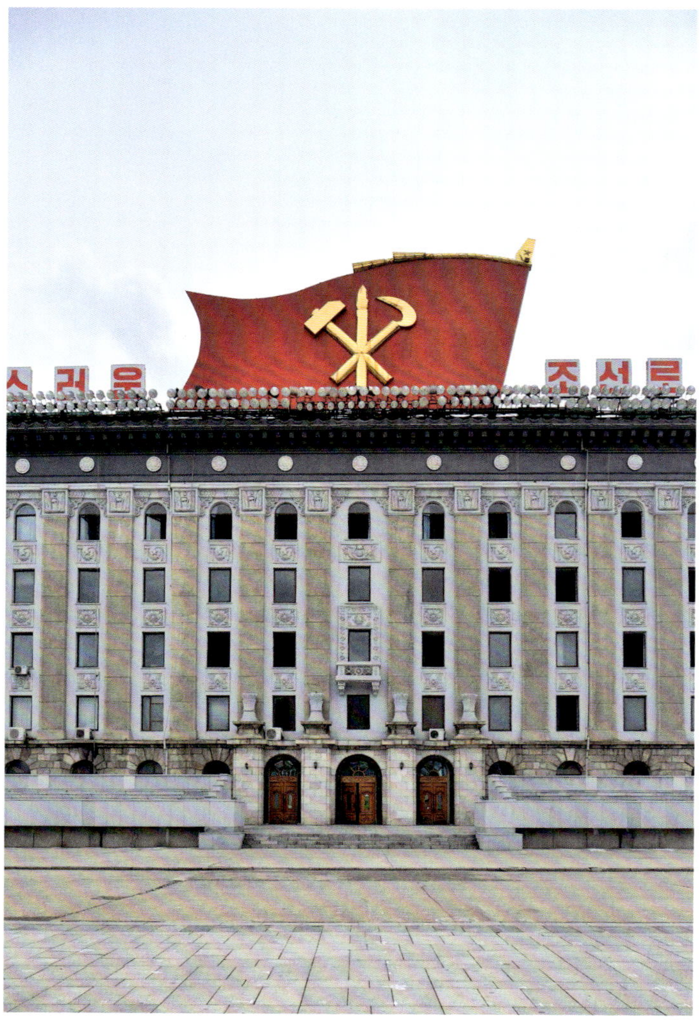

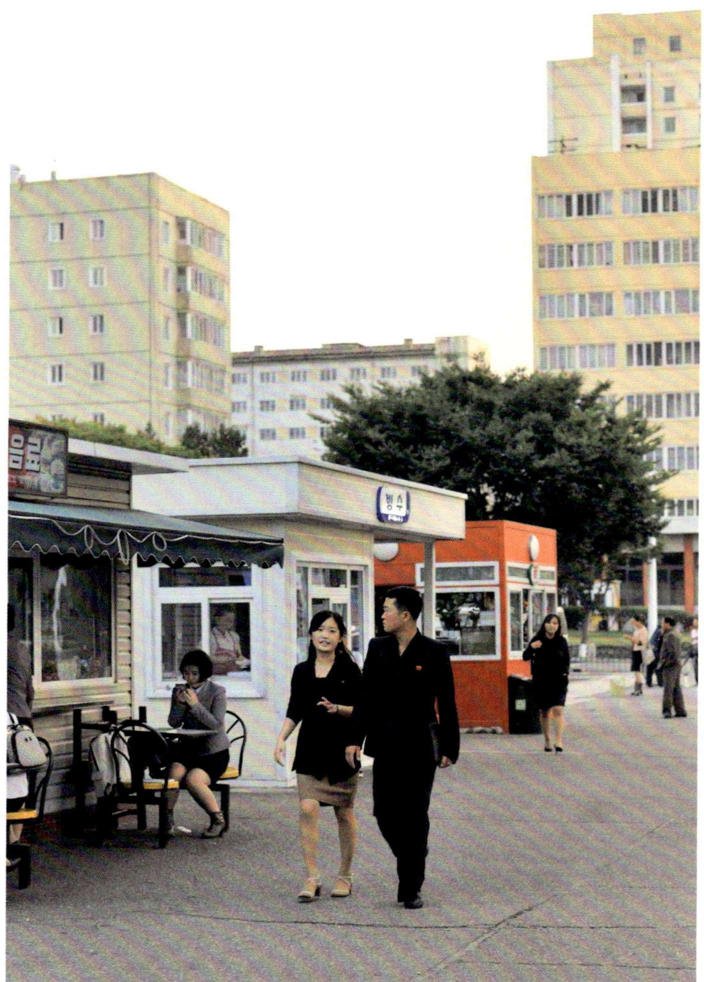

Pages 142–143 Candy-hued apartment buildings add a sense of playfulness to the extreme uniformity of Pyongyang's gridded streets. For architecture aficionados, specialised trips focusing on the capital's Communist and Soviet-style buildings can also be booked.

entire mountain ranges appear, then a gigantic globe rolls in surrounded by tens of thousands of entertainers: young gymnasts in tight-fitting costumes and elegant ladies in sumptuous ball gowns or army uniforms. Those not engaged in perfectly choreographed dances swoop through the air on zip wires or are catapulted from cannons. The message is that North Korea is not just big, strong, and proud, but is also a place of peace, harmony, prosperity, hope, and happiness. The sense of mass unity is complete. A beautiful way to experience daily life in Pyongyang is to ride the underground — the deepest metro system in the world, which also doubles as a nuclear shelter. Marvel at the artwork in its striking stations and mingle with school children and workers on their daily commutes. The city also has brewpubs, a pizzeria with a proper wood-fired oven, a zoo, and a fairground, along with a university library and a music school buzzing with nationalist zeal. There's even a bowling alley with American-made bowling balls. It feels like visiting any other city, although, of course, one is never without a guide. The most remarkable sights are the mausoleum of Kim Il Sung and Kim Jong Il, the father-son duo who led the country for three quarters of a century. Monuments, squares, and statues are all fascinating, but in the end, it's the human connection that makes a visit to the DPRK so special. On weekends and national holidays foreign travellers can easily mingle with locals when families have long picnics in the parks of Pyongyang. On Liberation Day — August 15 — the whole of Pyongyang seems to gather for a day off in places like Moranbong Park picnicking, dancing, singing, and listening to live music played in the pagodas. Families gather on the grass in the shade of trees, women wear spectacular dresses, and men drink soju (distilled rice) and happily share it with foreign visitors. Some sing and dance, while others simply relax in the grass. Elsewhere, parents teach their children to skate or ride bikes, or enjoy the rides at an amusement park like Kaeson Youth Park. Despite what many might think, these happy campers are not actors but ordinary people just enjoying a day off. Like people everywhere, they just seek a pleasant life. The Demilitarised Zone or the DMZ — the fragile border between North and South Korea — lies only 168 kilometres (104 mi.) from Pyongyang, connected via the Reunification Highway. A trip to the backcountry also offers a glimpse of rural village life among green rice paddies. Around four newly built ski resorts (Masik being the fanciest) with 25 kilometres (16 mi.) of pistes await visitors in winter, reaching altitudes of almost 1,100 metres (3,600 ft.). Ski passes cost 30 Euro a day and there's even internet, though access is frustratingly limited. Even though trips to North Korea are organised, guided, and restricted, they allow visitors to draw their own conclusions about the Hermit Kingdom. As Michael Palin concludes in his 2018 documentary after a fascinating visit of the DPRK: "For now, it stays a closed country but events are moving fast and we can only hope that positive change is coming to the people of North Korea.'

Creature Comforts

Lodging in the DPRK is straightforward. In Pyongyang, the Yanggakdo Hotel, a Western-style three-star behemoth with 700+ rooms, offers several bars and shops, along with a swimming pool, a bowling alley, a casino, karaoke facilities, and a revolving restaurant on the top floor. The Yanggakdo has air conditioning, internet access, and foreign TV. Local beer is 0,50 Euro per bottle.

Go Now

Some might ask if a visit to the DPRK is ethical, while others would say that a country shouldn't be judged until it has been explored. Even the most seasoned traveller will find a new perspective after visiting this secretive nation. Tourism offers scarce income for ordinary North Koreans, whose own horizons are widened as a result of exchanges with foreigners — bringing down the wall for both sides.

How to Go

Travelling to the DPRK isn't particularly complicated or dangerous, and once there, it's even relatively affordable. Expect to pay around 1,500 Euro per person for a five-day trip, including transfers, accommodation, most visits, and meals. A visa costs around 50 Euro and is easily obtained through the tour operator. International travellers can join a group or book an individual trip, but either way they will be escorted by guides. All visits, by plane or train, mainly start from Beijing.

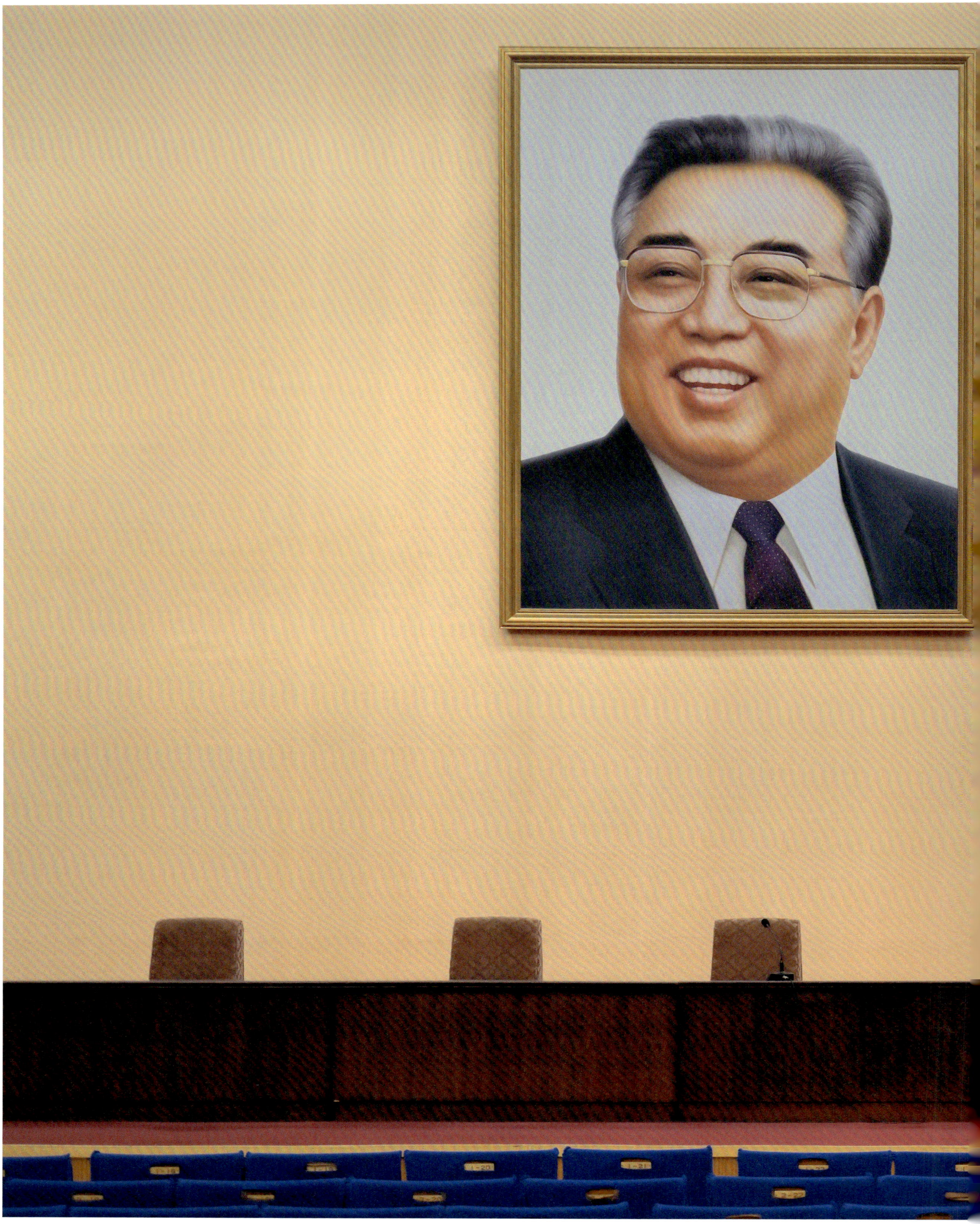

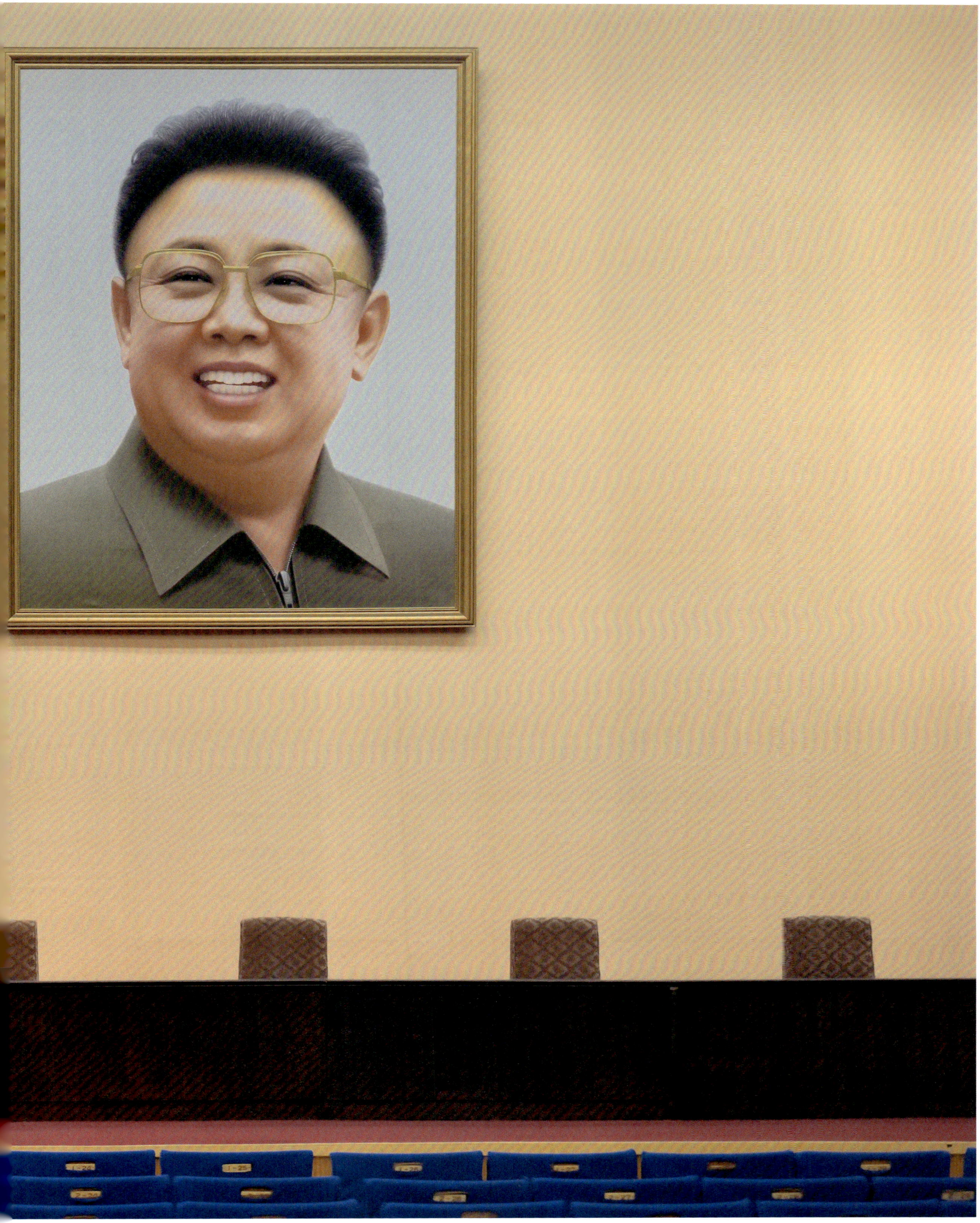

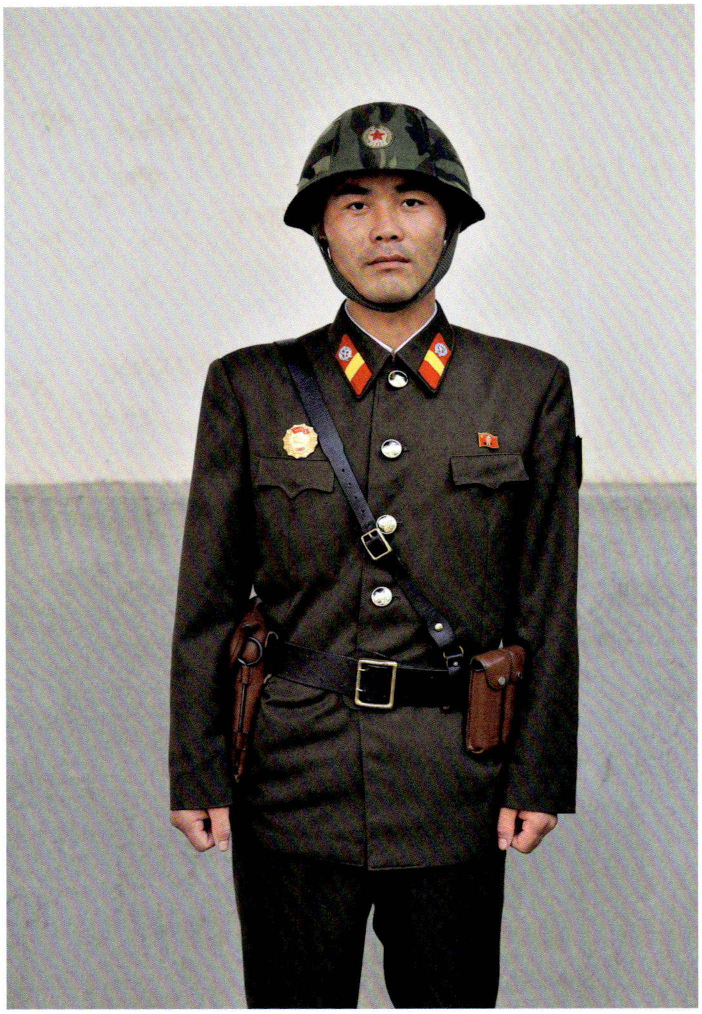

Opposite The 250-kilometre- (160 mi.) DMZ (Demilitarised Zone) between North and South Korea is one of the most heavily guarded borders in the world. The DMZ is a two-kilometre- (1.25 mi.) wide buffer, stretching coast to coast across the peninsula, lined on both sides with razor wire, heavy armaments, and tank traps. The long stretch of DMZ have been no man's land for more than 60 years, where wildlife has flourished undisturbed. The only location where visitors can temporarily cross the border on foot is marked by two blue buildings.

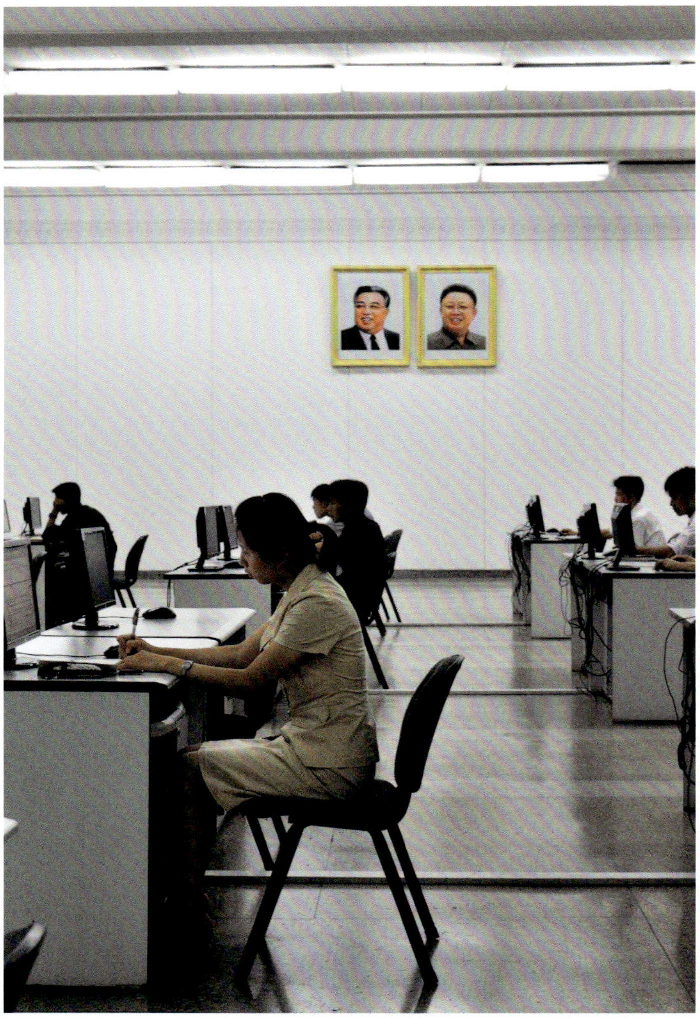

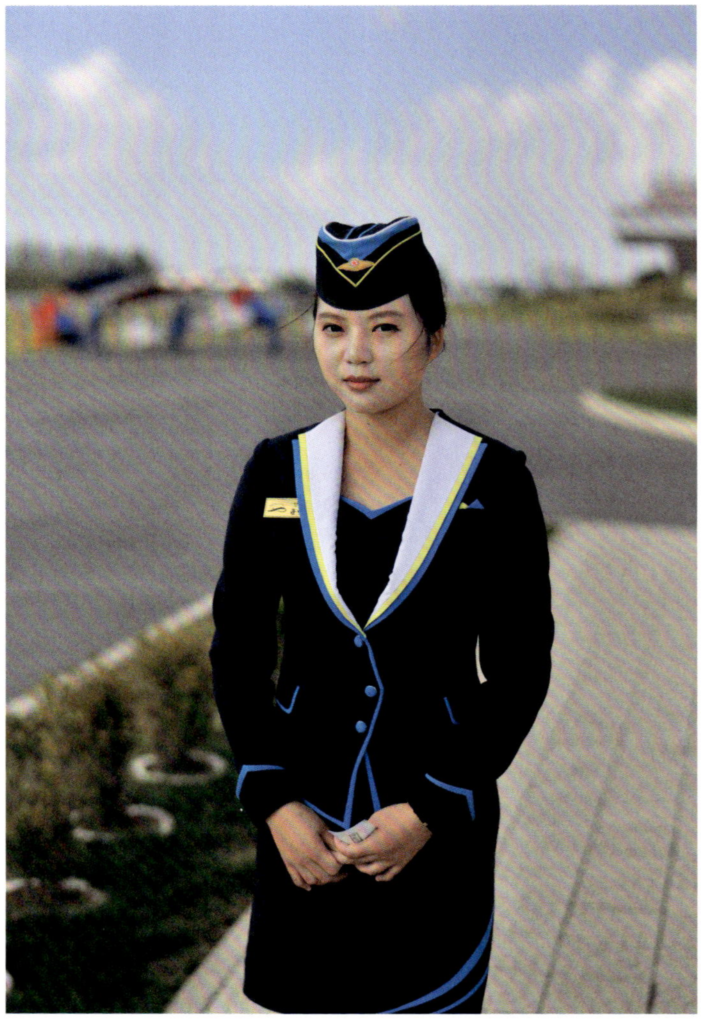

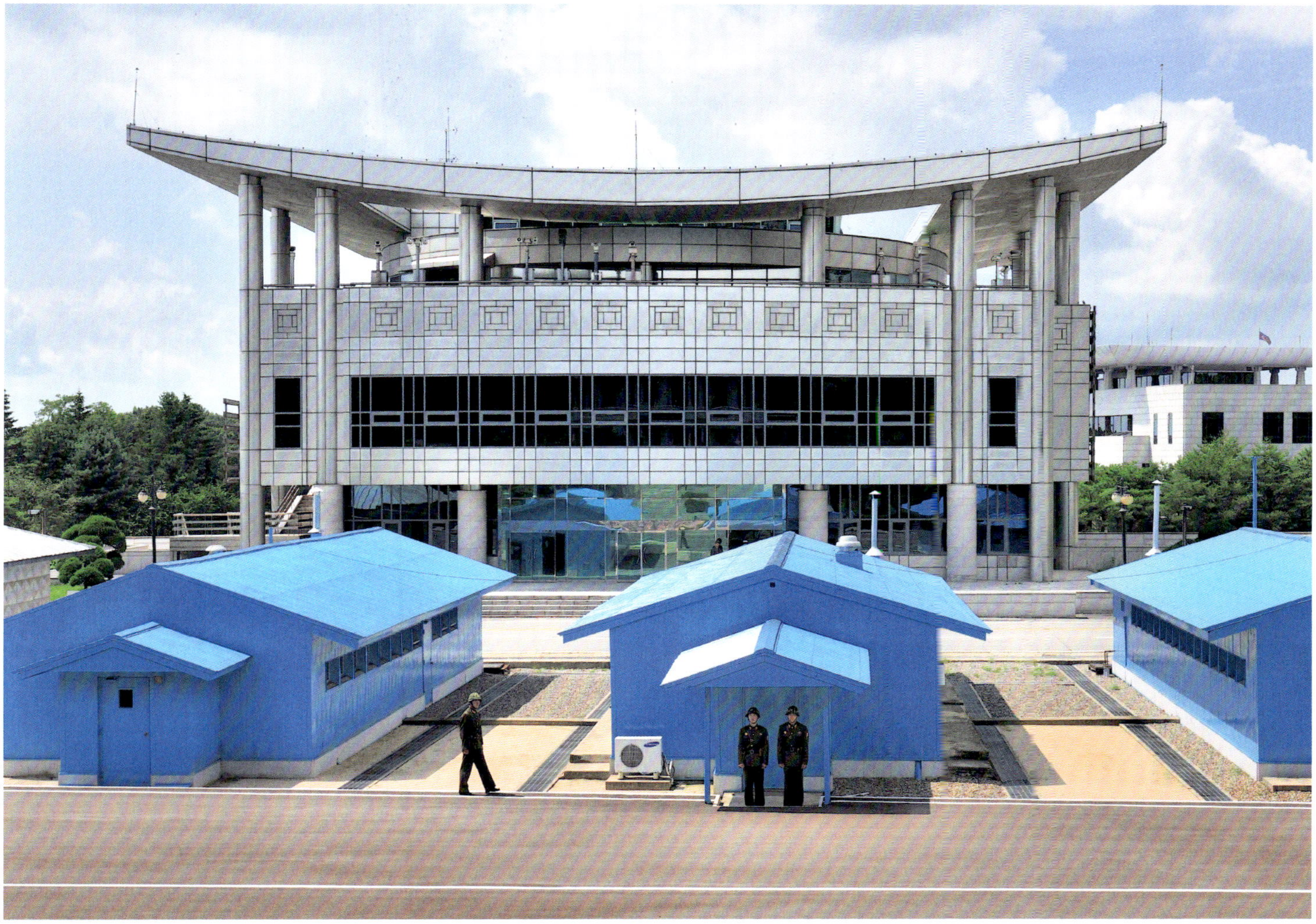

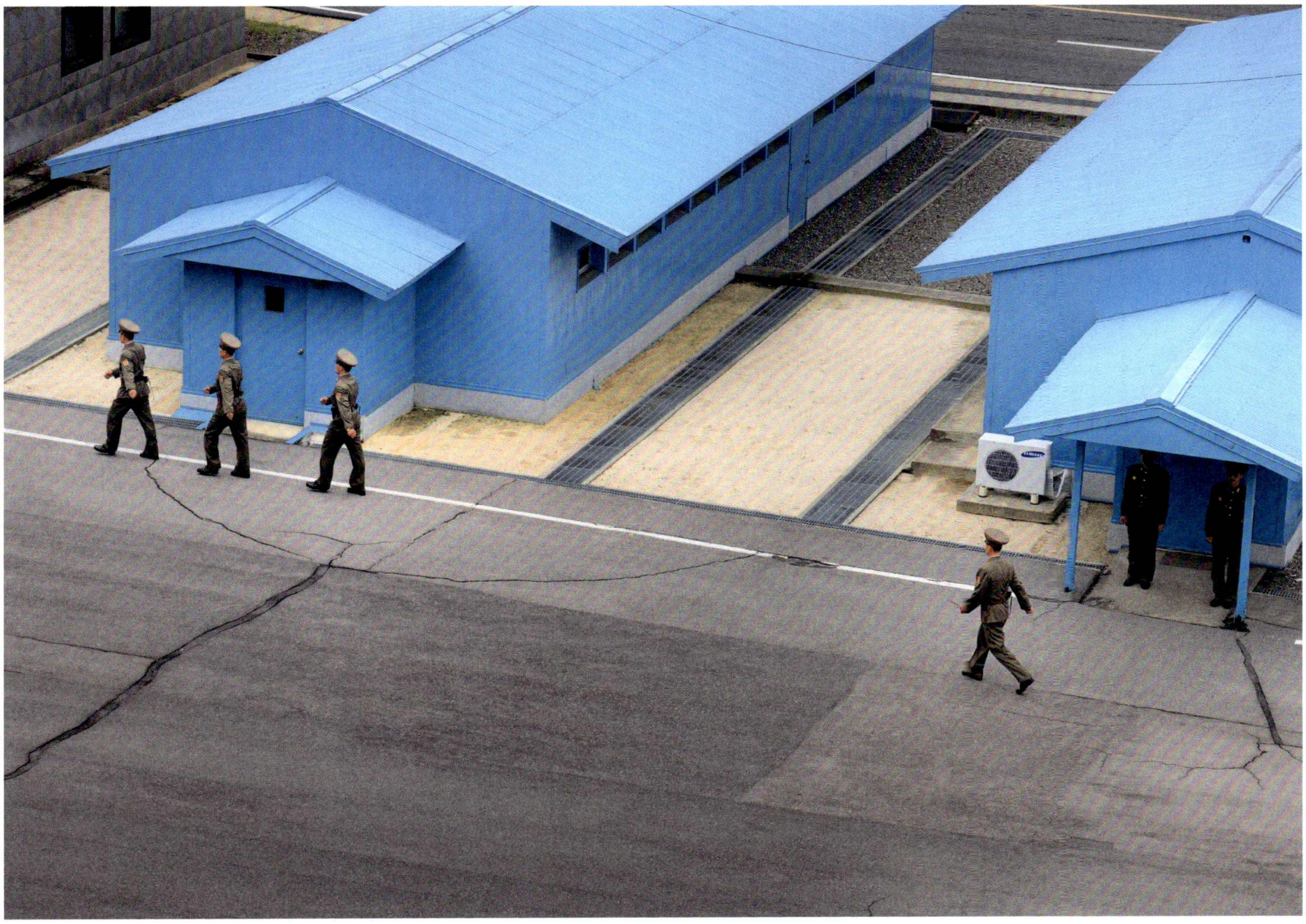

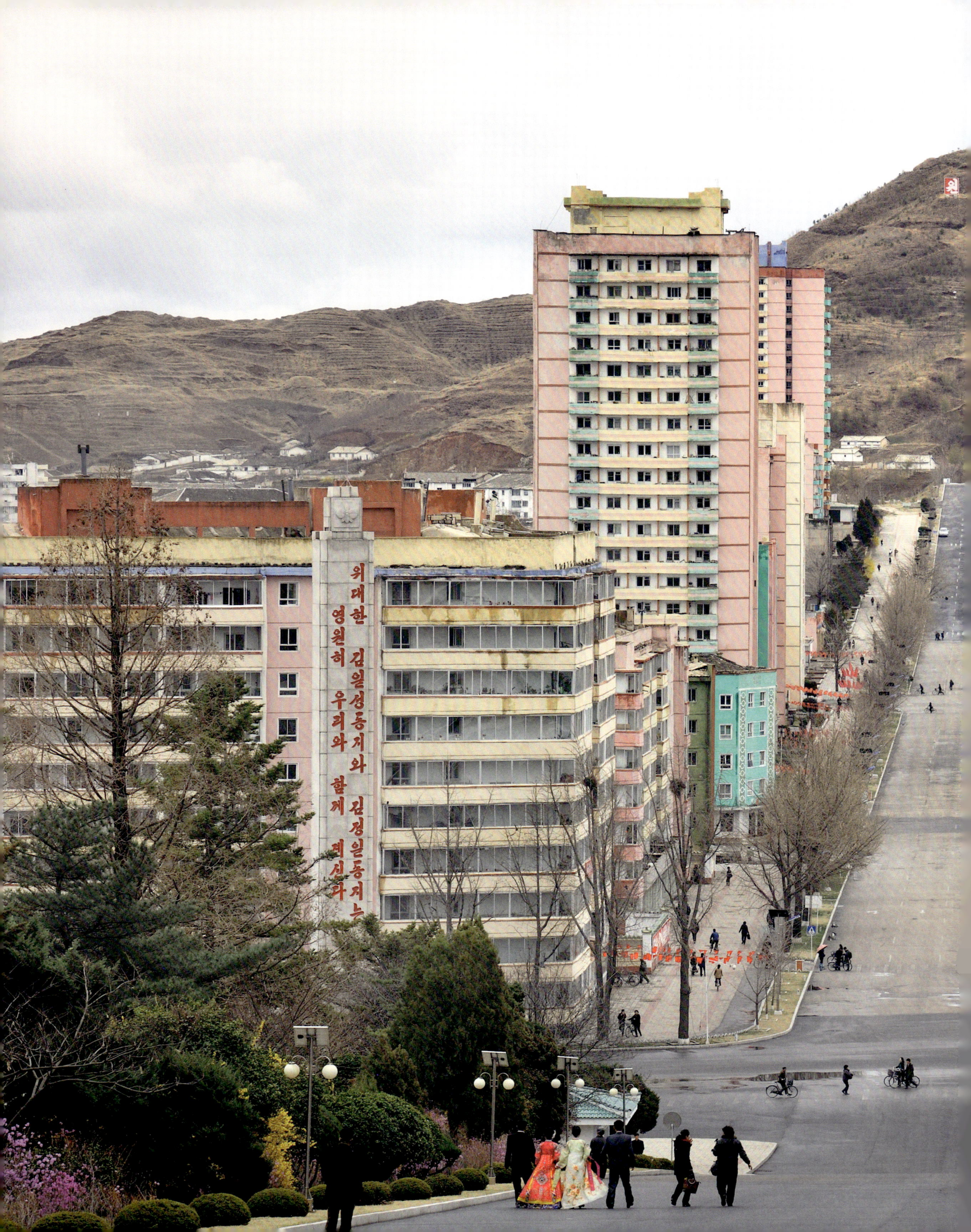
위대한 김일성동지와 김정일동지는
영원히 우리와 함께 계시다

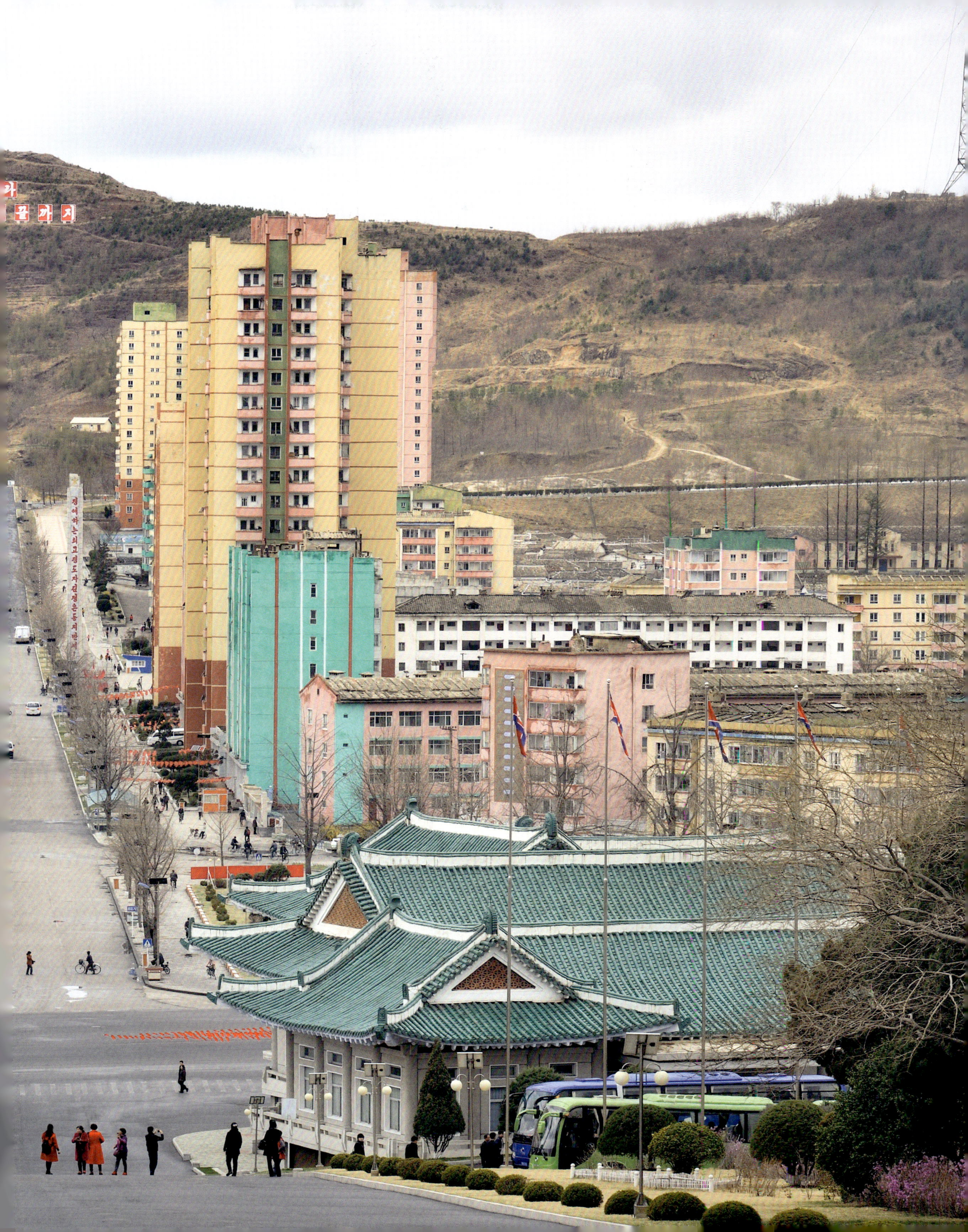
라
끝까지
경애하는 최고령도자 김정은동지 만세

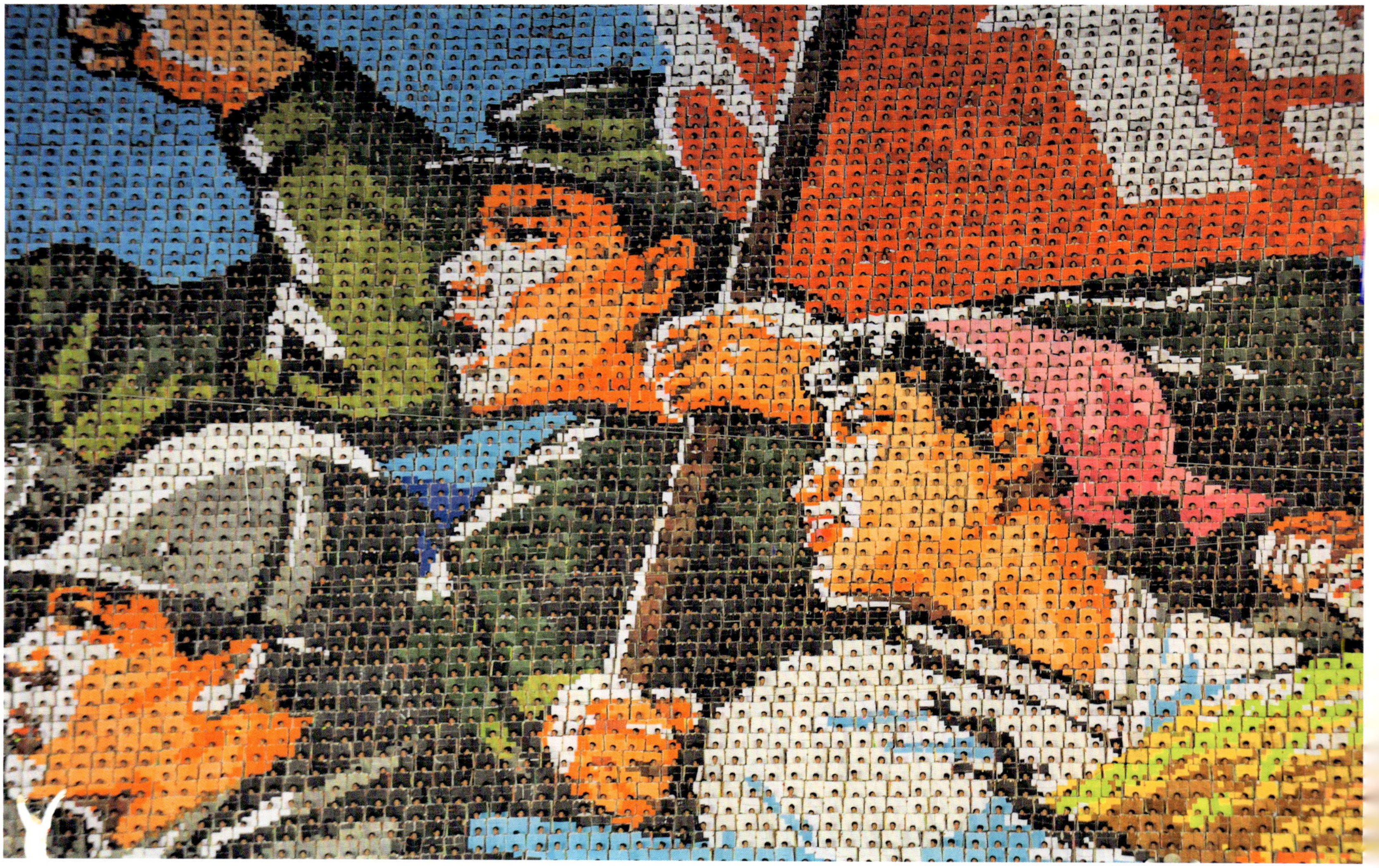

Above The Arirang Mass Games is an artistic performance of extravagant proportions. Featuring some 100,000 participants and drawing spectators from all over the world, the event was first held in 2002 and has since returned almost every year, although occasionally it has been cancelled without explanation. After a few years of absence, the Games returned in 2018, boasting a new theme and vision for the future, in celebration of the 70th anniversary of the DPRK. Foreign visitors can attend, and photography is allowed. The Mass Games are rooted in 19th-century Europe, where during the Industrial Revolution the human body came to be viewed as a machine. In this vision, humanity could be a well-functioning factory, a collective stronger than the sum of its parts. The earliest Mass Games were held in Germany and the Czech Republic, where large networks of gymnastics clubs were established. In regions that would later become part of the USSR, these mass gatherings became popular ways to celebrate socialism. In North Korea, the first of such events, titled 'The Era of the Workers' Party, took place in 1961.

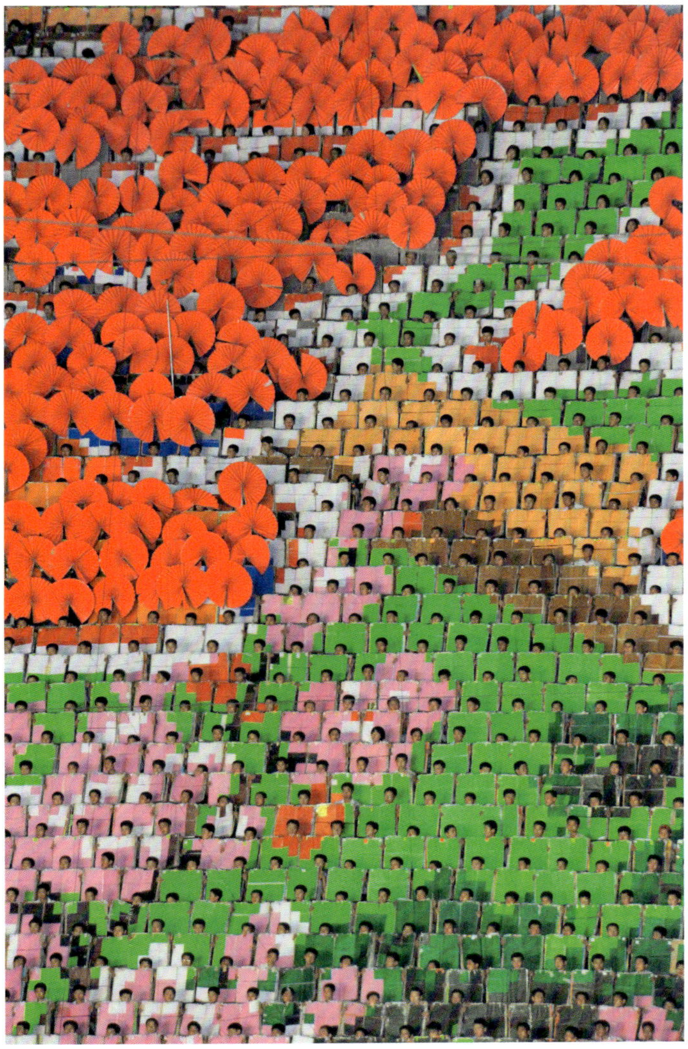

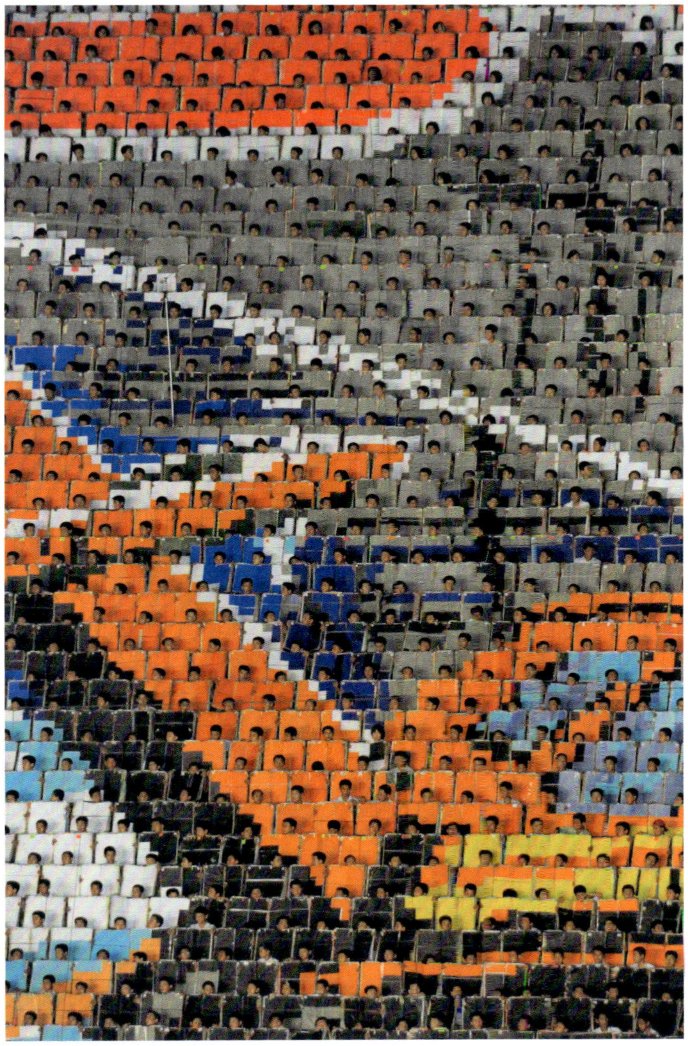

This page *I Want More*, a music video by British band Faithless, shows how hard the dedicated participants in the Arirang Mass Games train. The lyrics of the song ('more oneness, less categories, open hearts, no strategies') can be seen as describing the positive traits of North Korea. The clip is an excerpt from the British documentary *A State of Mind*, a rare glimpse into North Korean daily life. The filmmaker followed the lives of two North Korean schoolgirls as they train for the Mass Games. The film was made possible by Beijing-based Nick Bonner, who has been leading tourists and film crews into the DPRK for decades.

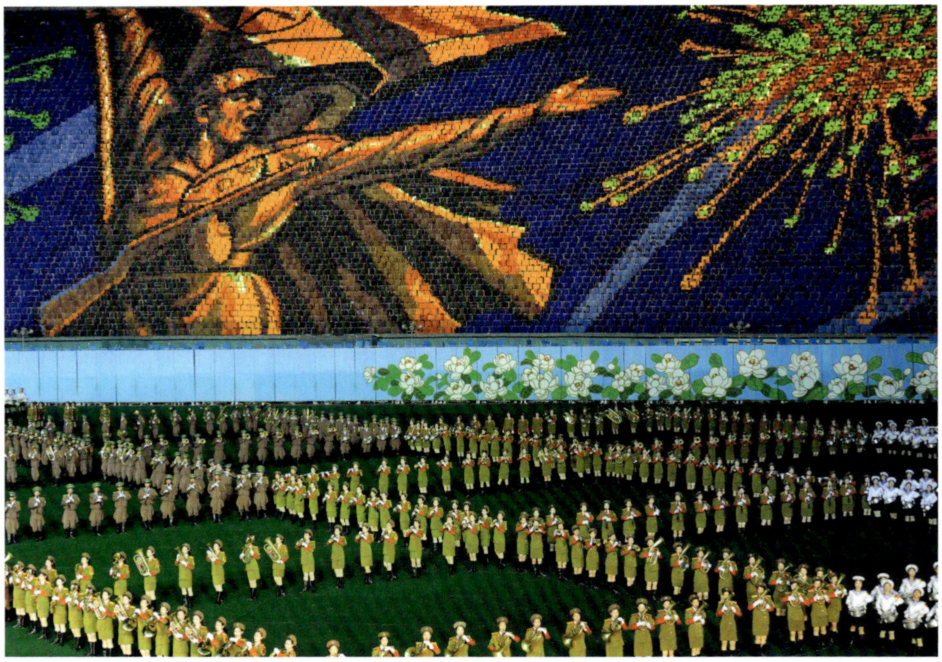

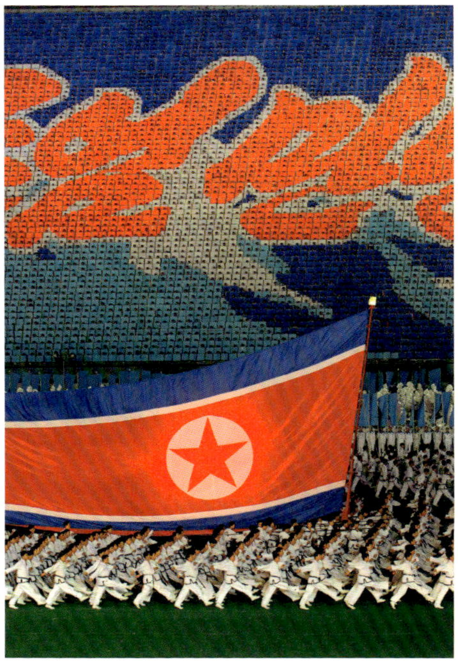

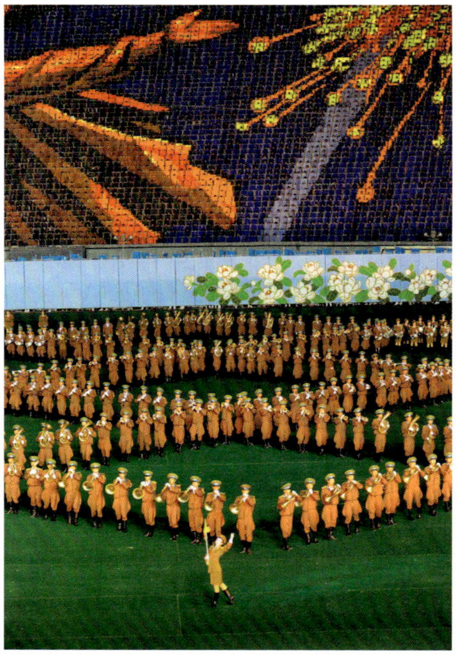

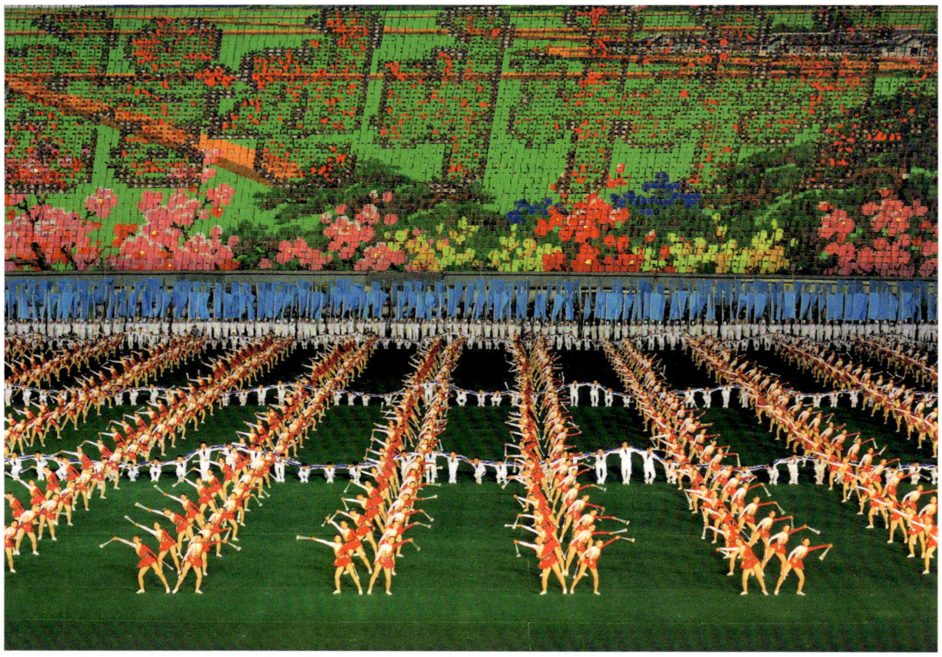

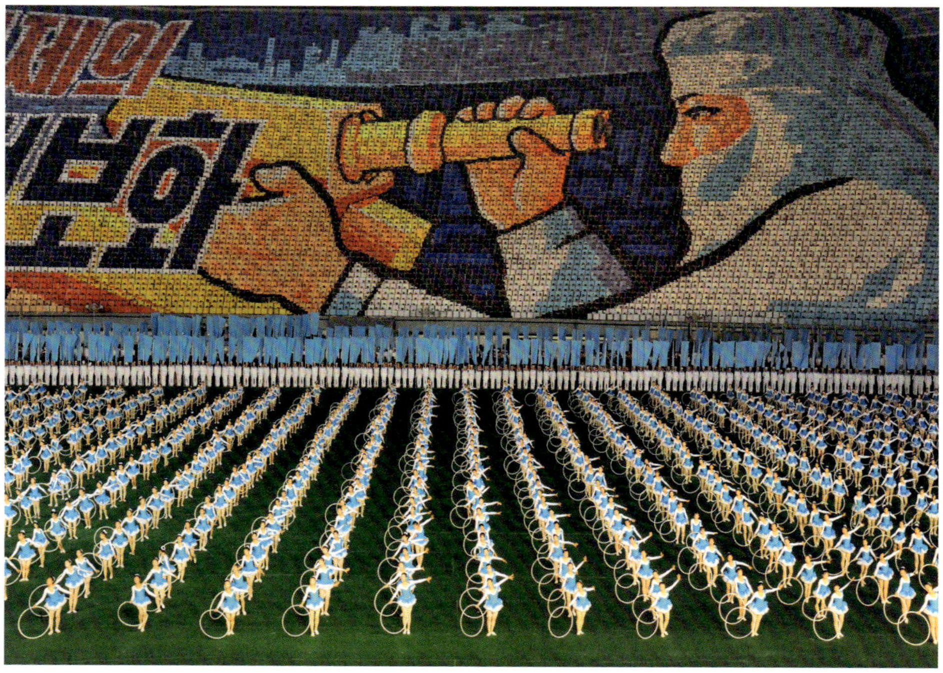

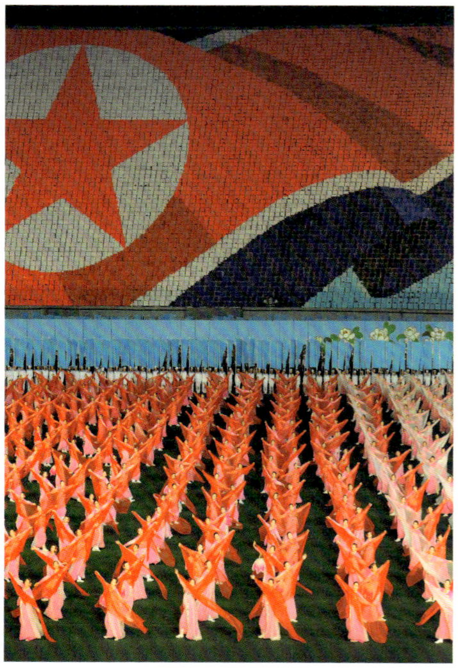

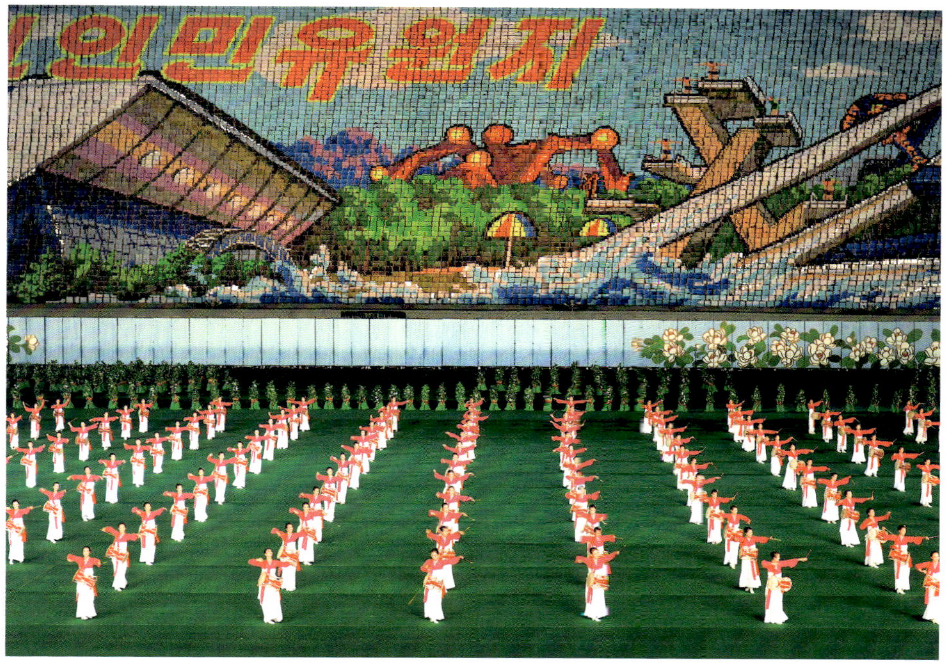
인민유원지

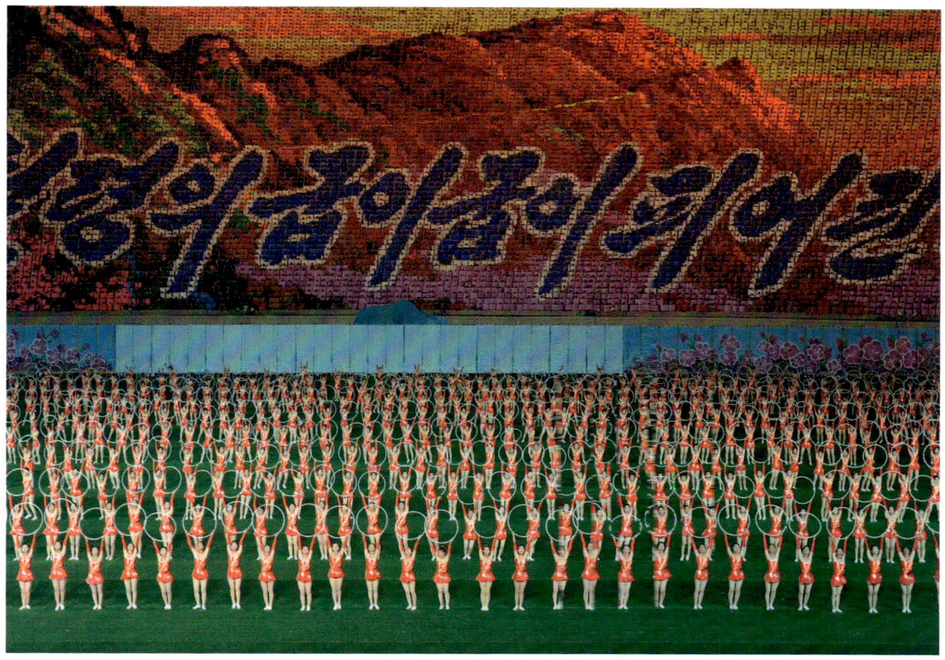

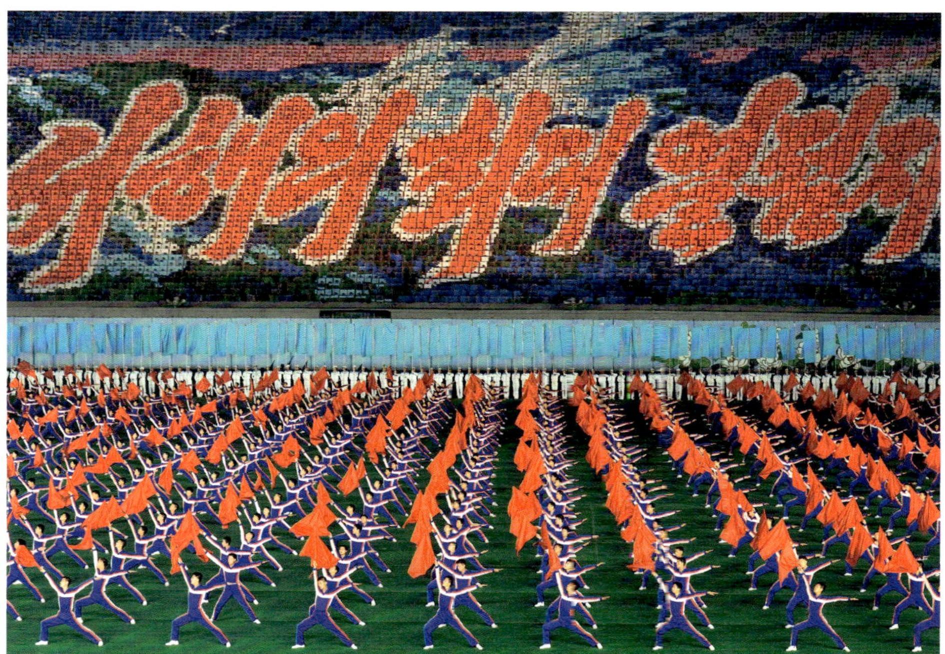

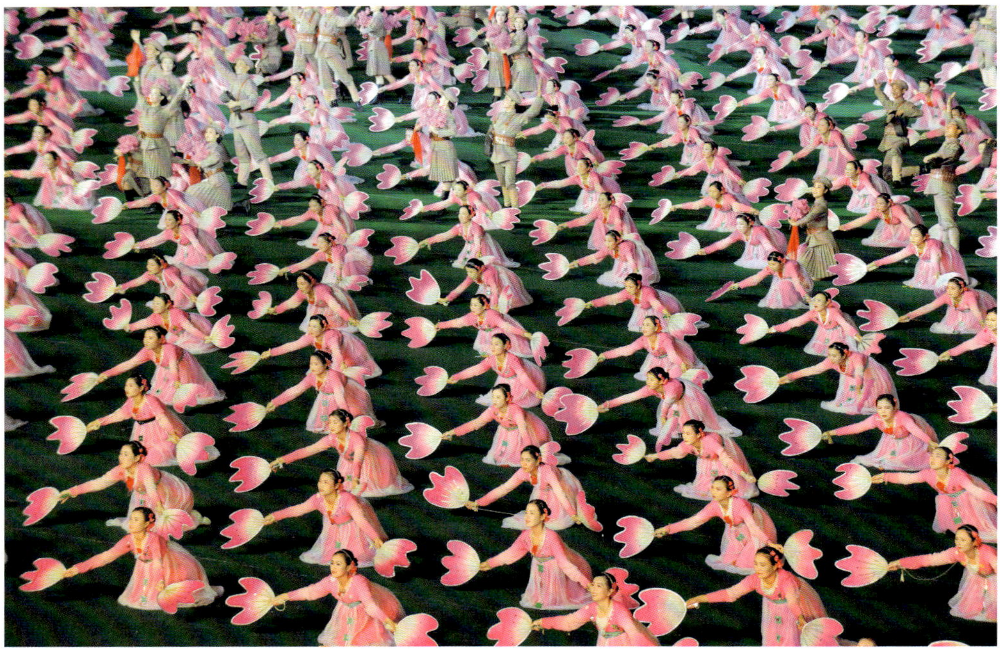

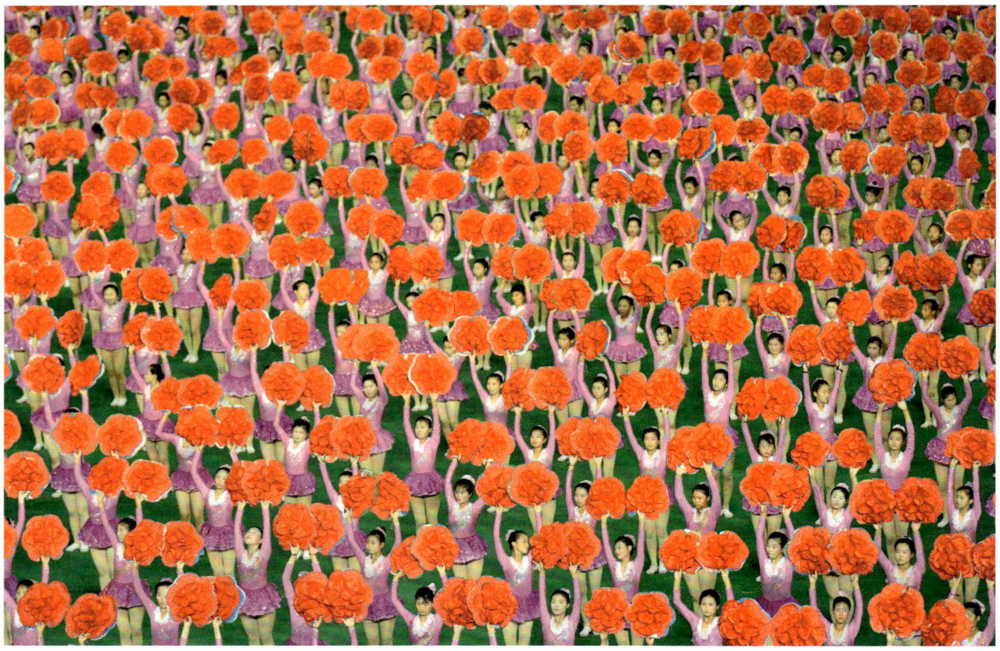

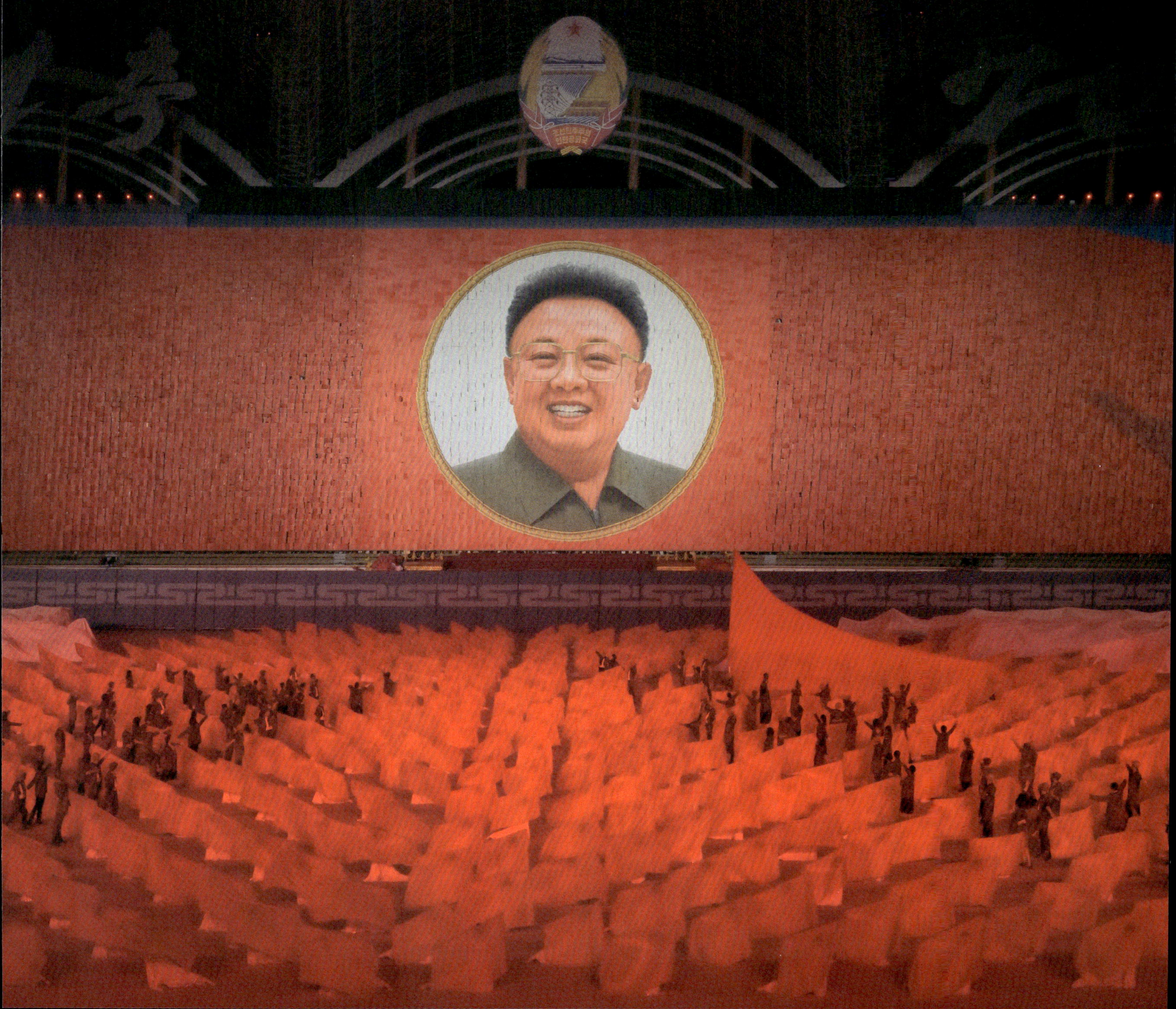

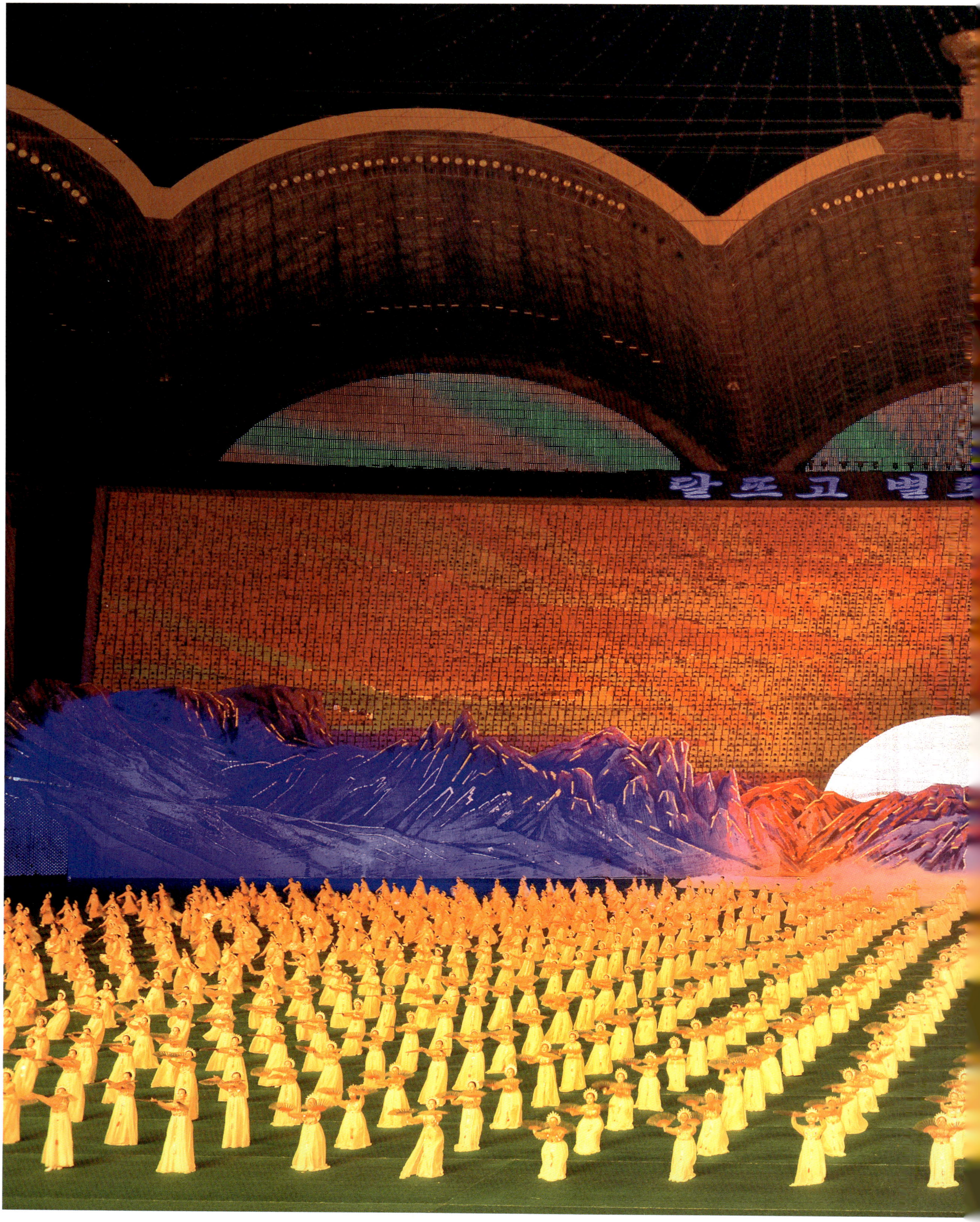
달뜨고 별

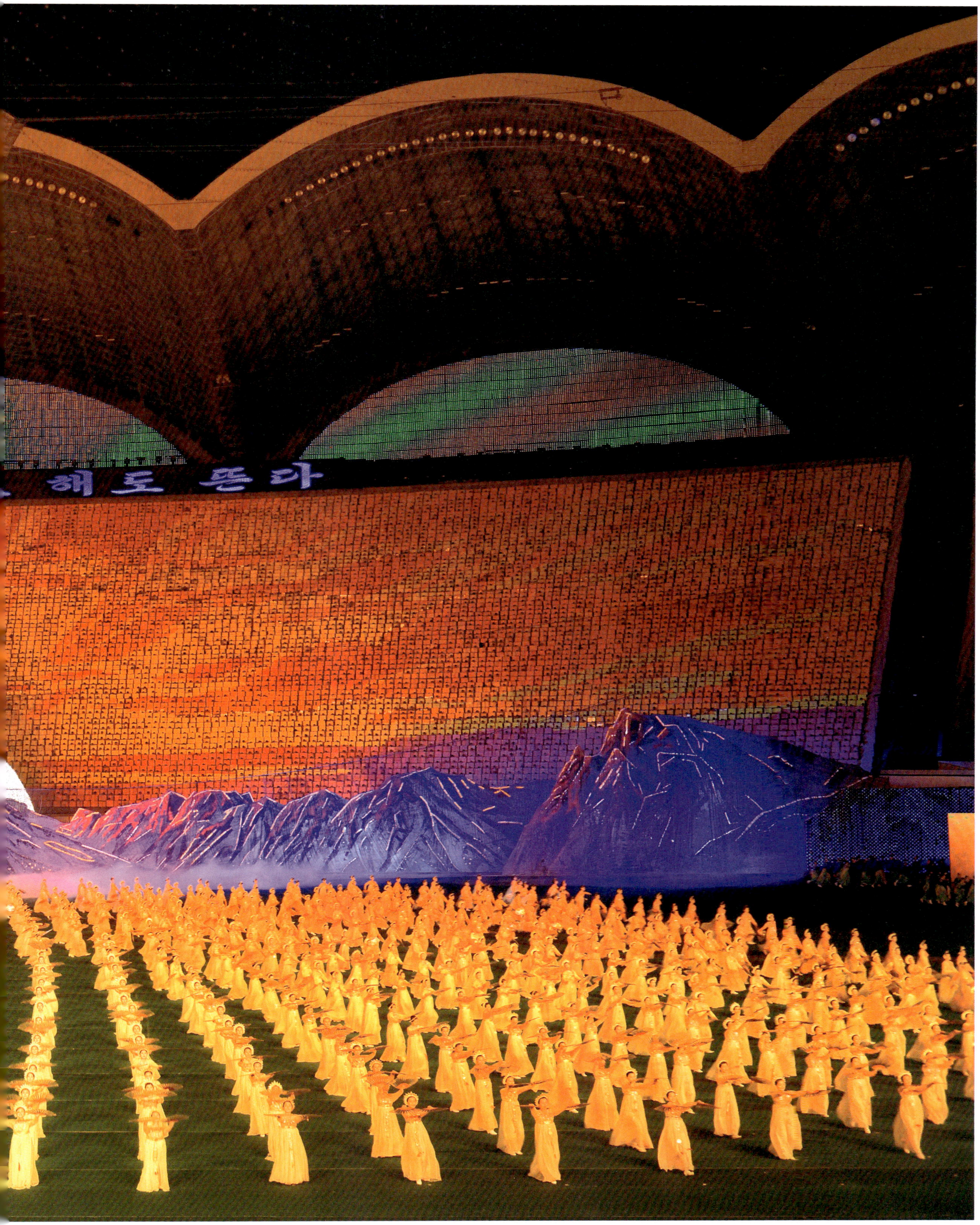
해 도 뜬 다

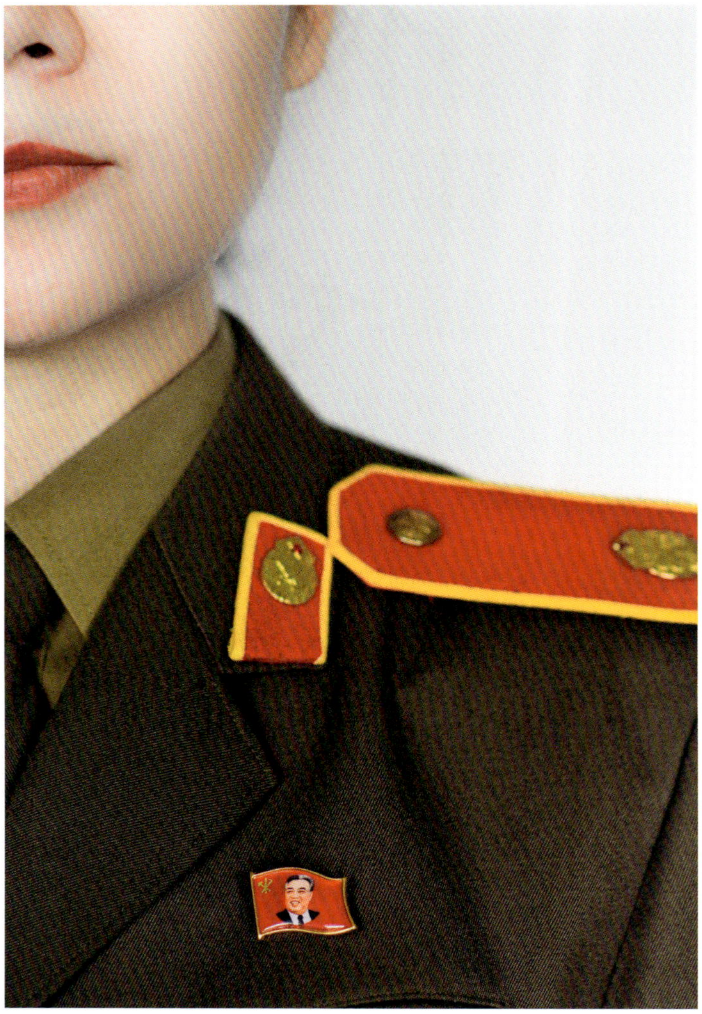

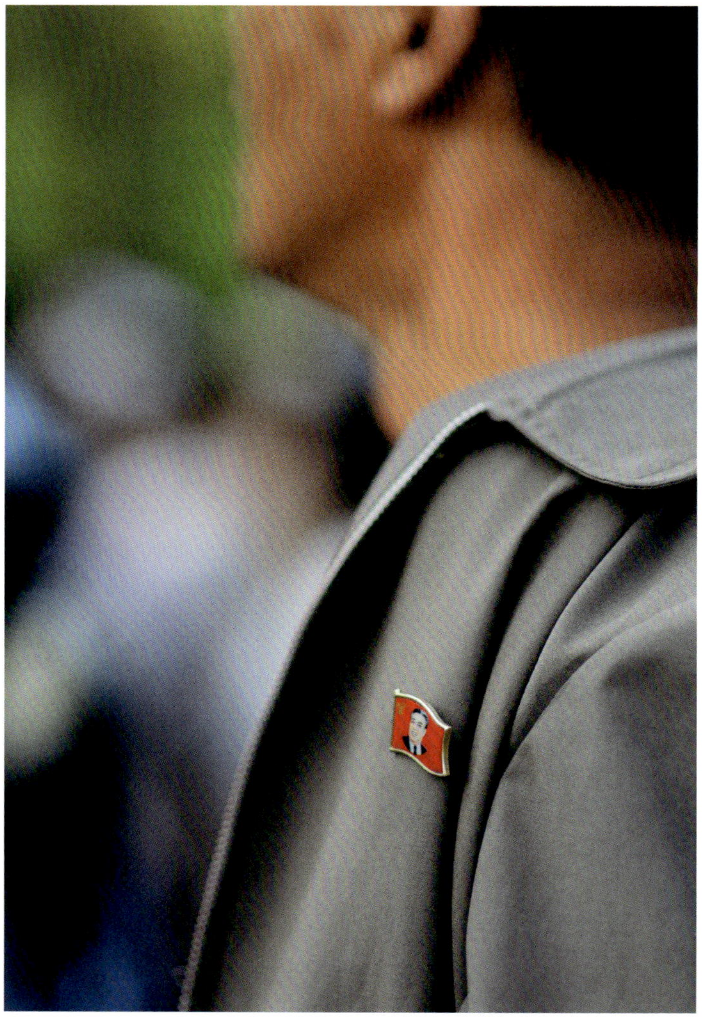

This spread Every North Korean over the age of 16 wears a enamel pin depicting the Great Leader Kim Il Sung, sometimes with the Dear Leader, his son Kim Jong Il. These lapels are strictly reserved for the people of the DPRK and cannot be obtained by foreigners. Always worn in full view, these pins are constant reminders of an extraordinary personality cult. The idea of children being 'the only kings of the country' is a core concept of North Korean propaganda. A prime example is Mangyongdae School Children's Palace, where students engage in afterschool activities like swimming and gym, but where there is also a library and a 2,000-seat theatre. The building, the largest of many buildings in North Korea known as 'palaces', is impressive inside and out, with 120 rooms and two wings extending from the central building, which are said to resemble the outstretched arms of a mother.

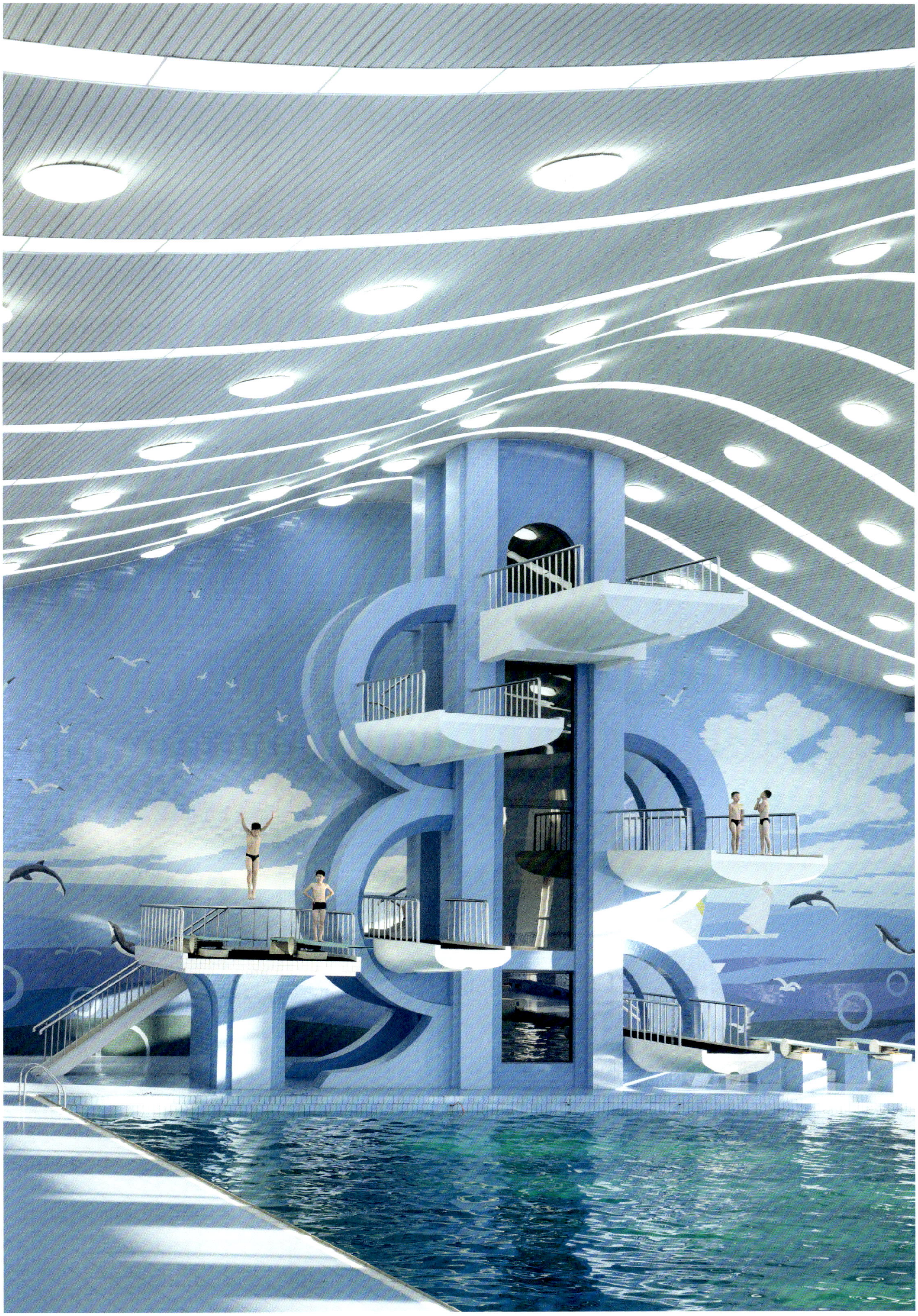

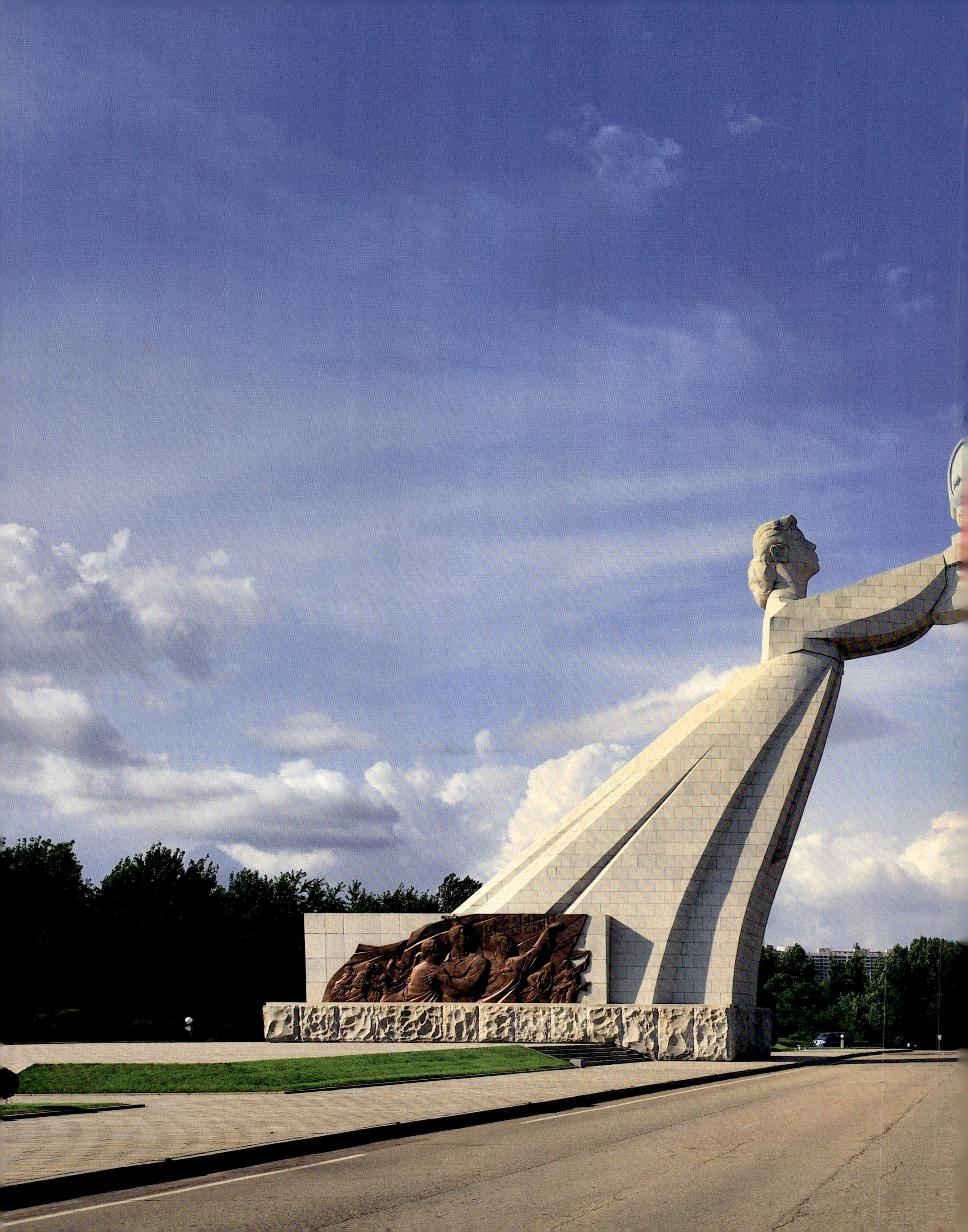

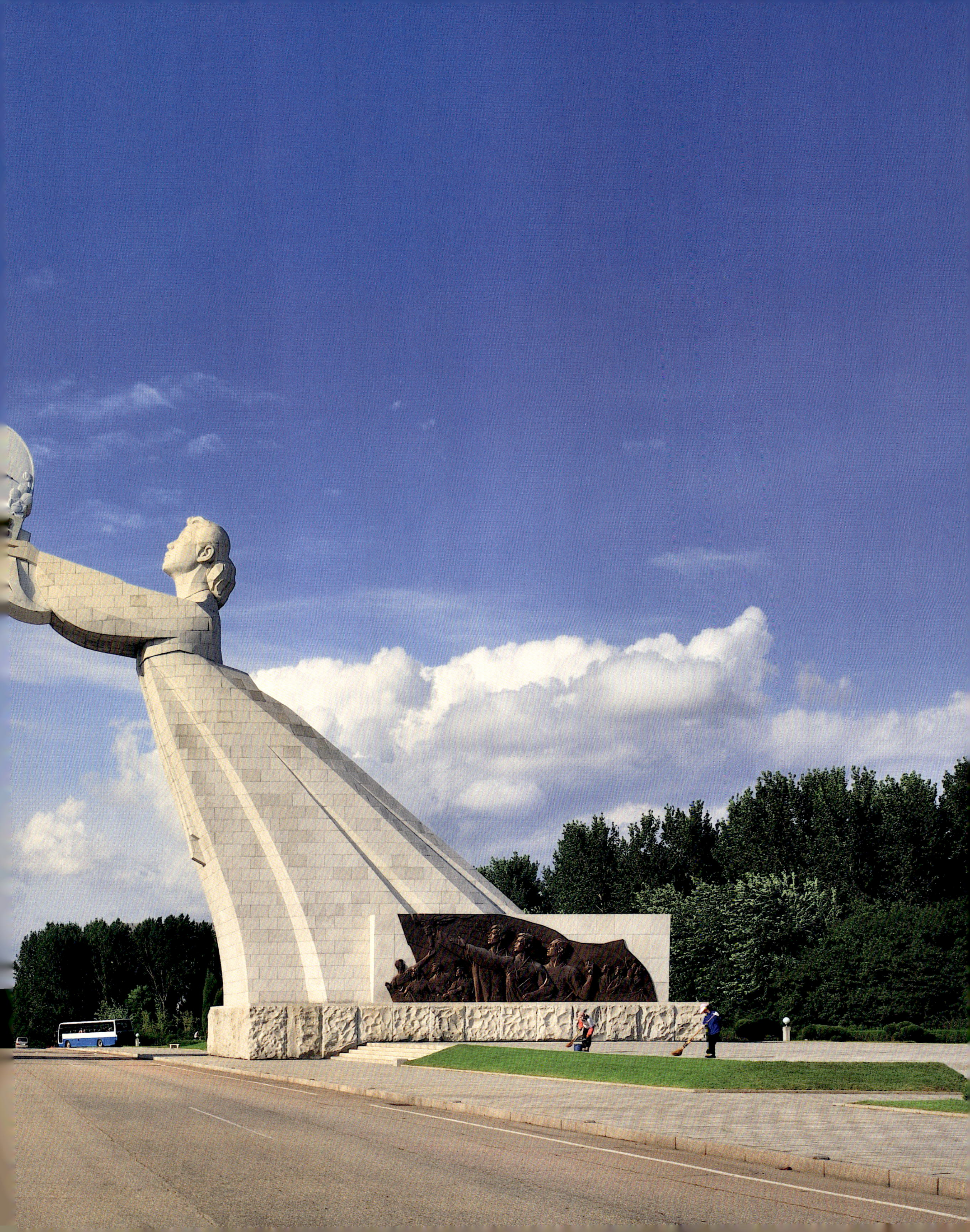

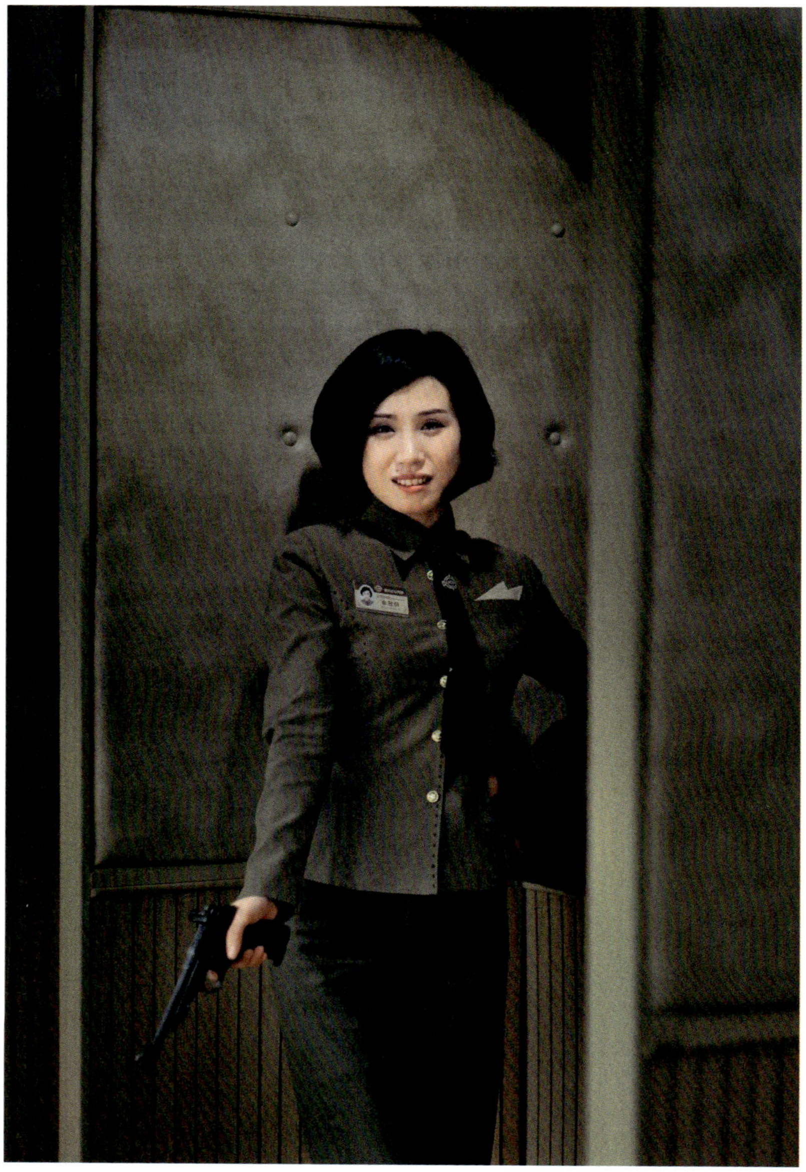

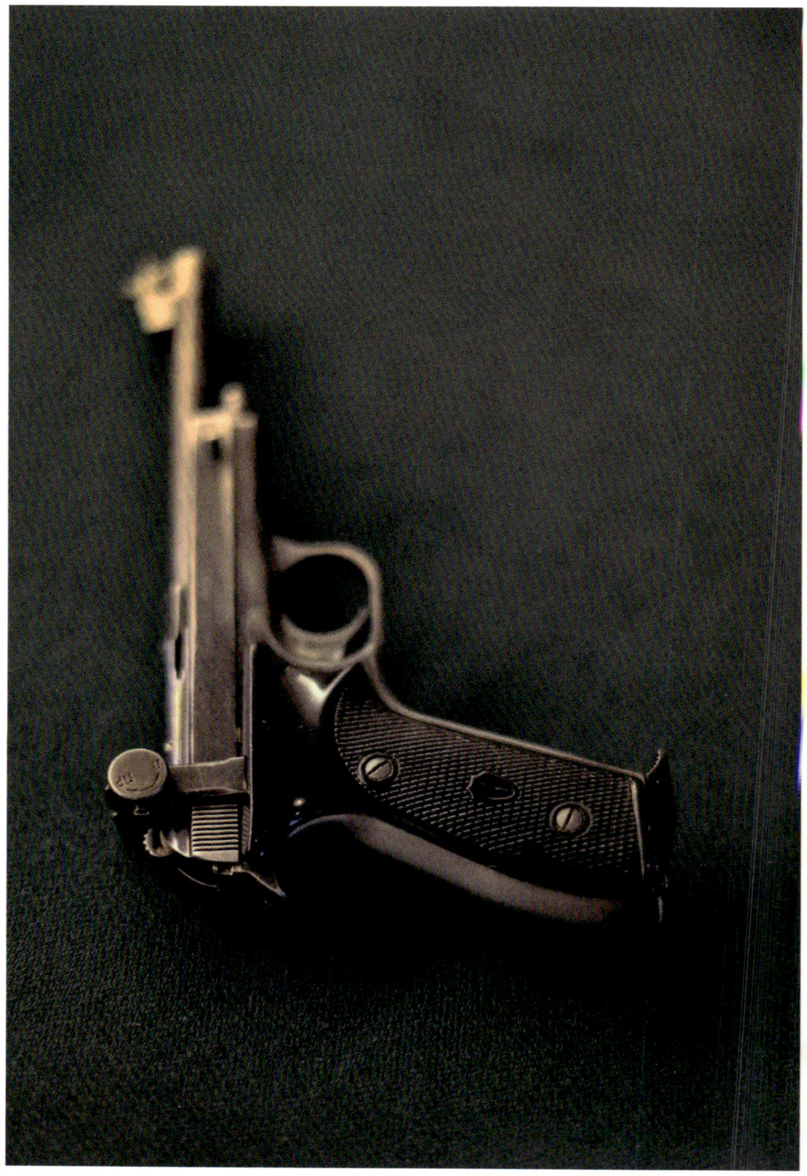

This page The Pyongyang shooting range, located in the city centre, is accessible to foreigners, who can buy bullets to shoot with 2.2 mm rifles or pistols. North Korean women, one a former Olympic sharpshooter, are among the instructors.

Previous spread South of Pyongyang, the Arch of Reunification towers 30 metres (98 ft.) above the highway towards the Demilitarised Zone on the border with South Korea. It was built to mark the visit of South Korean President Kim Dae-jung and implies that he agreed with the Three Charters of National Reunification plan that Kim Il Sung put forward.

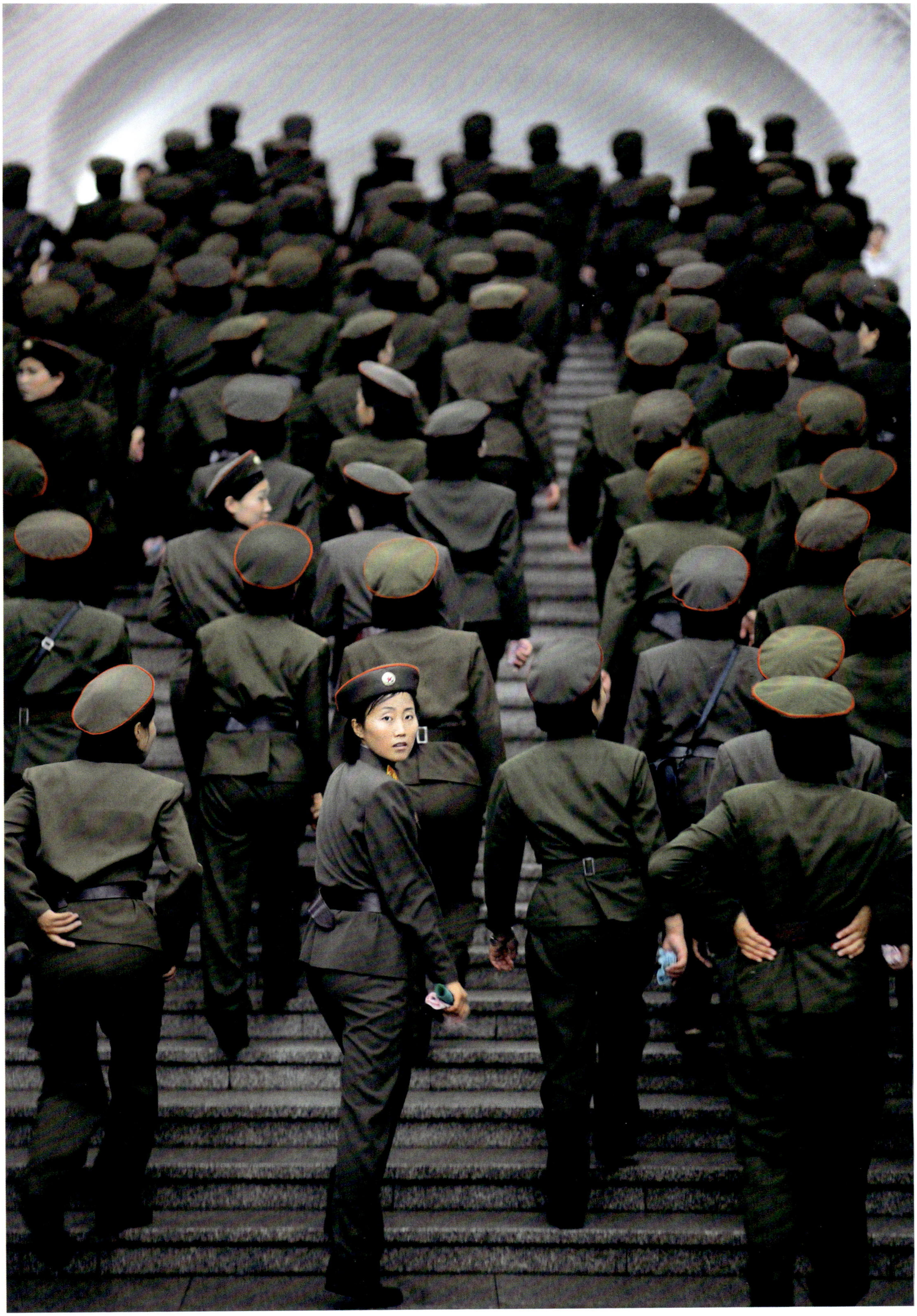

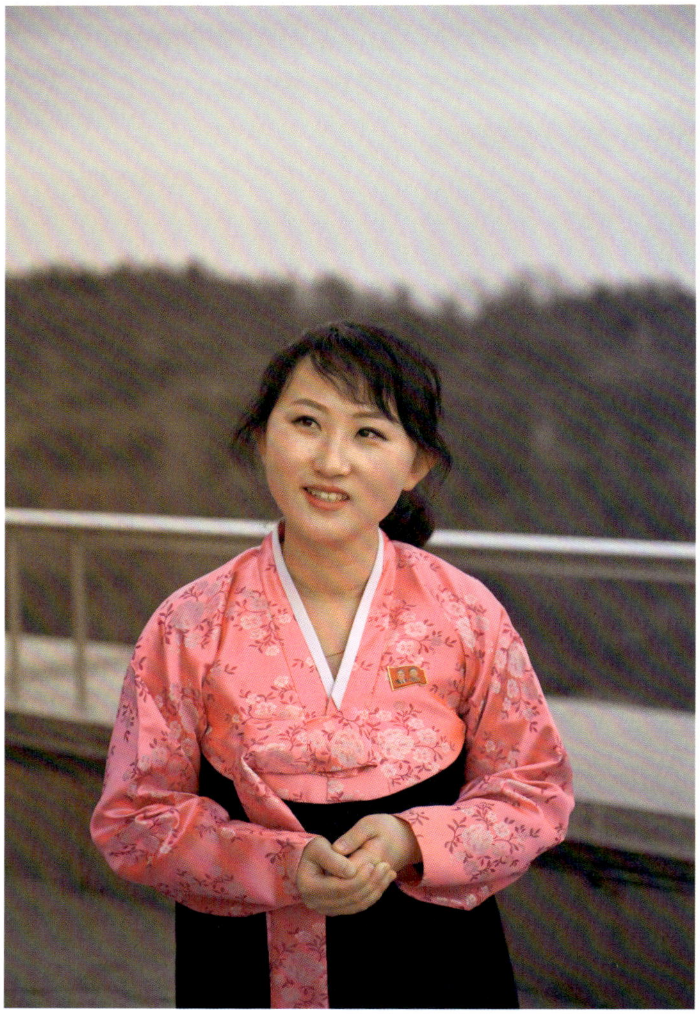

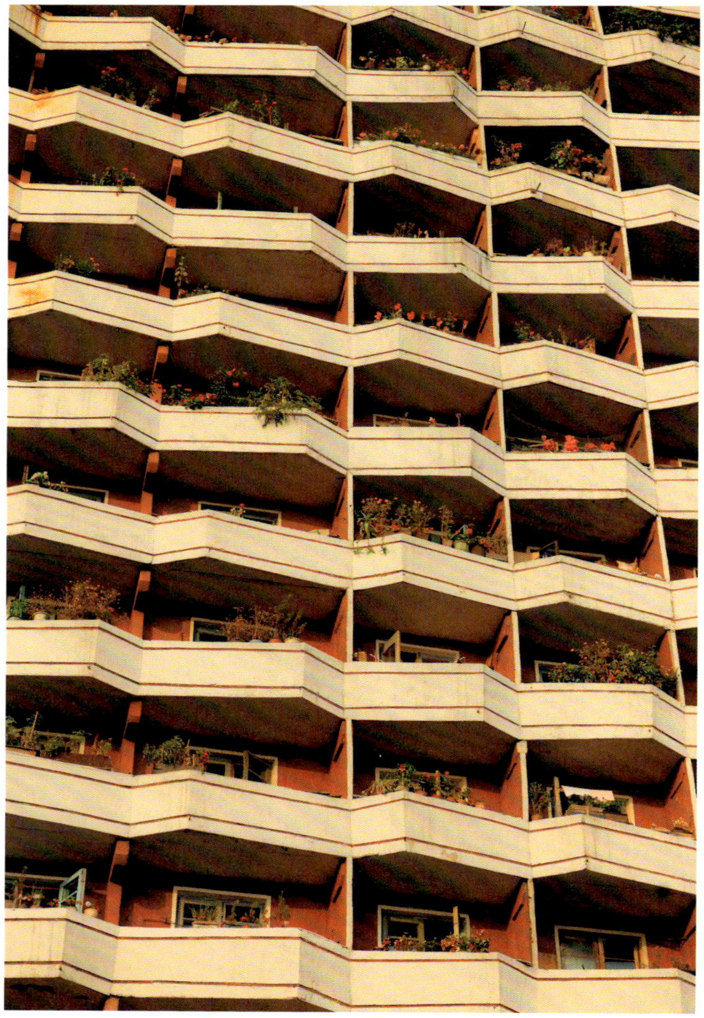

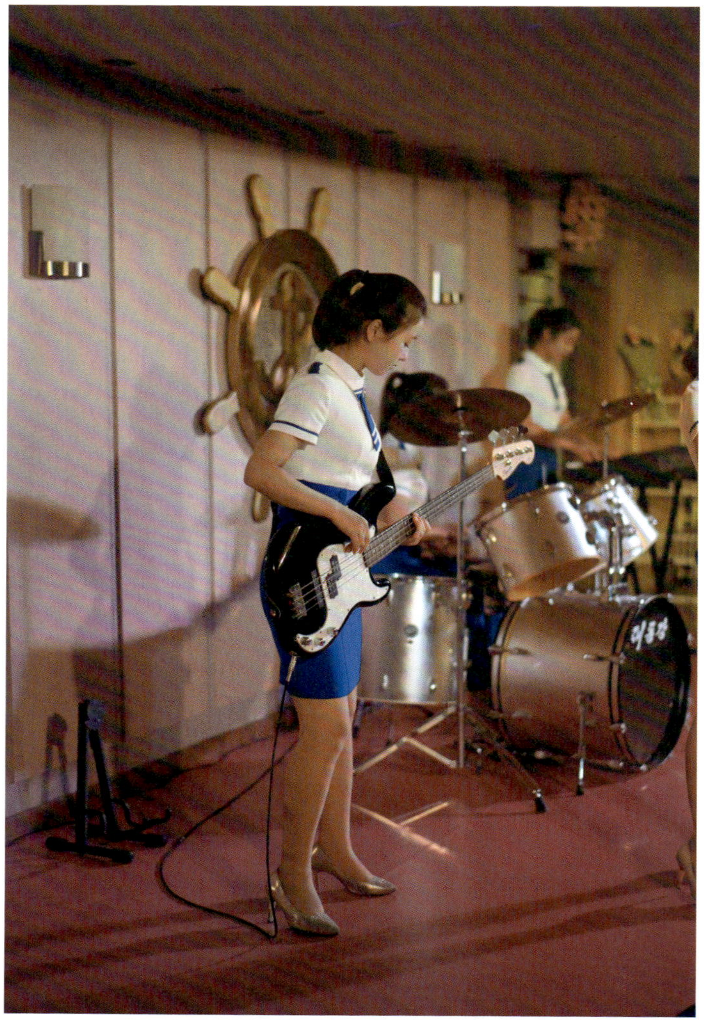

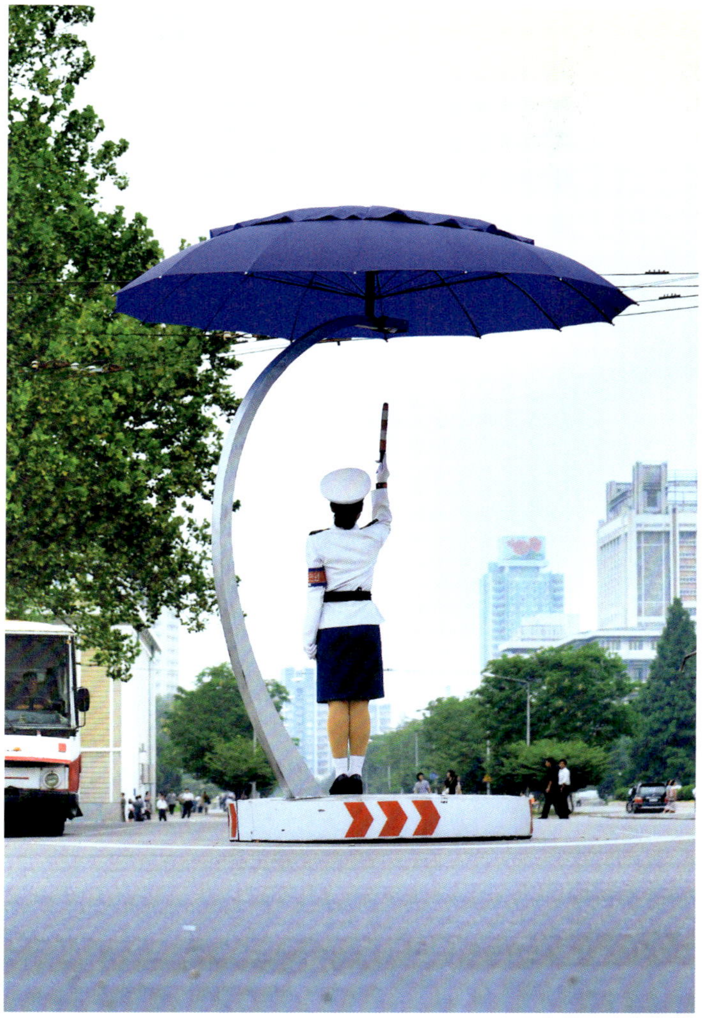
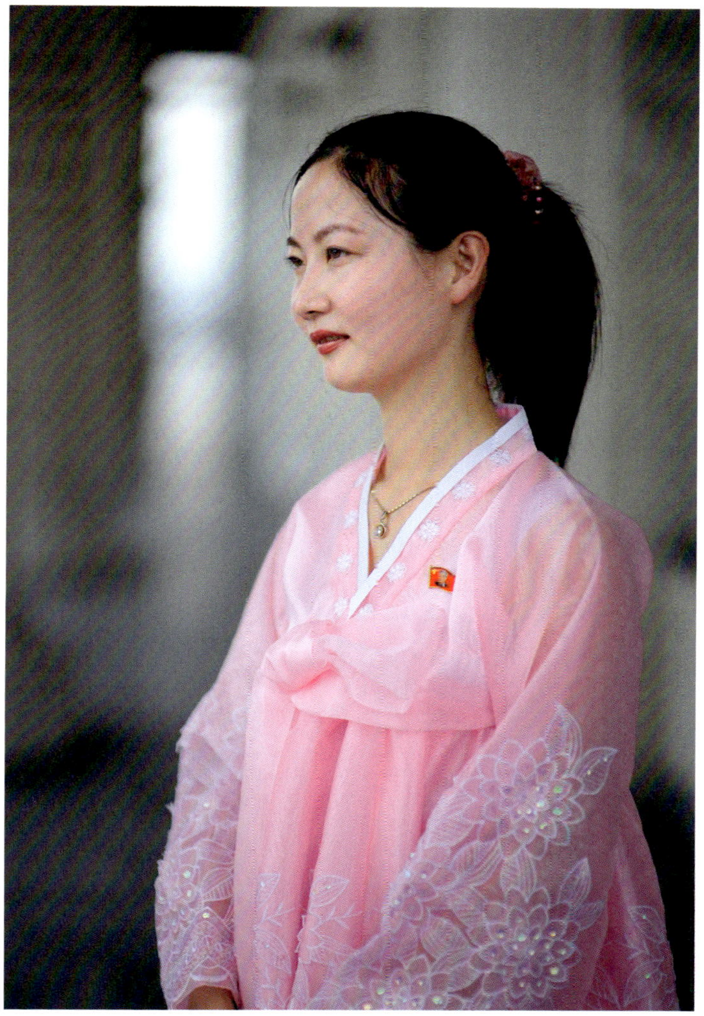
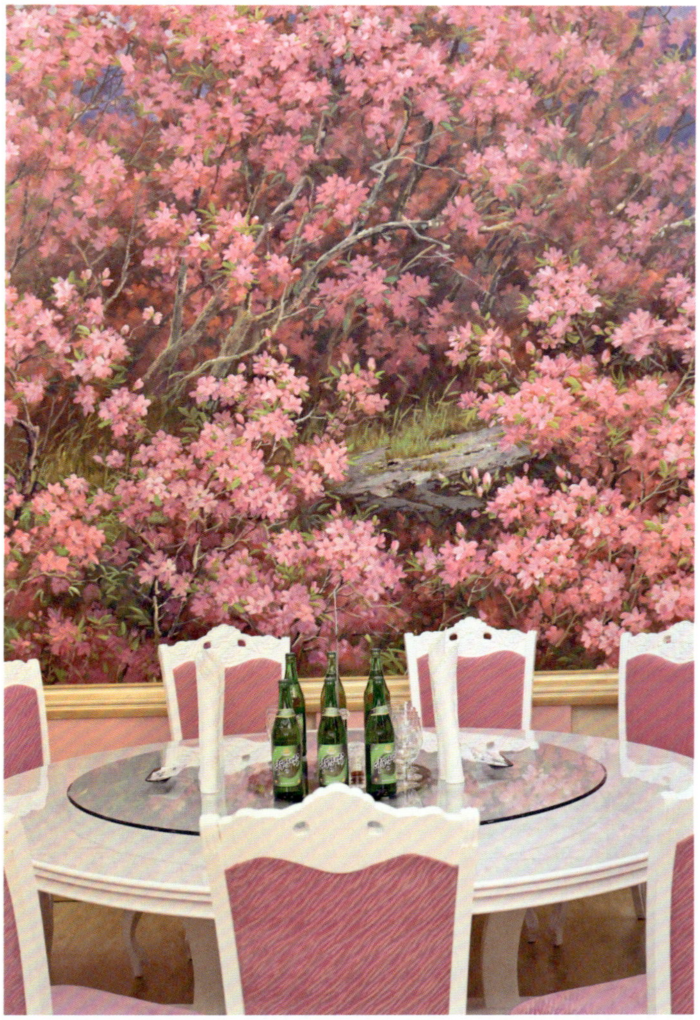
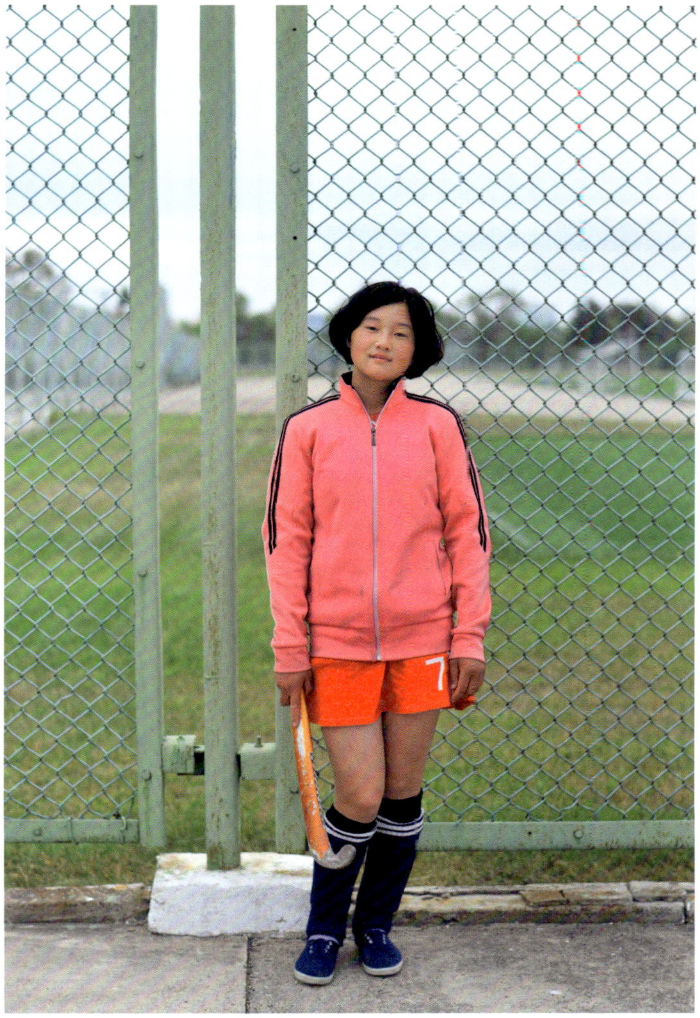

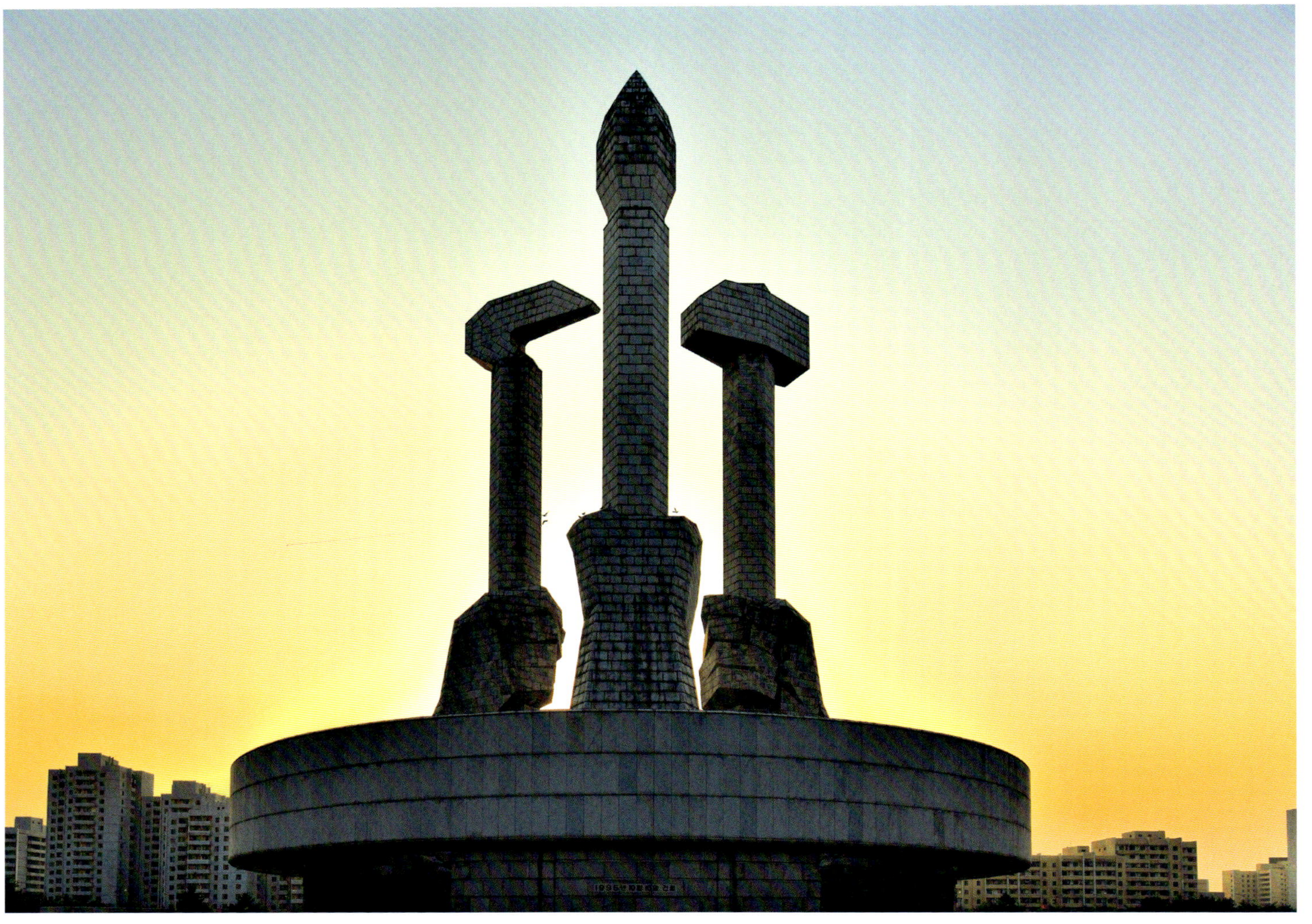

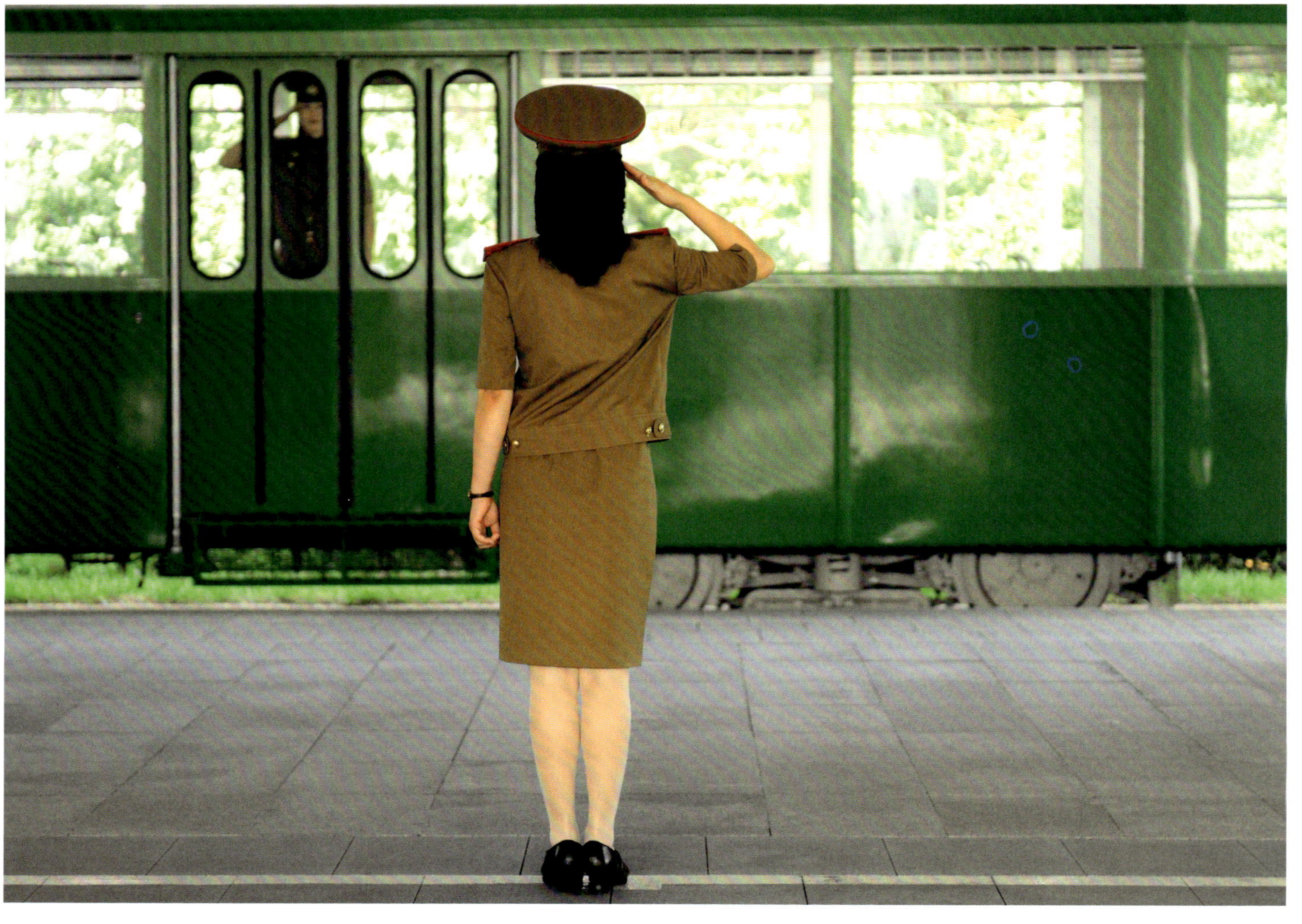

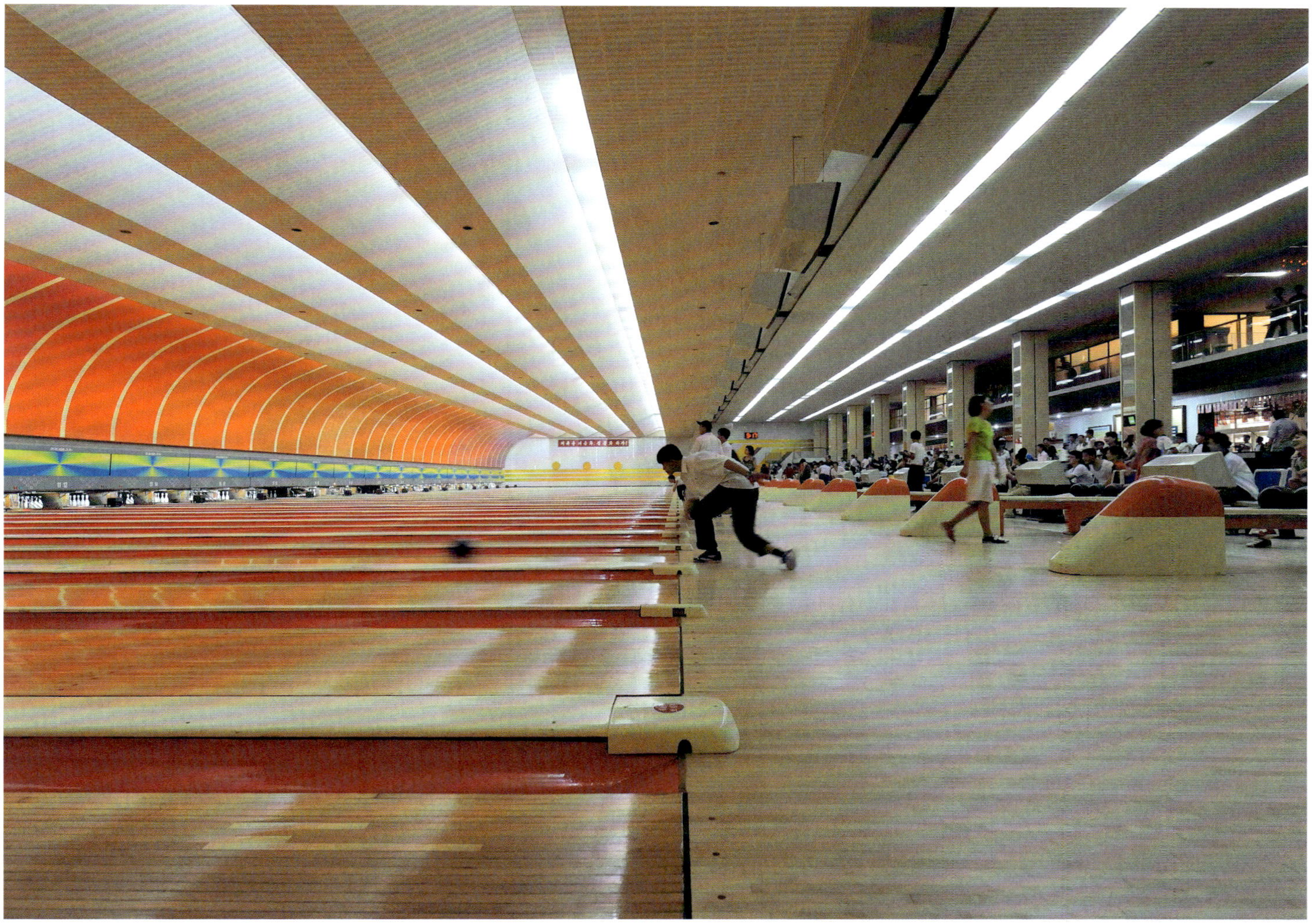

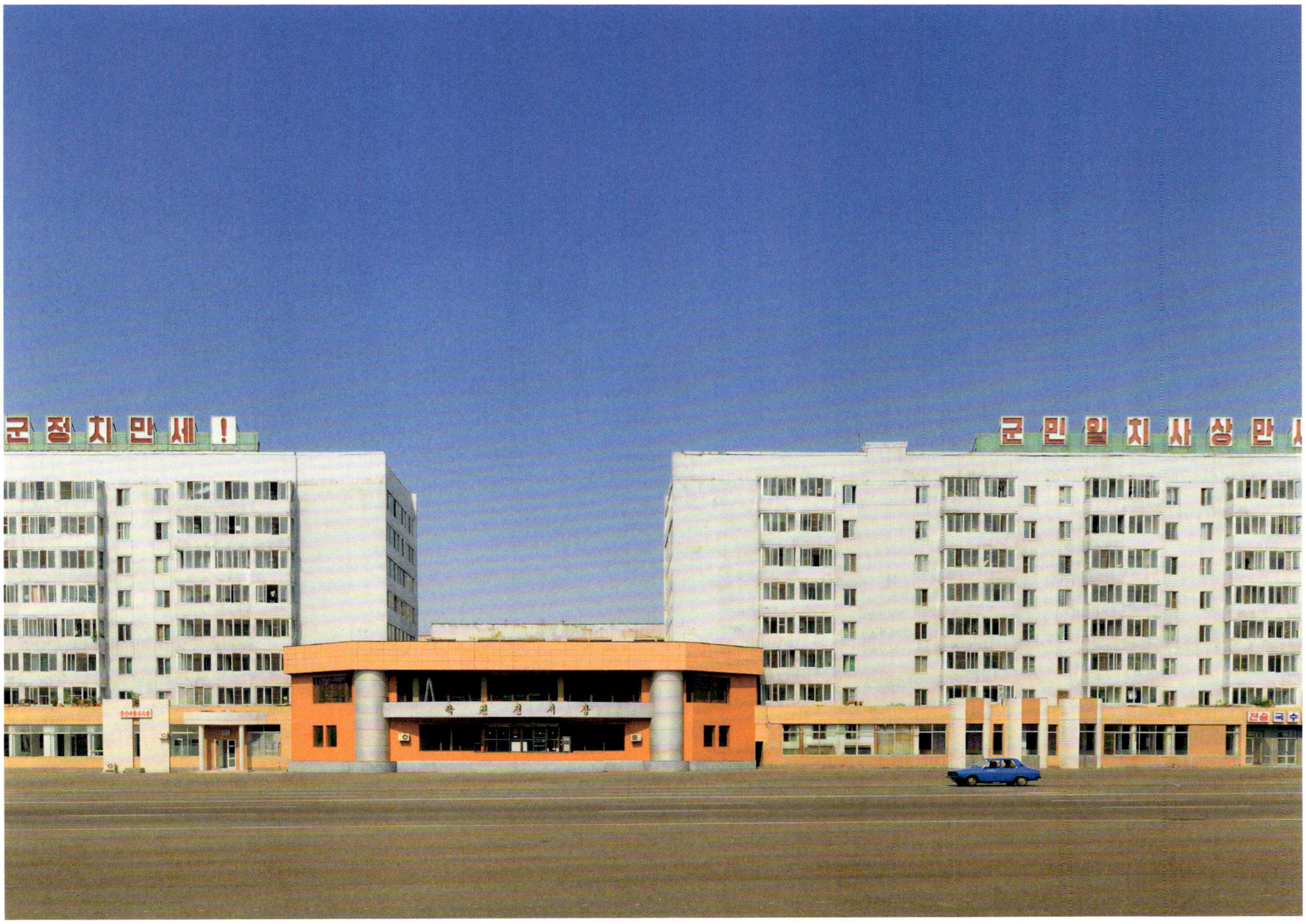

군정치만세!
군민일치사상만

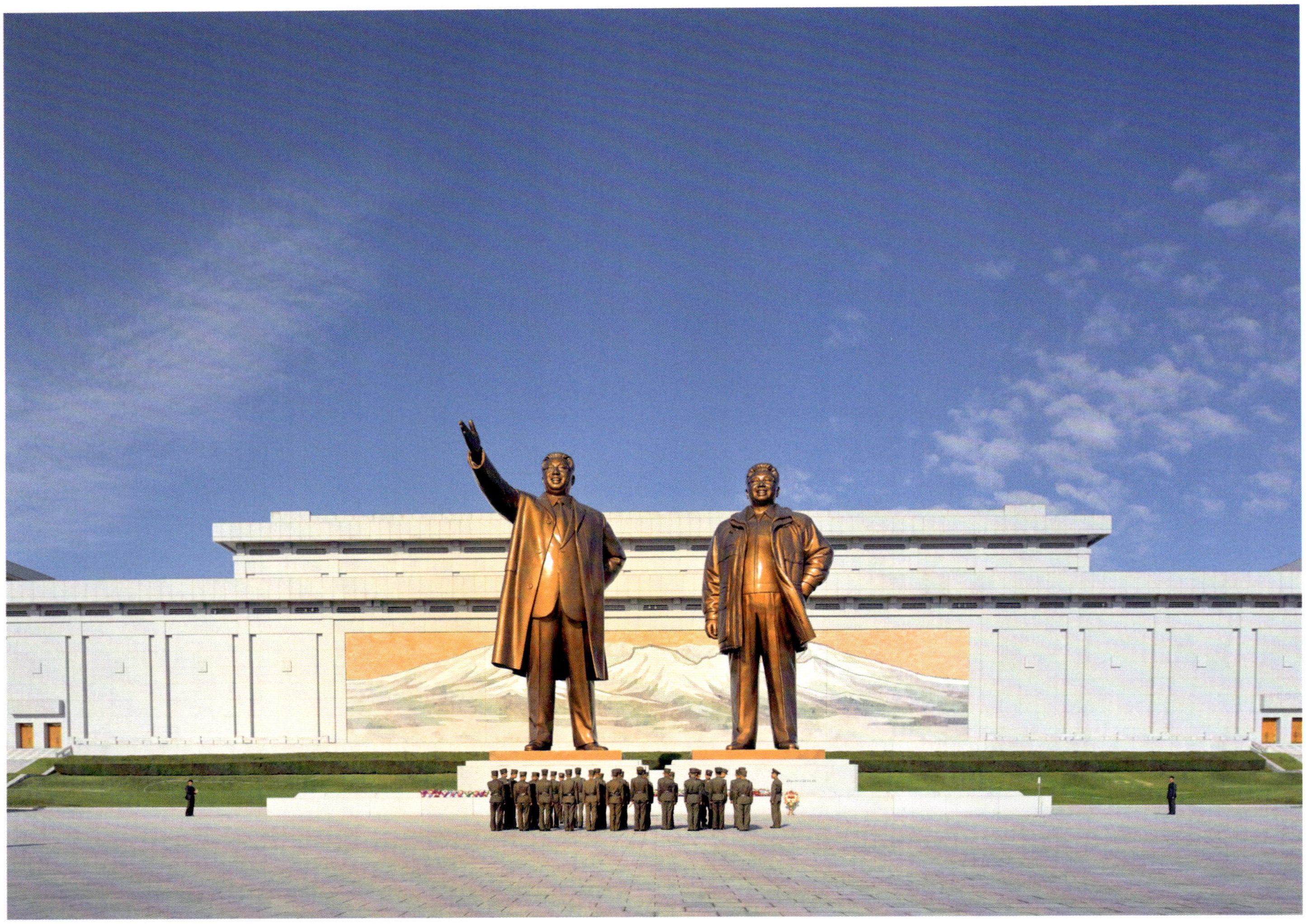

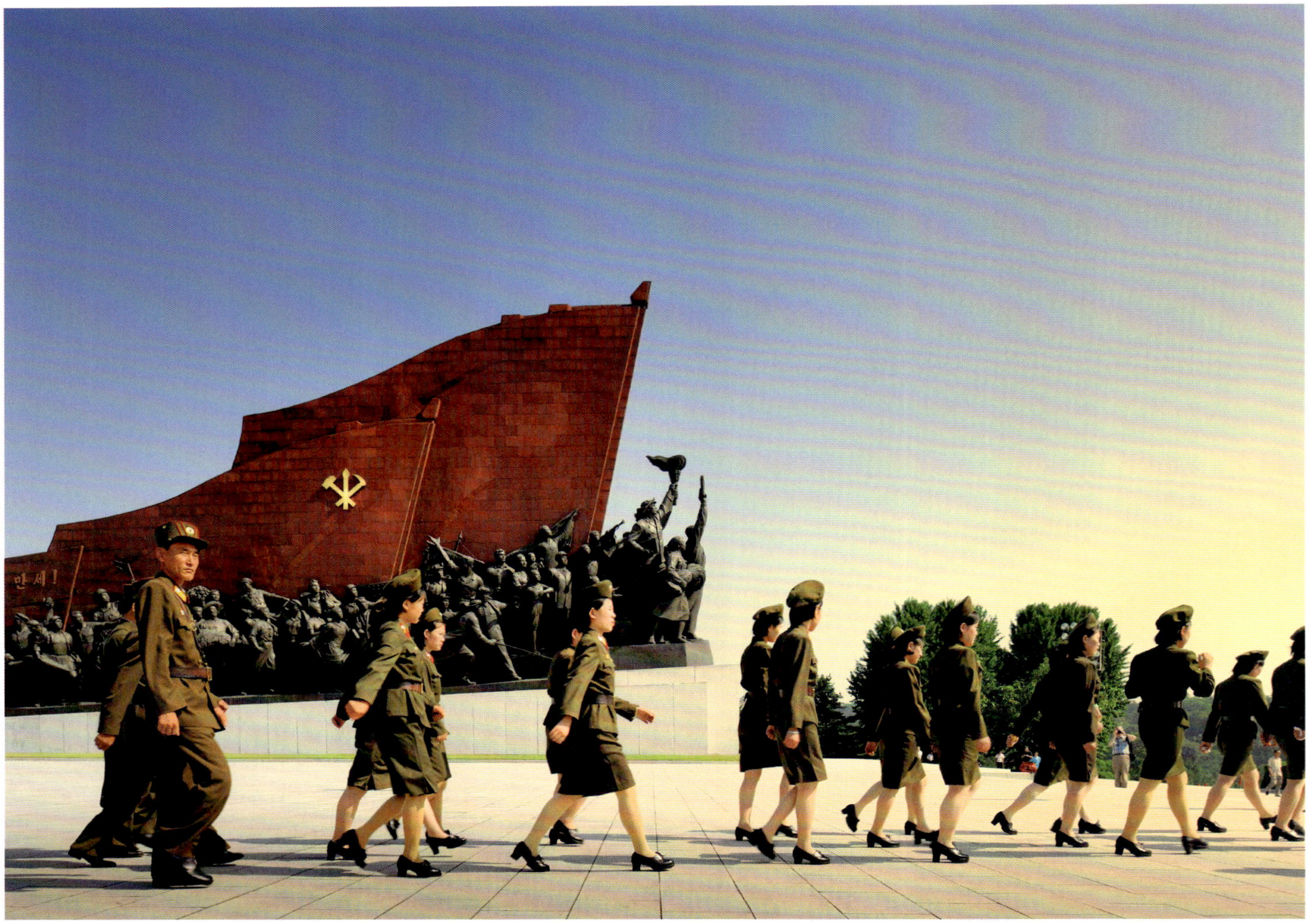
만세!

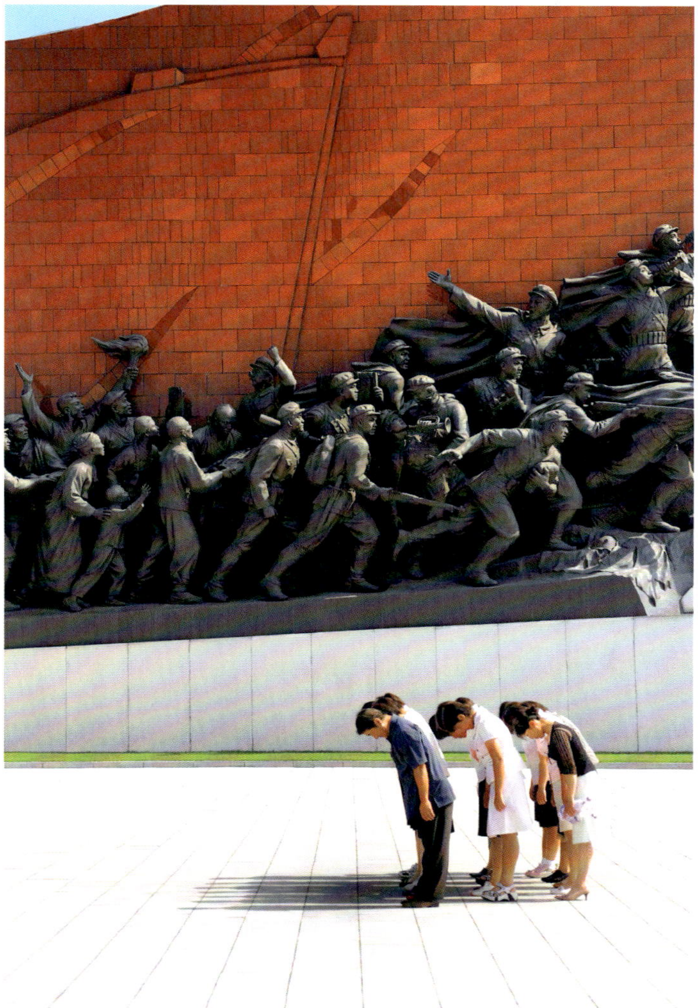

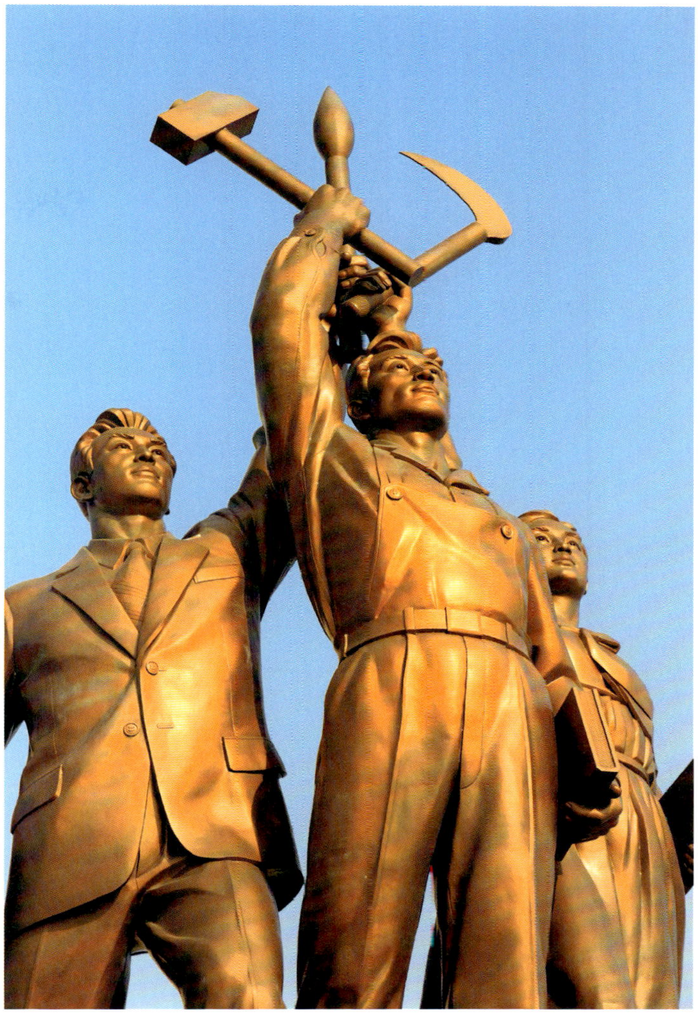

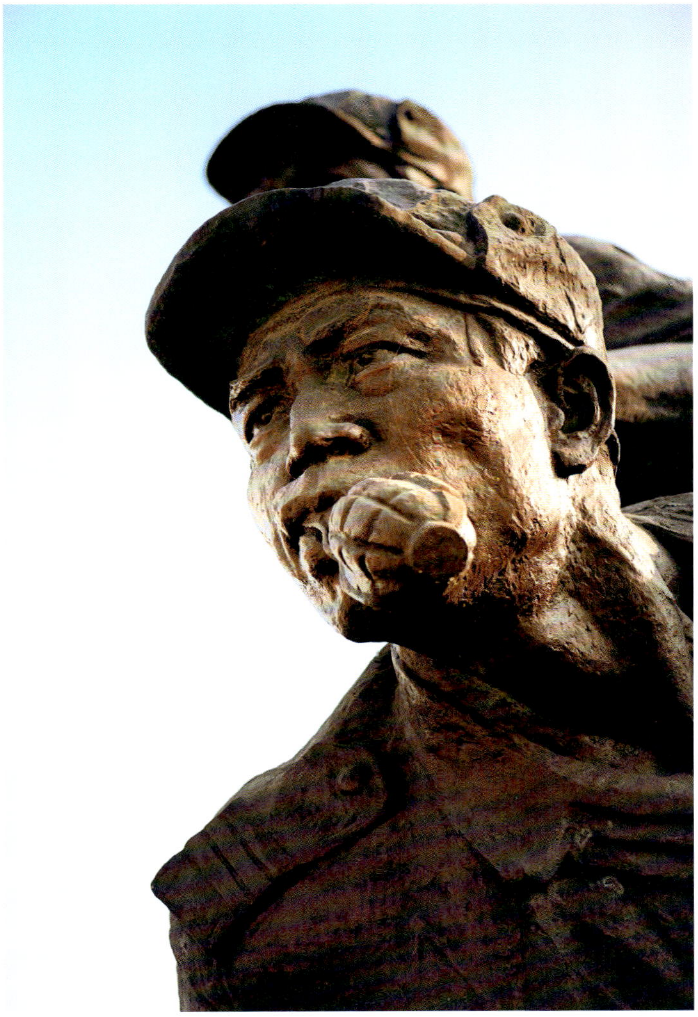

This spread Flanked by elaborate sculptures, two 22-metre (72 ft.) bronze statues of Kim Il Sung and Kim Jong Il tower over the centre of Pyongyang at Mansu Hill Grand Monument. Beyond, fabulous views stretch over Pyongyang.

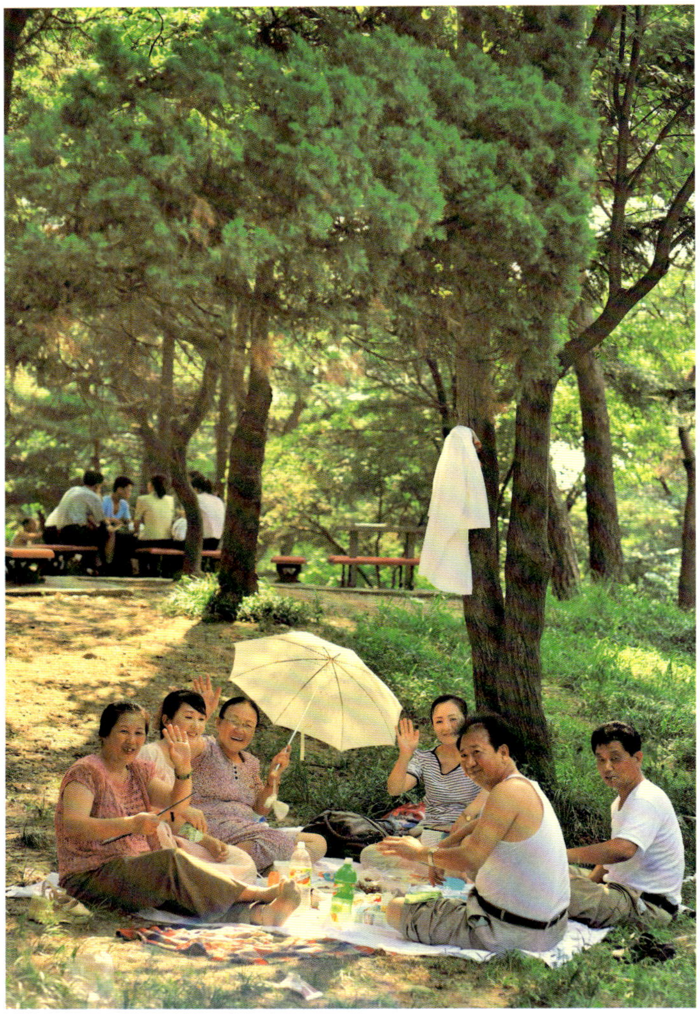
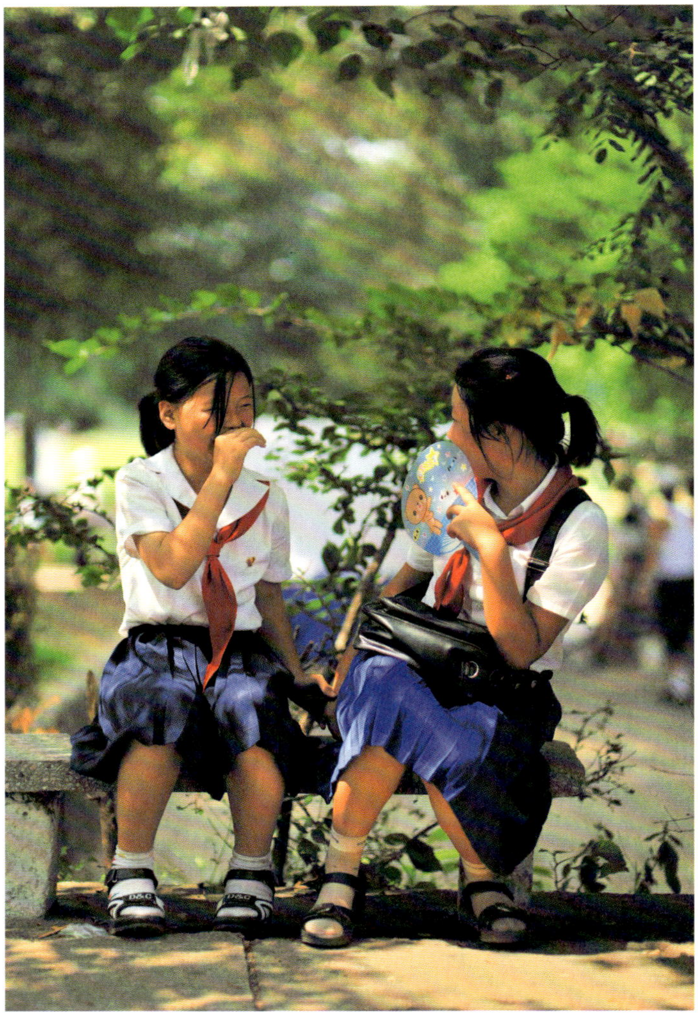
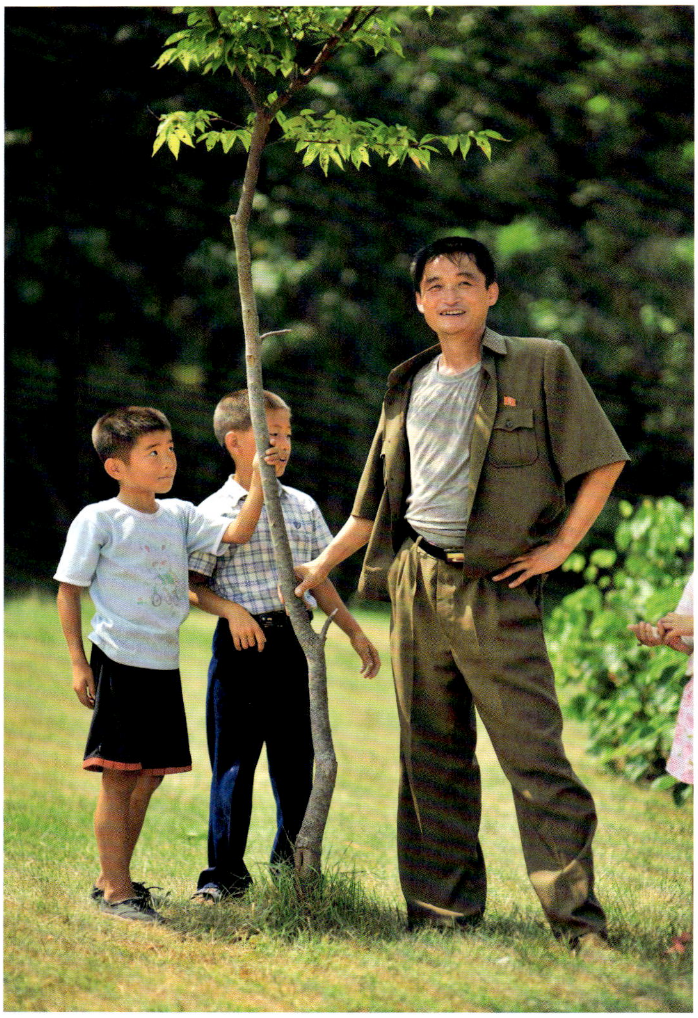
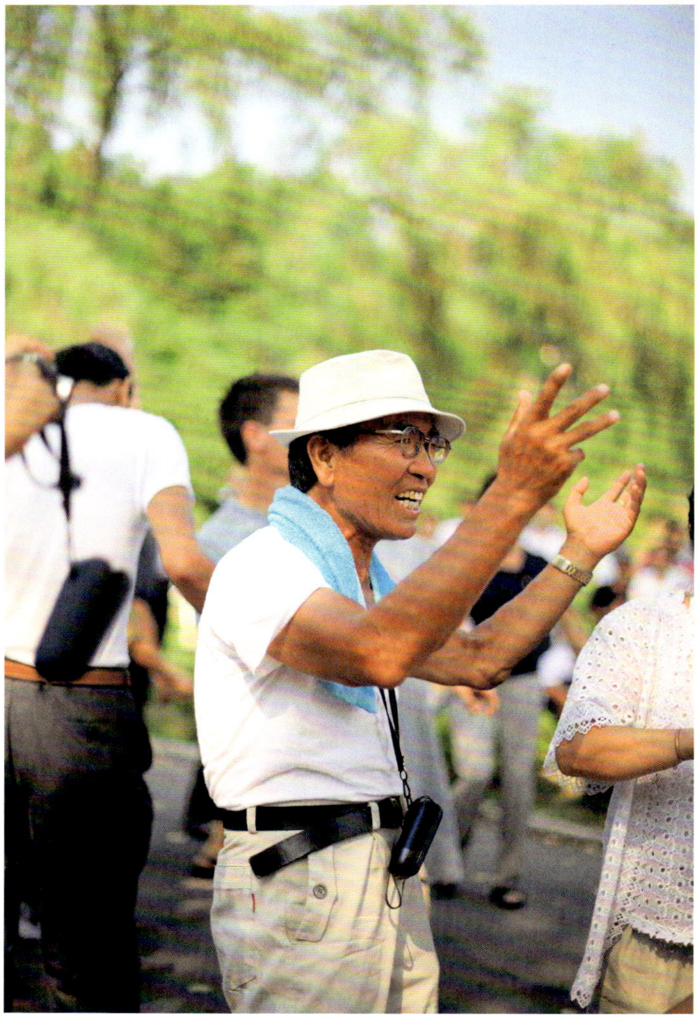

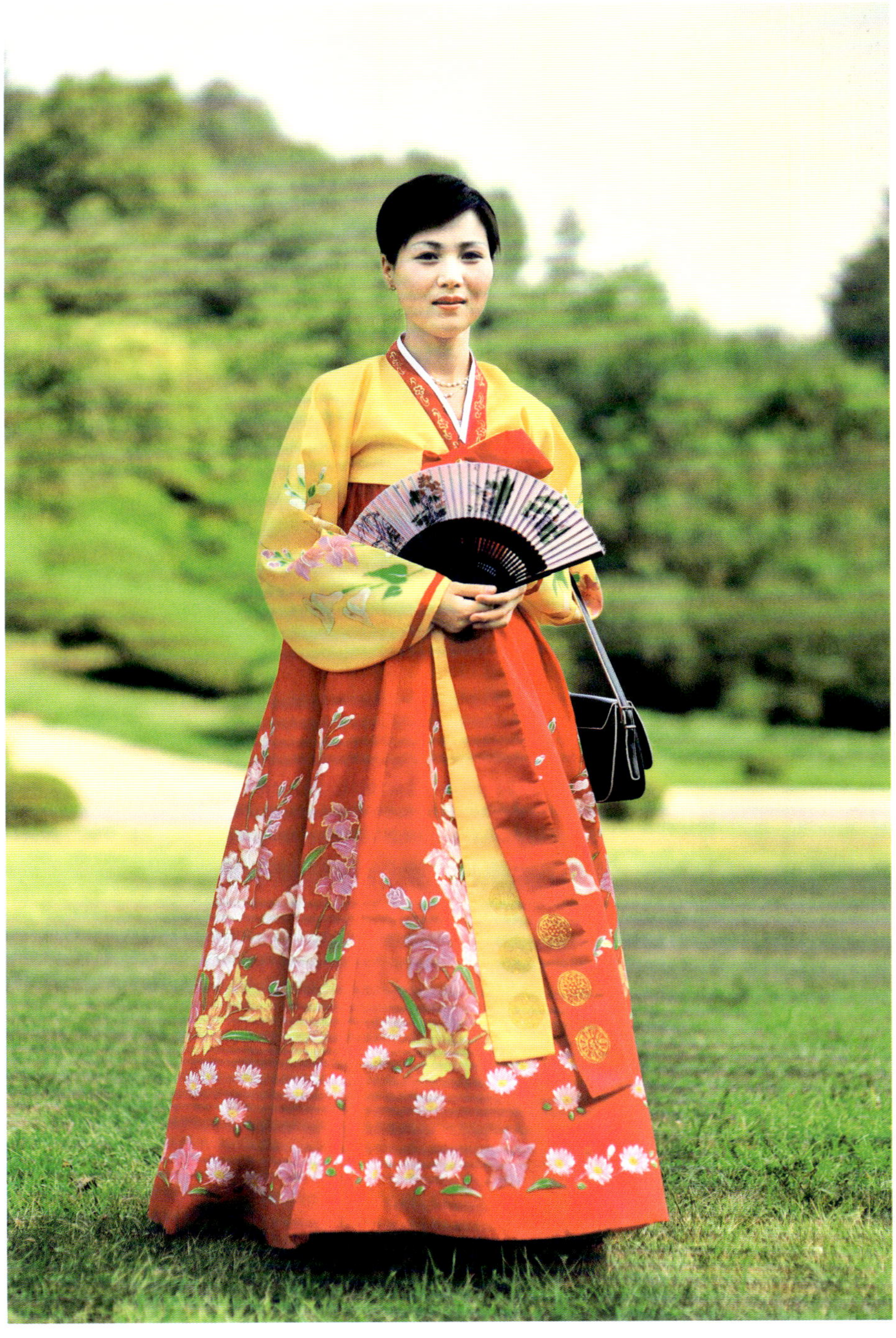

This spread Between the official visits to museums, monuments, and mausoleums, the traveller will gain a remarkable insight into the daily life of the North Korean. Every year on August 15, or Liberation Day, a lot of North Koreans flock to their parks to picnic, dance, sing, play... Families sit on the grass beneath the trees and most women wear their stunning Chosonot — the traditional dress — for the occasion.

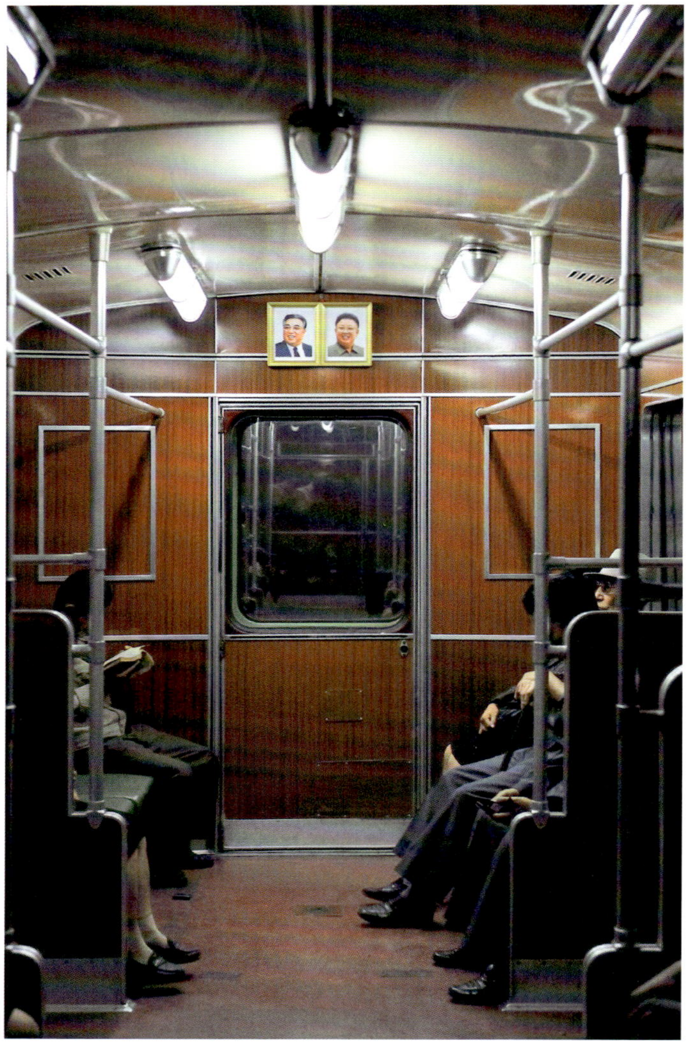

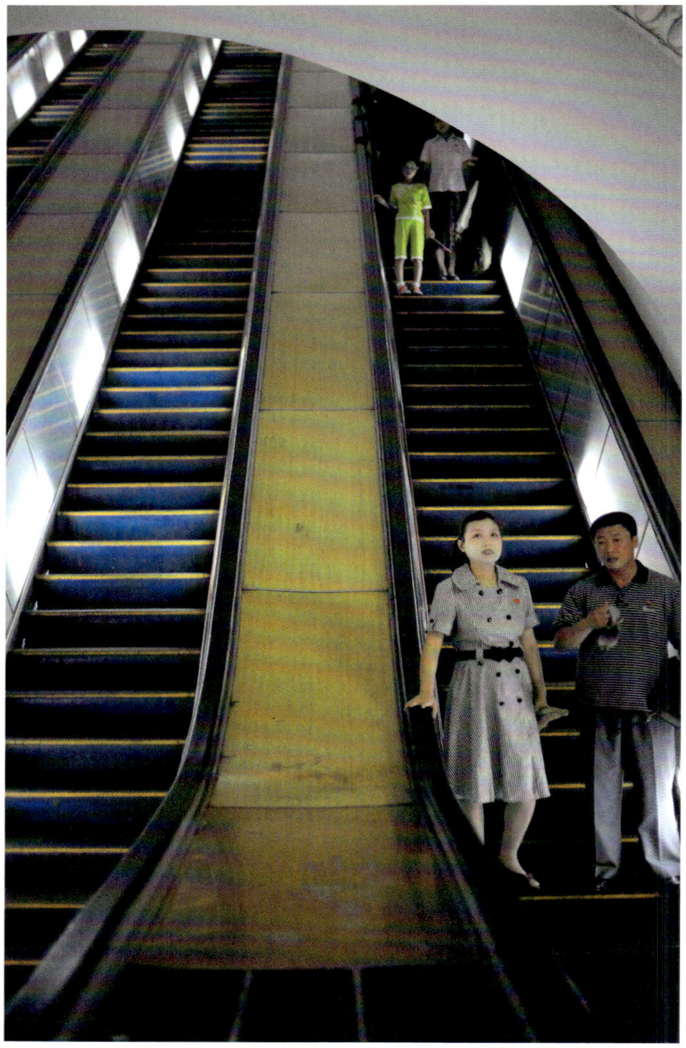

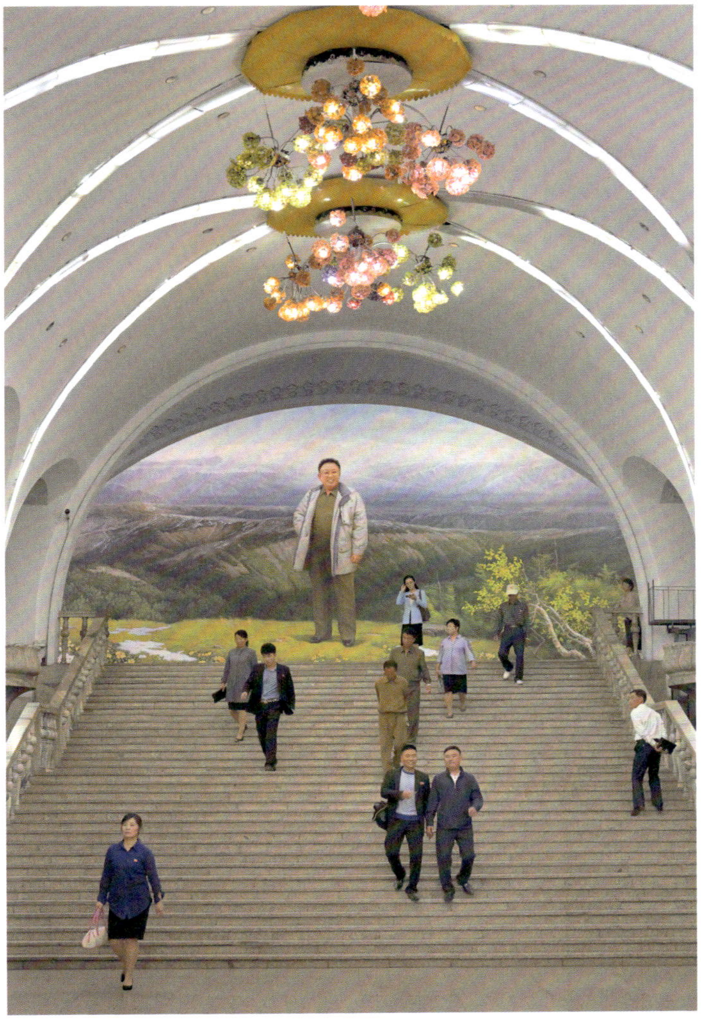

This spread Pyongyang's underground is one of the world's most striking and surreal transit systems. As if it were a museum, it includes stunning mosaics and enormous murals, as well as sculptural reliefs portraying North Korea's revolutionary history. Sixteen stations, each decorated along a different nationalistic and grandiose theme, are named to express ideology rather than location, such as Konsol (construction), Hwanggumbol (golden fields), Tongil (reunification), and Sungri (victory). Modelled after Moscow's, the system is said to be the deepest in the world; from street level it takes four minutes to descend 110 metres (360 ft.) to the stations. With heavy metal doors at the entrances, the stations could also be used as bomb shelters. This is especially evident at Yonggwang and Kaeson stations. Foreign visitors may ride the underground alongside commuters on their way to work or school, experiencing old German D-type carriages that have been discontinued elsewhere in the world. Although typically only some stations are open to foreigners, certain tour companies have obtained permission to explore them all.

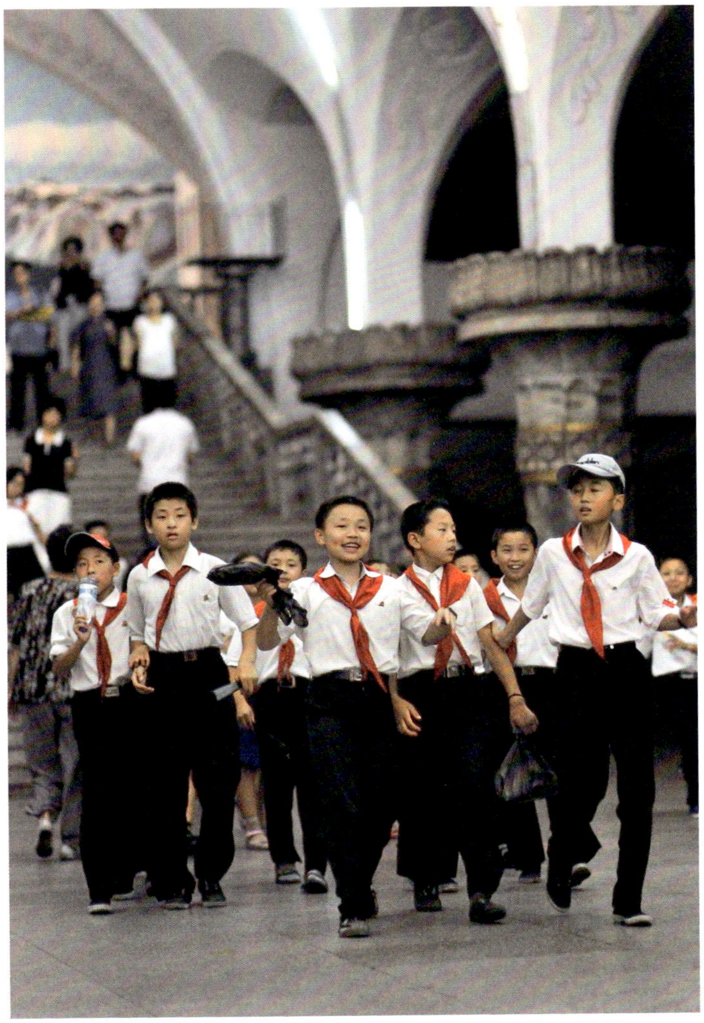

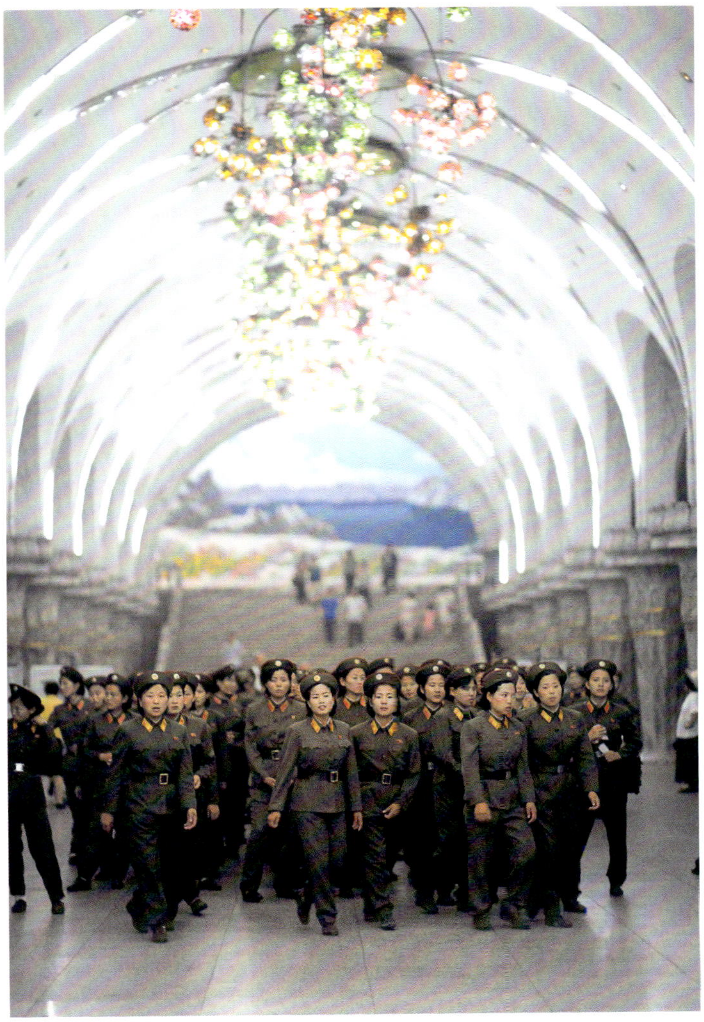

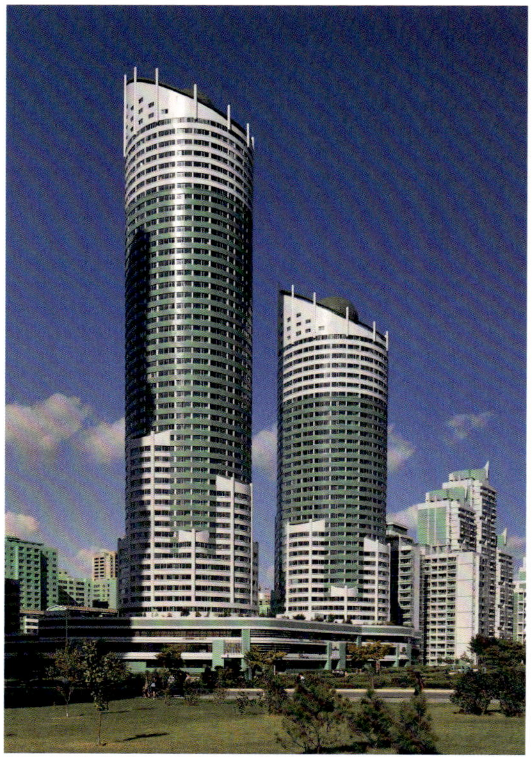
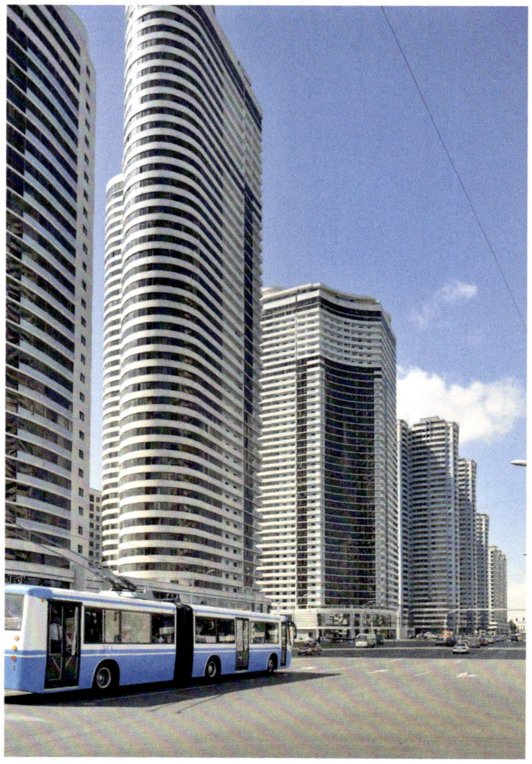
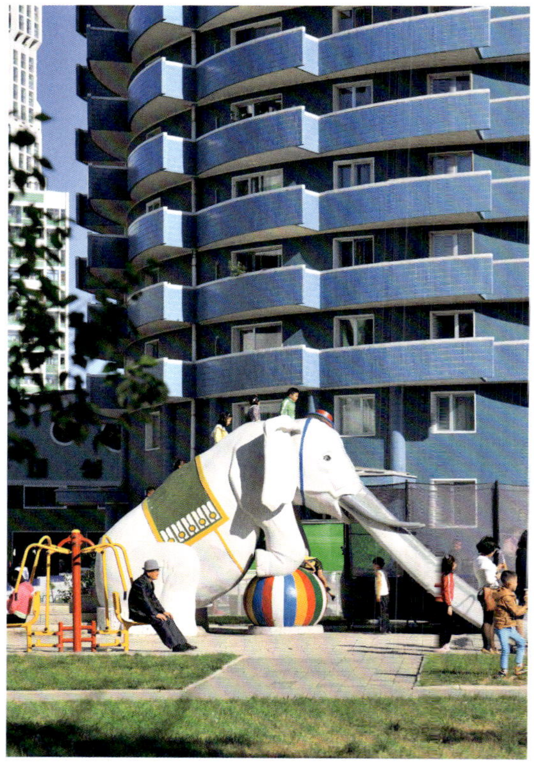
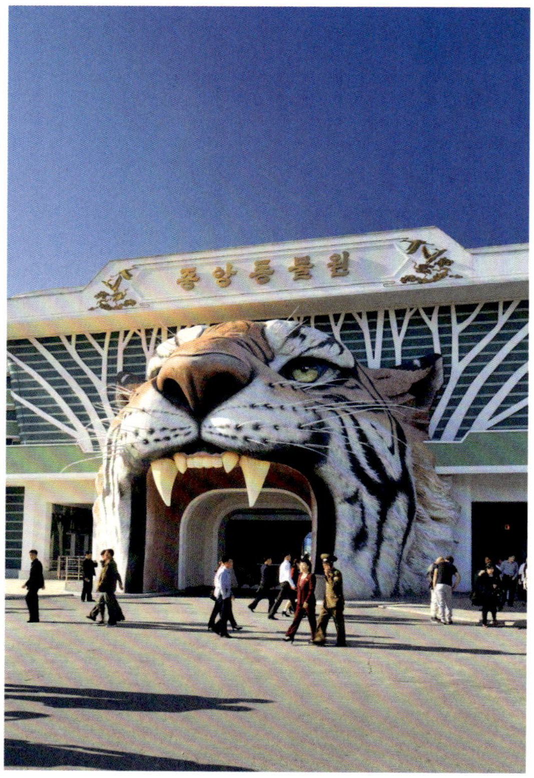

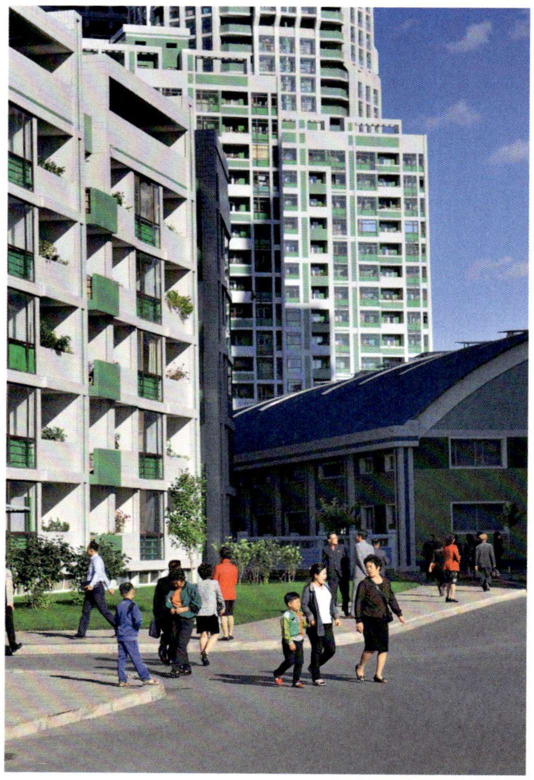
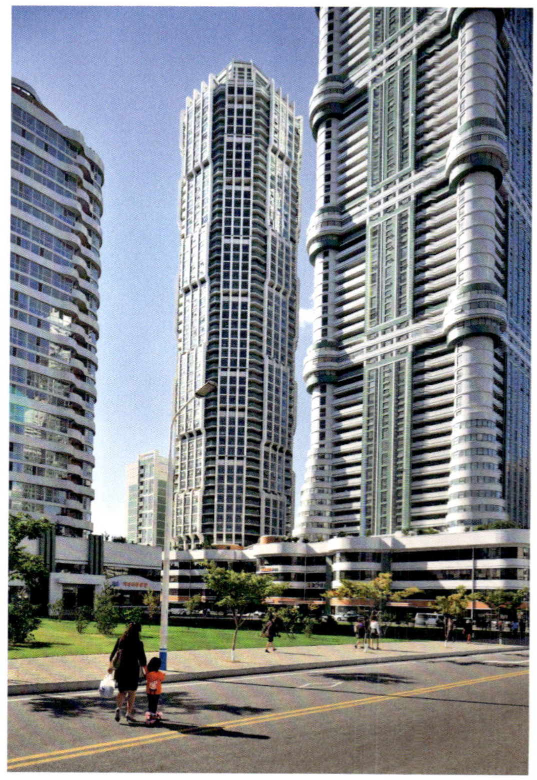

This spread The newest part of the city, nicknamed Pyonghattan by some, is a brand-new collection of high-rises and shiny apartment blocks along Mirae Science Street. The happy few who can afford to live here have made their fortunes in commerce. There are coffee bars, fancy restaurants, gyms, and supermarkets stocked with foreign food.

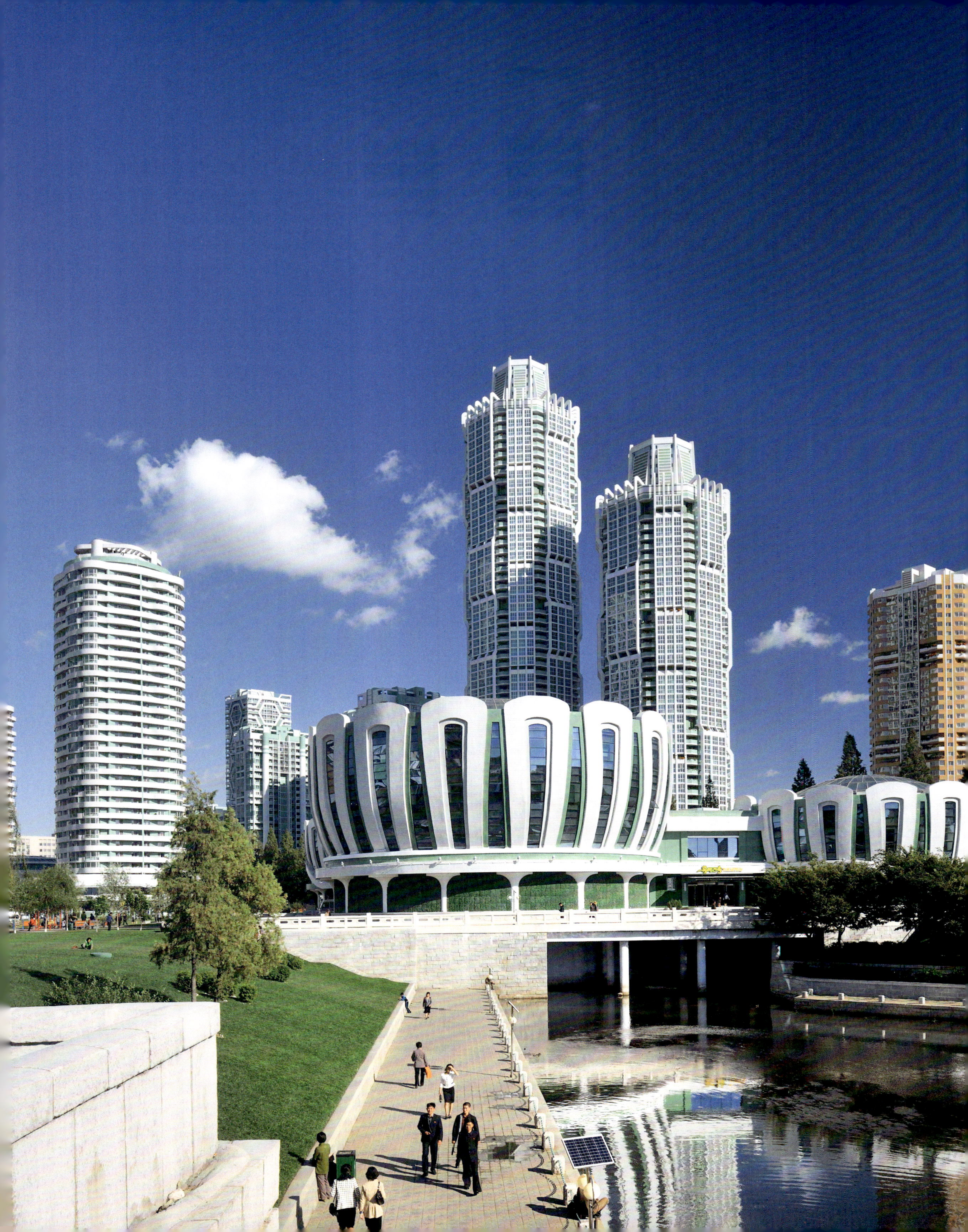

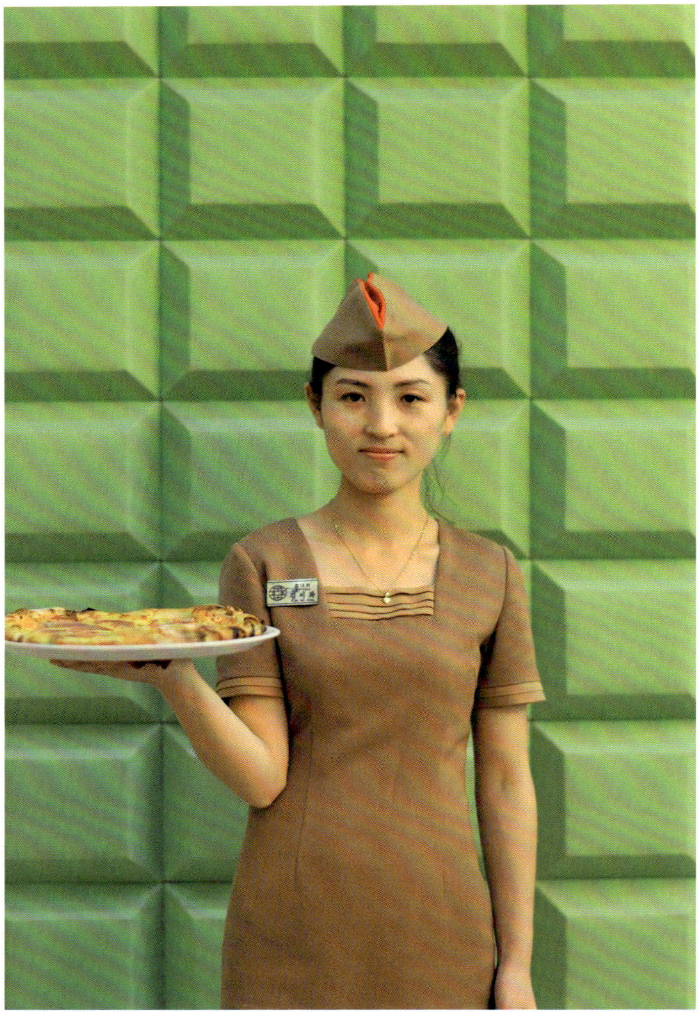

This spread Some trends from the outside world have managed to slip through North Korea's tight borders. In the late 1990s, Kim Jong Il invited a team of Italian chefs to Pyongyang to teach army officers how to make pizza and pasta. Then, in 2008, the Dear Leader sent Korean chefs to Naples and Rome to learn more. The first pizzeria in Pyongyang opened on Kwangbok Street. Some pizzas on the menu come with imported mozzarella, others with kimchi. All are baked in an imported wood-fired oven. Inevitably in the DPRK, dinner is followed by karaoke, even after trying the kimchi pizza. Bowling alleys, beauty salons, barber shops, ice-cream parlours and fast-food joints — Pyongyang has it all. Hamburgers, chicken burgers, vegetarian burgers — with or without cheese? — are served at seven (mostly Singaporean) fast-food restaurants. For the best french fries in town, head to Taedonggang Brewery #3, where a choice of seven beers is served on tap, with exposed brick walls and long wooden tables reminiscent of a German bierhaus. The fries come with ketchup and mayonnaise, just like in any other eatery around the world.

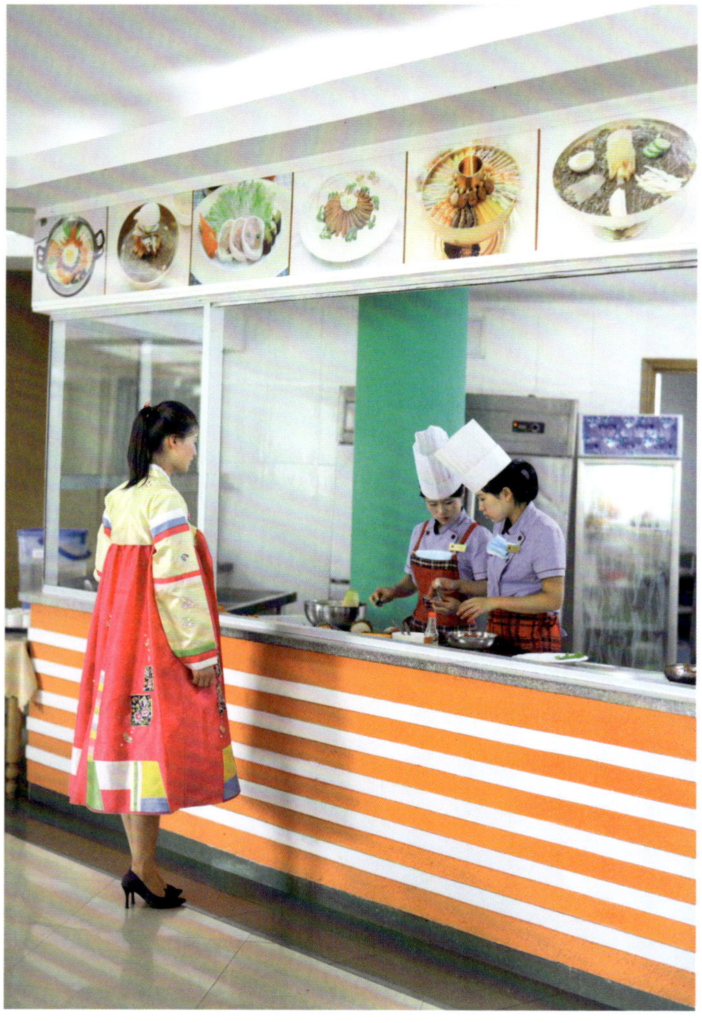

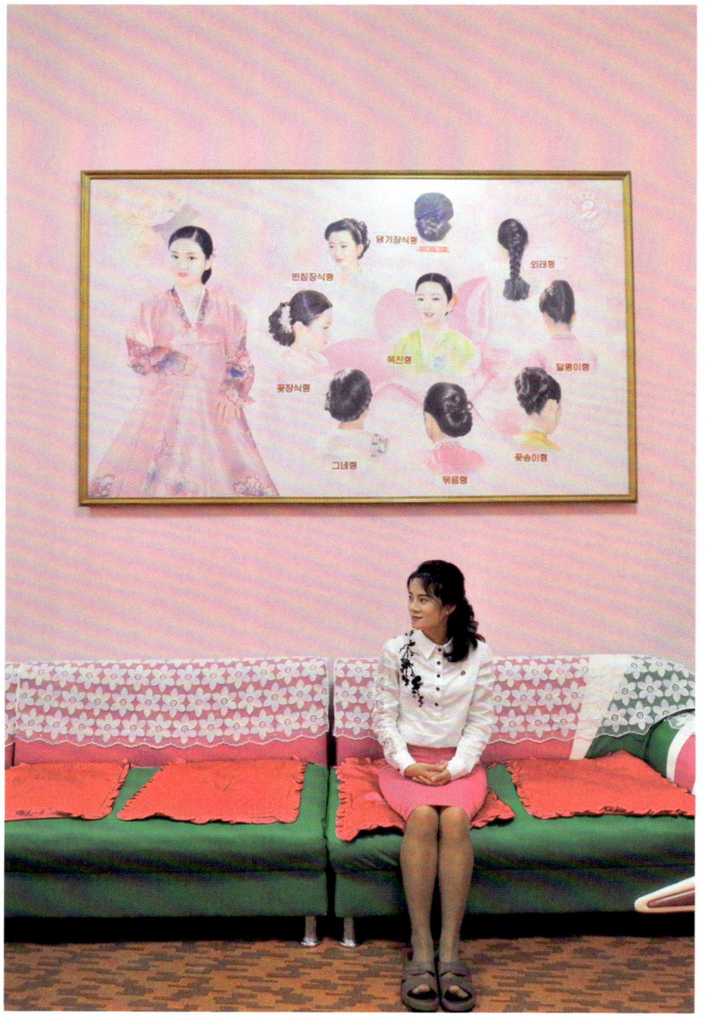

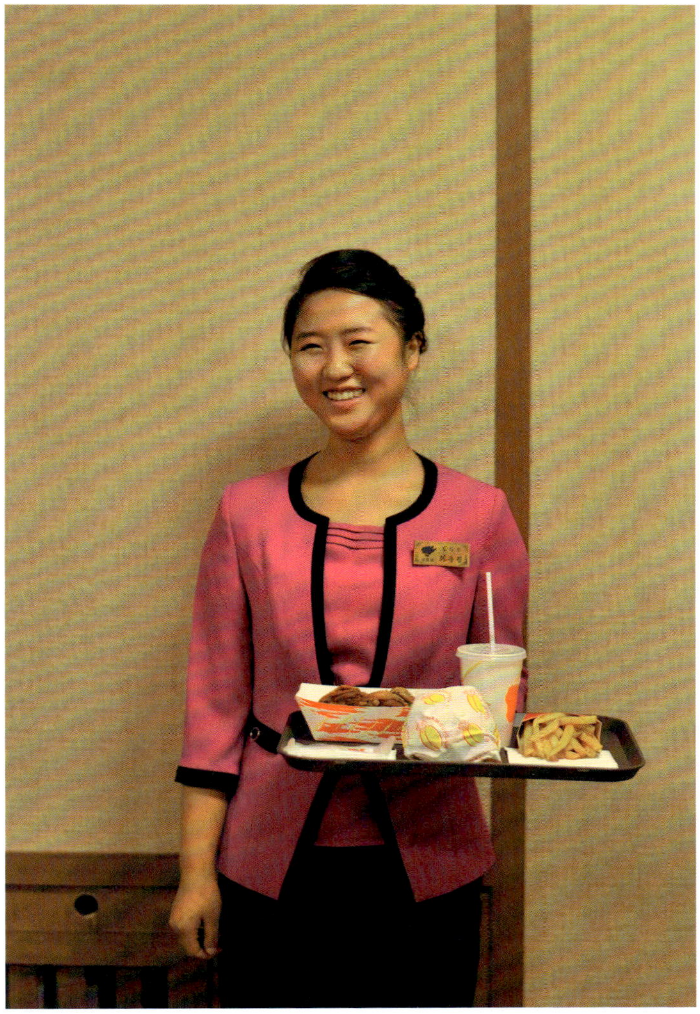

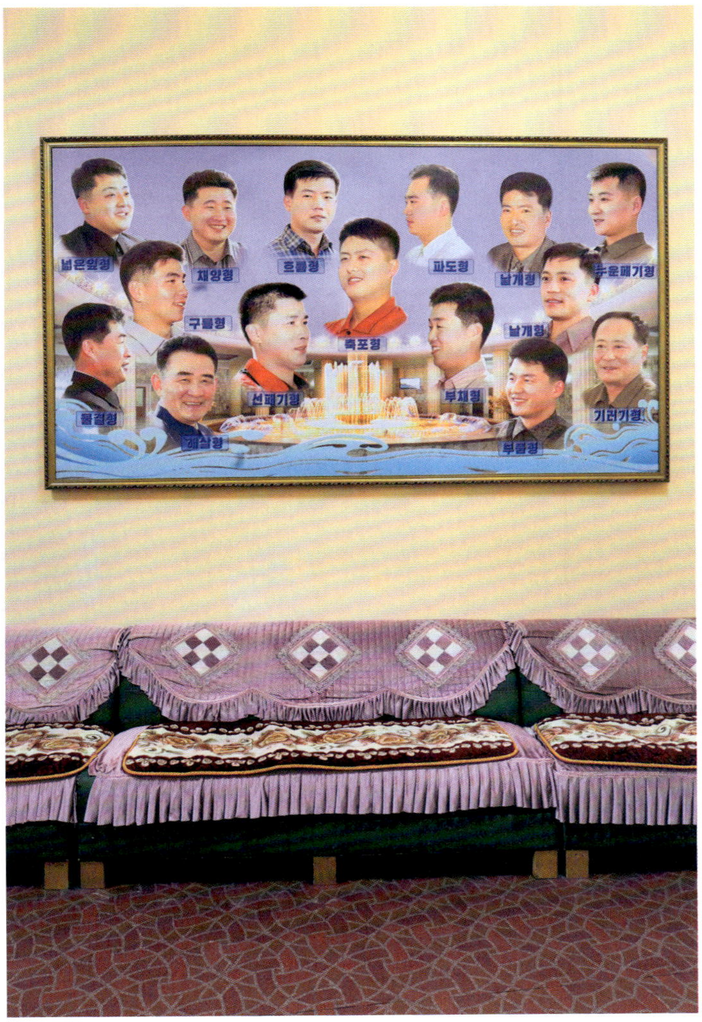

넓은잎형
채양형
흐름형
파도형
날개형
구름형
축포형
날개형
부채형
물결형
해살형
기러기형

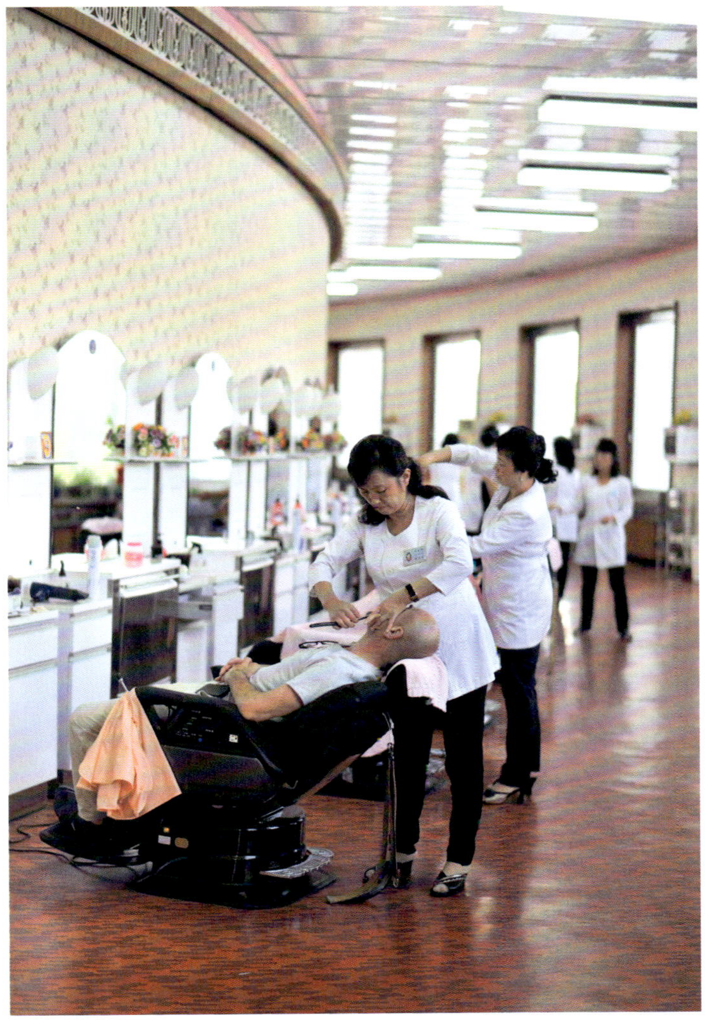

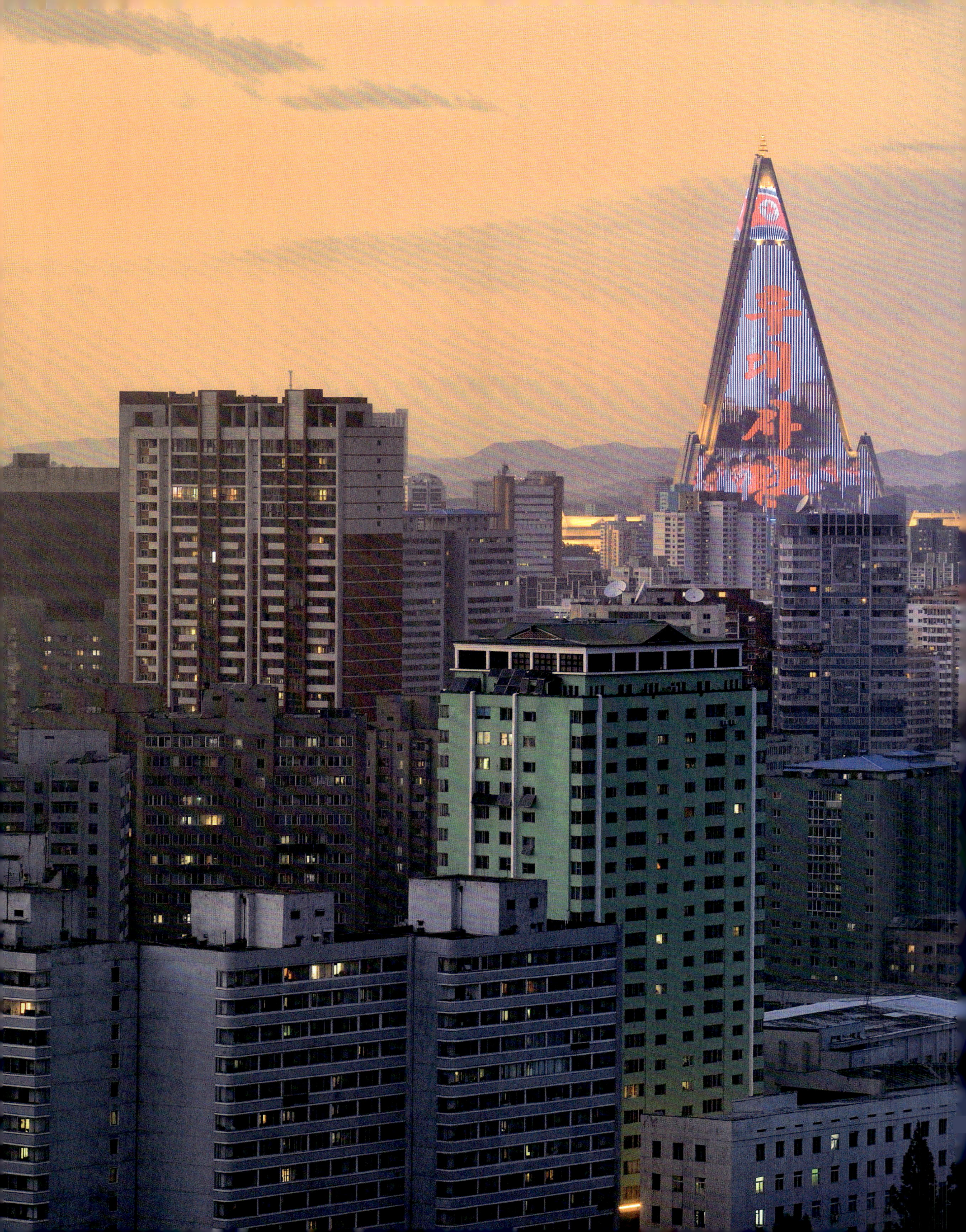

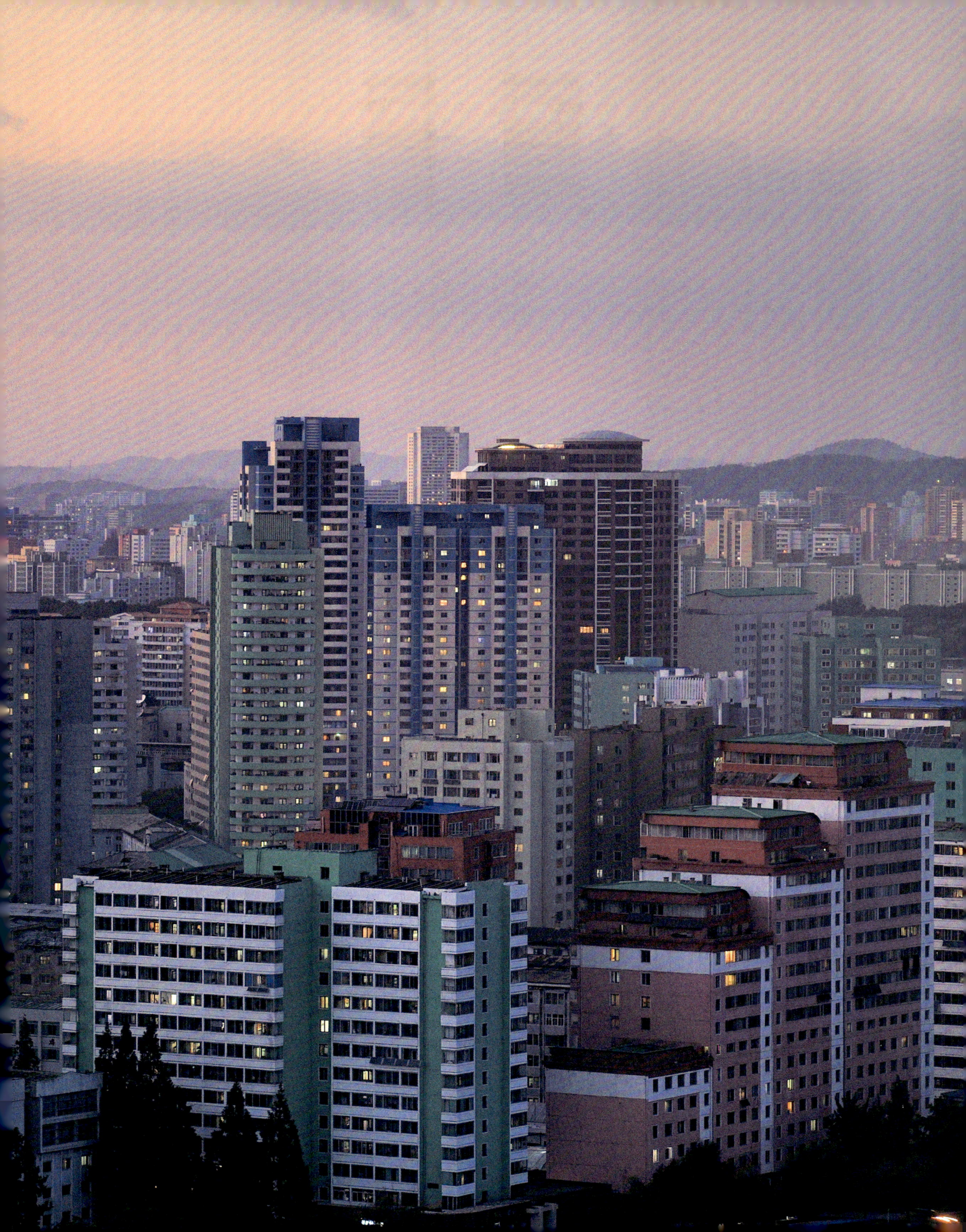

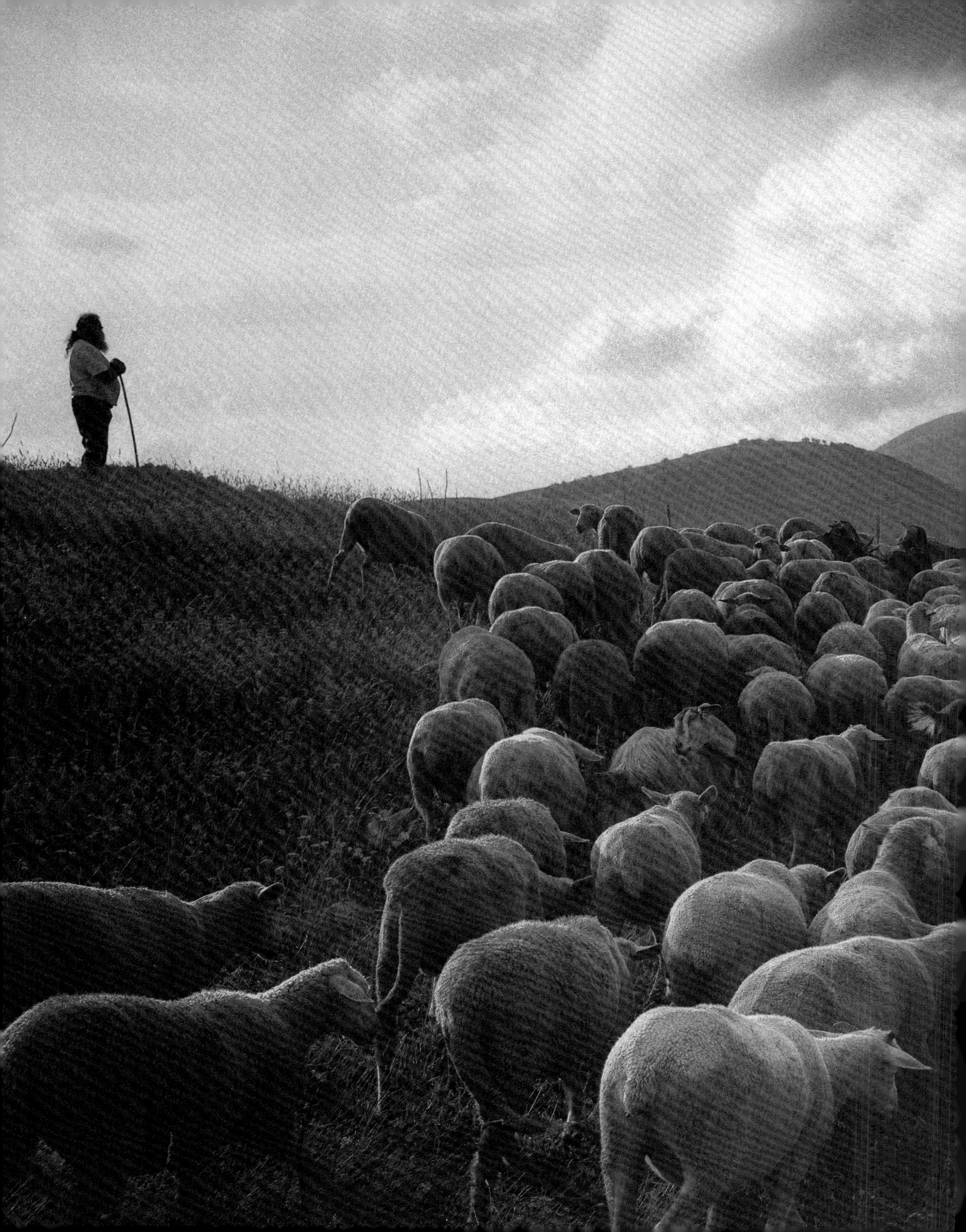

CHAPTER 6

A SLOW-PACED RAMBLE INTO THE HEART OF ABRUZZO

THE SEASONAL MIGRATION OF SHEEP AND GOATS

Italy

41°59'41.92" N
13°48'29.56" E

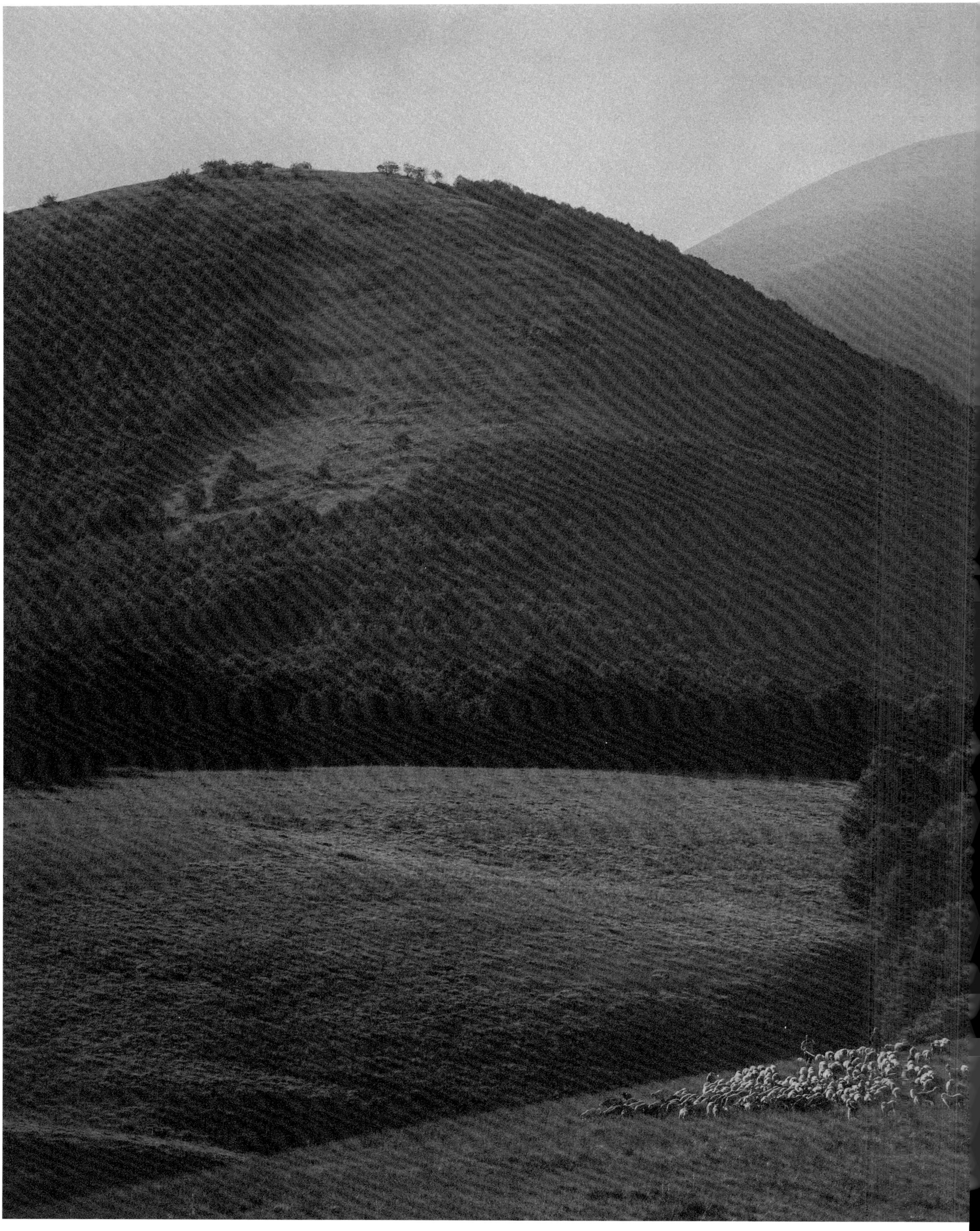

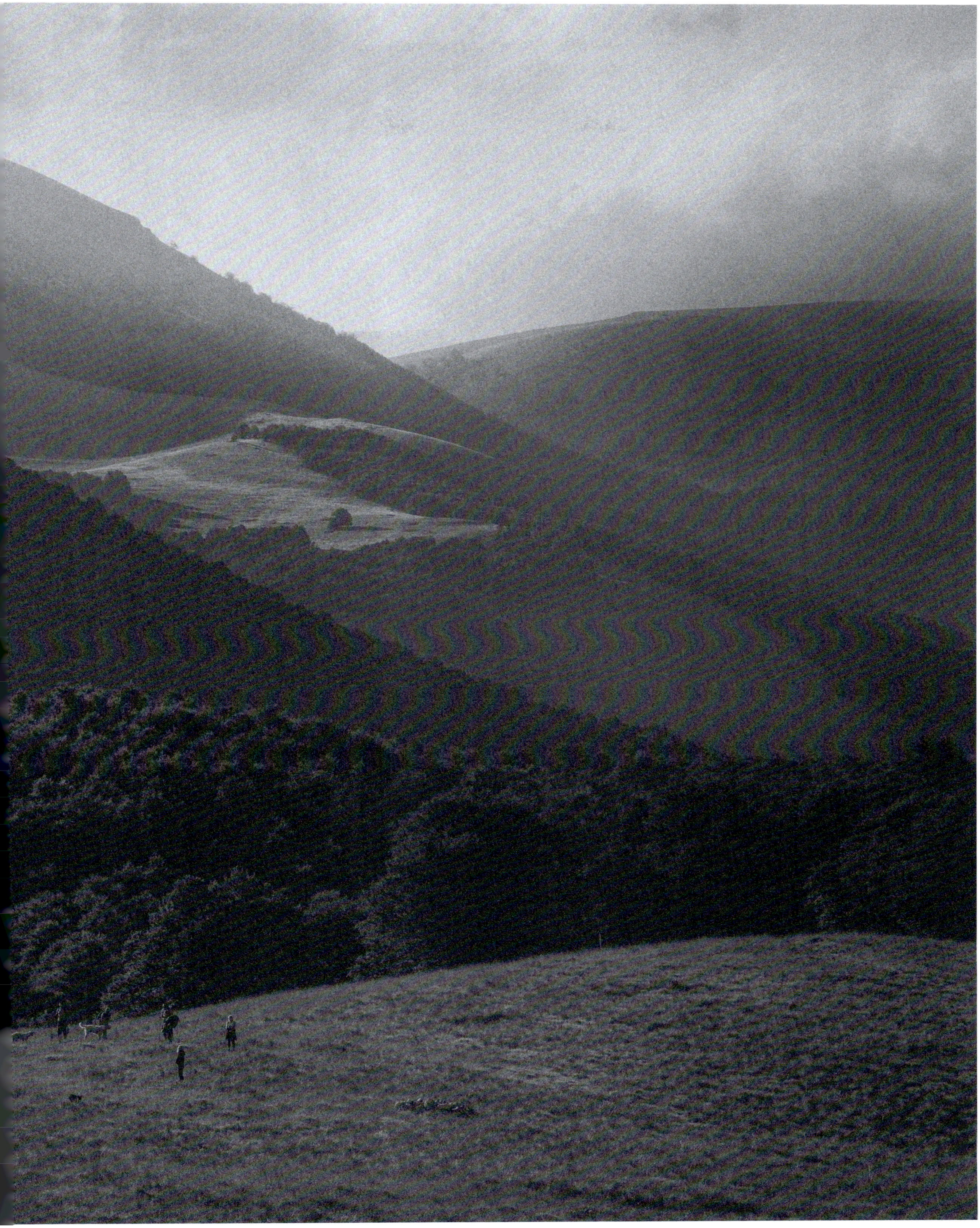

A Slow-Paced Ramble into the Heart of Abruzzo

The seasonal migration of sheep and goats

Country:	**Italy**
Best time to travel:	**Spring and autumn**
Closest airport:	**Rome**
Price tag:	**€**
Style of travel:	**A quiet hike in pursuit of tradition and heritage**

Introduction

For at least 3,000 years, the transumanza — a tradition of moving flocks of sheep and herds of goats between mountains and pastures — has been practised along the rocky slopes of Abruzzo. It captures the soul of this remote, sparsely populated, and mountainous territory, which is between the more famous regions of Tuscany and Puglia. Abruzzo is one of the last places in Europe where the migration of humans and domestic animals continues to define the local character. In its medieval villages, nearly everything one does or consumes is somehow linked to the transumanza, including the strong wines poured generously during long, joyful lunches. The connection between the people and seemingly pristine landscape that surrounds them is rooted in tradition. Designated a place of 'Intangible Cultural Heritage of Humanity' by UNESCO in 2019, the transumanza perpetuates deep, historic bonds between the people and their animals and ecosystems. Twice every year, in spring and autumn, thousands of sheep and goats are herded by their collective owners along with horses, dogs, and, occasionally, their families. Outsiders who wish to experience the soul of the transumanza can join in the tradition, which entails a long weekend of walking with 300 to 400 sheep and goats, slowly moving towards the green pastures of the Piano delle Cinque Miglia, a grazing area near the village of Roccaraso. On its way to one of Europe's most isolated regions, the route crosses the Sagittario Gorges, while also passing through immaculately preserved hill towns, including the charming village of Scanno. All departures are from the hamlet of Anversa degli Abruzzi in Gole del Sagittario, a stunning nature reserve in the L'Aquila region. Here around 250 villagers live frozen in time, still following the ancient, slow, simple customs of a forgotten world. Among them is Nunzio Marcelli. A seemingly bygone character with his majestic beard, piercing eyes, and strong hands, Mr Marcelli studied economics at the prestigious Sapienza Università di Roma but spends his days working the land and herding his sheep, fighting a quixotic battle against the intensive farming methods which threaten the age-old traditions and culture of Abruzzo. His farm, La Porta dei Parchi, is open to visitors who want to learn more about the traditional ways of farming. It's also a bed and breakfast serving simple and tasty alfresco lunches to diners seated at wooden tables looking out over spectacular views of the countryside. There is a shop selling organic produce, from fresh ricotta cheese to aged variations, honey, and a whole range of homemade meat such as the salamella rinascimentale, a sausage made from pork liver, honey, cooked grape must, and orange peel. The farm, which is organic, has more than 1,300 sheep and goats, and a dozen or so workers from nearby as well as from Macedonia and even Guinea-Bissau. To ensure the continued existence of real Abruzzese agricultural life, Mr Marcelli welcomes curious travellers to walk the transumanza but also to join 'Adopt a Sheep', an initiative that supports villagers' traditional ways of herding. In return for an annual fee, adoptive 'parents' receive a selection of Abruzzo's finest organic cheeses, meat, woollen socks and other items once a year. There is nothing industrial about anything Mr Marcelli produces. Members of his programme might receive an artisanal salami made from lamb and goat; Brigantaccio, the local pecorino, made from raw sheep's milk; or a ricotta cheese lightly smoked with juniper. Mr Marcelli's philosophy, 'produce less but better', has many admirers, especially in the United States, where *The New York Times* gave 'Adopt a Sheep' a glowing write-up.

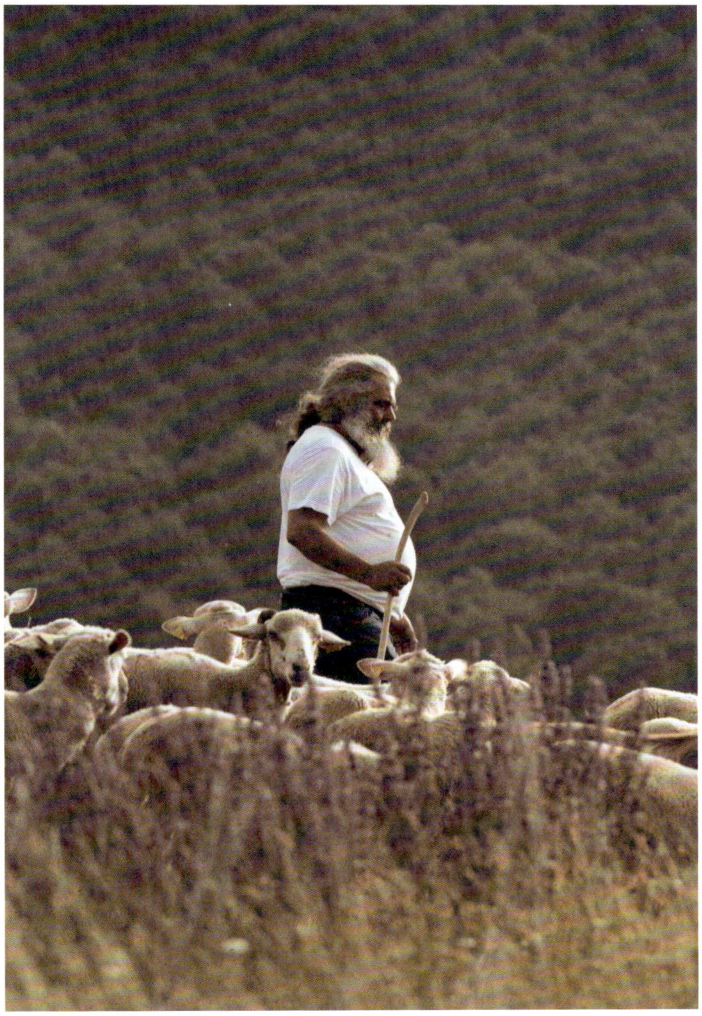
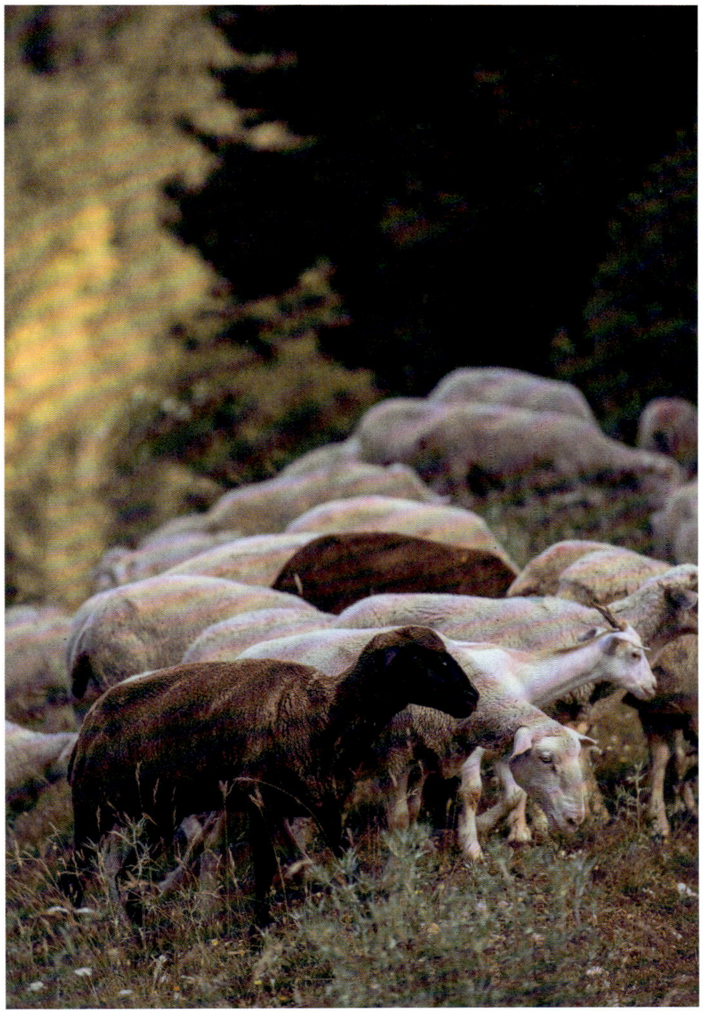
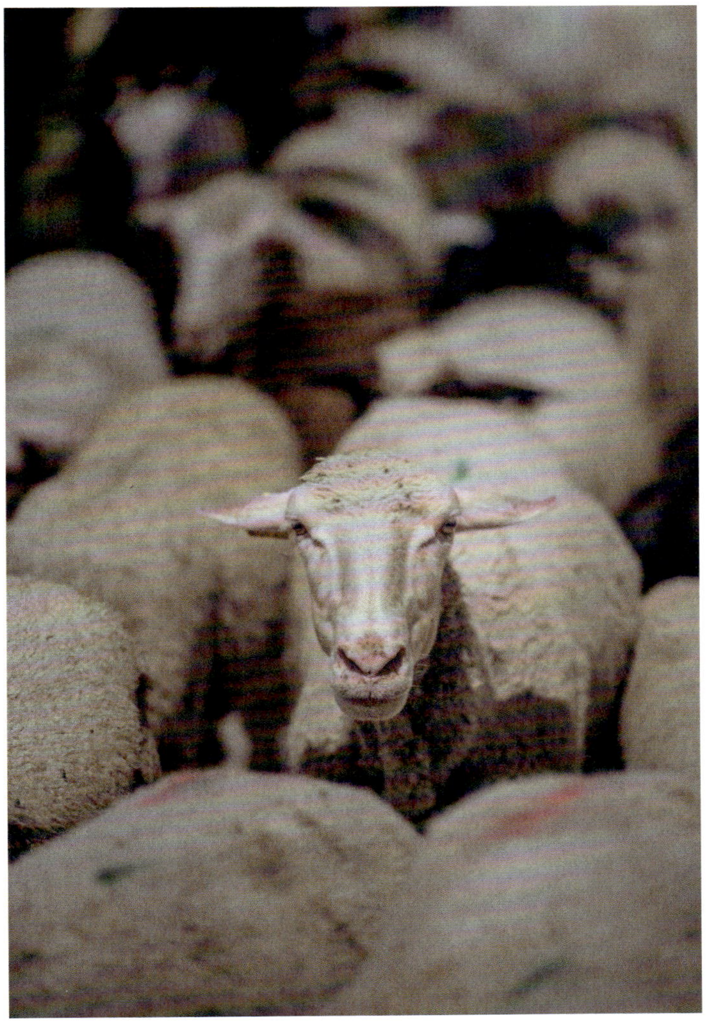
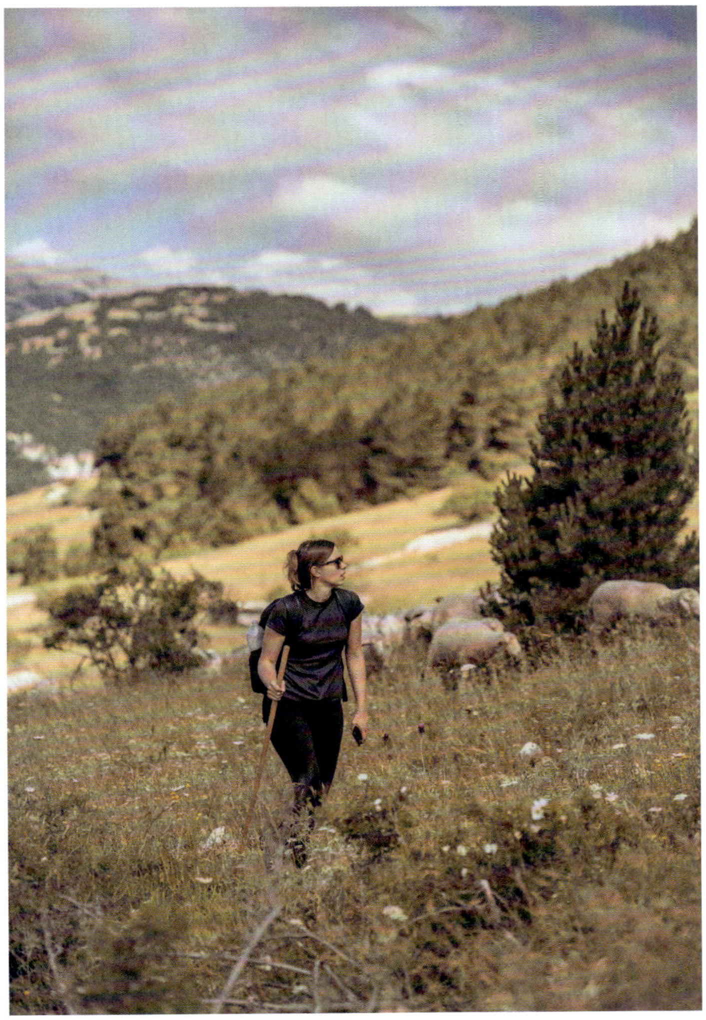

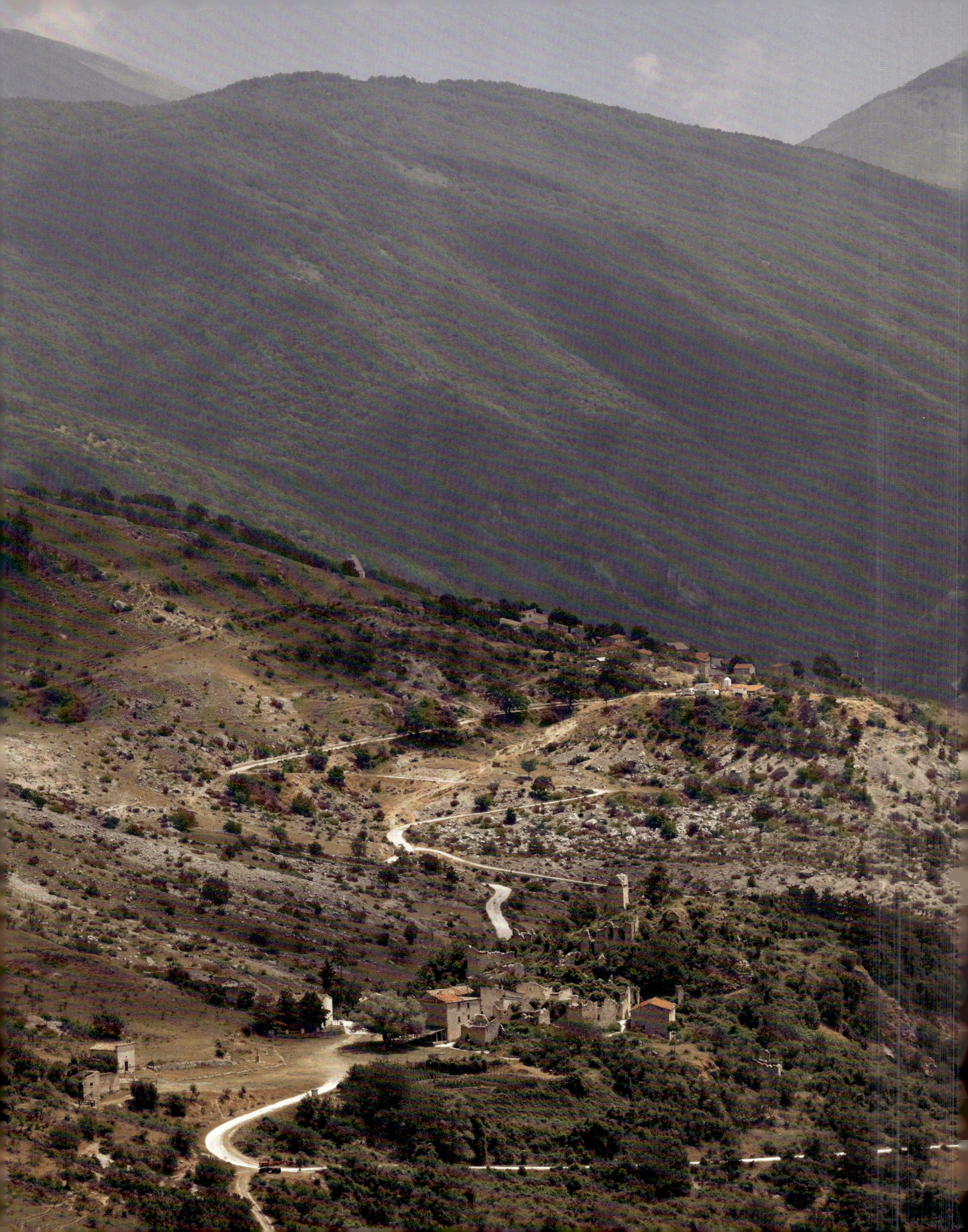

Opposite The goats and sheep of the transumanza once followed tratturi, ancient paths that were part of an immense network of trails along which churches, taverns, and bridges were built. These natural highways could be up to 111 metres (364 ft.) wide, and up to 245 kilometres (152 mi.) in length, but only a few are still accessible for the current flock. Today, only a few remain; most have been swallowed up by roads, villages, and private land. Along the way, strategically placed resting points called riposi would occasionally appear, providing shade and water to shepherds and their flocks. Nunzio Marcelli would like to see them become UNESCO World Heritage Sites, allowing travellers to discover Abruzzo and its traditions while preserving the tratturi's historic value.

The Views

Anversa, an archetypal Abruzzese village of hillside medieval houses set in a dramatic landscape, is the perfect point of departure. Locals meet several times a day in the village bar for coffee or an aperitivo. Older residents with opinions discuss the daily news and events, while kids play in the piazza where everybody seems to know everyone else. Surrounded by endless pastures and the snow-covered peaks of the Apennines, it is the ideal setting for the transumanza. The spectacle unfolds early in the morning, when shepherds open the gates of local farms, releasing undulating rivers of sheep. Assisted by clever maremmano sheepdogs and centuries of folk wisdom, the shepherds know exactly what their flocks need and how to protect them. Horses and donkeys are brought along to serve as pack animals. The shepherds who help out Nunzio Marcelli are members of a Romanian family who live on his farm. The father is a master cheesemaker whose ricotta, pecorino, scamorze, and caciottine are perfect for the walk. Nature is still unspoiled, and civilisation feels far away. Travellers may just as well leave their phones at home: on a three-day walk, there is no mobile signal, and internet access in the villages is severely limited. Sheep and goats in search of greener pastures, moving as a single uniform, fluffy mass, set the daily rhythm, while humans follow behind. The effect is meditative. The first day is spent crossing the wild landscape of the Sagittario Gorges. Over 450 hectares (1,100 ac.) in size, the reserve reaches altitudes of almost 1,500 metres (5,000 ft.) and is home to an extraordinary biodiversity that includes golden eagles, wolves, and bears. Next on the trail is the small village of Castrovalva. Set above gorges, it appears on paintings by M.C. Escher, who was inspired by its inaccessible vastness. At an altitude of 1,100 metres (3,600 ft.) is Lago di Scanno, a heart-shaped lake created 2,000 years ago when a landslide from the mountain above created a natural dam. The village next to it was born from a typical pastoral economy. A matriarchal culture existed here as the men spent most of their time away from the village, tending to their flocks, leaving the women in charge. To this day, elderly ladies wear the heavy woollen costumes typical of Abruzzo. The last stop is a wide, beautiful pasture on the Piano delle Cinque Miglia. Here the pure air is filled with nothing but the sounds of sheep bells, horses' hooves, and branches snapping underfoot. In this bountiful landscape, far from city life, the sheep will stop, eat, and rest until October. Then, the transumanza will move in the opposite direction, back to the farms in time for winter, continuing its endless loop.

Creature Comforts

From humble bed and breakfasts to simple shepherd huts, joining the transumanza means three days of basic lodging. However, the absence of modern comforts is generously compensated by home-cooked meals and plenty of Abruzzese wine, sometimes accompanied by guitar playing and folk singing. All for less then 300 Euro.

Go Now

The transumanza is recommended for everyone with a good level of fitness, and even inexperienced hikers are encouraged. Each day the trail covers around 15 to 20 kilometres (9 to 12 mi.). The long, slow journey through a remote region while accompanied by long-time inhabitants can bring heart and mind into a Zen-like state of calm.

How to Go

Despite its remote character, this quiet region of Italy is easily reachable with a flight to Rome and a scenic 90-minute drive (123 kilometres/76 miles) through lush valleys and medieval villages perched on craggy, snow-covered mountains.

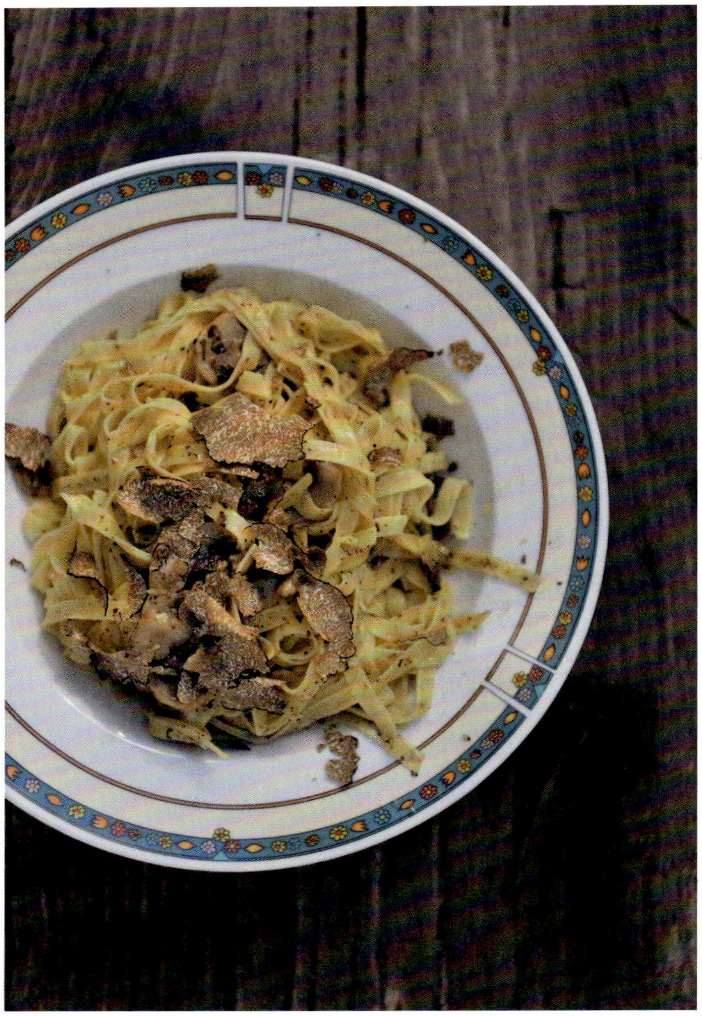
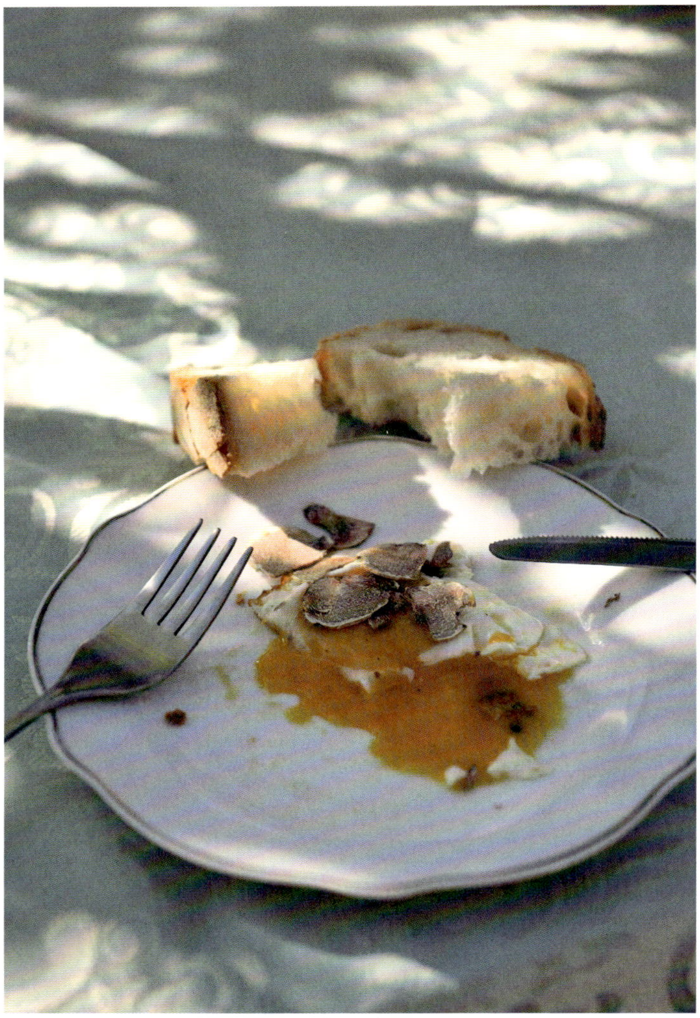
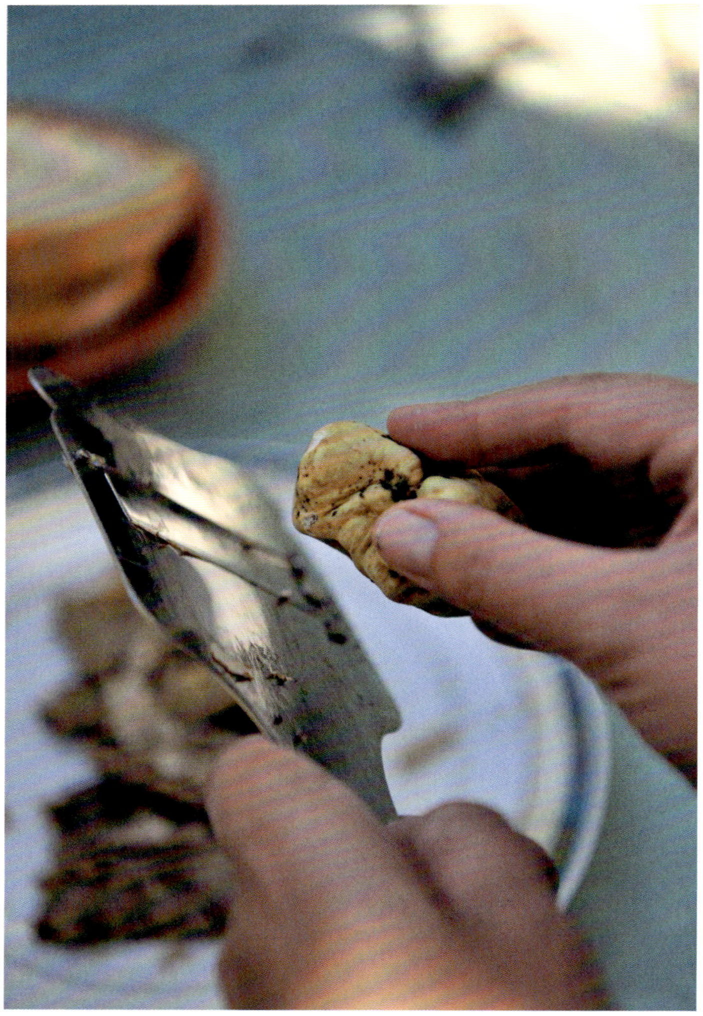
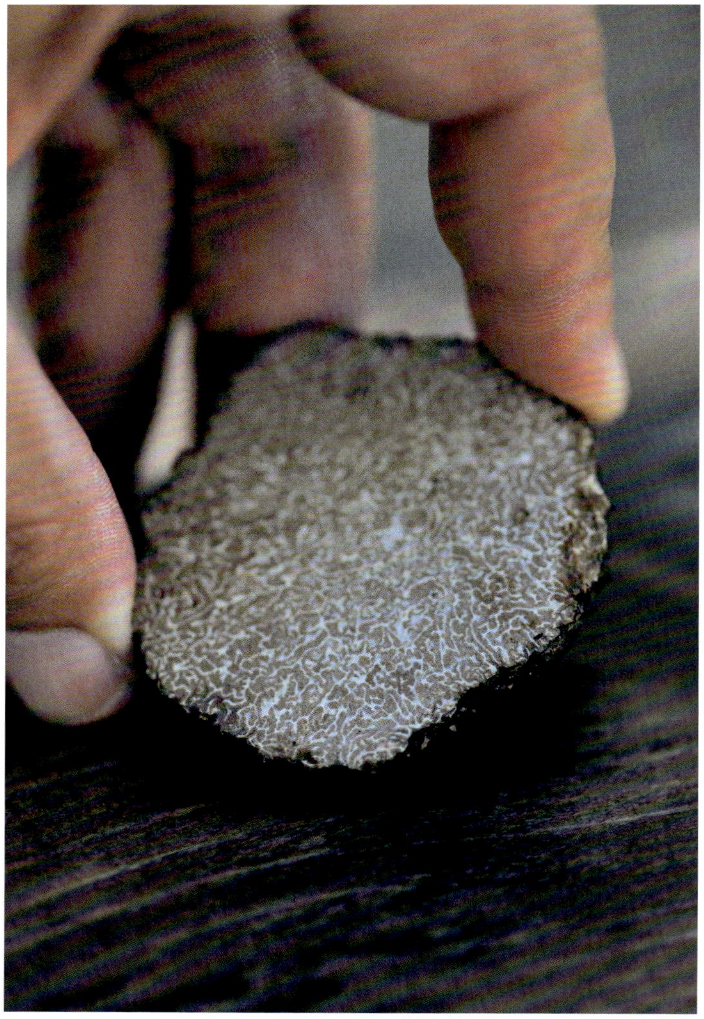

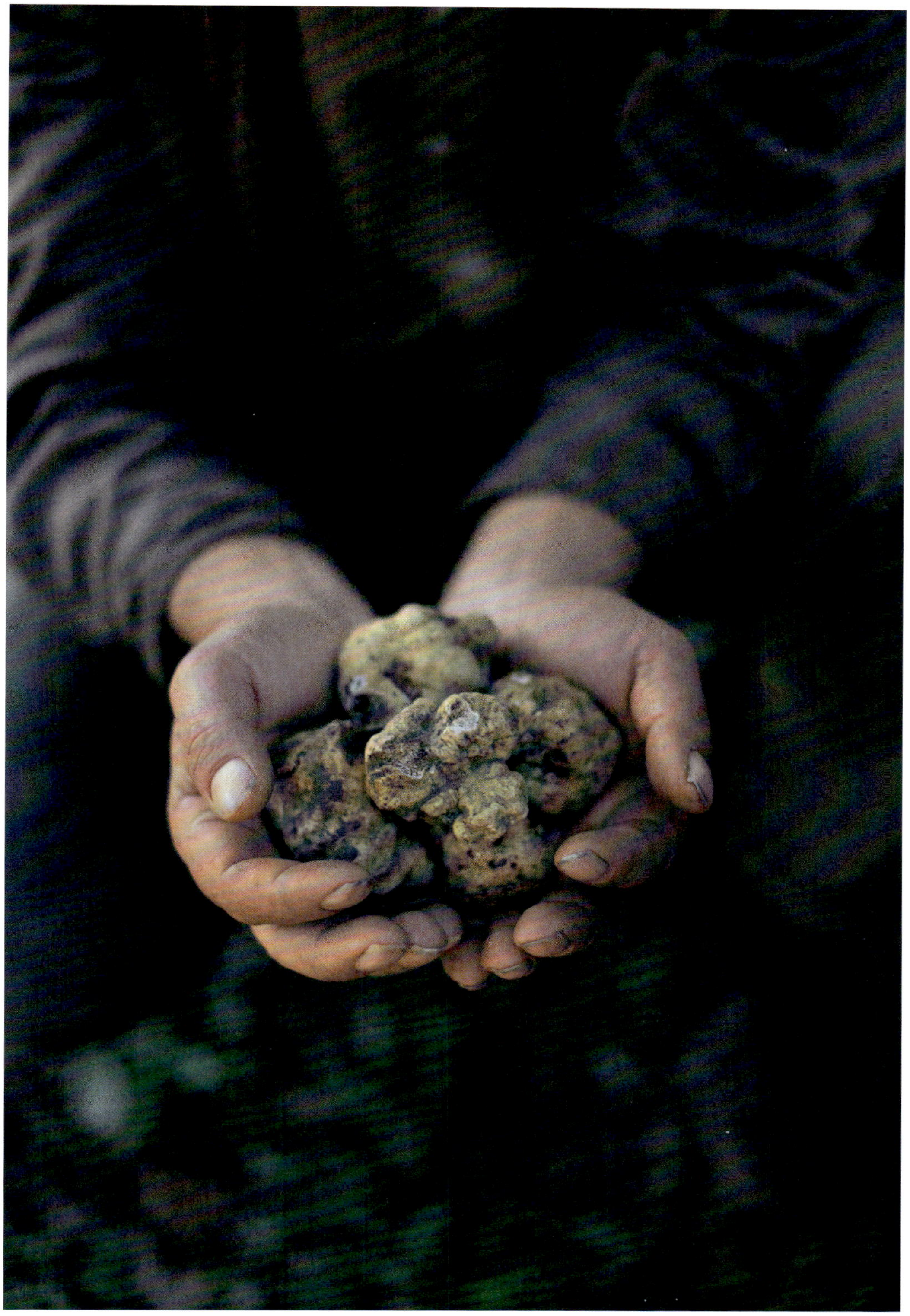

This spread Mushrooms and truffles found along the way enrich the cuisine of the transumanza, which also includes handmade fettuccine, homemade bread, salami, and no shortage of ricotta. White truffles account for 15 per cent of truffles harvested in Abruzzo from early October to late November. The mountain climate and olive groves, oaks, and hazelnut trees provide ideal conditions for truffle growing.

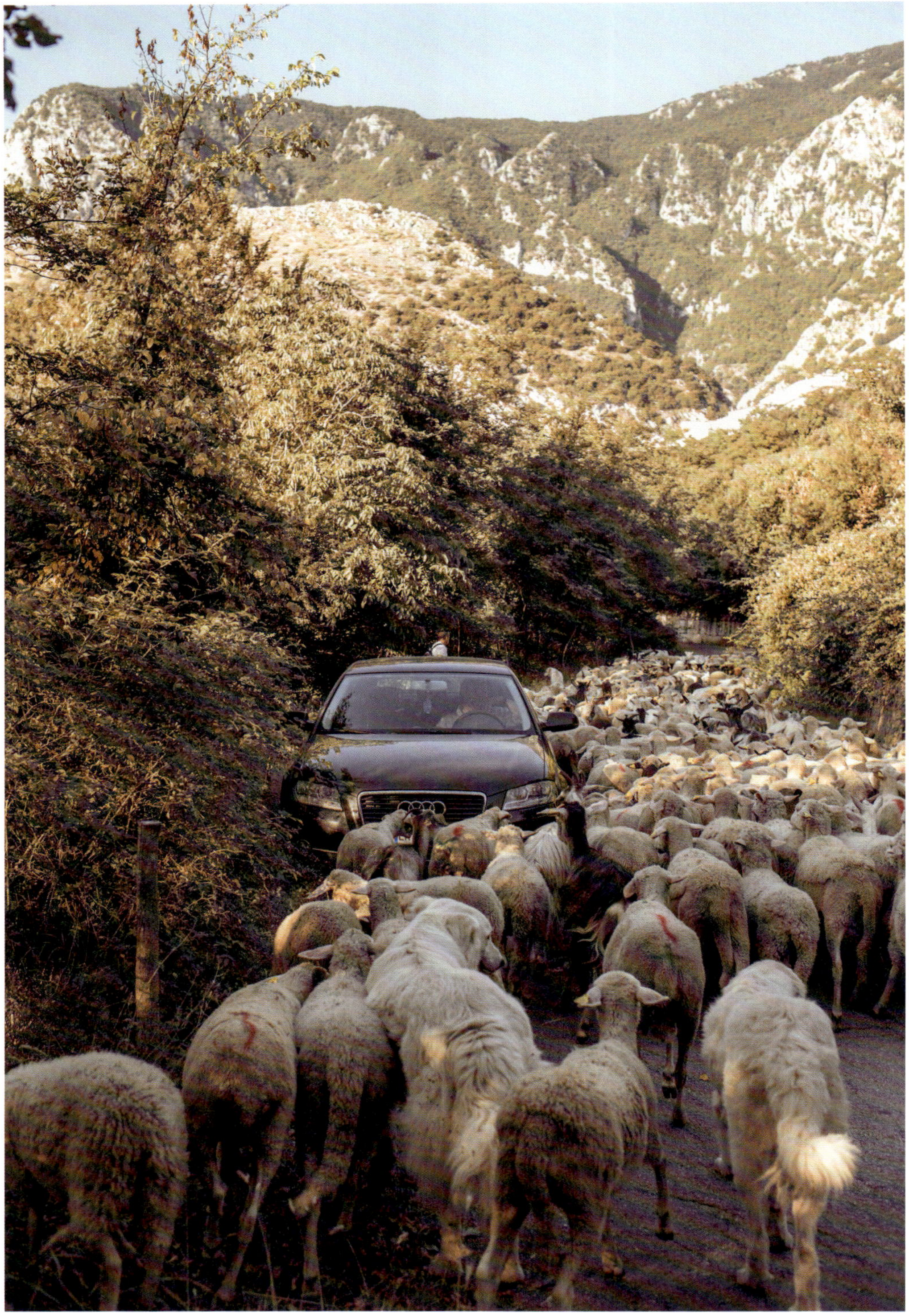

This spread En route to higher pastures, shepherds bring horses and donkeys to carry personal items, food, and plenty of Abruzzese wine. The Abruzzese shepherds have been doing this for thousands of years. Transumanza, or transhumance, comes from the Latin 'trans' for 'across' and 'humus' for 'earth'. This tradition is a way of life in Abruzzo.

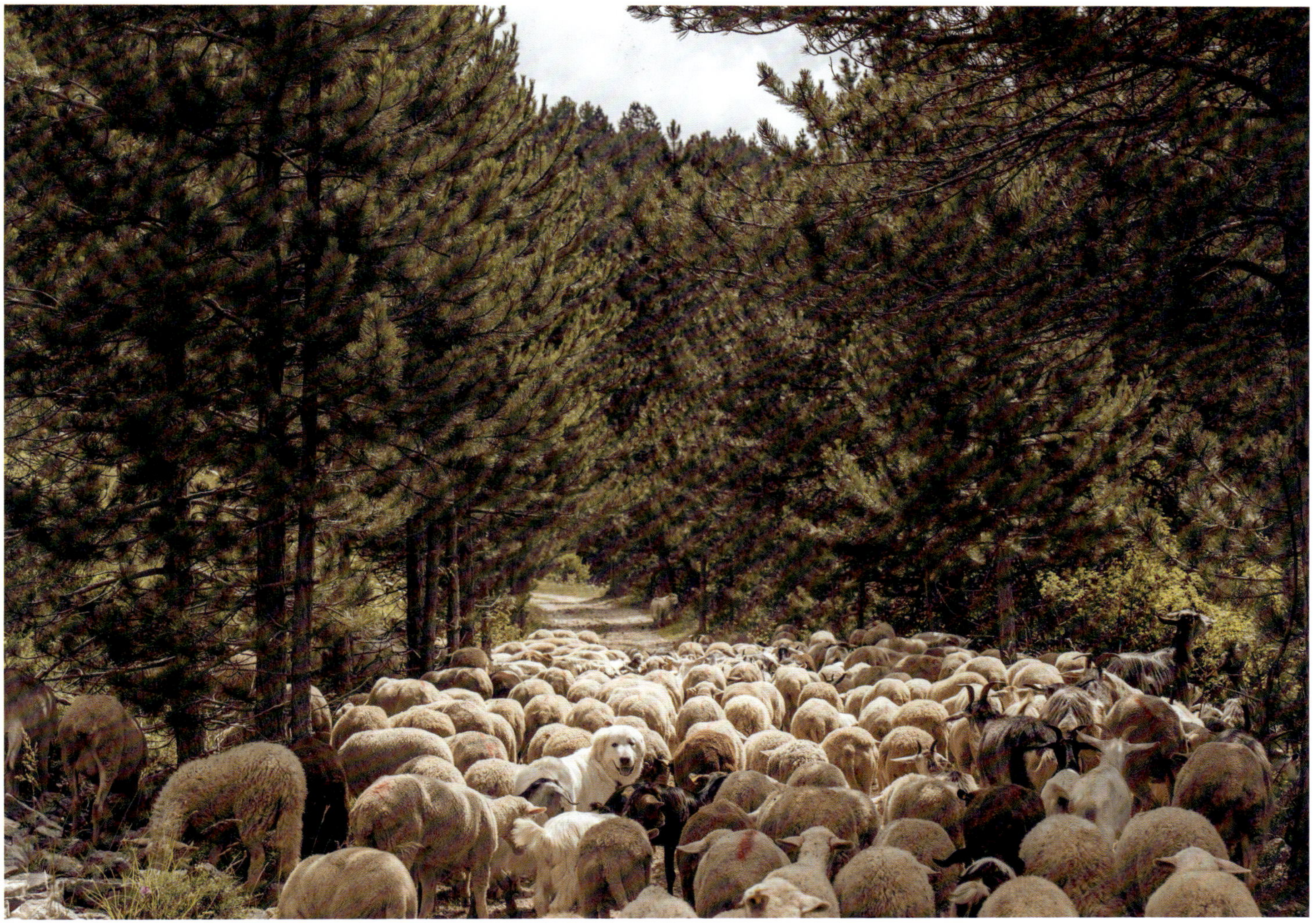

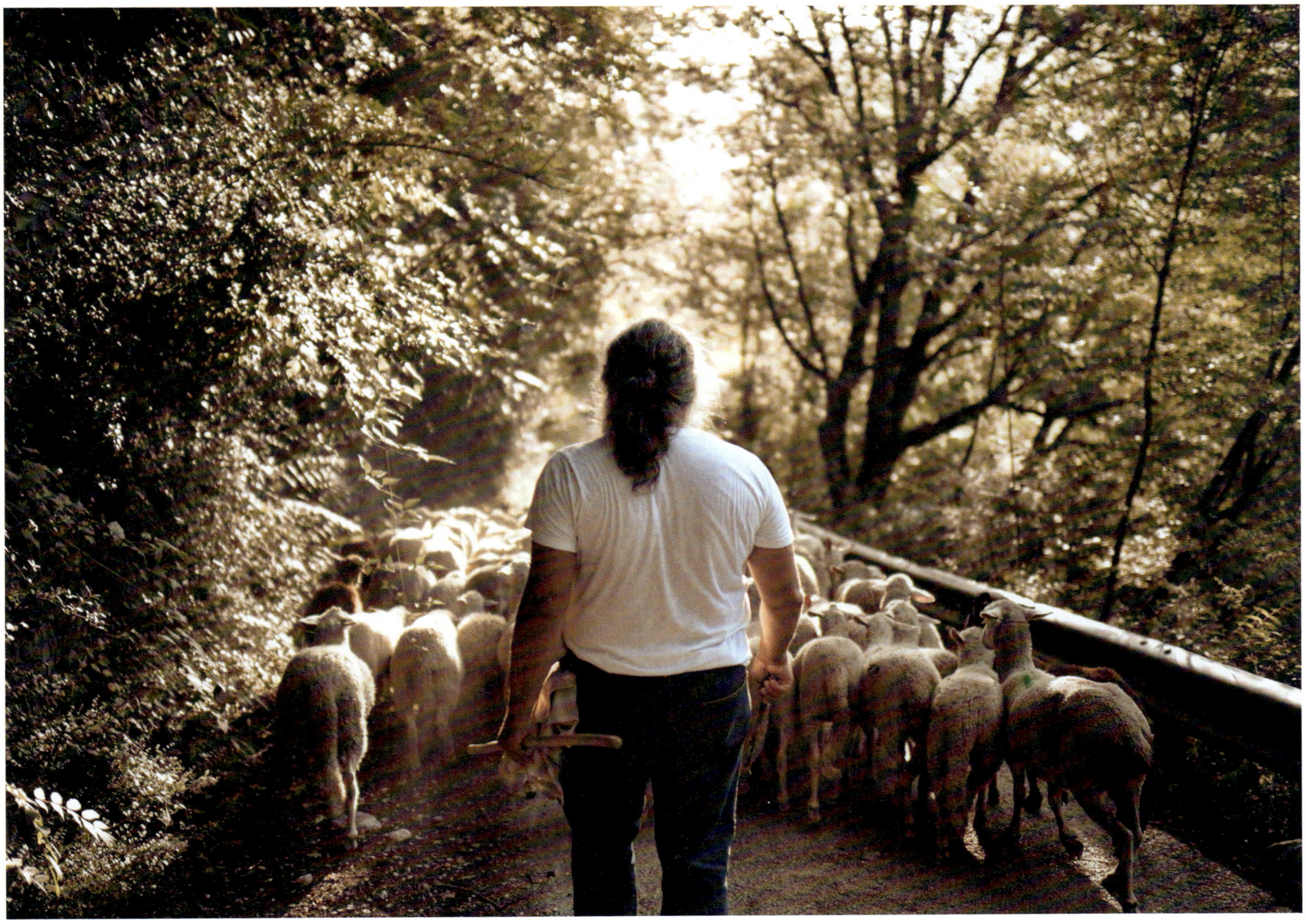

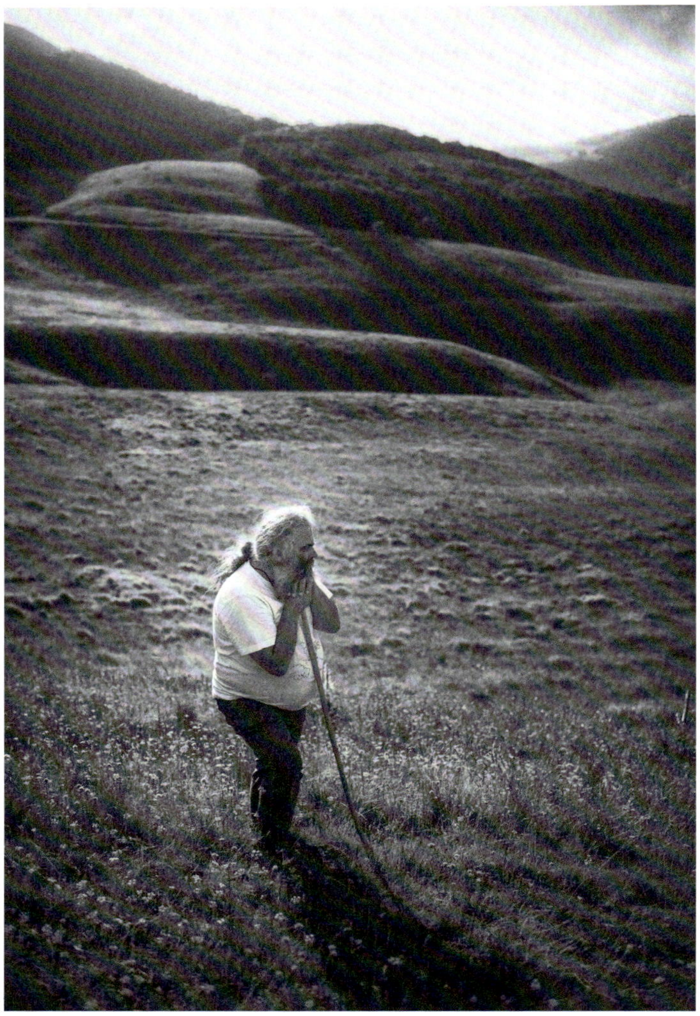

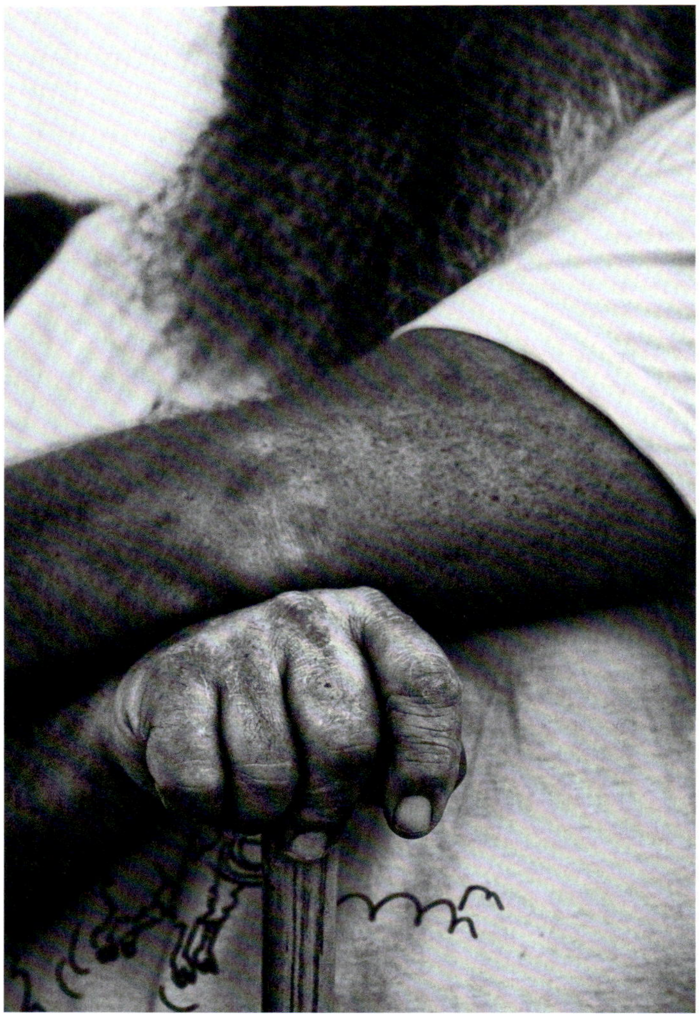

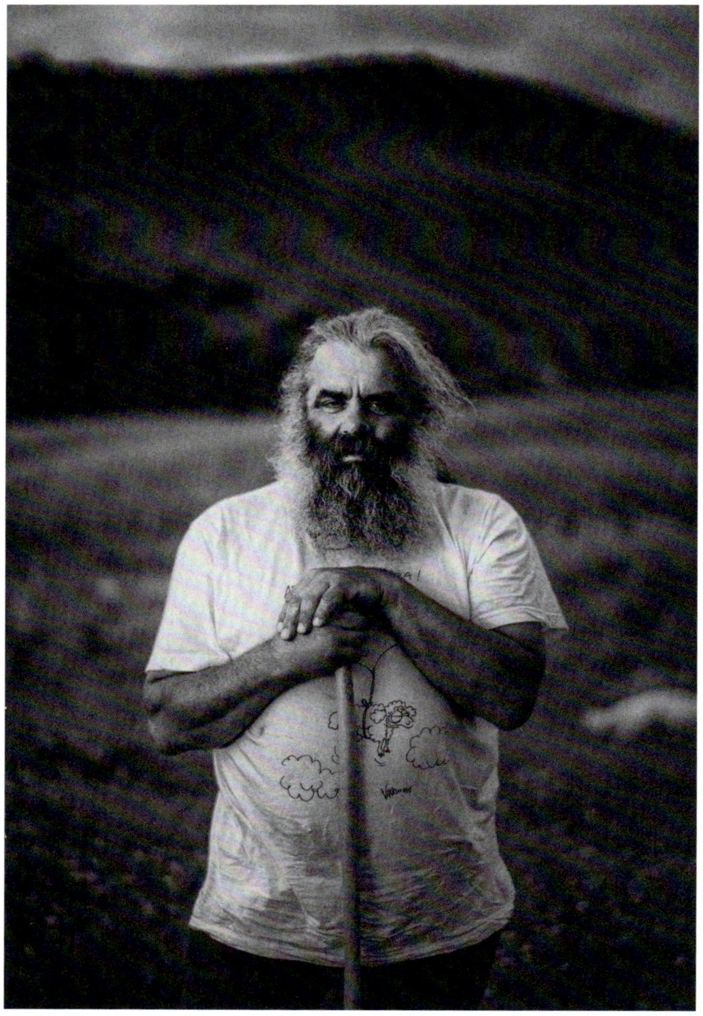

This spread Even nowadays when walking the transumanza, the Abruzzese landscape stays wild and pure with vast patches of remote nature. The sheep and goats are guided by the shepherds and the Pastore Maremmano Abruzzese, the tall and smart Maremma sheepdog used for many centuries as a flock guardian during the transumanza.

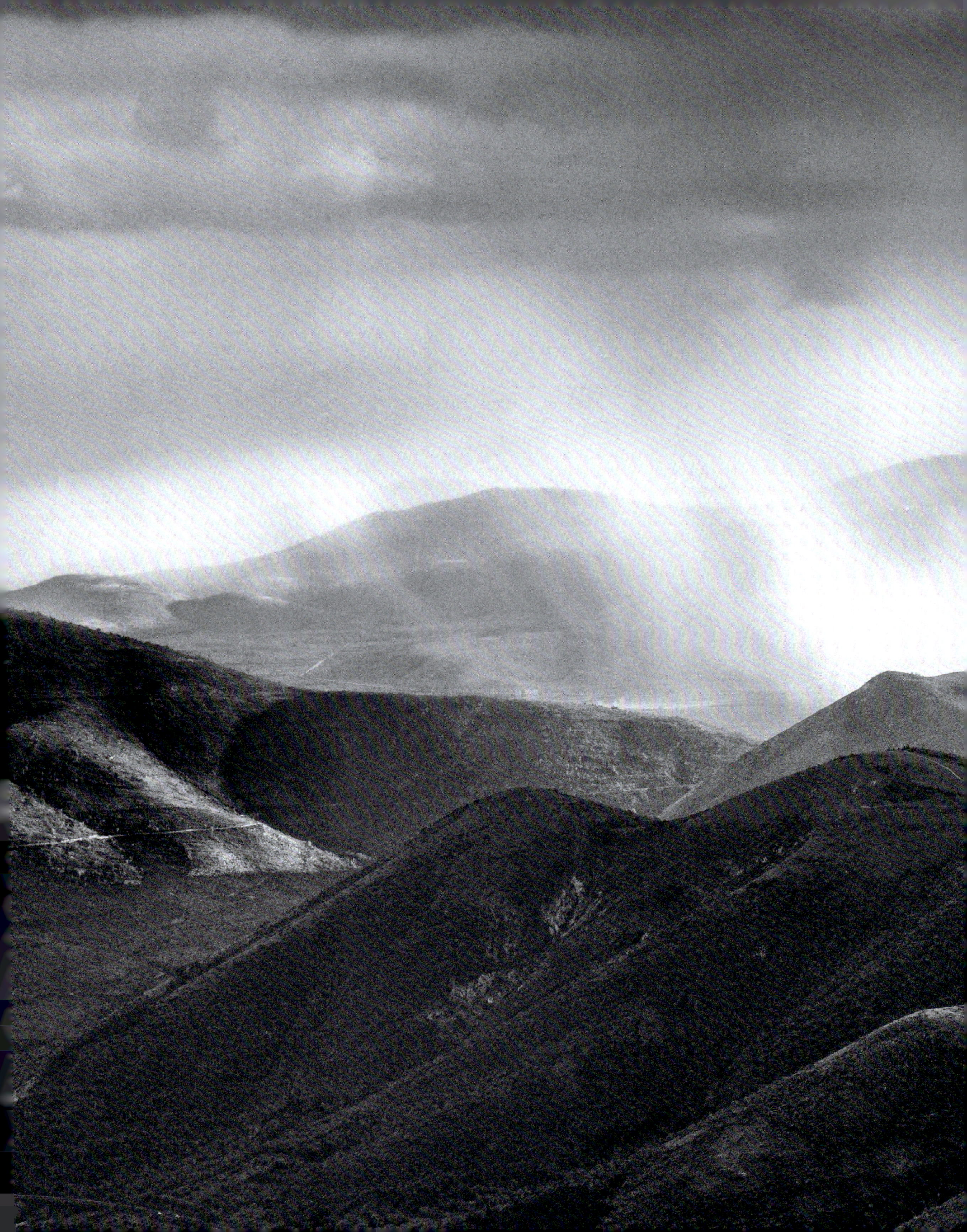

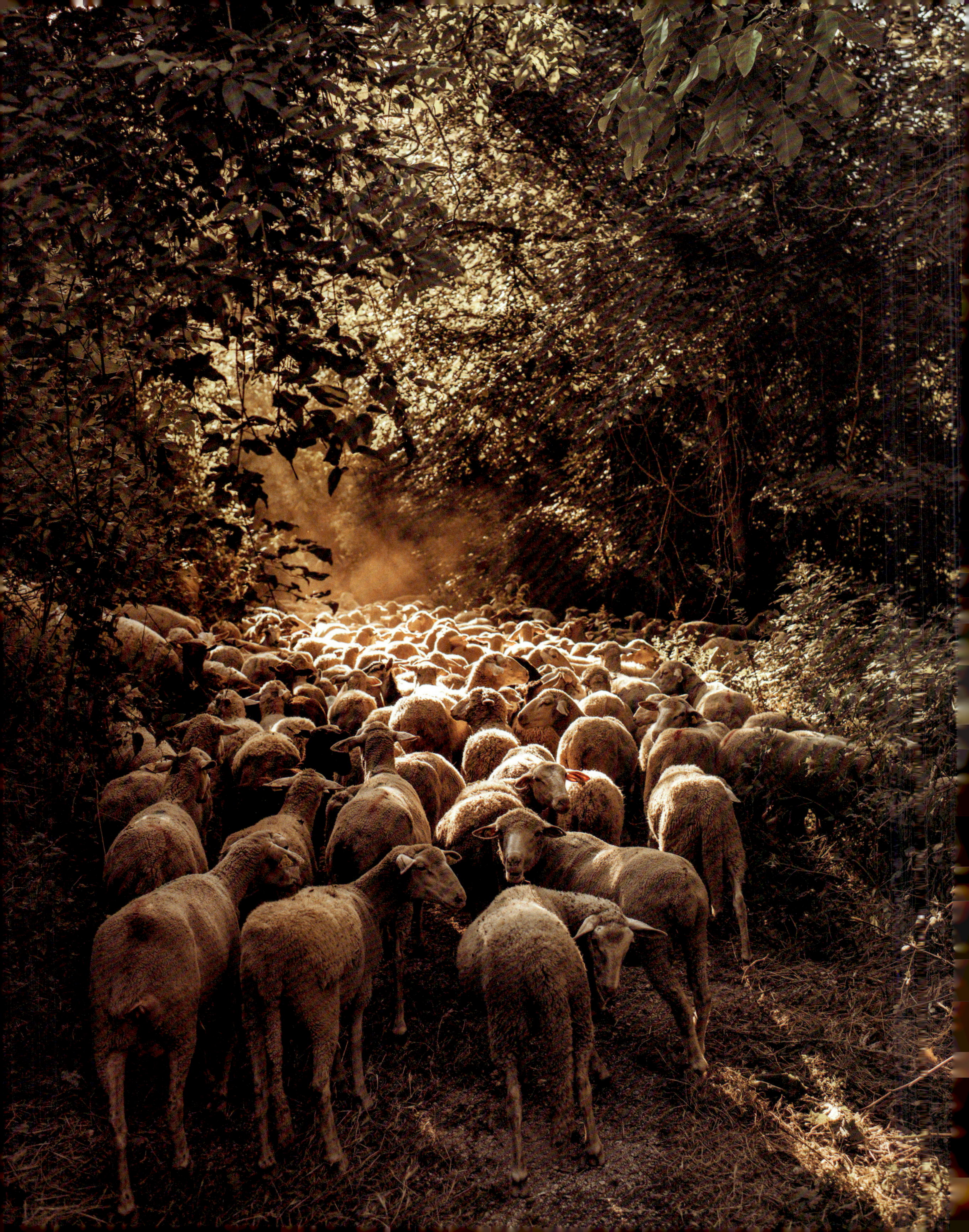

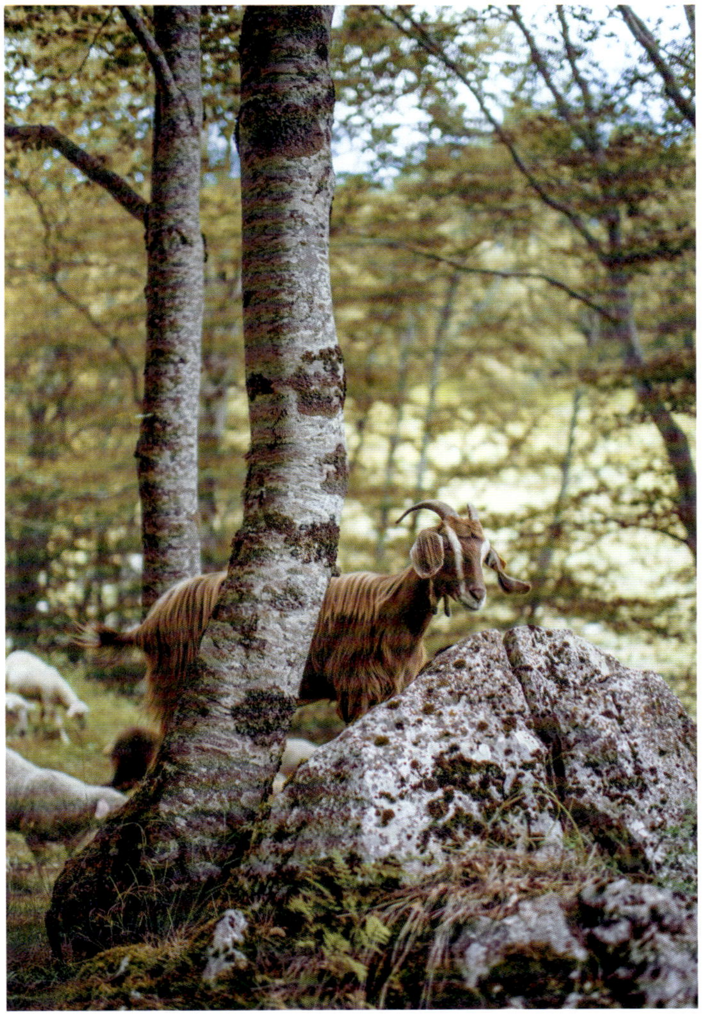
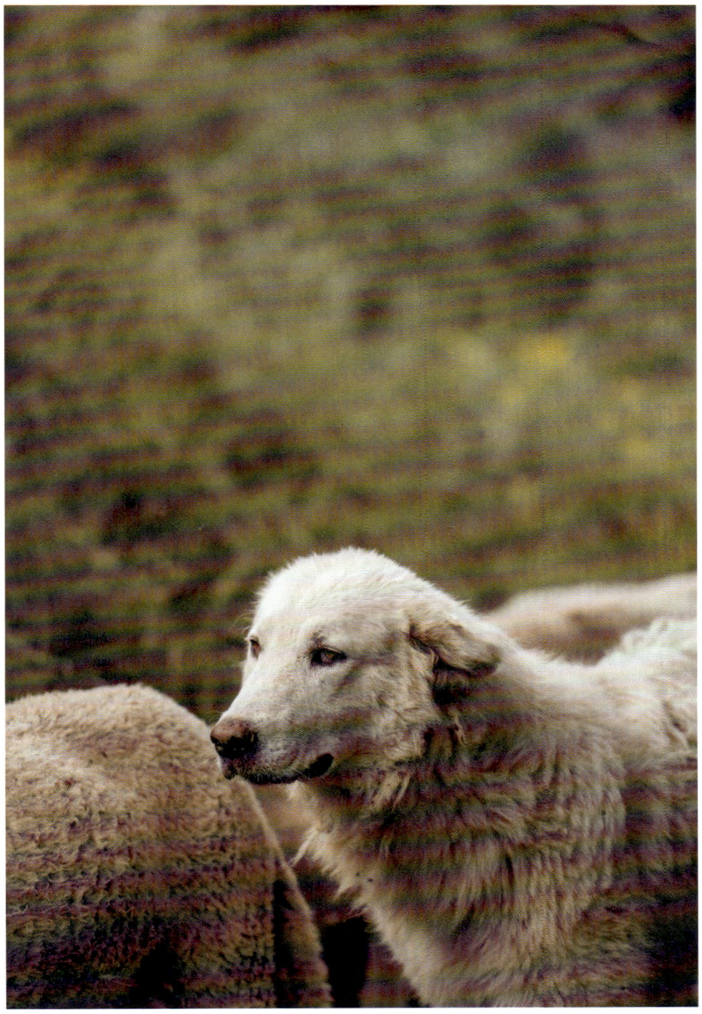
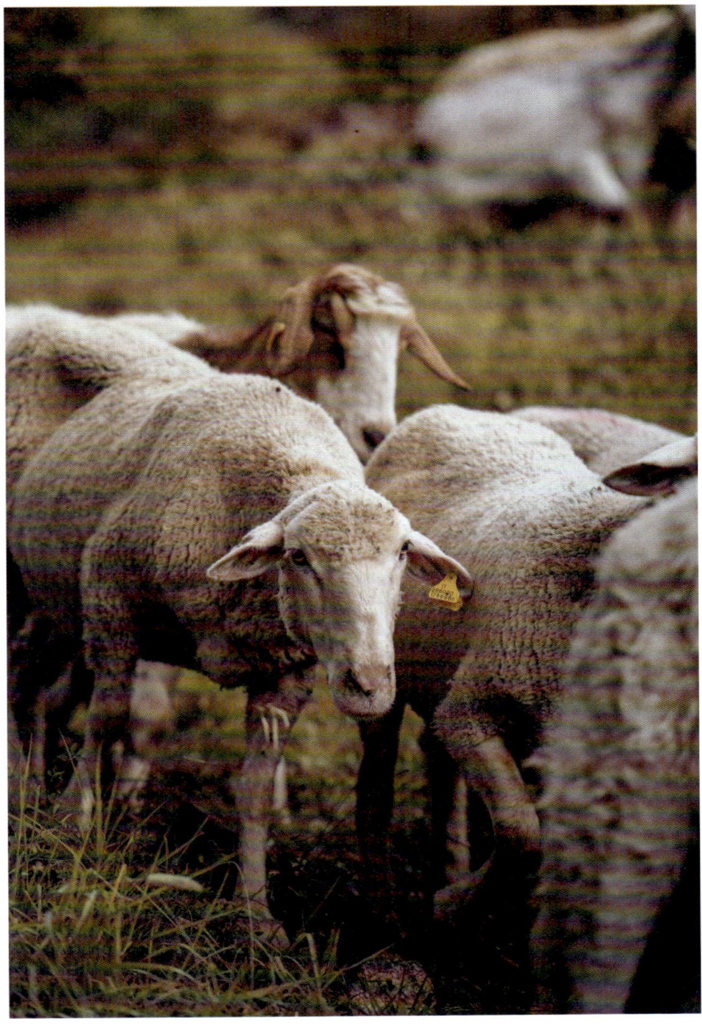
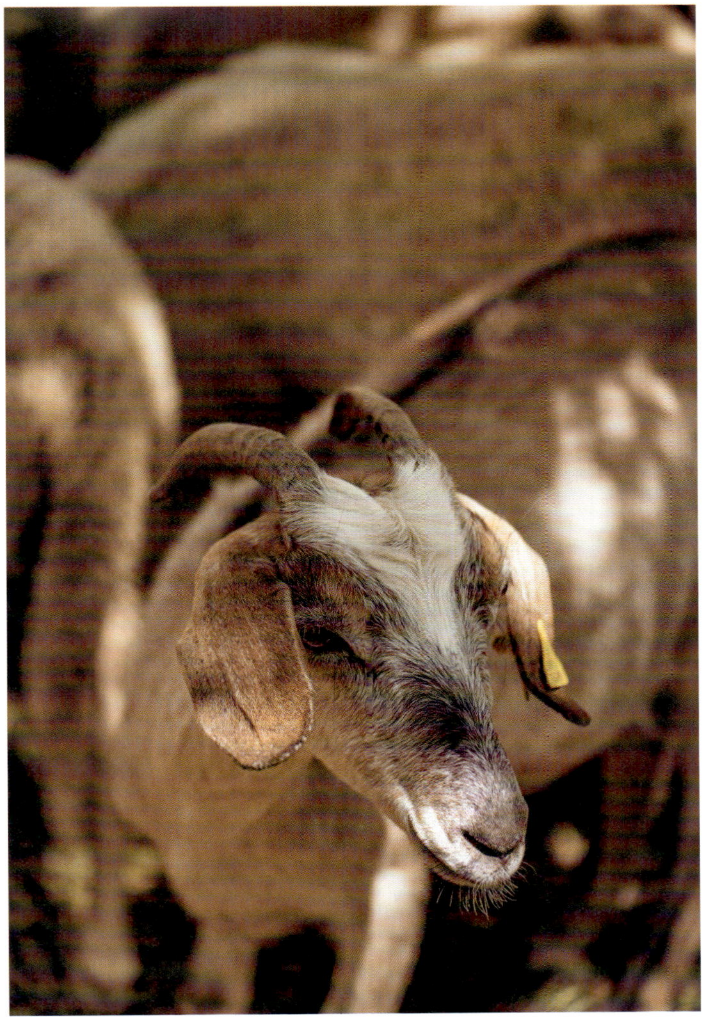

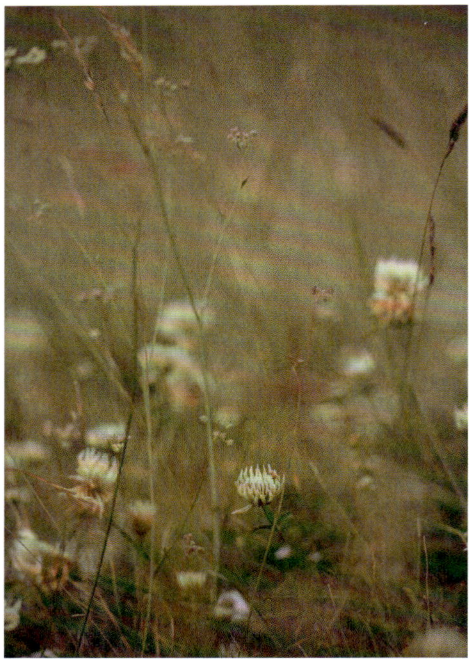
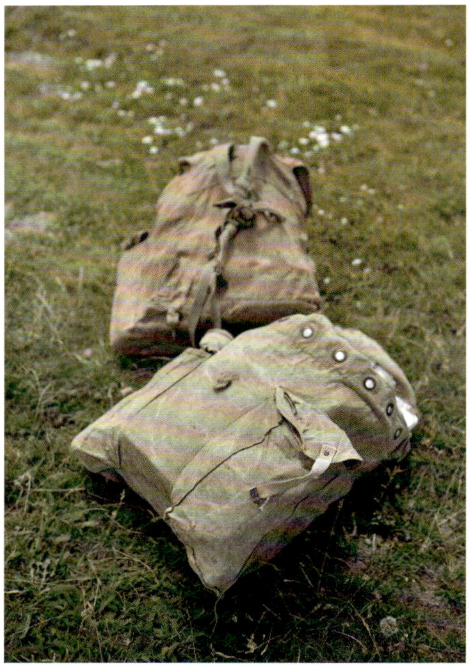
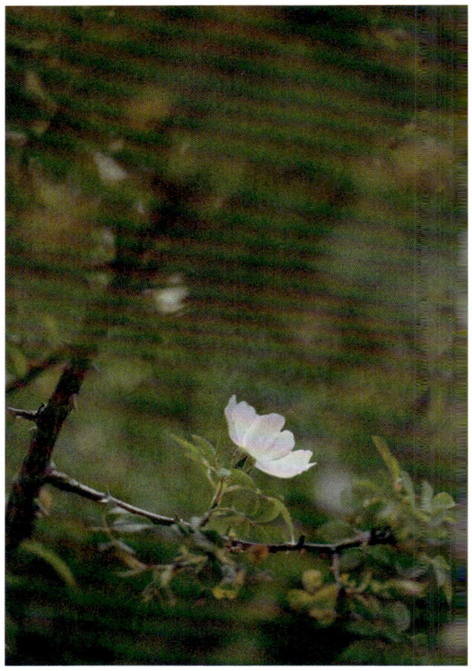
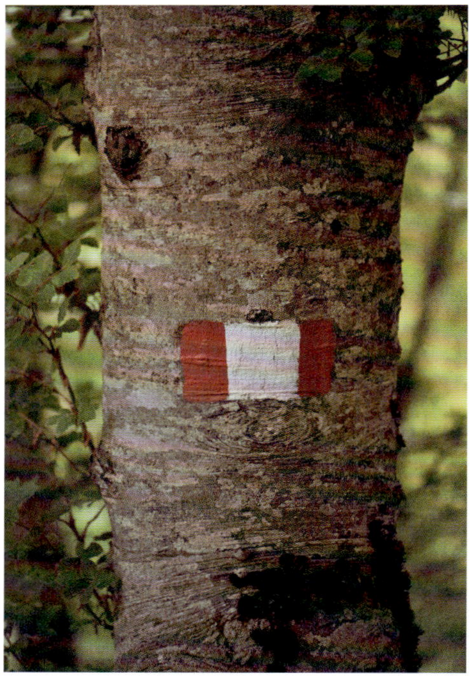
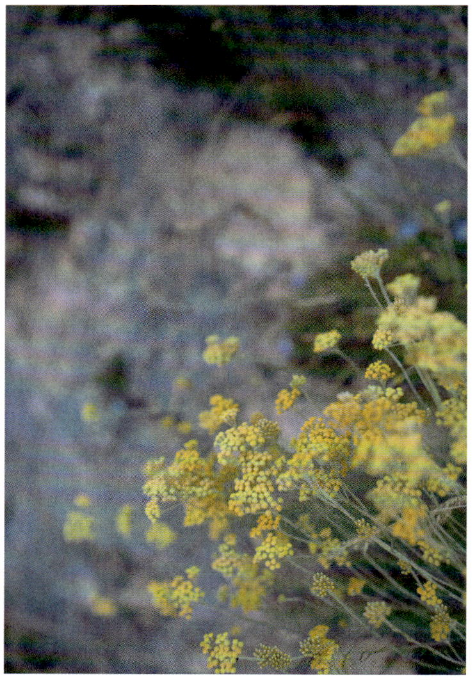
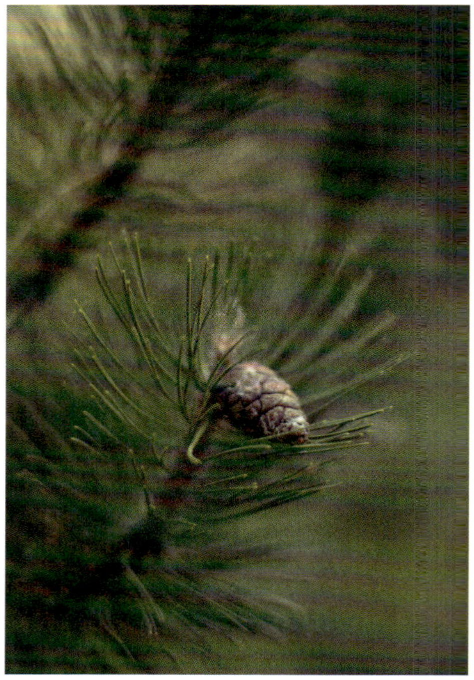

Opposite Goats can digest a wide range of herbs and produce a rich, high-quality milk that is safe to drink for people who are lactose intolerant.

Previous spread Elsewhere in Europe, wolves and bears were eradicated because of the threat they posed to farmers. Not so in Abruzzo, where an ancient alliance between shepherds and their sheepdogs ensures the safety of the flocks, while allowing these majestic predators to continue roaming in silent mountains and forests.
A carefully chosen manzir (castrated ram) is tasked with leading the flock.

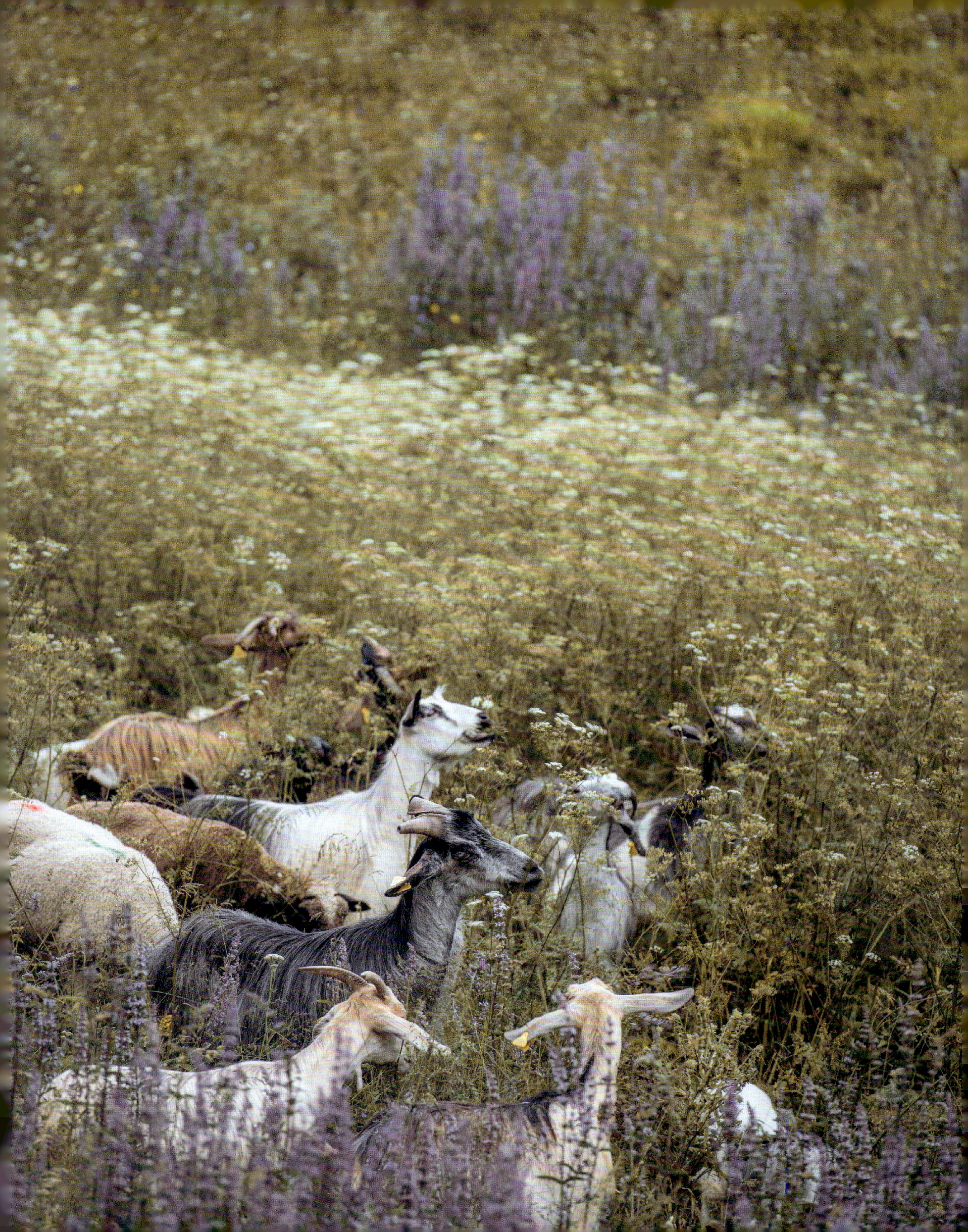

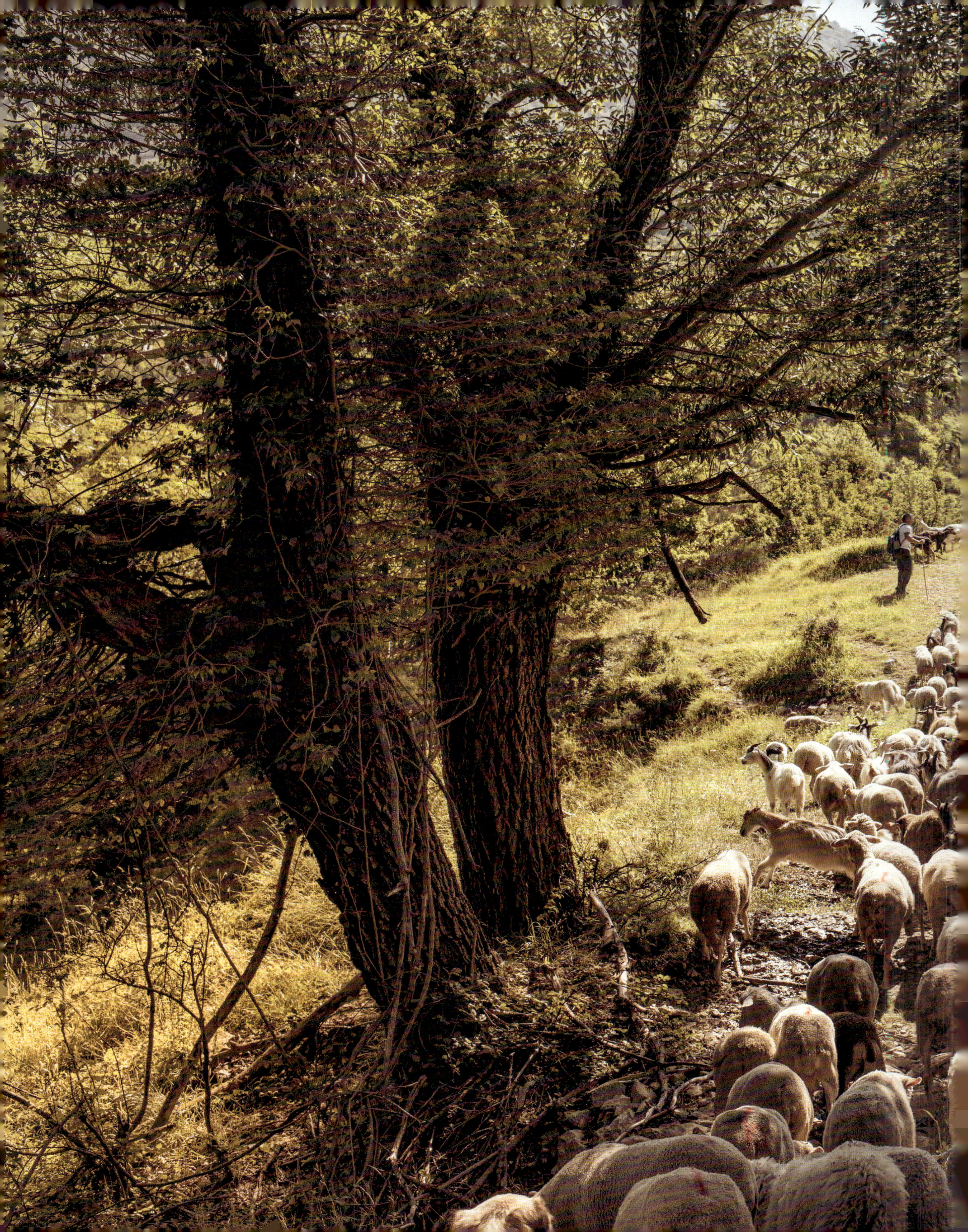

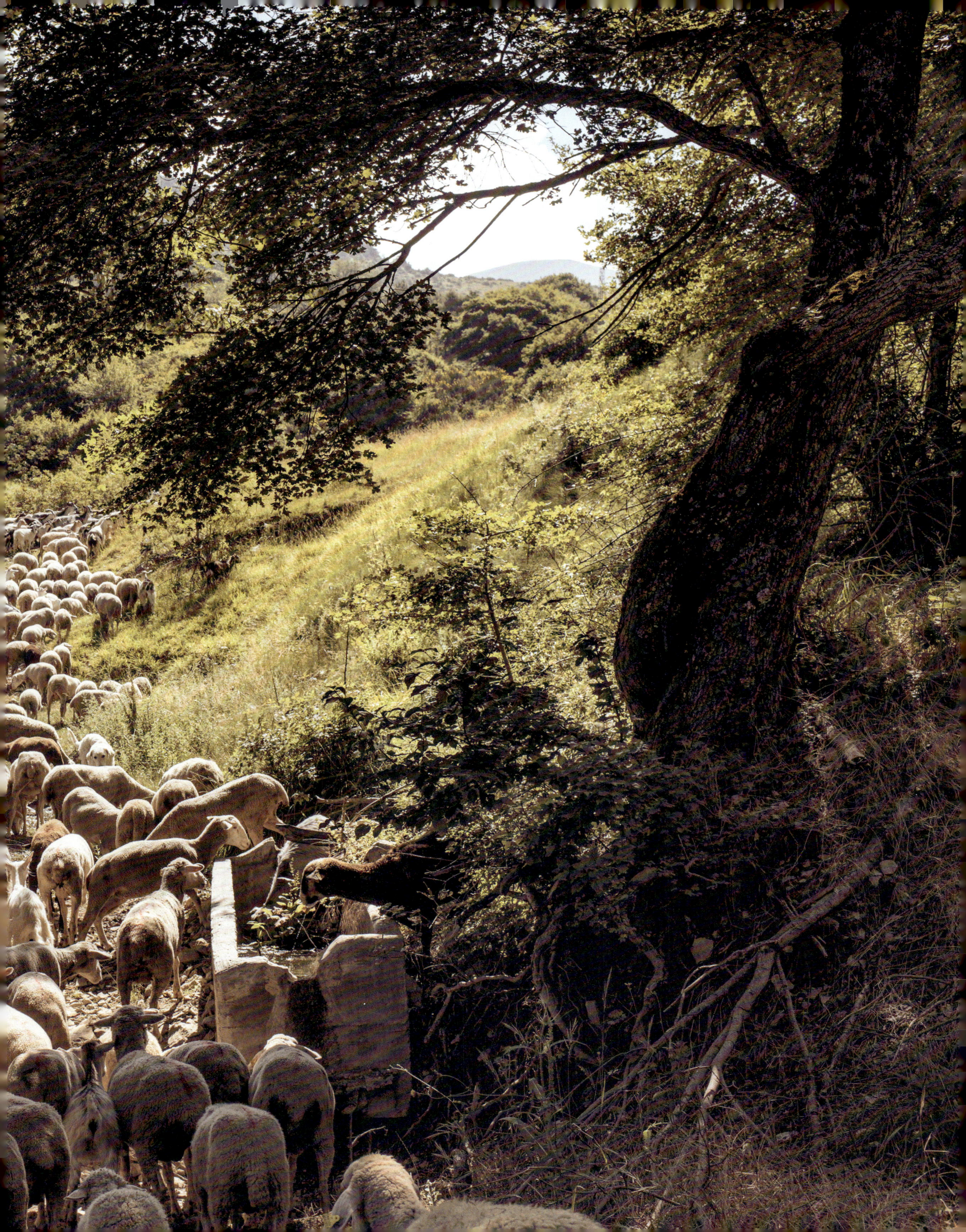

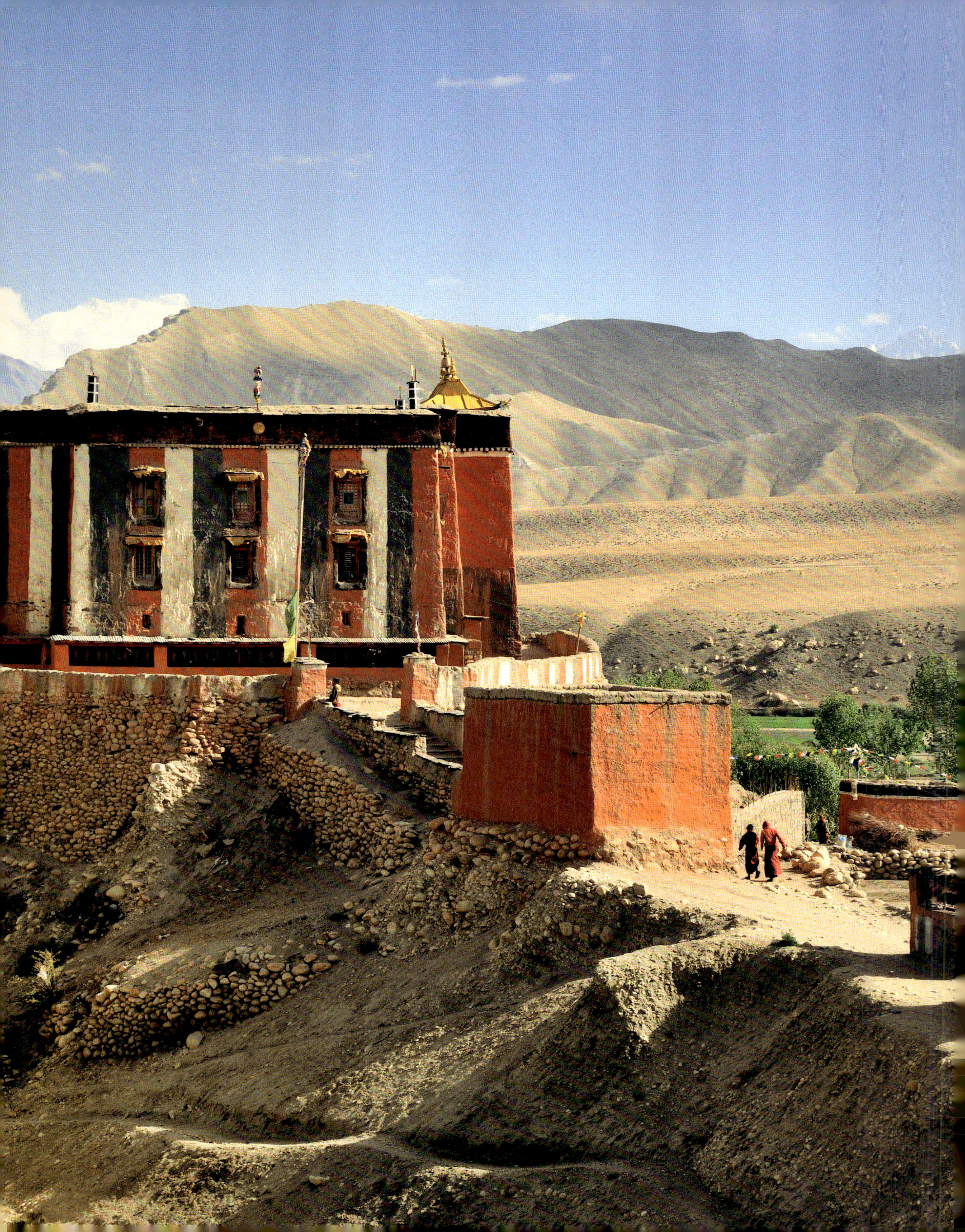

CHAPTER 7

IMMERSIVE TRAVEL THROUGH THE LOST KINGDOM OF LO

A HIMALAYAN SANCTUARY

Nepal

29°05'32.73" N
83°56'01.59" E

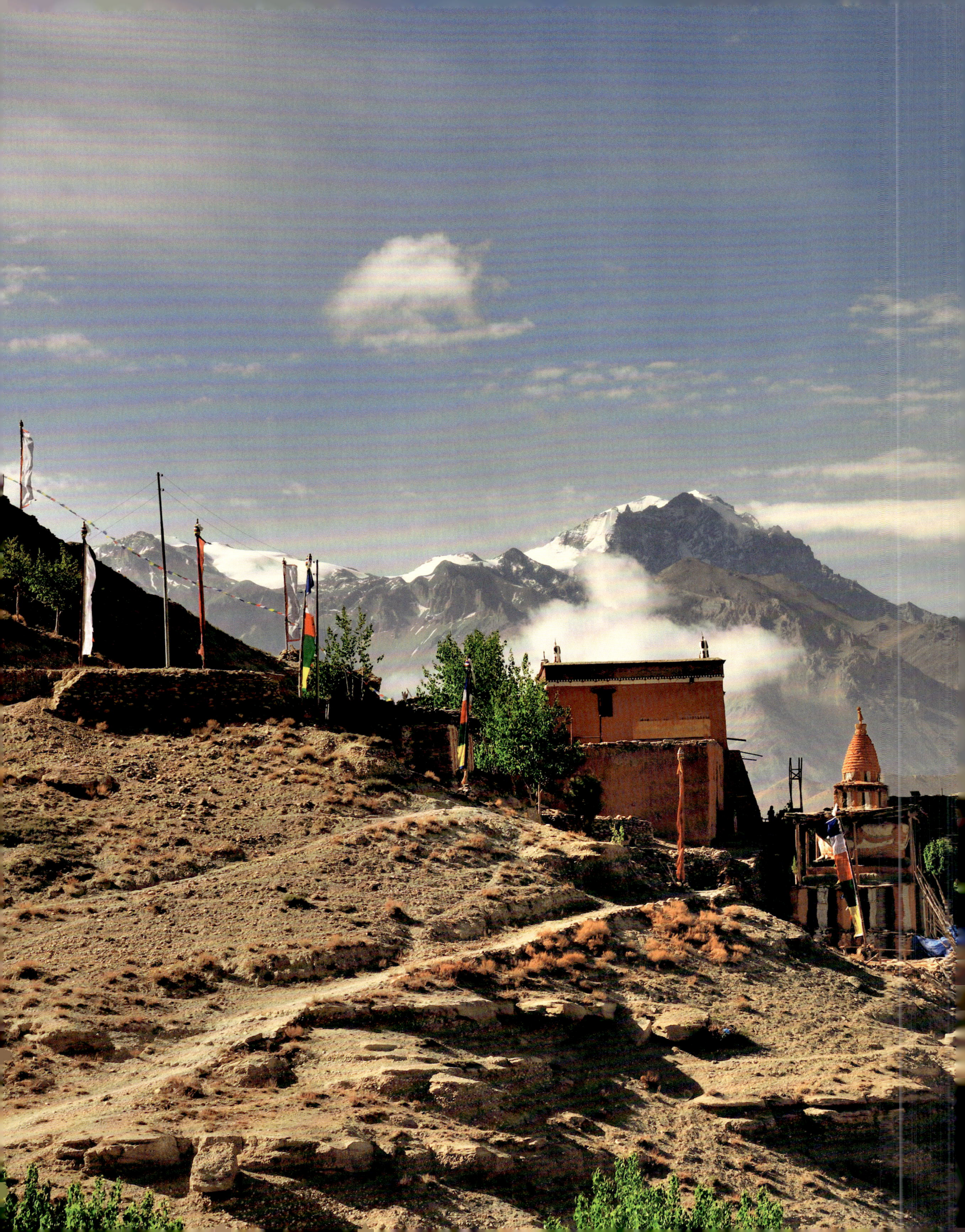

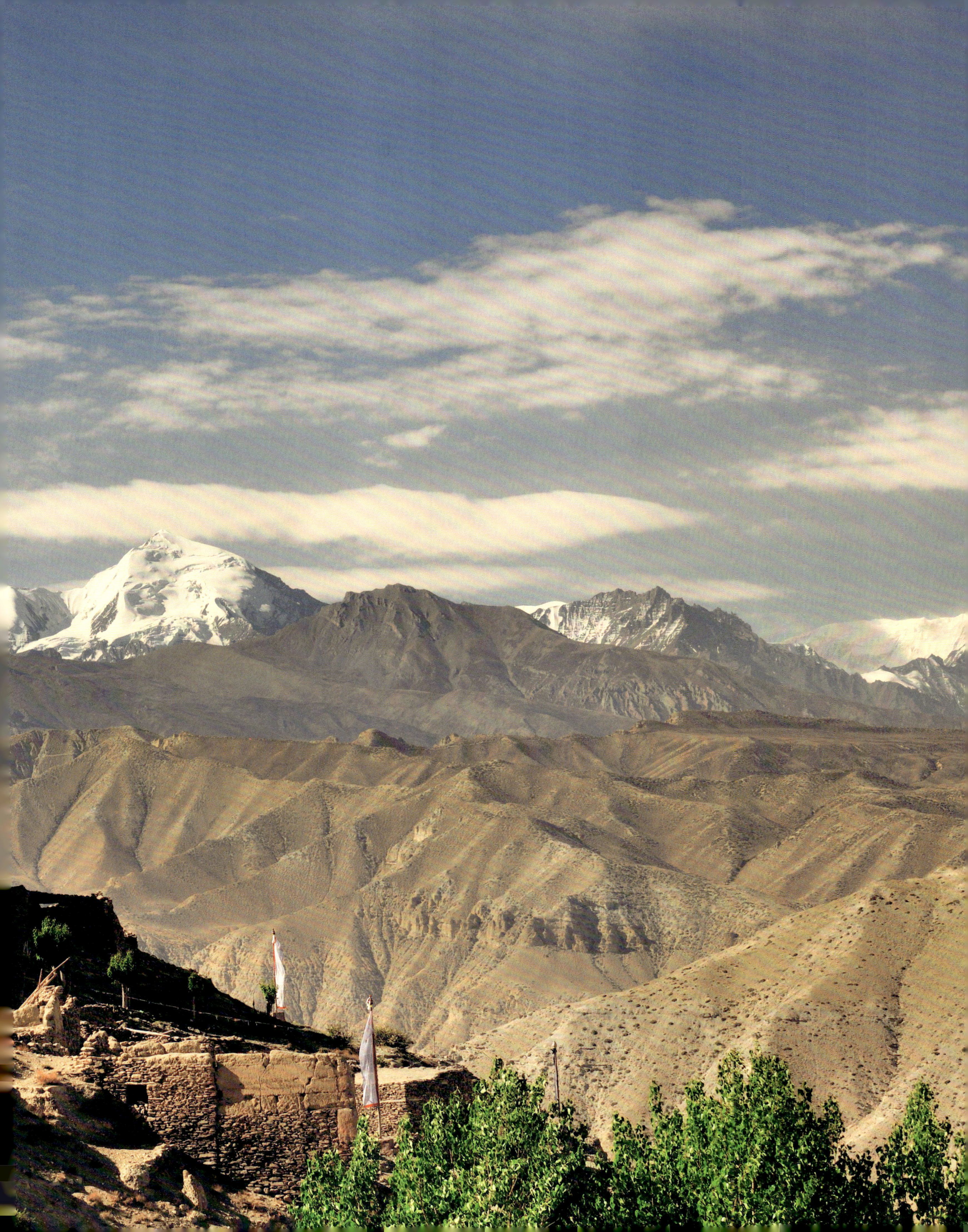

Immersive Travel Through the Lost Kingdom of Lo

A Himalayan sanctuary

Country:	**Nepal**
Best time to travel:	**May, June, September, and October**
Closest airport:	**Jomsom**
Price tag:	**€€**
Style of travel:	**Intrepid and bareboned**

Introduction

Upper Mustang is a spectacle of pristine and picturesque landscapes where rivers swirl around mighty Himalayan peaks and life in chalk-white villages moves slowly. Don't expect to meet other travellers. This isolated region of northern Nepal, which includes the former Kingdom of Lo, is culturally and linguistically tied to neighbouring Tibet. With little influence from beyond its borders almost every local still speaks the traditional Tibetic languages. Until 1992, Upper Mustang was a demilitarised area, off limits to outsiders. Hidden behind 4,000-metre- (13,120 ft.) mountain passes and peppered with colourful stupas, it remains one of Asia's best-kept secrets. As the crow flies, Upper Mustang is only a few dozen kilometres (couple dozen miles) north of Nepal's immensely popular Annapurna circuit, yet it receives only a fraction of its visitors. To trek here is considered a rare privilege. This uncharted region borders Tibet but lies a world away from this overrun hot spot. The Himalayas, a natural barrier, have preserved it for centuries, and until recently, the territory was only accessible on foot, by horseback, or by helicopter. Only after 2008, with the fall of Jigme Dorje Palbar Bista, the unofficial King of Mustang, was a dirt road built along an ancient salt-trade route to the diminutive capital of Lo Manthang. When the road opened in 2011, the region finally started to draw visitors, and now anyone with an appetite for adventure can explore and stay.

The Views

The gateway to Upper Mustang is the village of Kagbeni, where 10-day access permits can be obtained for a fee. From there, the road runs north to Lo Manthang, where the king once resided in an extraordinary palace built from mud, stone, and wood. Described by UNESCO as a 'medieval earthen walled city', Lo Manthang is on the organisation's 'tentative' list. The road from Kagbeni into Upper Mustang zigzags and spirals along steep slopes and gushing rivers, crossing tiny, Eden-like villages dwarfed by towering, snowcapped mountains. Nights are spent in charming teahouses in villages like Chhusang, where views stretch over endless barley fields and yak butter tea is served to warm up guests after a freezing night. In the days of the Kingdom of Lo, the official border was marked by the village of Ghiling, a gorgeous splash of fluorescent green surrounded by a dusty, bone-dry landscape with majestic Tibetan peaks on the horizon. Here, crystalline streams flow through an ingenious network of channels built to irrigate tall grasses, barley, greenish-yellow groves of birch and fruit trees. Horses drink from streams, goats wander freely, and yaks laze in the sun. Nearby, the cliffs above the Gandaki River at the village of Chele are dotted with cave dwellings dating back to the ninth century BCE, some hovering 50 metres (165 ft.) above the riverbank. Thirty kilometres (twenty some miles) north, the village of Ghemi is nestled in a fertile valley where horses graze next to a river. The Royal Mustang Guesthouse in Ghemi is owned by Raju Bista, a nephew of the former king, who now runs it with his wife, Tari. It's the perfect place to rest before venturing into an even more remote area beyond the 4,010 metre (13,160 ft.) Lo Manthang Nyi La Pass — the highest point of the trip. The road leads through an alien landscape of colourful rock formations ranging in colour from green to luminous yellow to Bordeaux red and fertile green. Topographically, the scale is difficult to fathom. If these valleys belong to the people, the mountains belong to the gods. In monasteries along the way novice monks study ancient scripts in the shadows and elderly locals spin prayer wheels. Ceilings are painted with intricate mandalas, and ceremonial tantric masks adorn the walls.

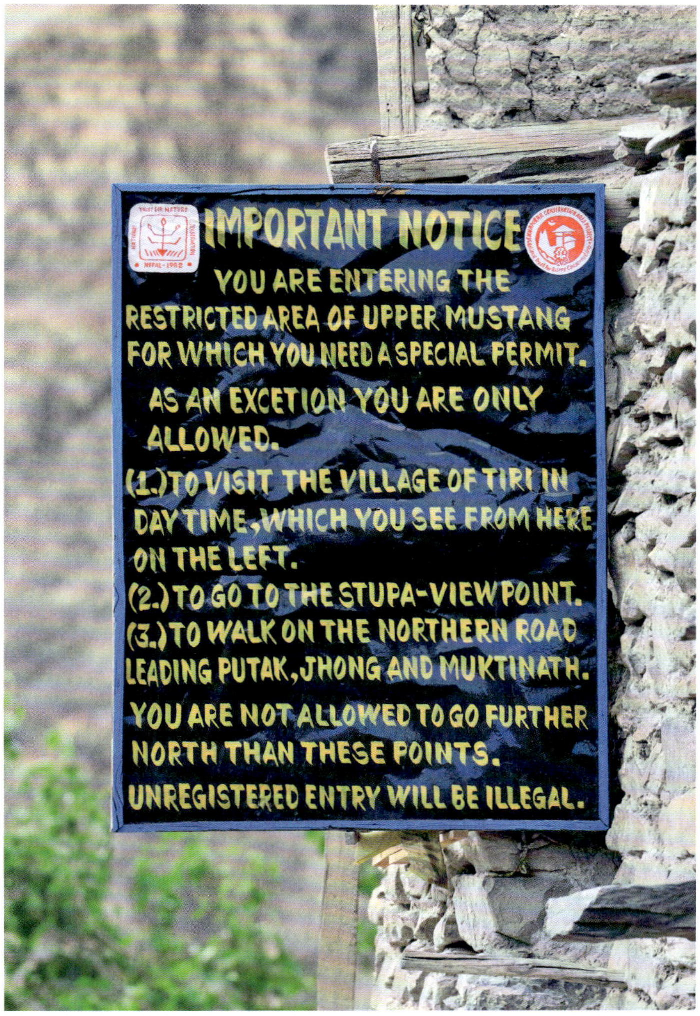
IMPORTANT NOTICE
YOU ARE ENTERING THE
RESTRICTED AREA OF UPPER MUSTANG
FOR WHICH YOU NEED A SPECIAL PERMIT.
AS AN EXCETION YOU ARE ONLY
ALLOWED.
(1.) TO VISIT THE VILLAGE OF TIRI IN
DAY TIME, WHICH YOU SEE FROM HERE
ON THE LEFT.
(2.) TO GO TO THE STUPA-VIEW POINT.
(3.) TO WALK ON THE NORTHERN ROAD
LEADING PUTAK, JHONG AND MUKTINATH.
YOU ARE NOT ALLOWED TO GO FURTHER
NORTH THAN THESE POINTS.
UNREGISTERED ENTRY WILL BE ILLEGAL.

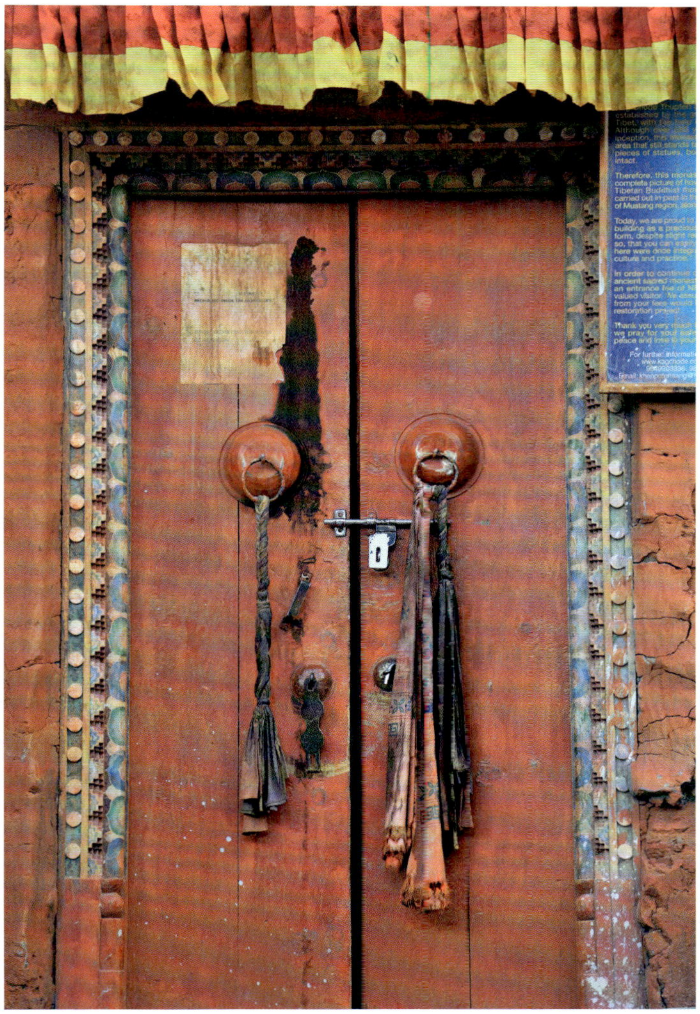

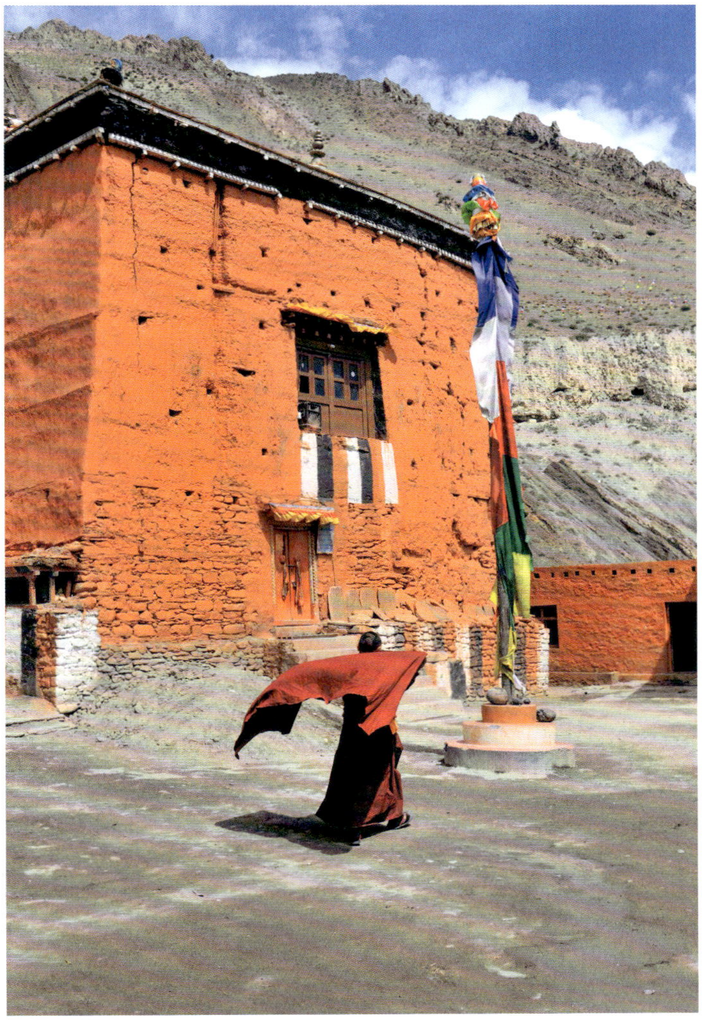

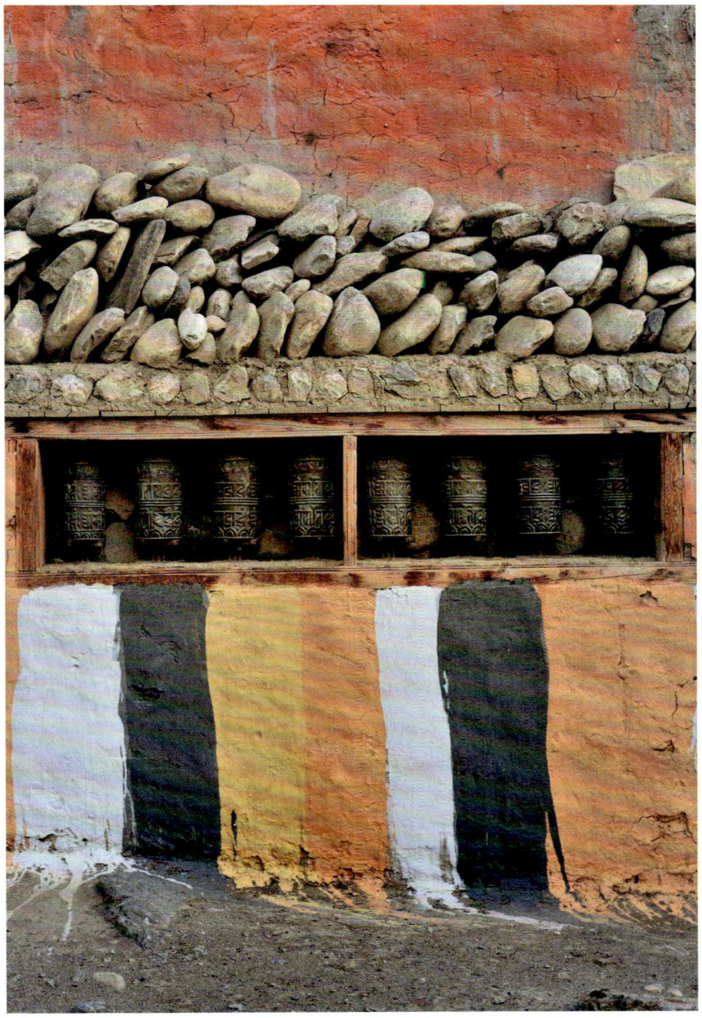

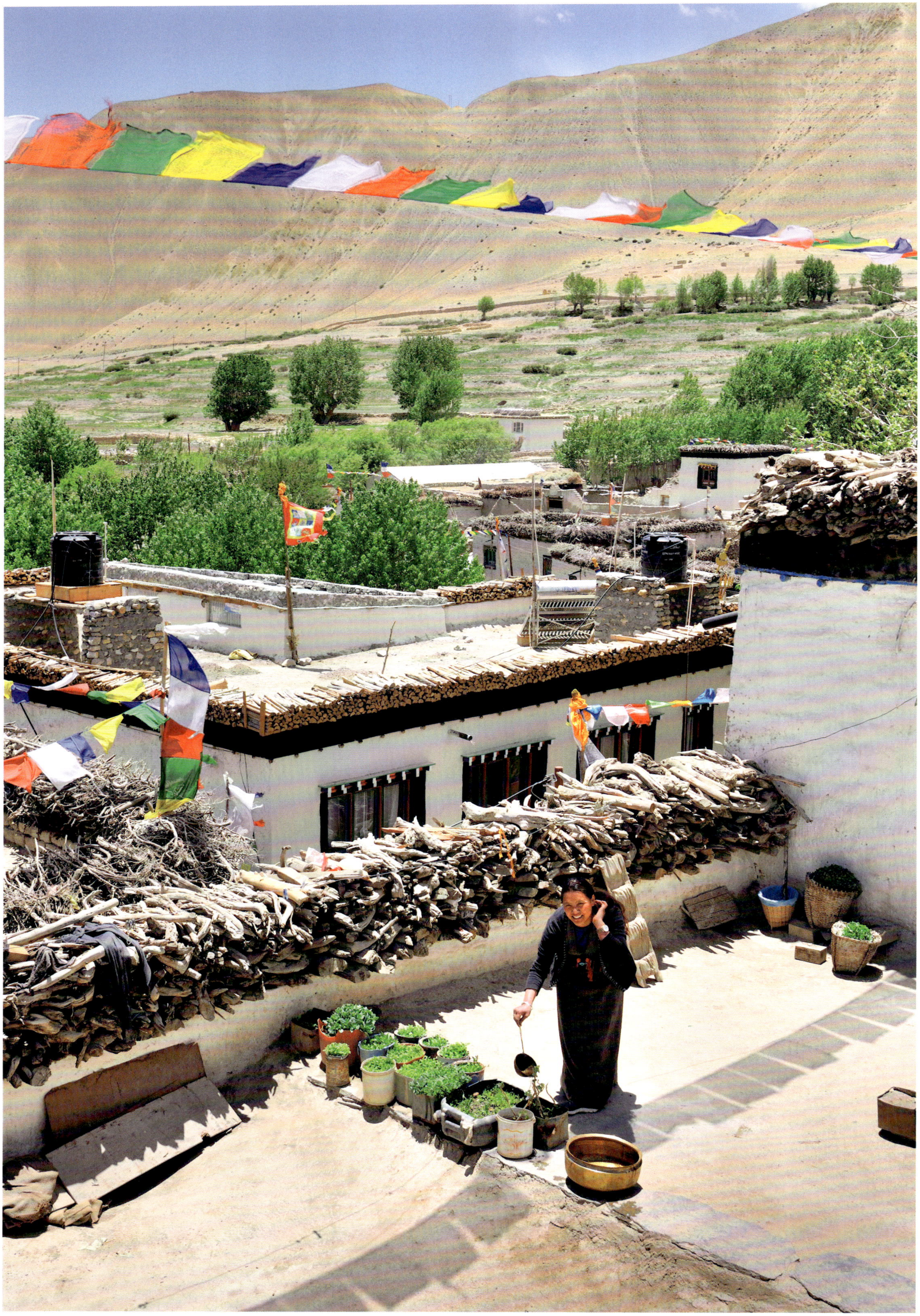

Opposite From the roof of the Royal Mustang Guesthouse in Ghemi, a pretty village of around 300 inhabitants, spectacular views stretch towards red and bluish mountains. The Royal Mustang Guesthouse is owned by the nephew of the former King of Mustang. The tiny hotel is hidden in two buildings, including an old family house with seven guest rooms and three large dining rooms, and a modern annexe boasting 10 elegantly decorated en-suite rooms with solar-powered showers. The guesthouse is the perfect base to feel how slow and peaceful life is in Ghemi and Upper Mustang in general.

Monks create mandalas of coloured sand with exacting care, only to dismantle them, symbolising the ephemeral nature of all things. Afterwards, the sand is scattered about the village to summon a fruitful harvest, offering insight into the ruminative rhythm and devotional nature of a Buddhist monastery. In villages lost in a towering landscape of snowcapped mountains, religion plays a central role. As it does in the region's capital a few hours away, at the end of the road. Lo Manthang has to be one of the most surreal cities on earth, dusty like the set of a western, tranquil and sacred like a monastery. Set on an immense mountain plateau and ringed by fields of buckwheat, it is a walled ensemble of whitewashed mud-brick houses and cobbled streets. The tallest structure here is the five-storey, 15th-century palace of the former king. Two of the city's four gompas (fortifications) are of a similar vintage. Here it is evident that Upper Mustang is one of the last vestiges of authentic Tibetan Buddhist culture, inextricably combined with pagan ceremonies called Bon-po. The city walls mark the edges of this compact city both physically and symbolically. The changri, as the walls are called, were made of pebbles, mortar, and red clay to defend the citizens and their cattle against invaders and winter winds. Today, for Lo Manthang's 2,000 inhabitants, the walls still function as a boundary between the sacred city and an outside world of demons and things unknown.

Creature Comforts

If you love bedding down in charming if basic teahouses, Upper Mustang is your place. Take the comfortable Royal Mustang Guesthouse in Ghemi, which has two beautiful buildings: the centuries-old teahouse that belongs to the Bista Royal family of Mustang, a 500-year-old dynasty. Raju Bista and his wife, Tari, also built a newer guesthouse next to the old dwellings. Spacious, cheery new bedrooms are brightened with Tibetan rugs, while en-suite bathrooms offer solar-heated water, and large windows look out onto majestic mountains. After sunset, when the village is silent, residents gather around colourfully painted wooden tables in the hotel's warm kitchen, where an all-female crew serves an evening meal and cold Tibetan beer. Don't miss an alfresco breakfast of freshly baked buckwheat pancakes washed down with sea buckthorn juice with views over the Himalayas at the New Asia Trekkers Hotel in Kagbeni. The hotel has simple but adequate en-suite rooms and a pleasant restaurant.

Go Now

Visiting Upper Mustang is like travelling back in time. The first gravel road between the dusty gateway town of Kagbeni and the enigmatic capital of Lo Manthang opened only in 2011, paving the way for intrepid visitors. Still, tourists must follow a 15th-century trade route on foot into what is perhaps the best-preserved example of traditional Tibetan life. Along with the new road, modernity has seeped in. China is eyeing Upper Mustang because of its natural resources and strategic location between Tibet and Kathmandu. Go now before the chirping of phones interrupts the monks' chants.

How to Go

From Kathmandu, the flight to the lakeside city of Pokhara takes 25 minutes, followed by another short flight to the small town of Jomsom in Mustang. Be mindful that flights are often delayed or cancelled due to poor visibility in the mountains. A bumpy 12-hour drive in a private 4x4 is an alternative, or a 30-minute heli-ride from Kathmandu to Lo Manthang. Taking a couple of days to acclimatise in beautiful Kagbeni is mandatory to adjust to the altitude. Pack a warm jacket and sleeping bag (there is no heating in the teahouses or inns) and dustproof bags for your gear. Don't forget about altitude sickness and prepare for intense sunlight during the day, followed by chilly nights.

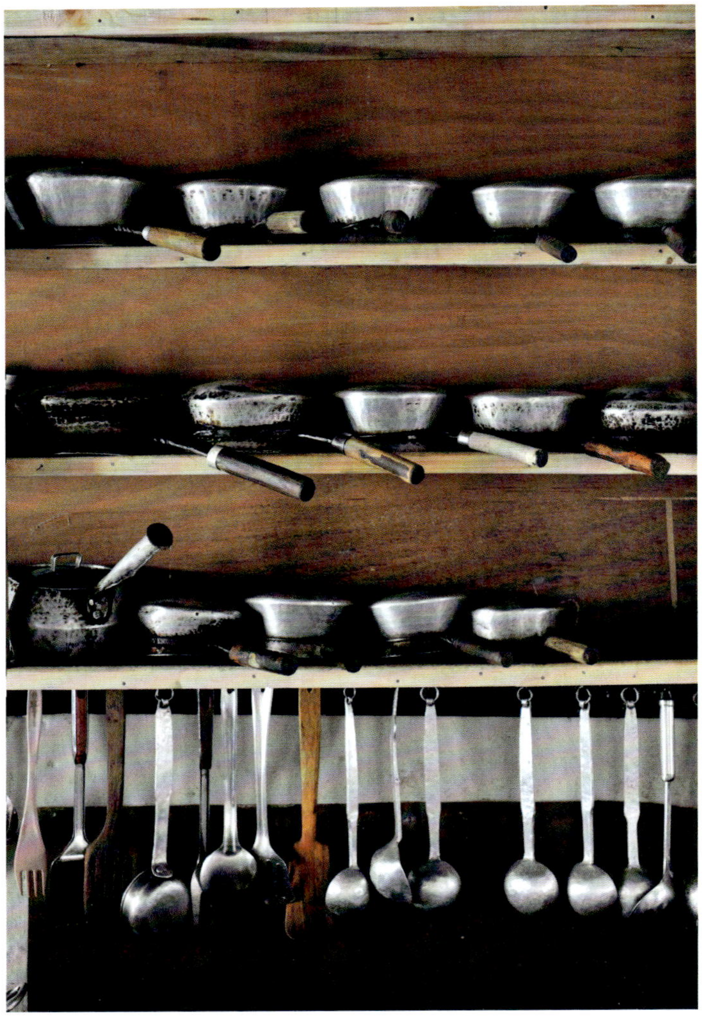

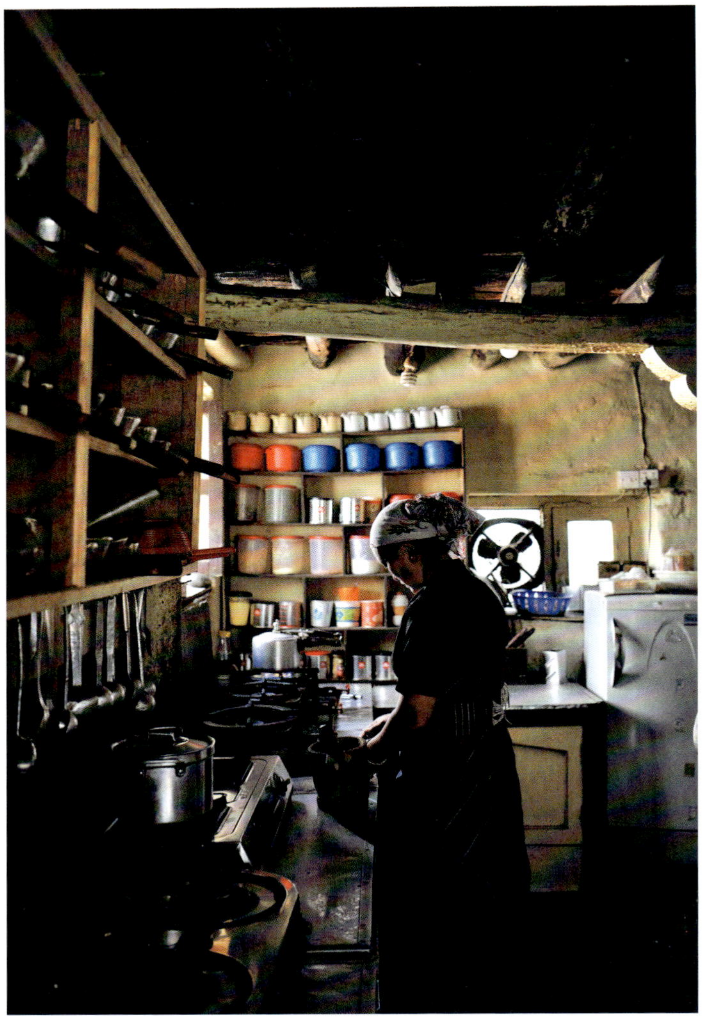

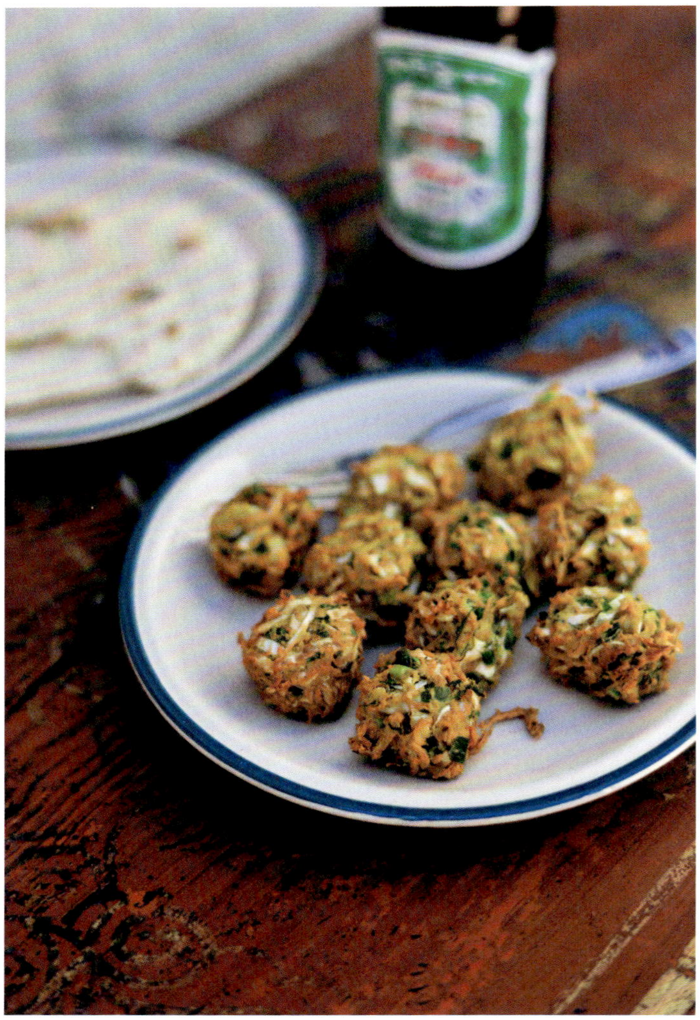

This spread In the kitchens, meals are prepared on blazing wood-burning stoves. The cuisine of Upper Mustang is deeply influenced by Tibetan culture: butter tea and buckwheat pancakes for breakfast; dumplings, noodles, yak cheese throughout the day. Guests and family members meet in the warmth of the kitchens to enjoy freshly prepared food along with bottles of Lhasa, a beer brewed in Tibet.

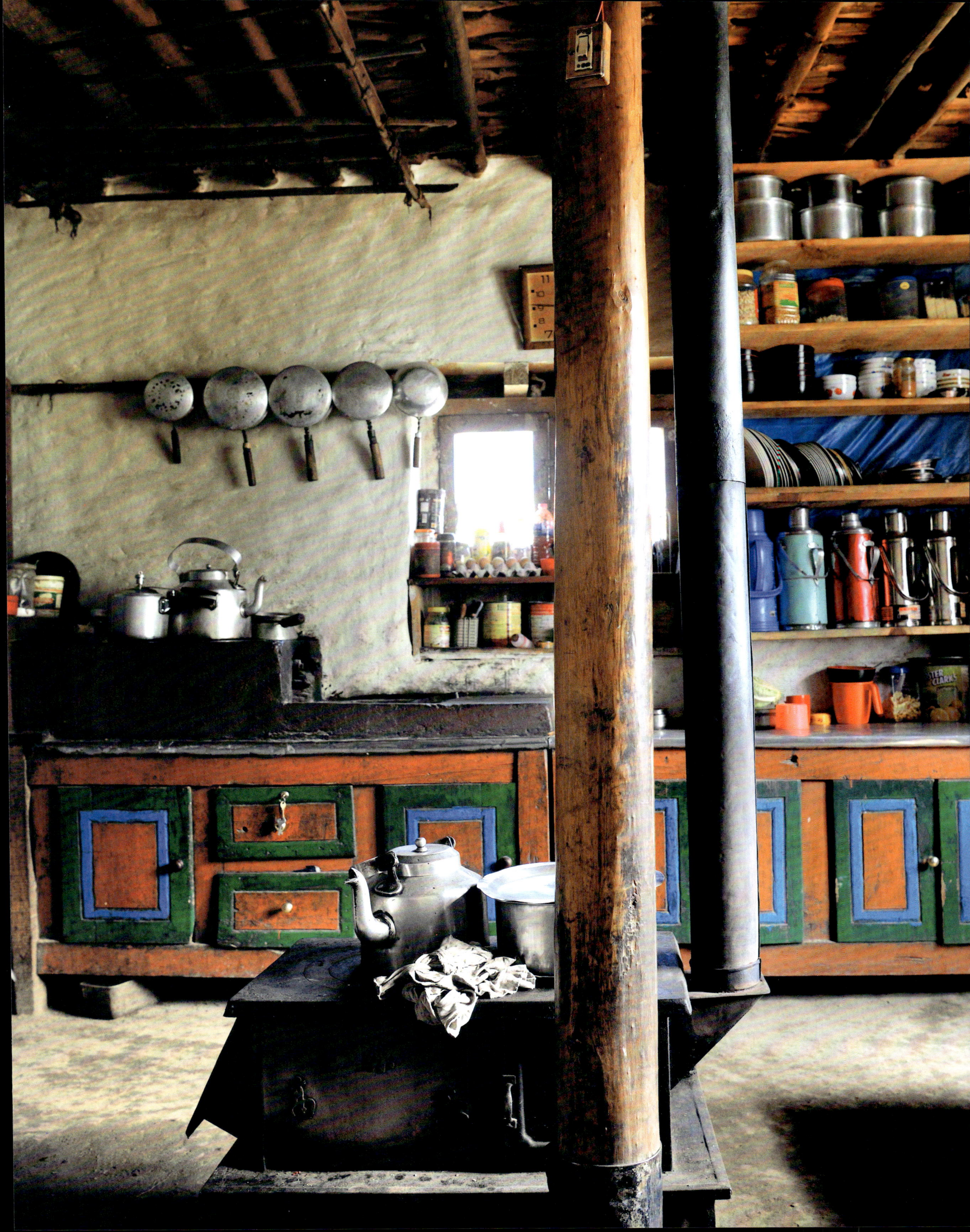

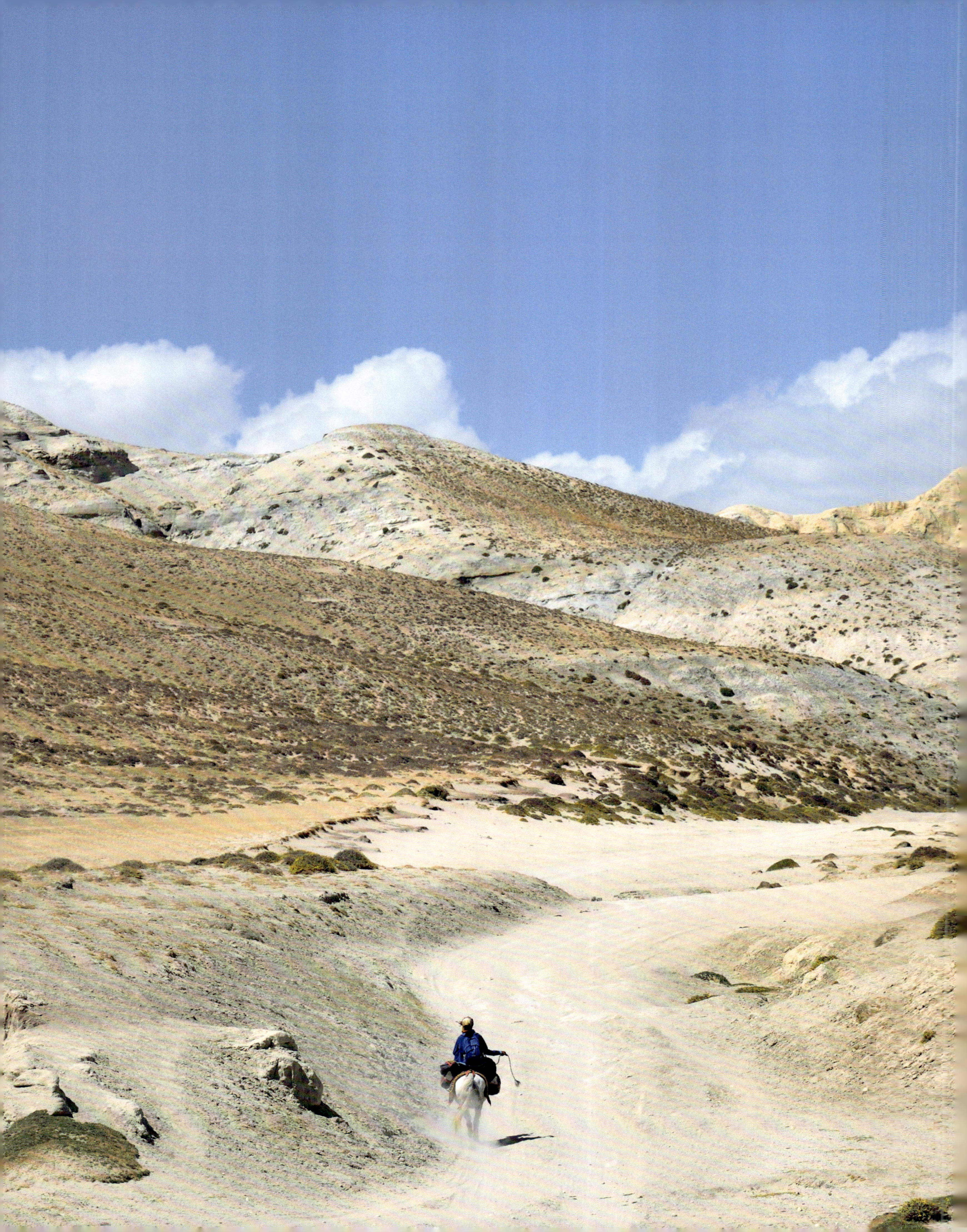

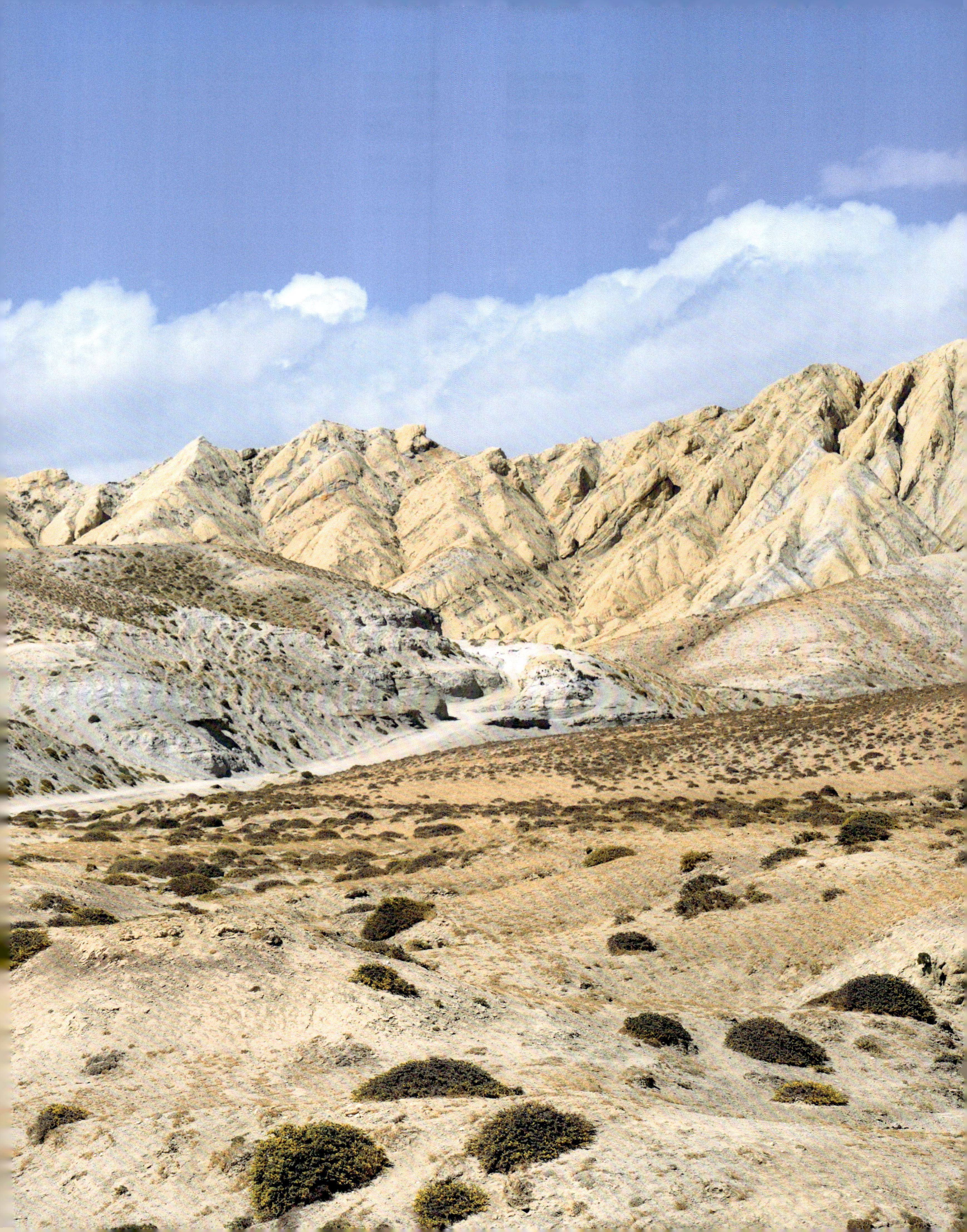

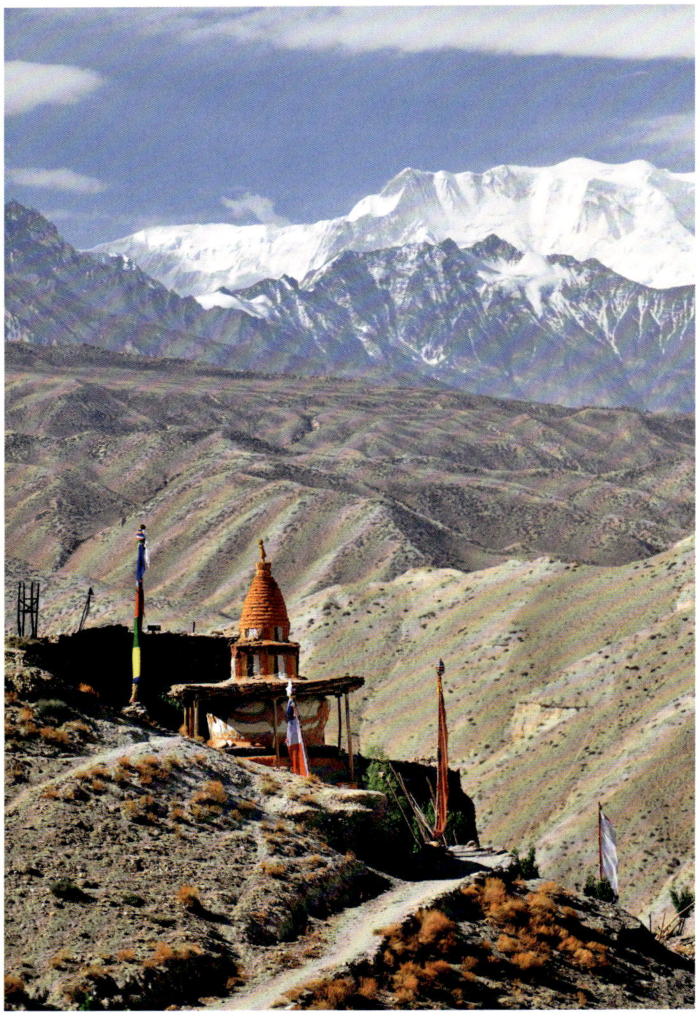

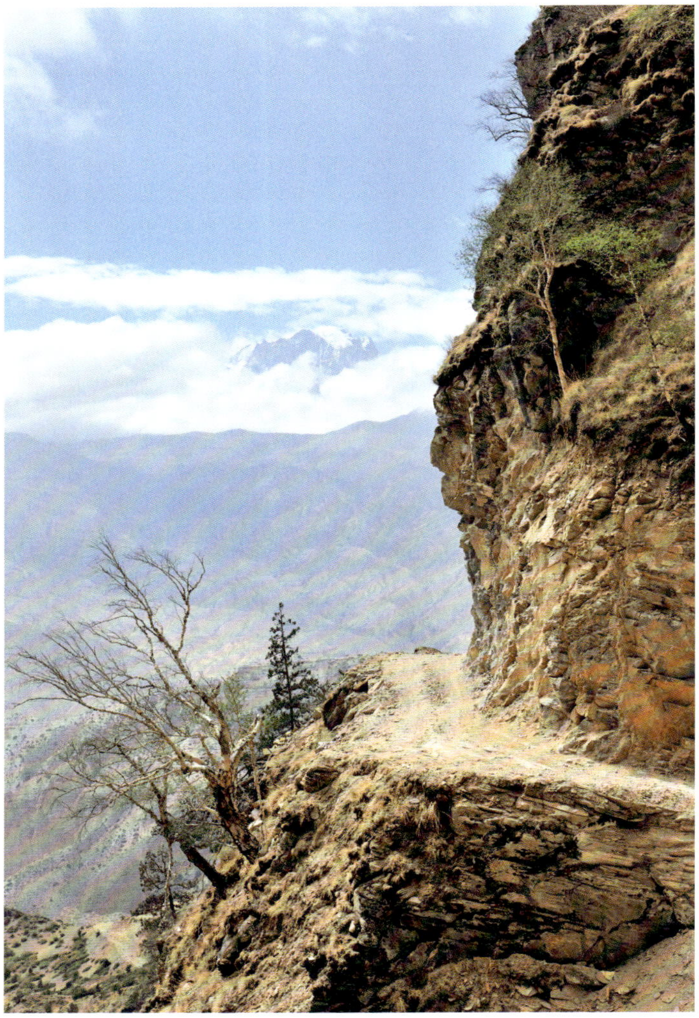

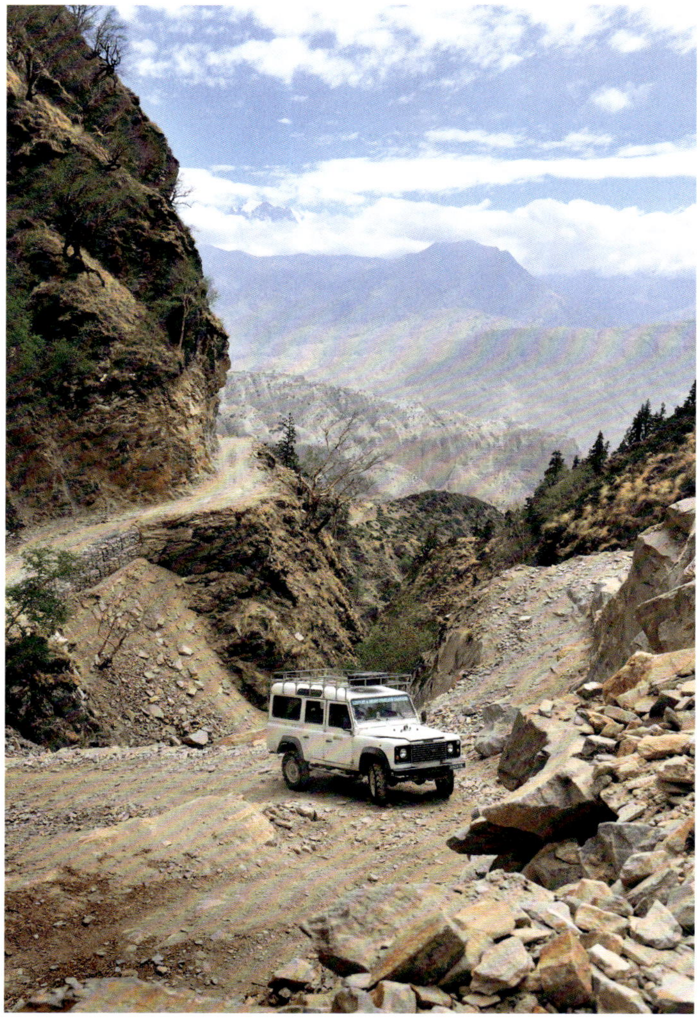

This spread Exploring Upper Mustang by 4x4 requires courage and a head for heights. Until a decade ago, when the first gravel road was constructed, the only way to travel in this part of the Himalayas was by horseback or on foot. The new road remains challenging, winding though high and arid mountain passes going up to 4,000 metres (13,125 ft.), with sharp bends and narrow pathways, descending into deep, lush valleys, and crossing riverbeds filled with beautiful pebbles and fossils.

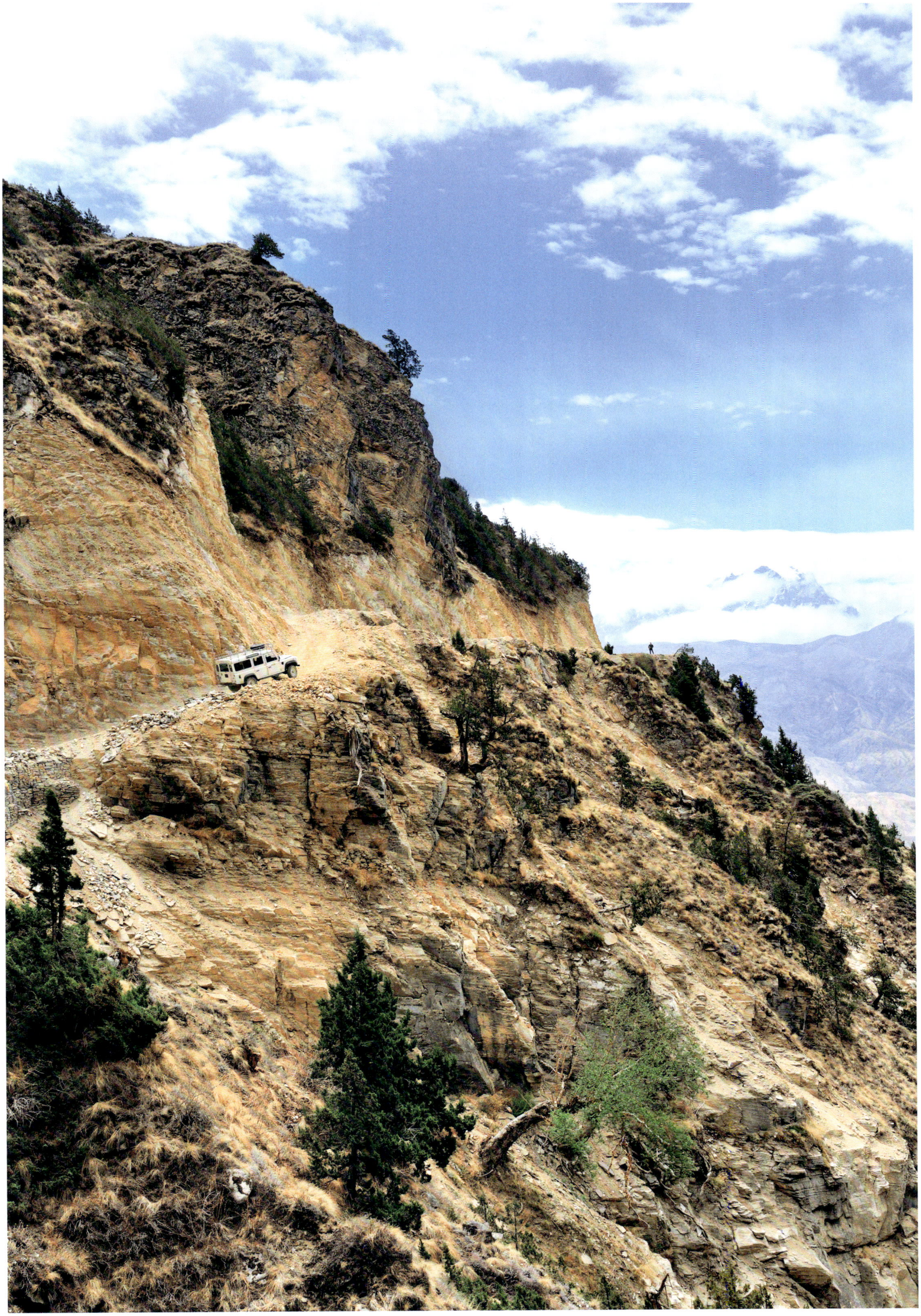

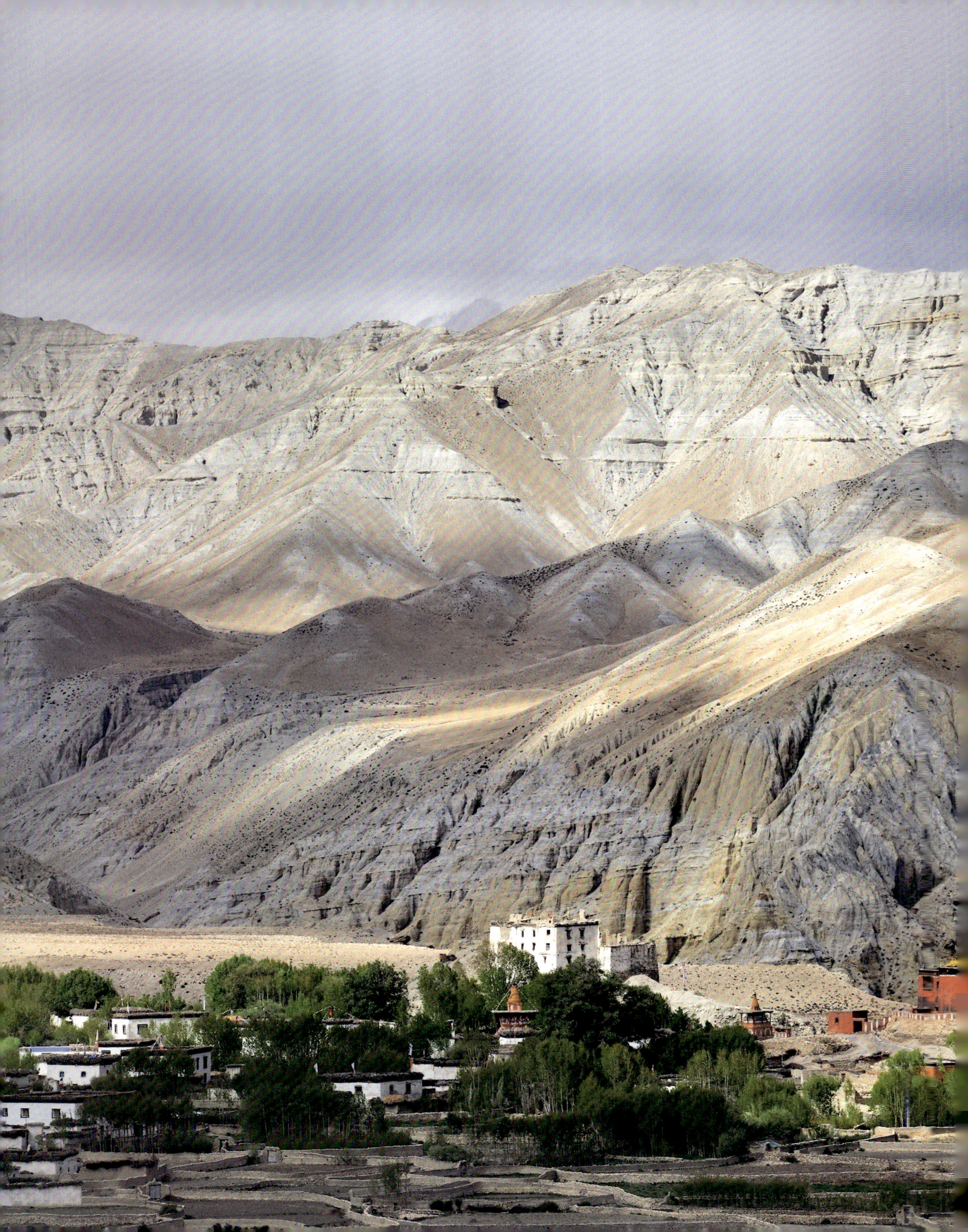

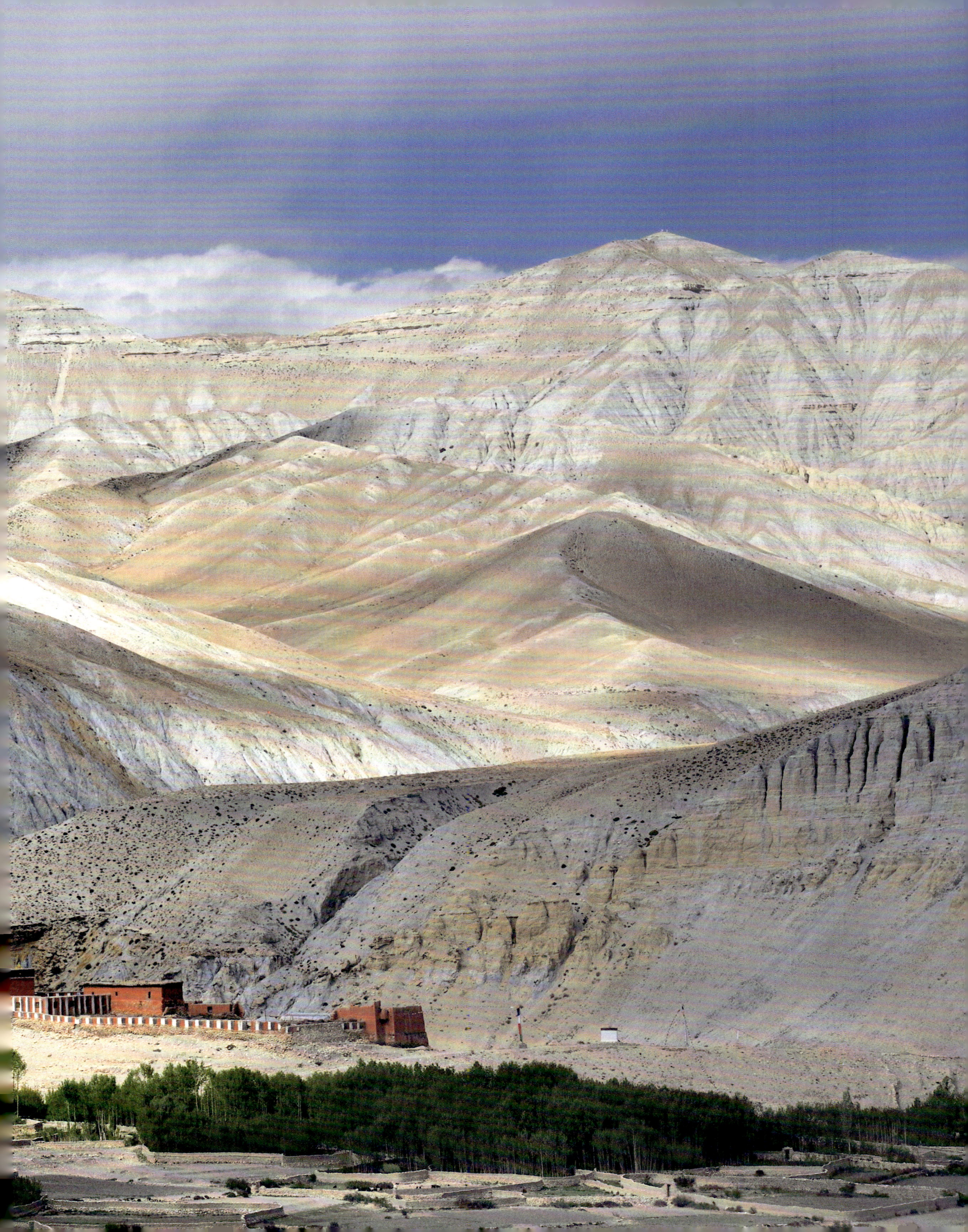

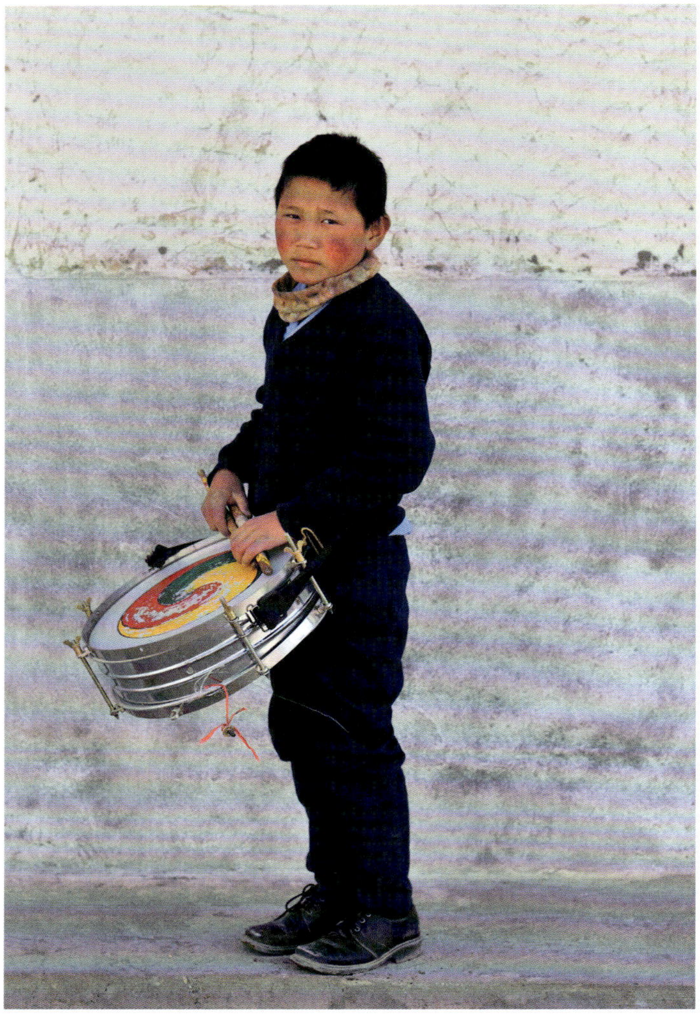

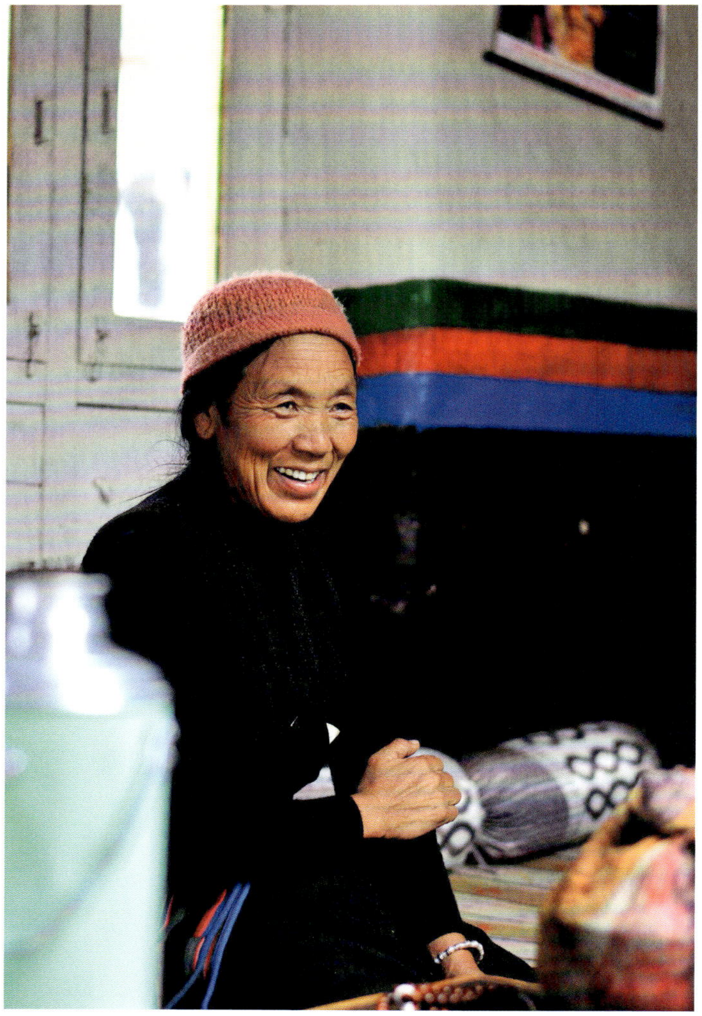

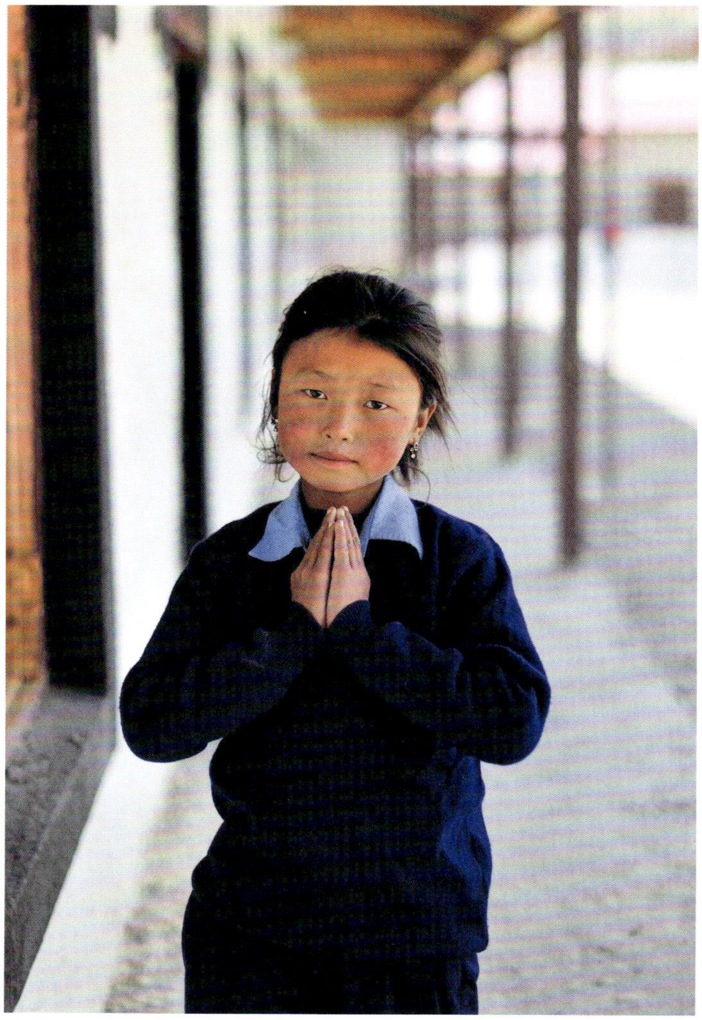

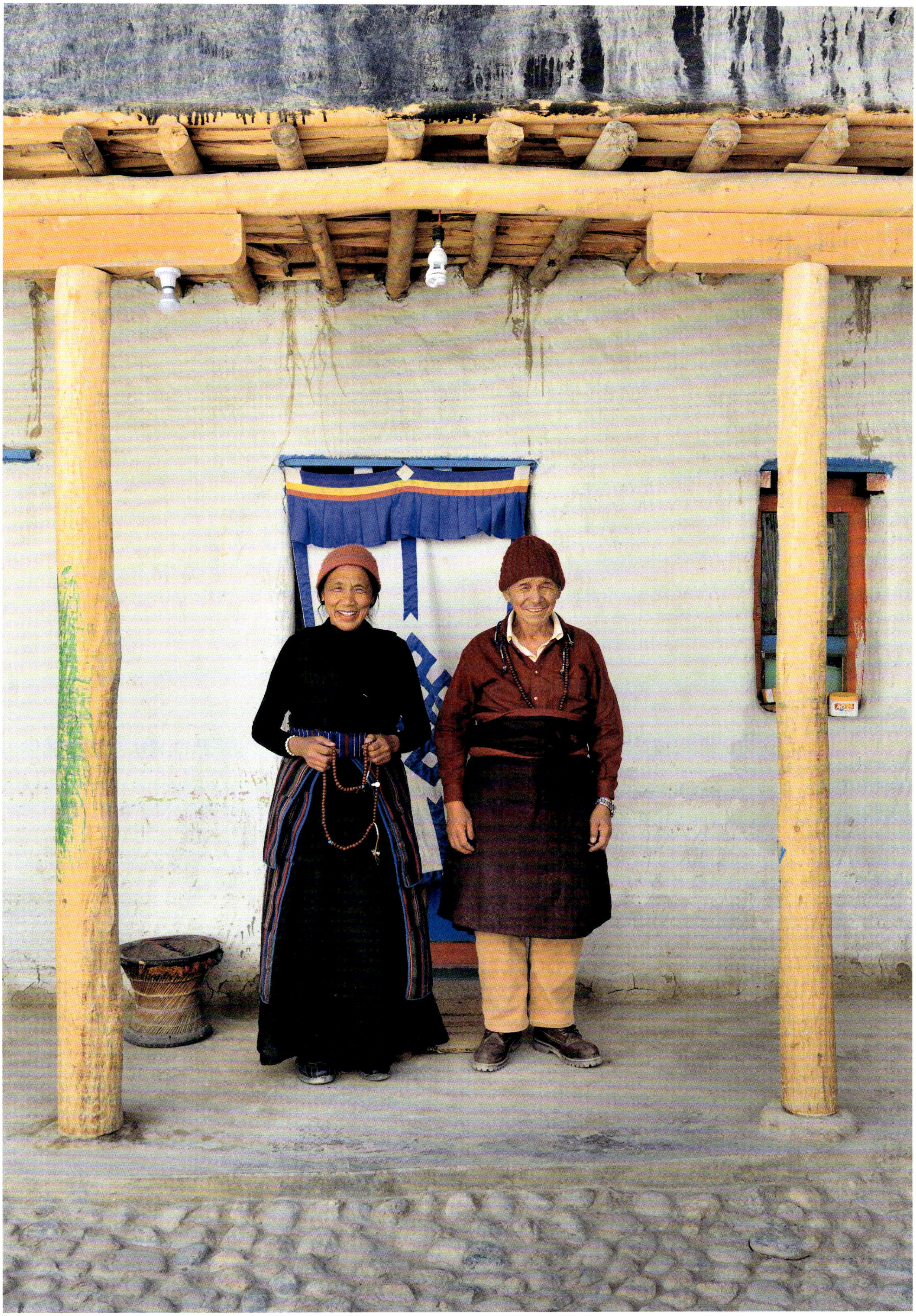

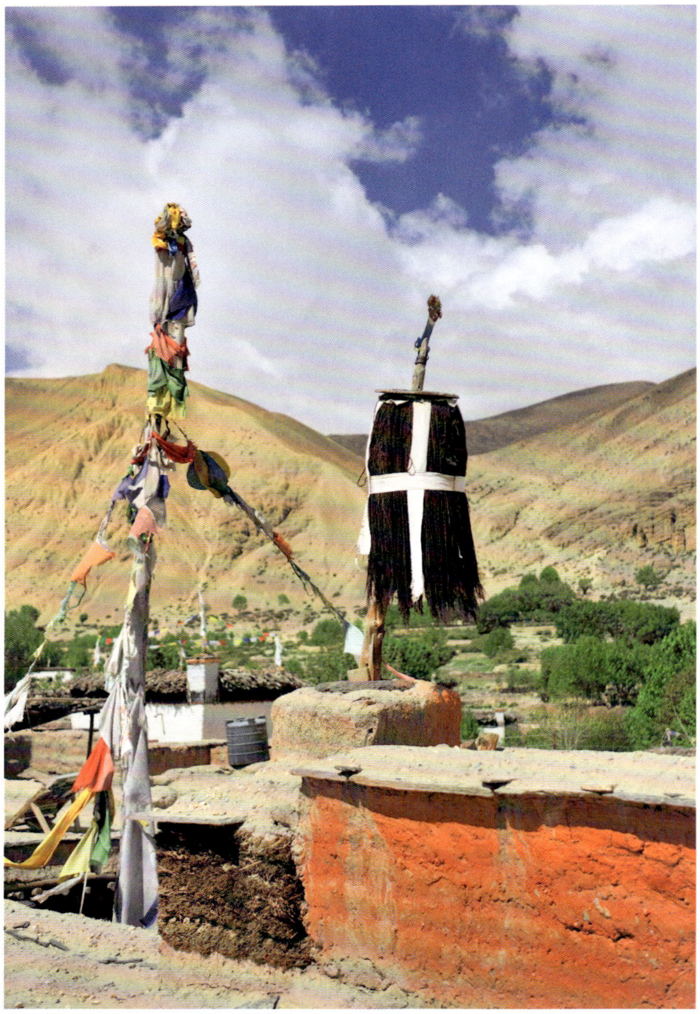
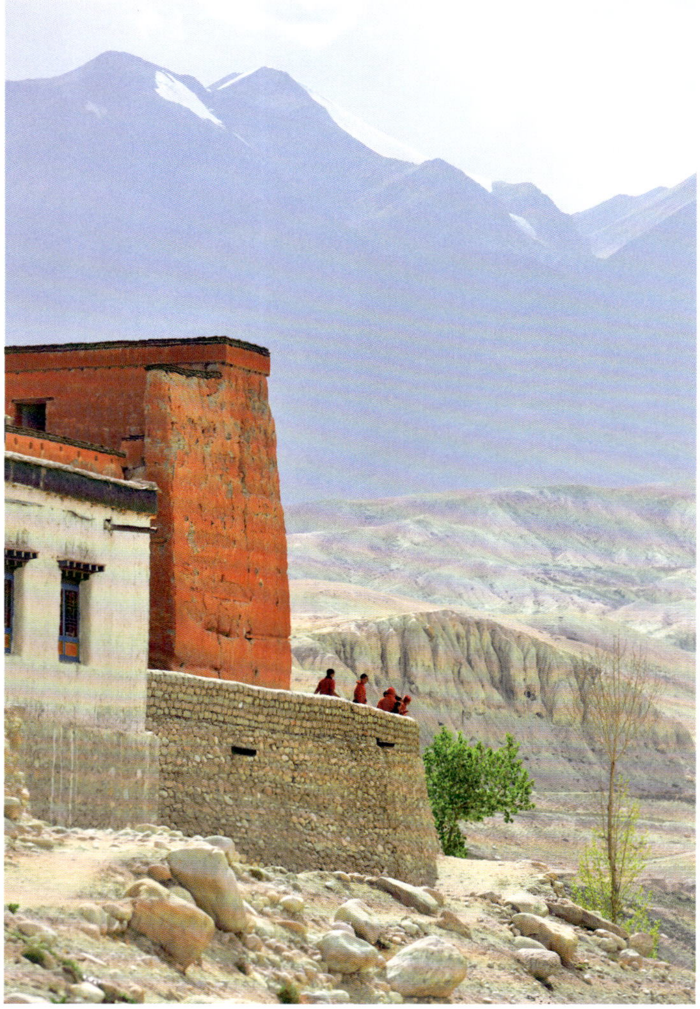
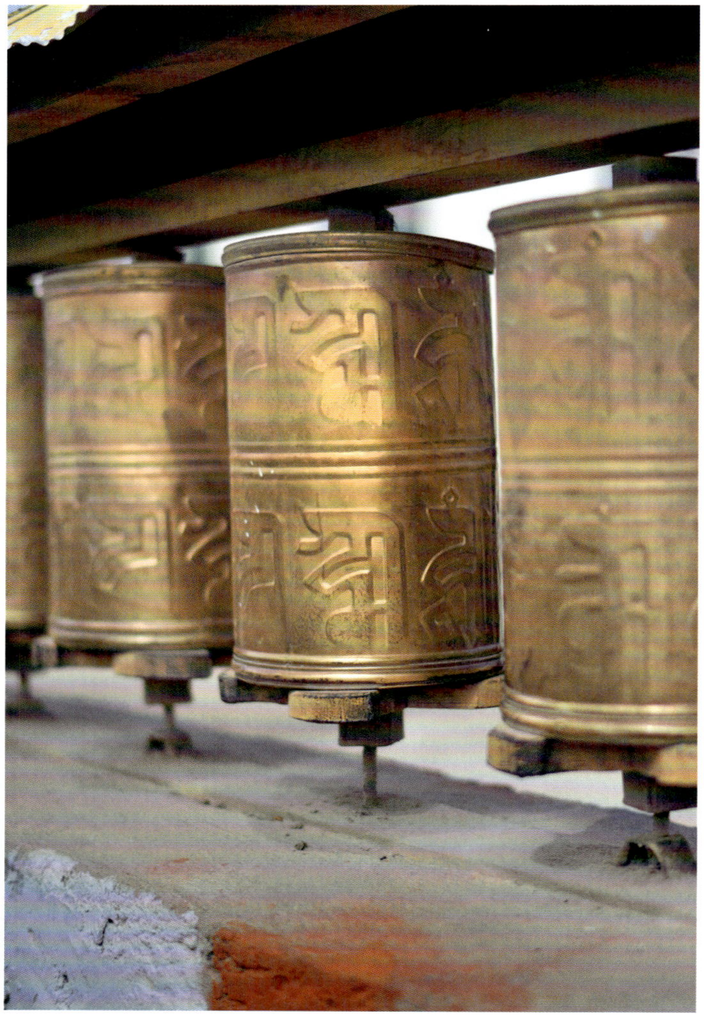
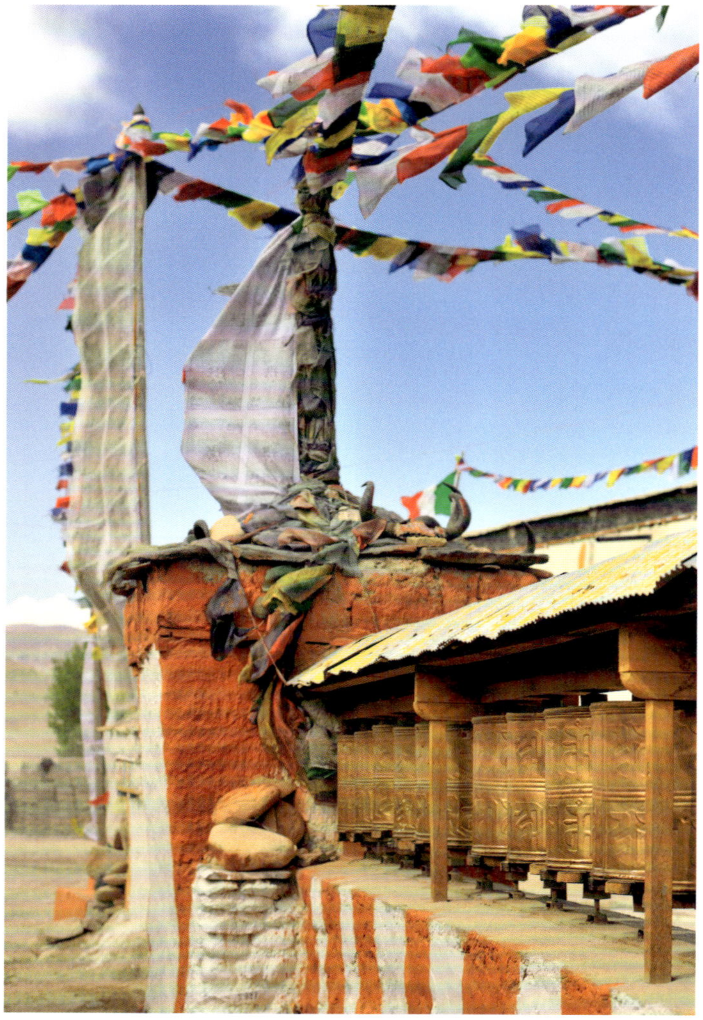

This spread In addition to temples and monasteries, royal palaces known as darbar are a vital part of the regional architecture, reflecting the economic and cultural heyday of the region in the 15th and 16th centuries. Although some were severely damaged by an earthquake in 2015, three darbar in the villages of Ghemi, Dhagmar, and Thingkar are being restored, thanks to a German research and funding project.

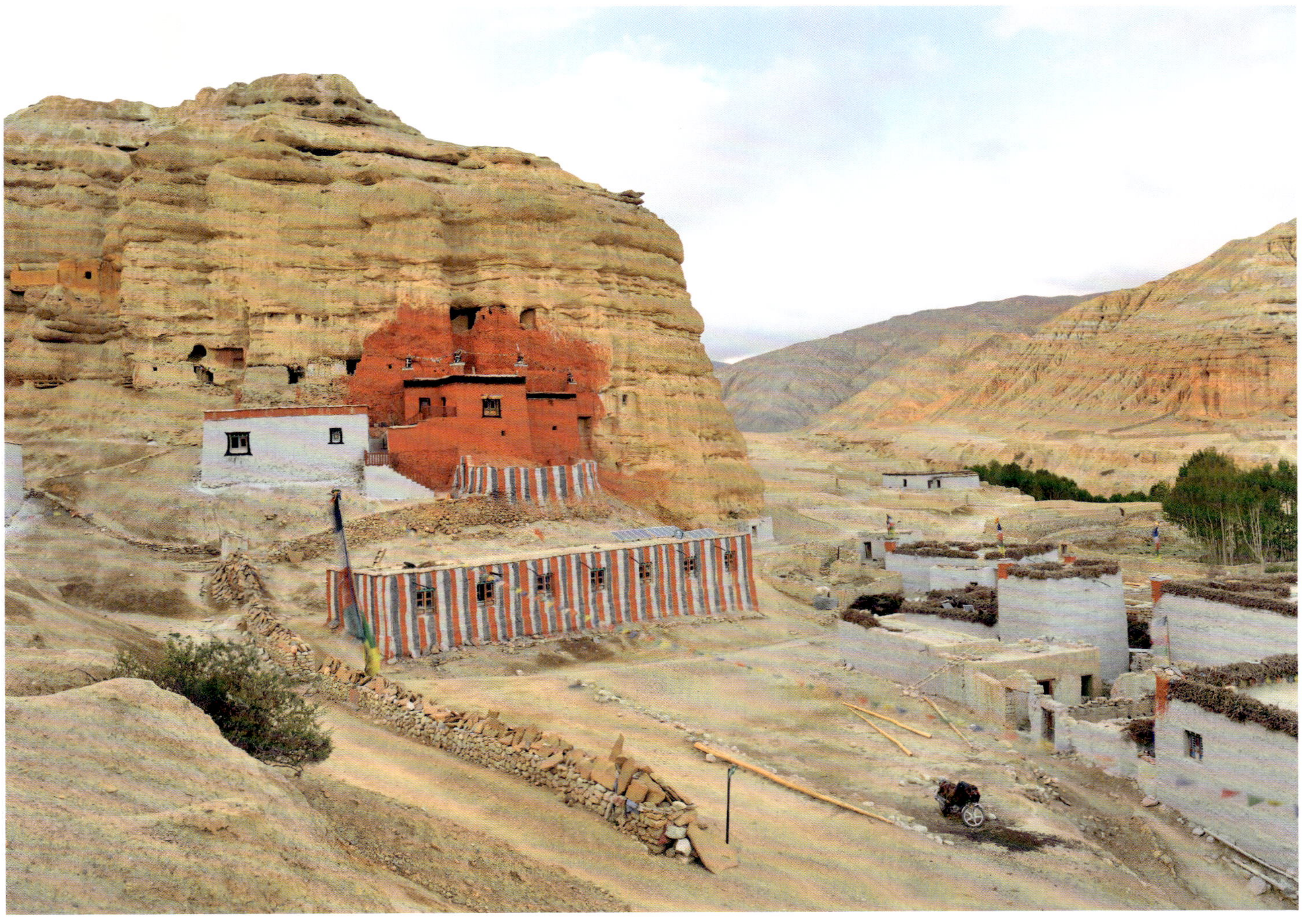

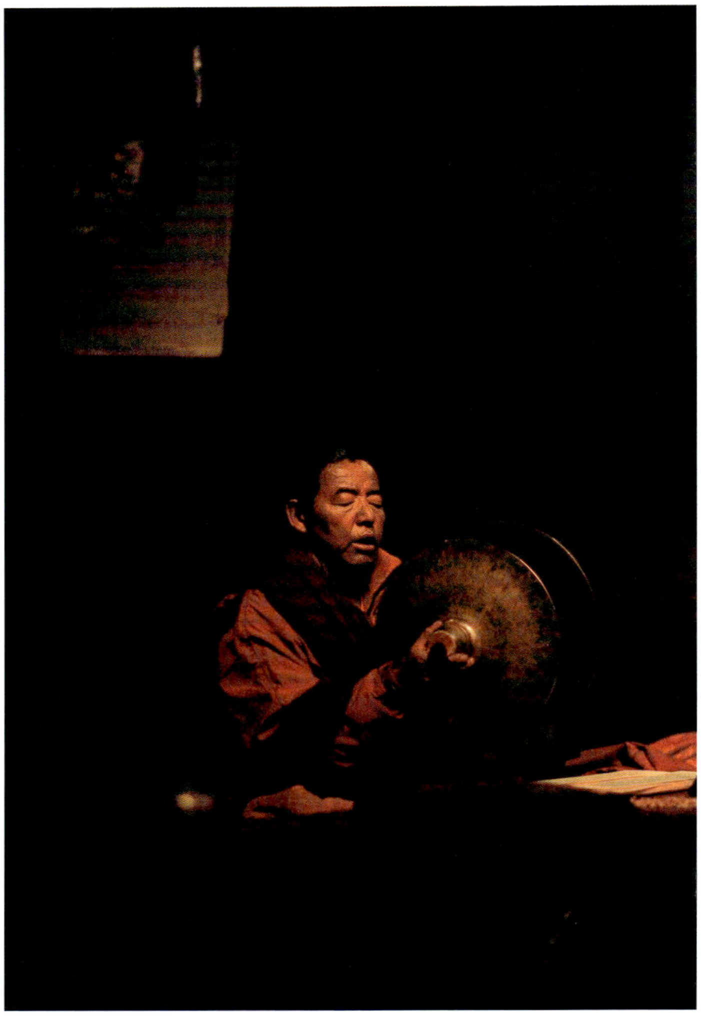
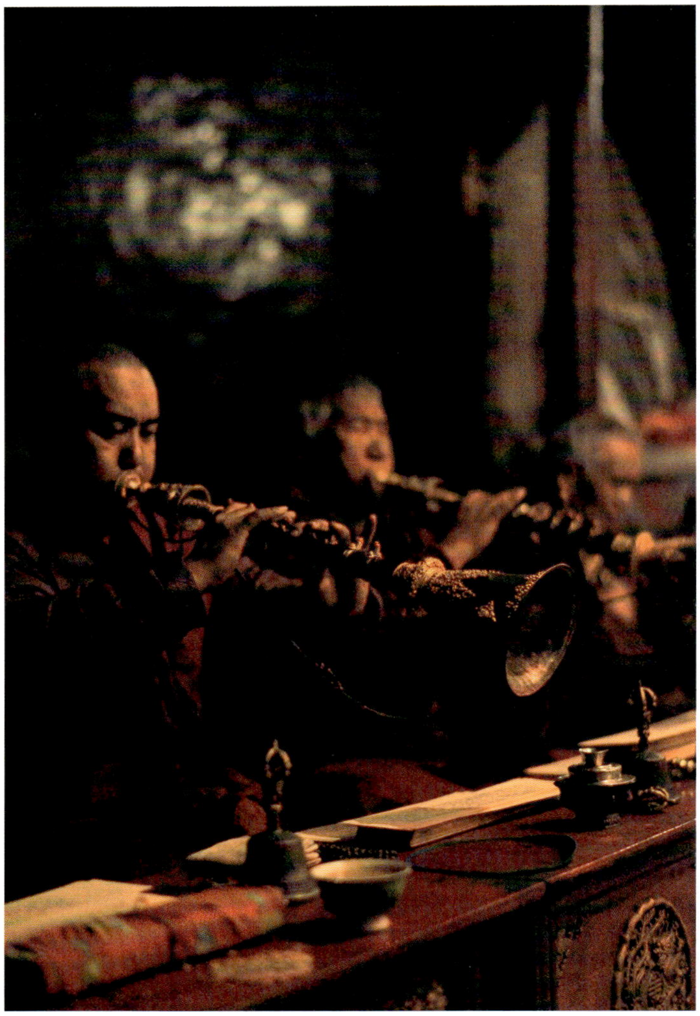
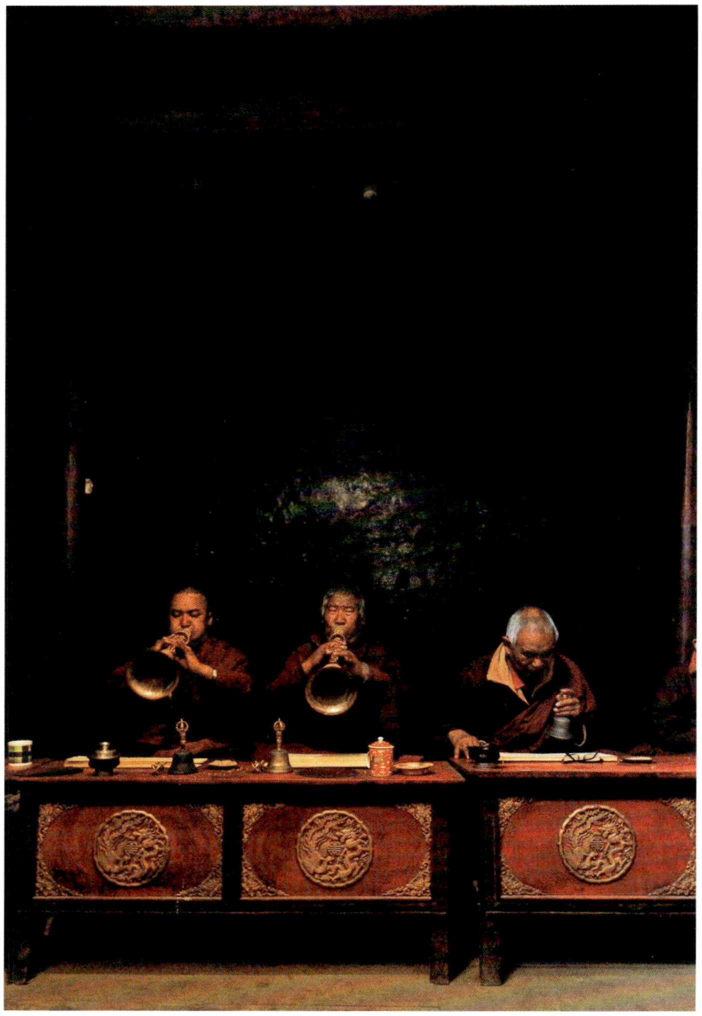
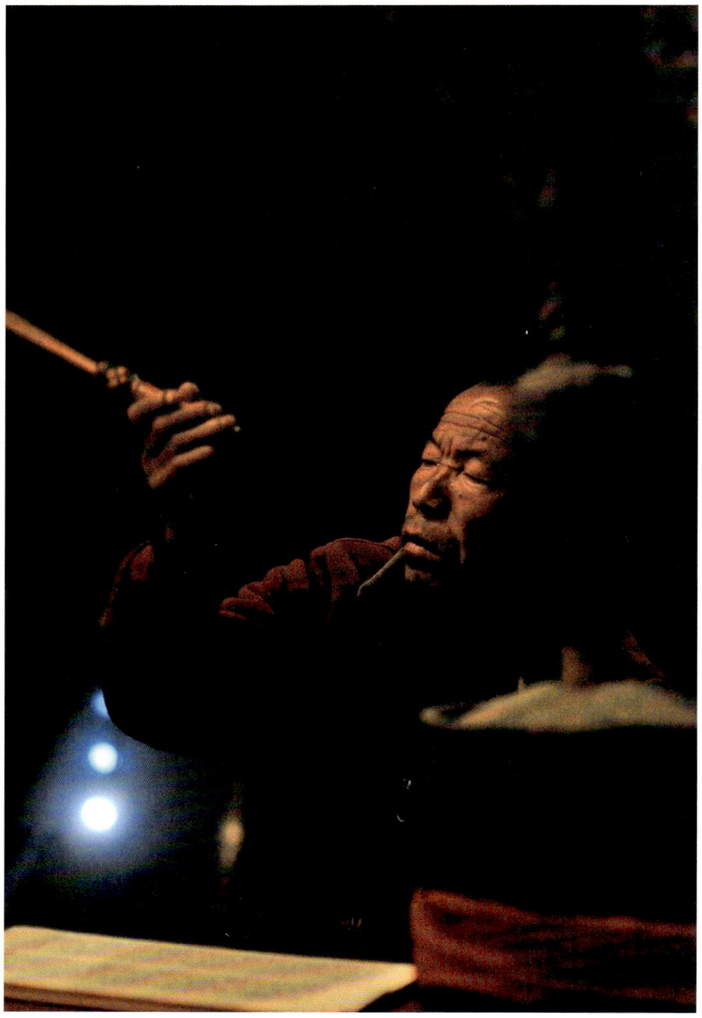

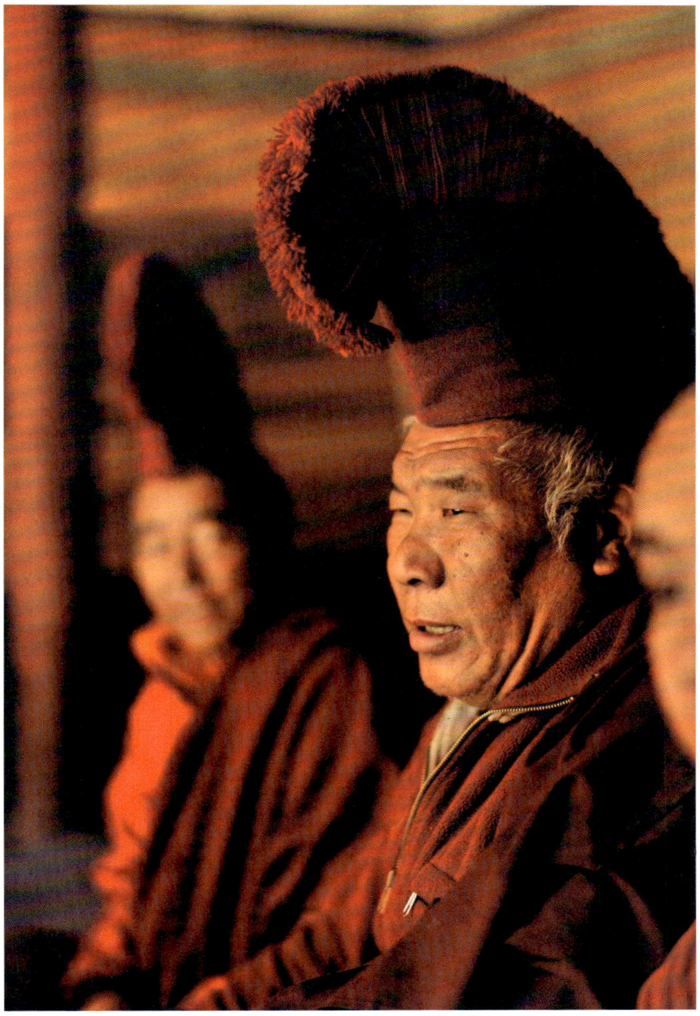

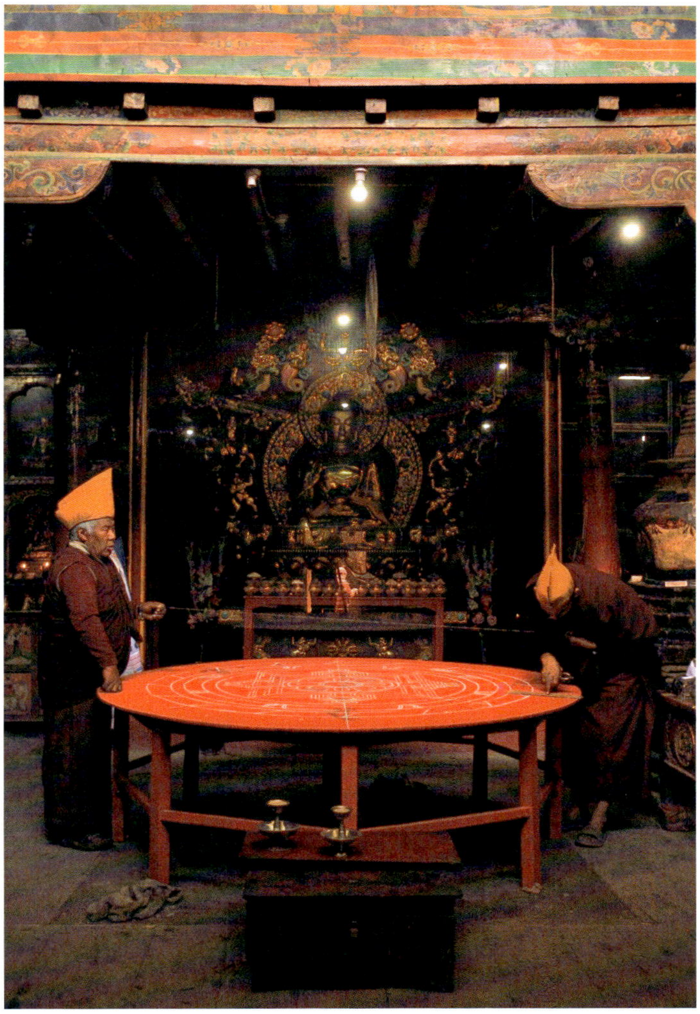

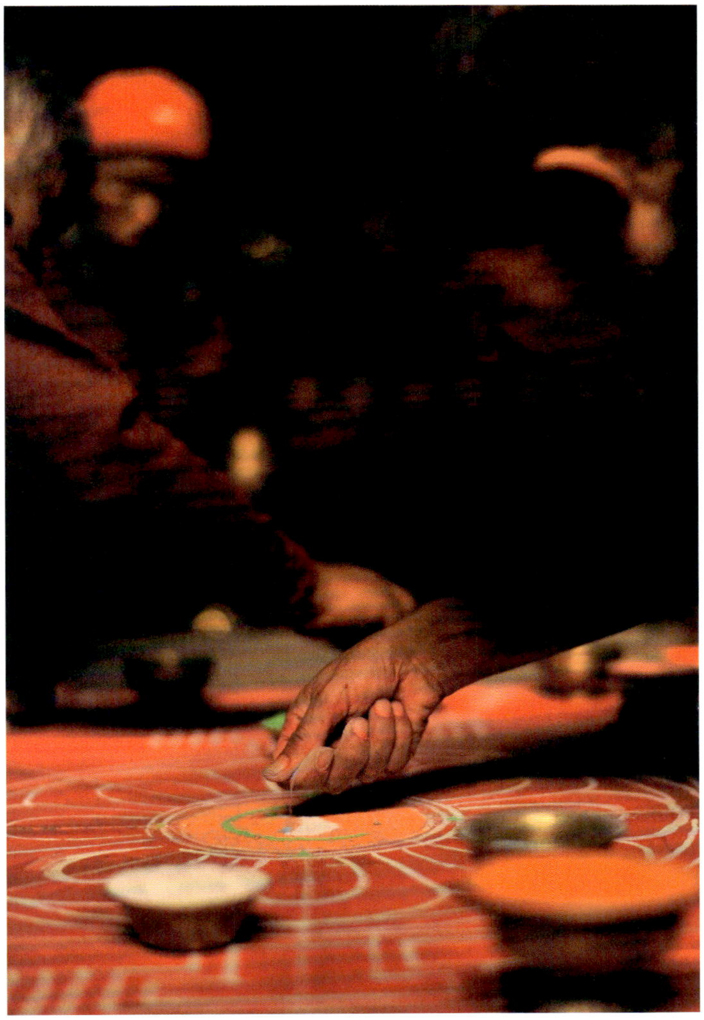

This spread At a religious ceremony called a puja, coloured salt or uncooked rice grains called akshata are distributed to the people of Ghemi in the hope of a prosperous harvest. Worship in Upper Mustang usually involves offerings of flowers or fruit to an image of a deity, or the creation of a mandala (symbolic diagram) of coloured salt or rice. In Hindu and Buddhist Tantrism, the mandala serves as an instrument of meditation, representing the universe, a vessel for the gods, and a collection of universal forces. The uncooked rice kernels represent prosperity, fertility, and bounty and, during pujas and other ceremonies, are thrown over devotees' heads. In Ghemi the mandala of coloured salt also symbolises the construction of something beautiful. Once accompanying ceremonies and viewings are finished, the mandala is deconstructed— in this case scattered in the river of a nearby village. This ritual symbolises the Buddhist doctrinal belief in the transitory nature of material life.

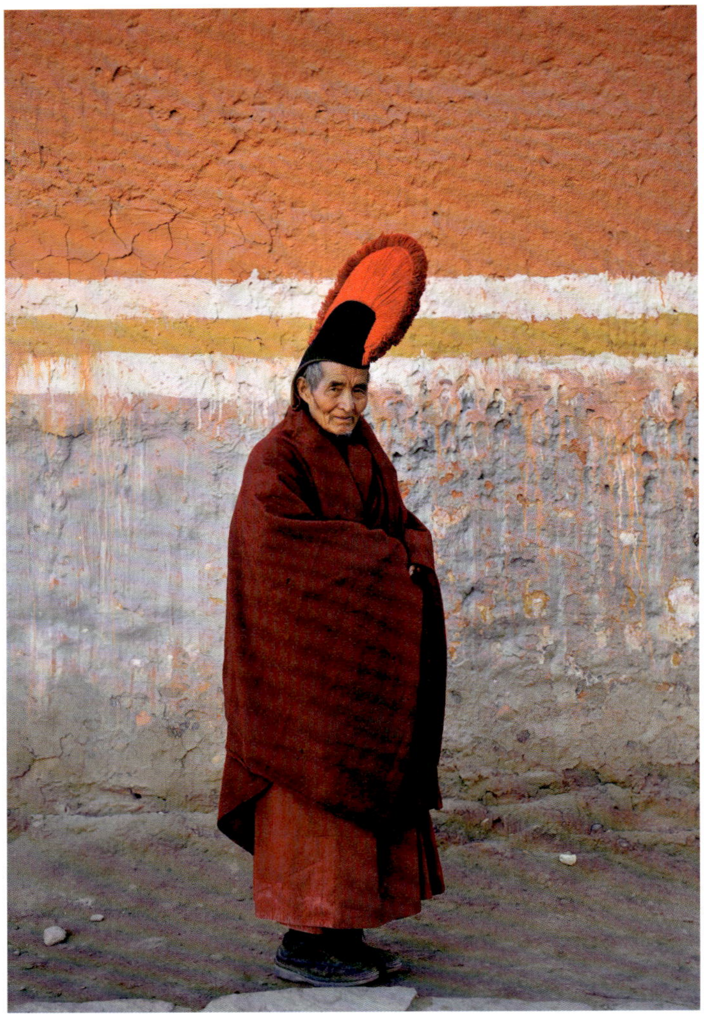
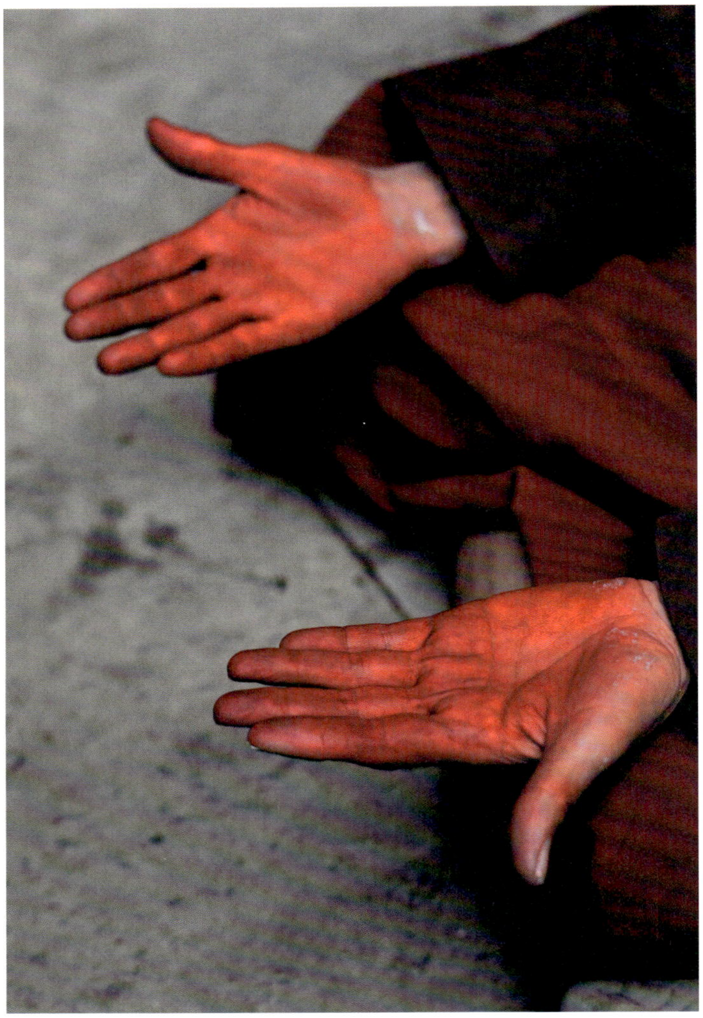
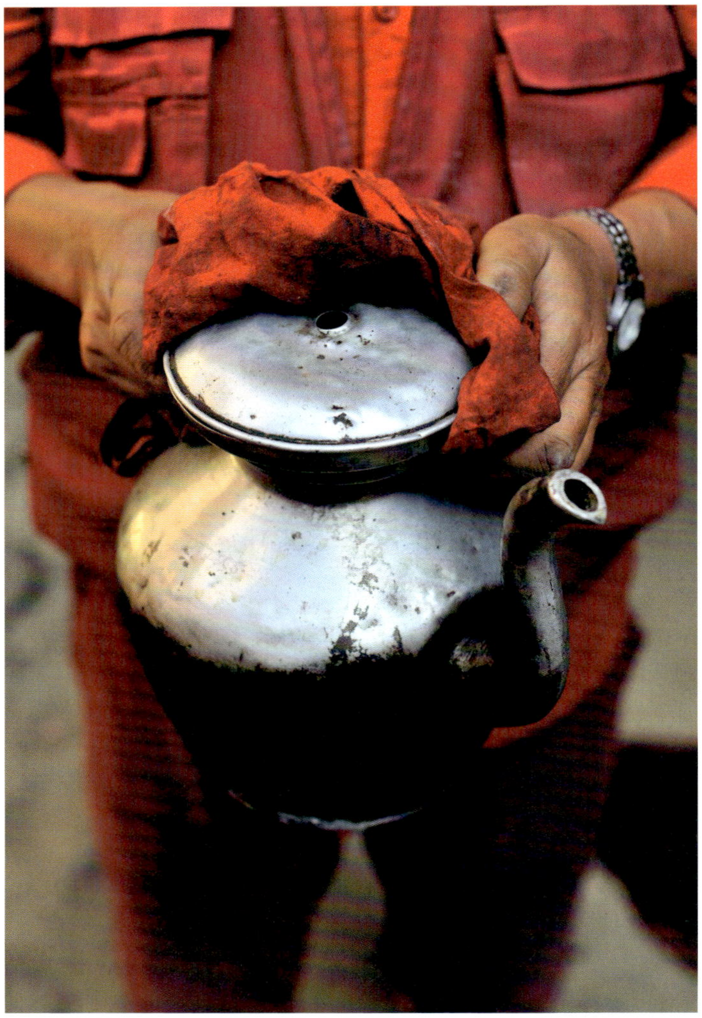
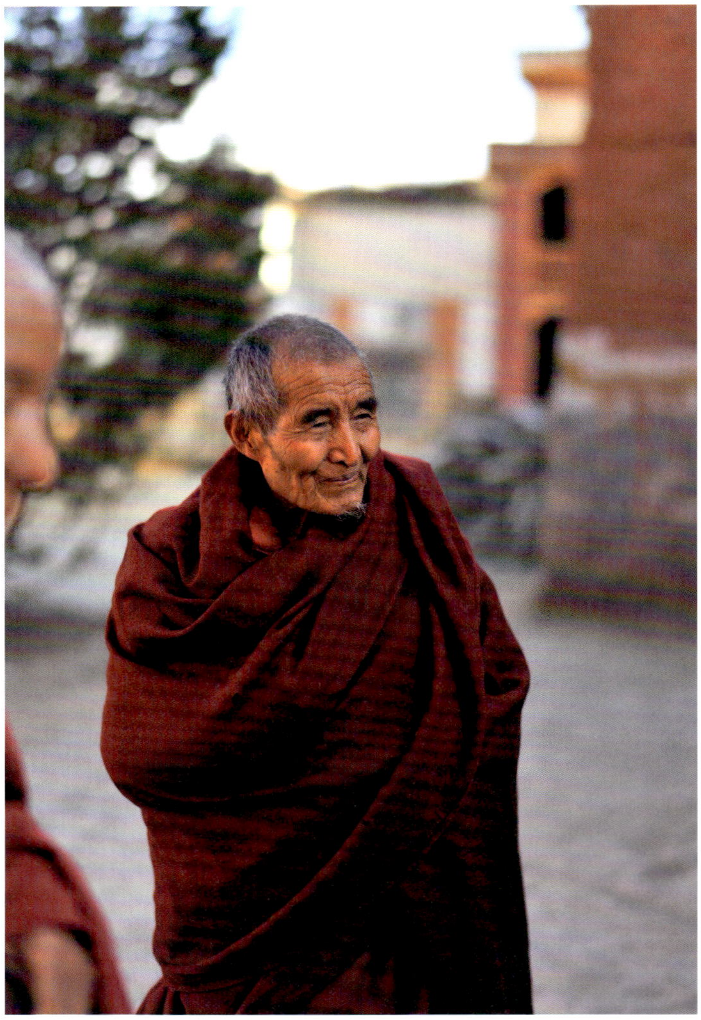

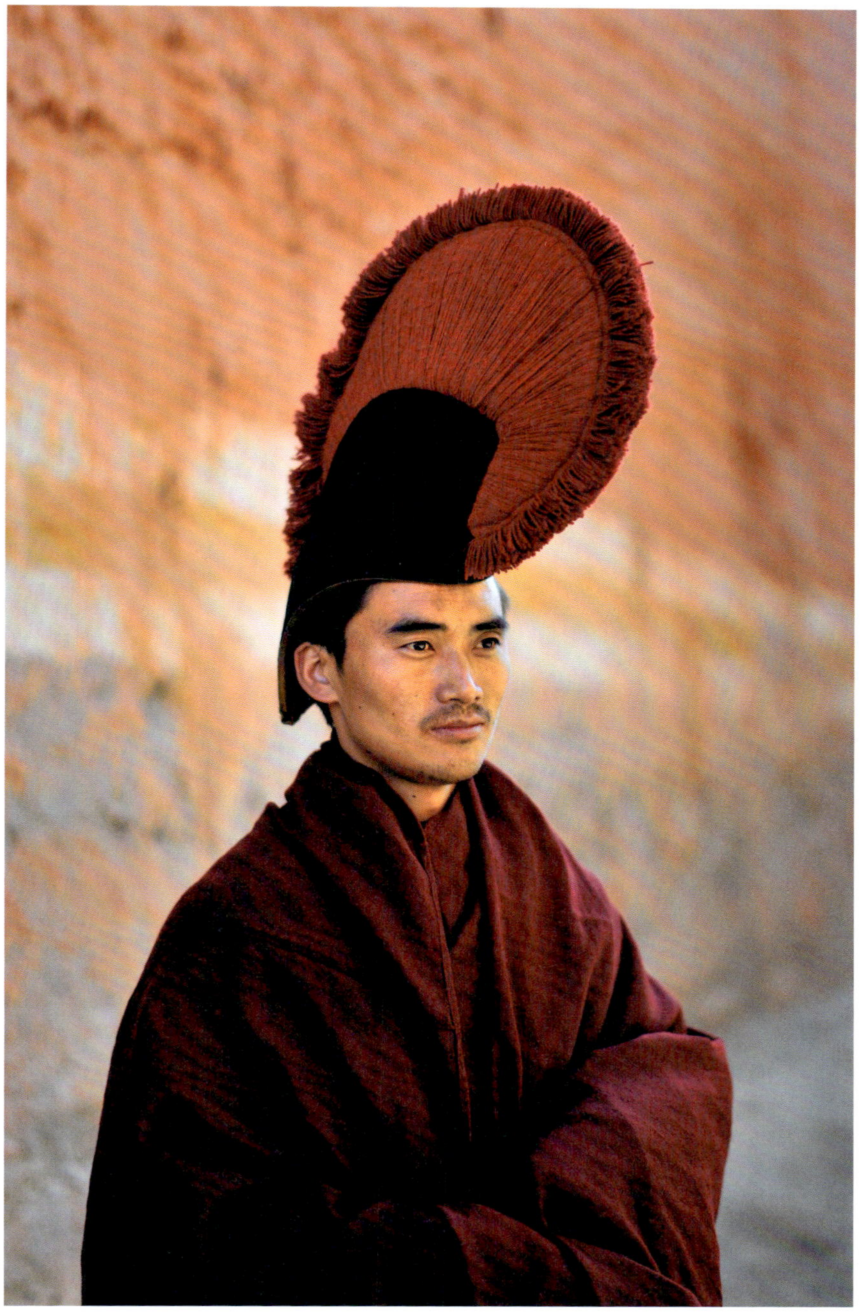

This spread Sakya Buddhism, one of four major schools of Tibetan Buddhism, is practised throughout the Lost Kingdom of Lo. Each village has monasteries, known as gompas, that are more than 800 years old and often still inhabited by monks. Although the people of Mustang live within the geographic boundaries of Nepal, their history is more closely linked with Tibetan religion, culture, and politics.

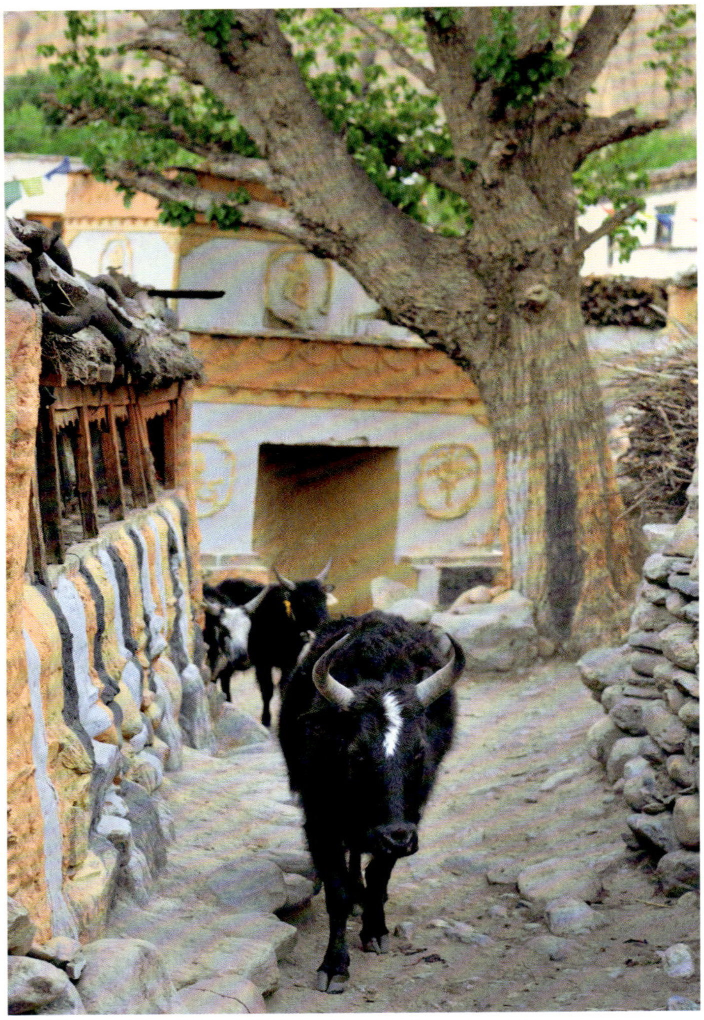

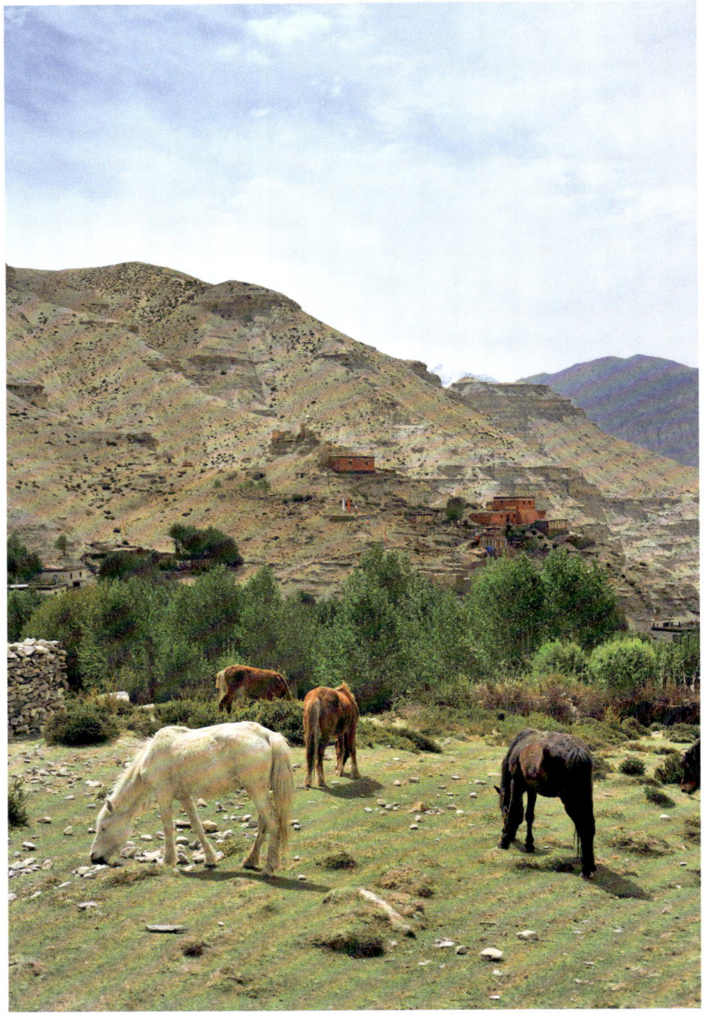

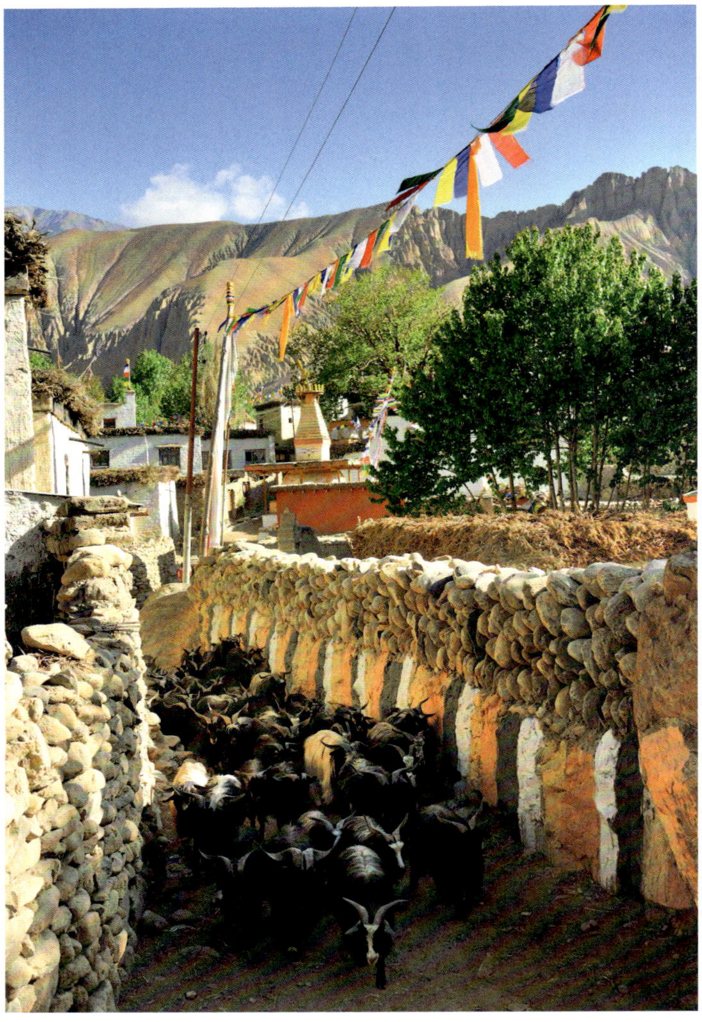

This spread The people of Mustang rely on an agro-pastoral system of trade and income. Water is scarce in this harsh climate; residents often build small channels to guide more water to their villages, irrigating fertile green zones that starkly contrast with their harsh surroundings, like little oases. Yaks, sheep, goats, horses, mules, donkeys, and dzos — hybrids between cows and yaks — graze freely in and around the villages. When dusk falls, the animals somehow find their way back home.

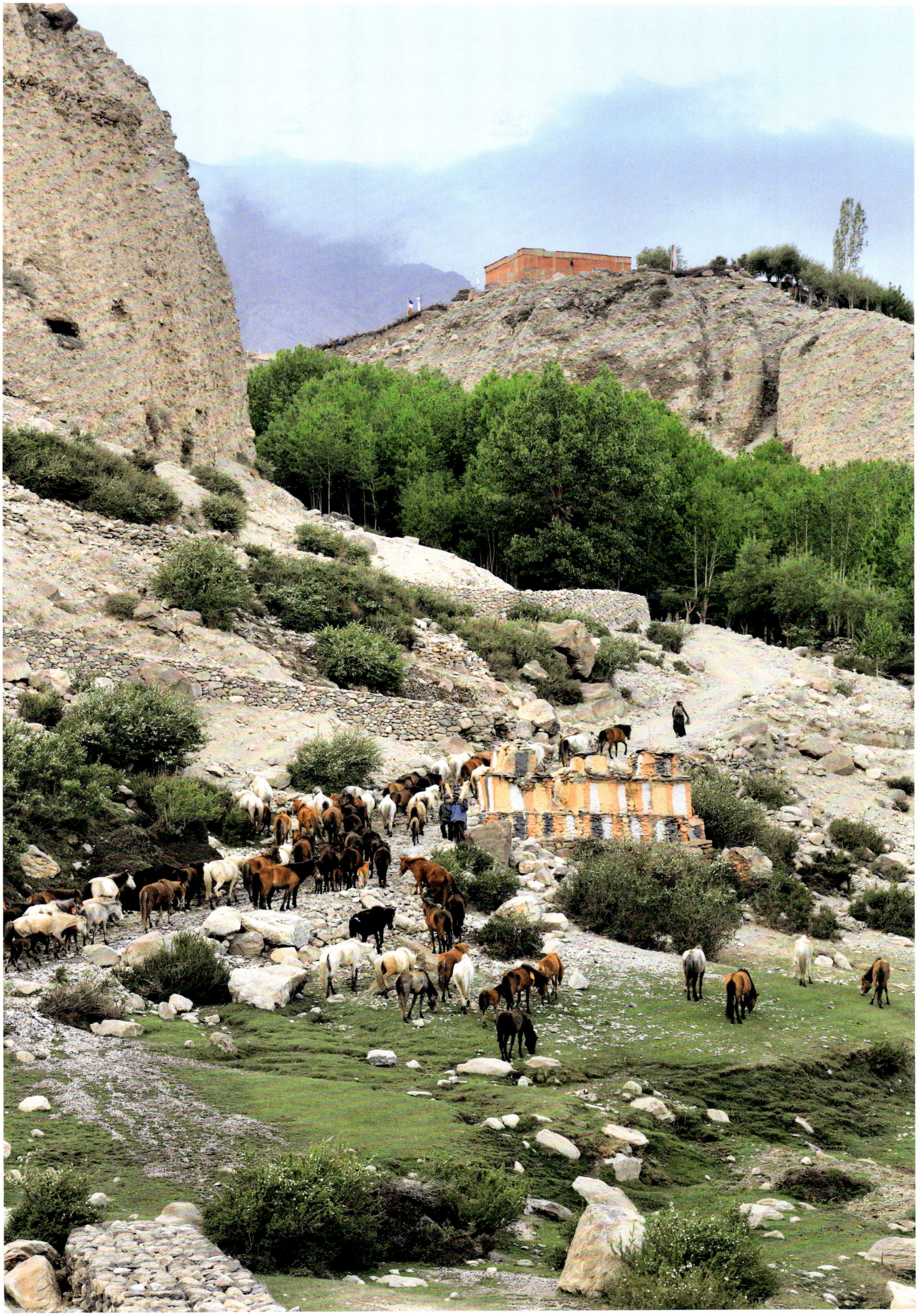

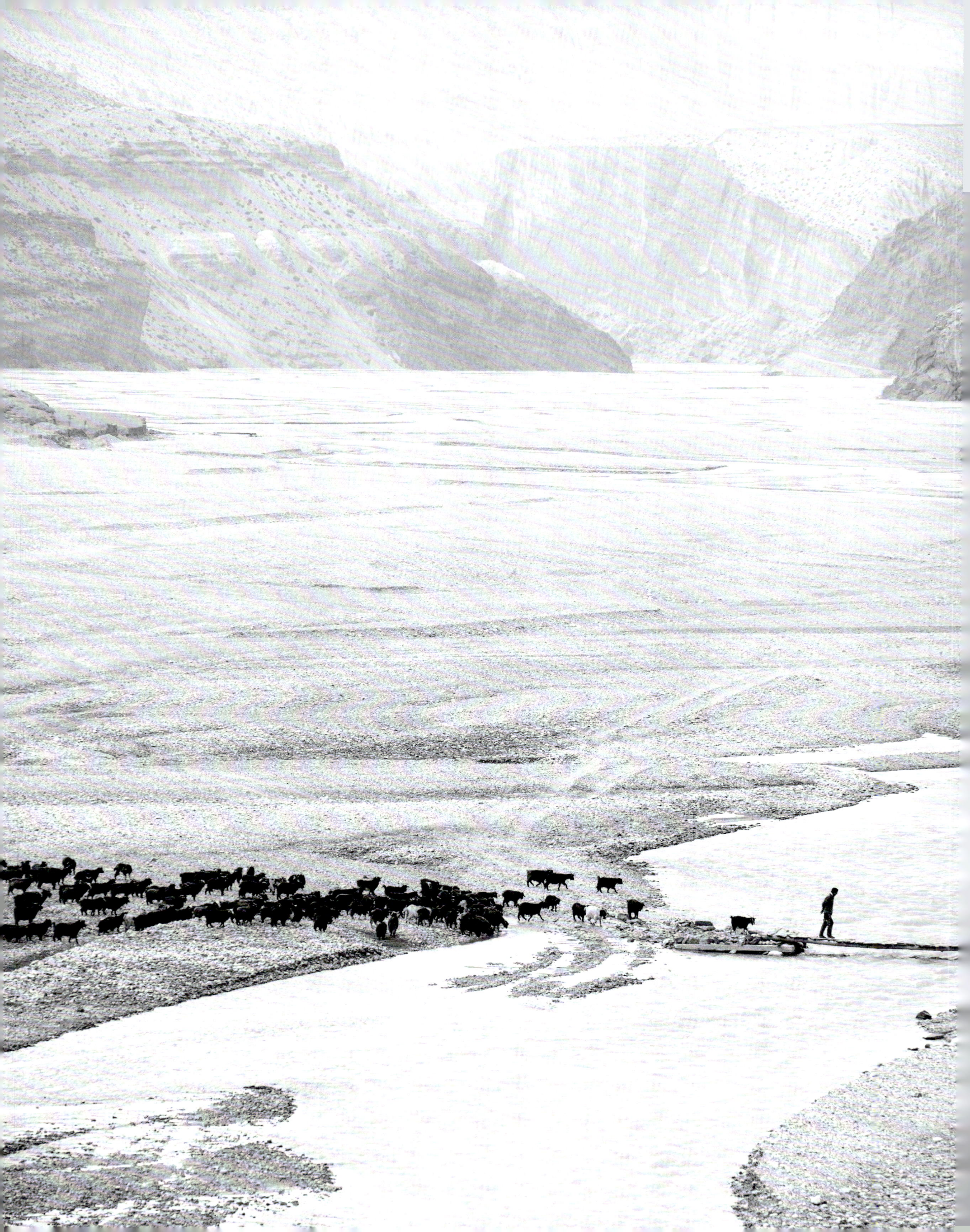

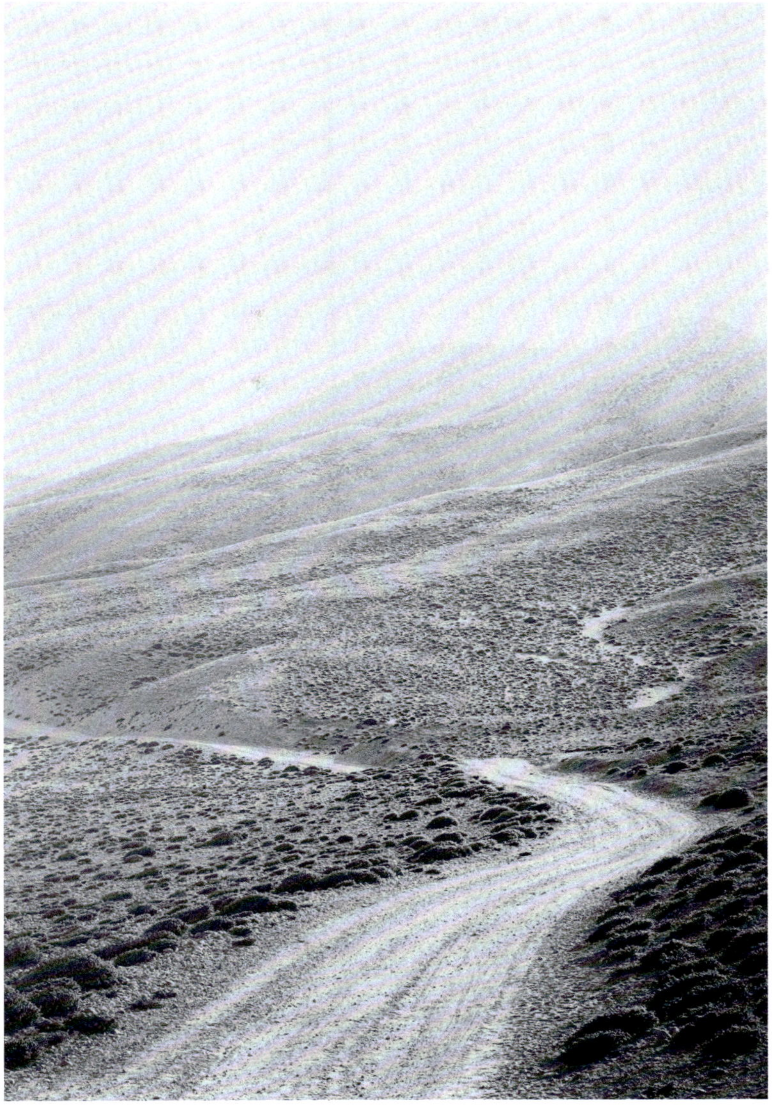

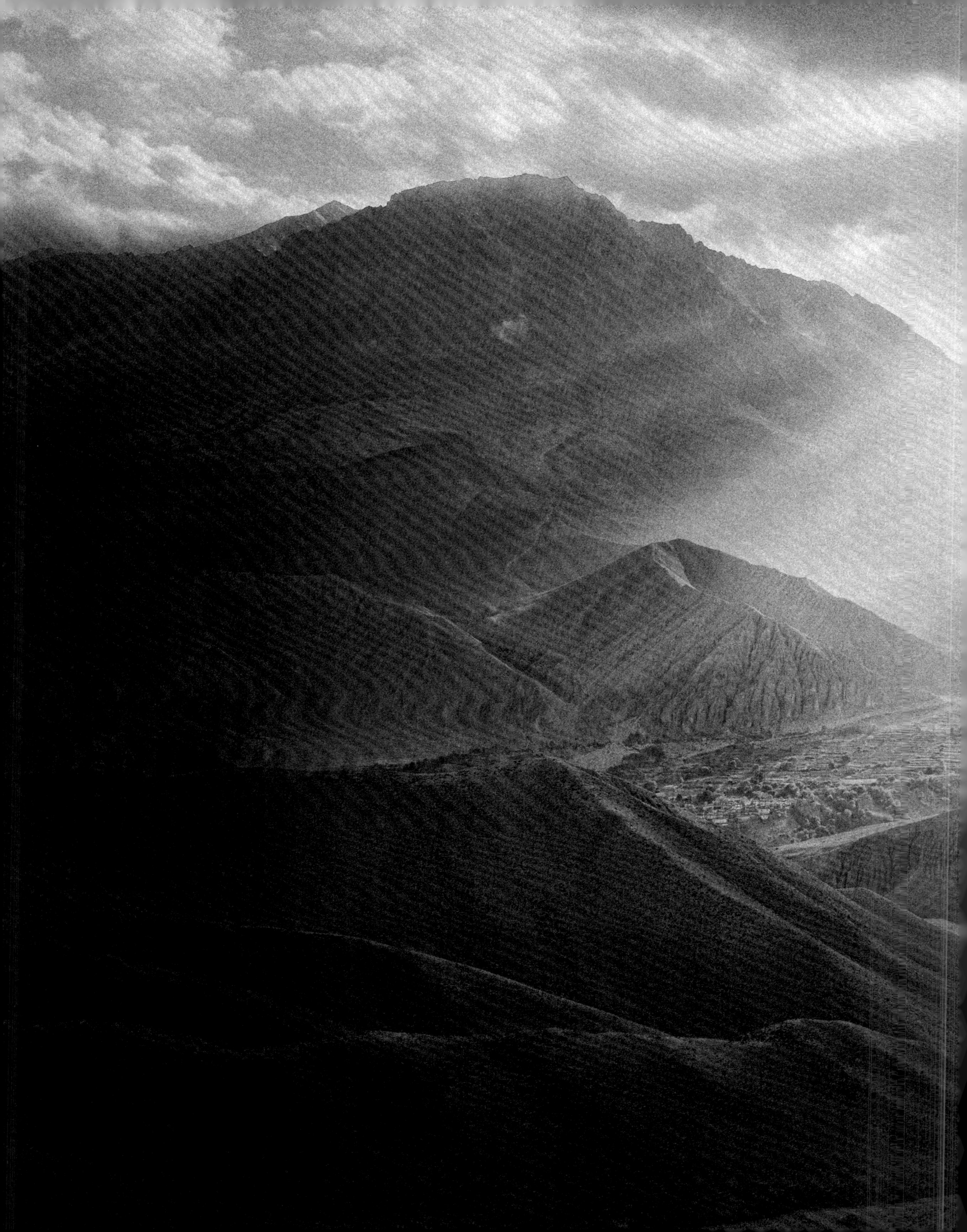

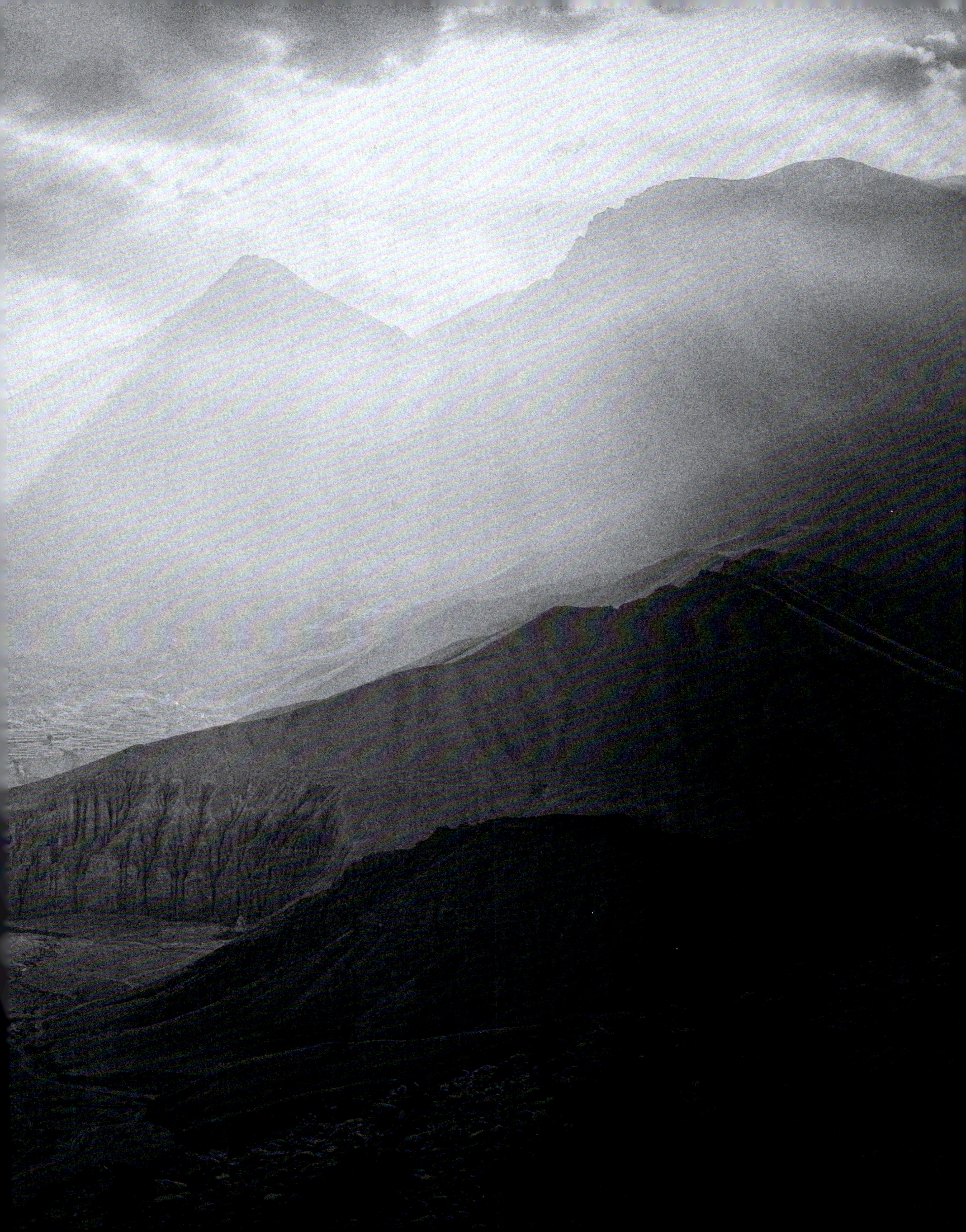

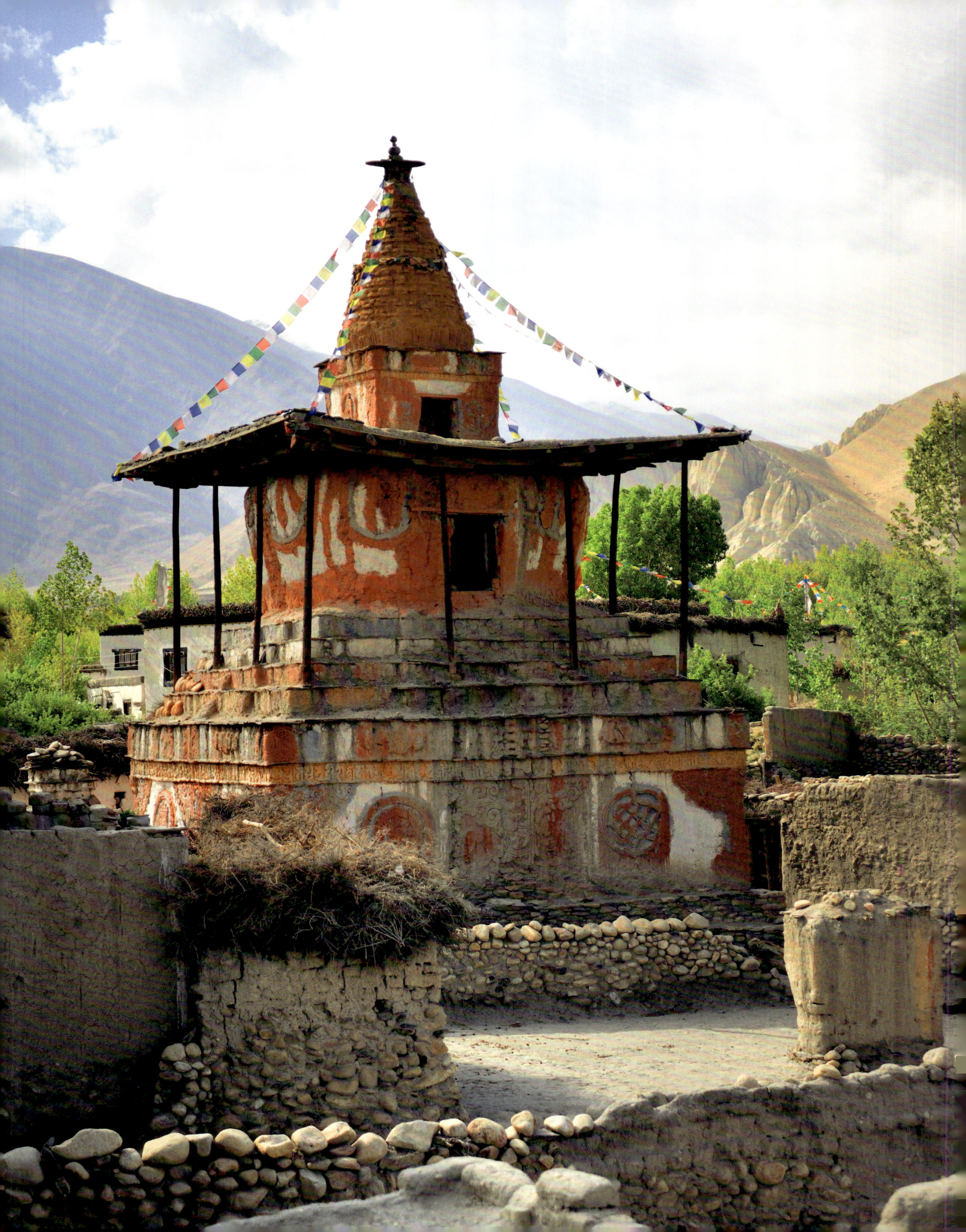

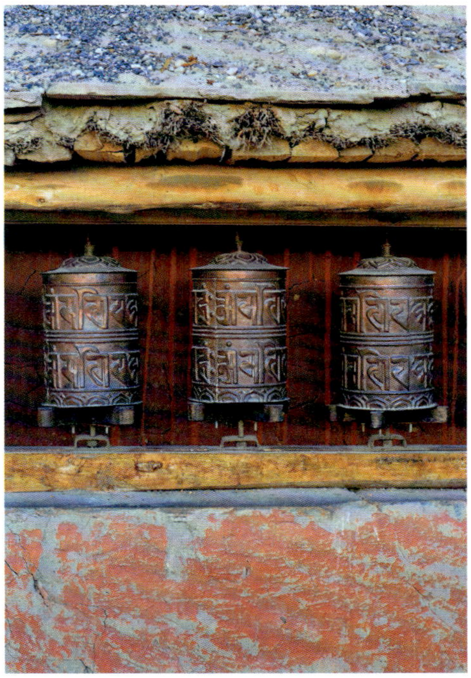
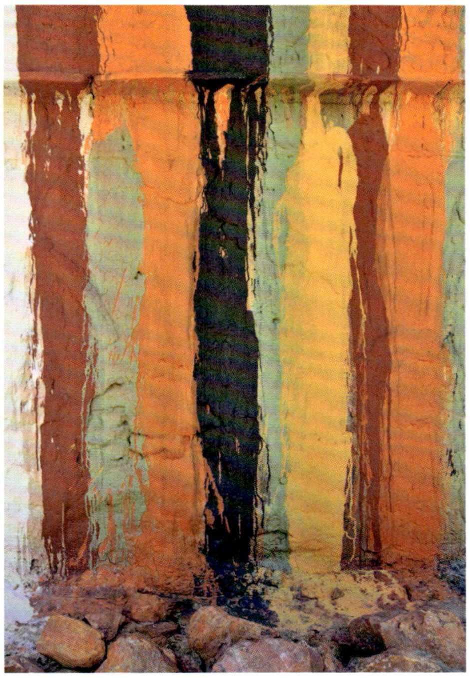
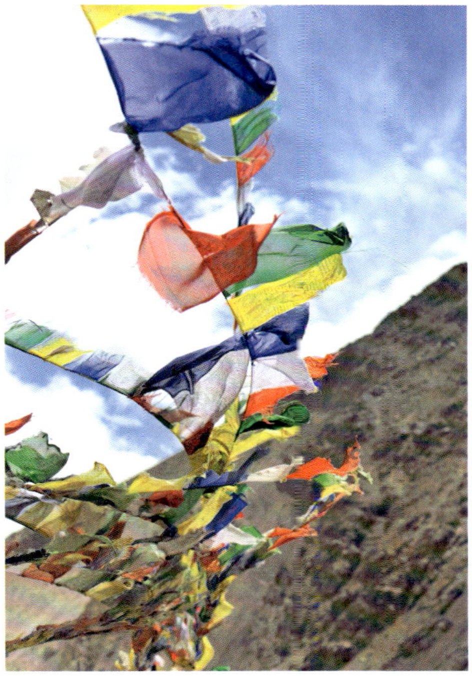
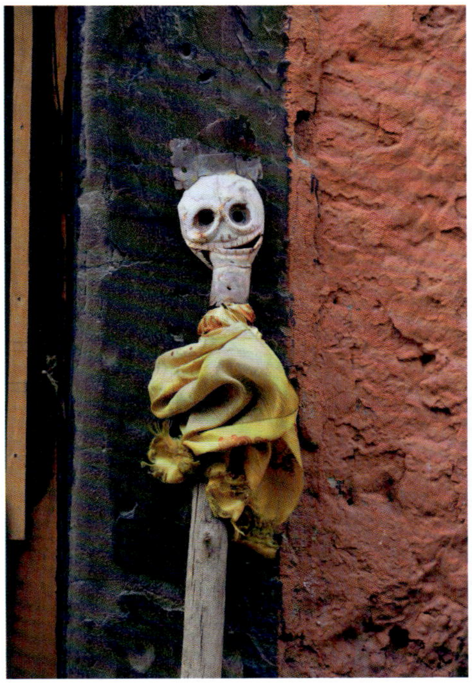
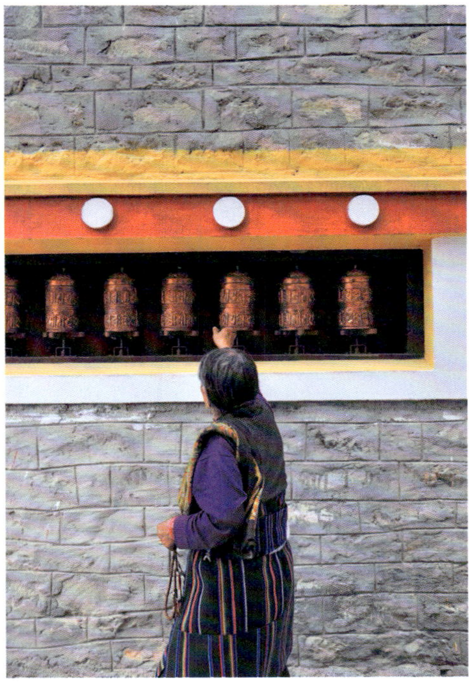
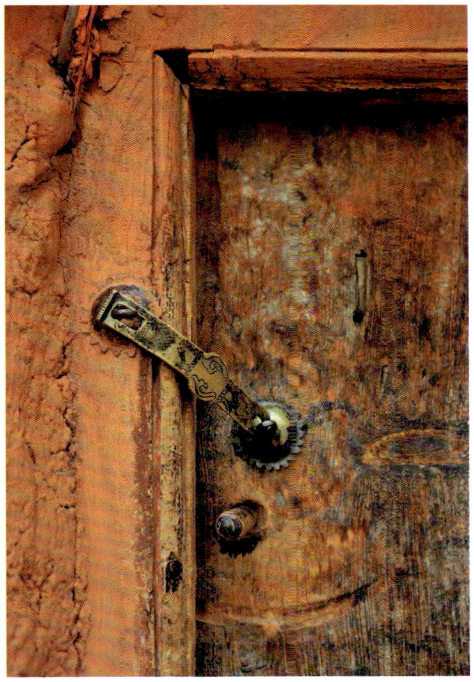

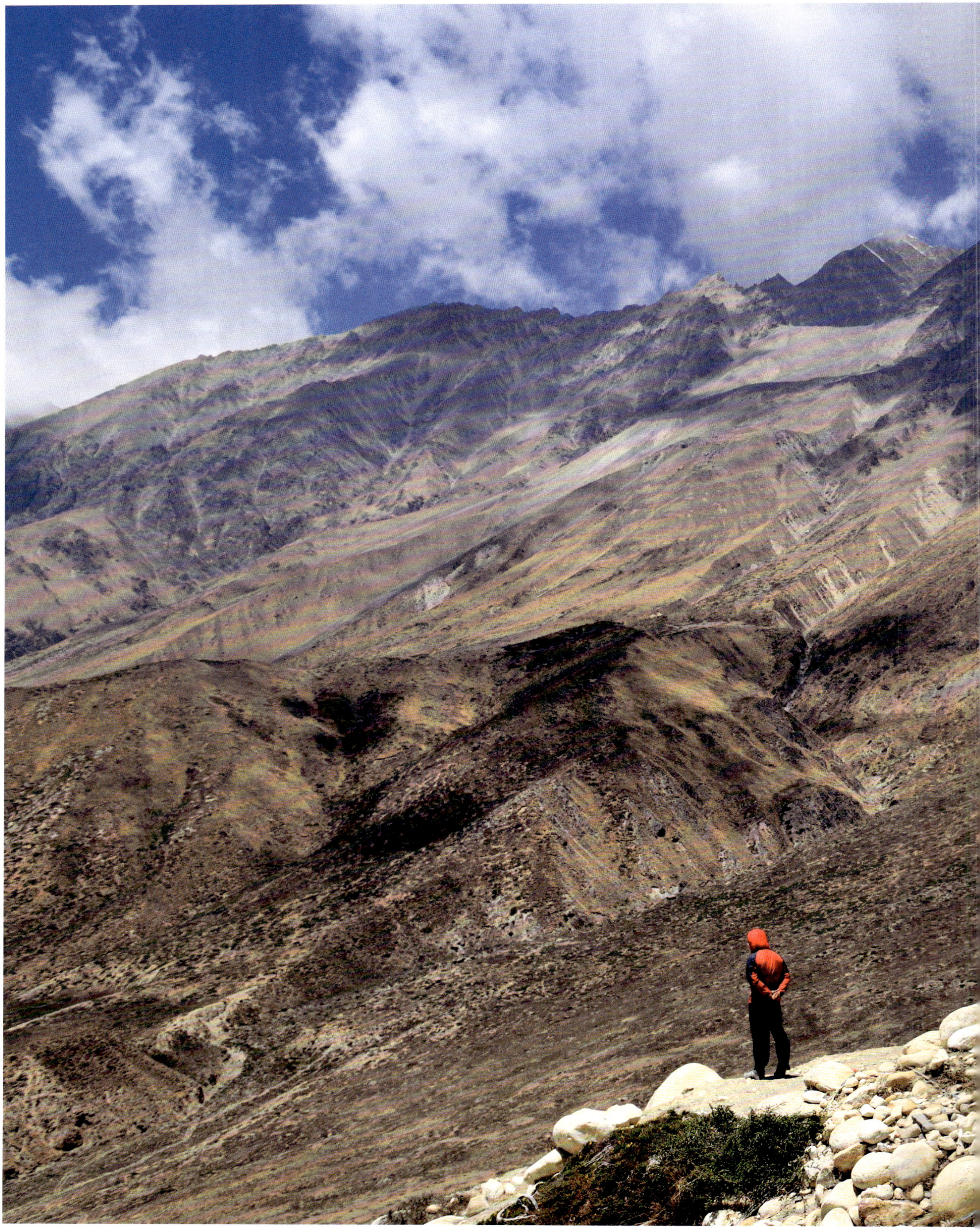

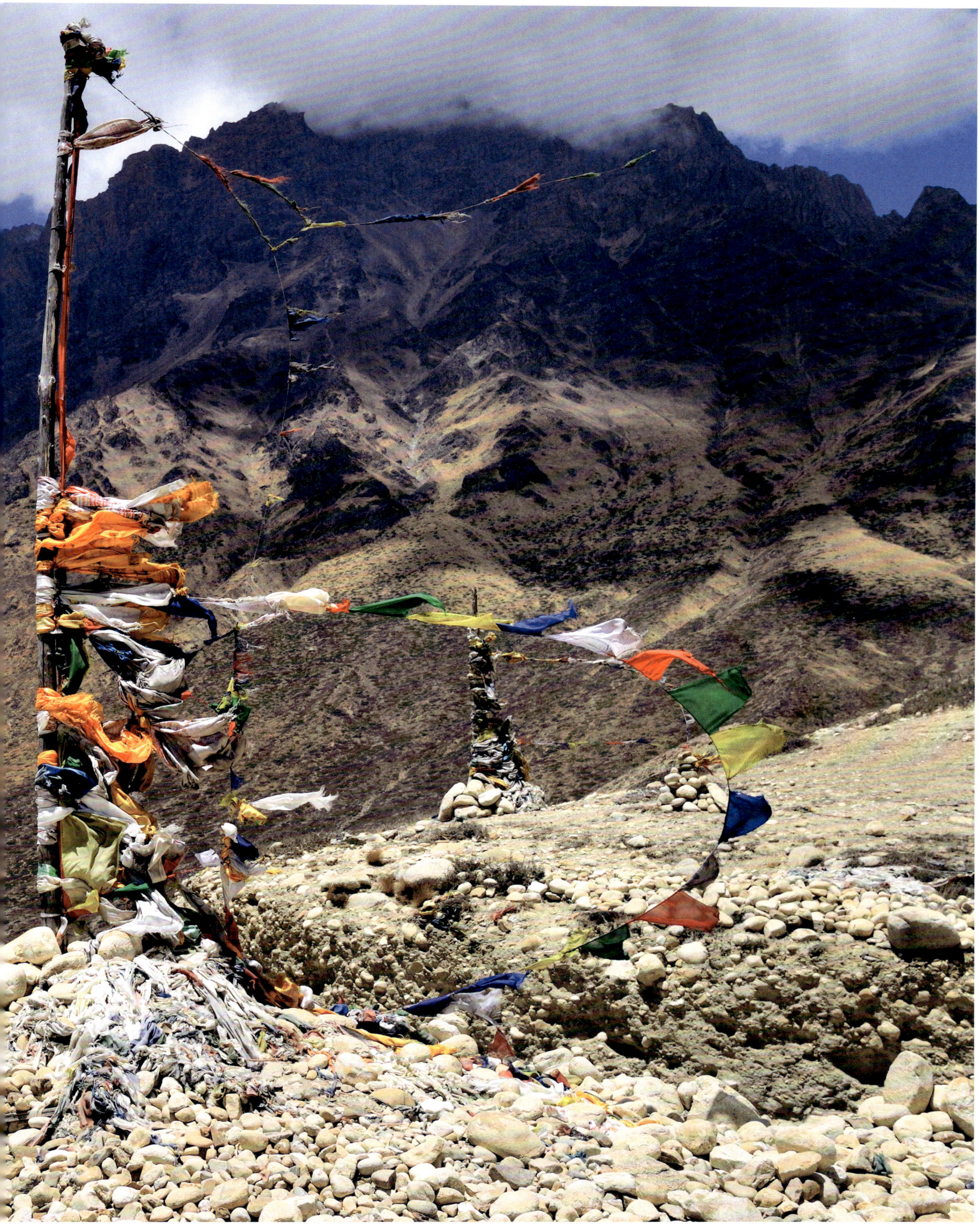

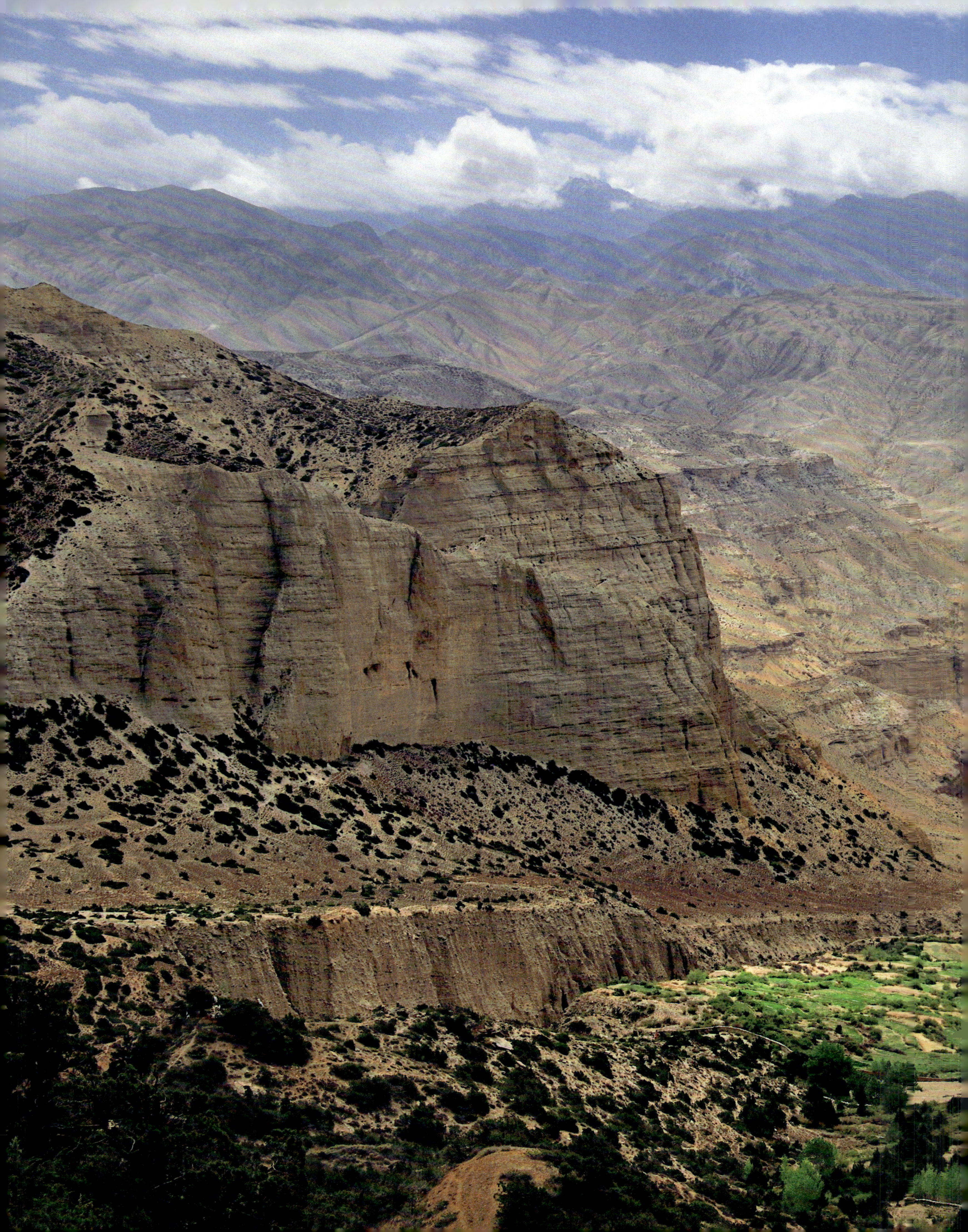

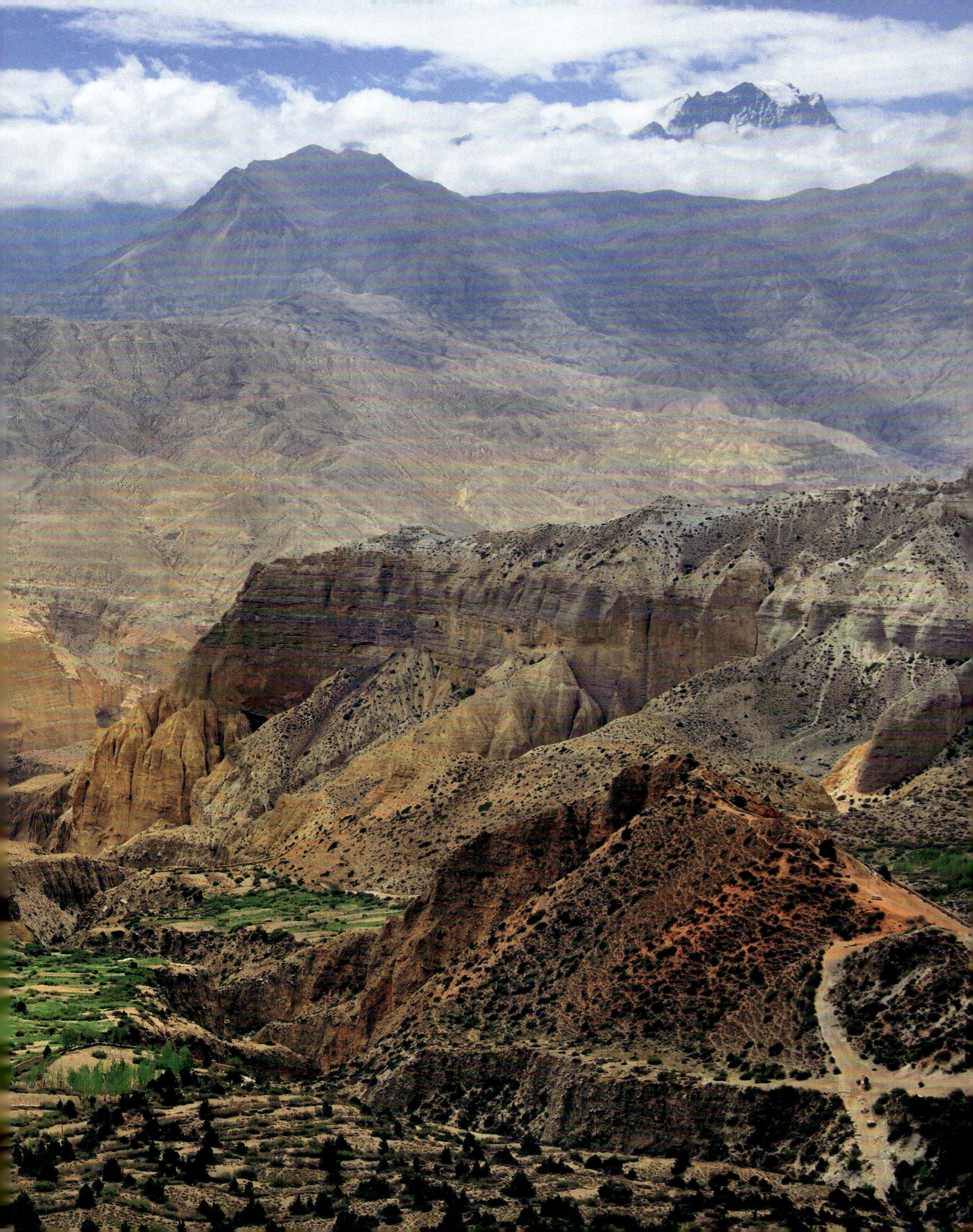

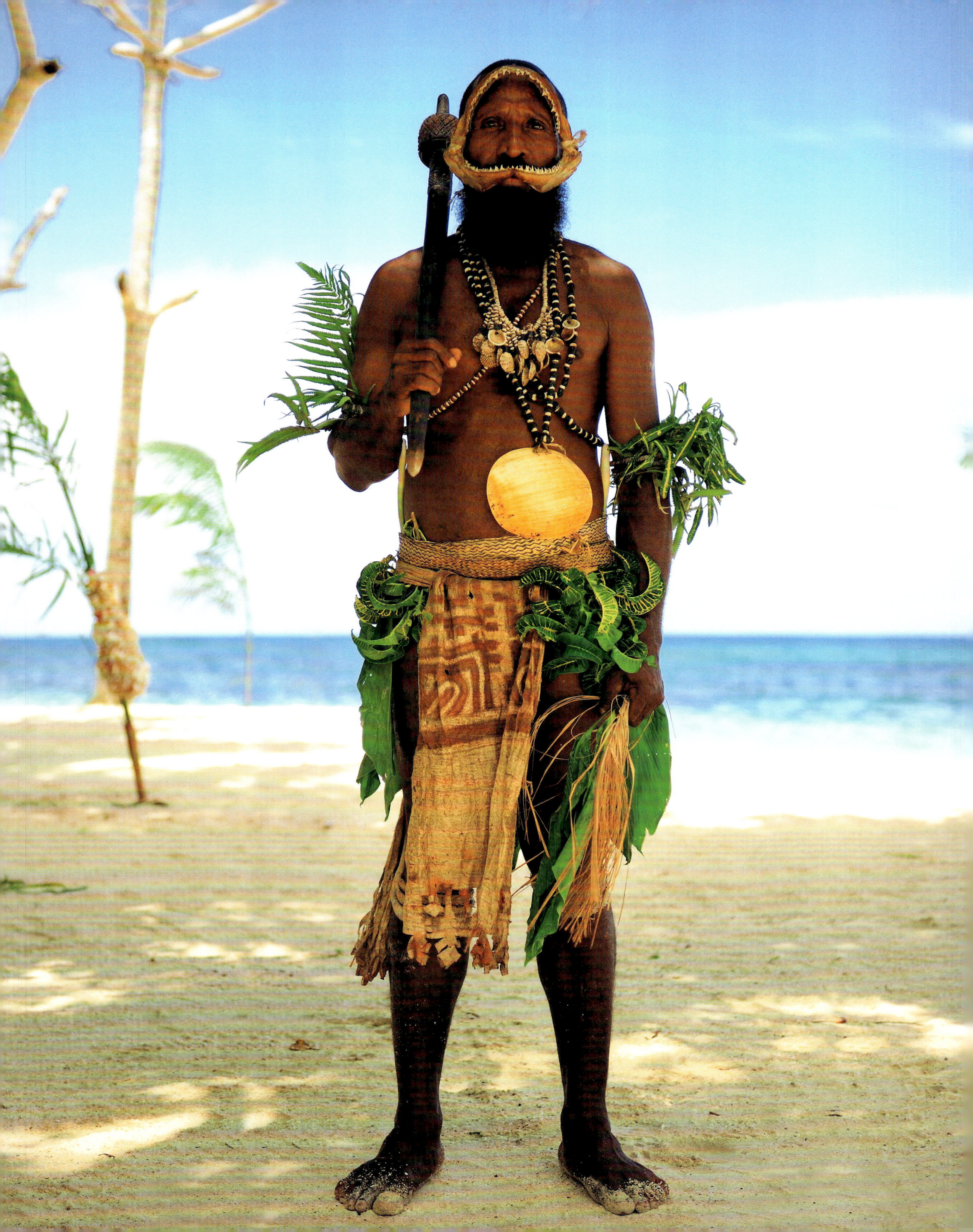

CHAPTER 8

A WILD EXPLORATION OF THE LAST TERRA INCOGNITA

INTO THE BIG UNKNOWN

Papua New Guinea

04°10'59.98" S
152°10'00.01" E

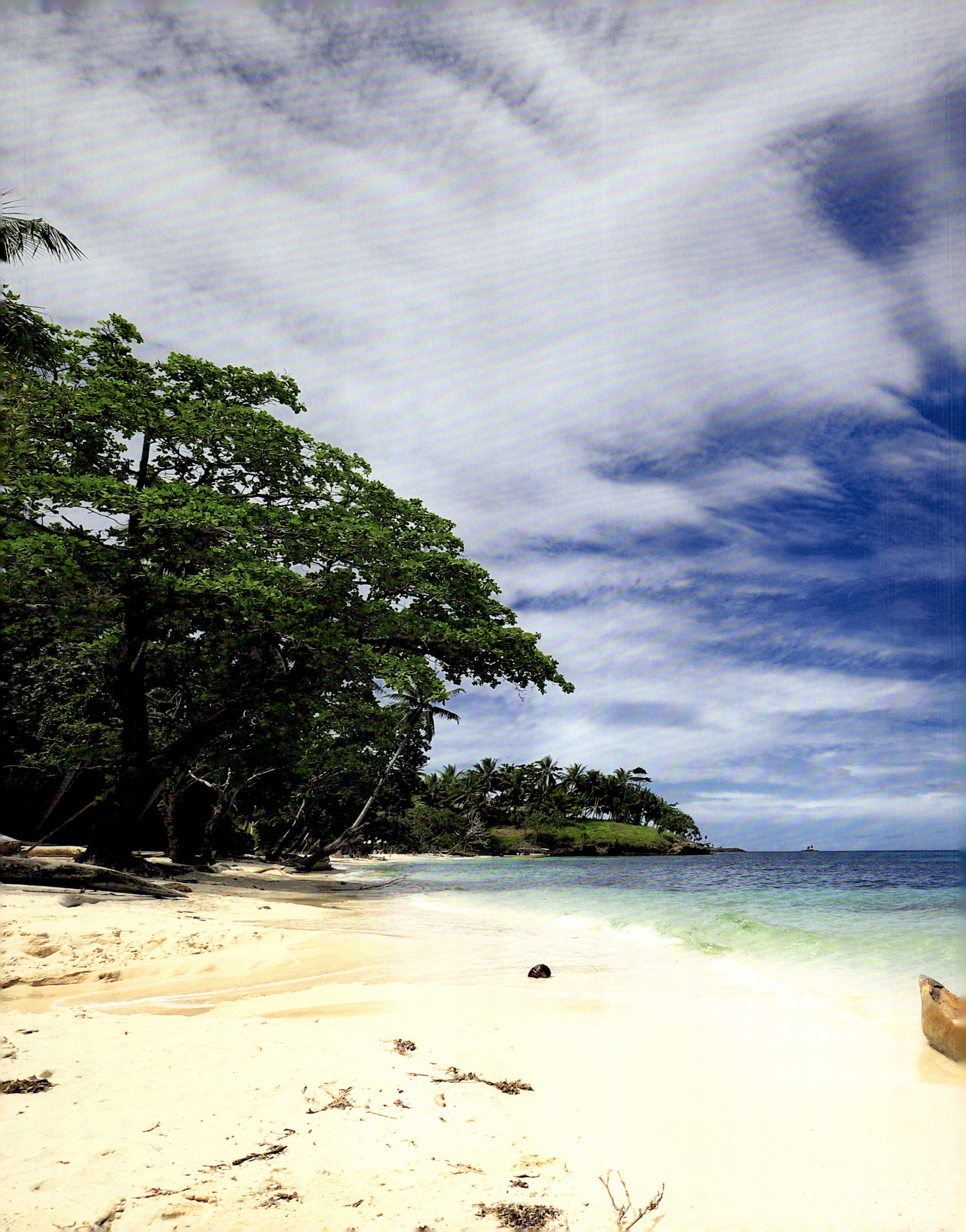

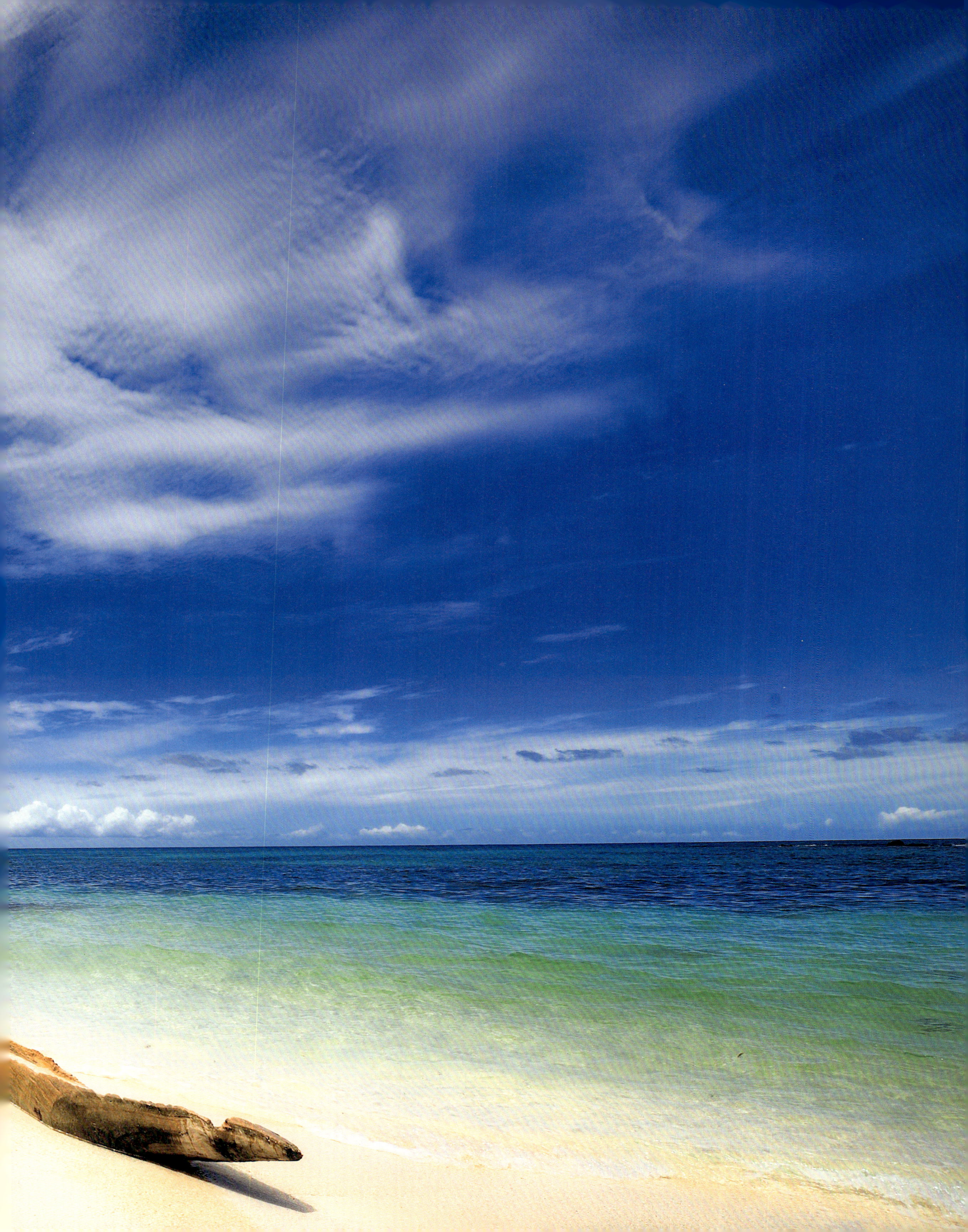

A Wild Exploration of the Last Terra Incognita

Into the big unknown

Country:	**Papua New Guinea**
Best time to travel:	**May, June, September, and October**
Closest airport:	**Port Moresby**
Price tag:	**€€€**
Style of travel:	**A wild and intrepid journey**

Introduction

Known as the most linguistically diverse country in the world and famous for its absolute wildness and rugged, natural beauty, Papua New Guinea (PNG) is the ultimate frontier. This part of Oceania, just a few hours' flight from Australia, offers travellers a world unlike any other, with more than 800 different tribes, each with its own language and customs. The land mass called Papua — the world's highest and second-largest island — is divided into a province of Indonesia called Papua and the country of Papua New Guinea. Flanked by active volcanoes and surrounded by a dreamlike, turquoise ocean dotted with atolls and coral reefs, it then morphs inland into unexplored rainforests and endless river deltas populated by outlandish fauna and flora. Papua has been described as the last paradise on earth, untouched by the modern world. A while ago, scientists even discovered a kind of 'lost world' high in the mountains of Papua New Guinea. Not a historical discovery but a sighting of an unknown bird species and butterfly, topped up with some exotic frogs and unrecorded plants. Even the Beck's petrel popped up after having been presumed extinct for years. Some call it too challenging, too obscure, too raw. Other travellers consider it the last paradise on earth. The truth lies hidden somewhere between a roaring volcano that refuses to go dormant and a silent and lush forest teeming with dazzling birds of paradise.

The Views

Vast and remote, PNG begs the traveller for patience, an open mind, and time for exploring. The Islands Region is a good example: in the province of West New Britain, it attracts scuba divers who come to explore magnificent World War II shipwrecks, while others come to hike its wild interior, home to tribes that were not exposed to the outside world until the 1960s. To the east, in the province of New Ireland, beaches like Lavongai, the spiritual home of tribes such as the Tumbuan and Malagan (known for their hallucinogenic rites to honour the dead), bring the world of Robinson Crusoe to life. Perhaps the most surreal attraction in the Islands Region is the former capital, Rabaul, a once colourful and pleasant city. In 1994 the eruption of two volcanoes, Tavurvur and Vulcan, buried Rabaul under thick ash. Although it killed only a handful of people, it left 50,000 more homeless. The growling, smoking Tavurvur remains active. After the eruption, Rabaul became something of a ghost town, a modern-day Pompeii, although many residents remain in the still-thick ash, proceeding as if nothing happened. The city is popular with visitors, who can stay at the Rabaul Hotel, which appears abandoned from outside but functions as normal within, complete with an English-style pub and Chinese restaurant. Whenever the wind blows in the 'wrong' direction, Rabaul receives a fresh coat of ash and must unbury itself once again. Rabaul sits on Blanche Bay, across which lies the city of Kopoko, where many former residents of Rabaul now live. From Kopoko there is a spectacular view of the smoking volcano. At night from the Rapopo Plantation Resort overlooking Tavurvur, the distant tectonic rumble sounds like a thunderstorm rolling in, and is sometimes accompanied by a red glow, blinking in the darkness like the eye of a beast. East New Britain is also home to the Baining tribe, known as 'the fire dancers' for their daring flame-walking rituals. To witness Baining fire dances, travellers can head into the New Britain hinterland on the McCarthy Trail, also known as the Rabaul Escape, a track formed by members of the Australian Lark Force in 1942 as they fled invading Japanese troops. These jungle treks, orchestrated in cooperation with local

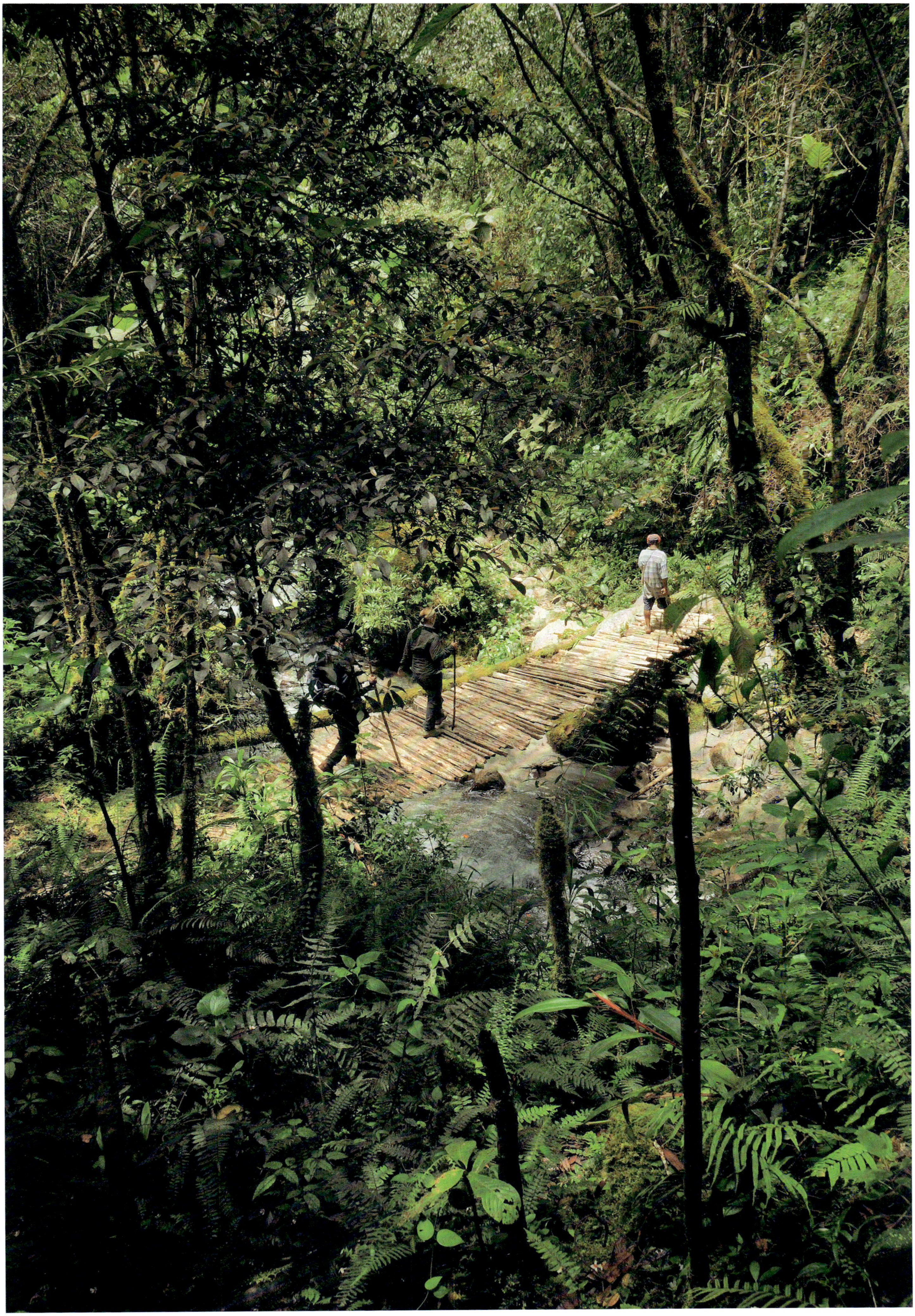

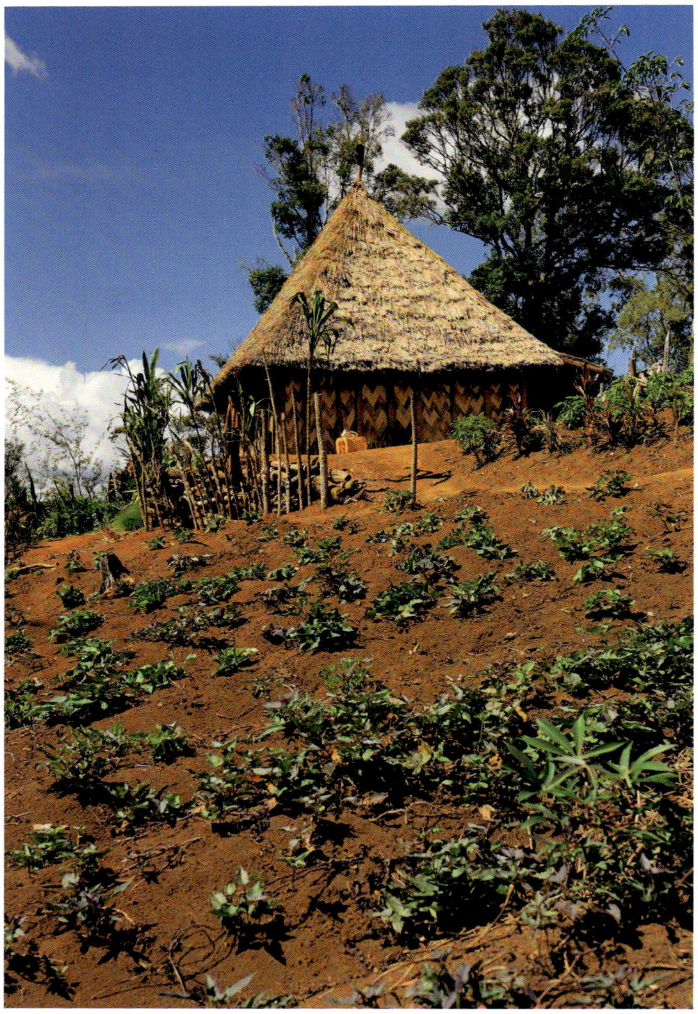

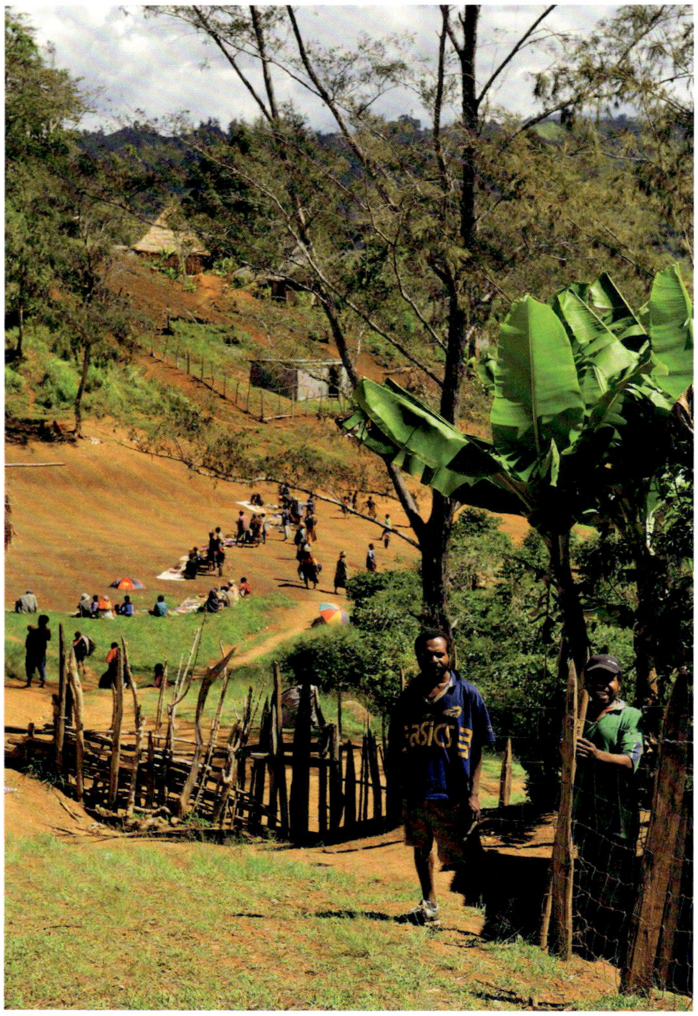

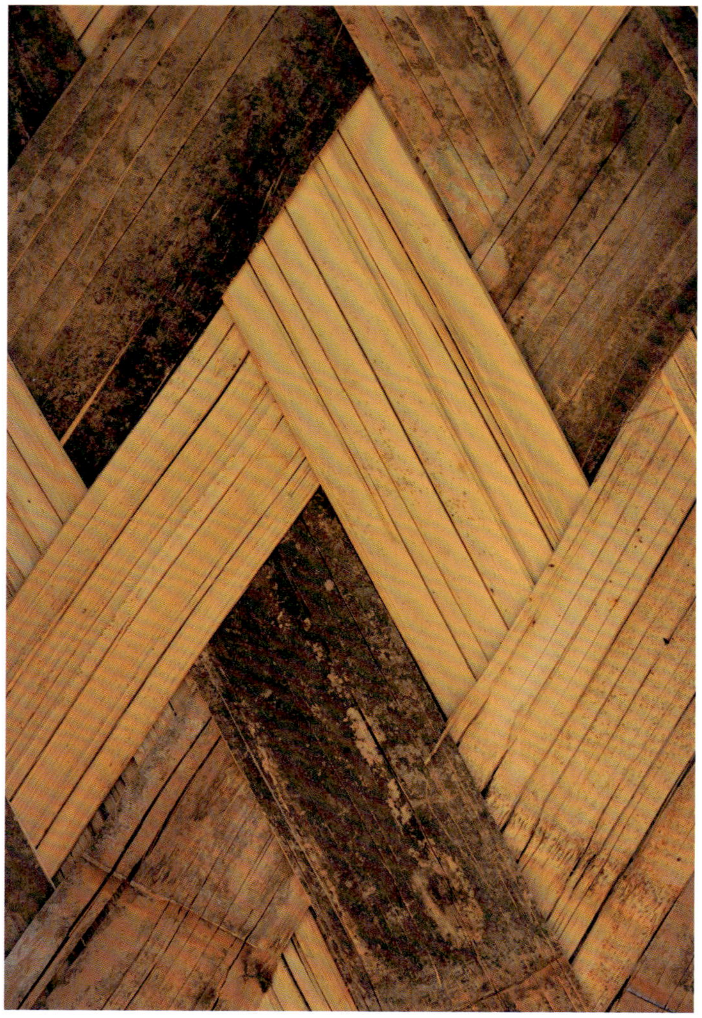

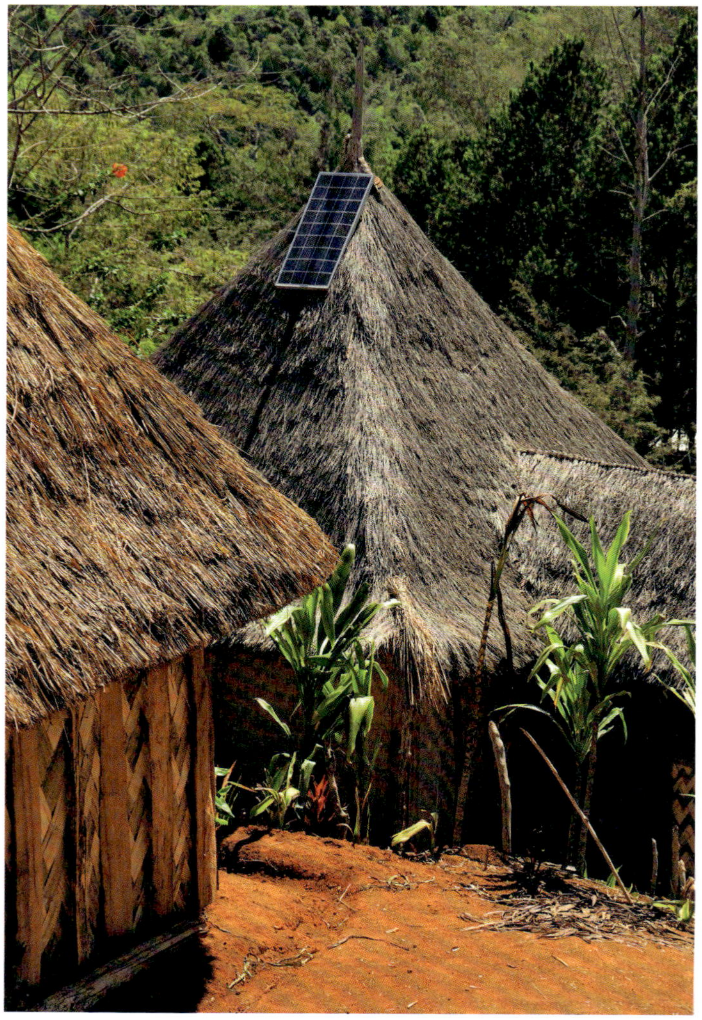

Opposite Straw-hut villages in the highlands between Goroka and Mount Hagen are sustainable thanks to traditional agriculture and powerful tribal loyalties. Despite their relaxed Melanesian attitudes, the villagers are self-sufficient and well organised, grow their own food, and build their dwellings with self-made tools. They hunt, fish, rear livestock, and cultivate vegetables in their perfectly manicured gardens.

residents, take four-to-five nights and cover a distance of 60 kilometres (37 mi.). Along the way, hikers stay in Baining and Pomio tribal villages, where, with luck, authentic fire rituals can be seen. The McCarthy Trail is a shorter, less touristed alternative to the more famous Kokoda Trail between Papua New Guinea's southern and northern coast. Another spectacular PNG hike is the Highland Highway, also known as the Okuk Highway, which wends its way through mountains between the two cities of Goroka and Mount Hagen. Here, straw-hut villages inhabited by 'mudmen' wearing frightening clay masks and long wooden finger tubes subsist on traditional agriculture and pig farming, and tribal loyalty remains powerful. After crossing the 2,478 metre (8,130 ft.) Daulo Pass, the Highland Highway leads to Simbu Province, high peaks, and narrow valleys etched with coffee plantations. The Western Highlands Province then begins with its capital, Mount Hagen, surrounded by more mountains and plantations, as well as one of the country's most fertile regions. Mount Hagen has a reputation for insecurity, especially at night, when raskols — bandits — prowl the streets. Wary travellers should stay outside the city at the comfortable Rondon Ridge Lodge, set on a mountain which towers above the valley. The charming lodge even offers electric blankets to offset the cold highland nights before a sunrise hike in pursuit of elusive birds of paradise. According to ornithologists, no place in the world offers better viewing than here in these remote highlands. And then there are the beaches of PNG—long, pristine, and lapped by the calm clear waters of the South Pacific, where divers claim that the visibility can be the same below water as it is above. At the Tufi Dive Resort seemingly countless war wrecks, including aircraft, lie within easy reach, strewn across the teeming seabed. Not far away, coastal clans such as the Amuioan use dugout canoes to venture deep into the mangroves, and visitors can observe traditional dancing on incandescent white beaches. The scene might seem surreal, almost dreamlike. Just like Papua New Guinea itself.

Creature Comforts

Luxurious hospitality should not be expected in PNG. Comfortable, yes, memorable, often, but five-star or even four, almost never. Tourism is still developing here. The best lodgings can be found at the Rapopo Beach Resort in Kapoko, the Bird of Paradise Hotel in Goroka, the Rondon Ridge Lodge in Mount Hagen, the Airways Hotel in Port Moresby, Tufi Dive Resort in Oro Province, and Loloata Island Resort in Port Moresby, which has its own private island.

Go Now

Regardless of its image, whether intimidating or merely exotic, Papua New Guinea is no ordinary destination. It is a place where you can still glimpse a way of life utterly unchanged by modernity. How it is experienced depends on the traveller's ability to appreciate an challenge in its own context: wild, remote, and untouched.

How to Go

Only a couple of airlines currently fly into Port Moresby Airport. The best connections are with Air Niugini, PNG's largest carrier, to Brisbane, Cairns, and Sydney in Australia, as well as Singapore, Hong Kong, Manila, and Tokyo. With a limited infrastructure featuring only a handful of roads (located mainly in the Highlands region), short domestic flights are the only practical way to get around this mountainous nation. Air Niugini flies to around two dozen destinations, and PNG Air operates a small fleet of turboprop planes (Dash 8) as well as smaller twin prop planes flying to around 20 often very remote, destinations. Be prepared for a dose of old-fashioned lack of infrastructure. Leave non-essential valuables at home and bring basics like a good solar flash lamp, mosquito net, and sturdy walking shoes. Good binoculars are also essential.

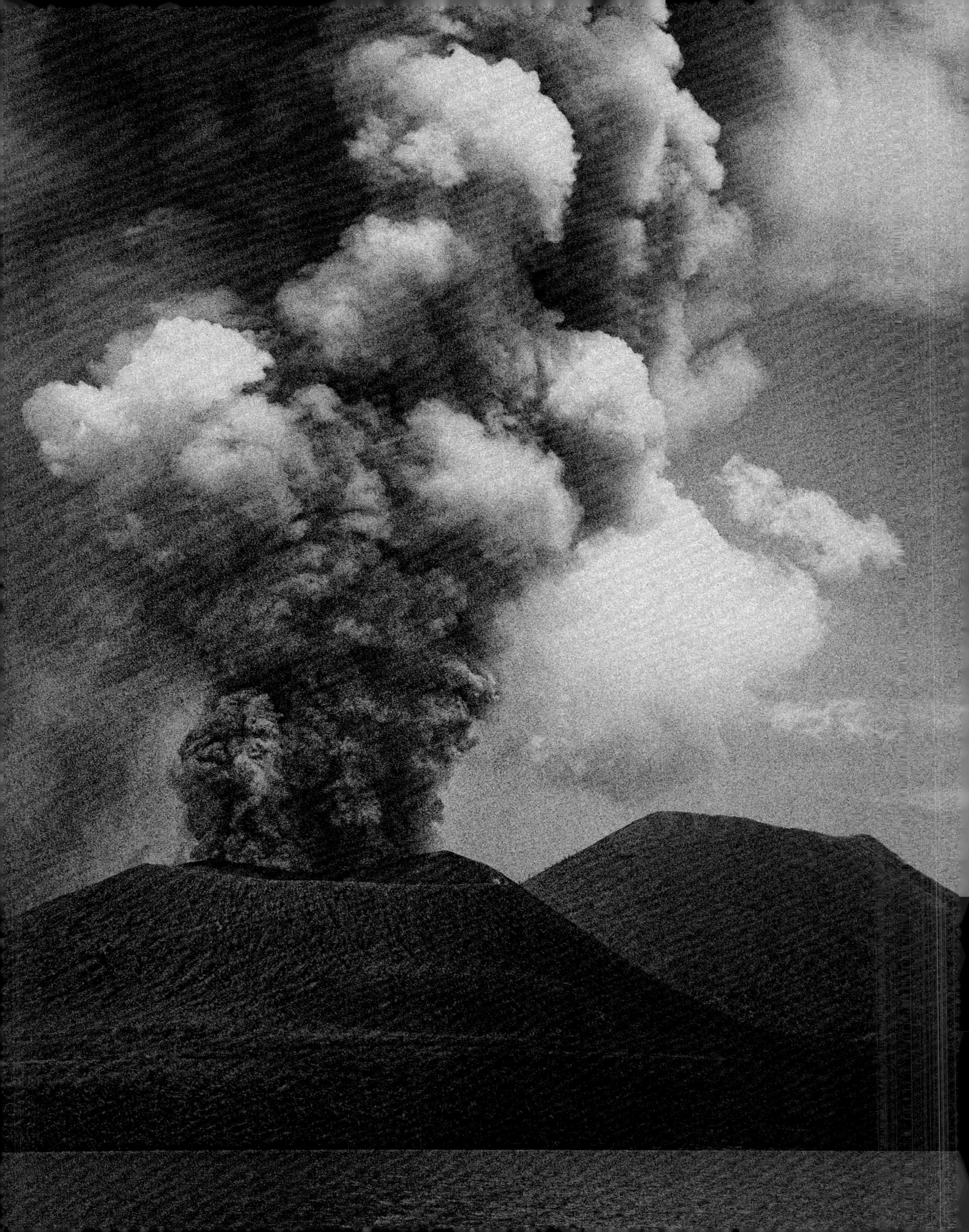

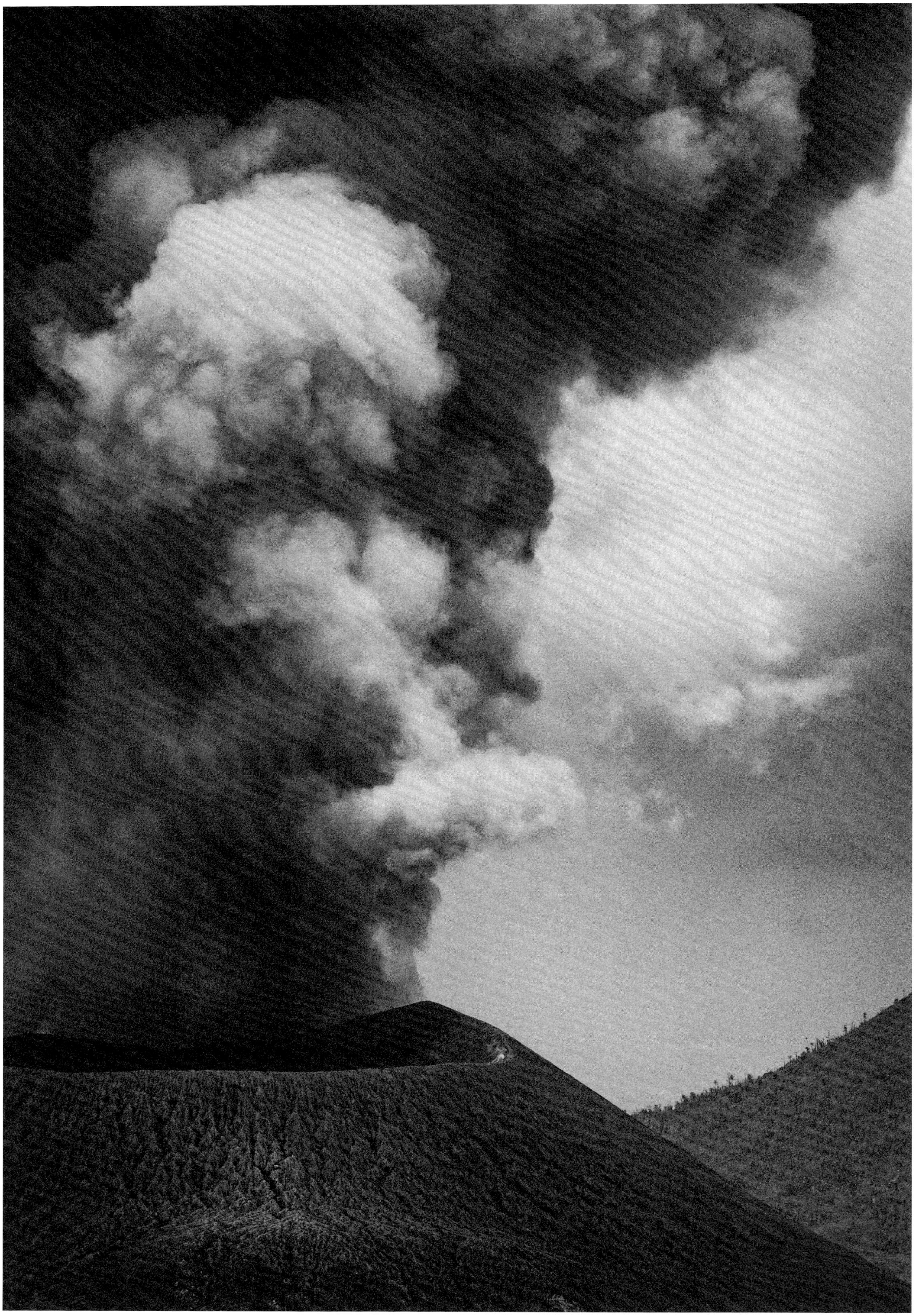

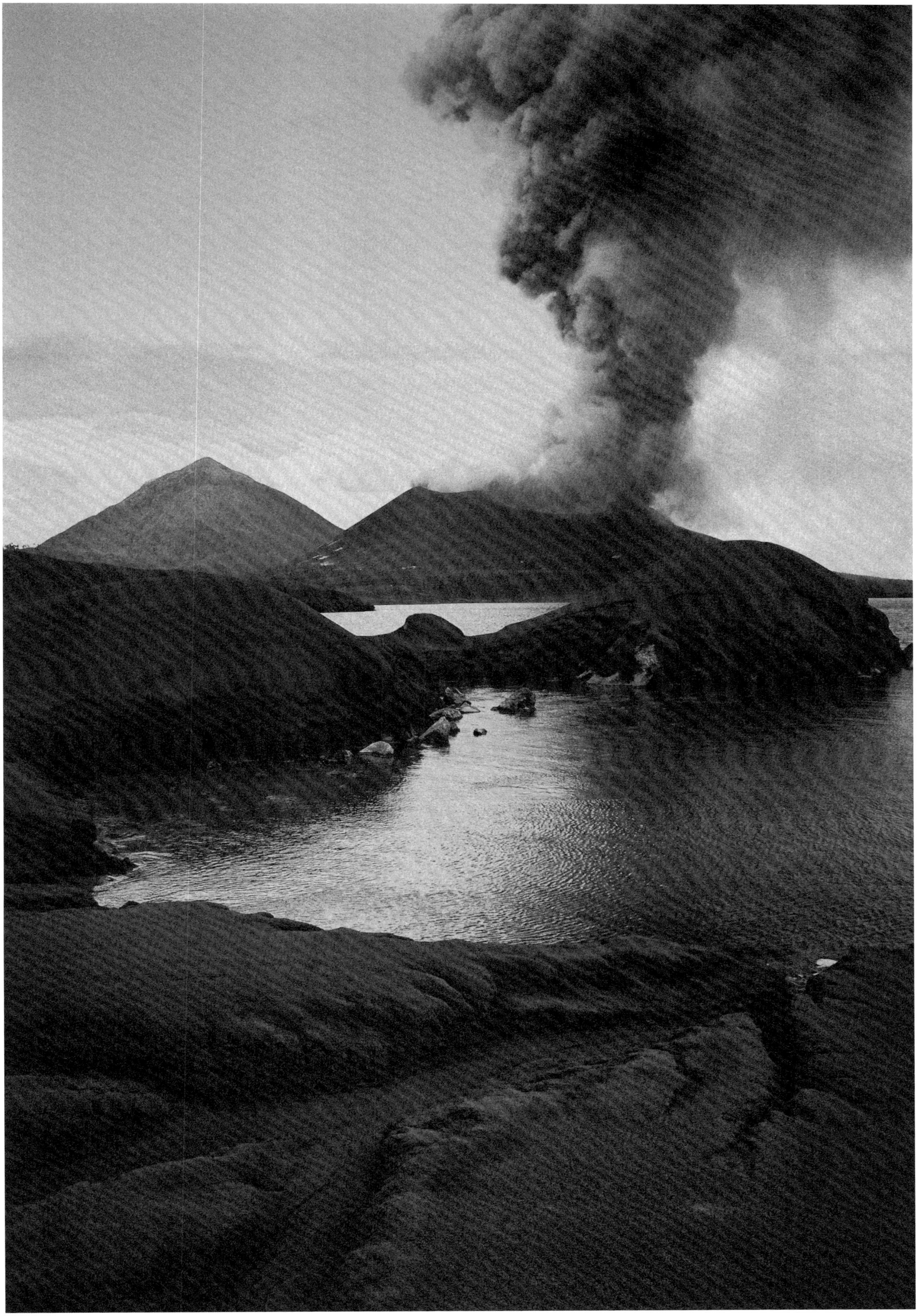

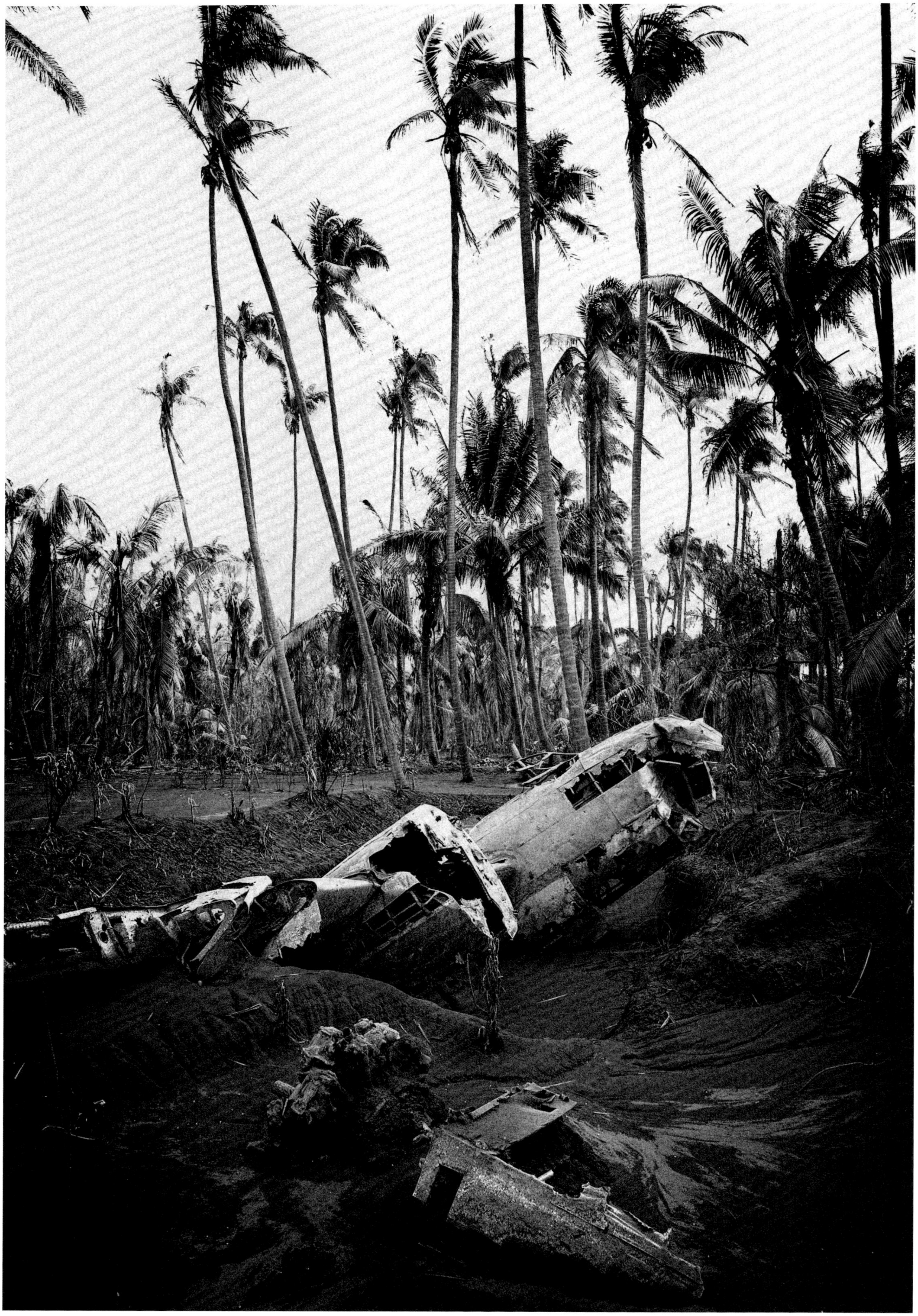

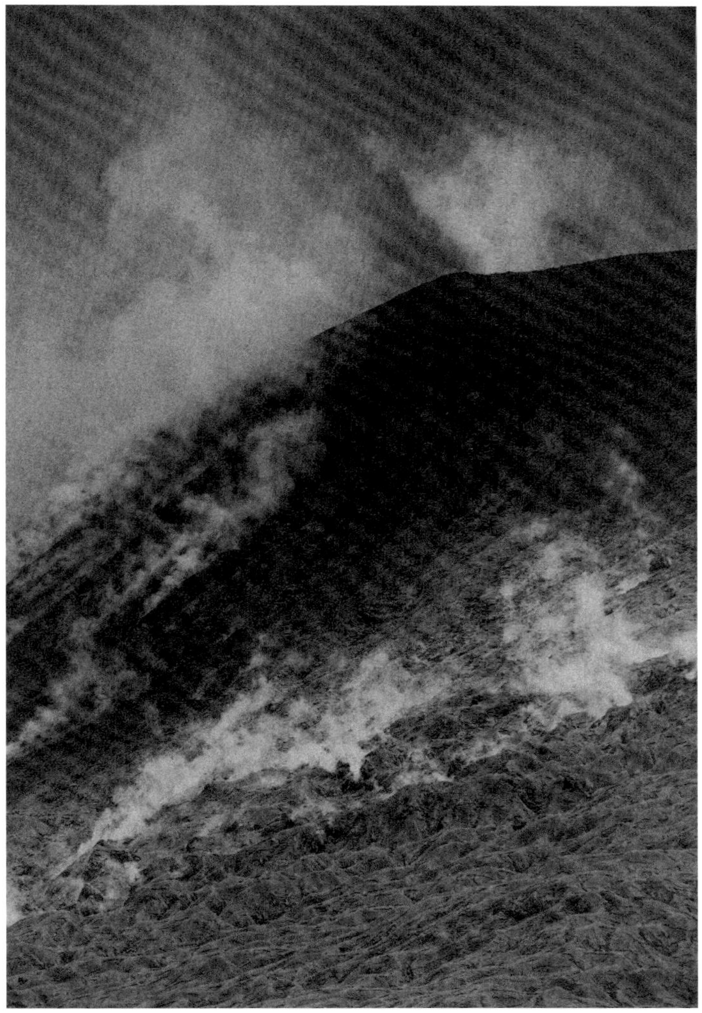

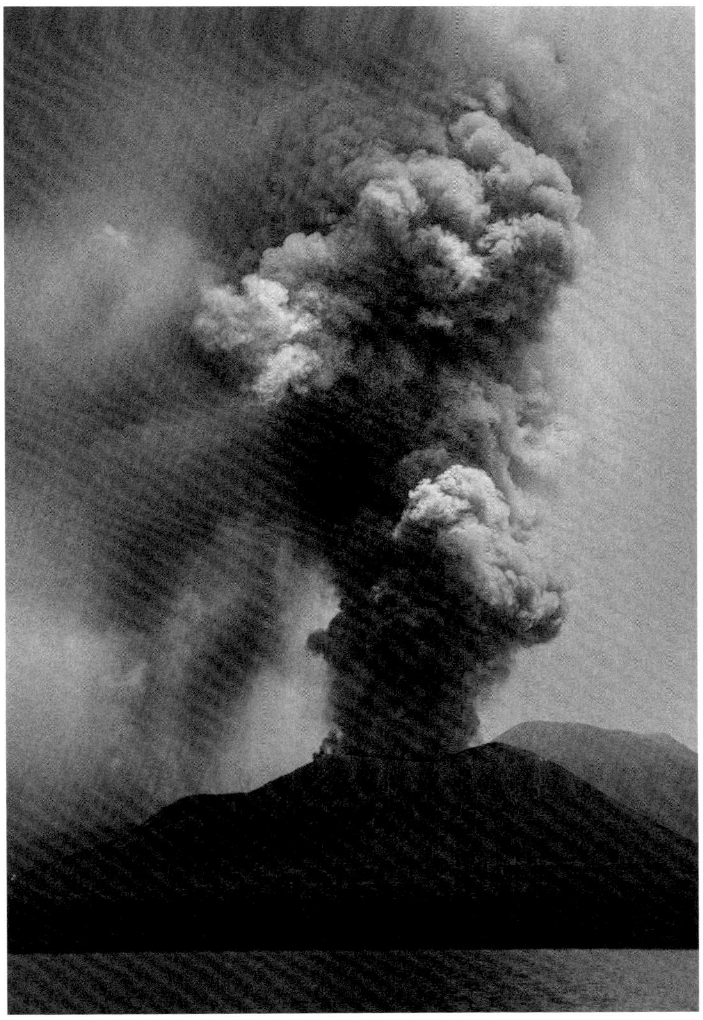

This spread Papua New Guinea has the most active volcanoes in the South West Pacific. The volcanoes form an arc from the north coast of New Guinea near the border with Indonesia to dreamy Bougainville Island in the east. Active and famous volcanoes of the island nation, including Manam, Karkar, Lamington, Langila, Ulawun, Rabaul, and Bagana, have produced some of the largest eruptions in recent millennia. Some have impressive calderas measuring over five kilometres (3 mi.) wide, while others are filled with water and form lakes, like Long Island, which measures 12 by 10 kilometres (7.5 by 6 mi.). The city of Rabaul, known as a 'Japanese Gibraltar', served as a naval and air base for the Imperial Japanese Army during the World War II. It is set between an active volcano and a picturesque bay. On land, ash-covered plane wrecks bear silent witness to desperate aerial battles and volcanic eruptions that followed. Many more lie hidden below the azure blue surface of the bay.

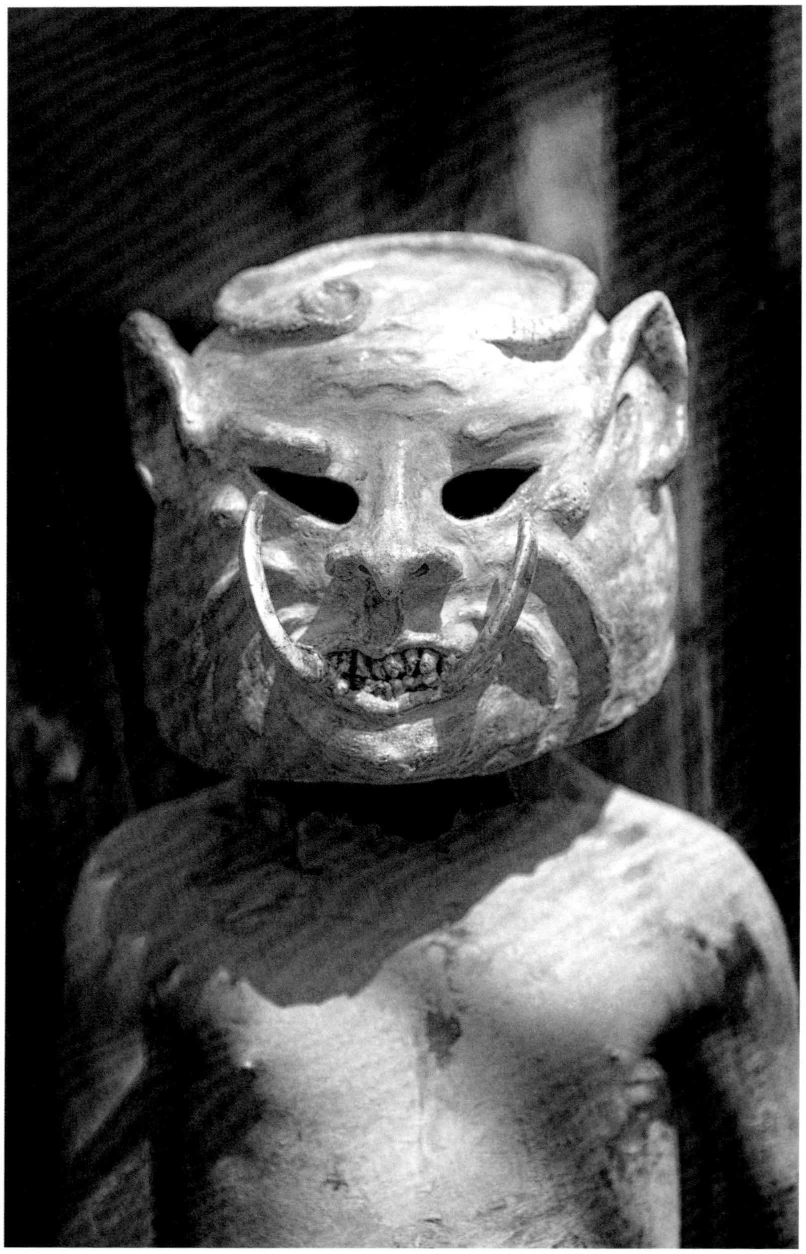

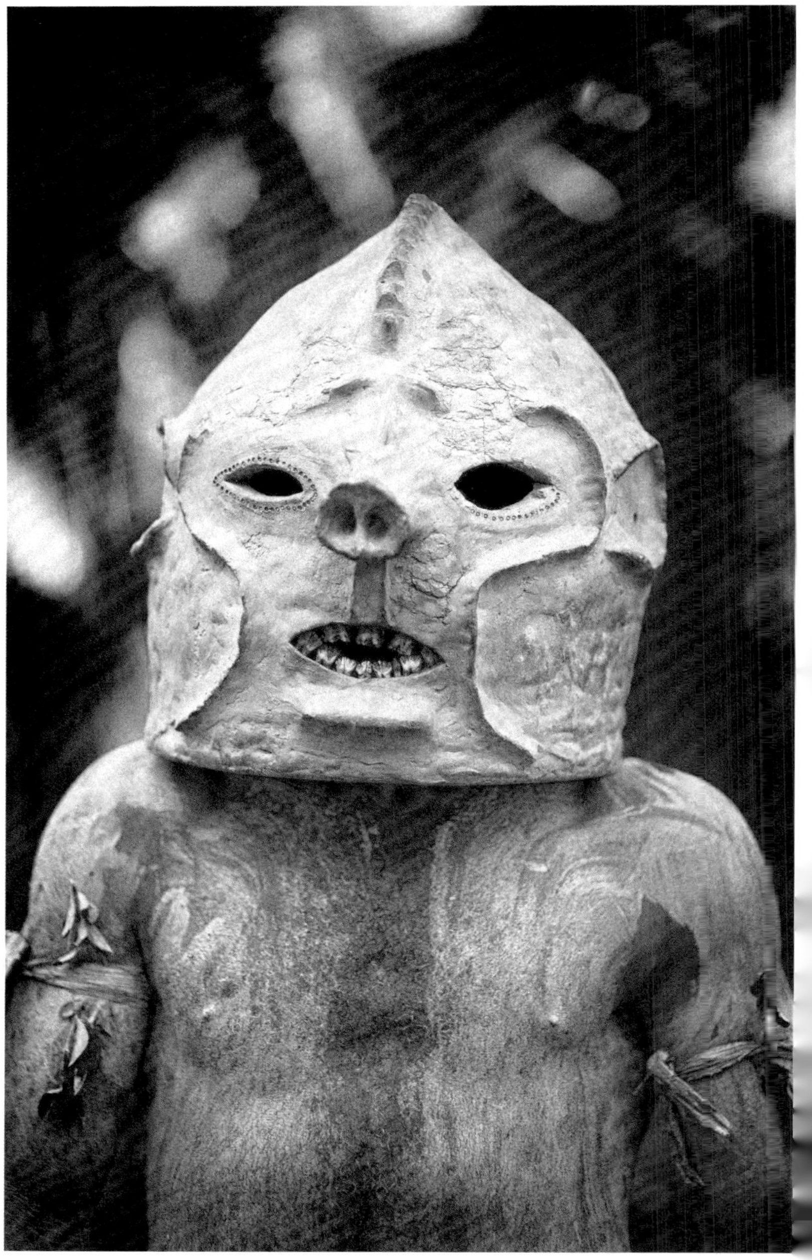

This spread The Asaro mudmen cover their bodies with white clay and wear heavy masks; bamboo claws cover their fingers. Nobody knows what inspired this tradition or when it started. According to one story, the Asaro once hid from an enemy clan on a muddy riverbank. As they emerged with their bodies covered in white clay, their rivals were frightened, believing only ghosts can be white.

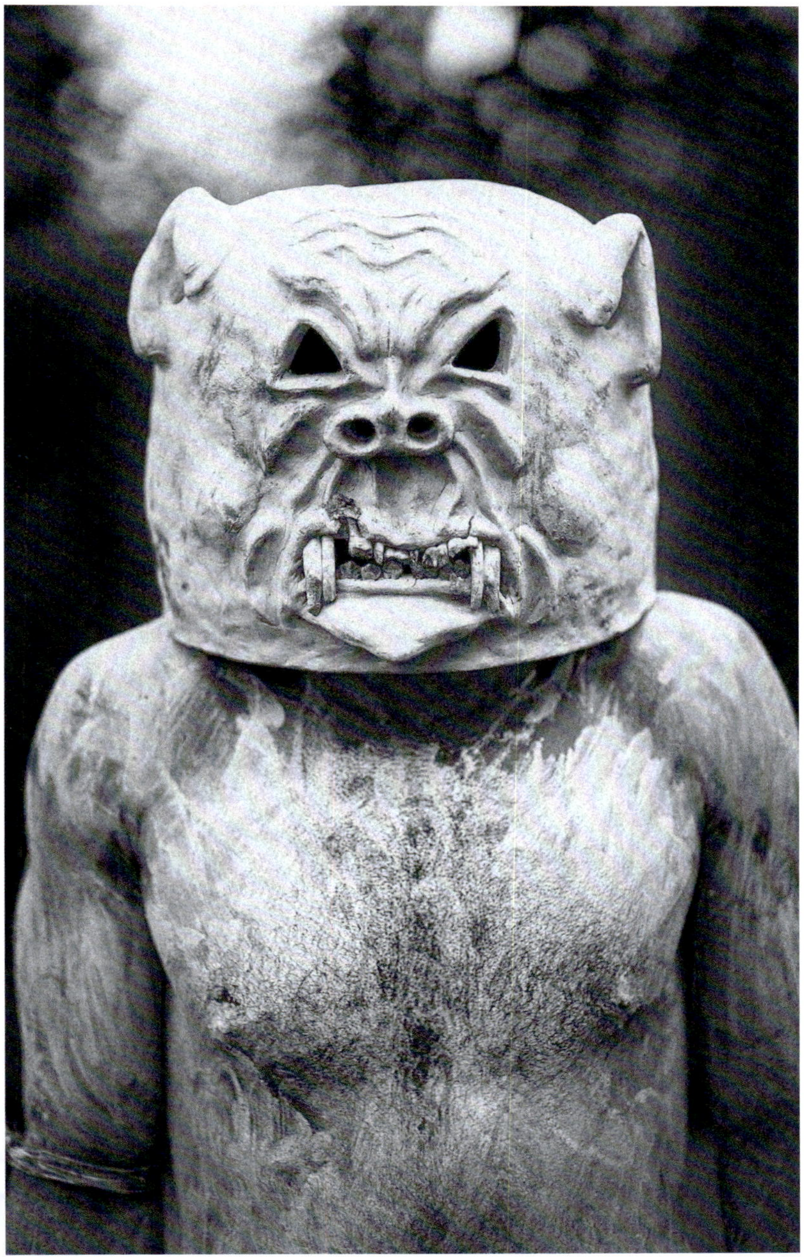
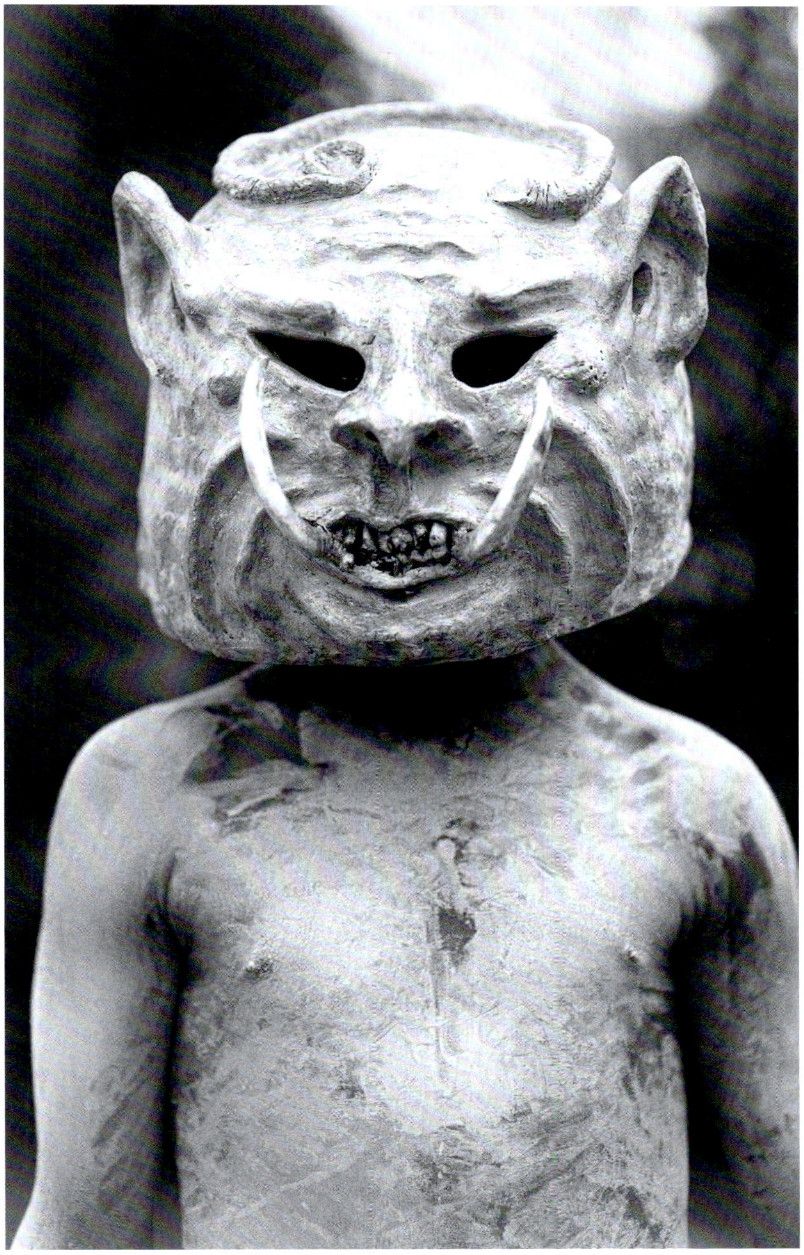

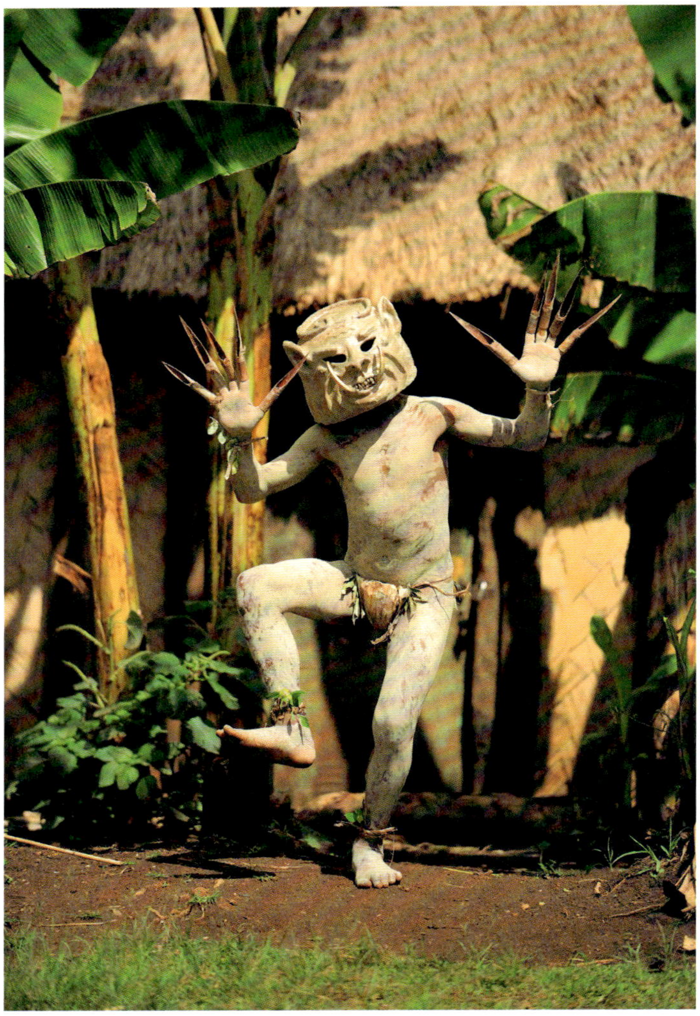

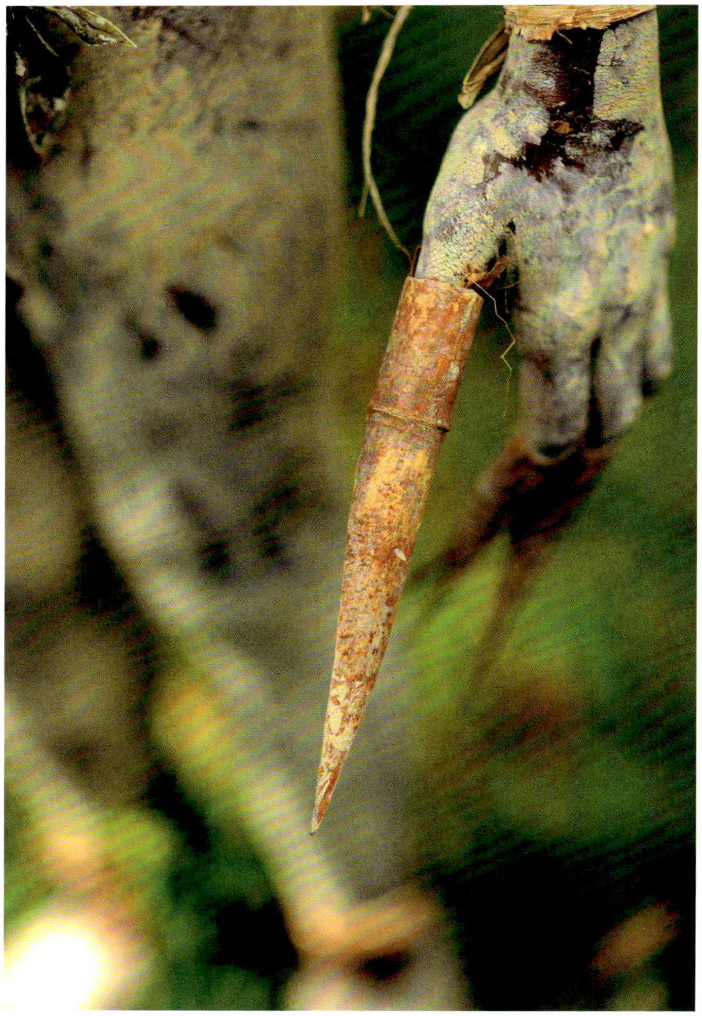

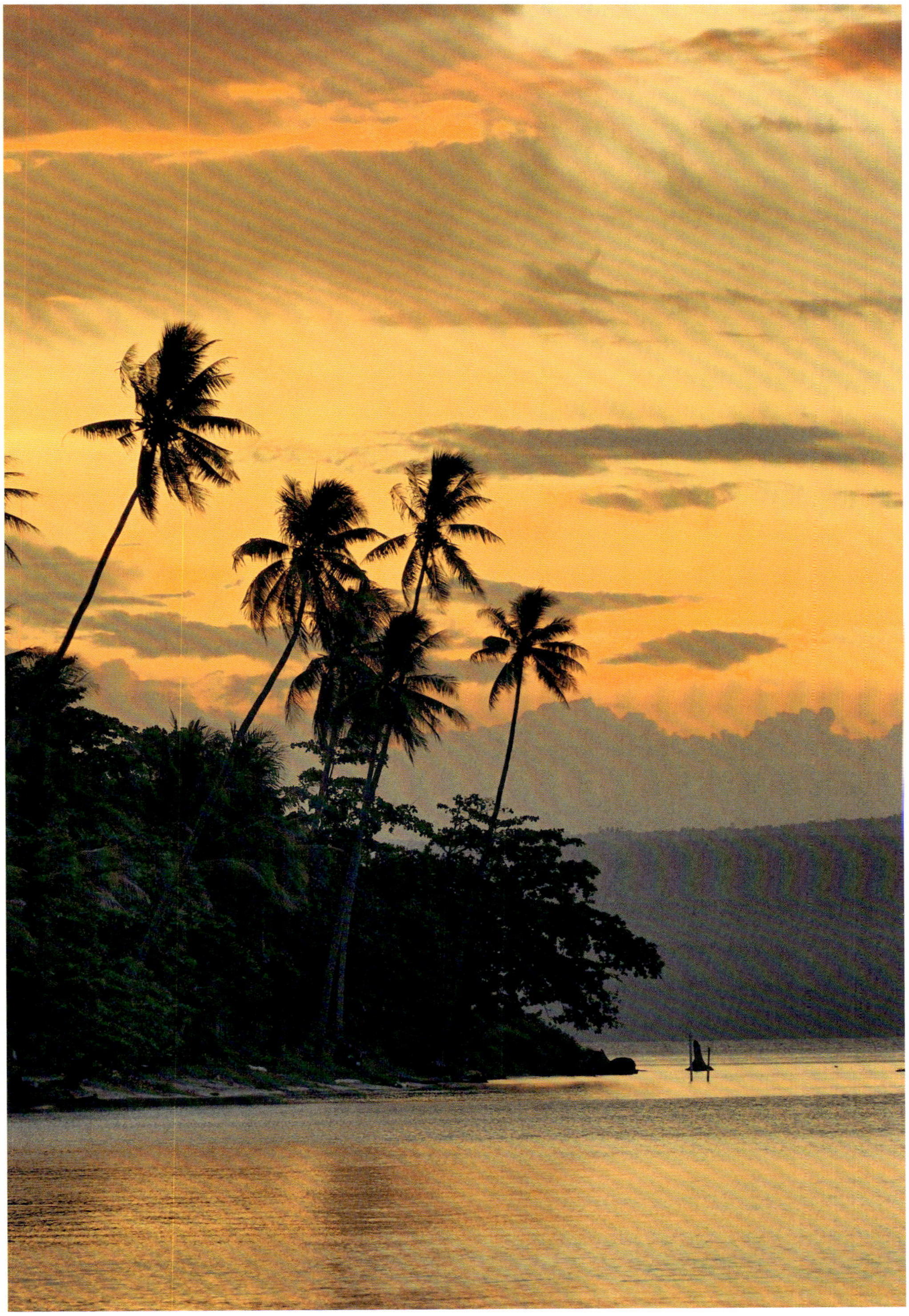

Above More than 600 islands dot the waters around Papua New Guinea. With white sandy beaches, shallow lagoons, and lush palm trees, they are reminiscent of Bora Bora or Tahiti, but pristine and wild, untouched by tourism. Some are perfect for divers and snorkellers, others are a surfer's paradise.

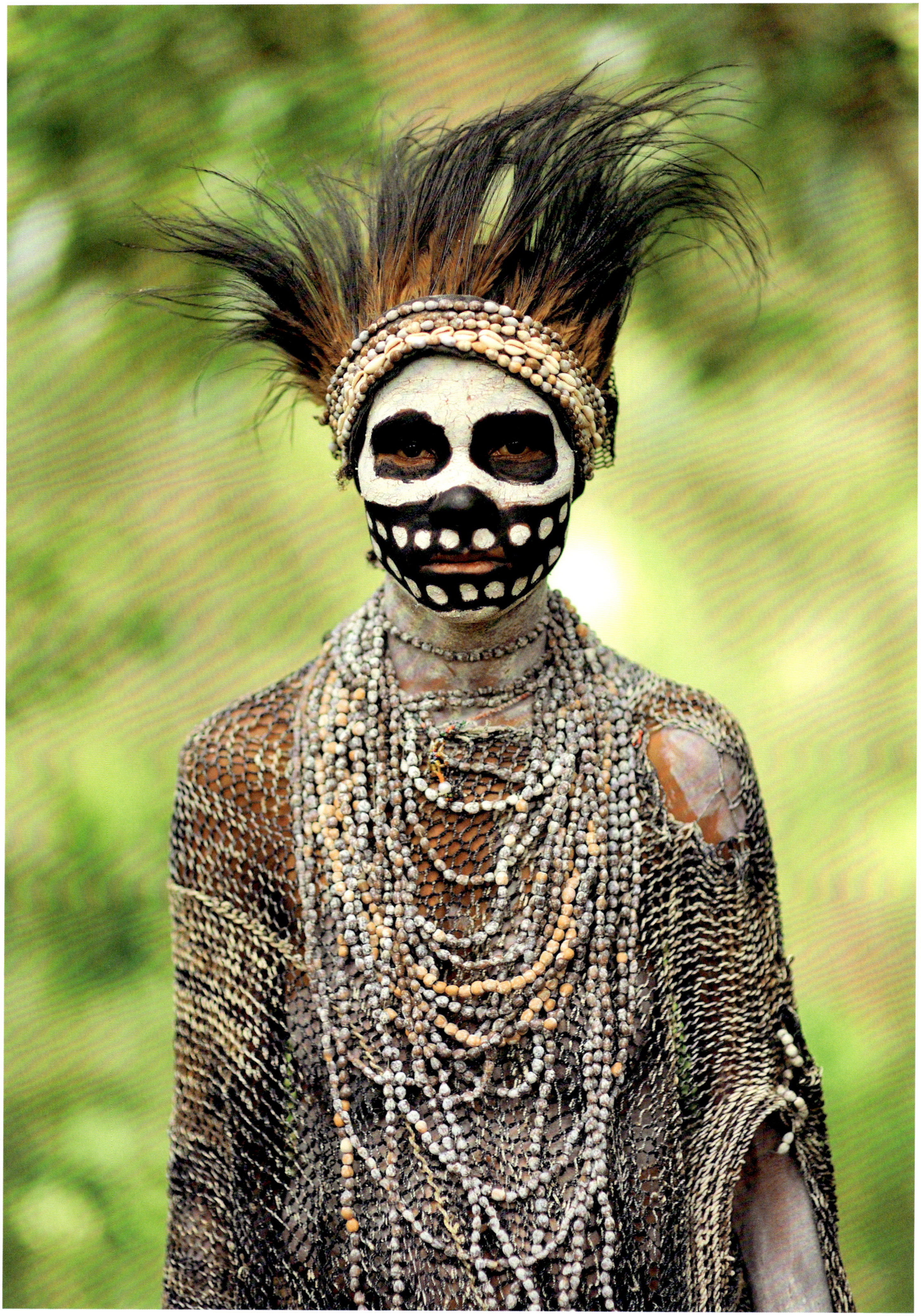

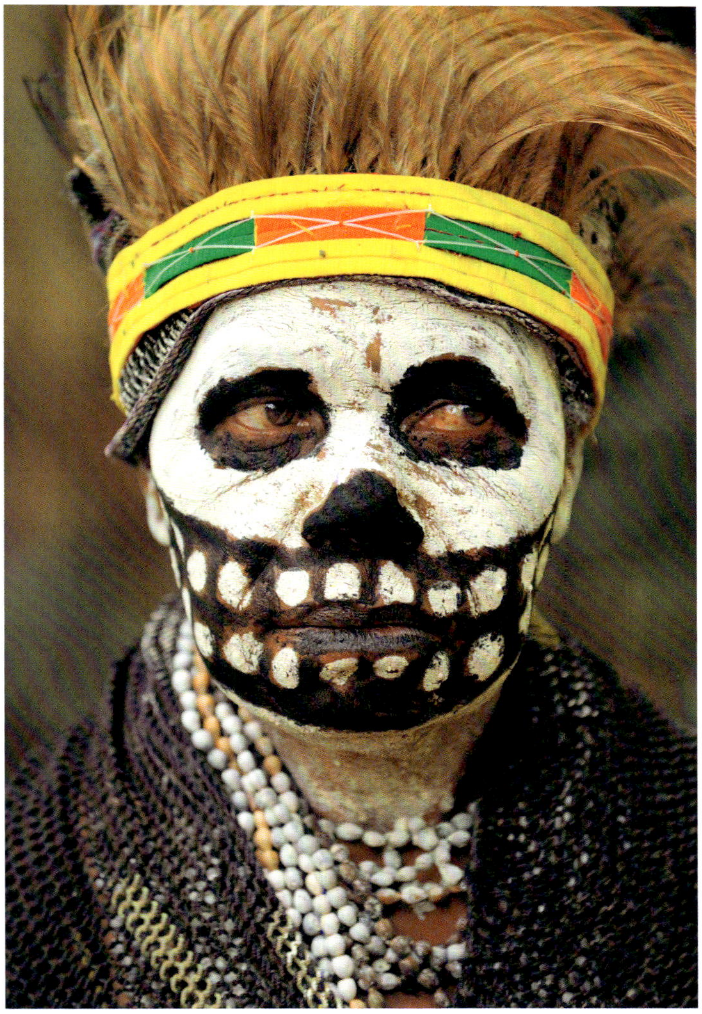

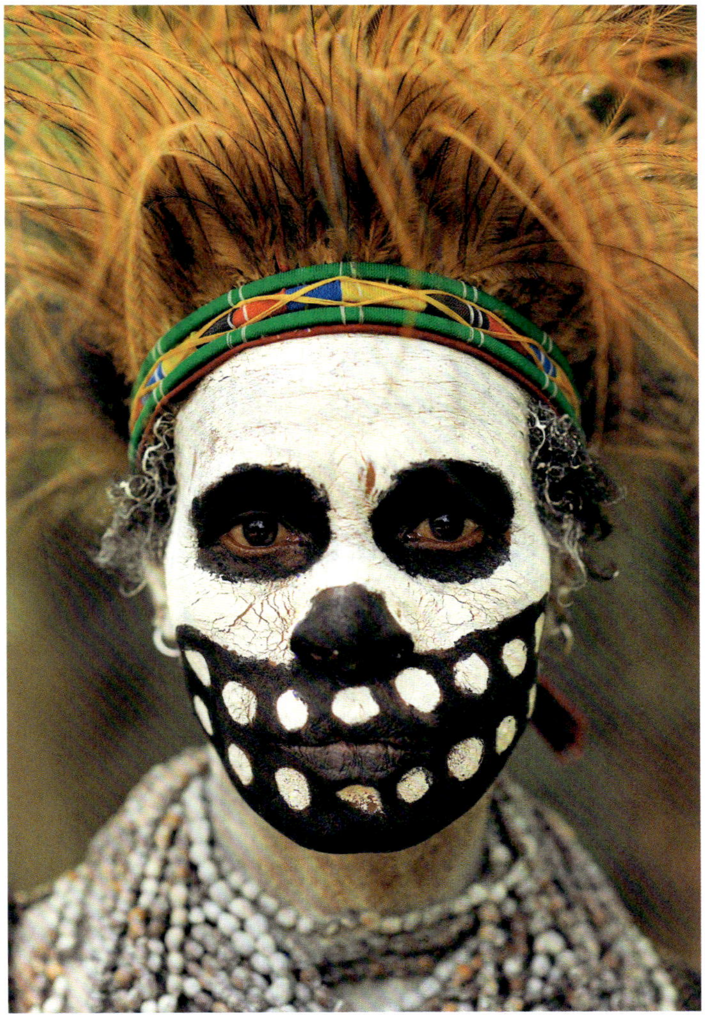

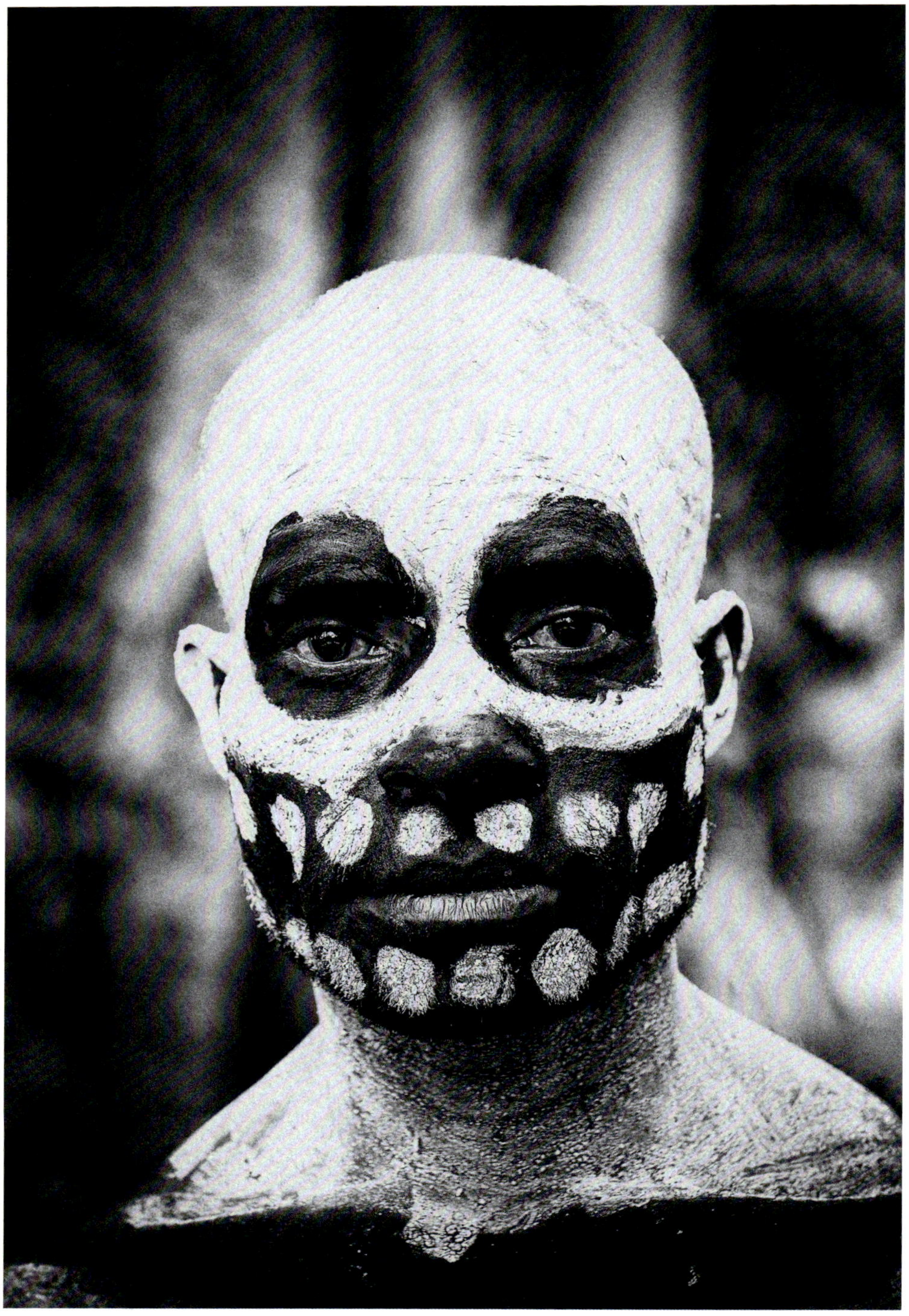

This spread Bugamo skeleton men terrify their foes with white body paint made from clay and ashes. Living in the highlands of Chimbu Province, their tribe was only discovered by explorers in the middle of the last century. Like most tribes in these parts, they are animists who believe that spirits reside in animals, plants, and in the objects that surround them. Evil spirits are called masalai, and puripuri witchcraft is practised.

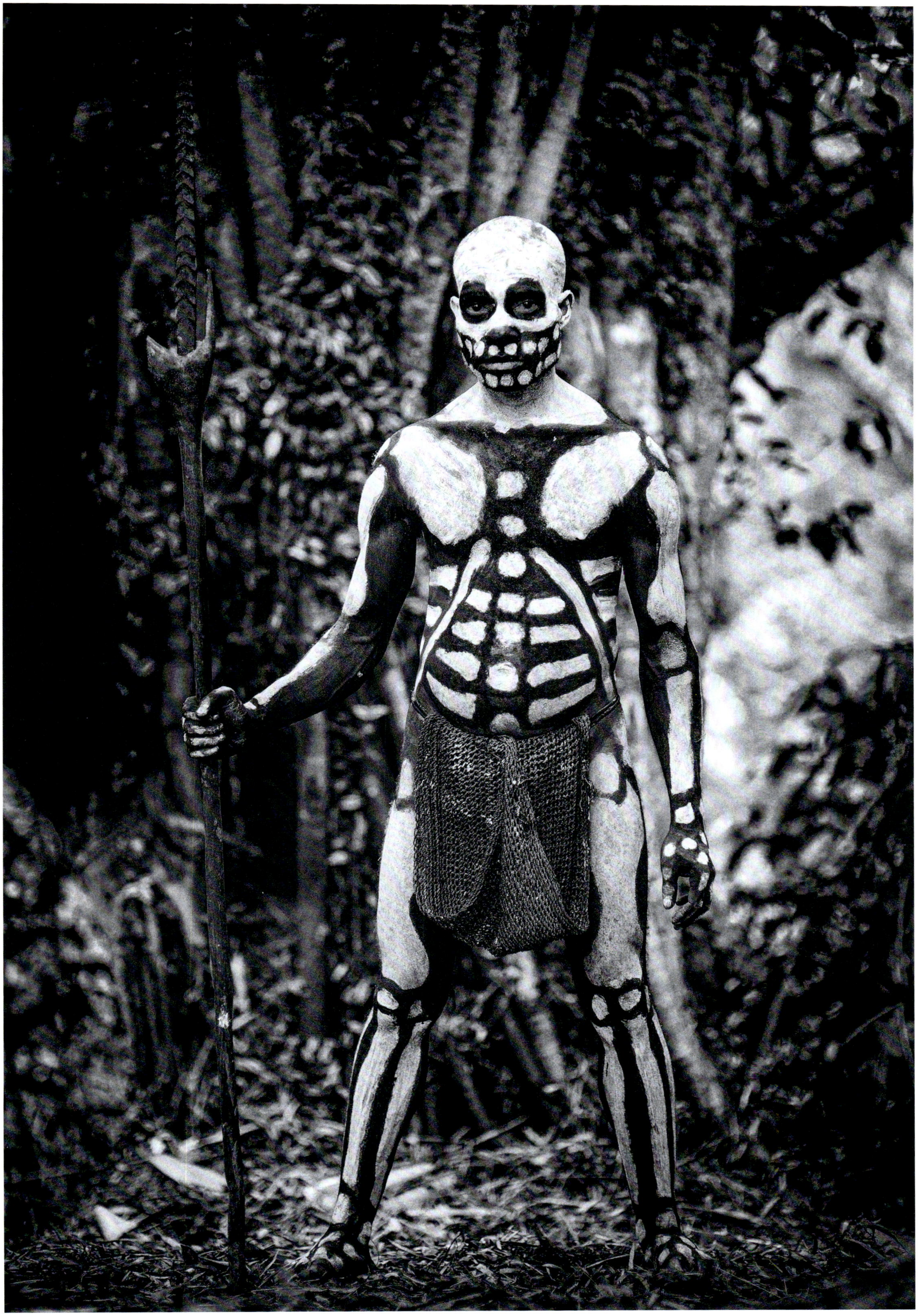

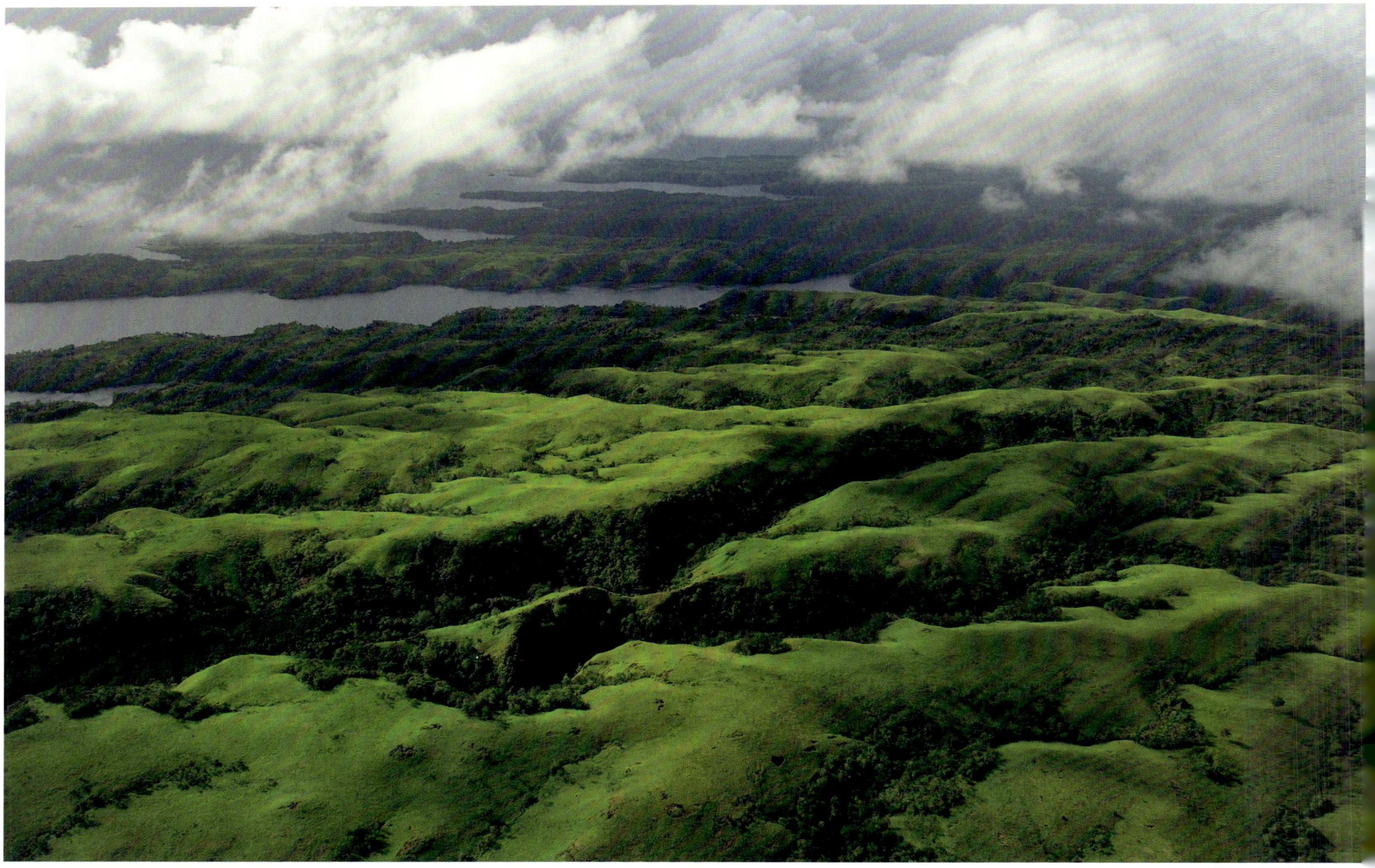

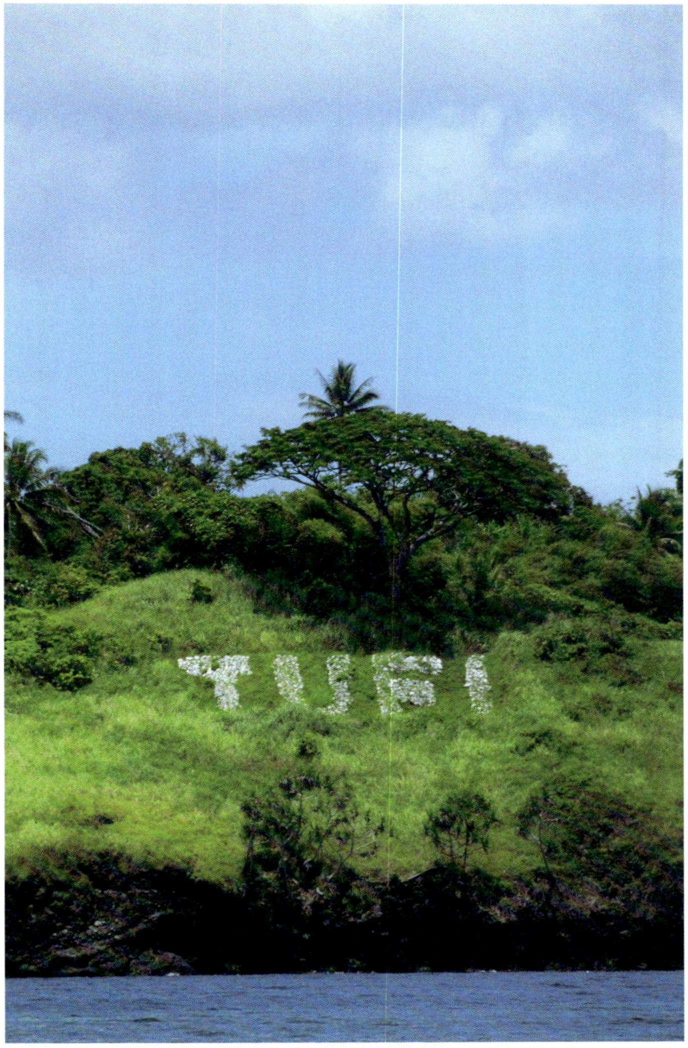

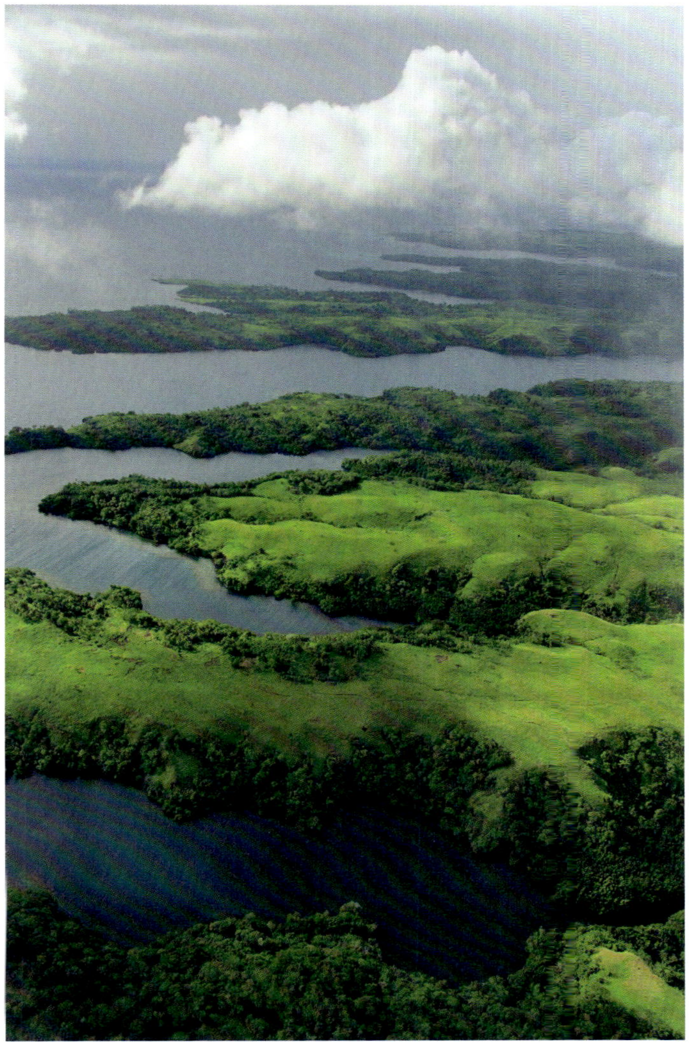

This spread Just an hour's flight from Port Moresby on the north-eastern coast, Cape Nelson has the most southerly fjord system in the world. Better known as rias, these fjords were shaped when enormous quantities of lava flowed into the Pacific Ocean. Seen from above, the peninsula is covered in lush vegetation reminiscent of the Amazon rainforest and surrounded by the blue of the Solomon Sea. This tropical and intensely wild setting makes it one of the most impressive regions in the country.

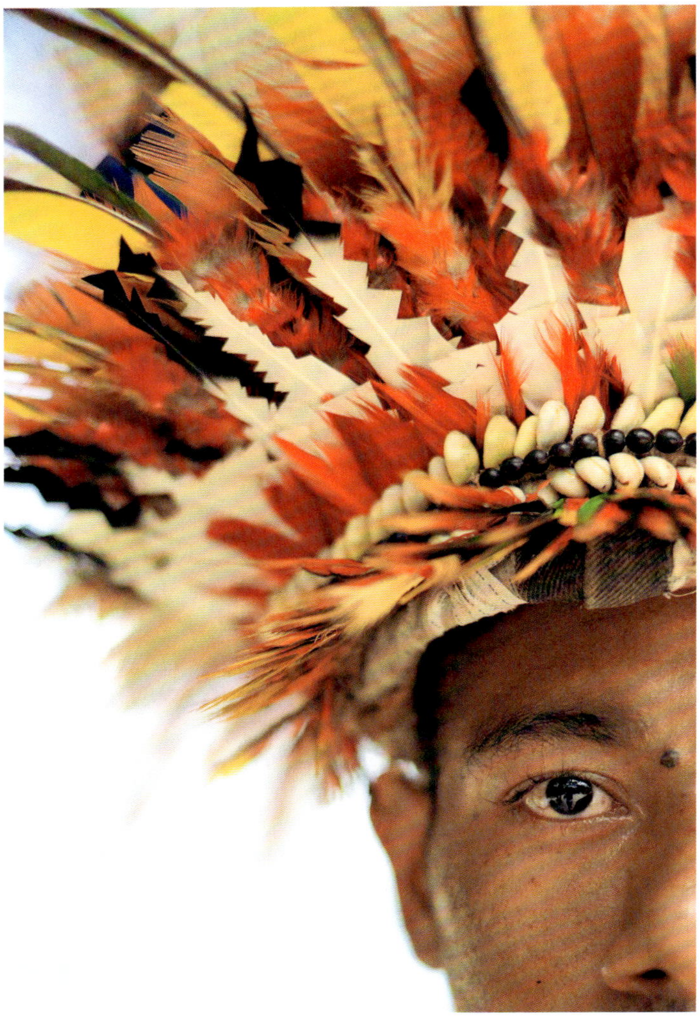
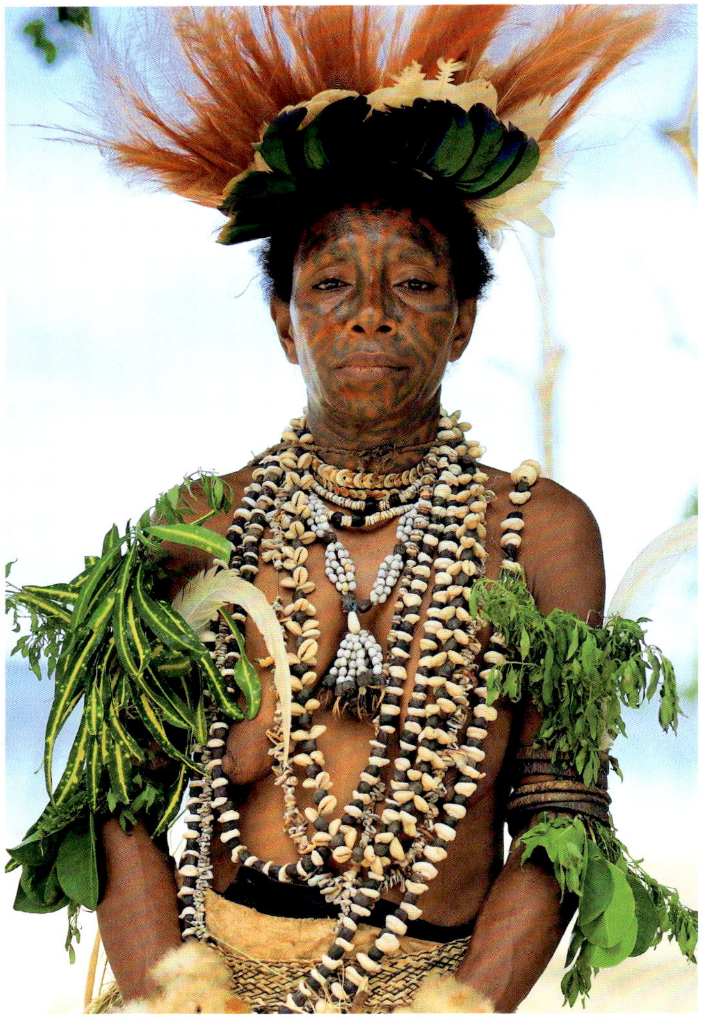
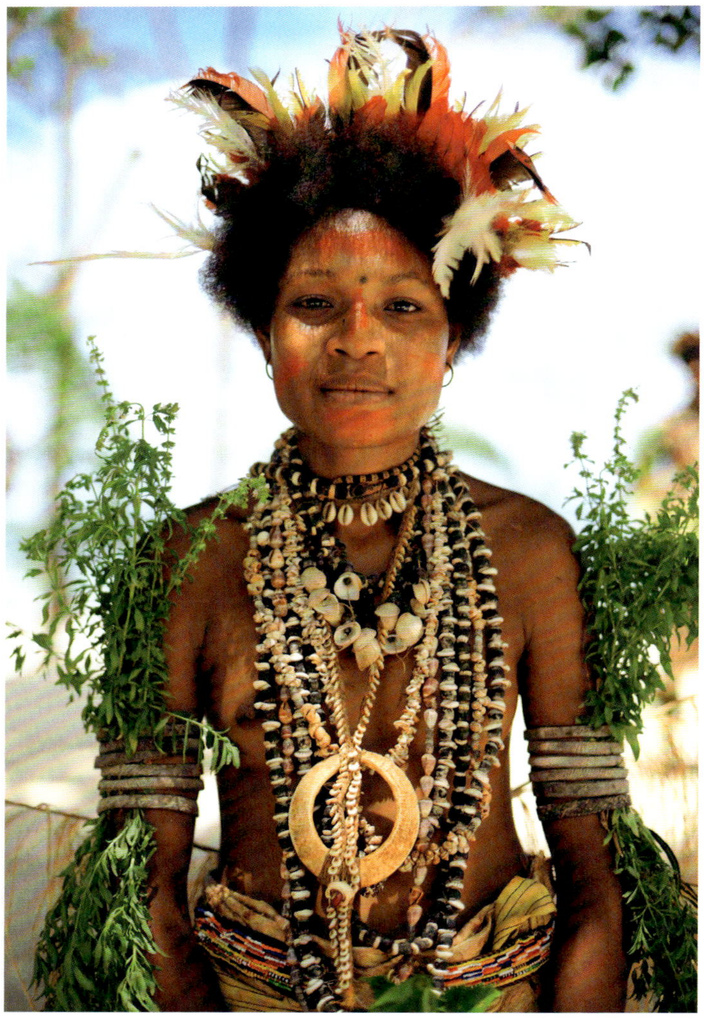
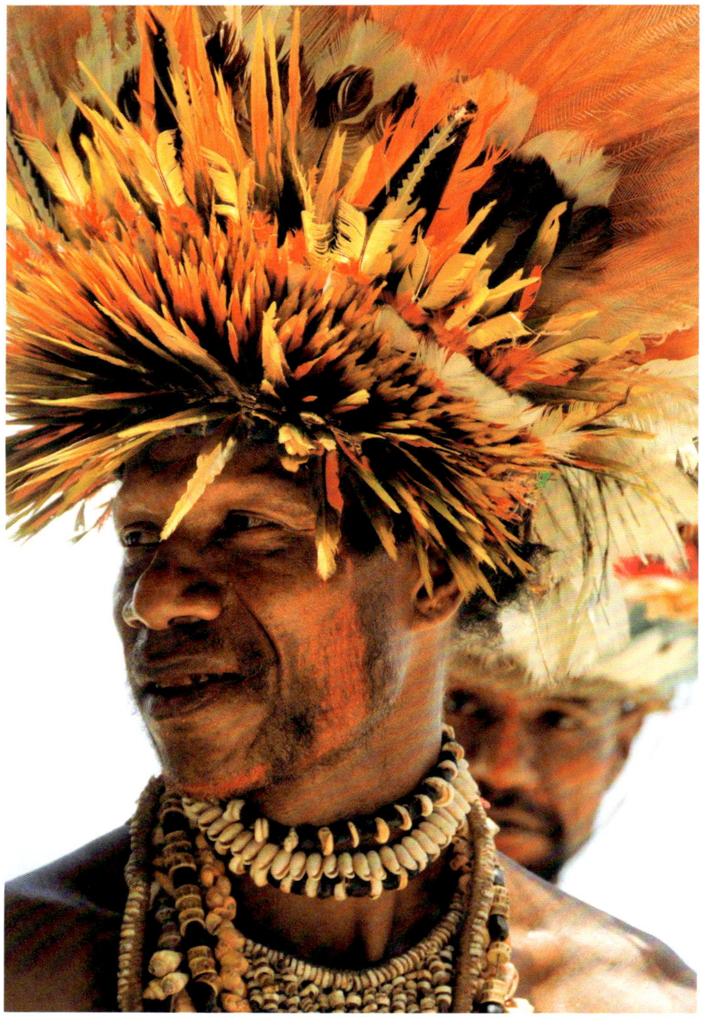

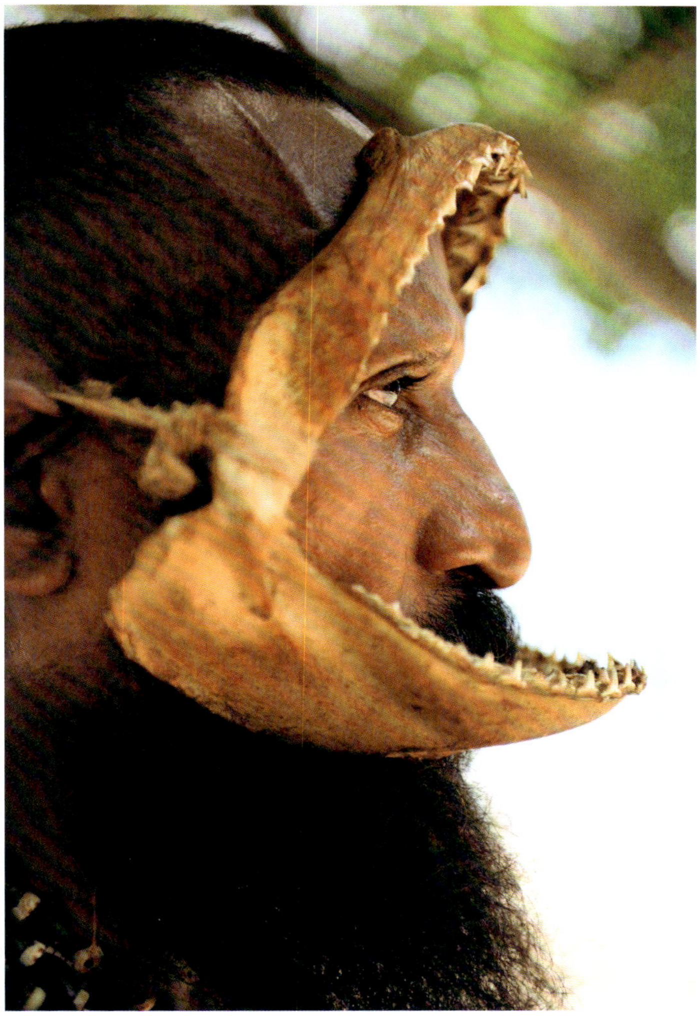

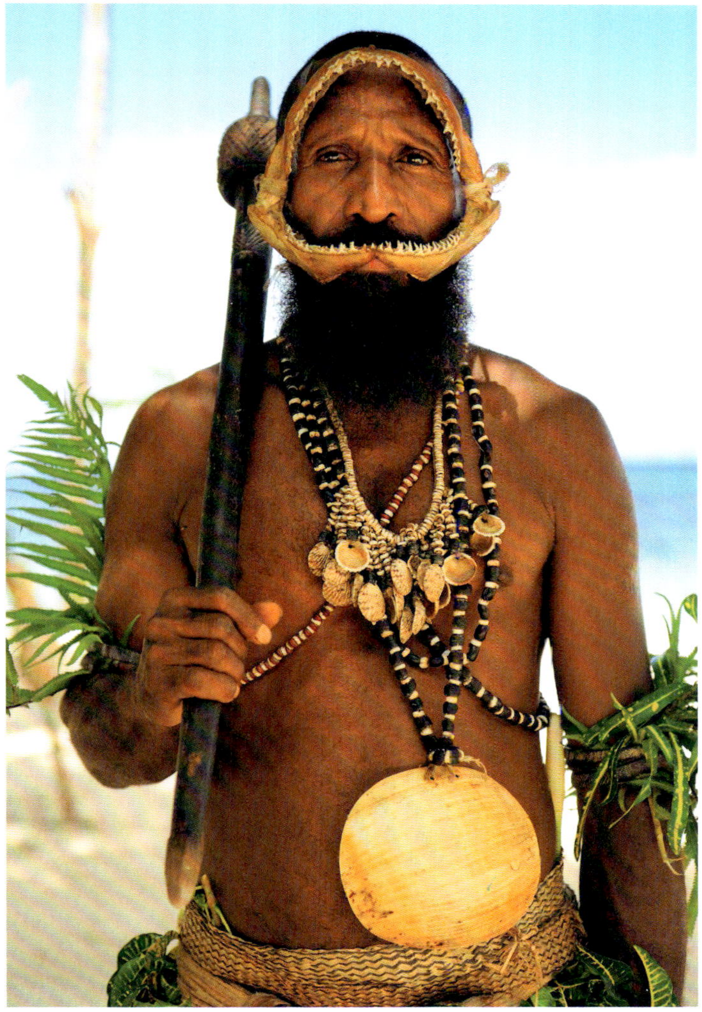

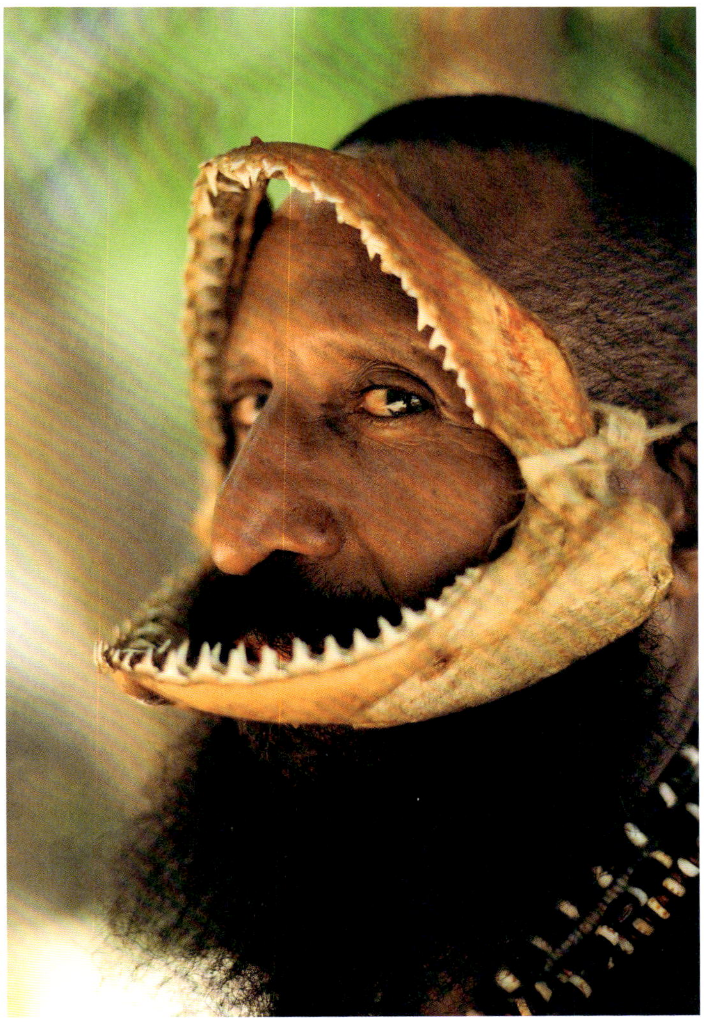

This spread On the beaches in the tropical mangroves near the town of Tufi, people of the Karafe tribe dwell in simple huts. Visitors arrive by dugout canoes over the calm waters of the McLaren Fjord. All festivities are celebrated with so-called sing-sing events that often happen on the beach and involve the beating of drums and hypnotic chanting. Karafe women wear shell necklaces and hibiscus flowers in their hair, while men wear impressive headdresses made of bird-of-paradise feathers. The chief wears an elaborate mother-of-pearl necklace, and the jaw of a shark on his head — valuable heirlooms passed from father to son.

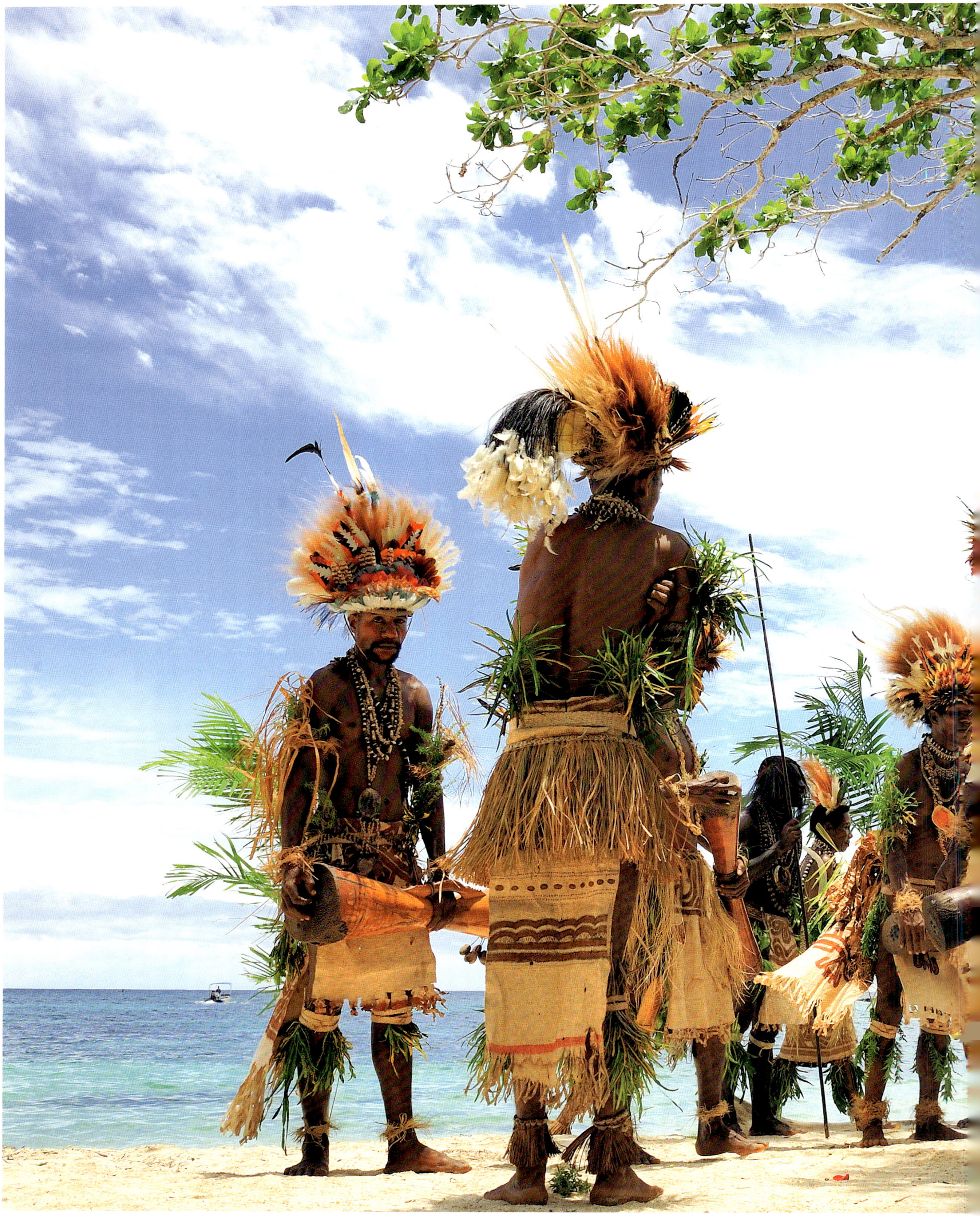

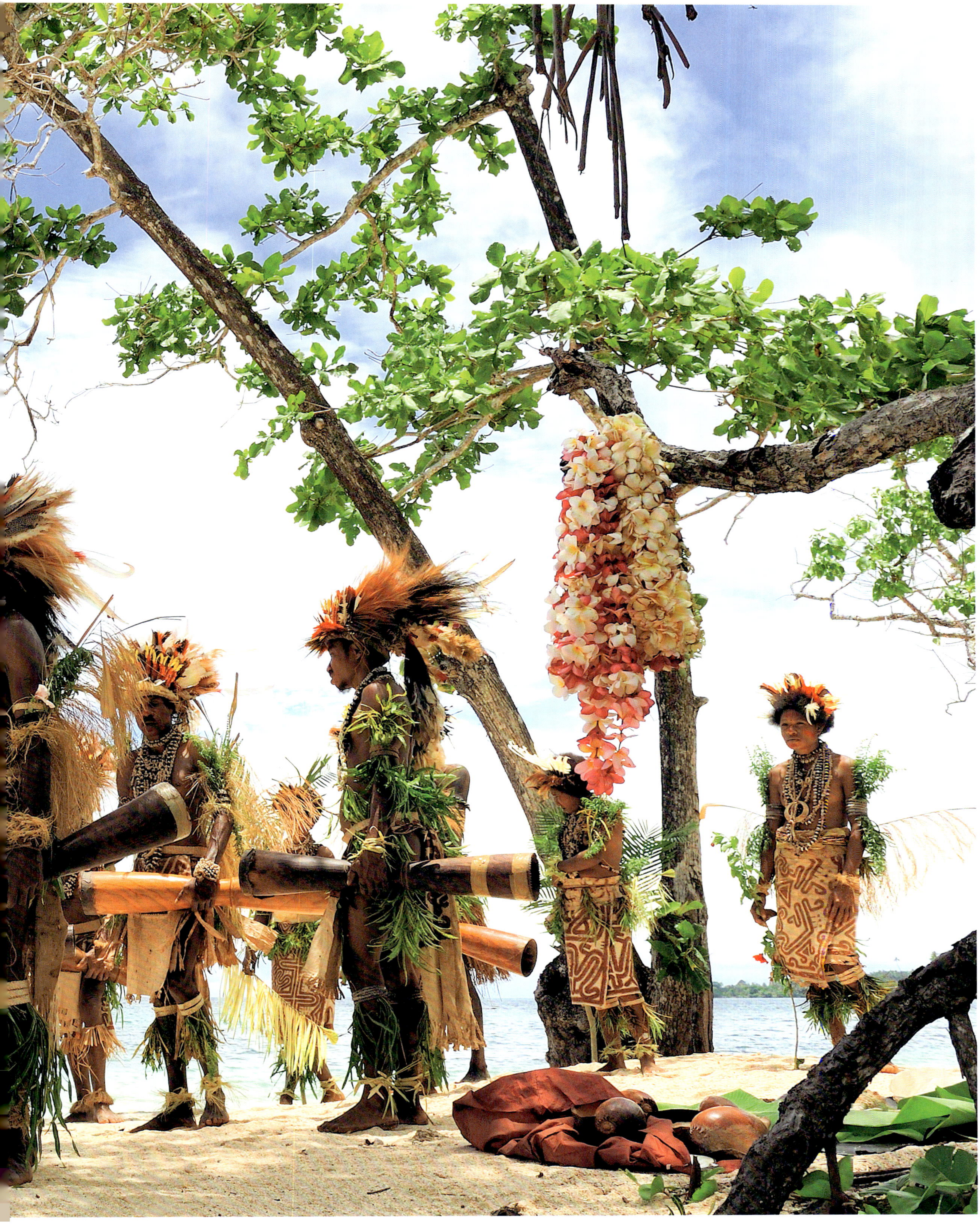

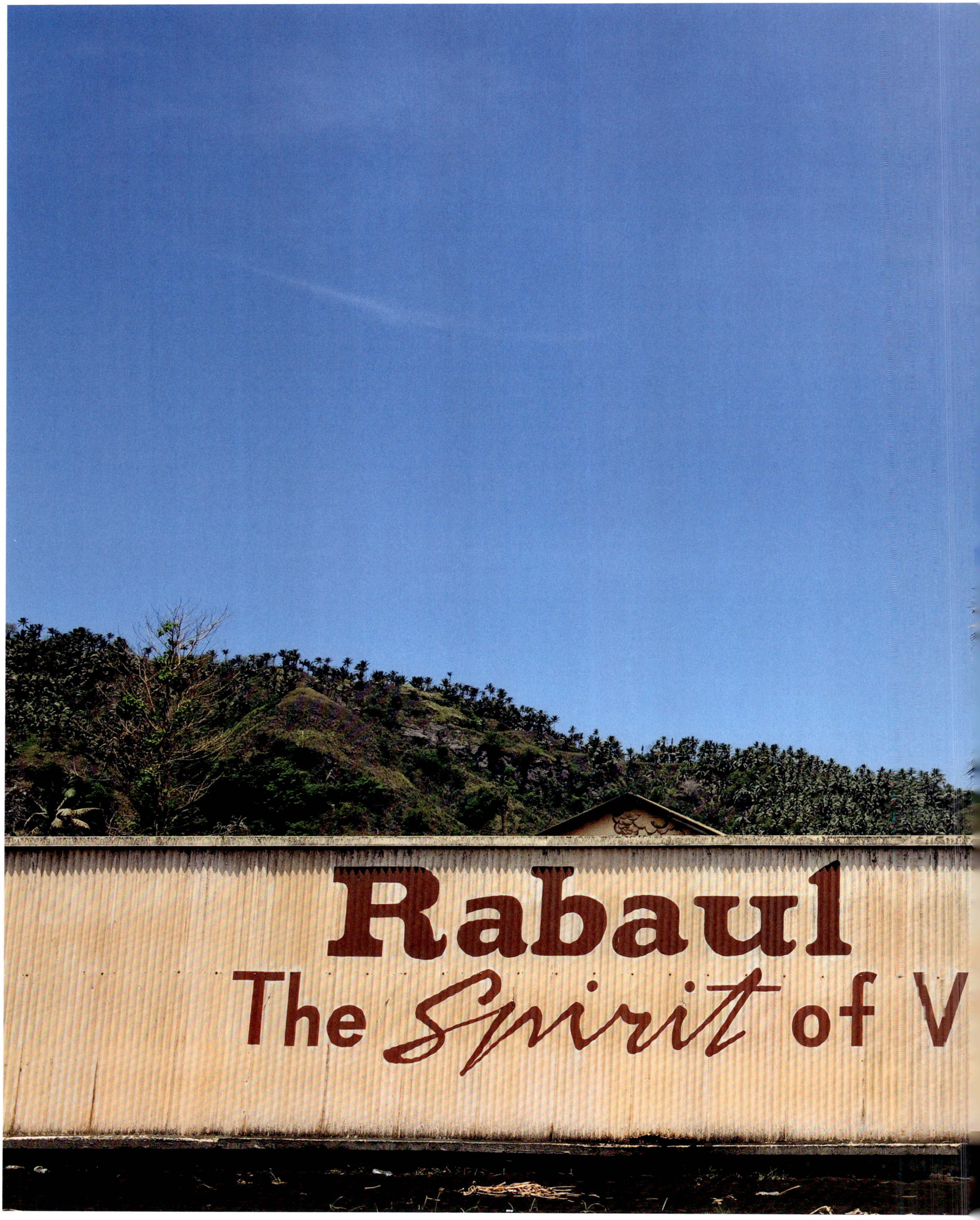
Rabaul
The Spirit of V

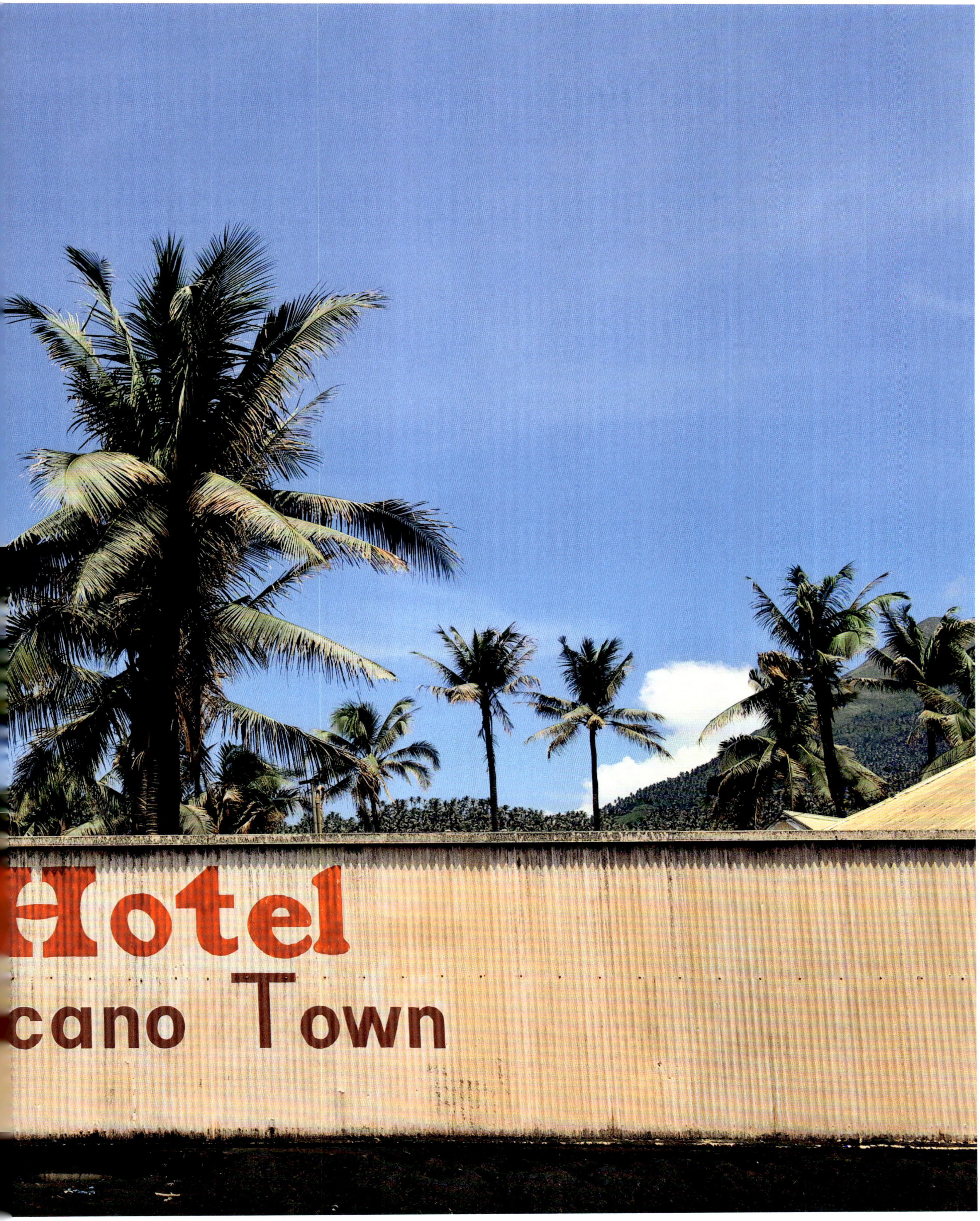
Hotel
cano Town

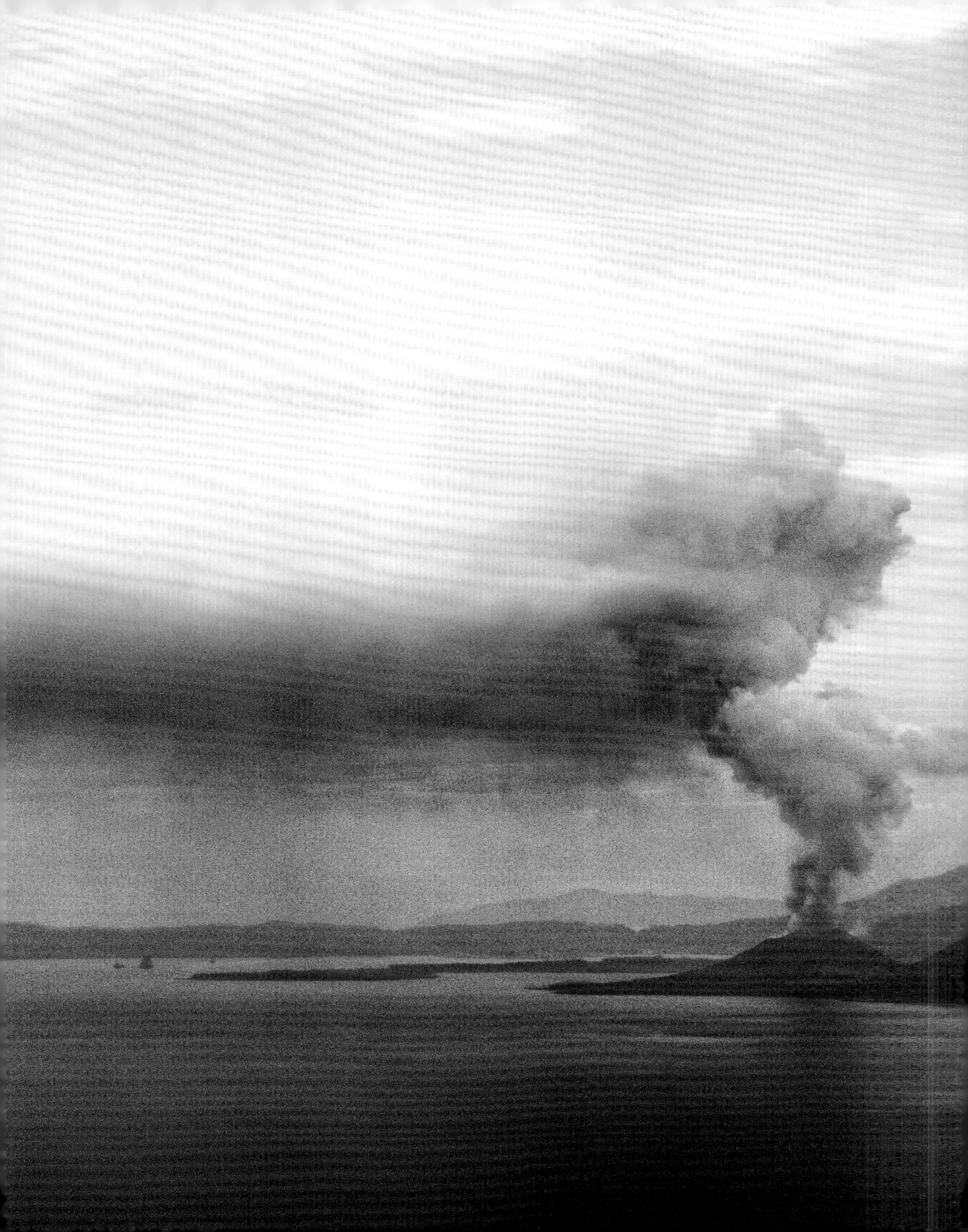

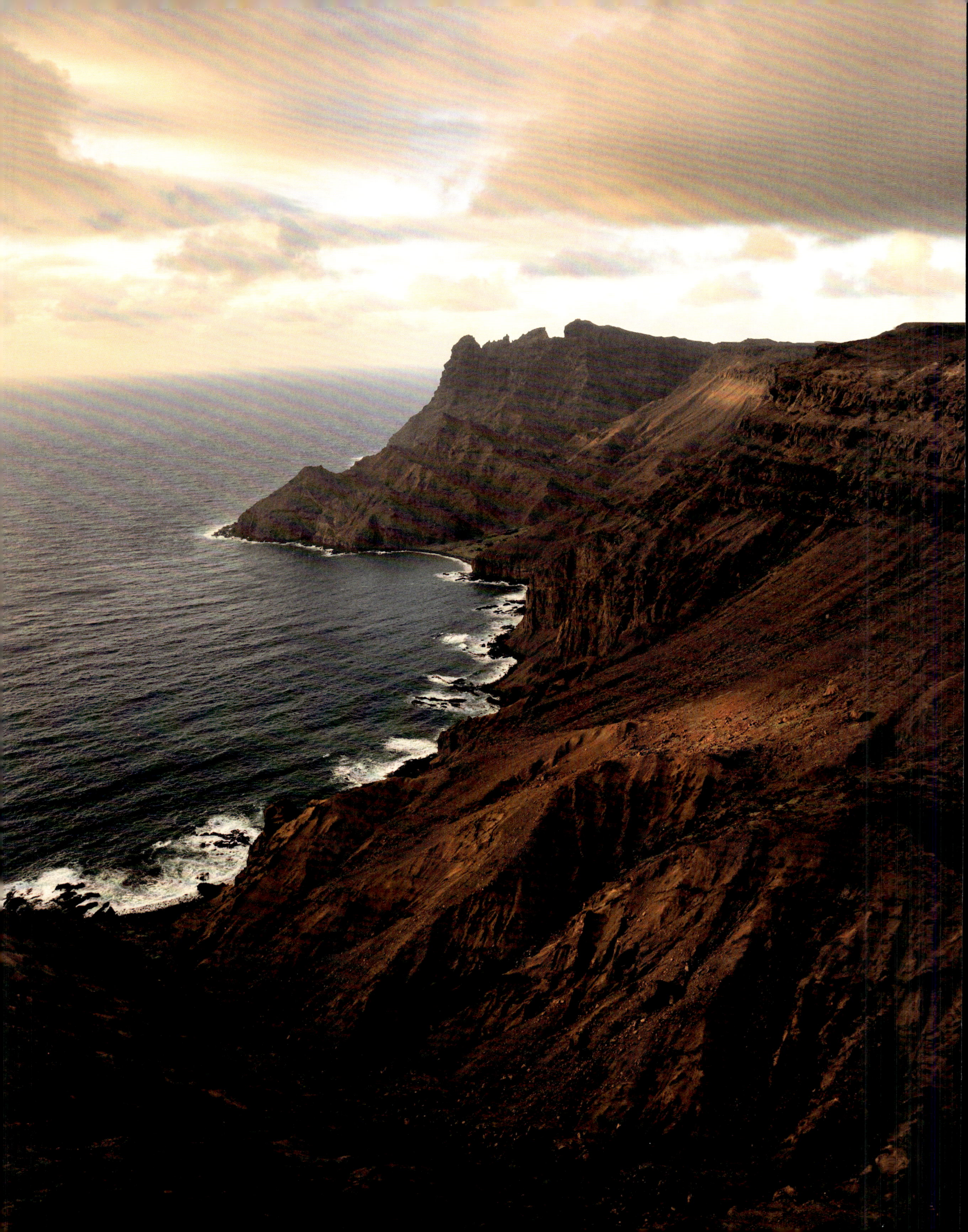

CHAPTER 9

ST HELENA, AN ATLANTIC ISLAND REFUGE

A WINDSWEPT ROCK

St Helena

15°55'34.13" S
05°43'01.59" W

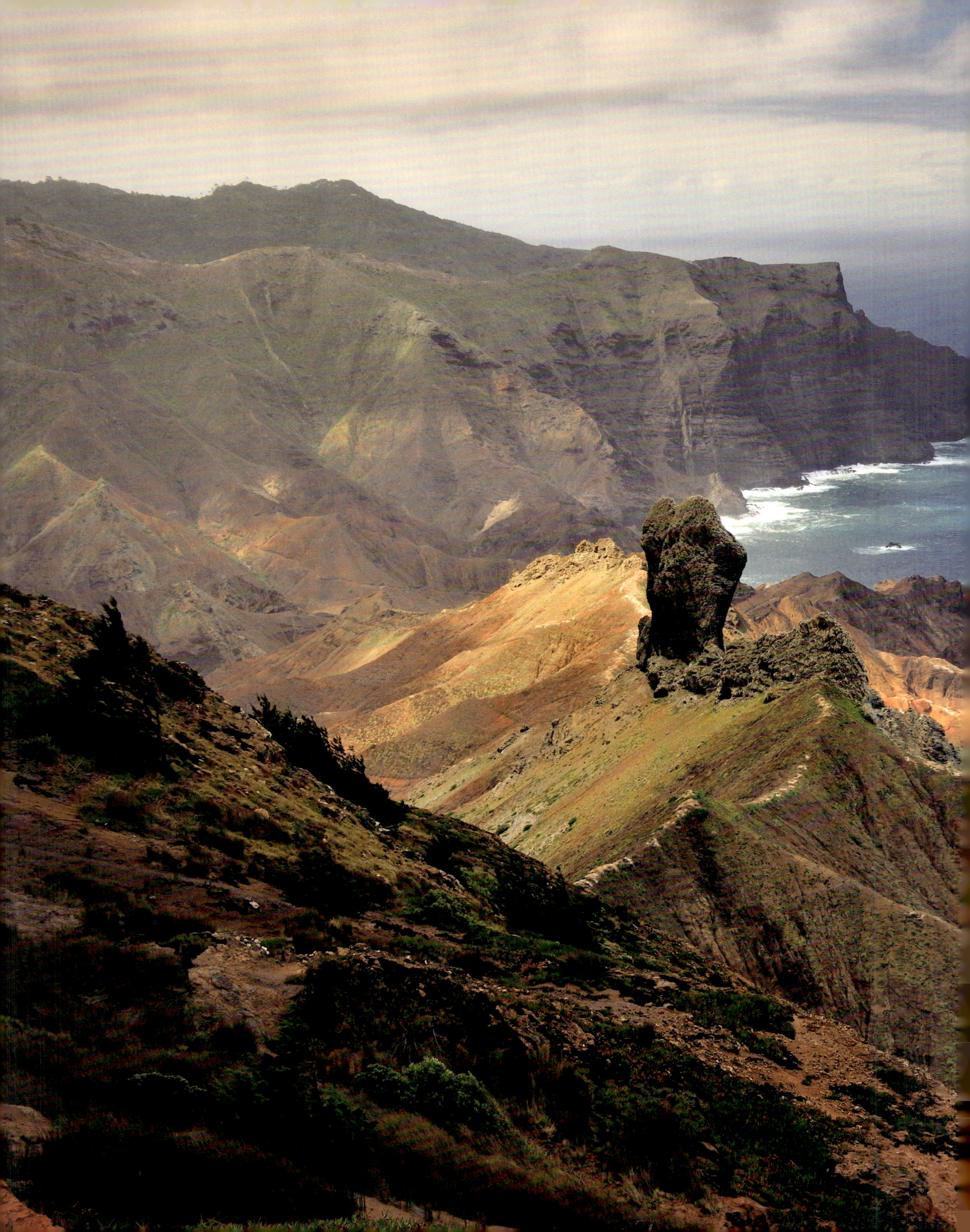

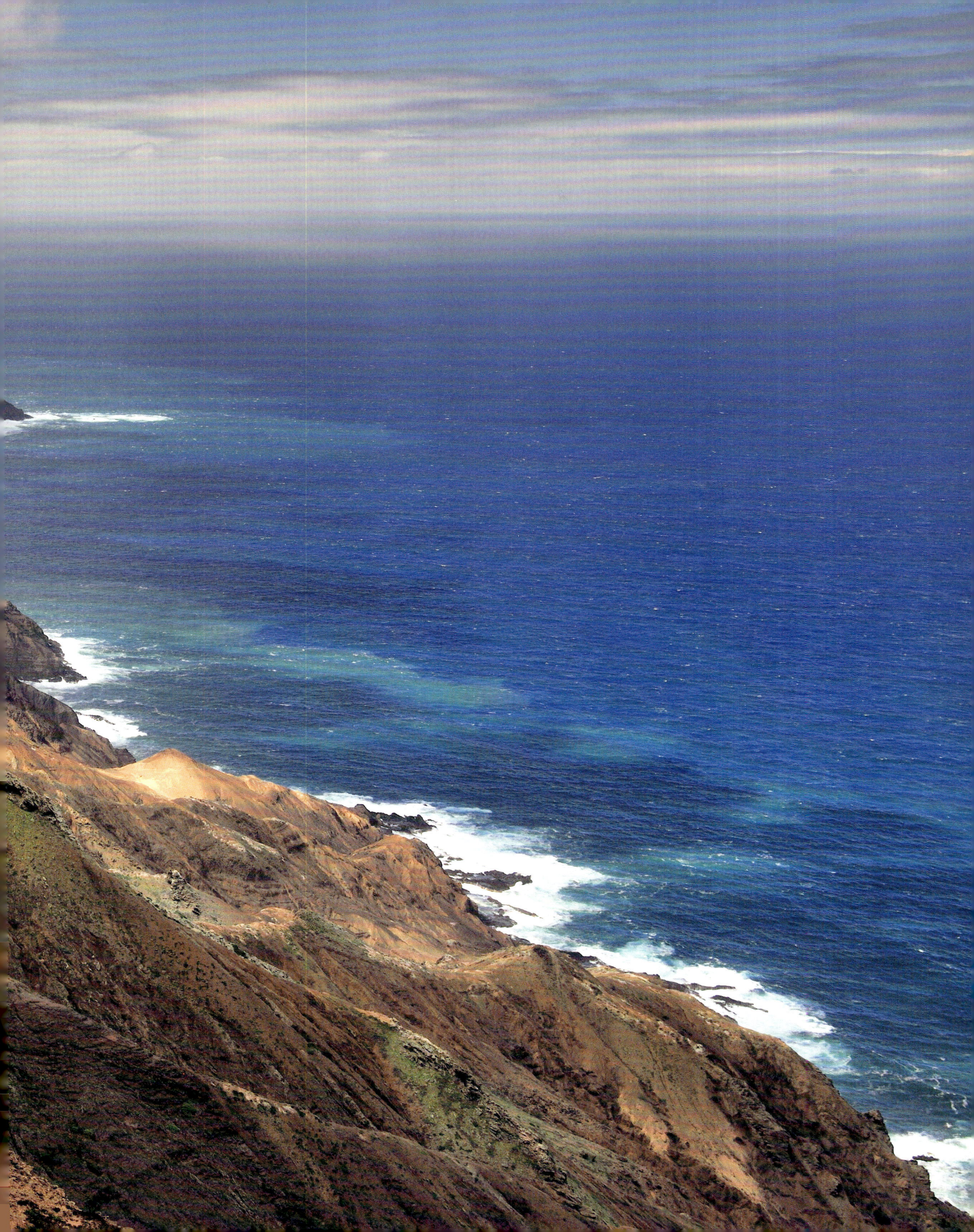

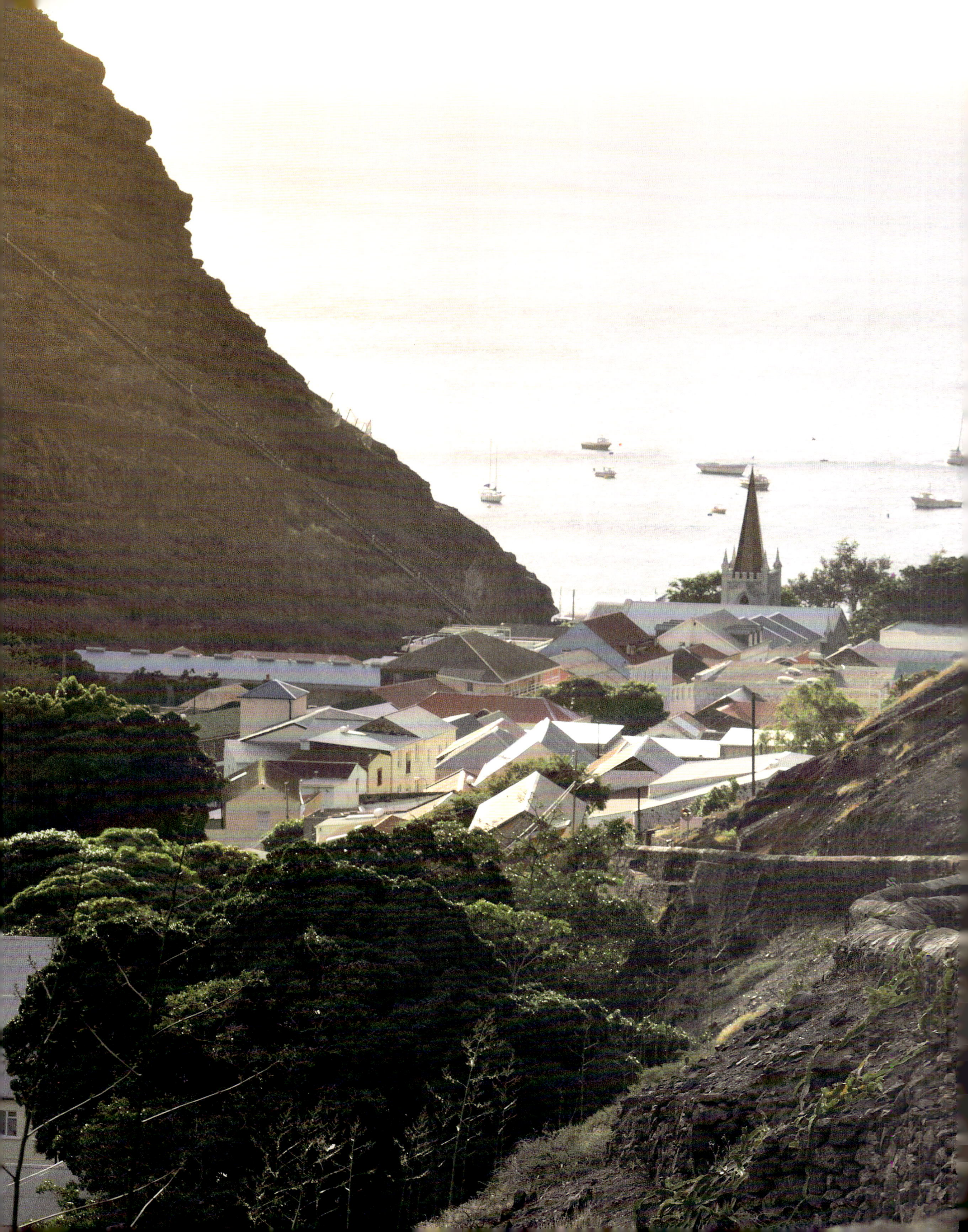

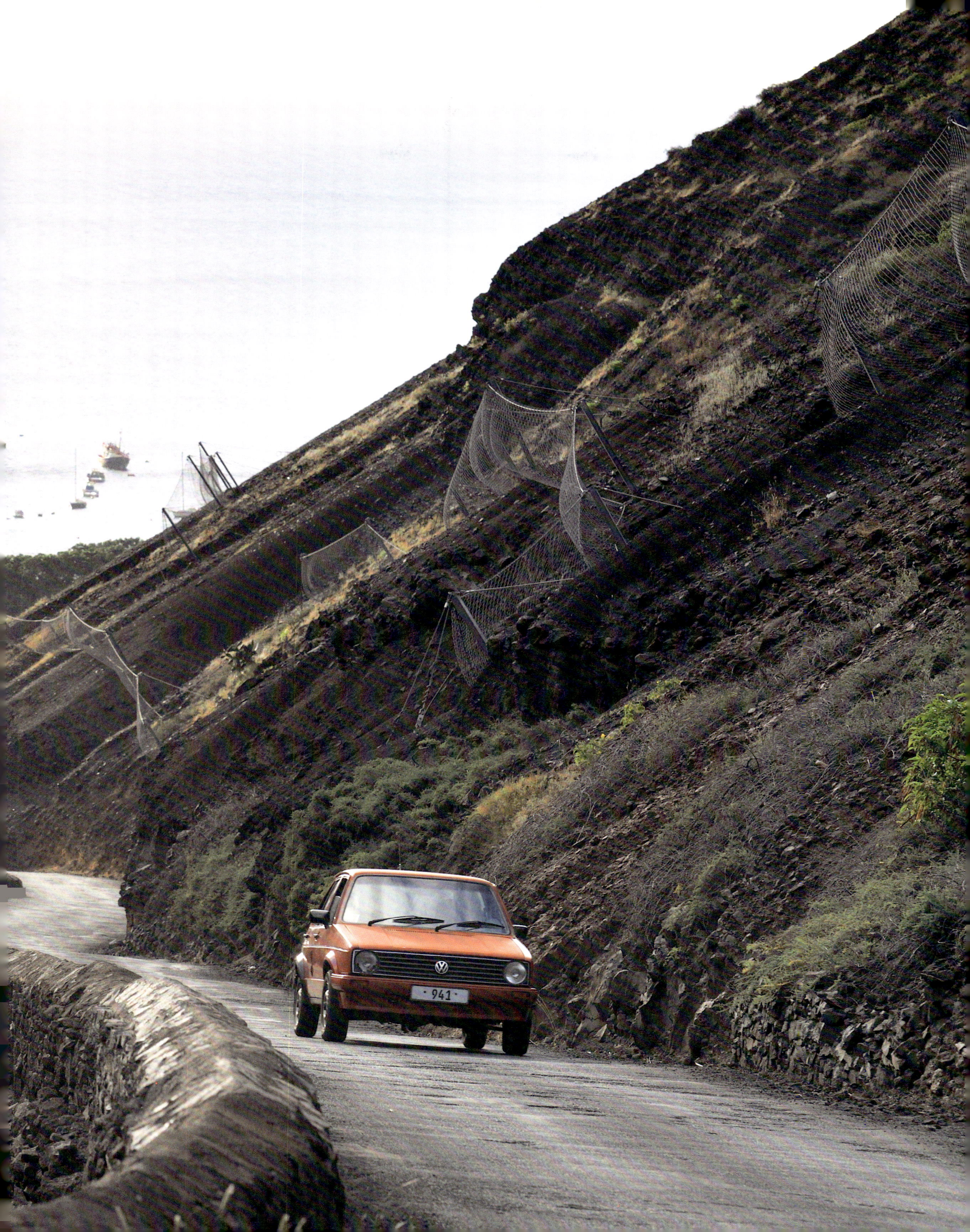
941

St Helena, an Atlantic Island Refuge

A windswept rock

Region:	**St Helena**
Best time to travel:	**November to March**
Closest airport:	**St Helena**
Price tag:	**€€€**
Style of travel:	**Slow island wanderings**

Introduction

Chiefly known as Napoleon's prison and not much else, St Helena remains one of the remotest inhabited islands in the world — a volcanic rock rising from the South Atlantic, a couple of hours by plane off the coast of Namibia and Angola. Farther west is Brazil. Nothing but the fierce and moody Atlantic Ocean surrounds this speck, bringing strong weather and even stronger winds. St Helena is one of the hardest places to reach on earth and only around 4,500 'Saints' live here permanently. Its character could be described as Scotland dressed in a Hawaiian shirt. It is a curious mix of extremes, where dusty, sub-Saharan, almost desert-like prairie contrasts with impenetrable Jurassic rainforests and pretty, well-manicured English-style gardens filled not only with roses but also voluptuous tropical flowers. A narrow road of only 138 kilometres (86 miles) swirls around and through the interior, transporting the driver from one extreme setting to another and offering a new perspective of the ocean around every bend. Whether it's a view of towering cliffs or pretty beaches, everywhere feels like the edge of the world. Landing on this craggy rock — a British colony since 1659 — is not easy. Until October 2017, the only way to get to here was on board the Royal Mail ship, *RMS St Helena*. The British postal ship arrived from Cape Town once every three weeks and the sea voyage took five days in one direction and another five days to return to mainland Africa. As well as bringing container-loads of supplies for the island, the ship also carried up to 100 passengers, often a mix of Saints and travellers. The *RMS St Helena* was no ordinary cruise. Passengers dressed up in their best outfits in the evening to dance to American country music — a favourite music style on the island — and to enjoy chic cocktail parties and formal dinners. The British government had been planning to build an airport on St Helena since the 1950s. When, in 2016, the 285 million GBP airport was opened, they discovered that high winds made it impossible for large commercial carriers to land or take off from the 1,550-metre- (5,085 ft.) long runway. In October 2017, after much controversy and years of (re)construction, the airport was finally reopened.

The Views

When the Airlink Embraer 190 aircraft touches down on this legendary runway, it feels like landing on a distant planet. The island's maximum driving speed of 48 kilometre per hour (30 mph) is understandable given the number of bends in the road. It's impossible to drive 'fast and furious'. The island might be far away from the rest of the world but exploring it feels like travelling to different parts of the planet. There are desert-island landscapes that recall Lanzarote, lush jungles reminiscent of Praslin in the Seychelles, and meadows where sheep graze on green grass with flax in the distance. Every drive brings scenic highlights including inactive volcanoes, lush valleys and tiny villages with colourful houses, and terraced fields where cattle graze. There's also the impressive Plantation House, the official home of the governor of St Helena since 1792. The English-style garden is home to a colonial-looking governor's mansion where giant tortoises live, imported from the Seychelles and leading a happy life here. Some 4,500 Saints continue their daily lives in a time capsule. Many parts of the island do not have a mobile-phone signal, much less internet, which remains impossibly slow and prohibitively expensive. As a consequence, instead of being engrossed in social-media feeds, Saints happily wave at passing cars and will stop for a chat at every opportunity. With cargo ships only arriving now and then, patience is a way of life. Grocery stores often run out of items such as walnuts or avocados, and sometimes even potatoes. Once, the

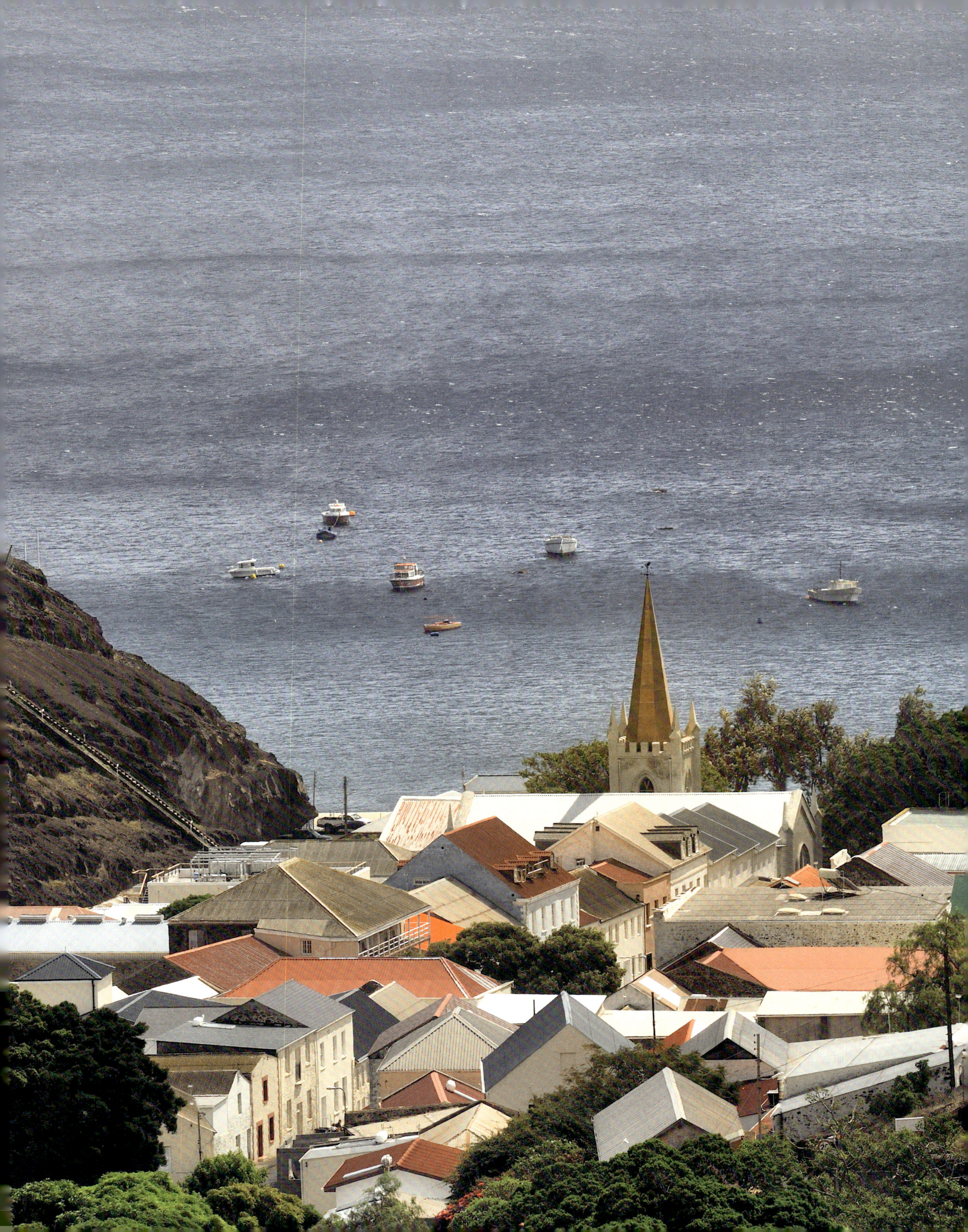

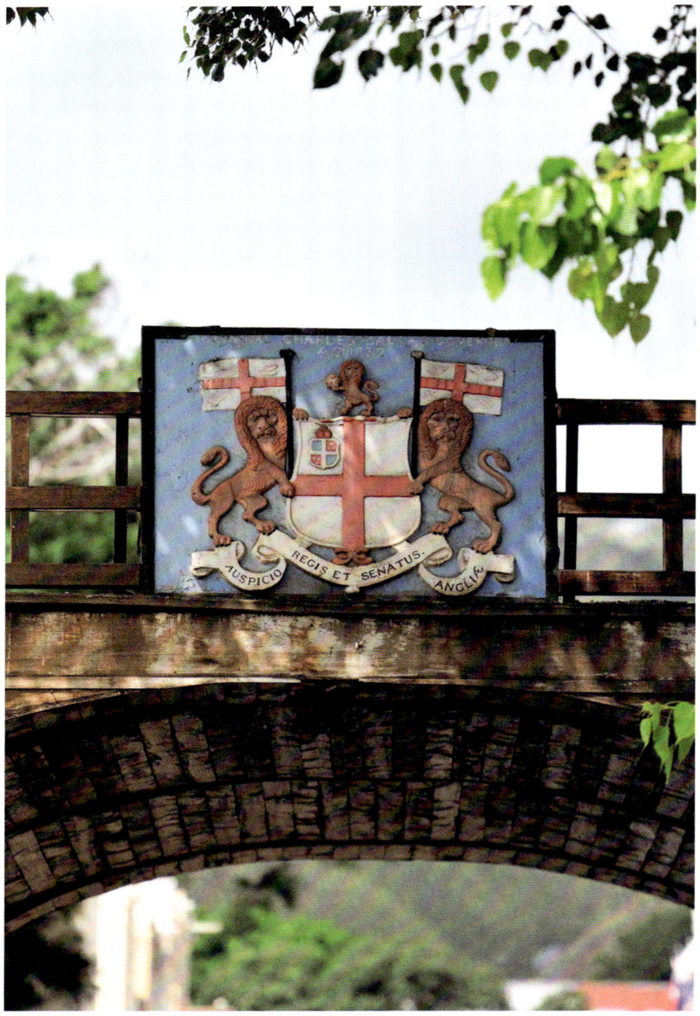
AUSPICIO REGIS ET SENATUS ANGLIÆ

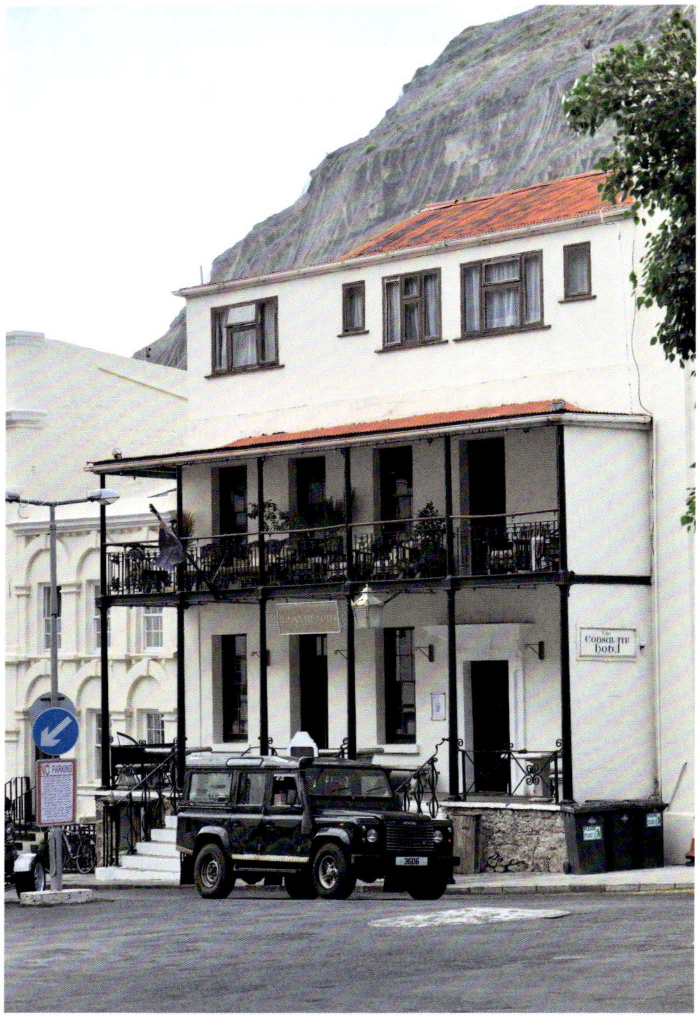
NO PARKING

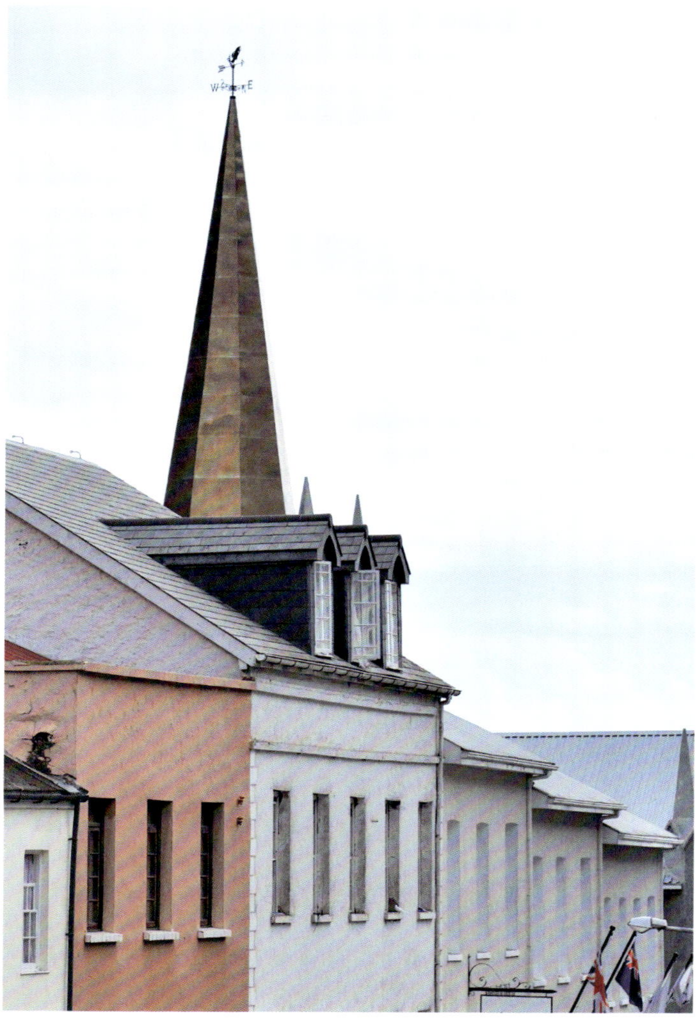

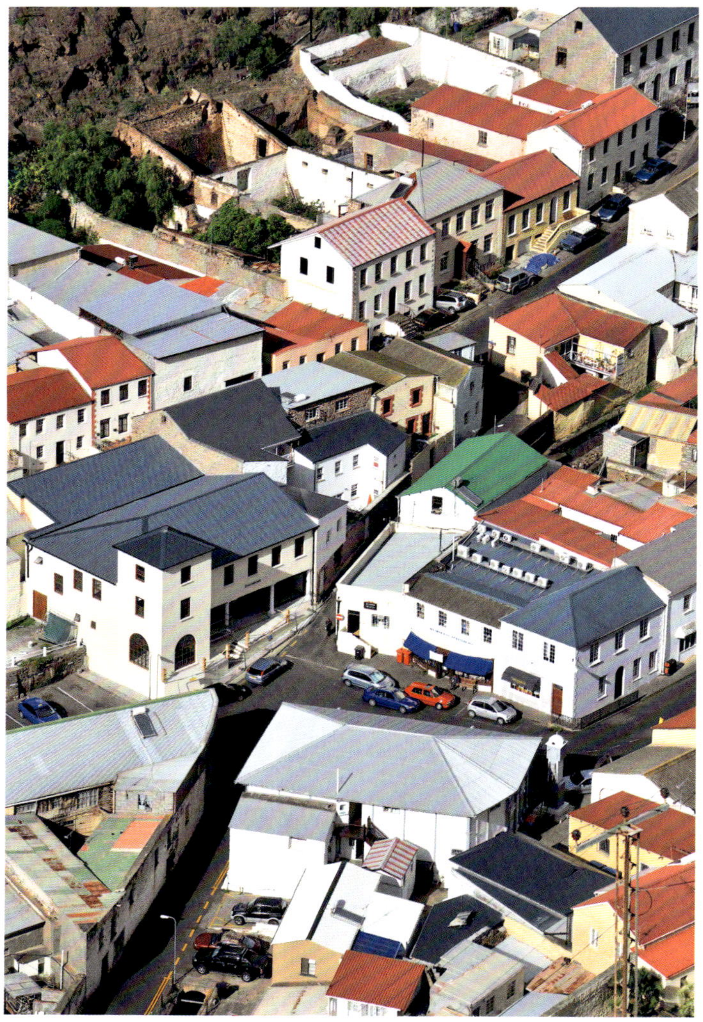

Opposite Isolation is St Helena's appeal. The pocket-sized capital, Jamestown, nestled between volcanic mountains and almost spilling over into the deep blue Atlantic, is probably one of the quirkiest towns in the world. One of its more curious features is the soaring 699 steps of Jacob's Ladder, a steep staircase leading from town up to Ladder Hill Fort.

island almost ran out of beer — a near calamity. At the same time, isolation is, and always has been, St Helena's greatest asset. If it hadn't been deemed suitably off the map to exile Napoleon after his defeat at Waterloo, the island might have been abandoned a long time ago. Even after centuries of abuse, the isolated ecosystem remains of great value to biologists and conservationists. Bees on St Helena have remained uniquely free of the diseases that have decimated populations around the world. Coffee shrubs on the local plantations can all be traced back to one arabica plant that arrived from Yemen, the aroma unchanged since it graced the tastebuds of Napoleon himself. St Helena is a walker's paradise. The network of 21 so-called 'post box walks', each up to 12 kilometres (7.5 mi.) long, leads through green valleys, meadows, cloud forests, and stone deserts. A few hair-raisingly steep walks follow the coastline, offering stunning views of wild and godforsaken bays. Thankfully, Diana's Peak, the highest point of the island, is an easy hike, its summit marked by a solitary, straight Norfolk pine that is said to have been planted there — for future use as a ship's mast — by the crew of Captain James Cook, who visited the island twice. One of the most scenic hikes is to Blue Point, walking along the cliffs, past ginestra bushes, flax, and juniper trees that are so battered by the wind that their highest branches touch the ground. Everything points northwest here, obeying the force of Mother Nature. Despite their submissive appearance, these plants are resistant to the fierce winds swirling around the island. When driving down to Jamestown, the capital of St Helena, the views are impressive. This small Georgian era town nestled at the foot of two mountains with the Atlantic in the distance, has shops and bars, a swimming pool, and a tiny shooting range next to a compact port, where boats and small ships bob on the water. There's always something going on in Jamestown, especially at the weekend. On Saturday evenings people shuffle around the dance floor at the Standard bar with disco lights crawling over the old walls, lip-syncing to the country songs that are so popular on St Helena. Or they head out to the oldest hotel in town. Since 2008, the Consulate Hotel has become a favourite meeting place for locals and visitors. This beautiful mid-18th-century building on Main Street has colonial wrought-iron balconies and a nostalgic décor filled with antiques and vintage knick-knacks. Locals come here for coffee — and lots of English tea — served with rich, homemade cakes or for pizza evenings with live music. Tourists bed down here for the homely ambience and friendliness of Hazel Wilmot, who runs the hotel. She calls the hotel a window to the world, with guests coming from all continents.

Creature Comforts
Choice is limited for an overnight stay on St Helena. There are countryside lodges where home-cooked dinners are served in a distinctly British setting. There is also the Mantis Hotel in Jamestown, the most luxurious option on the island (and the only spot with free, high-speed internet). Food is a casual affair on St Helena. From coffee shops in town — serving coffee grown on the island — to garden cafés and even a Thai restaurant, to home-cooked experiences like Wranghams. Here, Debbie Fantom serves her guests a set menu featuring organic produce from her garden in the shadow of the Peaks National Park and with views over Sandy Bay.

Go Now
Attracting travellers to the island since the postal ship stopped sailing is, according to one Saint, like fighting the strong wind that often holds St Helena in a grip. While in the past, the *RMS St Helena* brought a steady number of travellers who stayed at least one week, the majority of younger travellers are barely aware of the island's existence. Though St Helena's airport hasn't delivered on its promised boom in tourism, a submarine fibre-optic cable filled with the frenzied madness that is broadband internet could potentially have a far bigger impact. Despite arriving on the island in 2021, it won't be operational until early 2023. Under the radar and often under a thick blanket of clouds, St Helena is on the bucket lists of the occasional island hoppers, ambitious sailors, and those looking for solitude and something entirely unique.

How to Go
The only way to reach St Helena is via the fortnightly flight from Johannesburg OR Tambo Airport, leaving on Saturday morning, weather permitting. The six-hour flight includes a designated stop at Walvis Bay in Namibia for refuelling before continuing its journey to St Helena. On the second week of each month, the incoming flight continues on to Ascension Island, returning to St Helena on Sunday for its departure to Johannesburg OR Tambo. During peak season, December to February, there is a midweek flight operating on Tuesdays from Cape Town to St Helena, with the return flight leaving the same day.

This spread For a while, St Helena's 285 million GBP airport was dubbed 'the world's most useless'. The inaugural flight was postponed for over a year due to treacherous wind conditions since it is classified as a 'Category C' airport, the toughest in its tier. Only a dozen pilots in the world are qualified to fly into St Helena. Finally, in October 2017, the first commercial flight landed at the airport, seven years later than planned. In February 2018, the *RMS St Helena*, the island's much beloved lifeline to the African mainland, made its last voyage. Since then, the only scheduled service to the island is a fortnightly flight from Johannesburg.

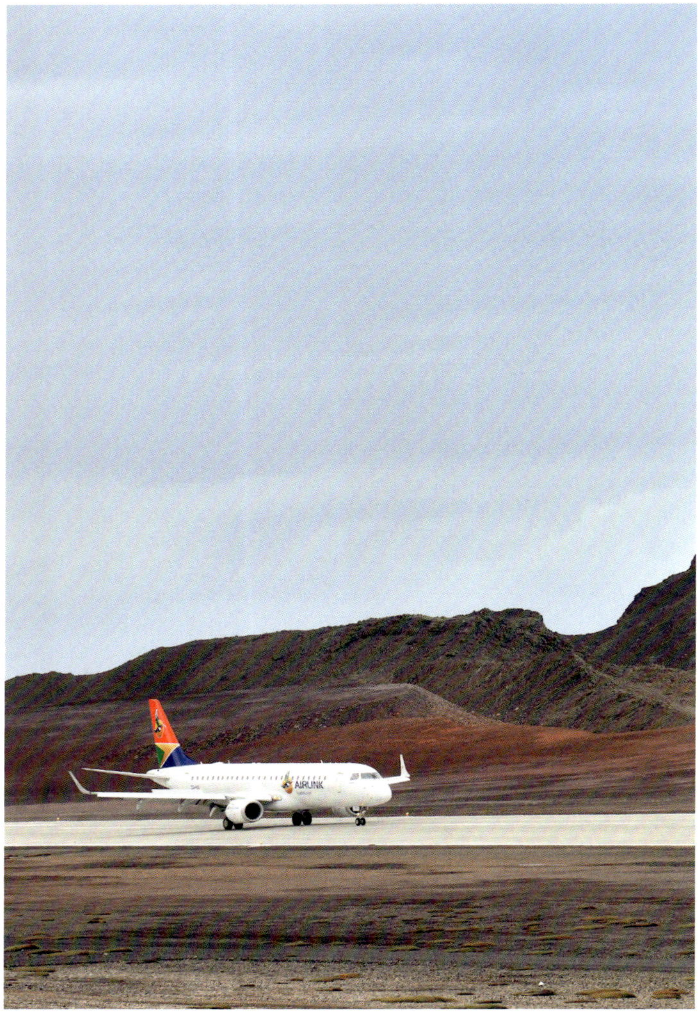

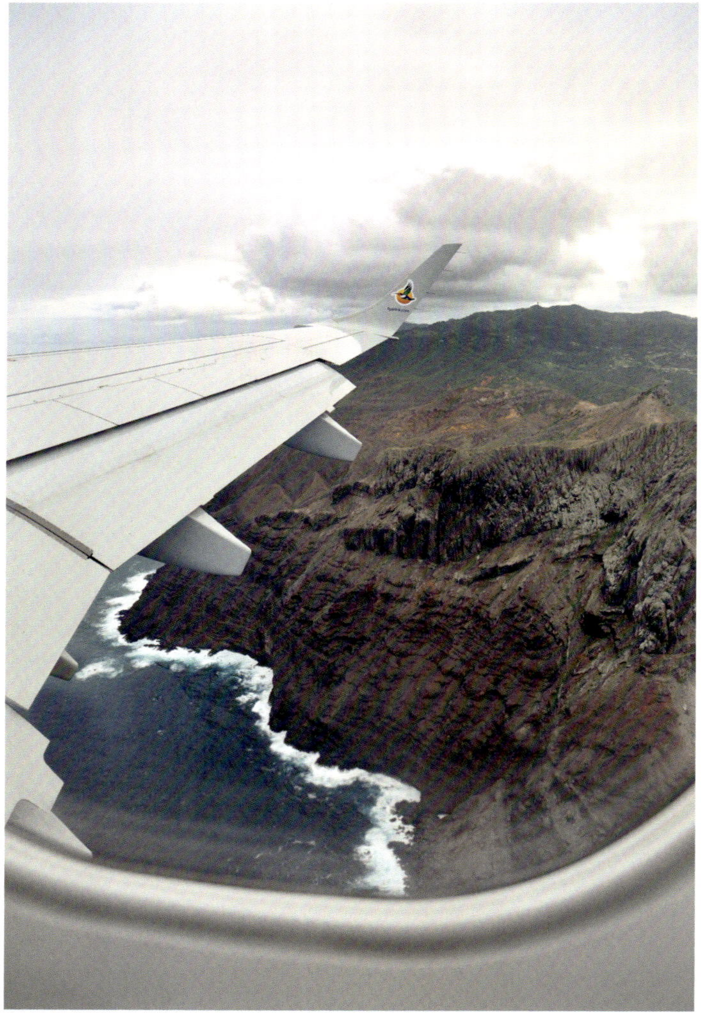

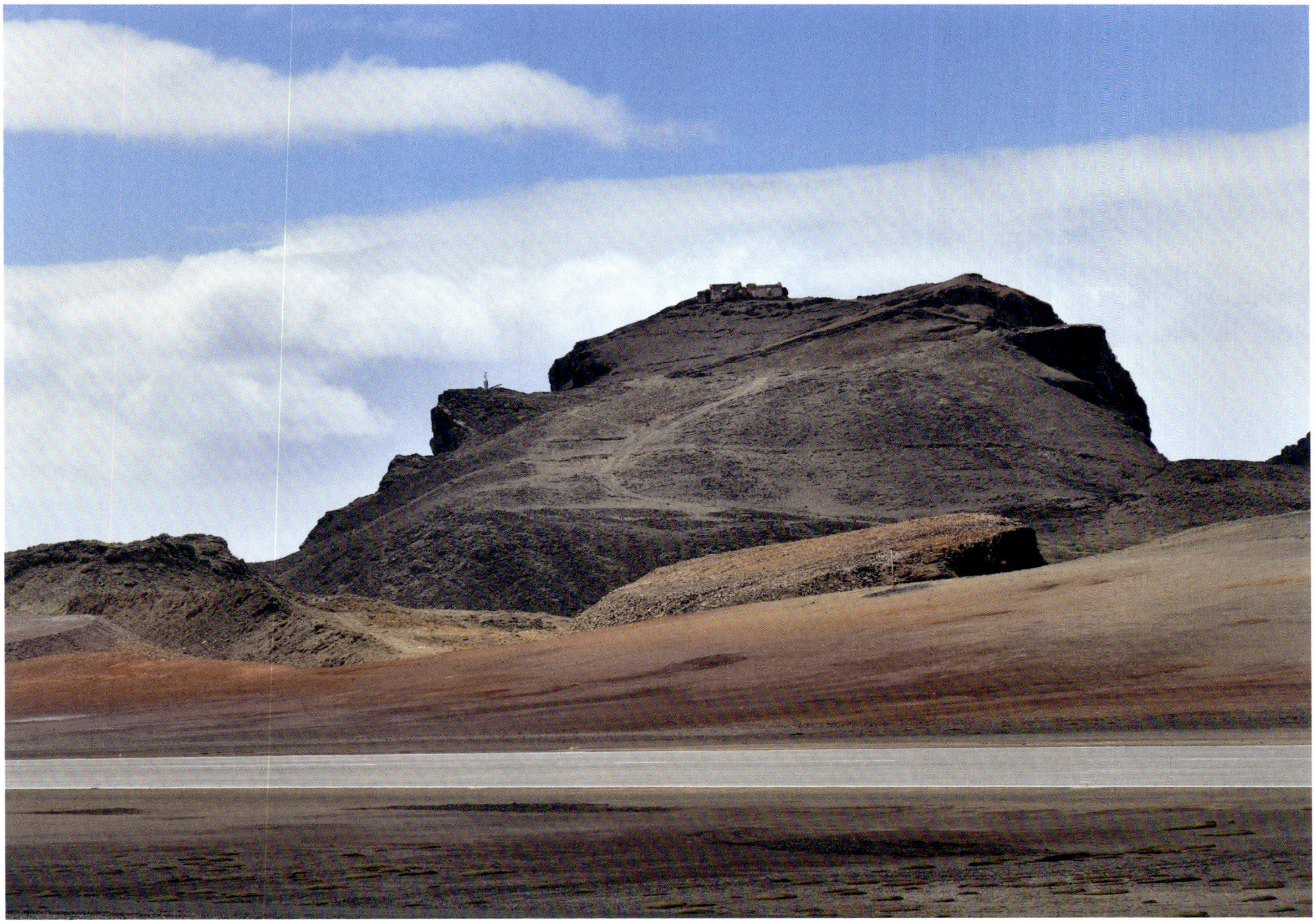

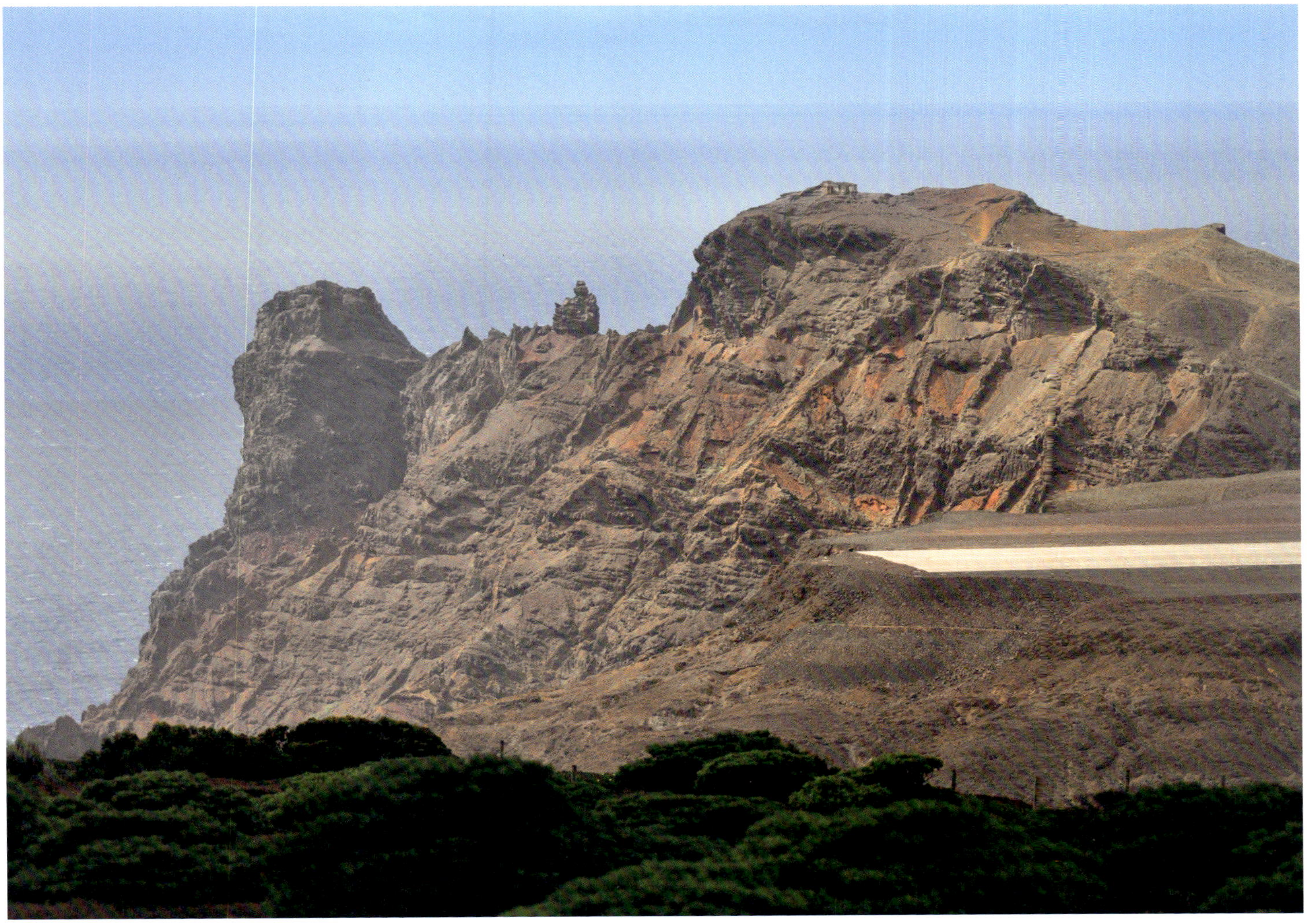

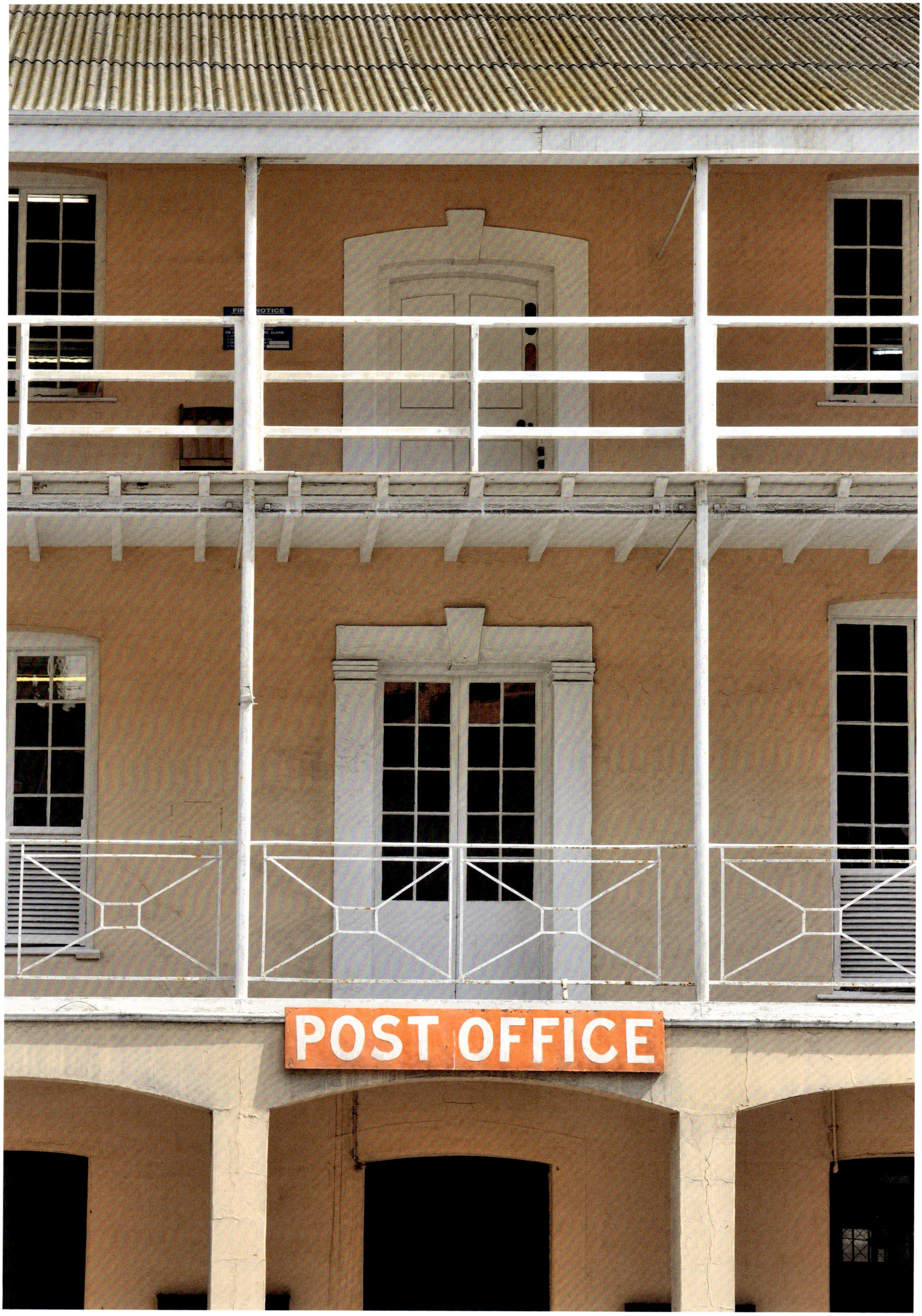
POST OFFICE

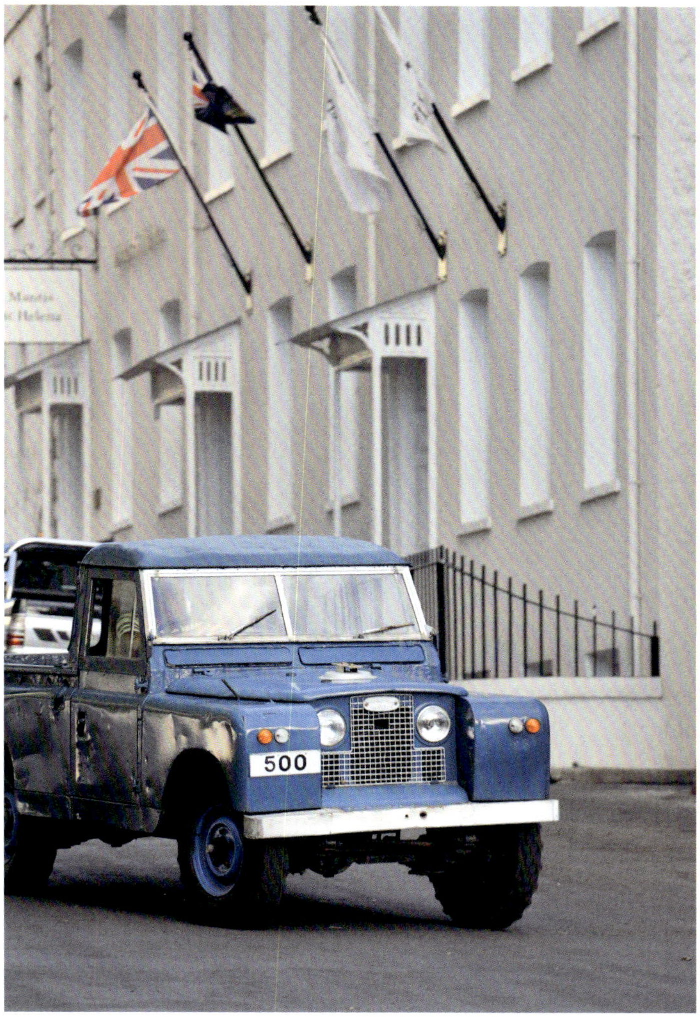

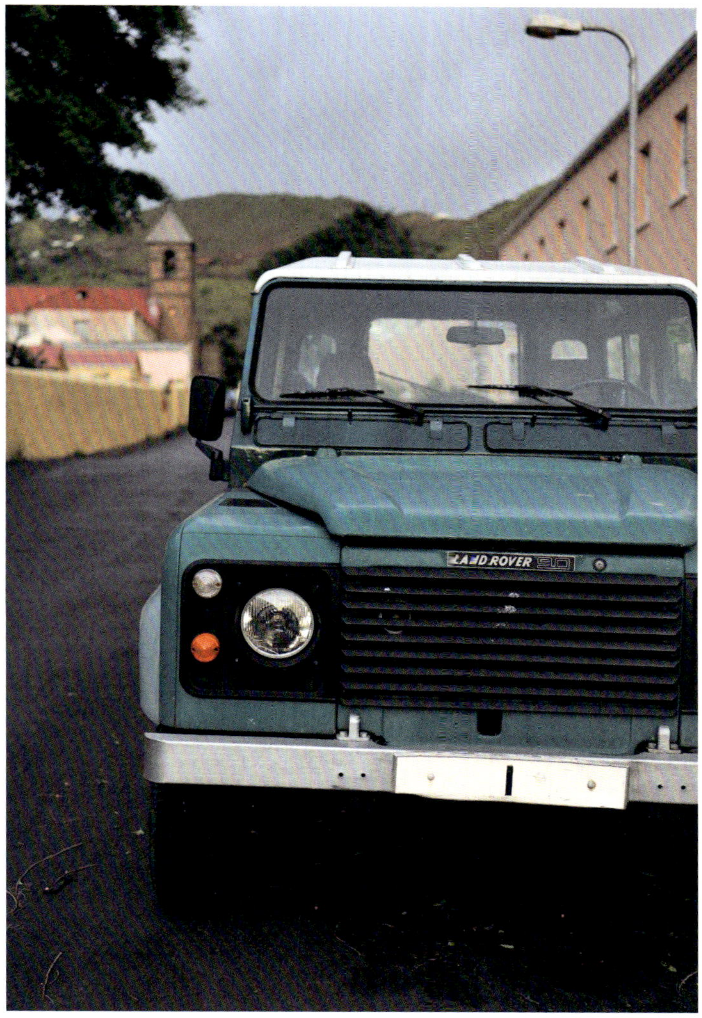

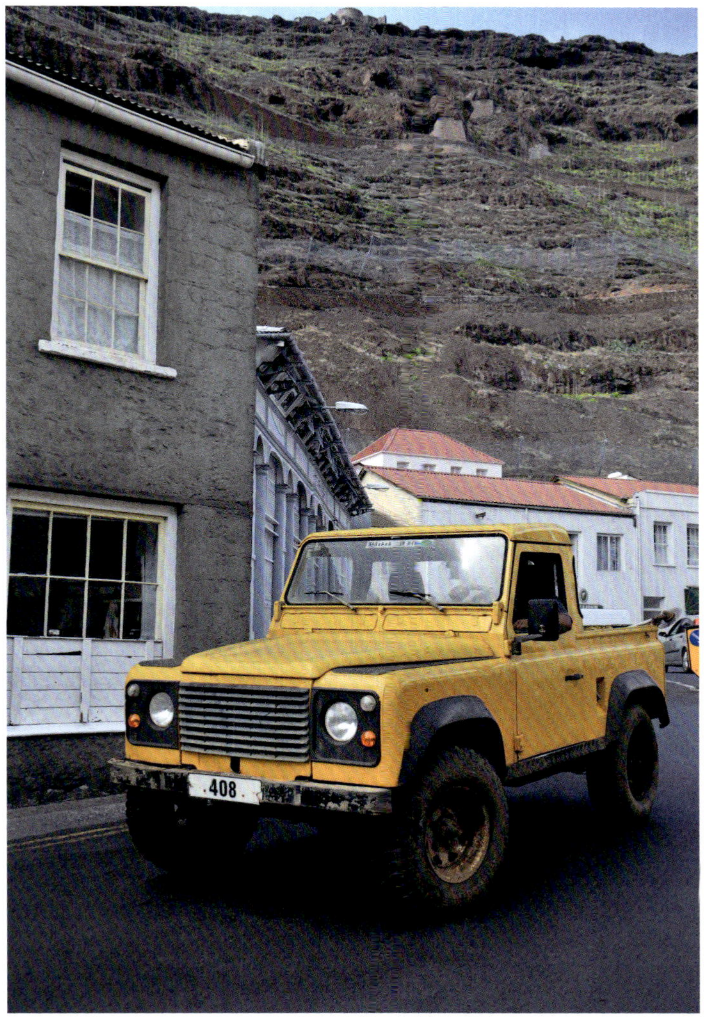

This spread Coming from such a remote and obscure territory, St Helena's stamps have always been much sought after by collectors. The biggest source of income for the post office on Main Street in Jamestown is its philatelic service. To meet the demand, a special Stamp Advisory Committee meets regularly to discuss new designs. In 2016, it issued a commemorative stamp celebrating the long-awaited opening of the island's first airport. Naturalist Charles Darwin, explorer Captain James Cook, novelist William Thackeray, and astronomer Edmond Halley all visited St Helena. Before the Suez Canal was built, 30 ships a day docked here, but now the island has only visiting ships. With only 4,500 residents, St Helena is a society of three-digit number plates and five-digit phone numbers. Those who were born here are known as Saints: a mix of European, African, and Asian ancestries including settlers, soldiers, and slaves who arrived since the second half of the 17th century. A general atmosphere of friendliness and conviviality is extended to anyone who has taken the trouble to get to the island.

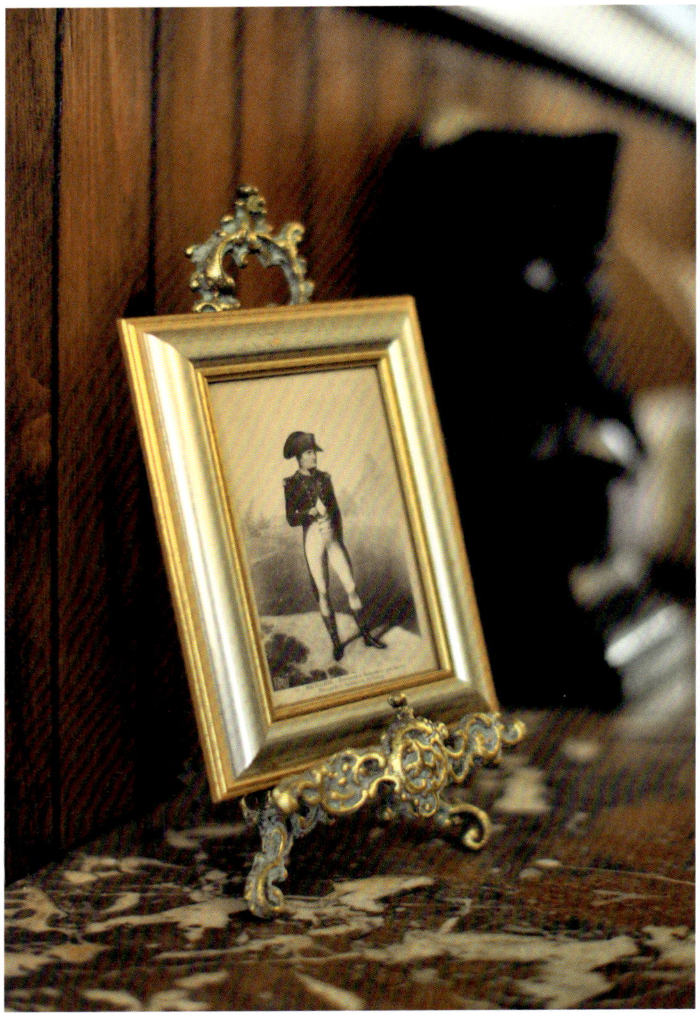

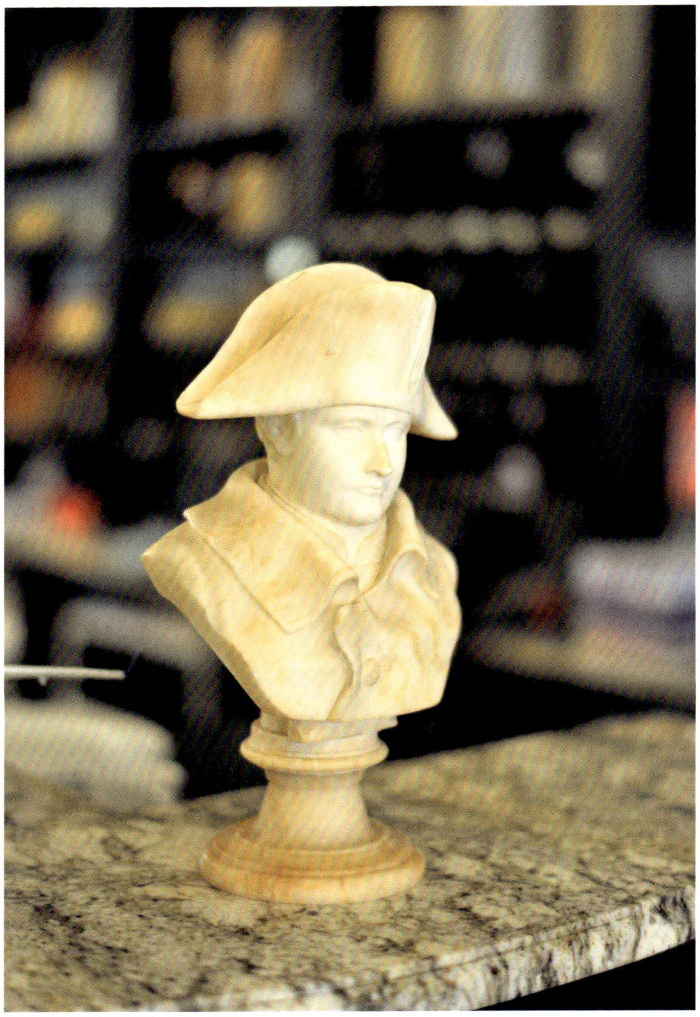

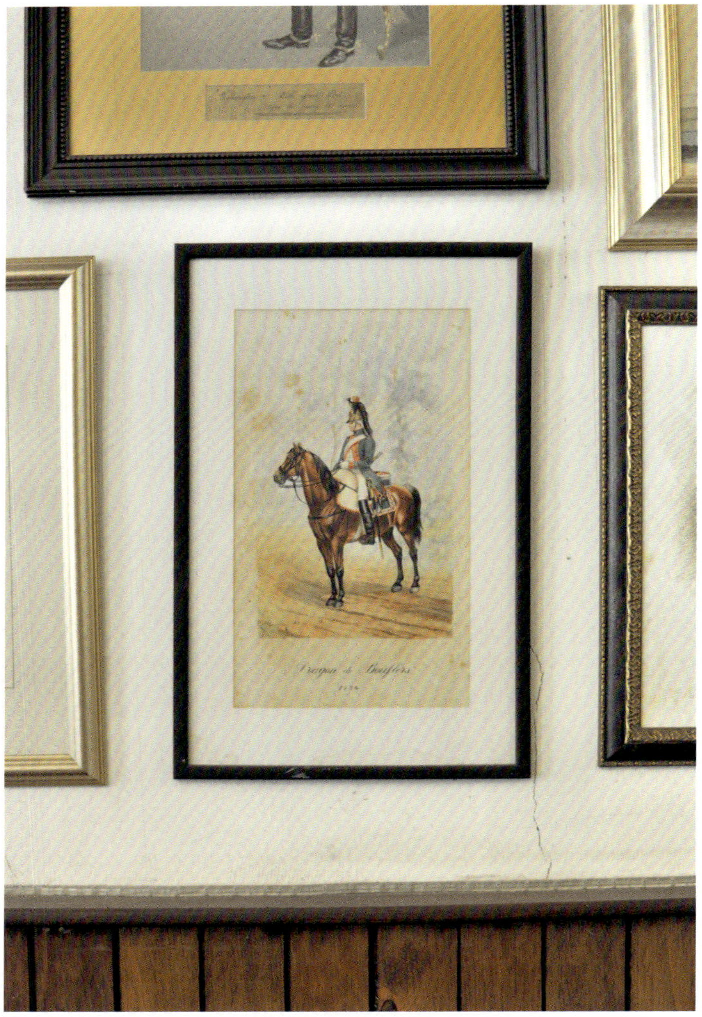

This spread The Consulate Hotel, owned by Hazel Wilmot, is one of the most beautiful buildings along the main street in Jamestown. The balcony on the first floor, with old wooden chairs in which to sit and read a book from the library and observe life in Jamestown, is strictly reserved for hotel guests. A sign on the reception desk sums up life on the island: "Some afternoons, mornings, or even whole days, we are not here at all," it warns, "but lately we have been constantly open, except when closed, although we should be here, unless we left early." Though the sign is likely an example of tongue-in-cheek island humour, it's hard to say for sure. The décor is eccentric and has a wealth of paintings and busts depicting the exiled emperor Napoleon, and beyond the wooden doors of the bar there is a pleasant covered patio where live music is played and pizza evenings are enjoyed.

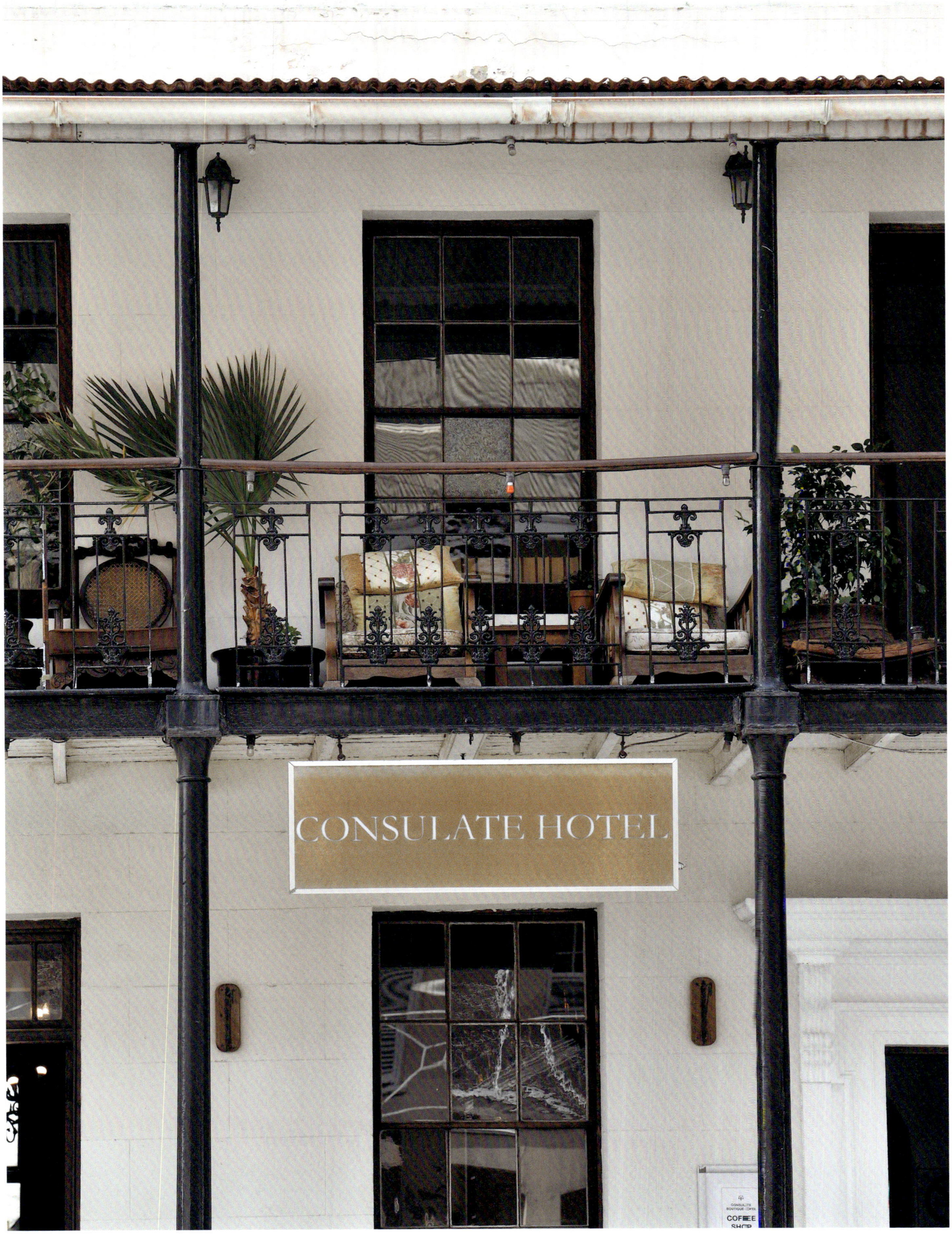
CONSULATE HOTEL

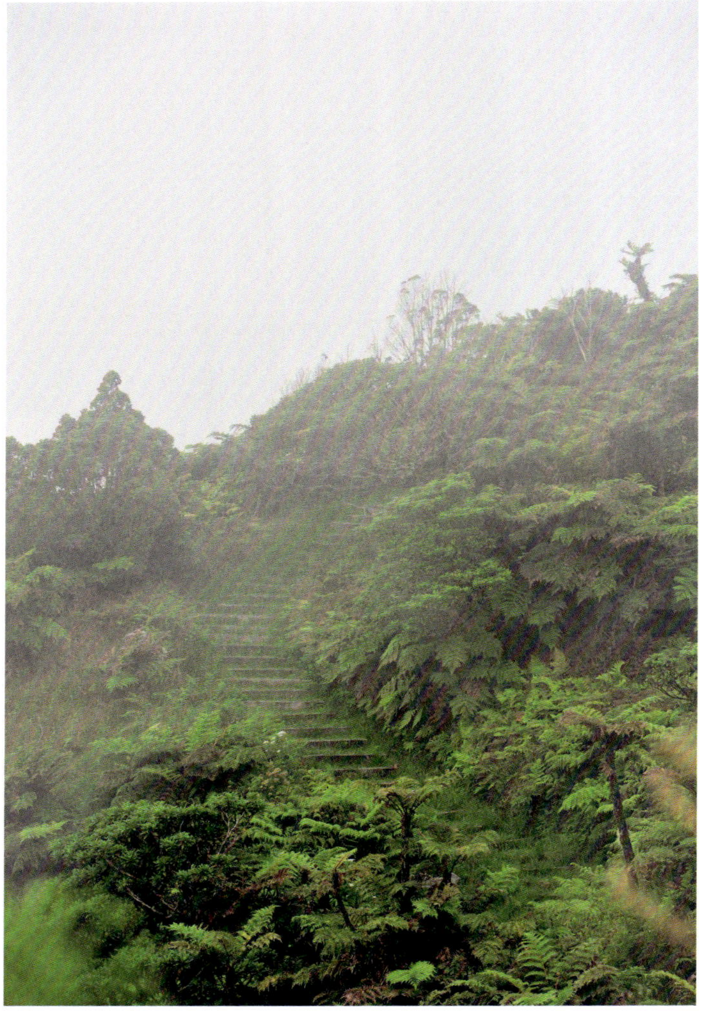
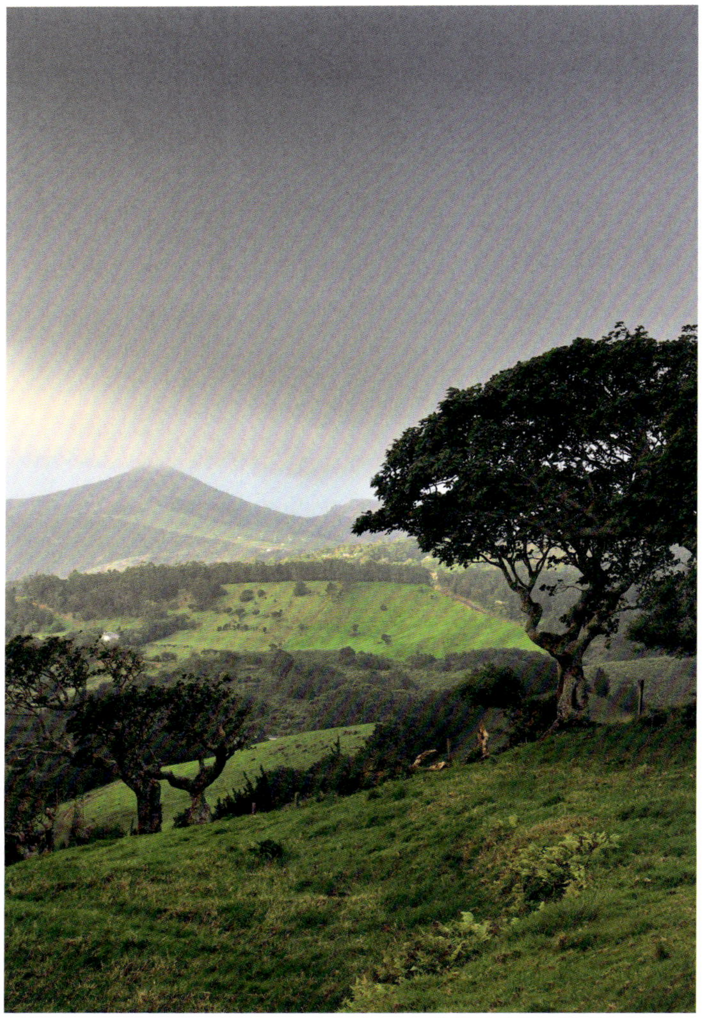
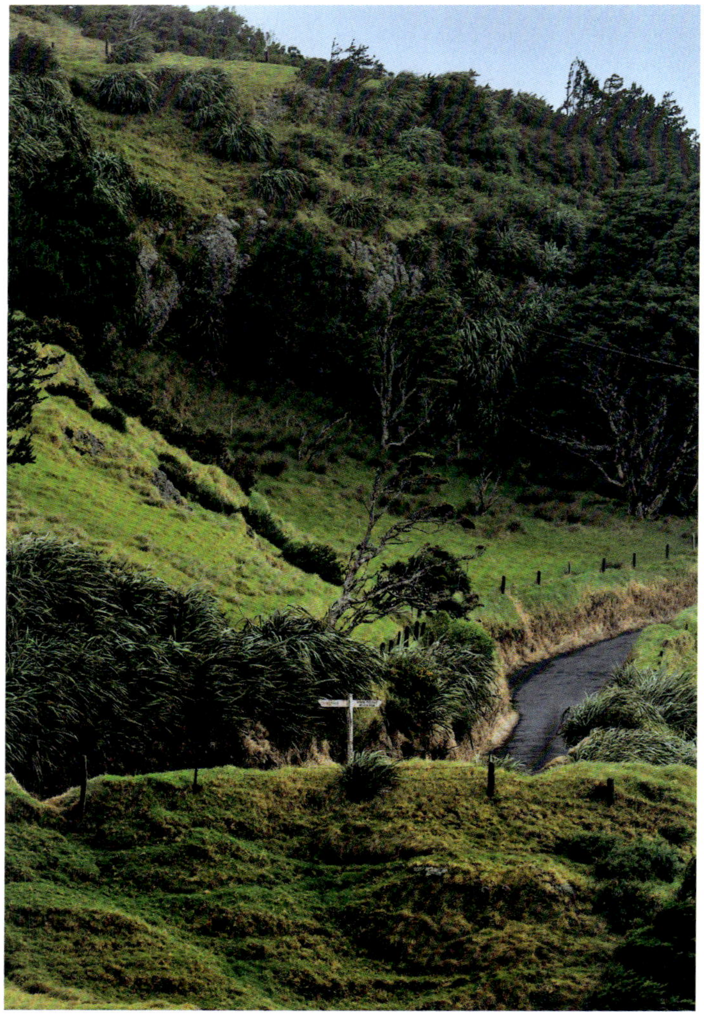
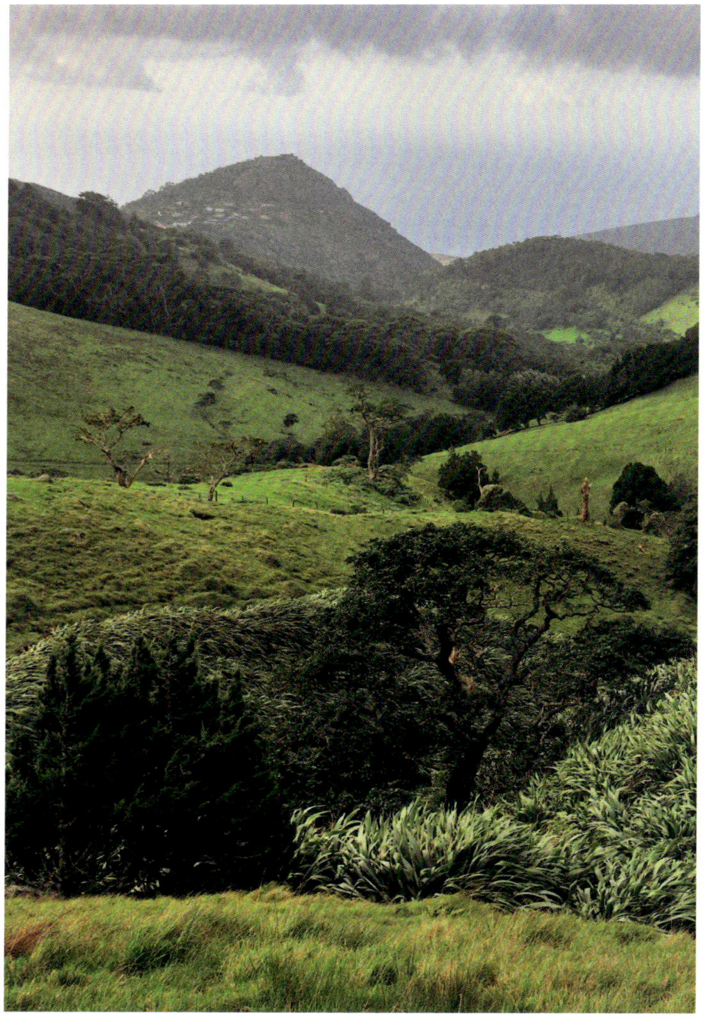

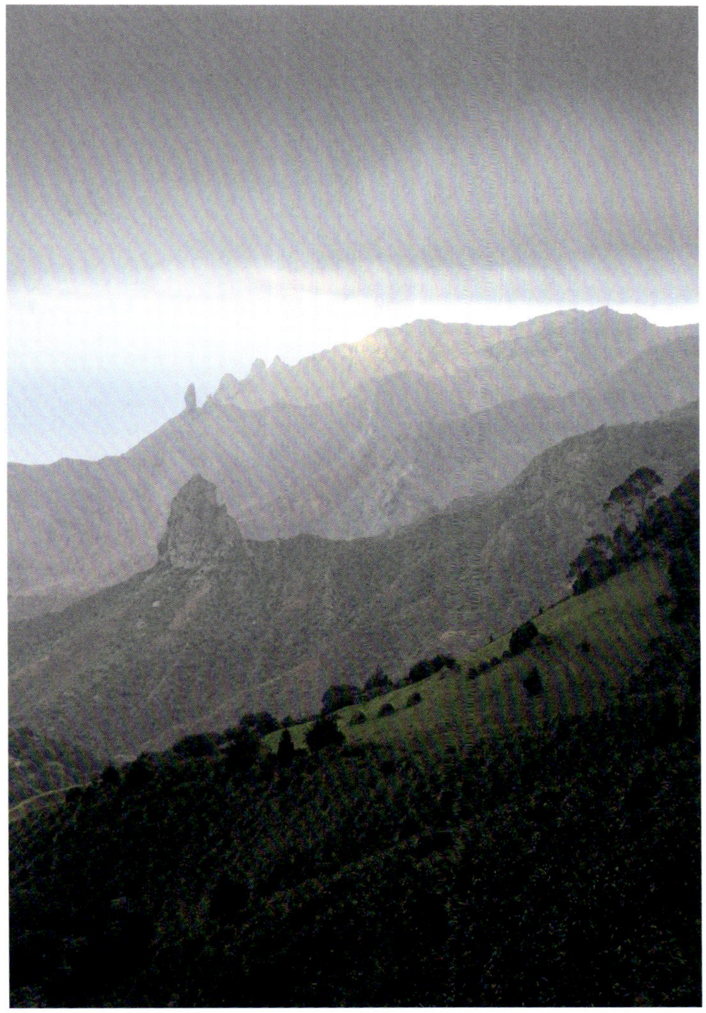

This spread St Helena's landscape changes around every bend of the narrow roads zipping through the island. One side might feature ocean-carved cliffs or black sandy beaches, the other sheer valleys shrouded in vast fields of flax or lavish forests filled with green ferns and often covered in a spooky tidal mist drifting in from the Atlantic. Hiking is one of the most popular island pursuits. One of the easiest and most scenic 'post box walks' leads to the summit of Diana's Peak, the highest point on St Helena. It's often shrouded in fog, but when the sky clears, dramatic views stretch over the green interior of the island towards two solitary rock pillars, known as Lot and Lot's Wife, towering over the volcanic beach of Sandy Bay.

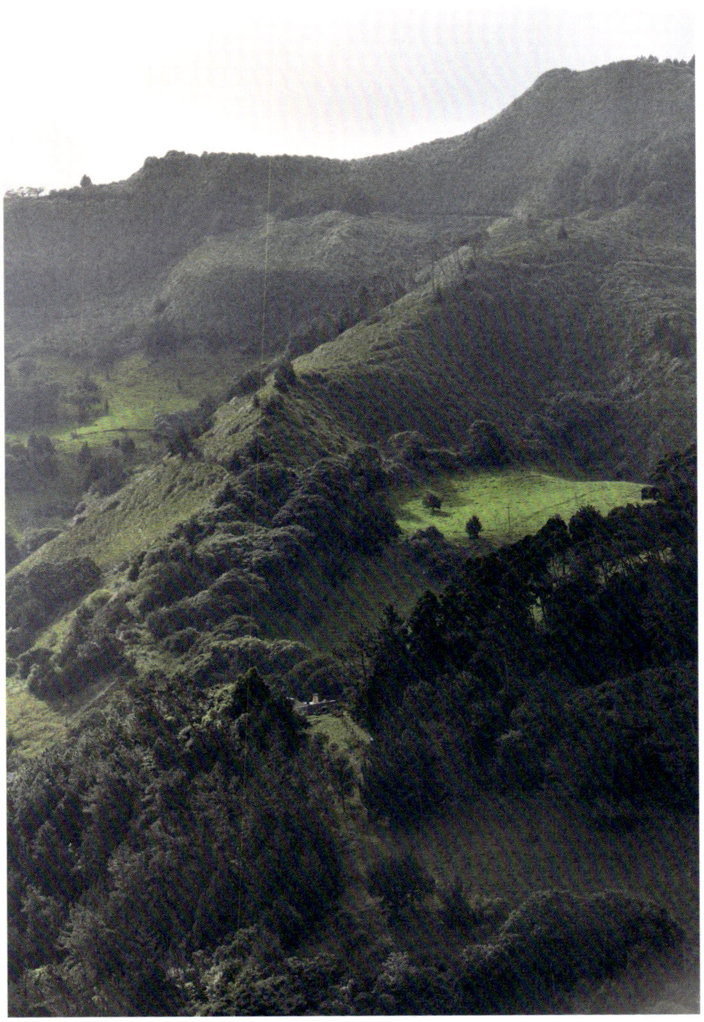

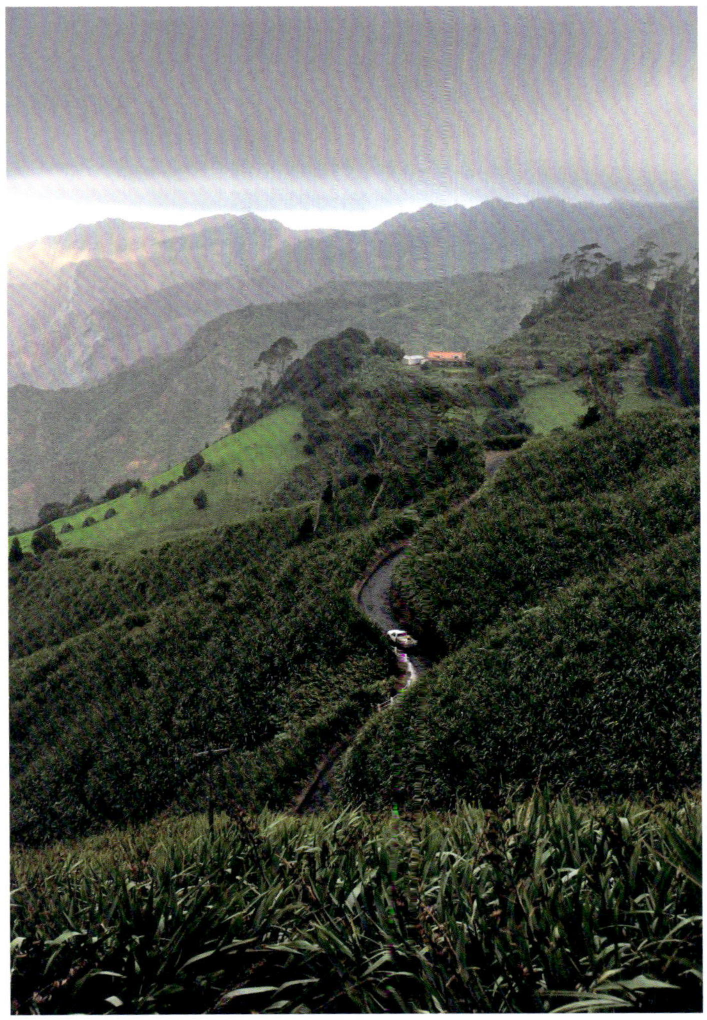

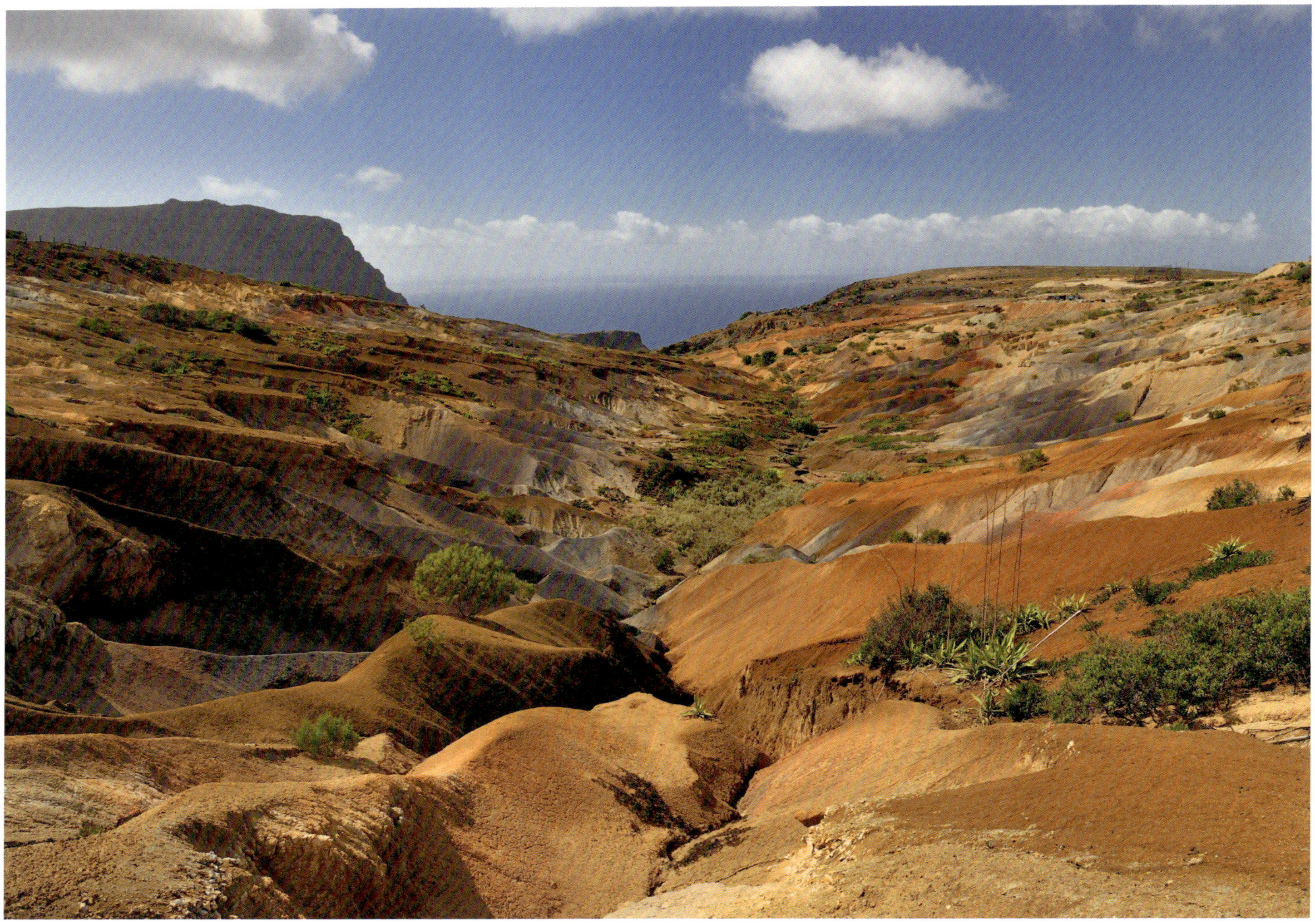

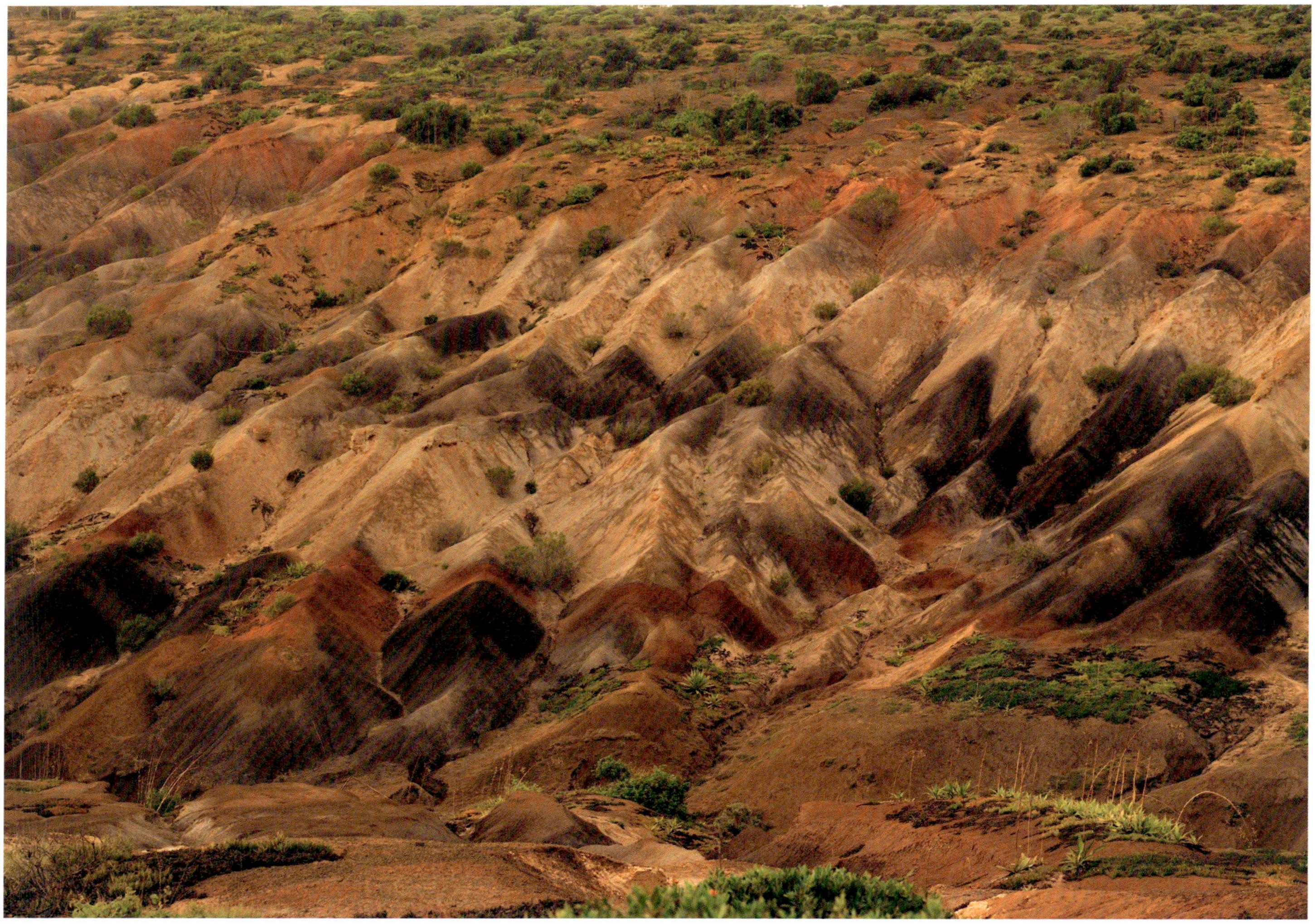

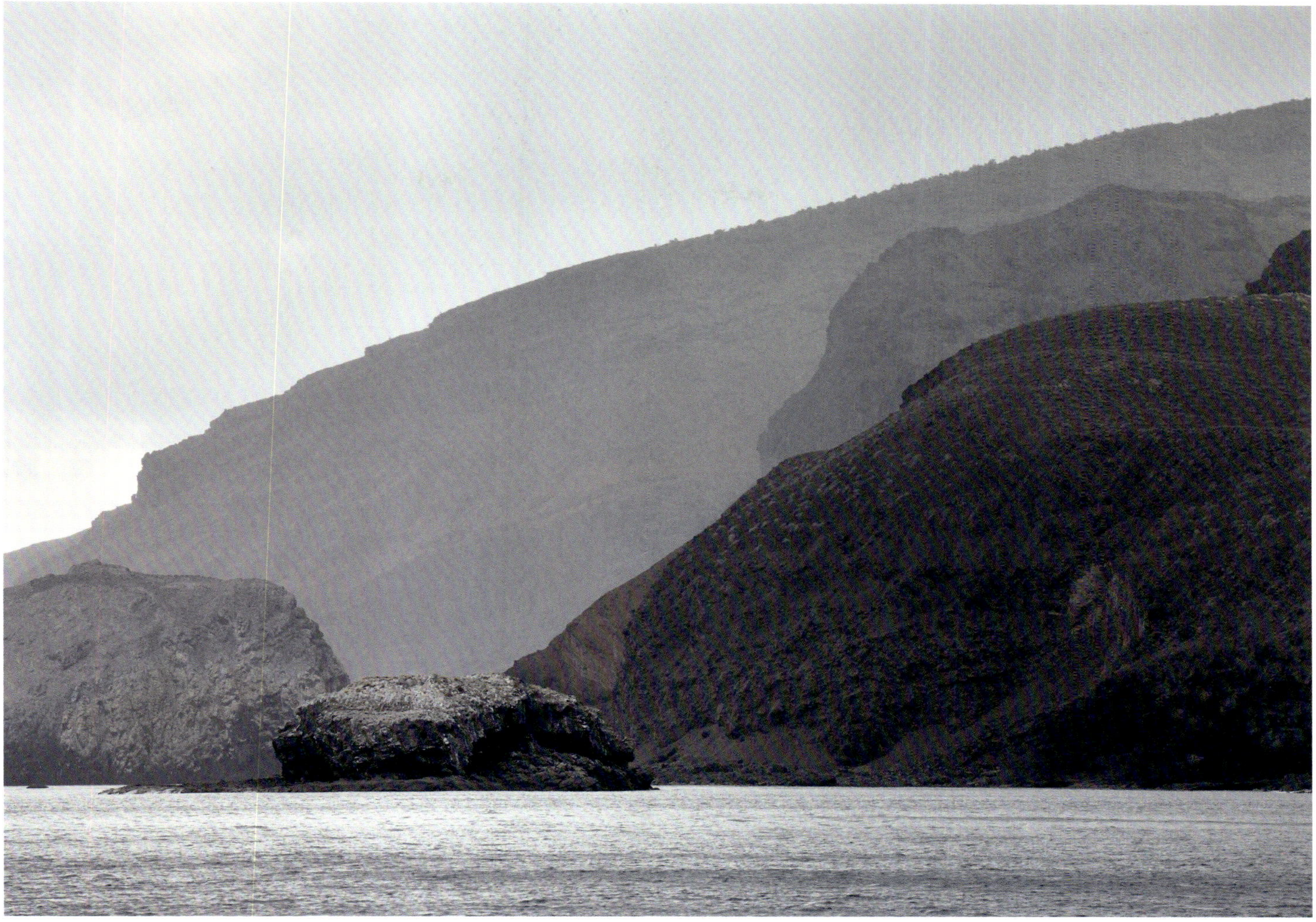

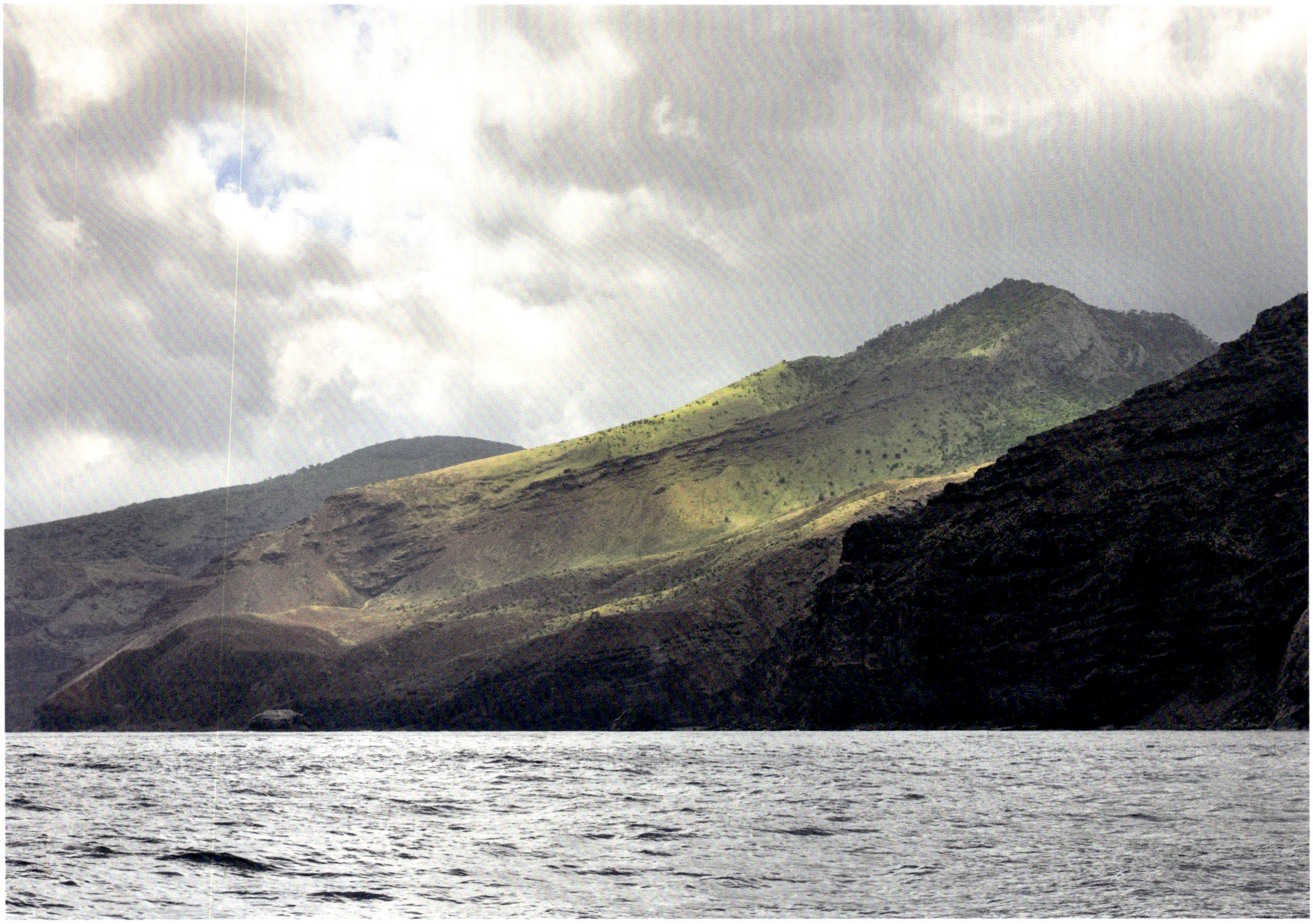

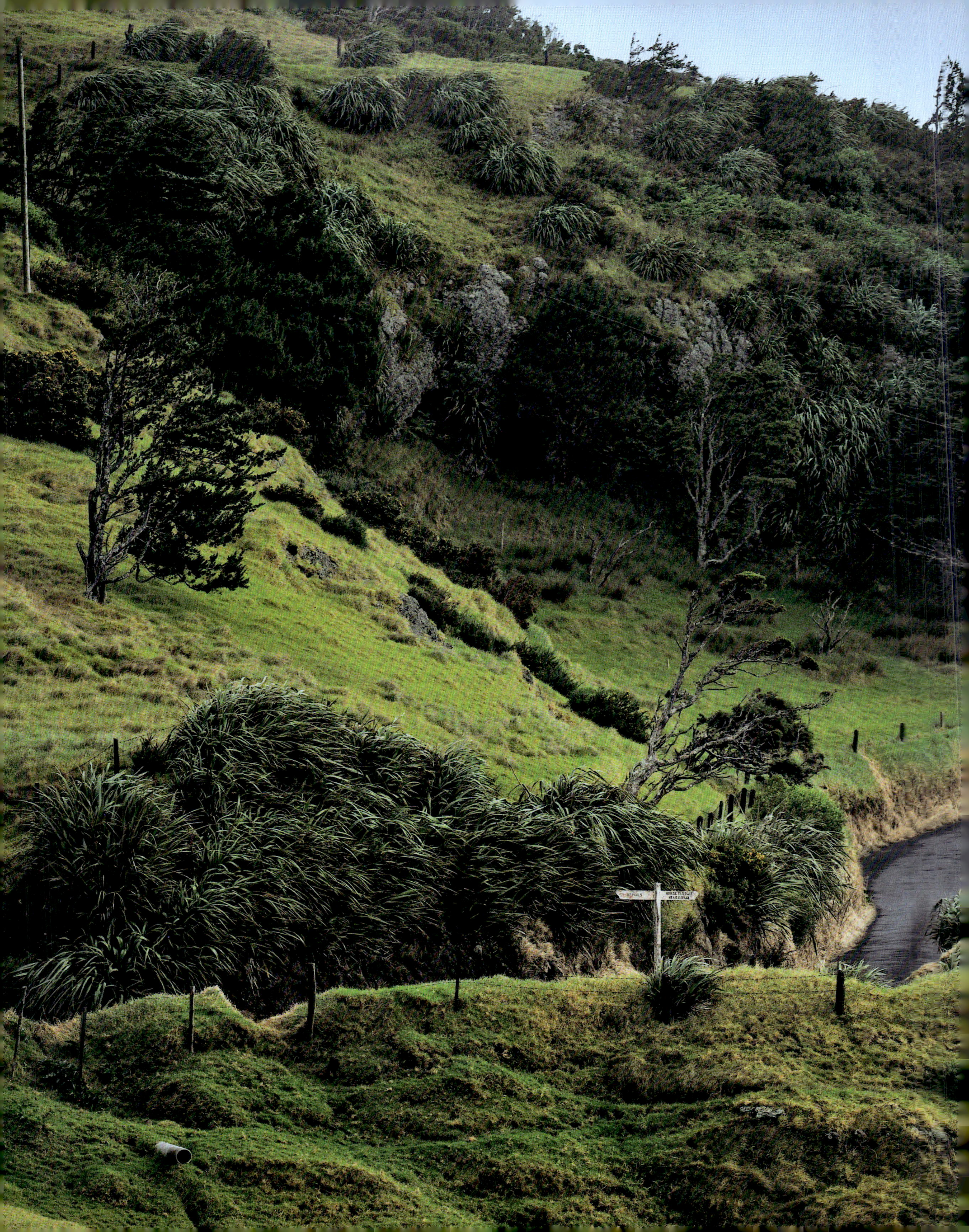
ST PAULS

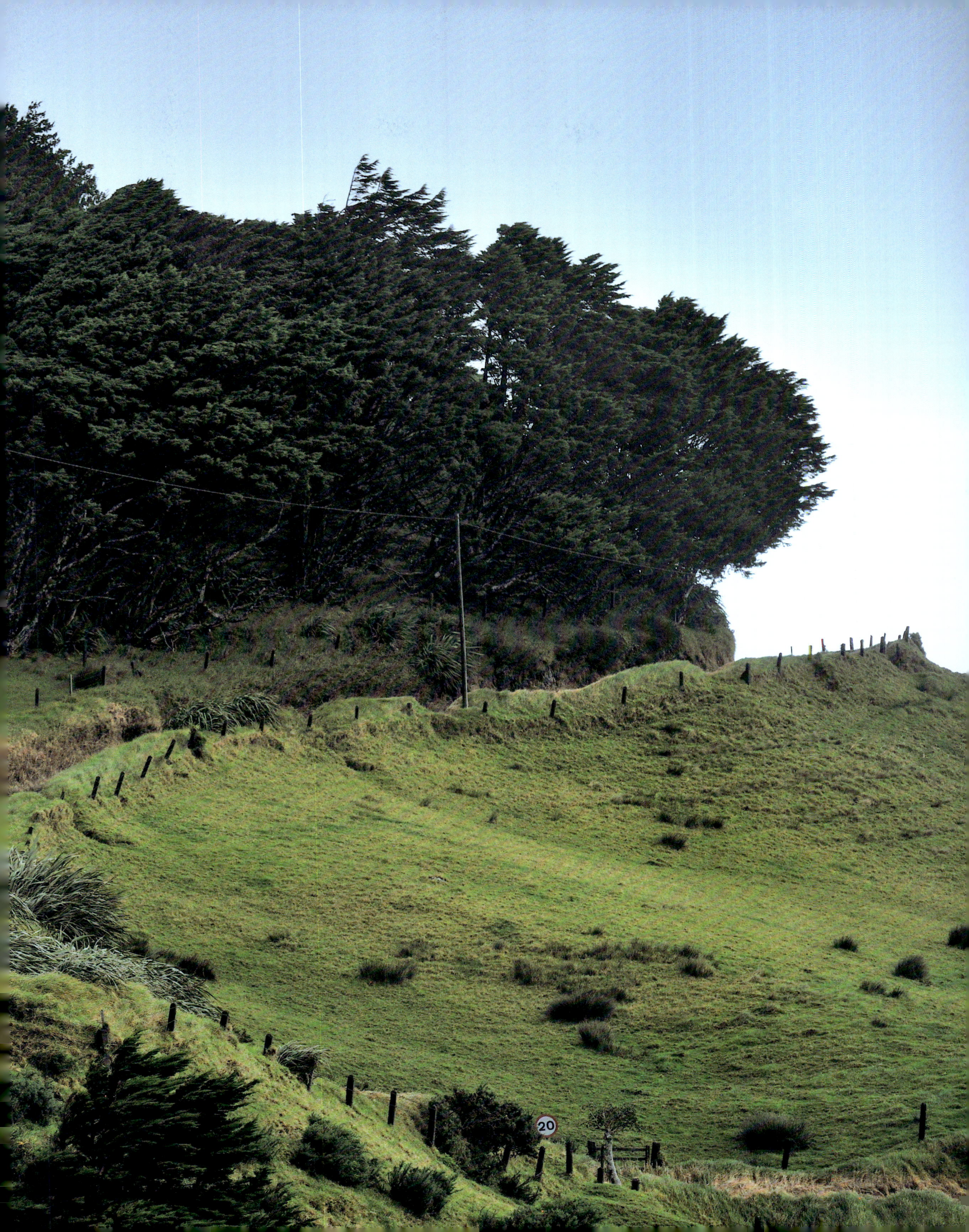
20

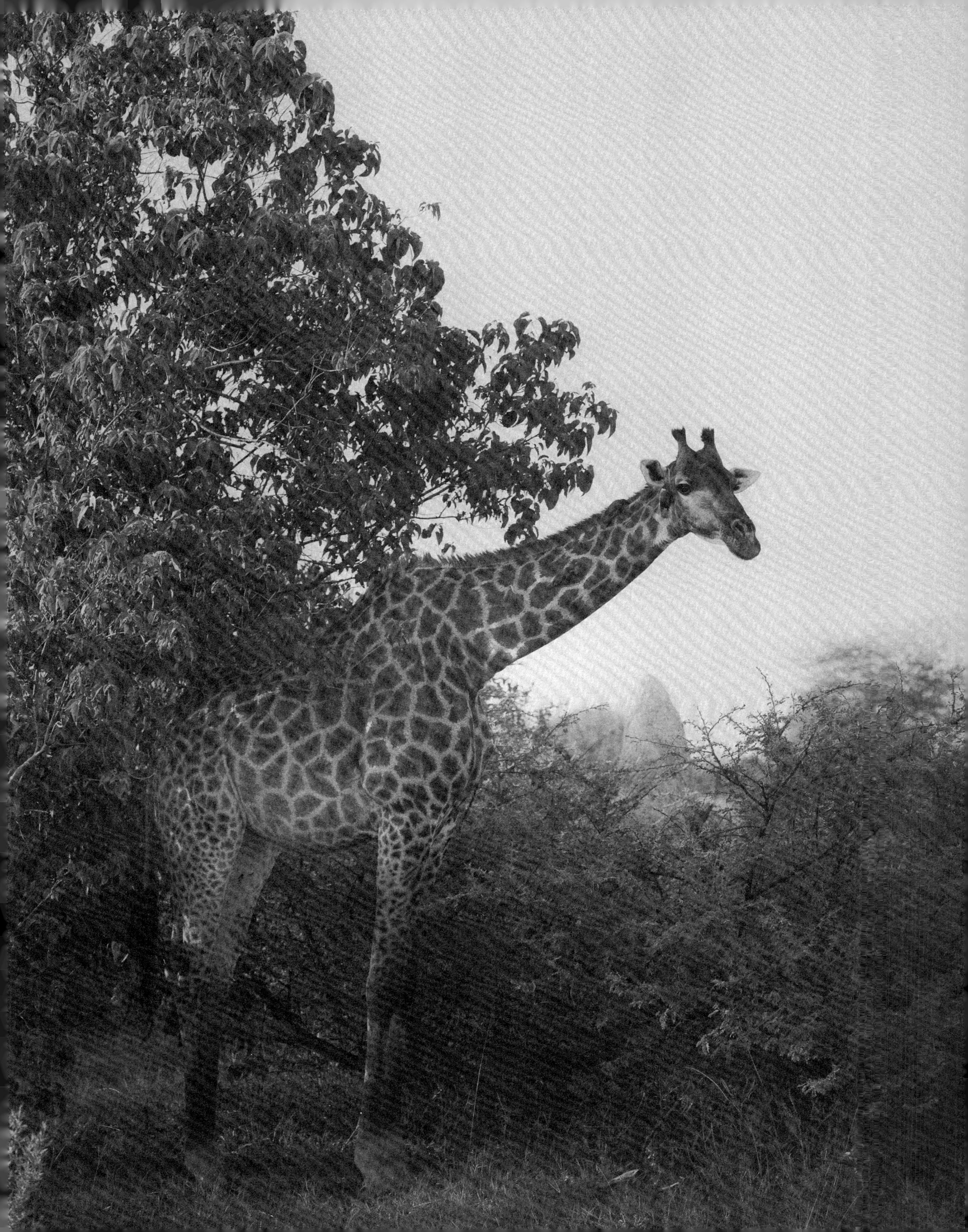

CHAPTER 10

OLD-SCHOOL AND CLASSIC SAFARI IN THE OKAVANGO DELTA

A MODERN UPDATE ON THE GOLDEN AGE OF EXPLORATION

Botswana

18°50'26.01" S
22°41'34.1" E

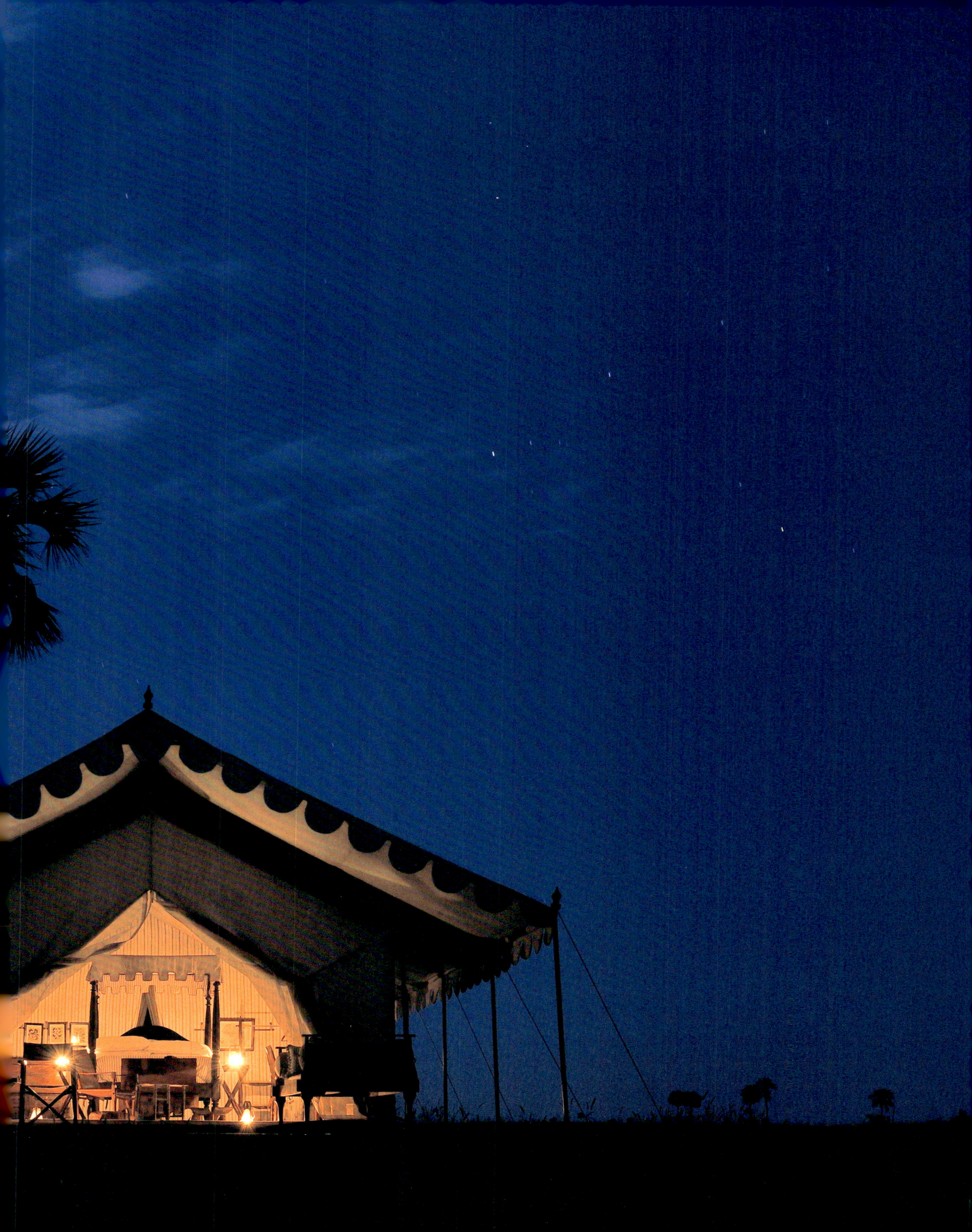

Old-School and Classic Safari in the Okavango Delta

A modern update on the golden age of exploration

Country:	**Botswana**
Best time to travel:	**Between April and September**
Closest airport:	**Maun**
Price tag:	**€€€€**
Style of travel:	**Old-school and elegant exploration**

Introduction

Seasoned travellers have discovered that the best way to experience a good old African safari is to travel with legendary safari guide Ralph Bousfield, who is considered one of the finest safari guides on the Mother Continent. Ralph has roots in Africa going back to the 17th century. His father, Jack, was the fourth generation of the family to live in Africa, a skilled tracker and hunter who killed over 50,000 crocodiles. Stylish and rugged, the 59-year-old bears the scars of his profession, including those of a snake attack and an almost fatal bout of blood poisoning. He has also survived nine plane crashes, although this hasn't stopped Ralph from flying all over the continent aboard a vintage red bush plane named A2 SAN — referring to the San Bushmen. According to Ralph, the key to flying in Africa is to know how to crash, or to do it with someone experienced. Even in the bush a touch of (British) humour and self-reflection do wonders. He even lost his father in one of those plane accidents. On his right arm Ralph wears a collection of amulet bracelets. On his left arm, a Rolex. Around his neck, a collection of talismans, neatly hidden in the skin of an antelope's scrotum. These are beads and rocks given to him by shamans and important tribesmen, along with a talisman from his friend and mentor, the late American photographer Peter Beard, who was a close friend for more than 25 years. A trained biologist, he was raised in the African wilderness, where he became a master tracker and learned the language of the Kalahari Bushmen, equipping him to escort celebrity clientele — from former American presidents and royalty, to filmmakers like the Coen brothers — into the remotest parts of the continent. The place where Ralph takes guests most often is the Okavango Delta, one of the oldest and most pristine ecosystems in the world. According to him, just by looking at ancient trees here, scientists can learn about evolution and how humankind progressed through history. Seventy per cent of the surface of the Okavango delta consists of pools, waterways, and swampland. Fed by run-off from mountains in Angola, hundreds of kilometres away, crystal-clear water runs though shallow channels and eventually evaporates in the Kalahari Desert. The Kalahari is one of those rare places where indigenous plants and animals have orchestrated their biological cycles in tandem with the seasonal rains and floods of the Okavango. To experience it during a private mobile safari is a privilege.

The Views

"Lose wonder and part of you dies," Ralph likes to say. In Moremi Game Reserve, a wildlife park in the Okavango (the park is three times the size of London), the pursuit of wonder means eschewing the more typical high-adrenaline pursuit of animals adopted by most safari operators in favour of a holistic and improvisational approach to wildlife encounters, without chases in roaring caravans of 4x4s. Driving slowly, early morning and late afternoon, is the preferred way, during the periods that Ralph calls 'the gentle times of day'. The Okavango is a labyrinth of underwater gardens for wildlife seeking refuge from the African dust. Whether they're swimming in a shallow lagoon or creeping along a trail at dawn, Ralph, ever alert, instructs his guests in a more gentle approach to wildlife viewing. At night guests camp on secluded islands where they sleep under nothing but diaphanous mosquito nets, with the diamond dust of the night sky framed by colossal mangosteens. Bedding down in the bush, bathed in the scent of wild sage, is a far cry from the world of the luxury 'tent' complete with air conditioning and internet. A mobile safari captures the essence of the 'safari' in its truest form, by taking

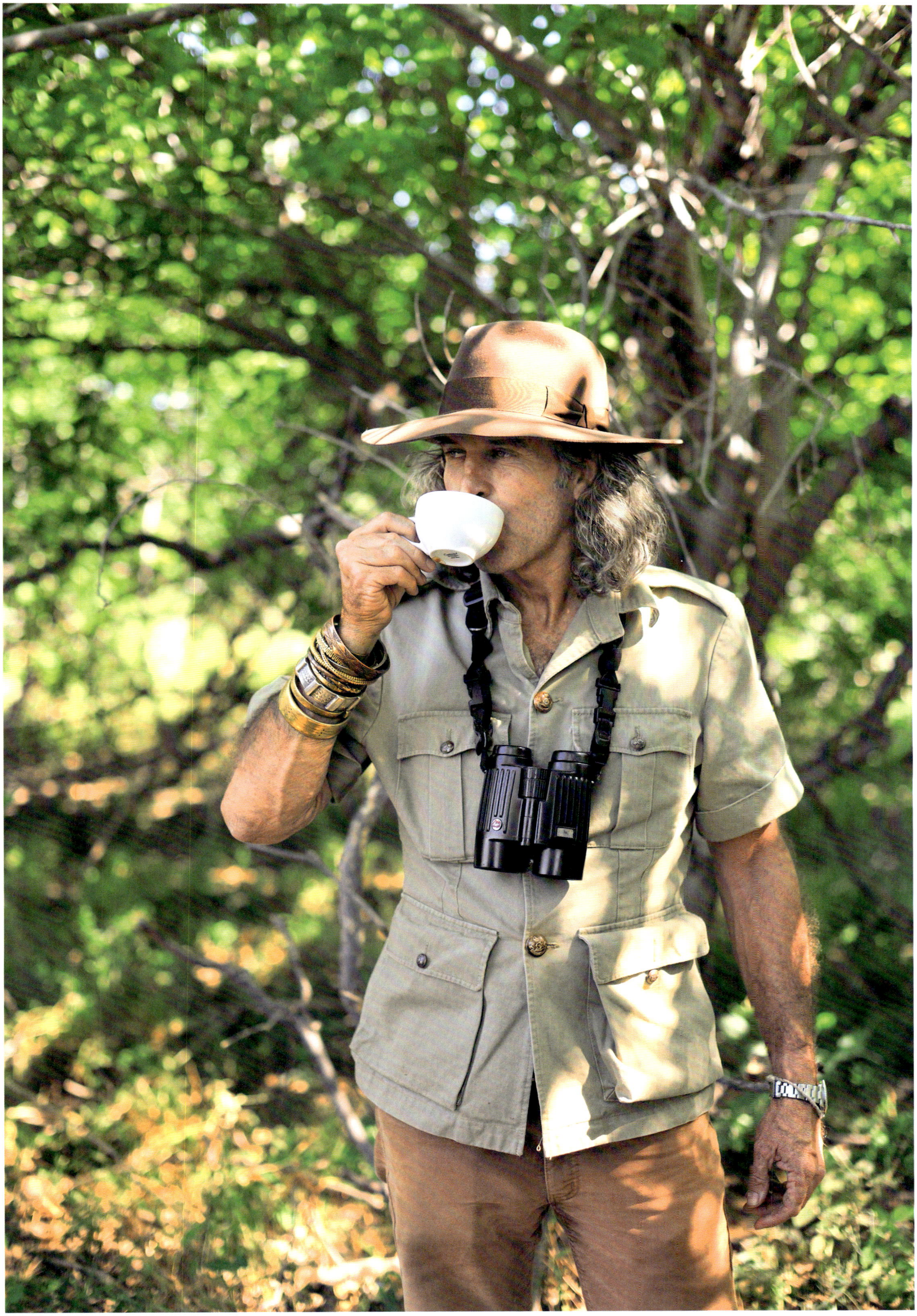

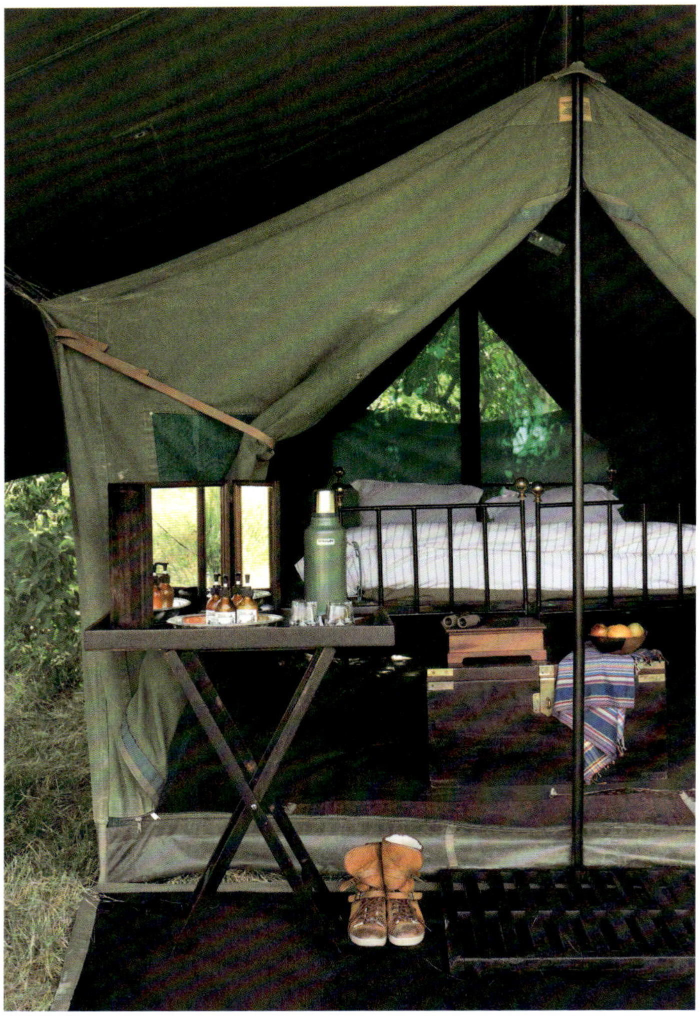

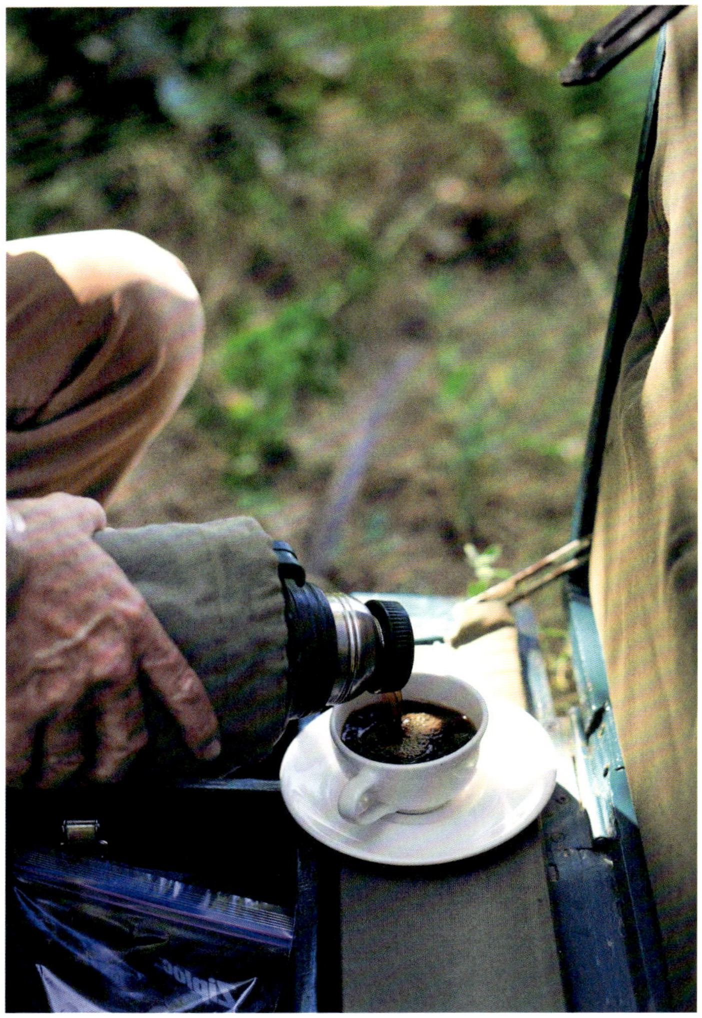

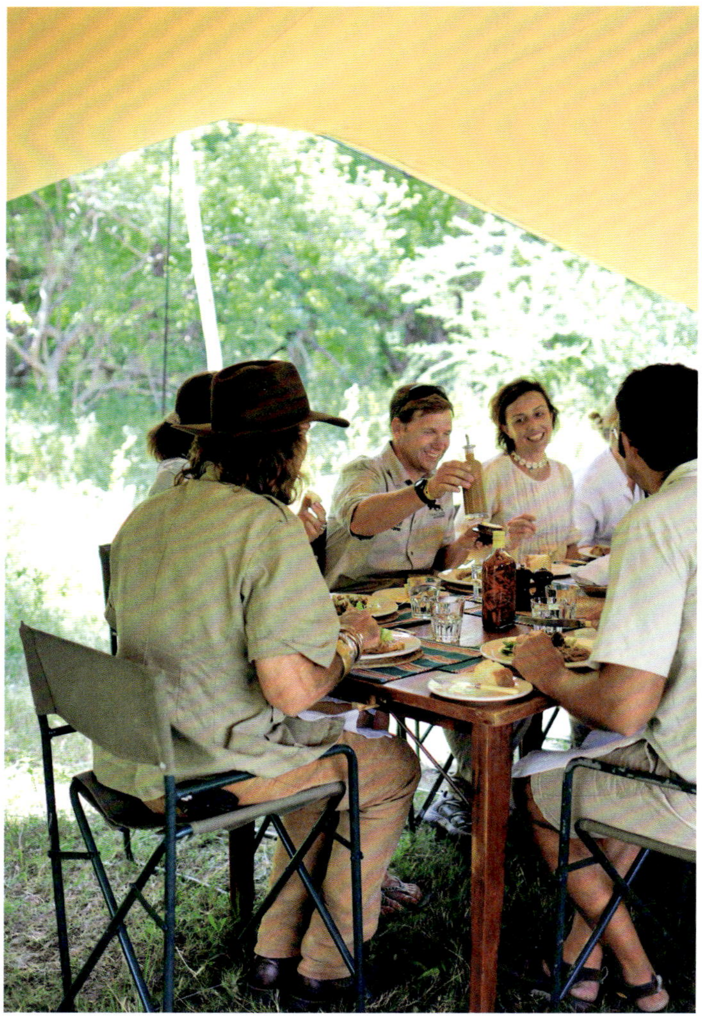

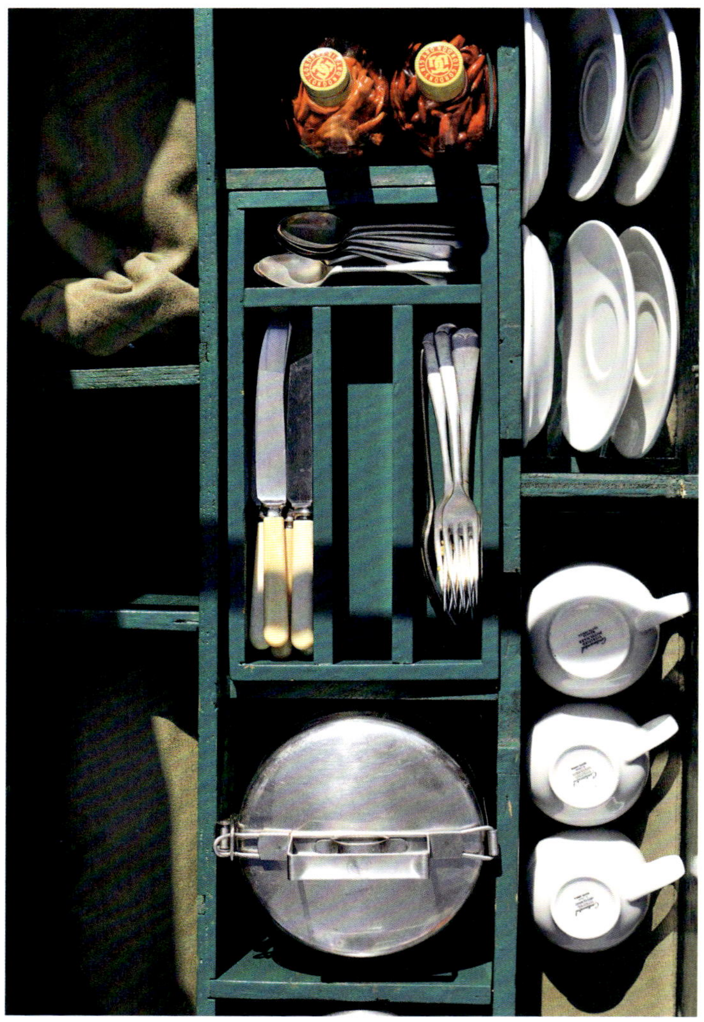

Opposite Lunch in the bush is served — an elegant and chic assortment of homemade treats packed in tiffin lunchboxes and served on delicate china, with vintage silverware and beautiful glasses. The elegant furniture in cast iron or wood is handmade in Ralph Bousfield's workshop in Botswana. Some curios or accessories in the mobile camp come from his extensive travels around the world.

its lead from expeditions carried out by early explorers and hunters. Ralph likes to stay long enough to fathom the soul of a place, to observe its fauna and flora, yet not long enough to disturb it nor have the animals get too used to the presence of people. The philosophy is to move on without a trace, leaving barely a footprint. Nor is Ralph hung up on securing glimpses of big predators. He's just as happy showing guests the lilac-breasted roller or the side-striped jackal, or pointing out a barely noticeable chameleon hiding in a tree at night. Creatures undetectable without his extraordinary tracking skills. Floating through the Okavango on a slow-moving boat, there is little to do except gaze over the gently flowing water, trying to spot wildlife in the lush foliage — crocodiles, perhaps, or the rare sitatunga and the famous lechwe, an aquatic antelope with elongated hooves. (The latter has a water-repellent secretion on its legs so it can move fast through water.) Driving with Ralph in one of the open safari 4x4s is no less entertaining, thanks to his theatrical storytelling, which derives from knowledge of the natural world learned from the Bushmen. For example, San Bushmen survival skills will be demonstrated via an introduction to hoodia, a succulent that suppresses appetite, allowing the Bushmen to cross long stretches of the Kalahari with little food. After the mobile safari in the Okavango Delta, Ralph might fly his guests over to the Kalahari Desert to one of his permanent safari camps on the Makgadikgadi Salt Pans. A mind-blowing landscape which often appears as a mirage can be experienced on horseback or quad bike, perhaps while fly camping on the shimmering salt pans to feel and hear the heavy and thick silence of this site. Once again Ralph exposes in a gentle and respectful way the soul of his land and its people, the San Bushmen. Instead of focusing on elephant and lion sightings, Ralph tries to tell the story of humankind which, after all, originated in Africa.

Creature Comforts

Ralph's mobile camps can be set up anywhere, even the Okavango wetlands. A typical camp consists of a cluster of safari tents with en-suite bucket showers and steaming hot water from an open fire, along with proper bush toilets and comfortable cast-iron beds. There is also a tented lounge with Moroccan carpets, plush cushions, and a huge table decked in linen and antique silverware. Paraffin lamps light up the dark African night. San Camp in the Kalahari Desert is a cluster of majestic white tents stocked with museum-worthy antiques and collectibles, from the heydays of Ralph's father. The communal spaces are classically stylish as well, with fine linen and delicate tableware used during long, relaxing lunches in between game drives and explorations. Another safari camp not to miss and also set up by Ralph's late father is Jack's Camp. Surrounded by the vast and pristine landscapes of the Makgadikgadi Pans and once again reminiscent of a traditional 1940s East African safari camp. With just eight tents, this is the most beautiful and intimate spot to sit on a private veranda and take a dip in the small plunge pool while overlooking the shimmering Makgadikgadi salt pans.

Go Now

Things change slowly in southern Africa and remote areas like the Okavango Delta. The modern world is, for the moment, still kept at bay by endless deserts and impassable deltas. Travellers who visit now will be able to connect to its unique flora and fauna, but also with the region's ancient cultures and traditions.

How to Go

The gateway to the Okavango Delta is Maun International Airport in the sleepy town of Maun. From there it is a 30-minute flight by bush aeroplane. Mobile camping, including full board, alcoholic beverages, and guiding is priced at 2,000 Euro per person per night, with a minimum of four guests. Mr Bousfield charges 2,800 Euro per day and per group for his guiding services.

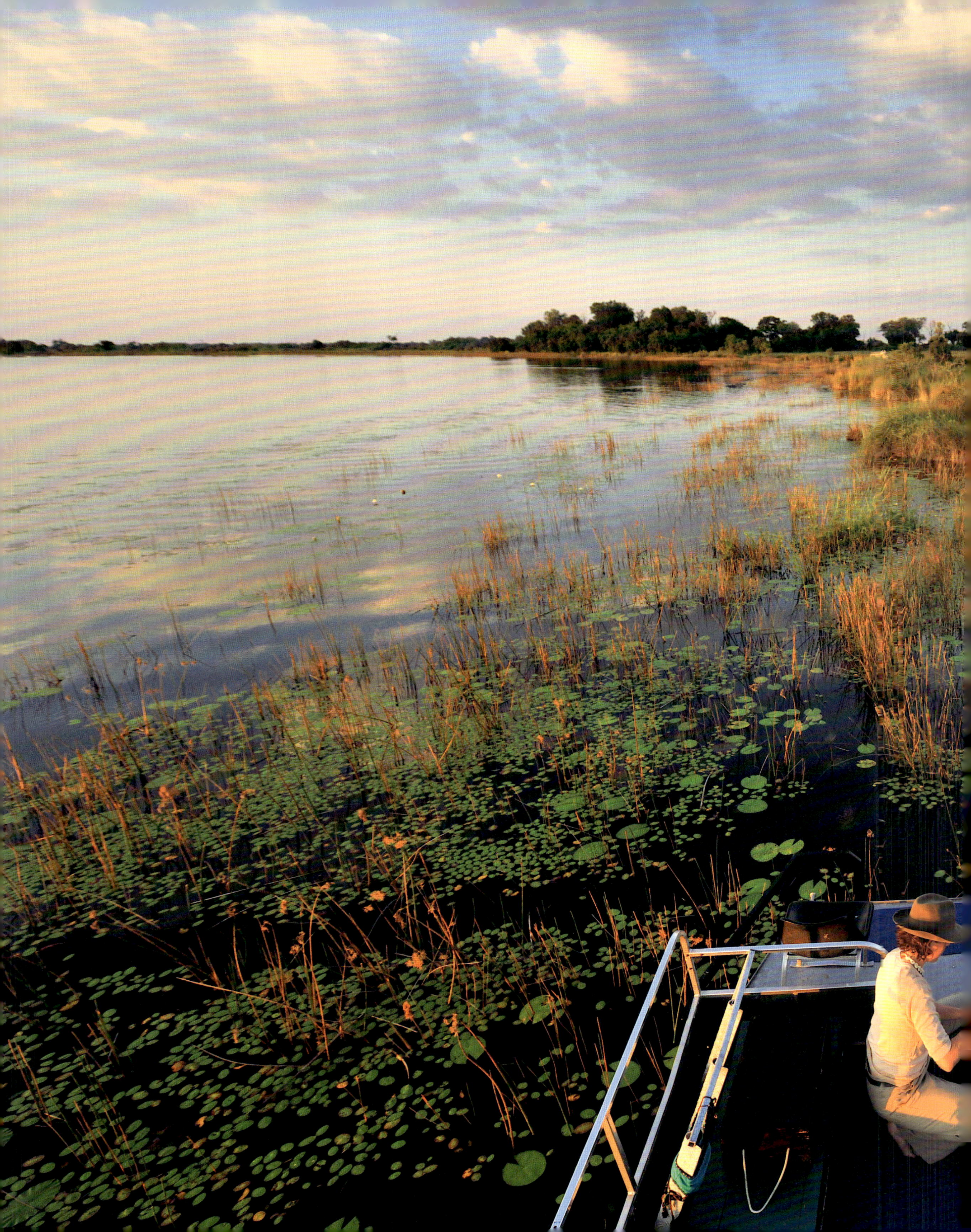

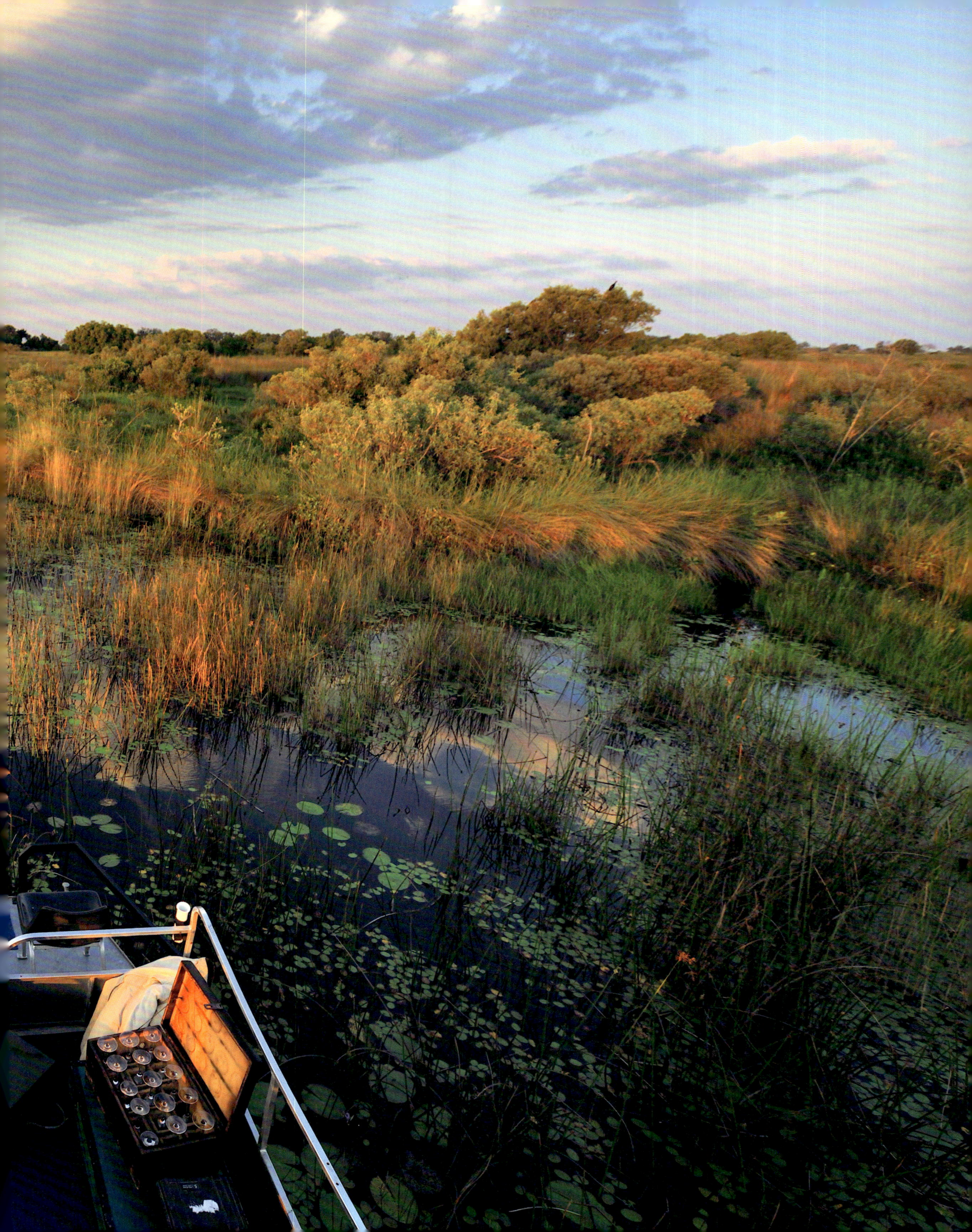

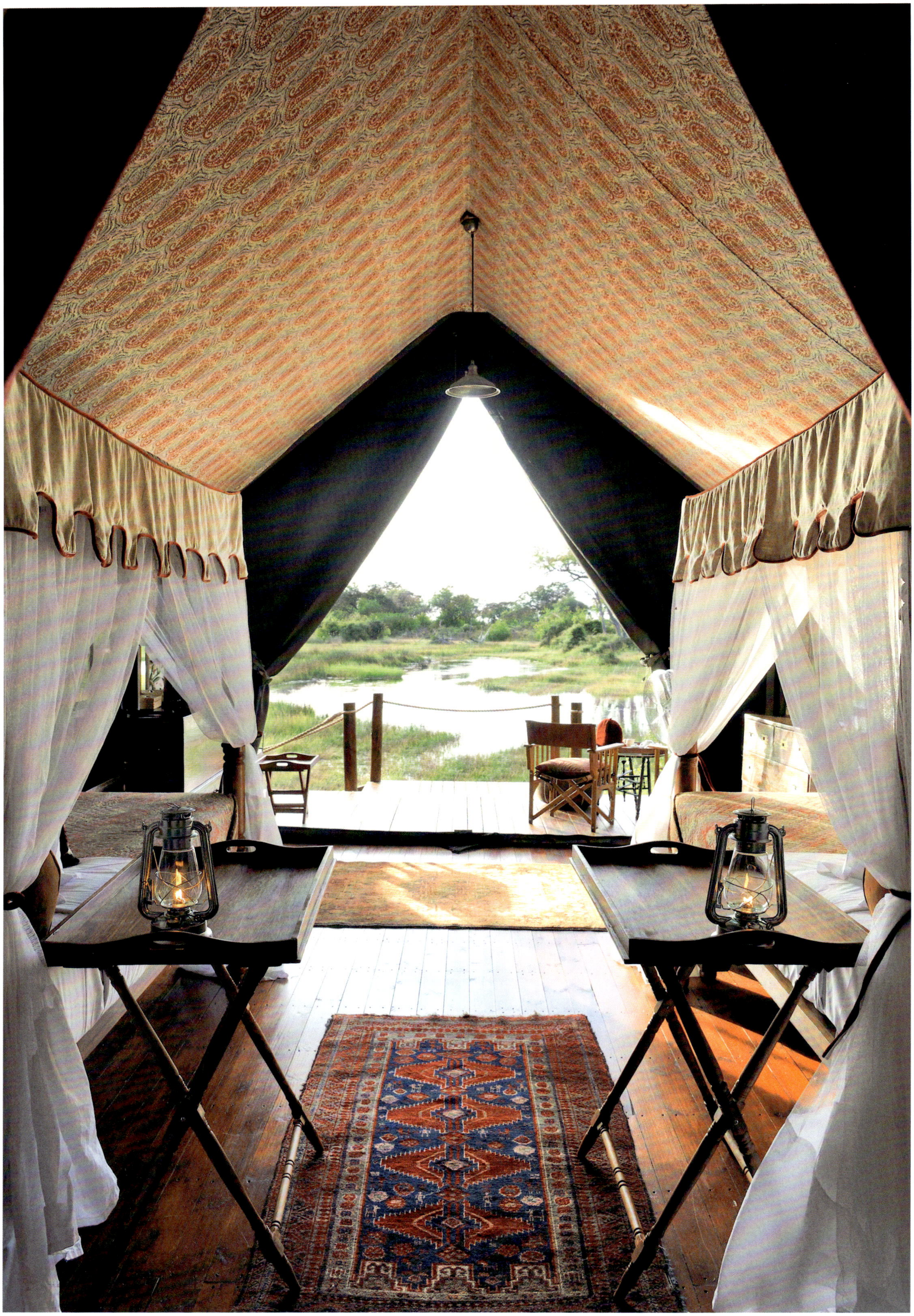

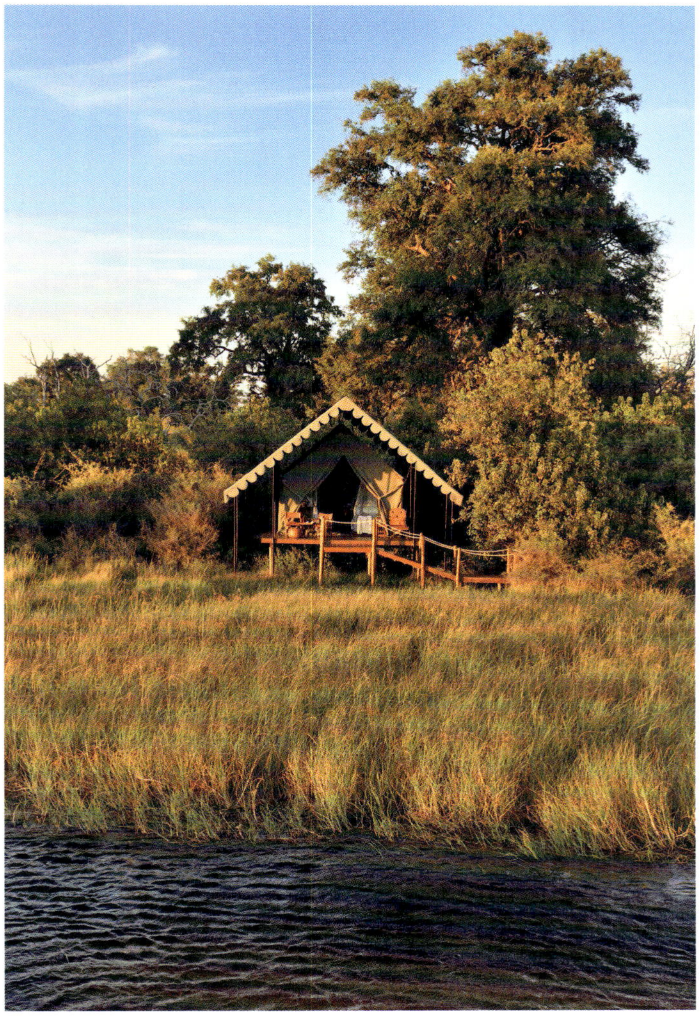
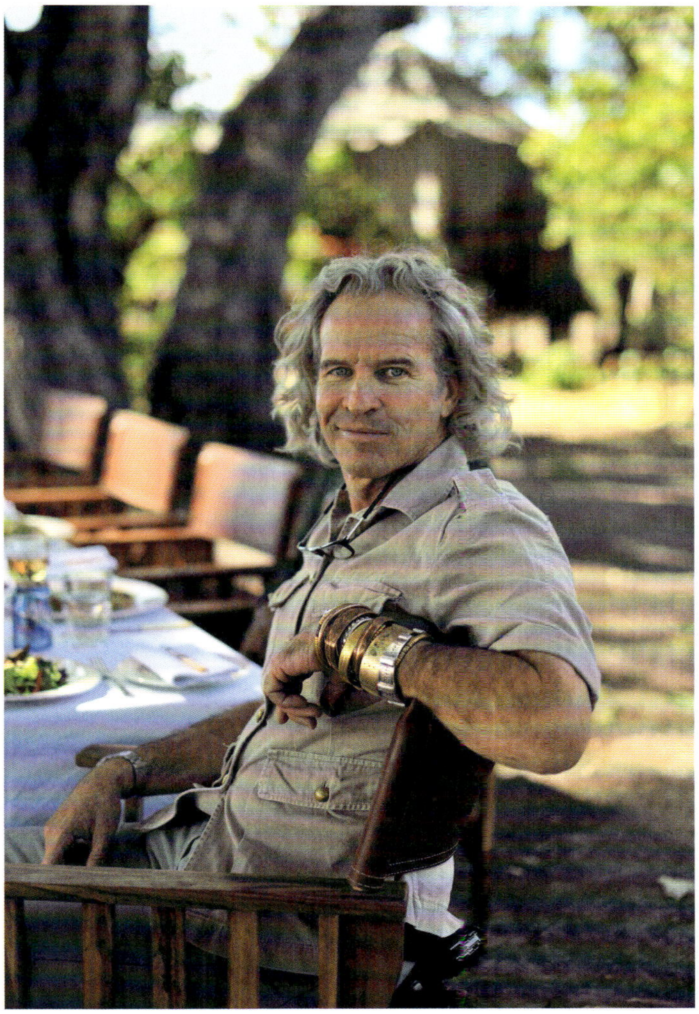
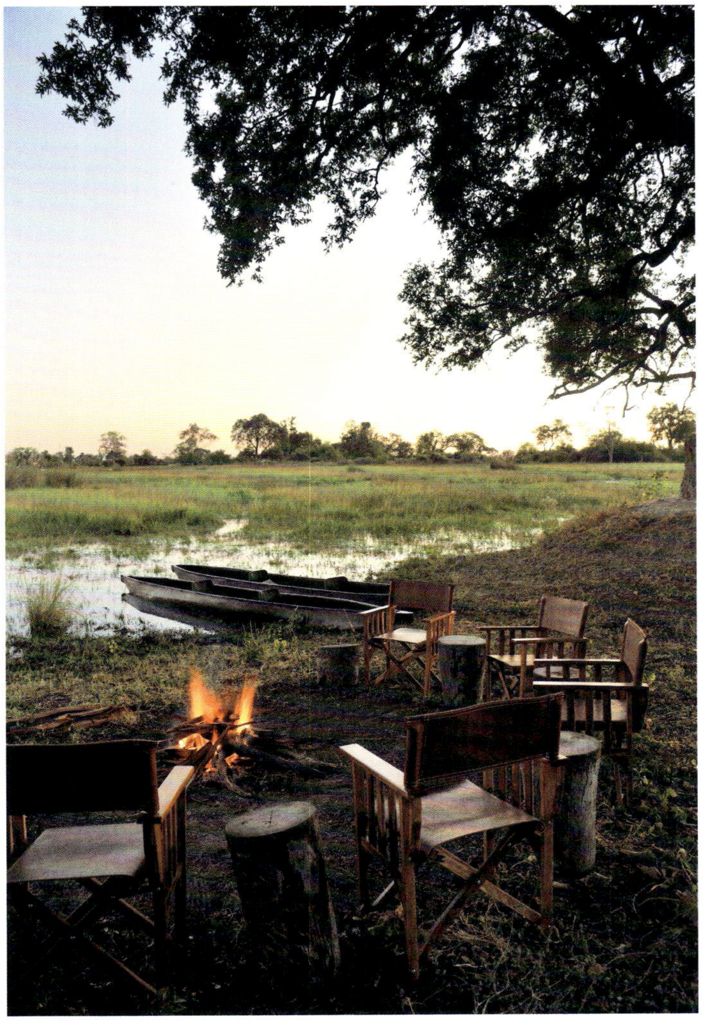
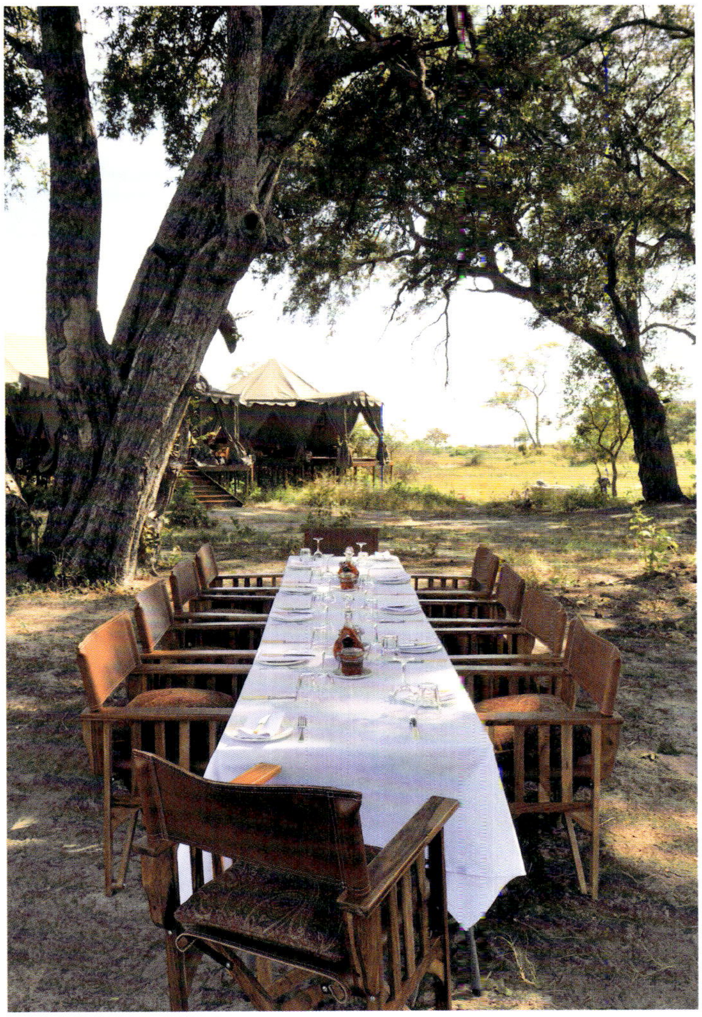

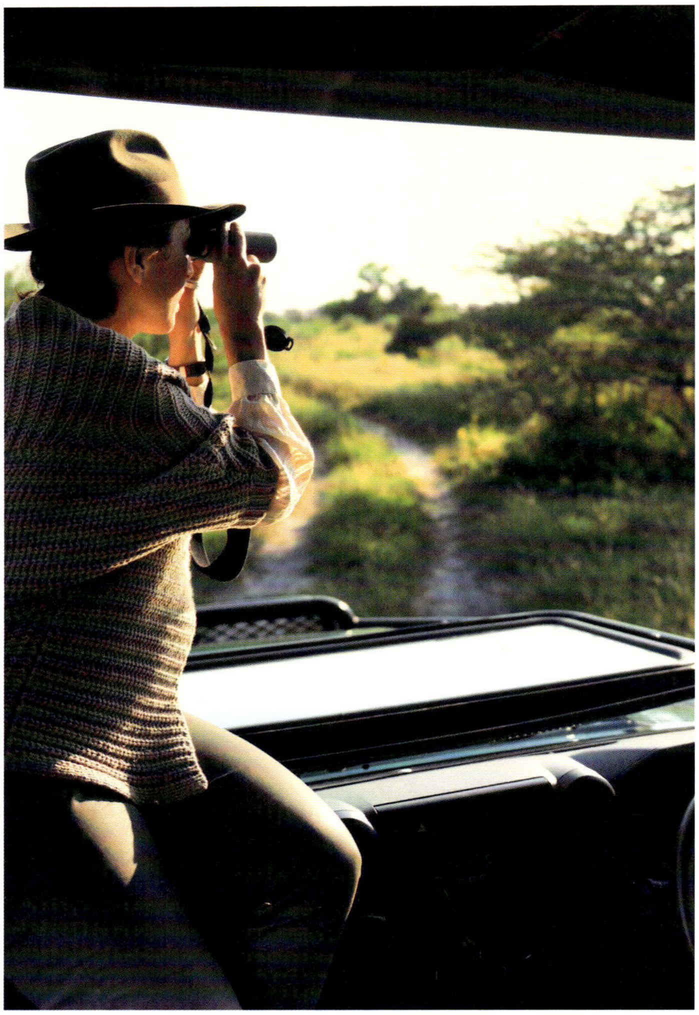

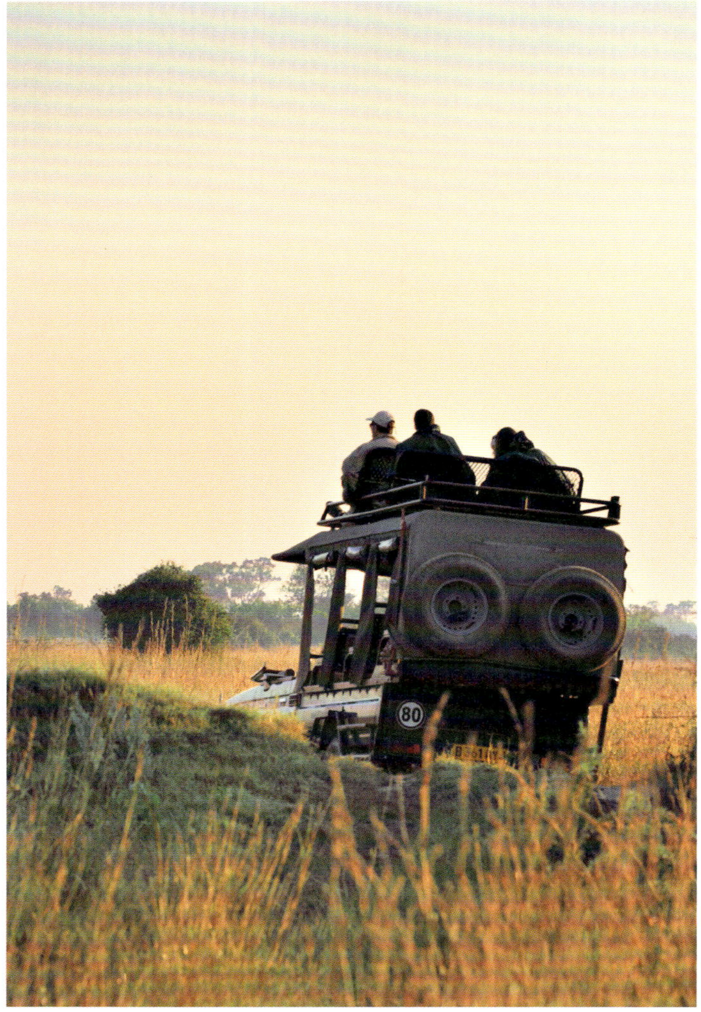
80

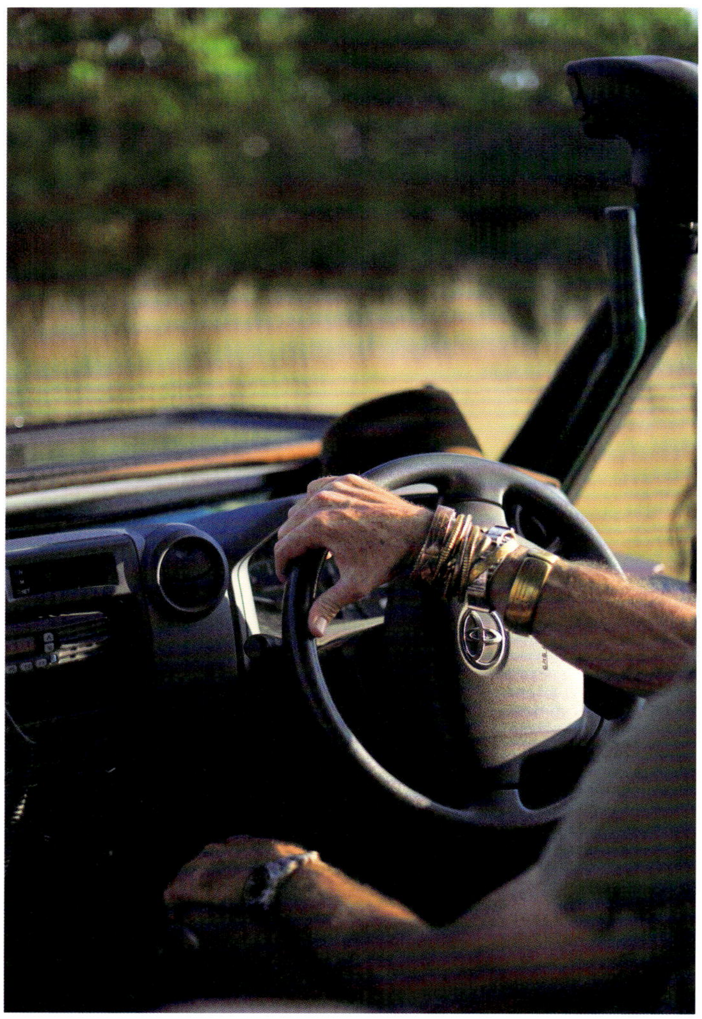

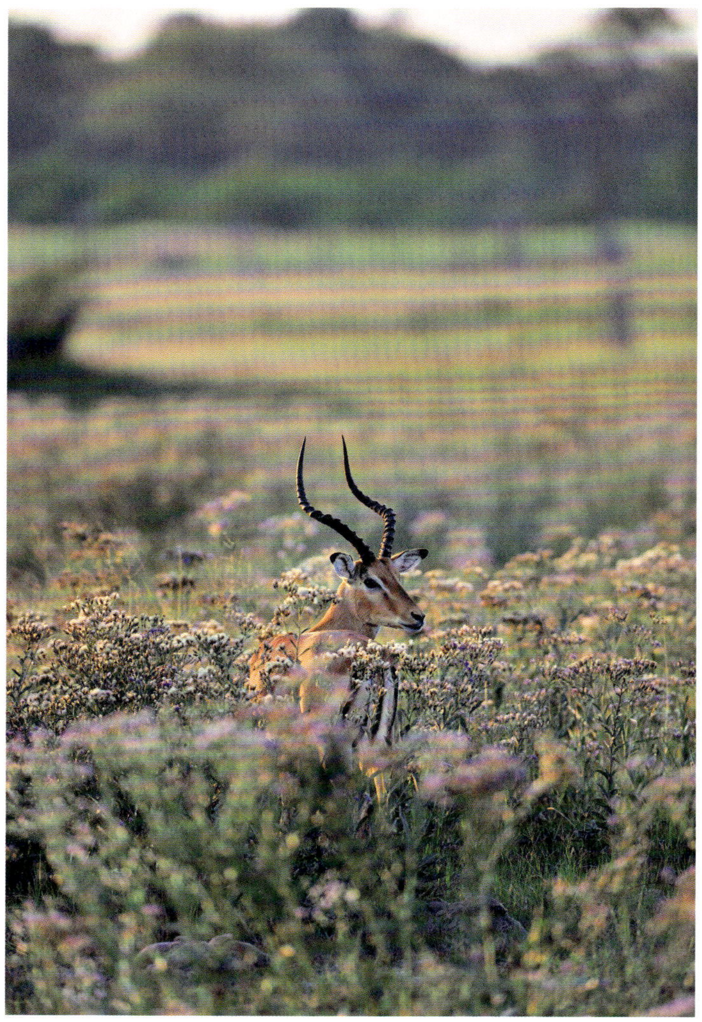

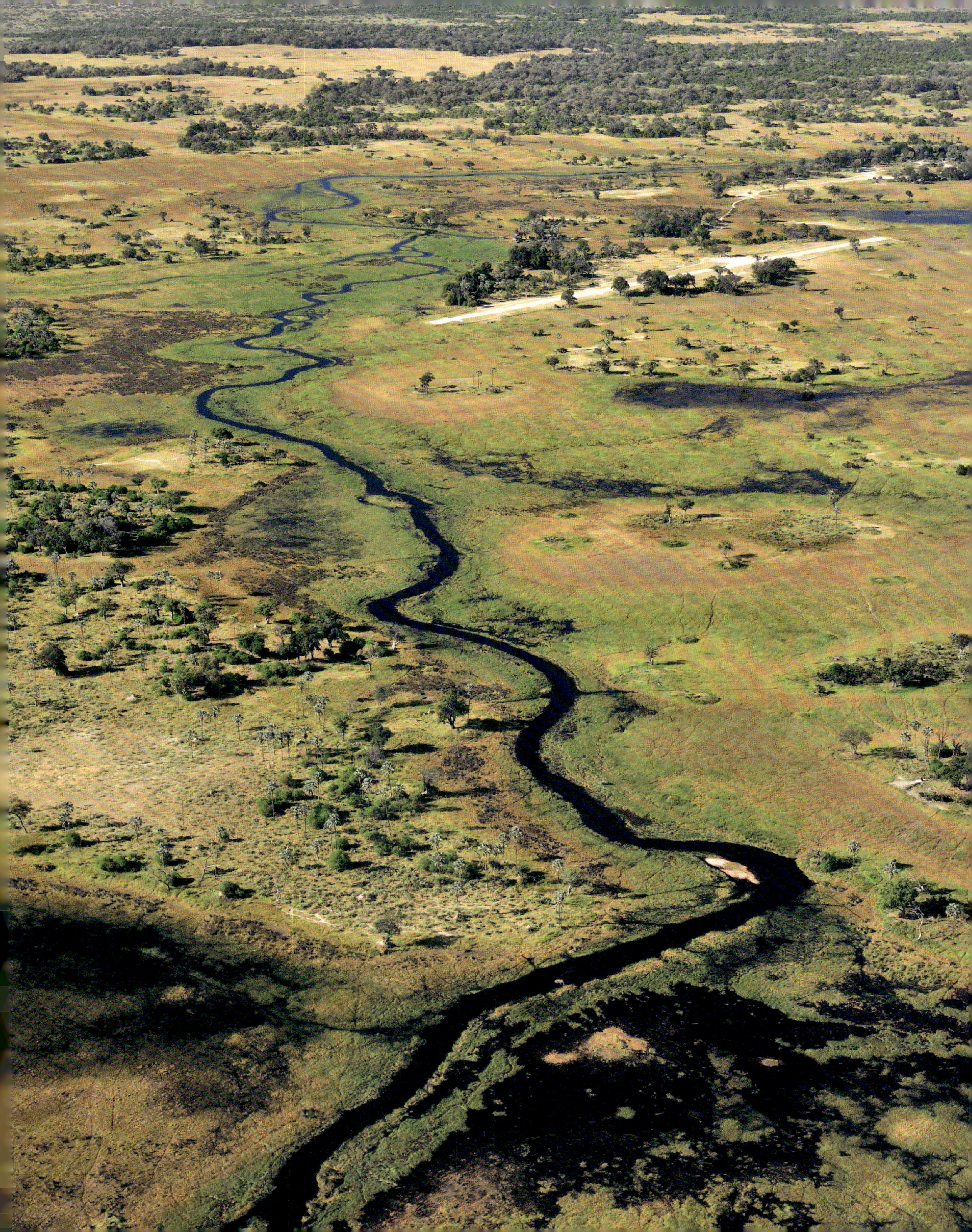

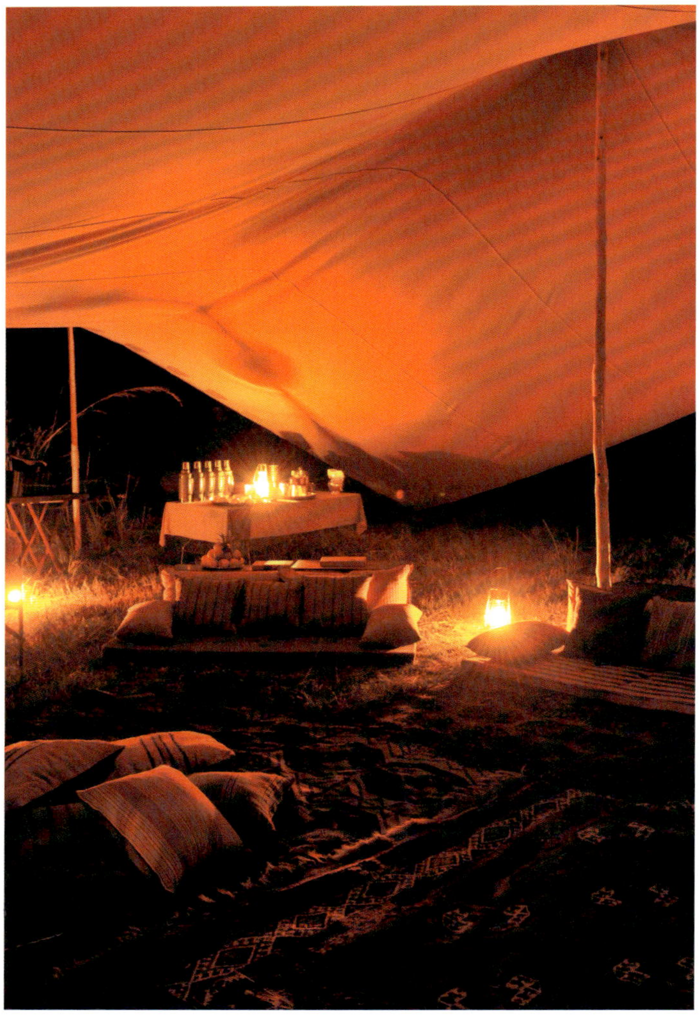
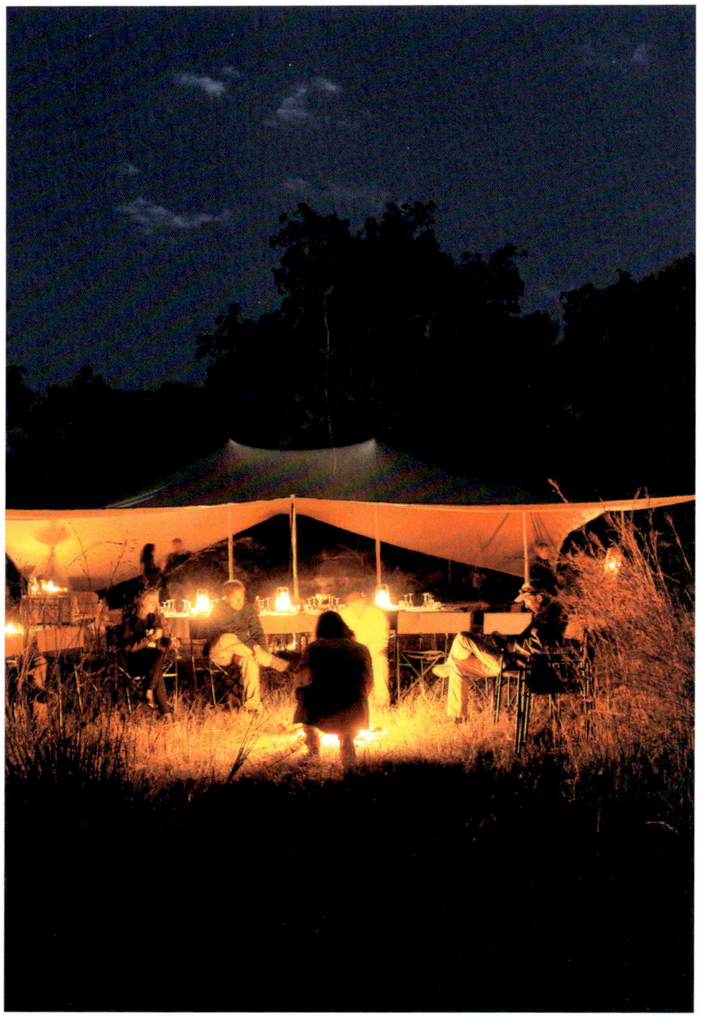
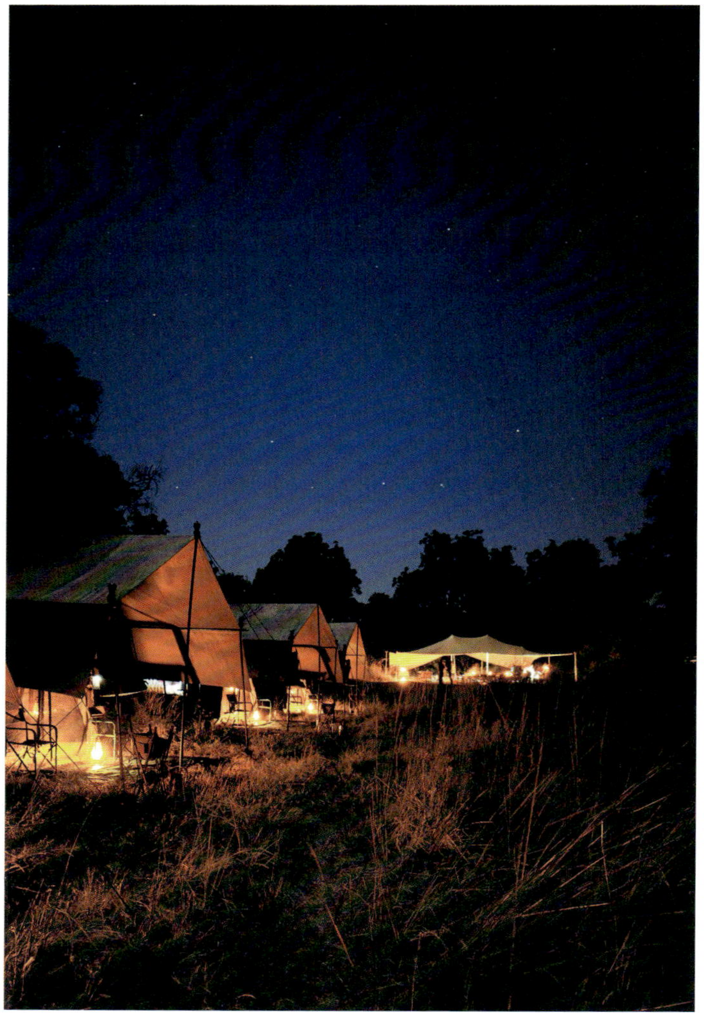
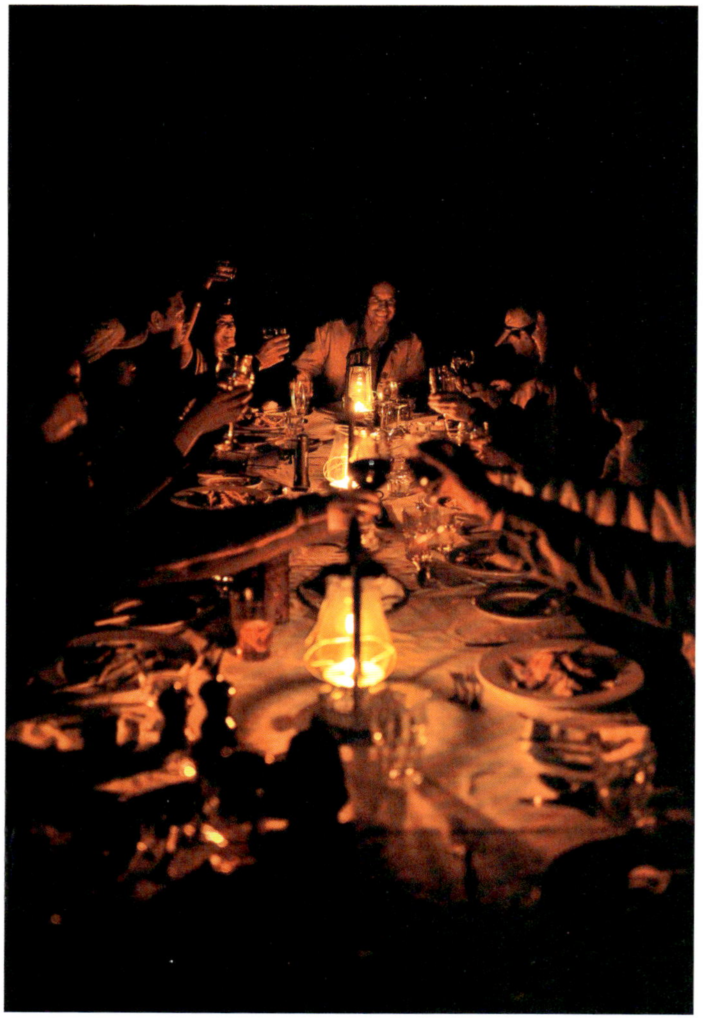

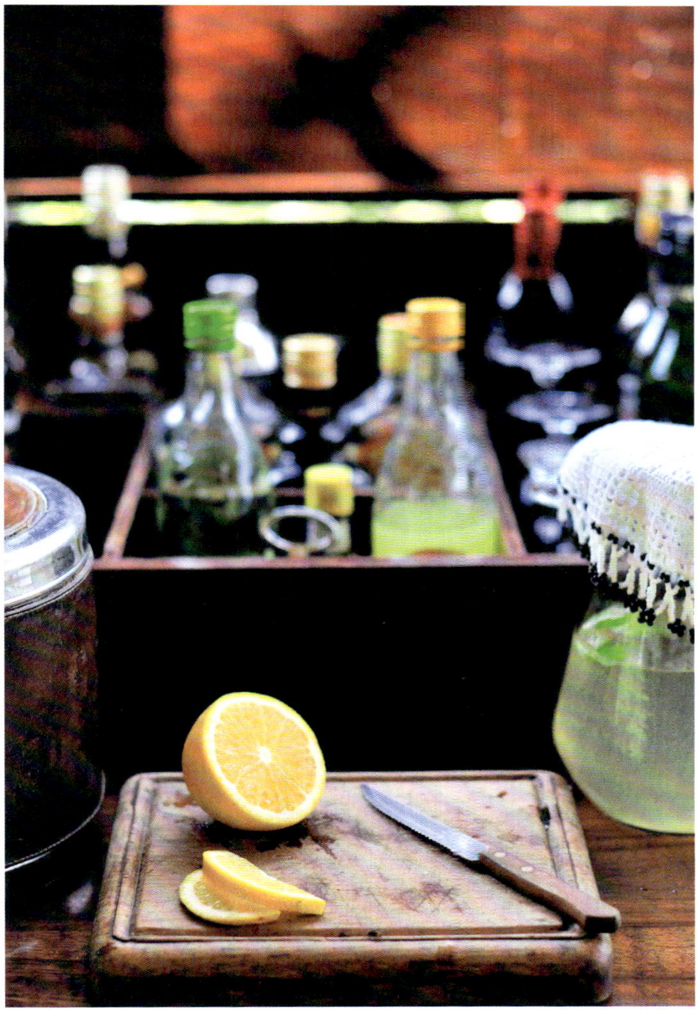

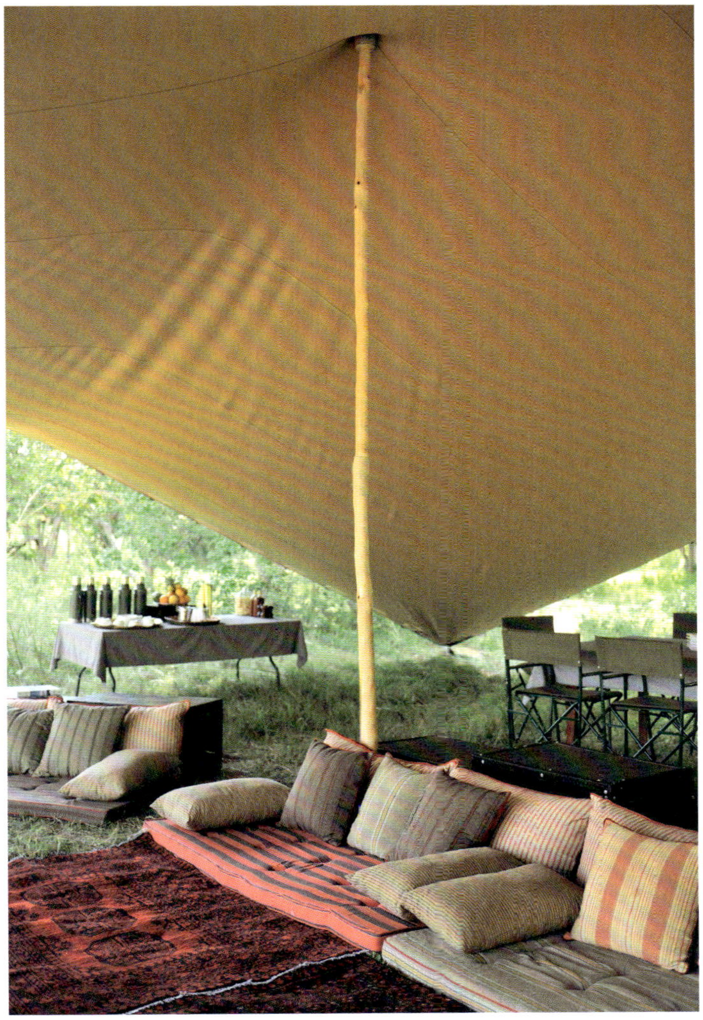

Previous spread One of the largest and least spoiled wetlands in the world, the Okavango Delta is home to over 1,500 plant and bird species, an abundance of reptiles and fish, and more than 130 mammal species, including the world's largest remaining population of elephants. More than 130,000 animals gather here in search of fresh water and lush vegetation, constantly reshaping the Delta's channel system with their movements.

This spread A mobile safari isn't just about spying on wild animals or experiencing the serenity of the delta. After nightfall, travellers connect or reconnect over classic gin and tonics, sharing travel — or life — stories around a roaring campfire.

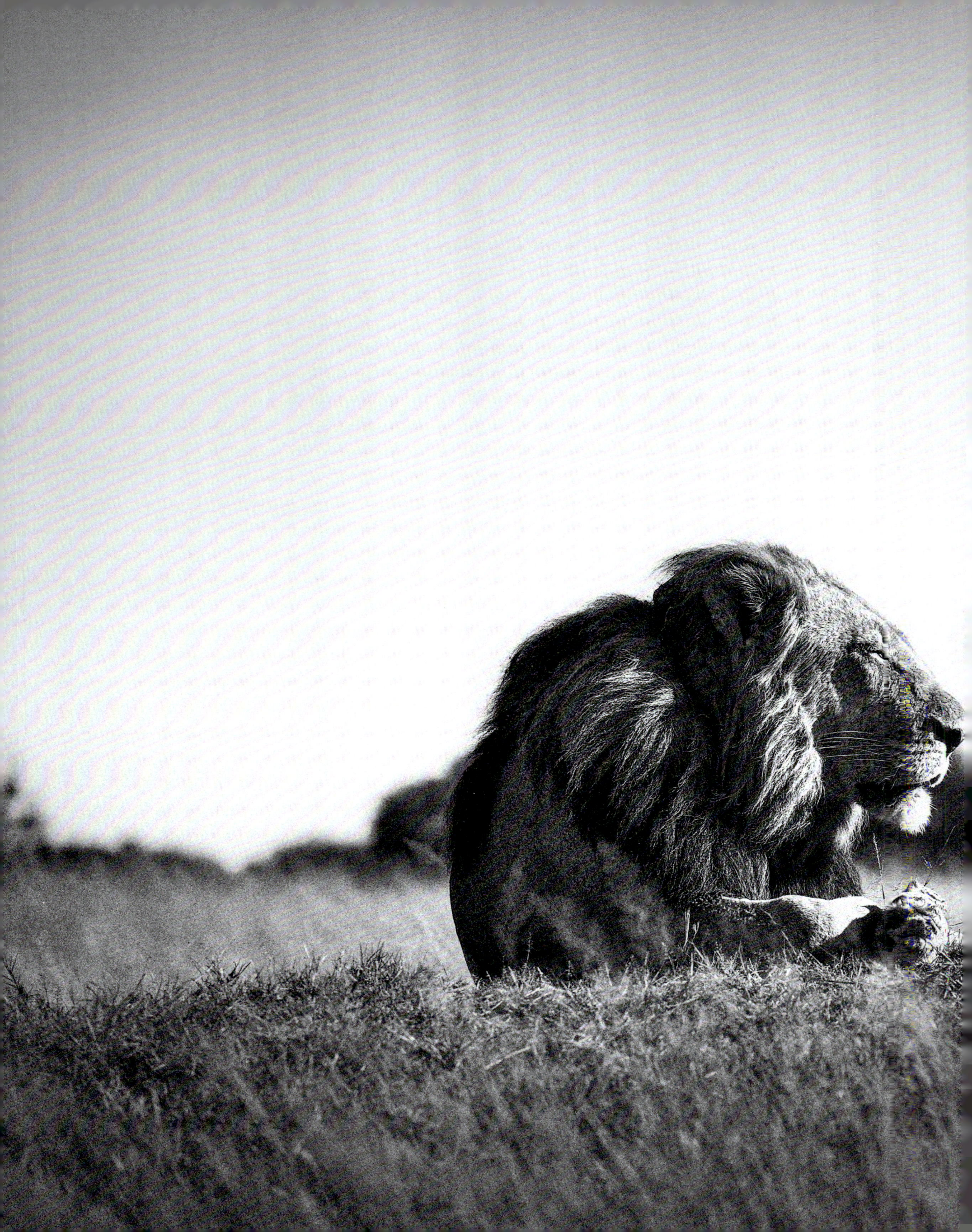

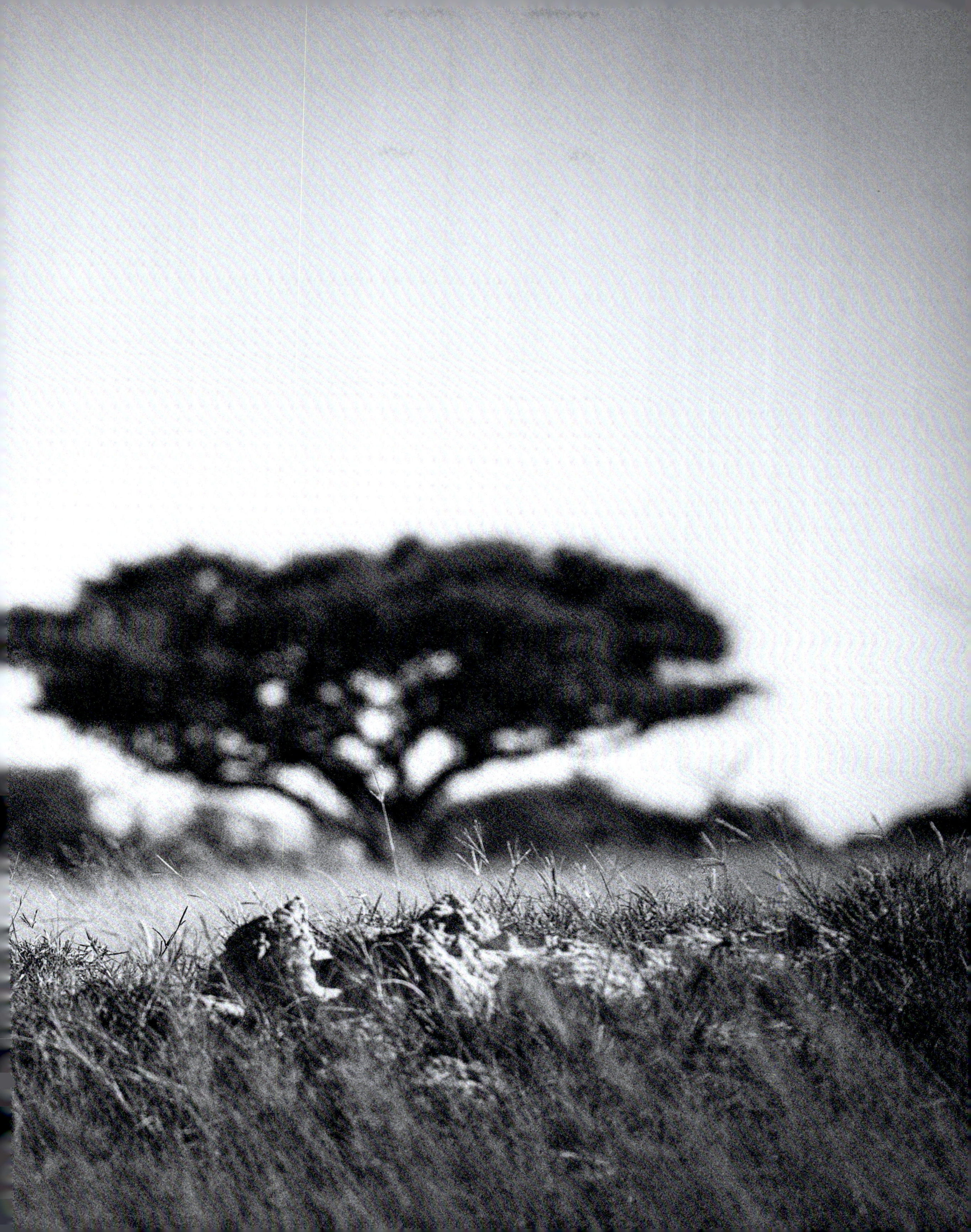

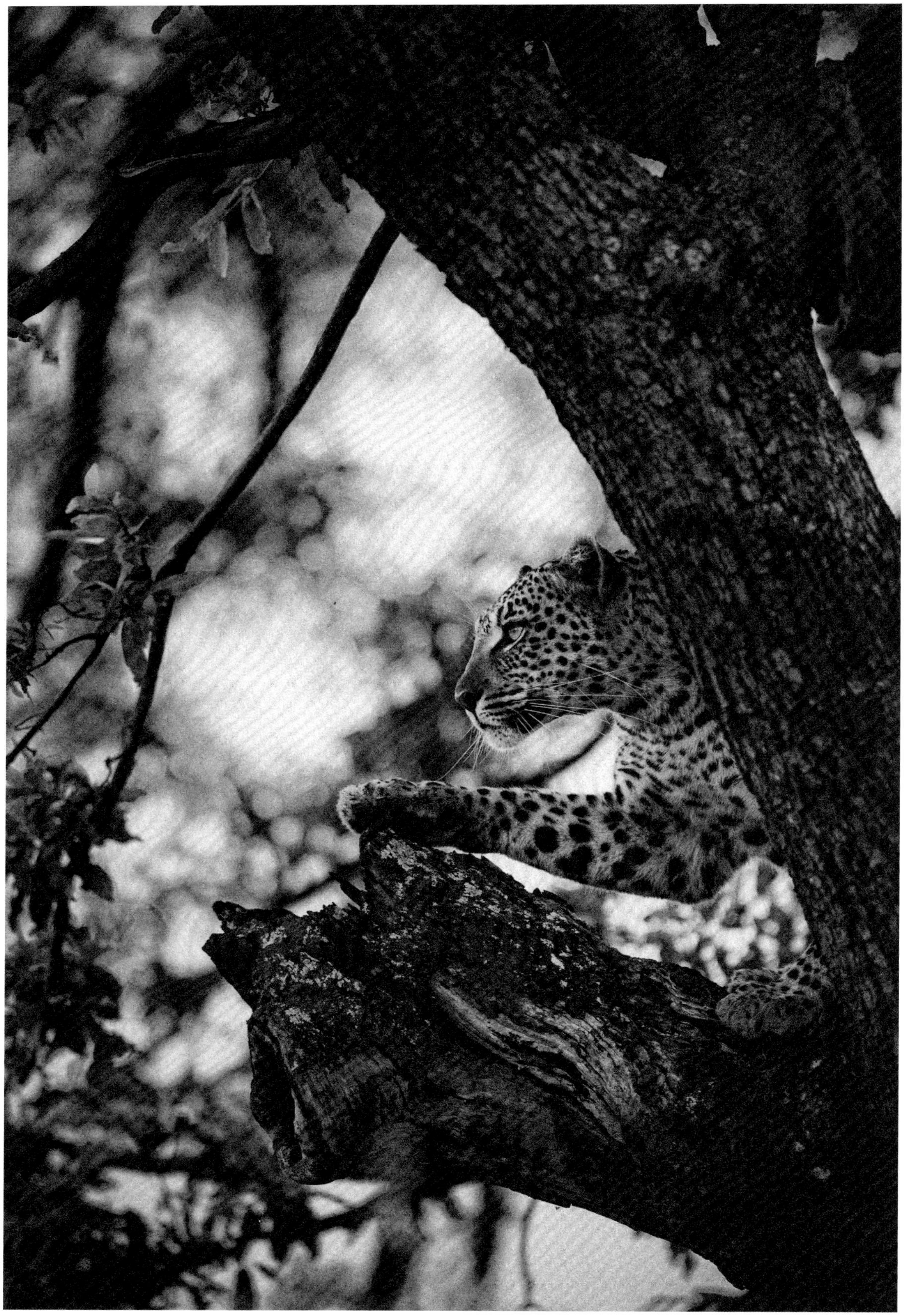

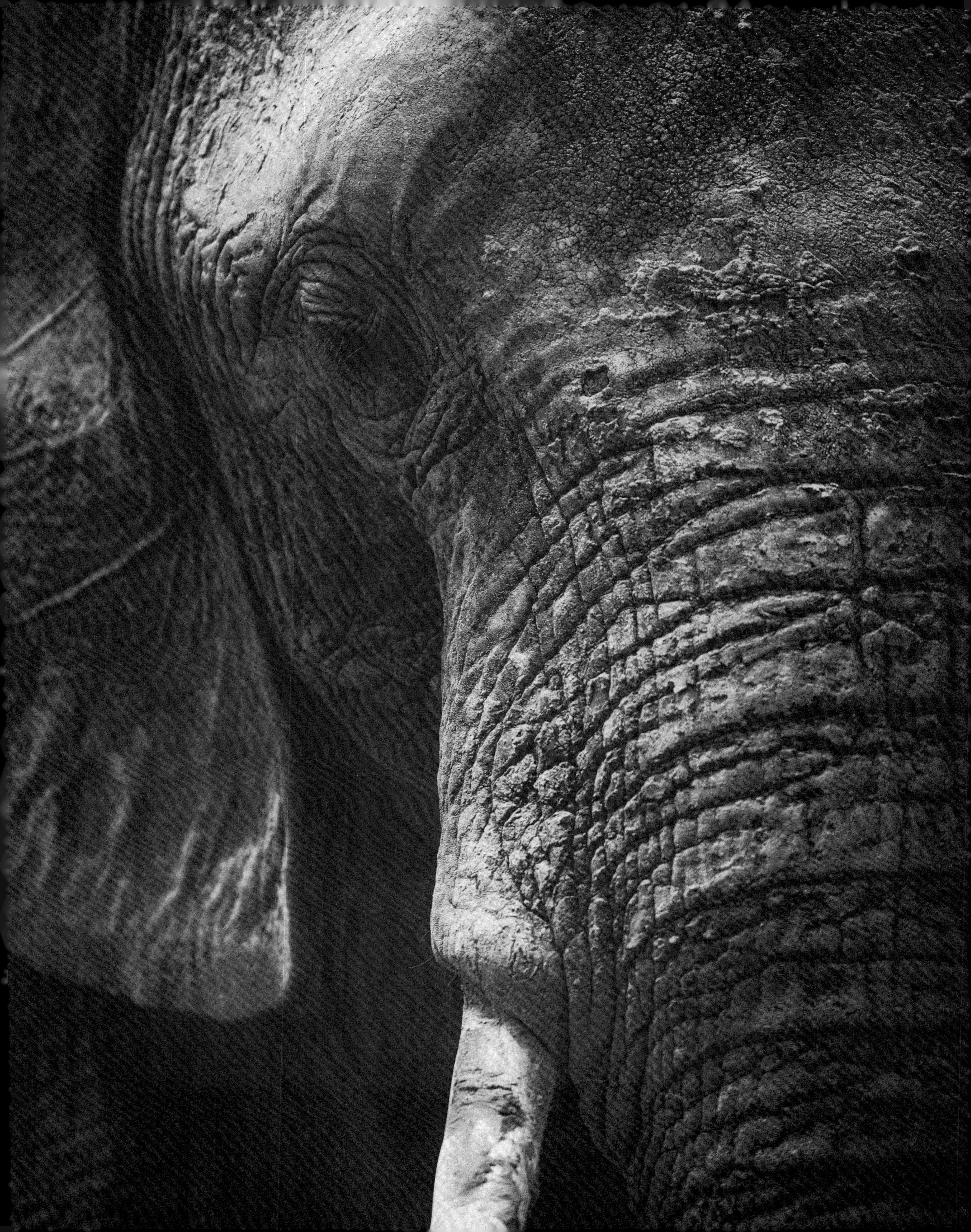

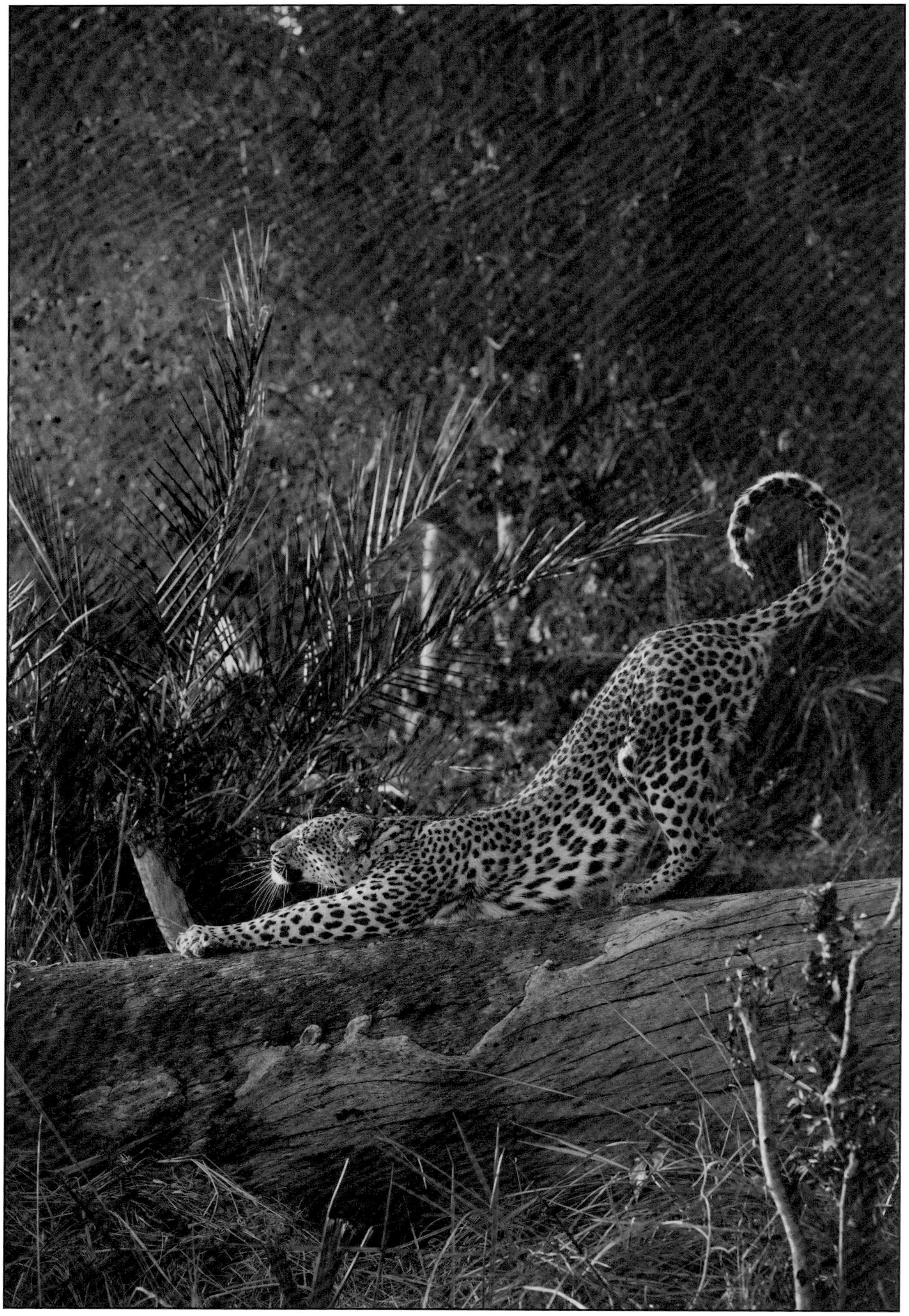

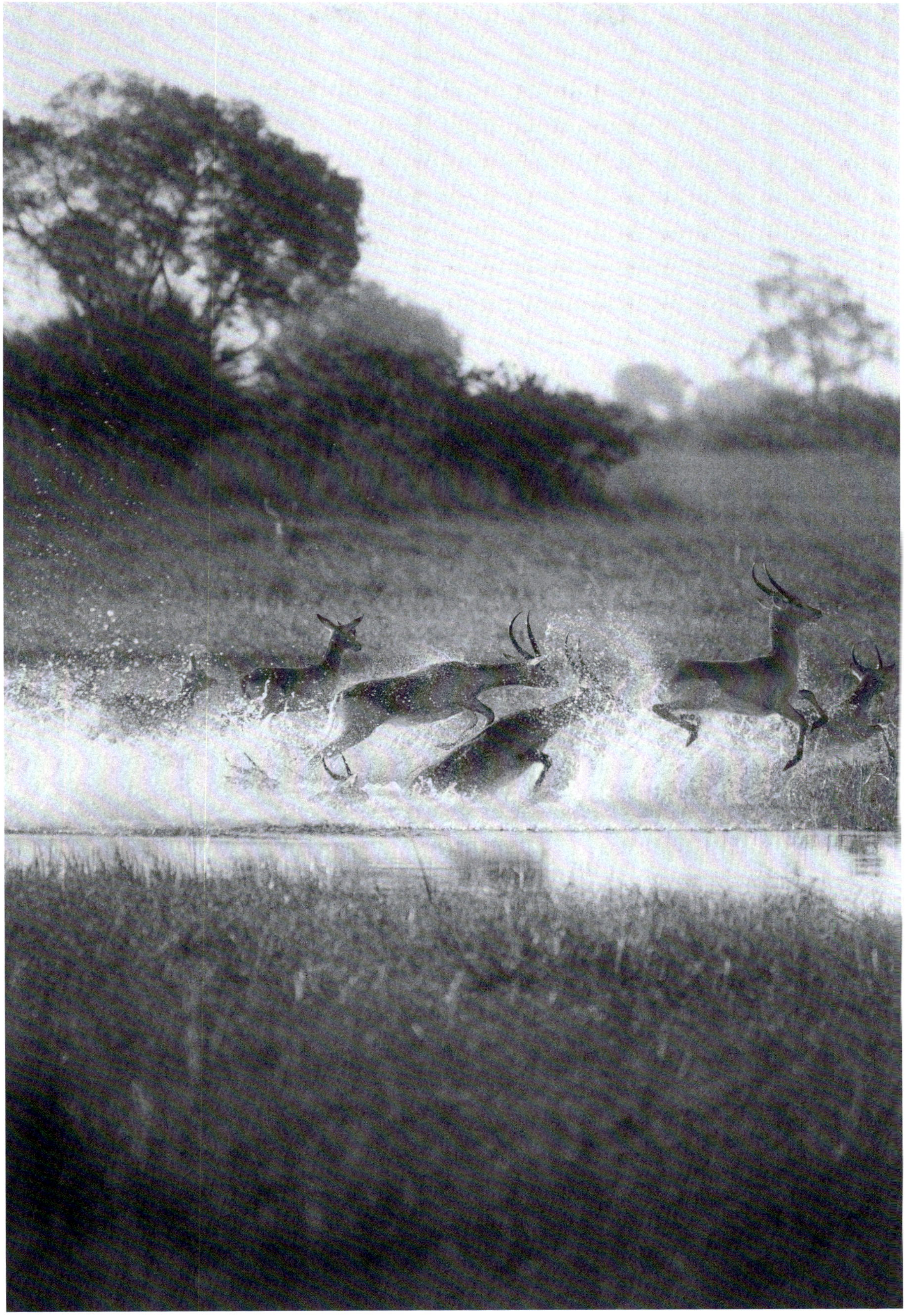

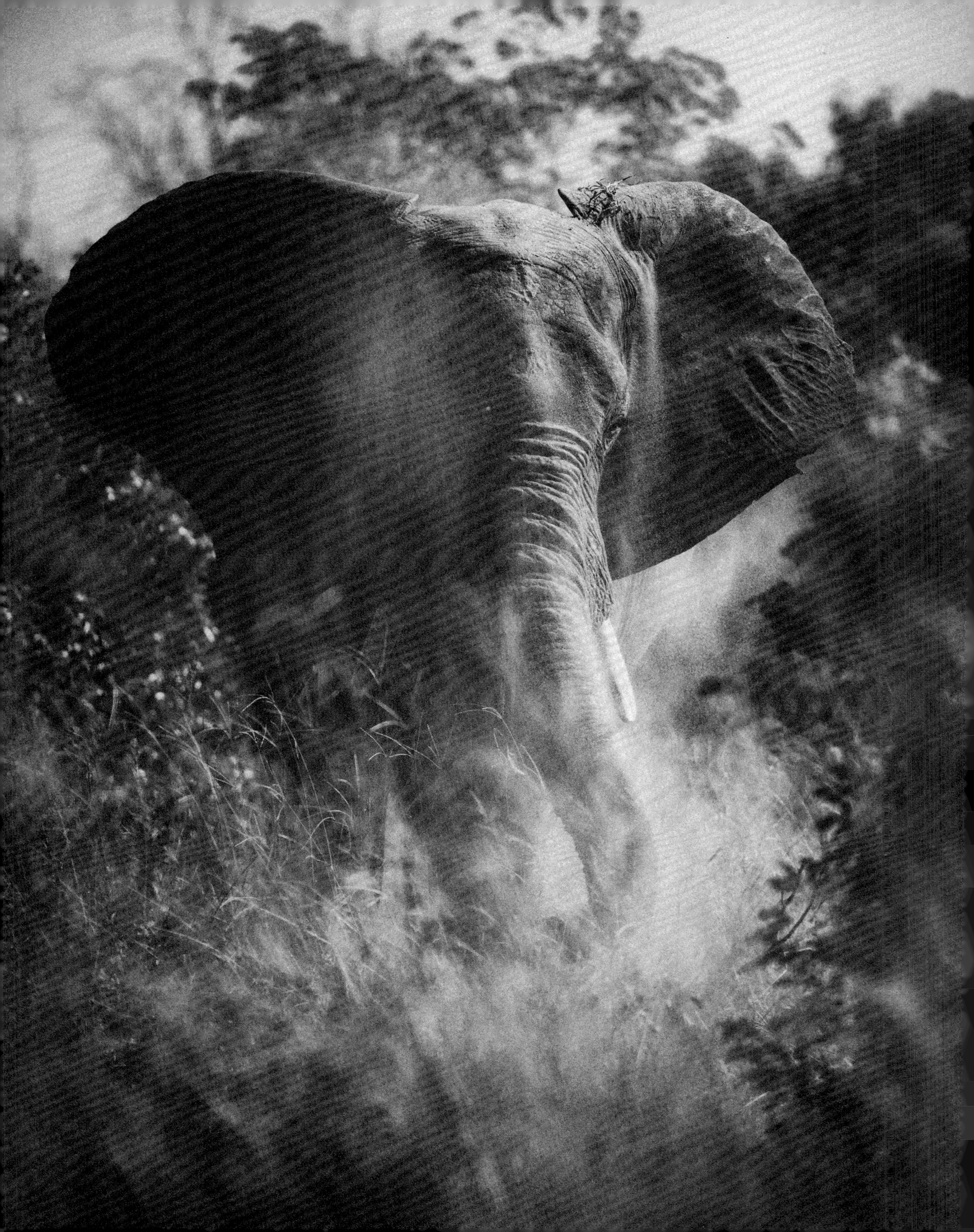

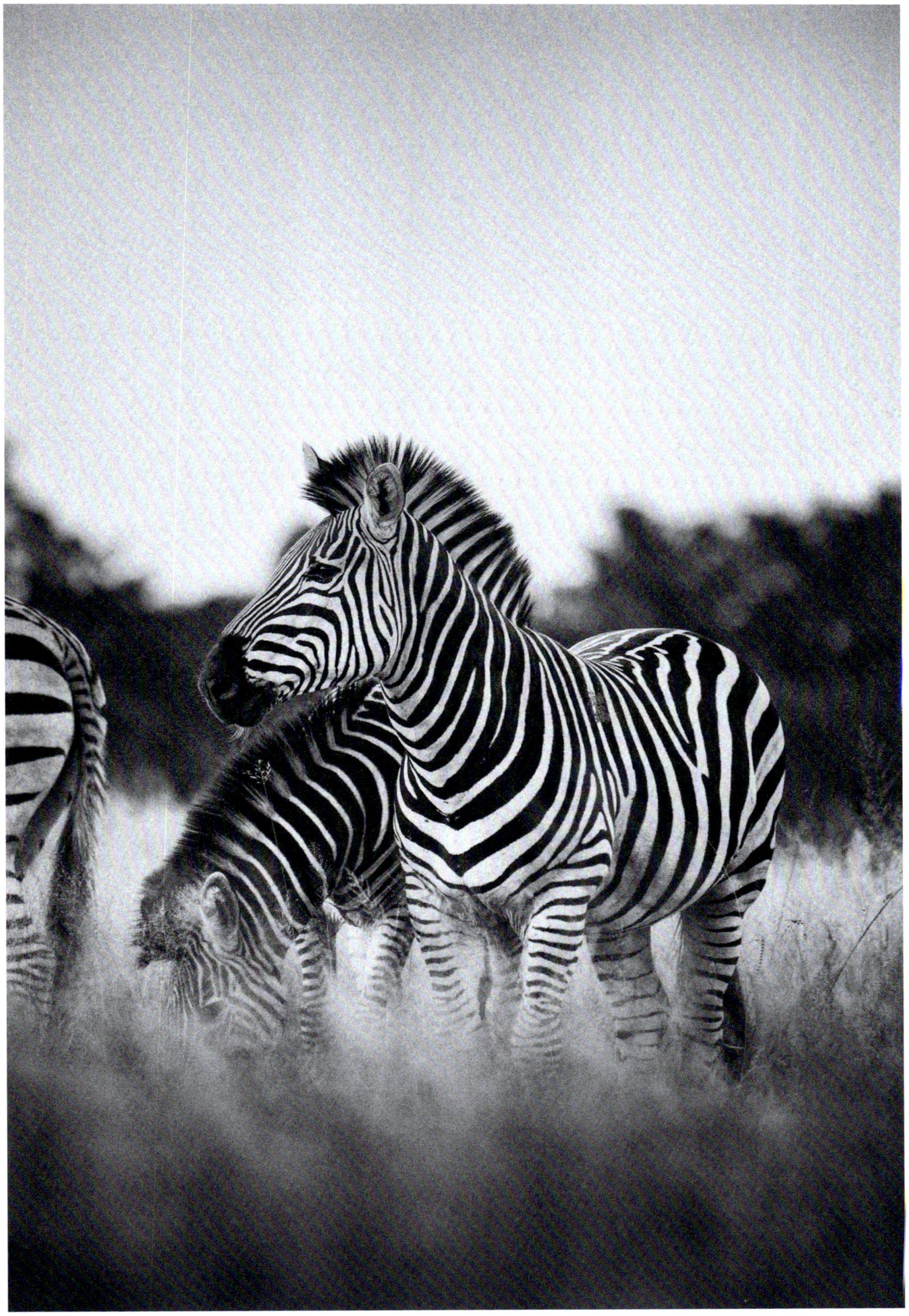

This spread Wildlife migrations are some of the greatest spectacles on earth. One of the largest of its kind happens twice a year in Botswana, when 25,000 zebra traverse a distance of 300 kilometres (185 mi.) between the Okavango Delta and the Makgadikgadi Pans.

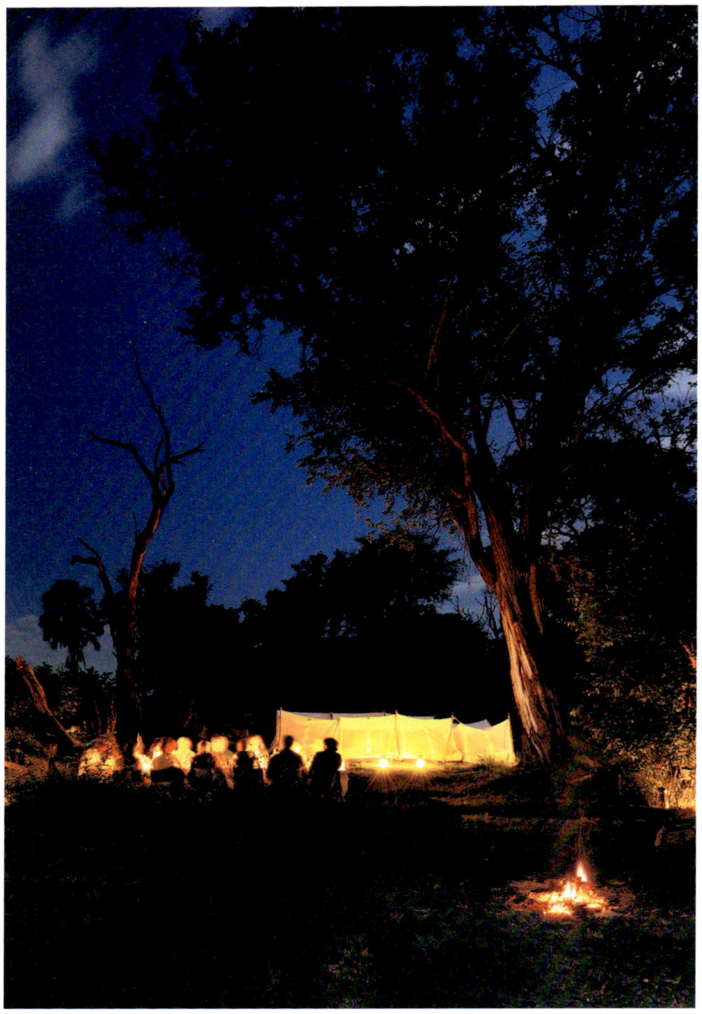

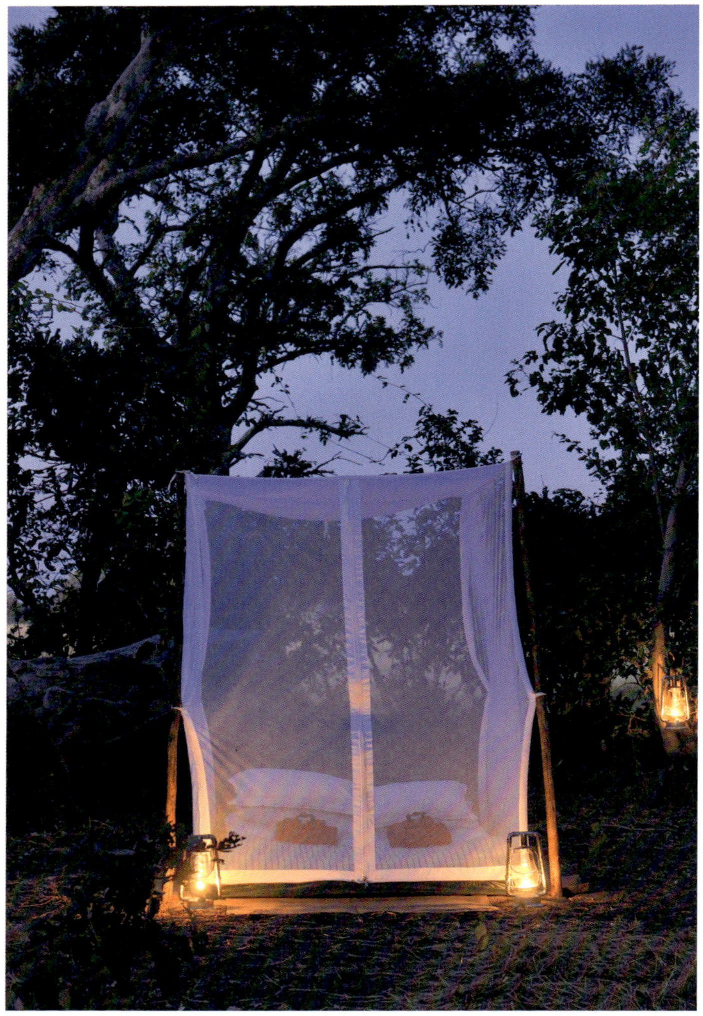

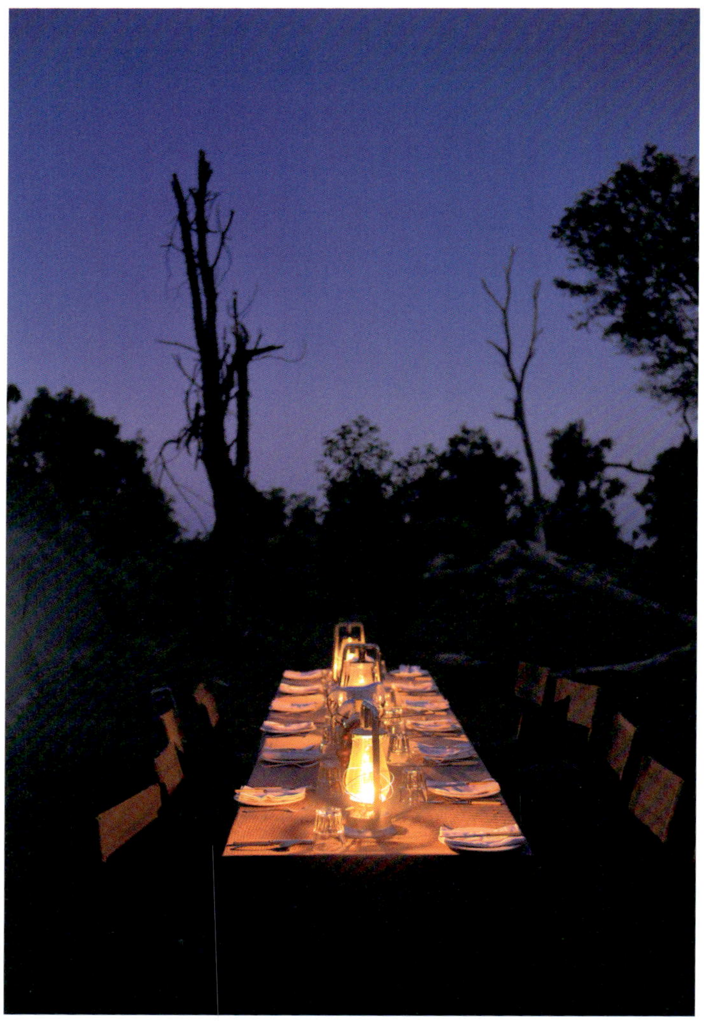

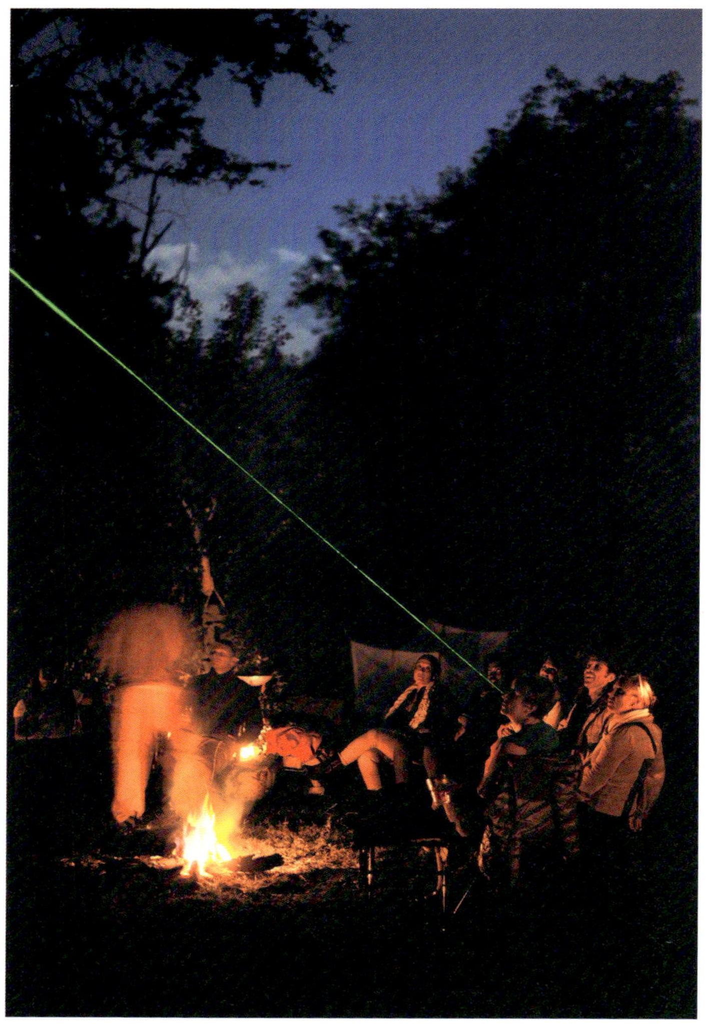

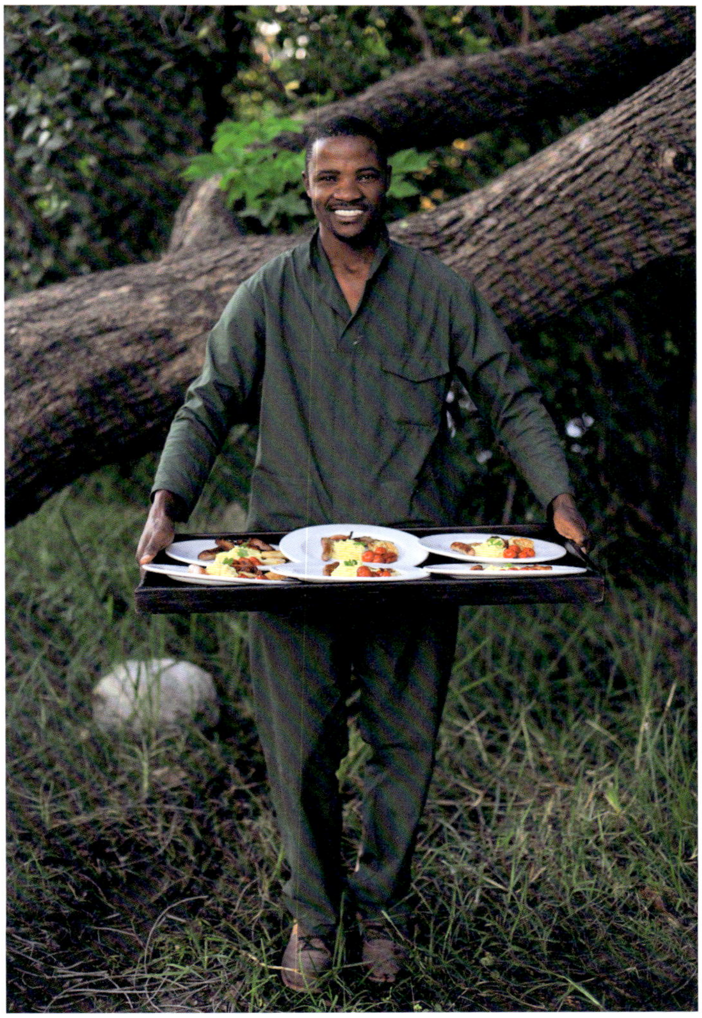

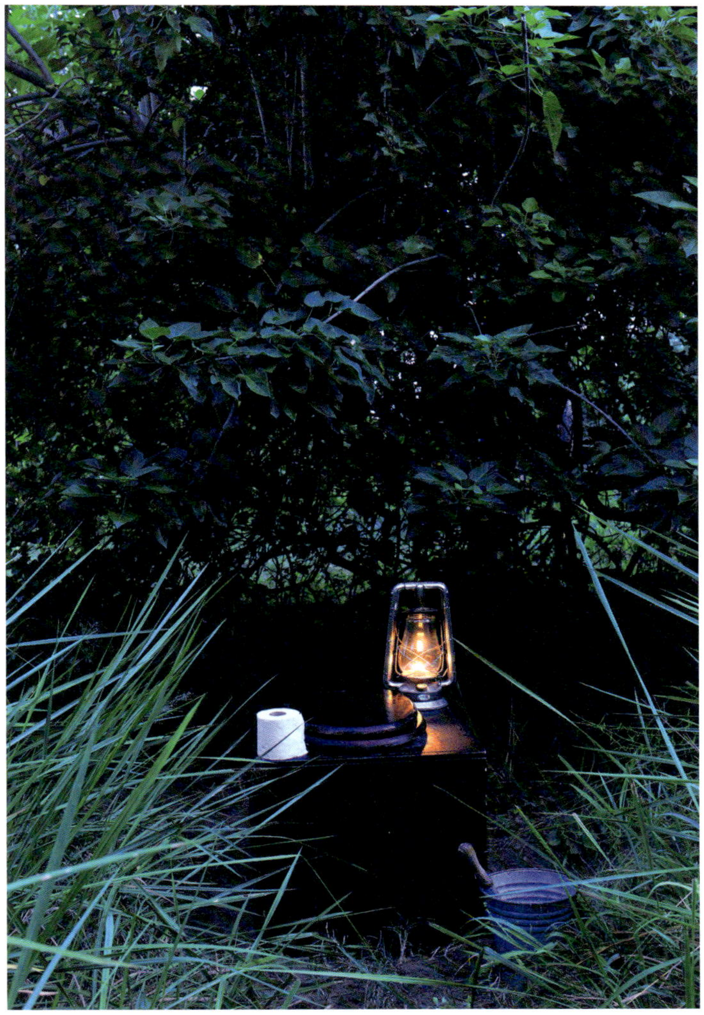

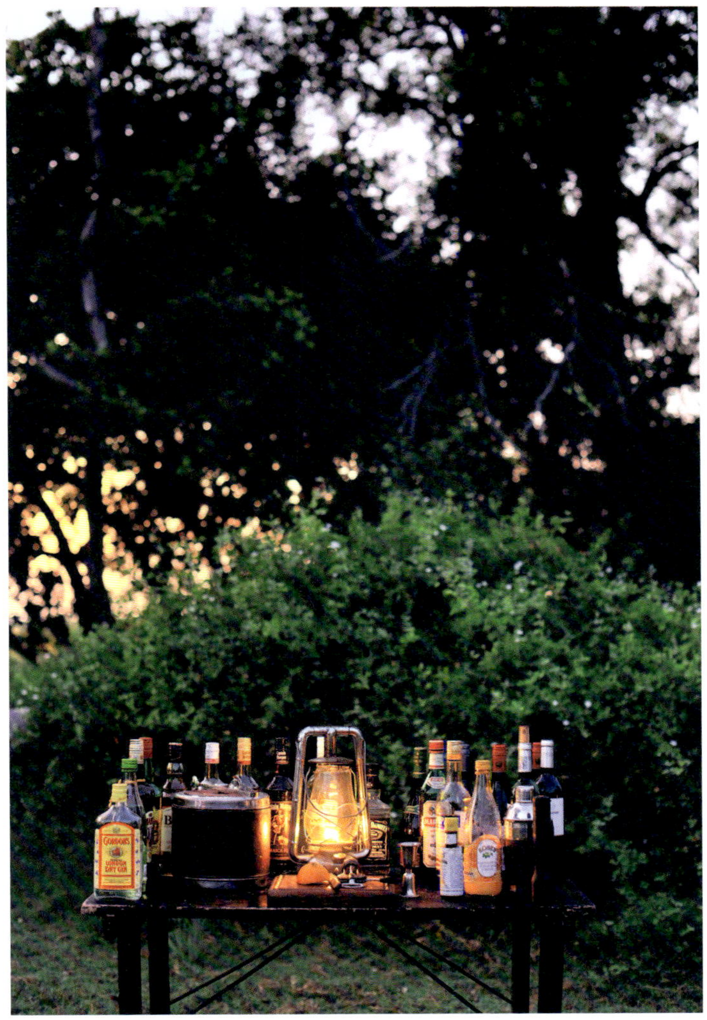

This spread Fly camping on small islands in the delta, surrounded by intricate watery channels filled with delicate lilies and reeds, is a highlight of Ralph Bousfield's mobile safaris. Lazy evenings are for long dinners and stargazing, while crisp mornings are spent enjoying hearty breakfasts prepared over a wood fire.

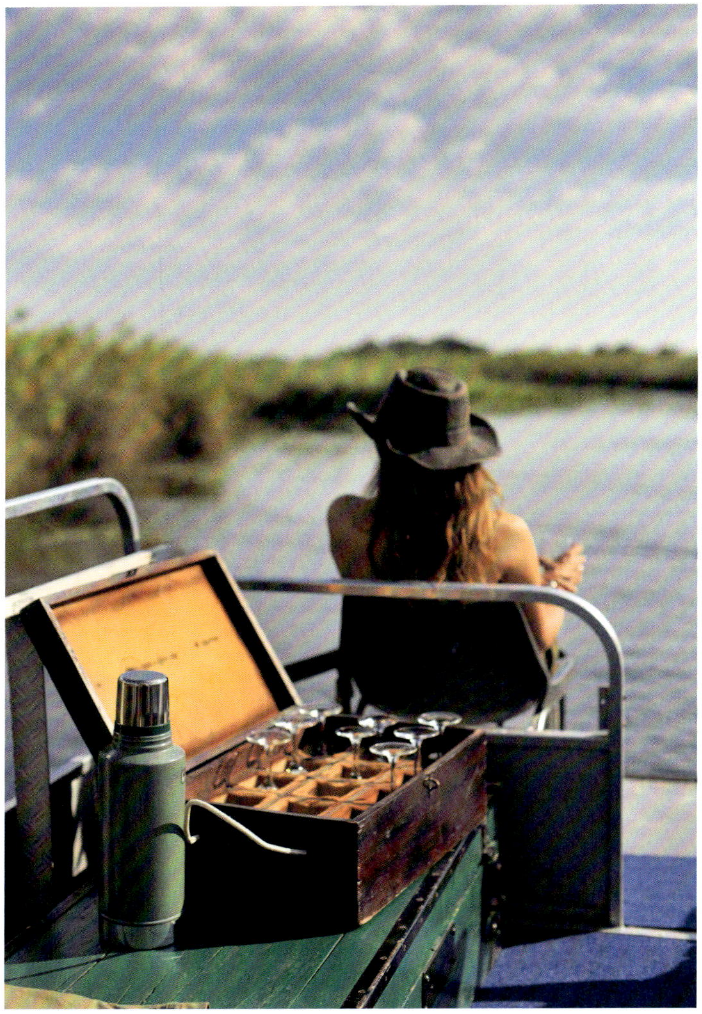

This spread Floods and dry spells alternate in a never-ending loop in the Okavango Delta. On a leisurely small-boat safari, and safe from crocodiles and hippopotamuses, travellers find shallow places to swim in exceptionally clear water. There is also the opportunity to float through the reedy labyrinth in a slim mokoros, a traditional canoe of the delta.

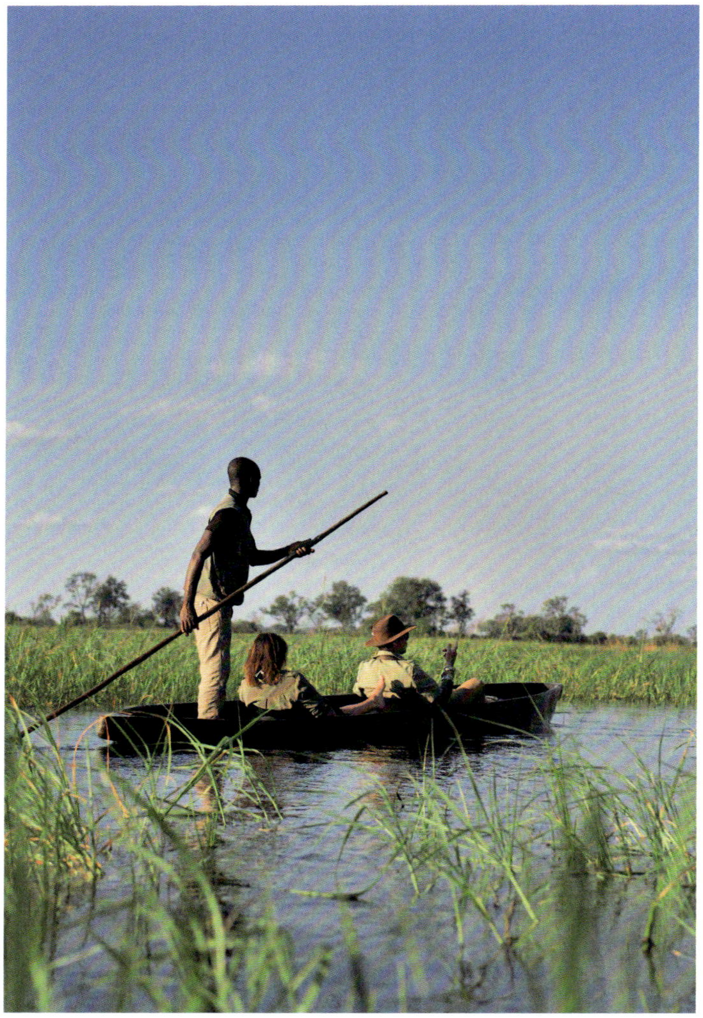

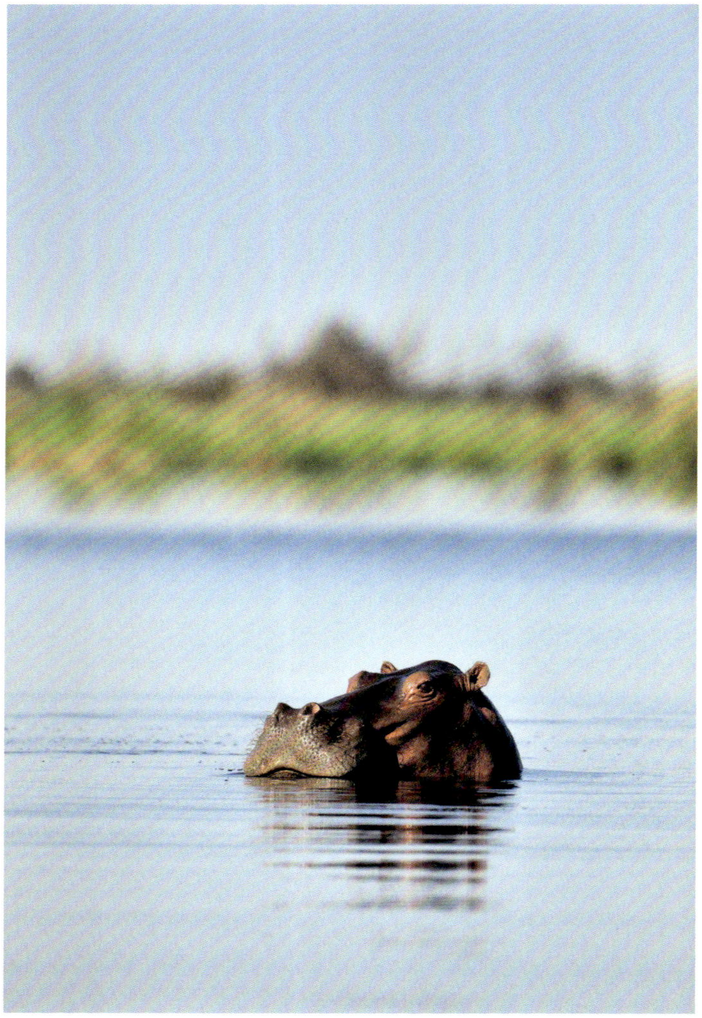

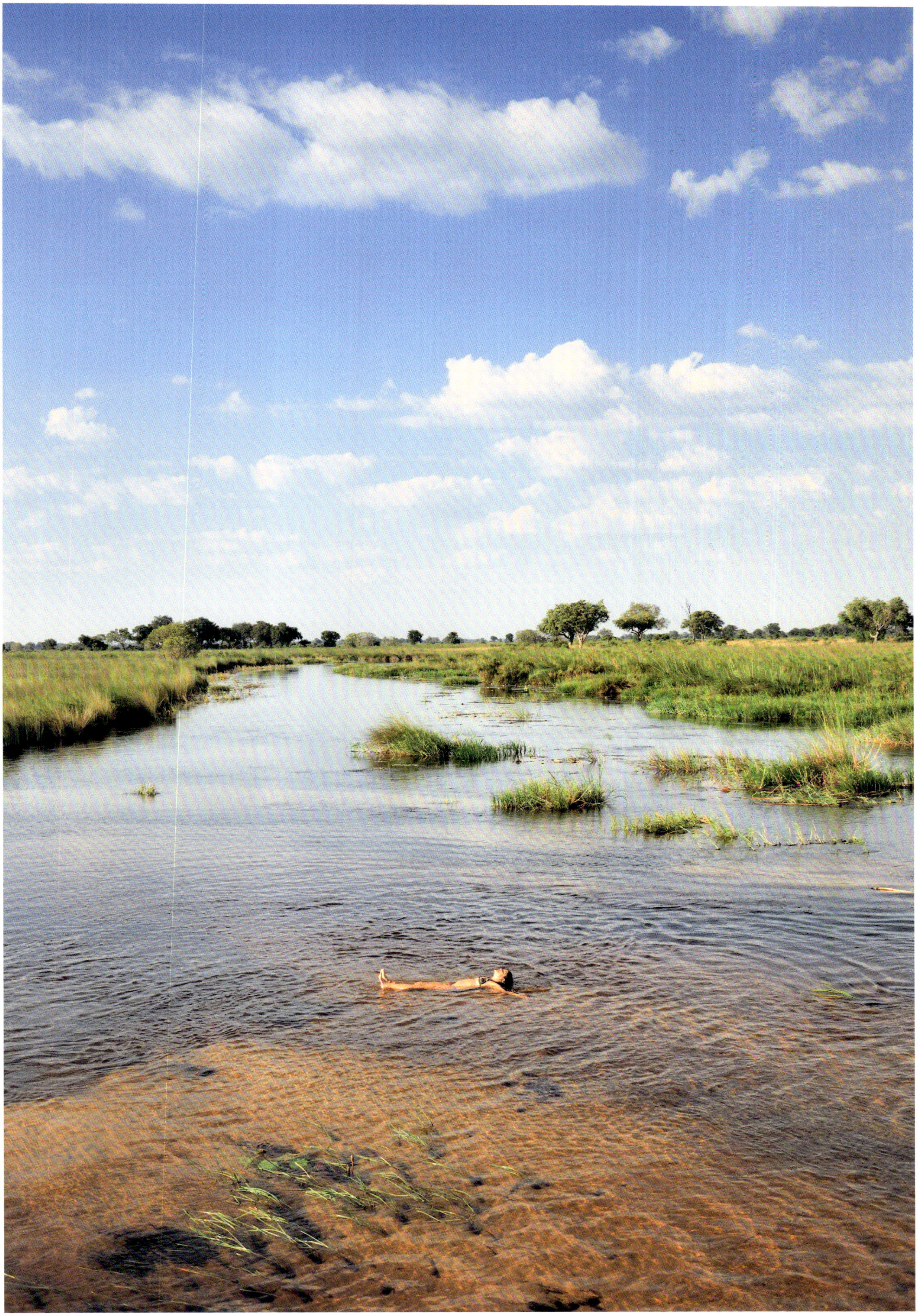

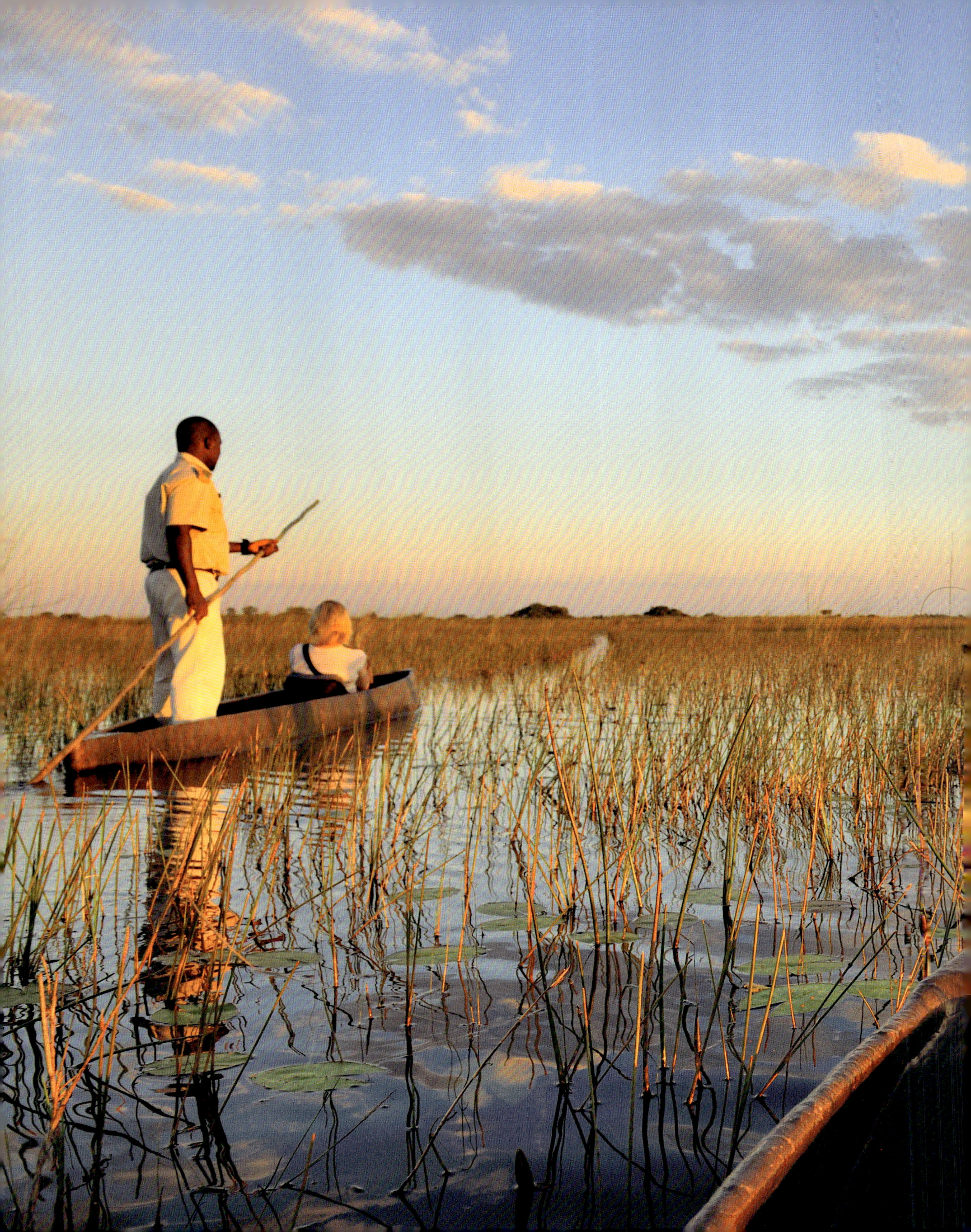

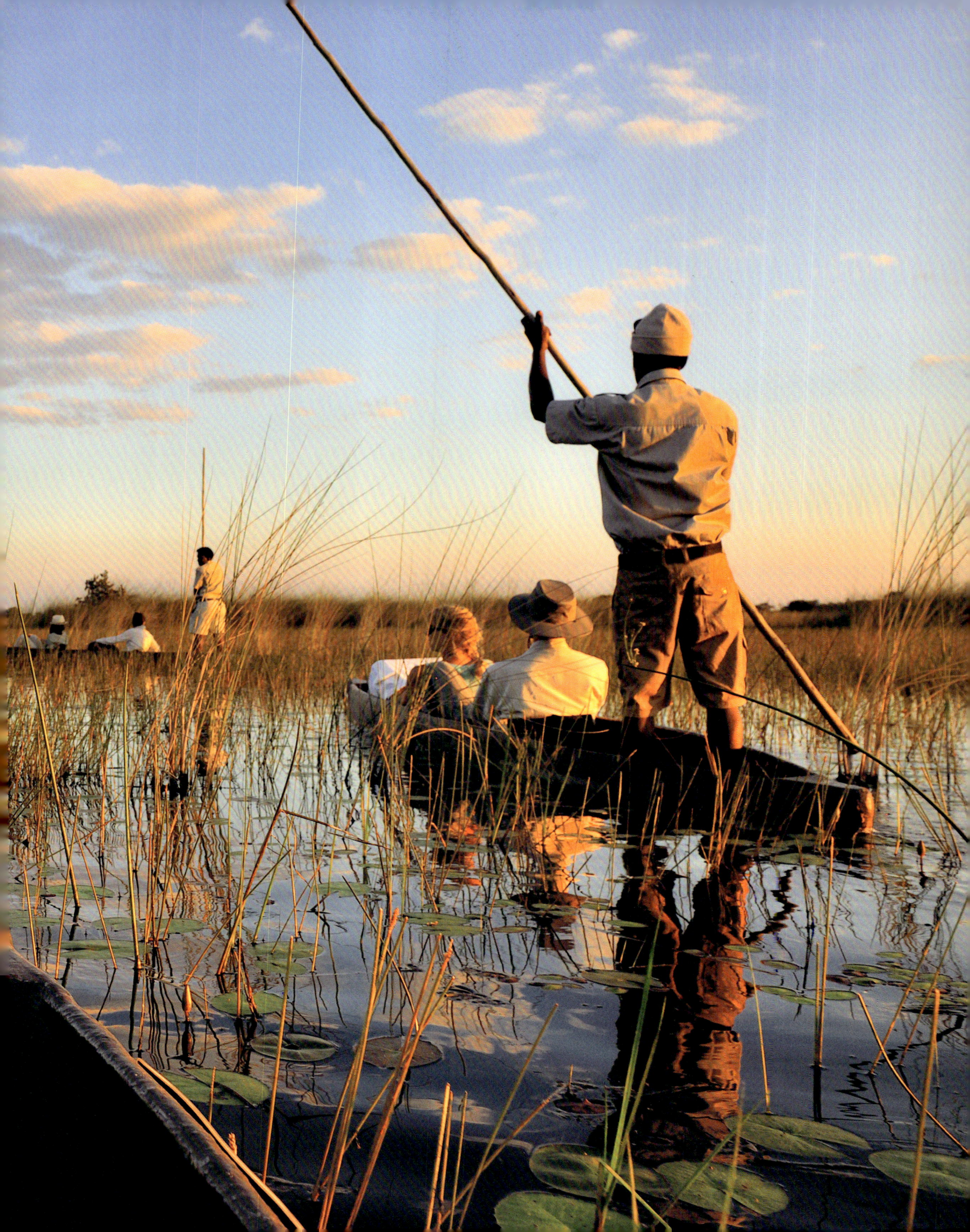

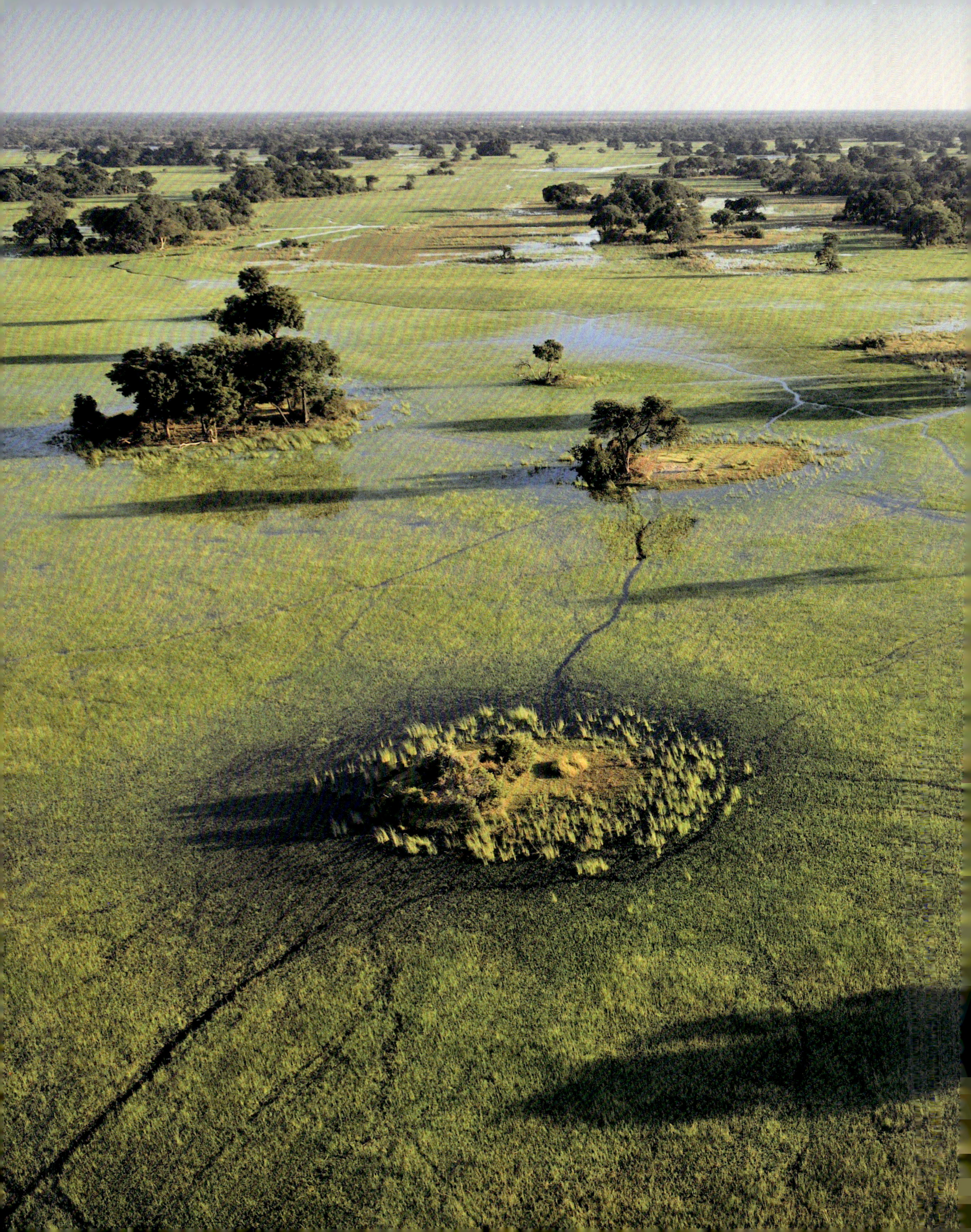

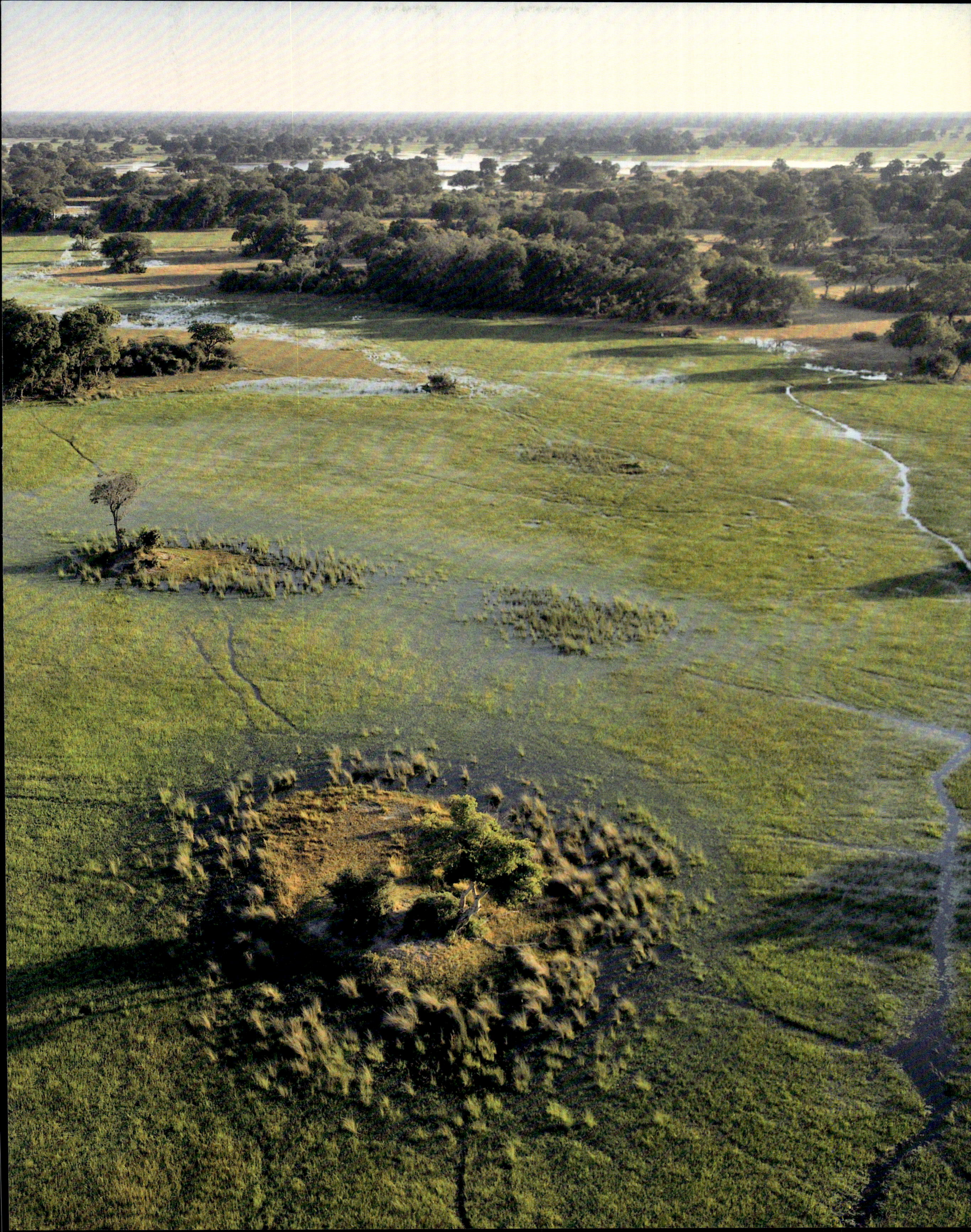

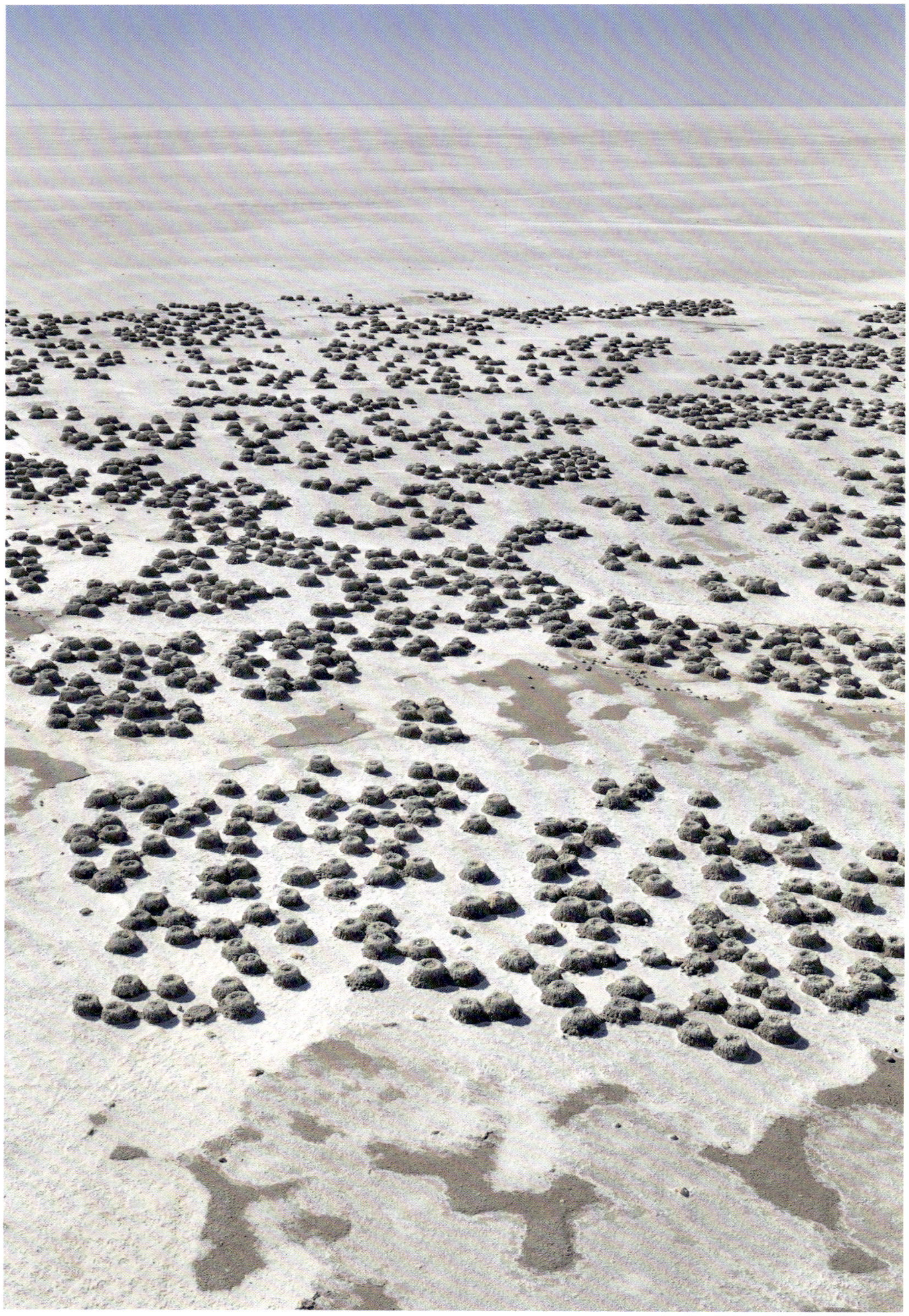

This spread While flying with the helicopter over the great Makgadikgadi Salt Pans there is a chance to observe the gigantic breeding colonies of greater and lesser flamingos on Sua Pan. The abandoned nests form an arty pattern spread throughout the vastness of the salt pan.

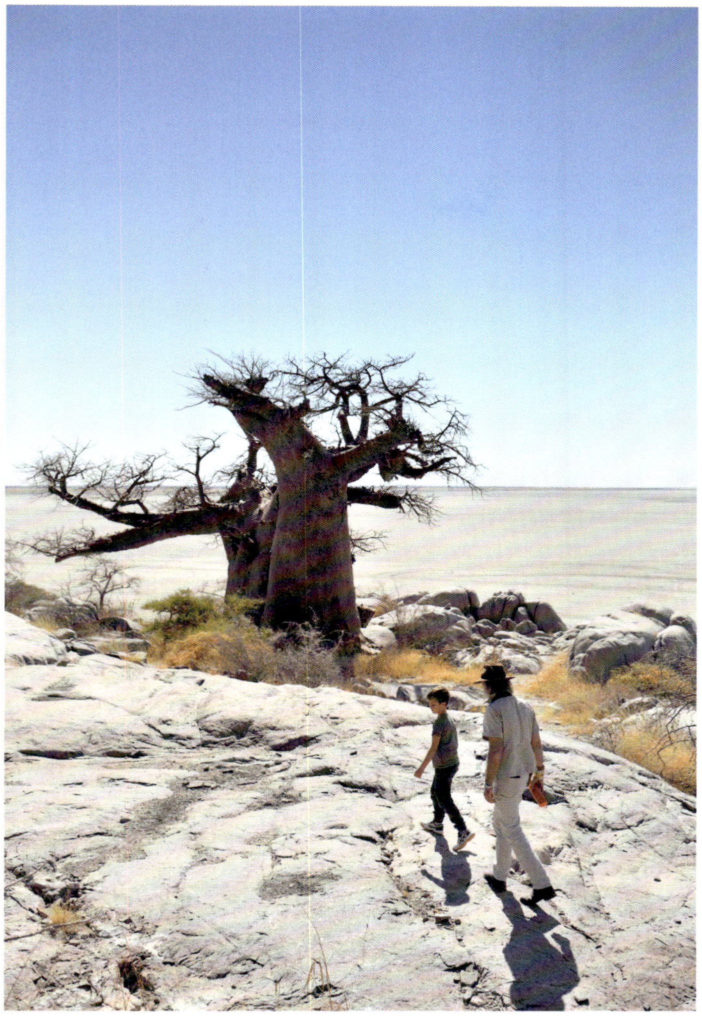

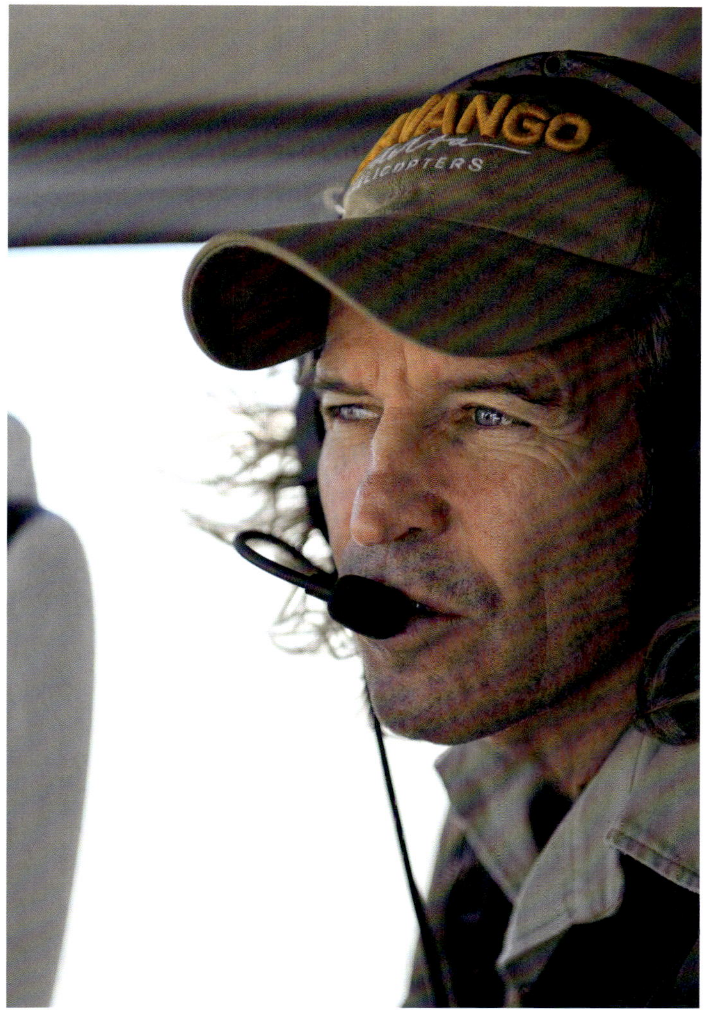

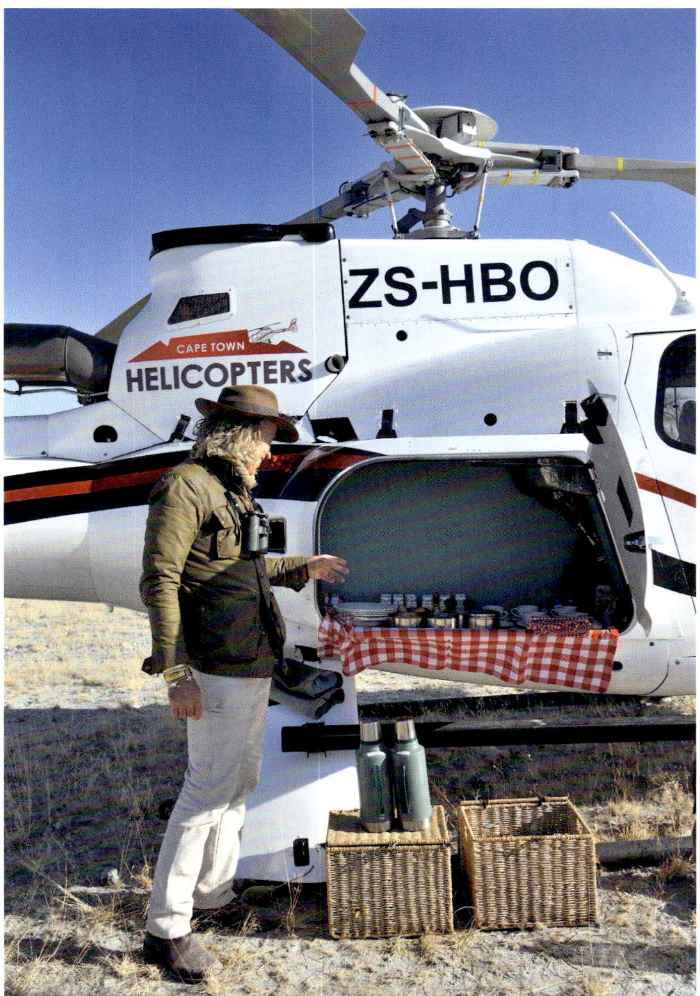
ZS-HBO
CAPE TOWN
HELICOPTERS

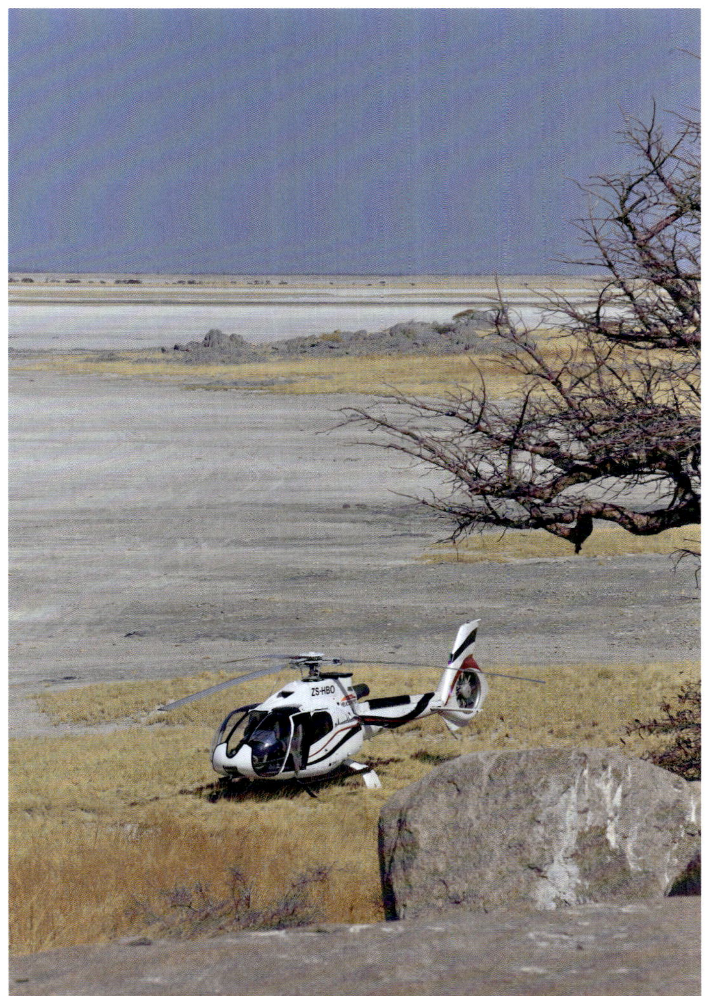

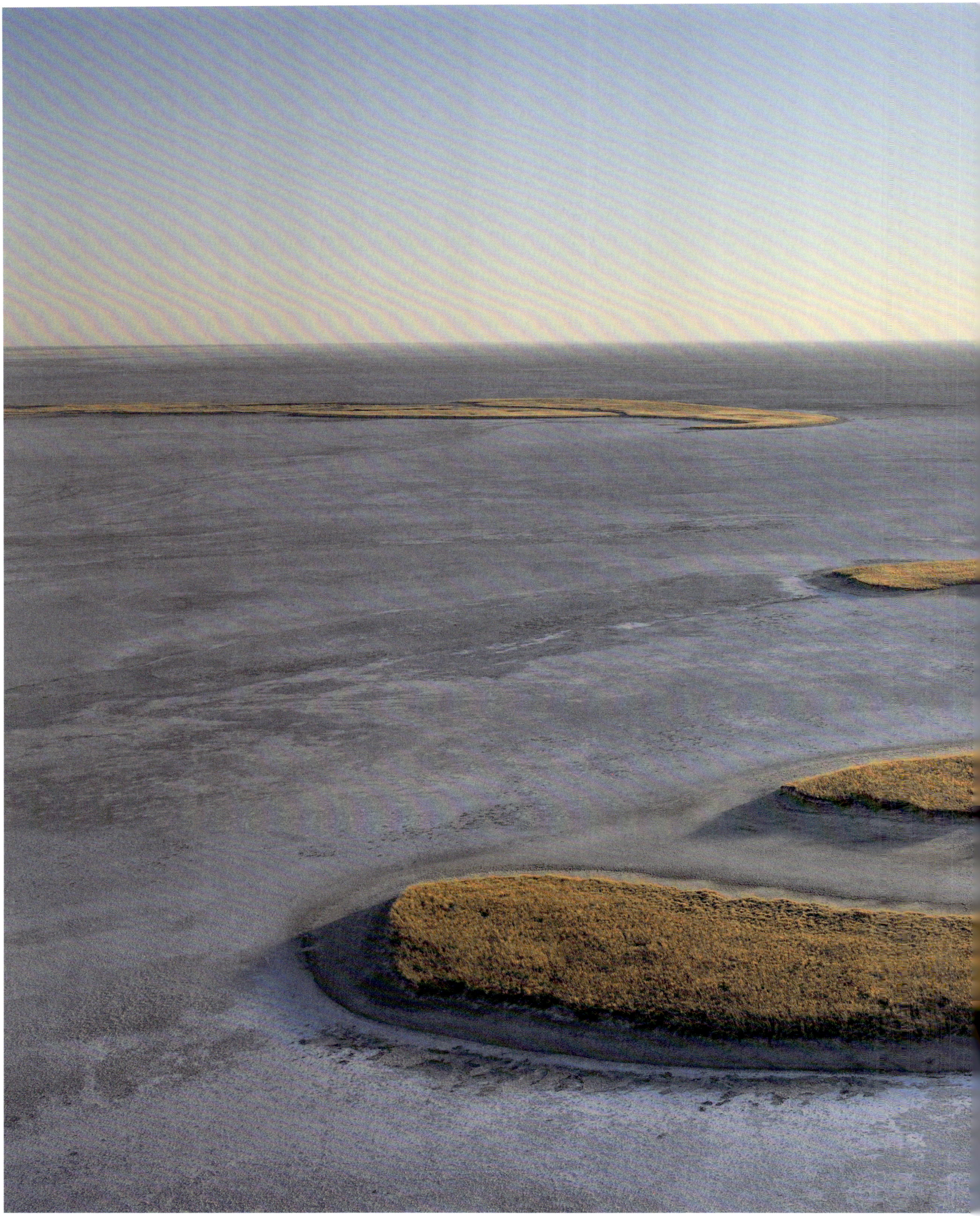

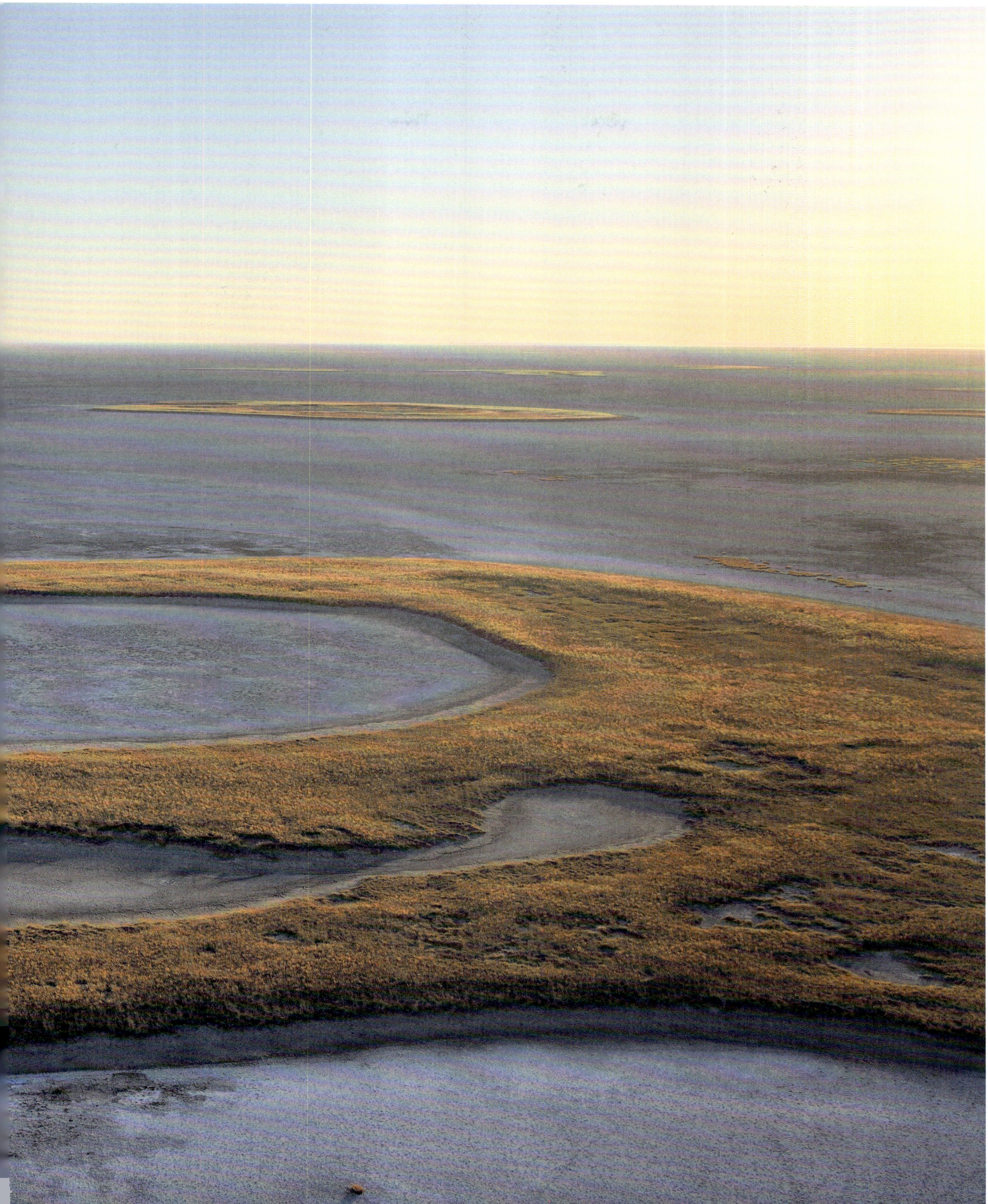

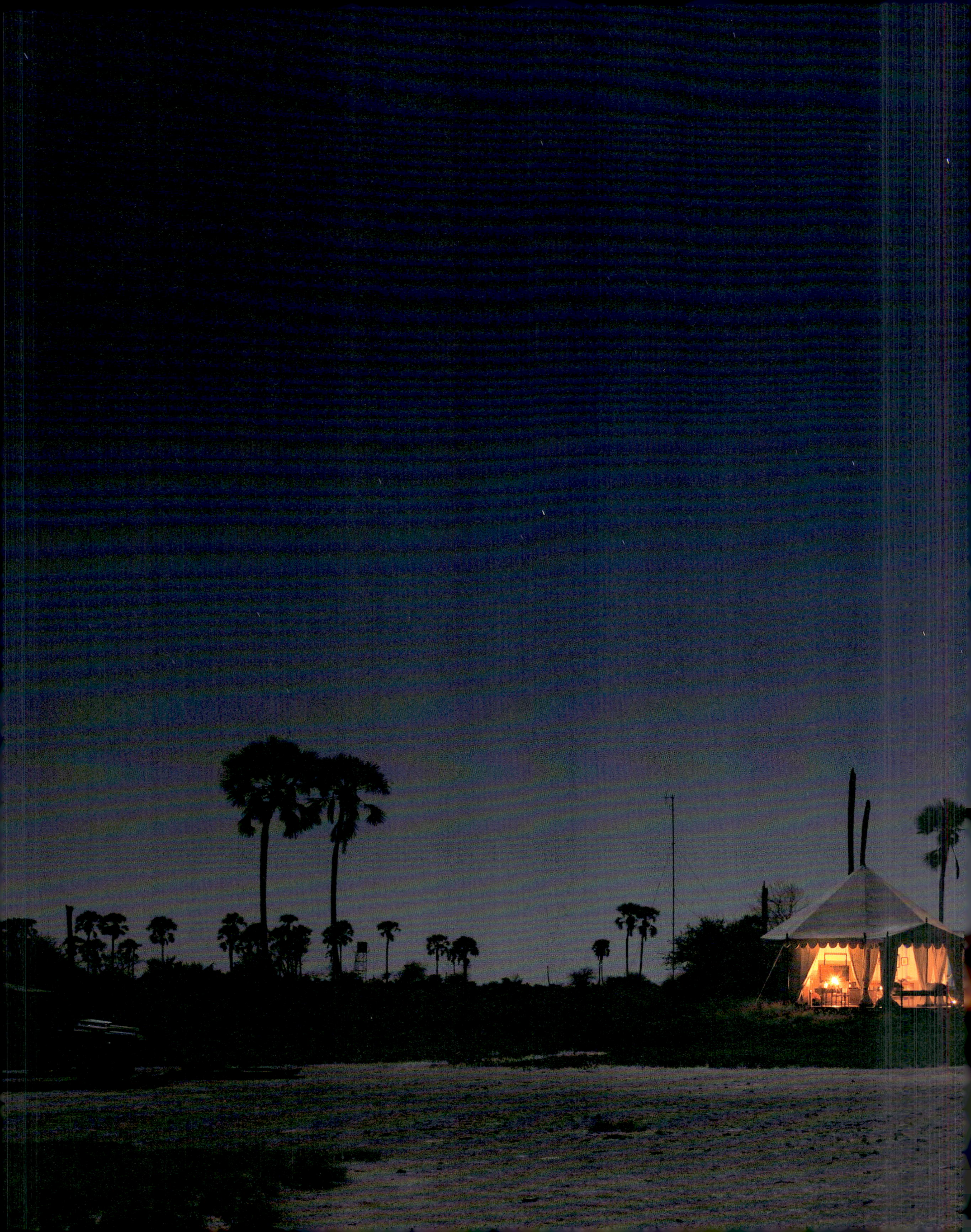

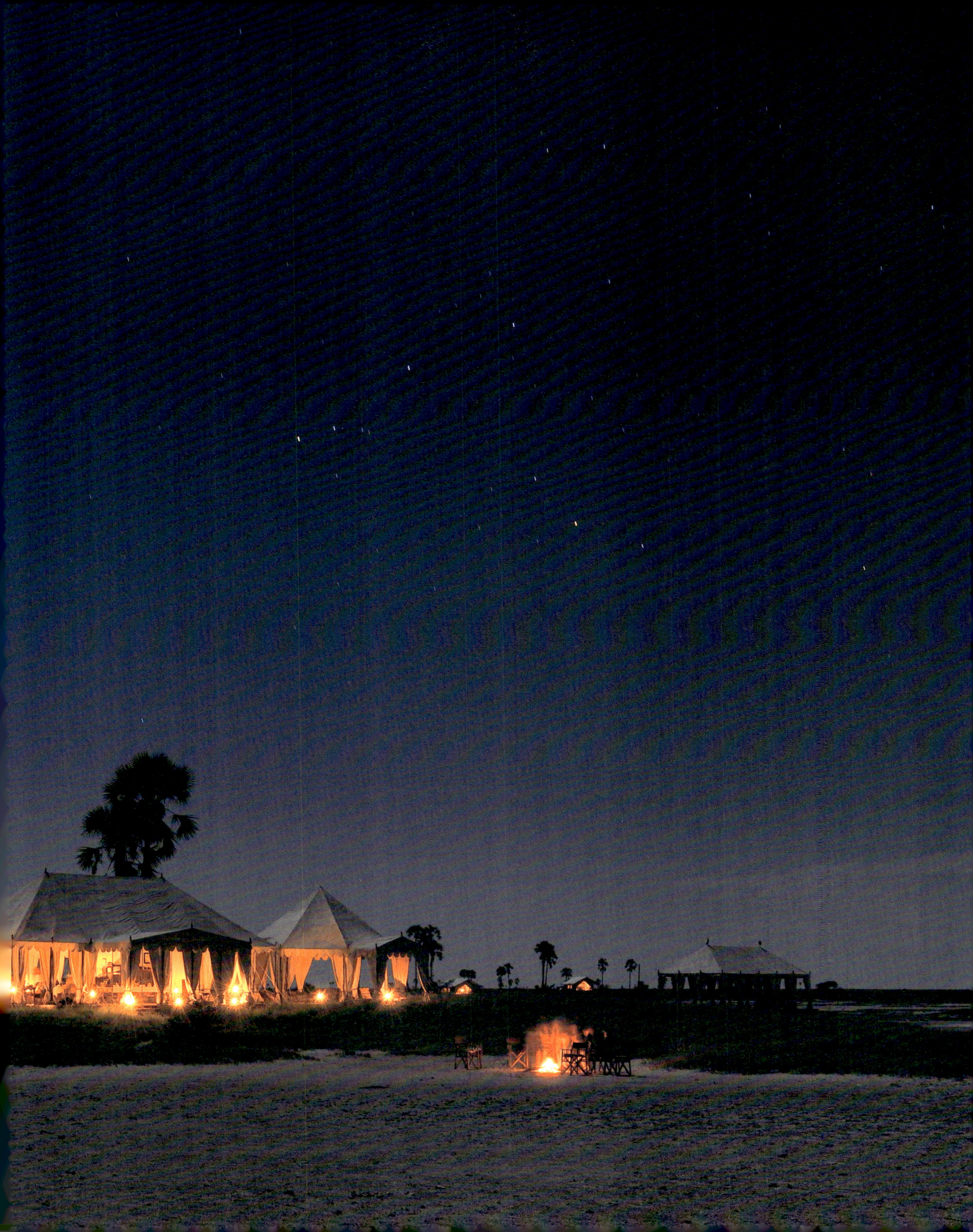

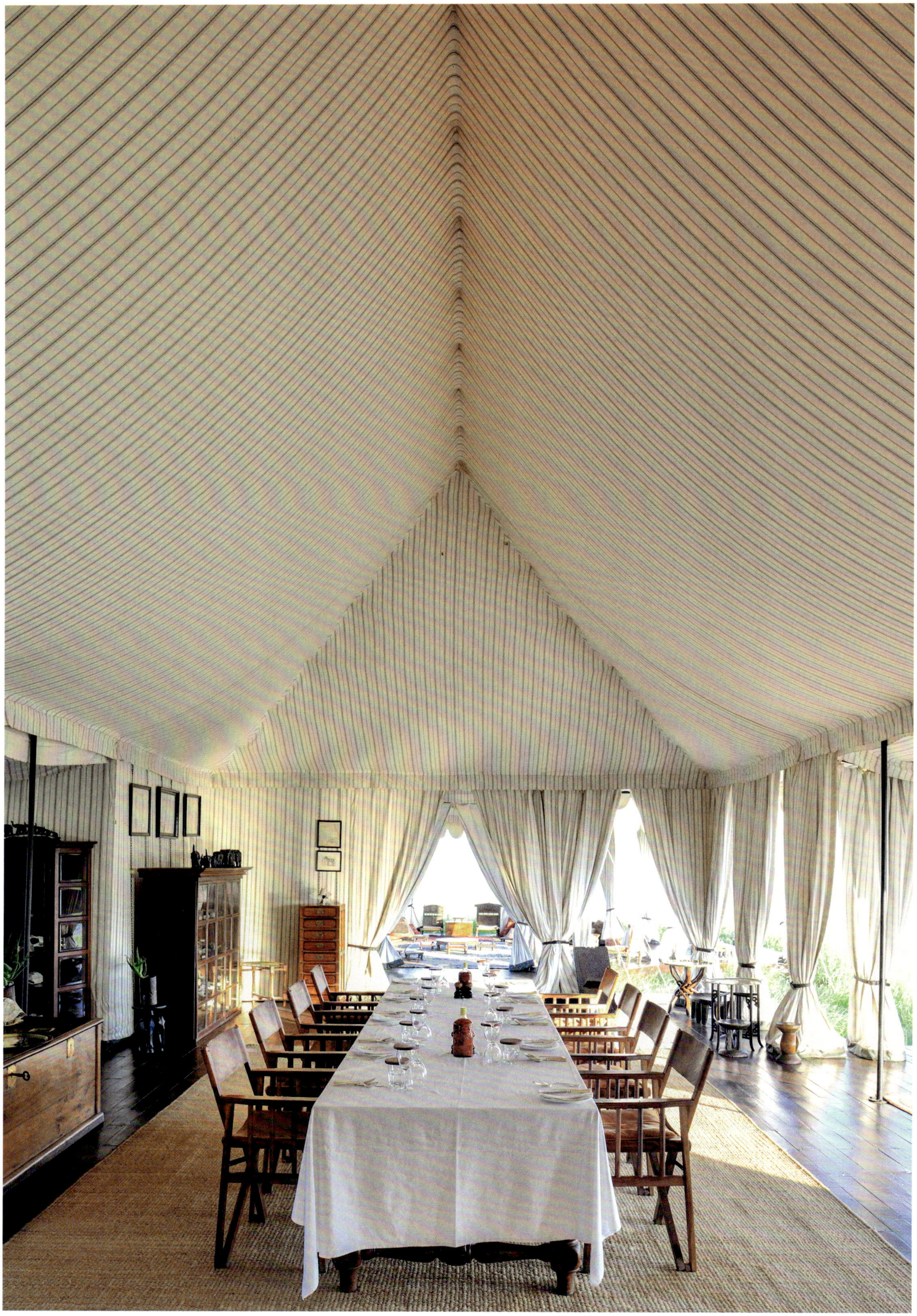

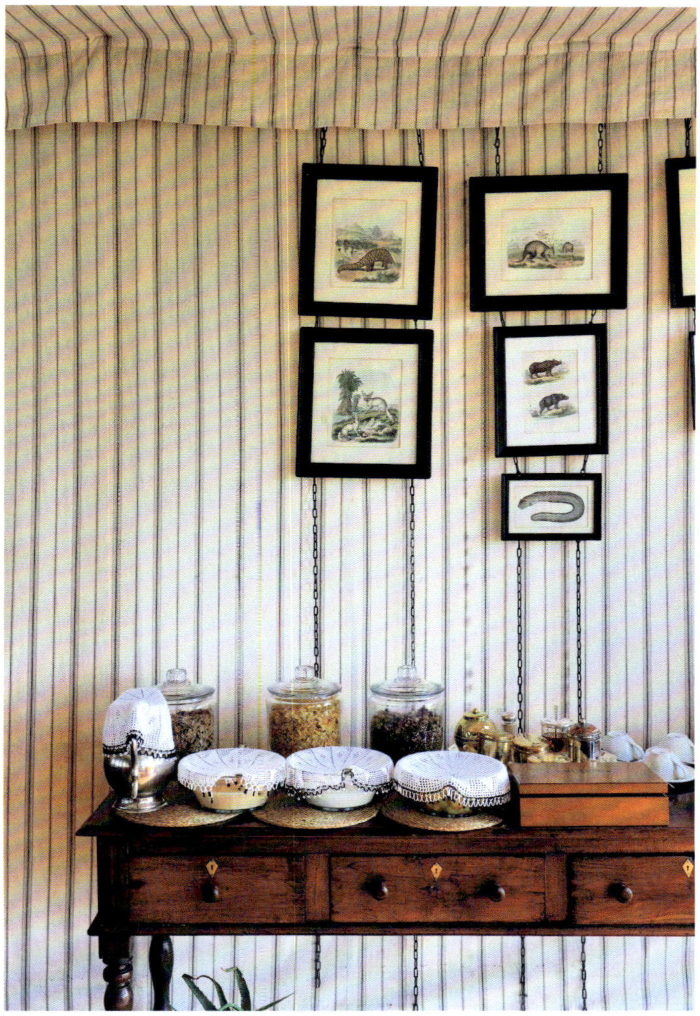

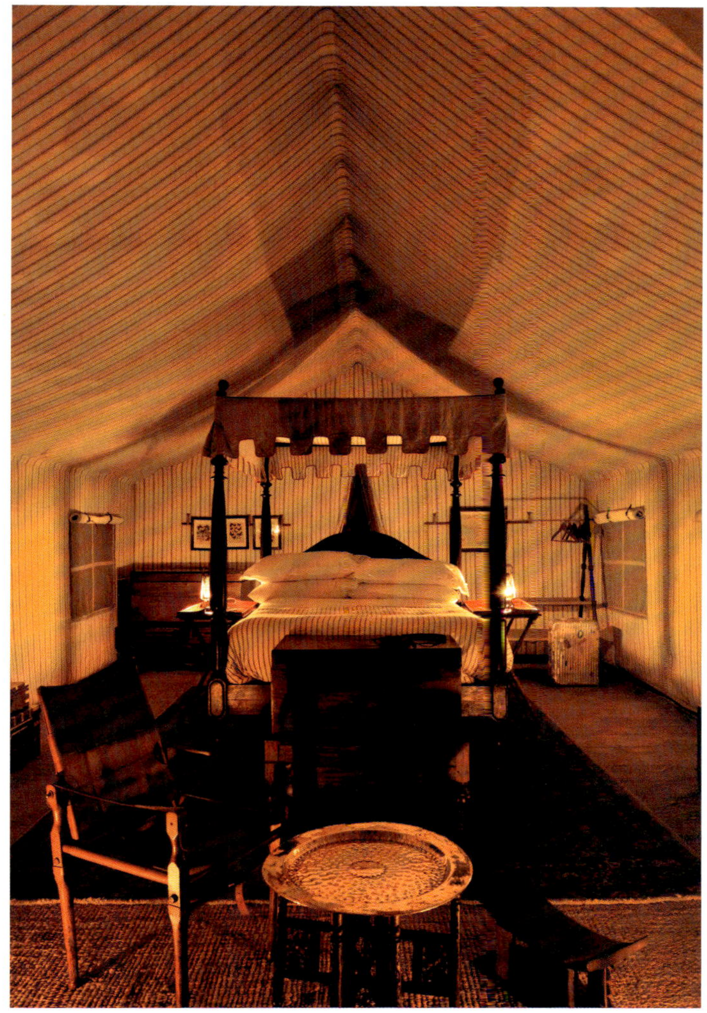

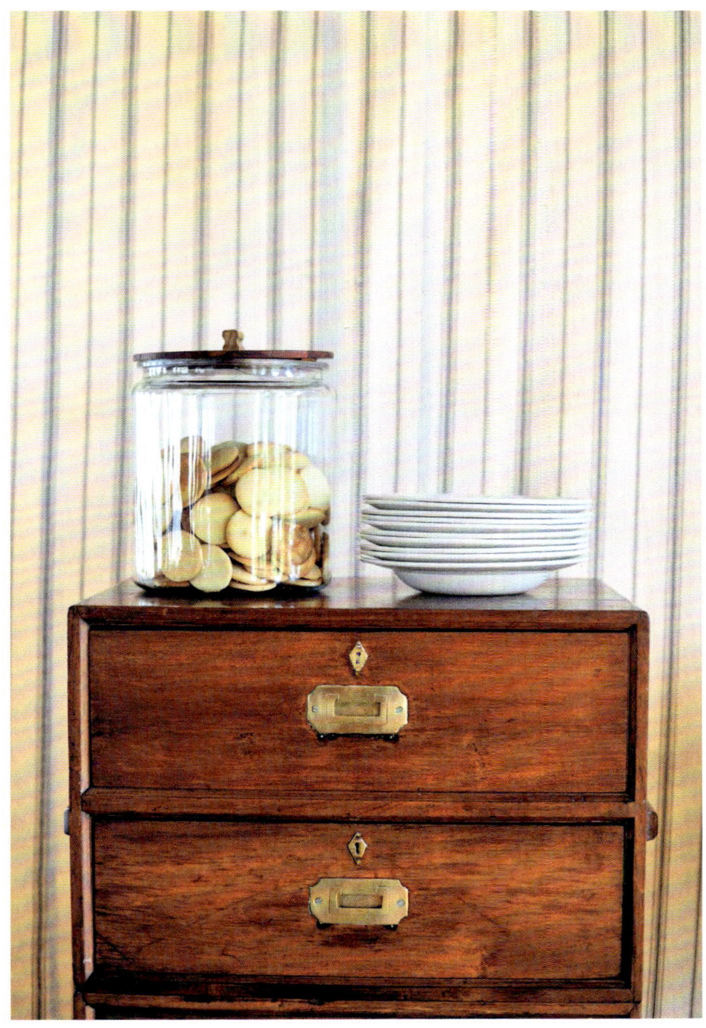

This spread Appearing like a mirage in the otherworldly Makgadikgadi Pans, Ralph Bousfield's San Camp is an intimate assembly of seven canvas tents with wooden decks furnished with antiques and lush fabrics. Air conditioning and Wi-Fi are absent; instead, elegant oil lamps and chic four-poster beds create an ambience of old-school African romance. At San Camp, Ju'hoansi Bushmen offer insights into their culture and survival tactics in this remote part of Africa. The intriguing landscapes of Makgadikgadi Pans National Park encompass dry savannah and stunning salt flats that are among the largest in the world.

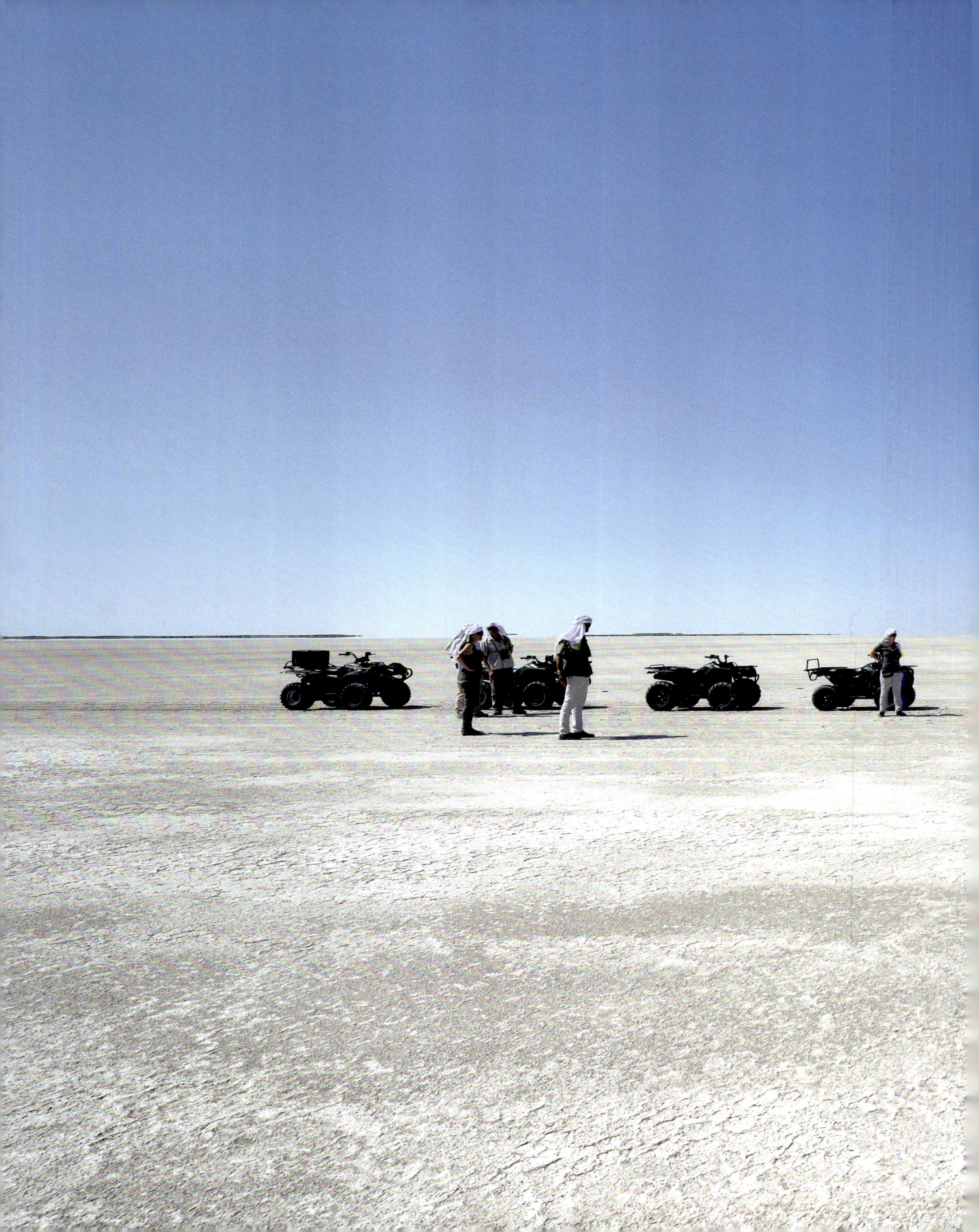

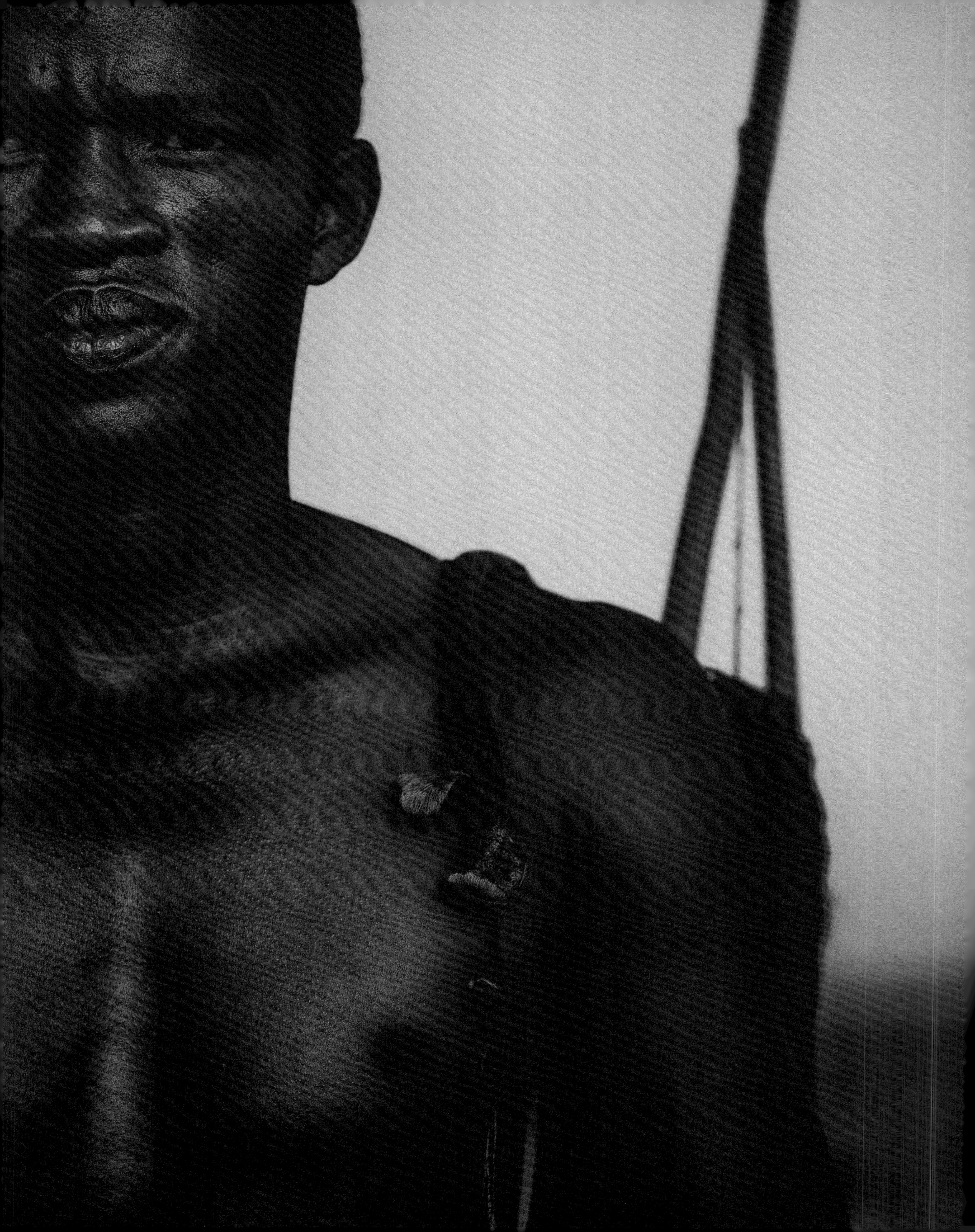

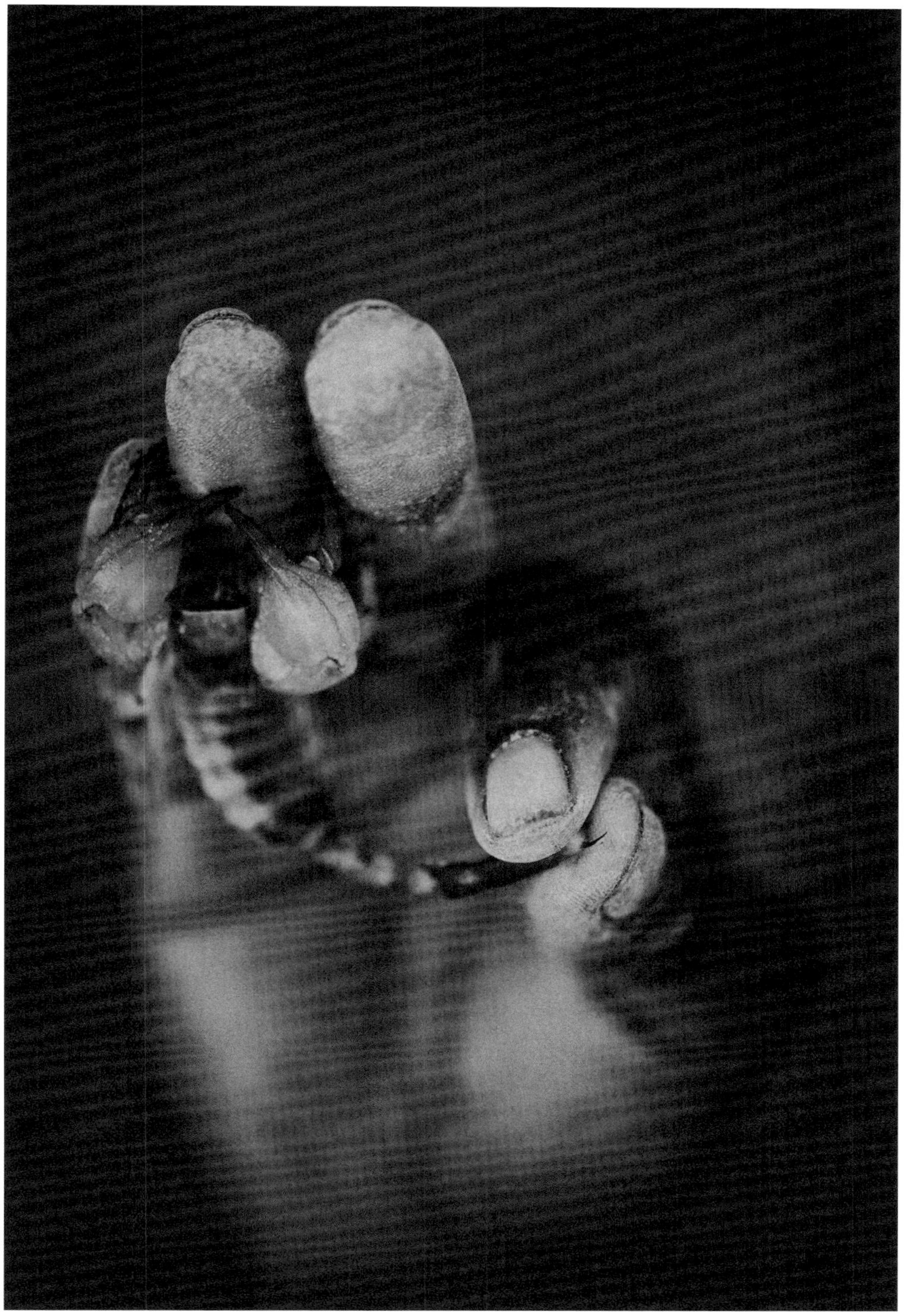

This spread The San Bushmen douse their arrow shafts with a concoction of animal venoms and plant poisons when they go out hunting. A particularly deadly combination uses snake or scorpion venom, combined with euphorbia sap. This gluey substance is put on the shaft, just behind the metal or bone tip of the spear. The San often use the tracker beetle to put it close to a scorpion burrow. The little black beetle then pushes and shoves the obstruction (scorpion) out of its way to the surface where the Bushman is waiting to catch it.

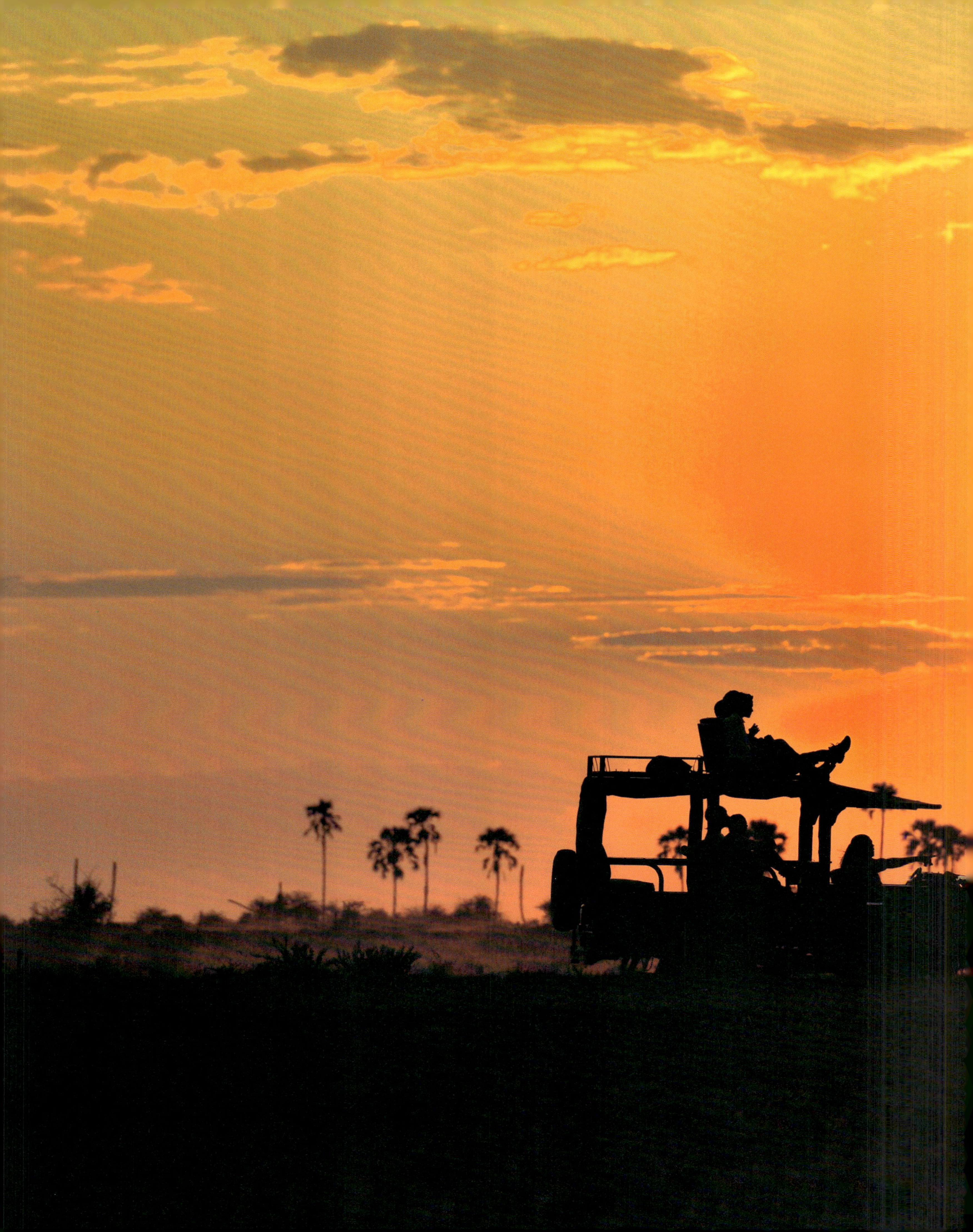

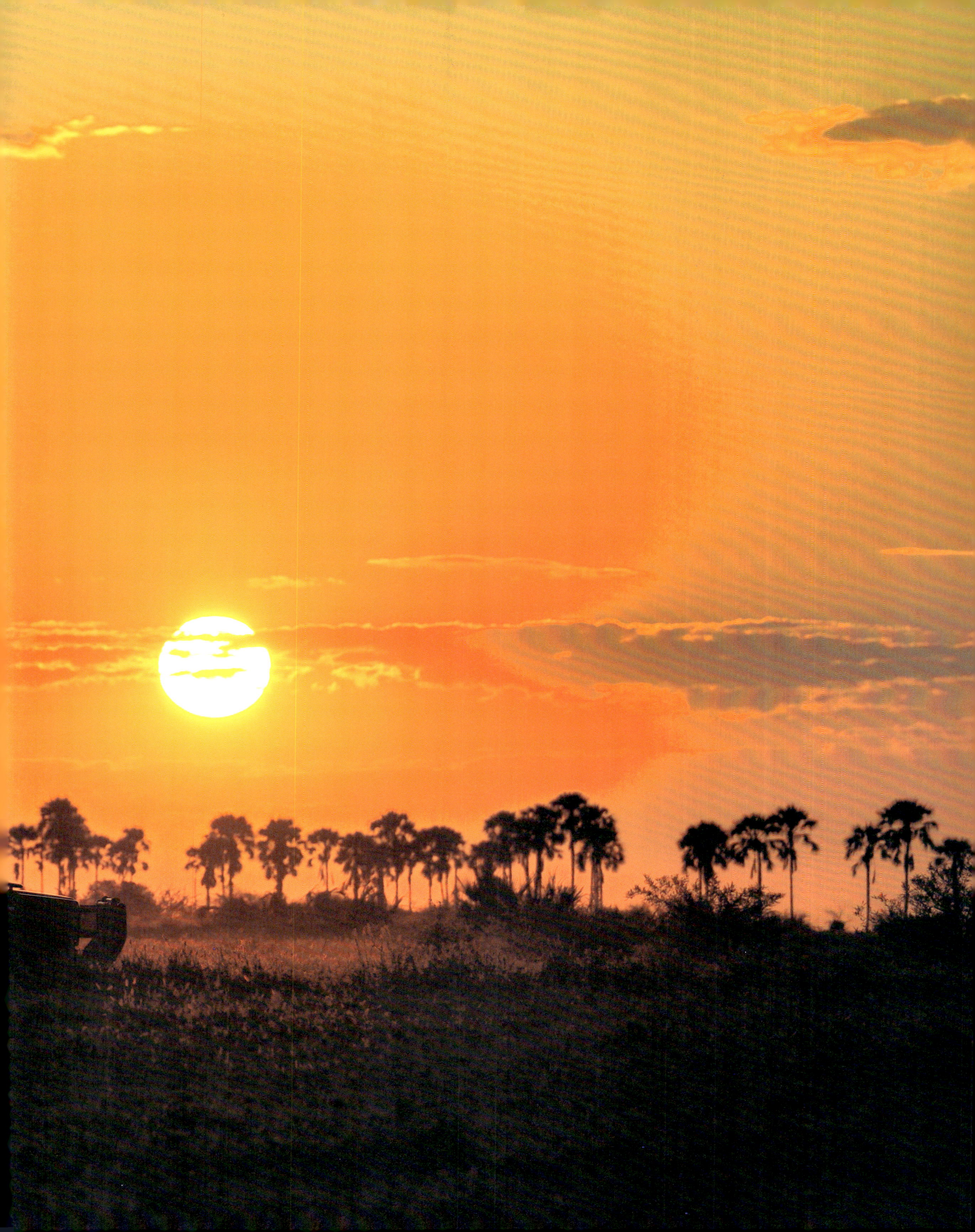

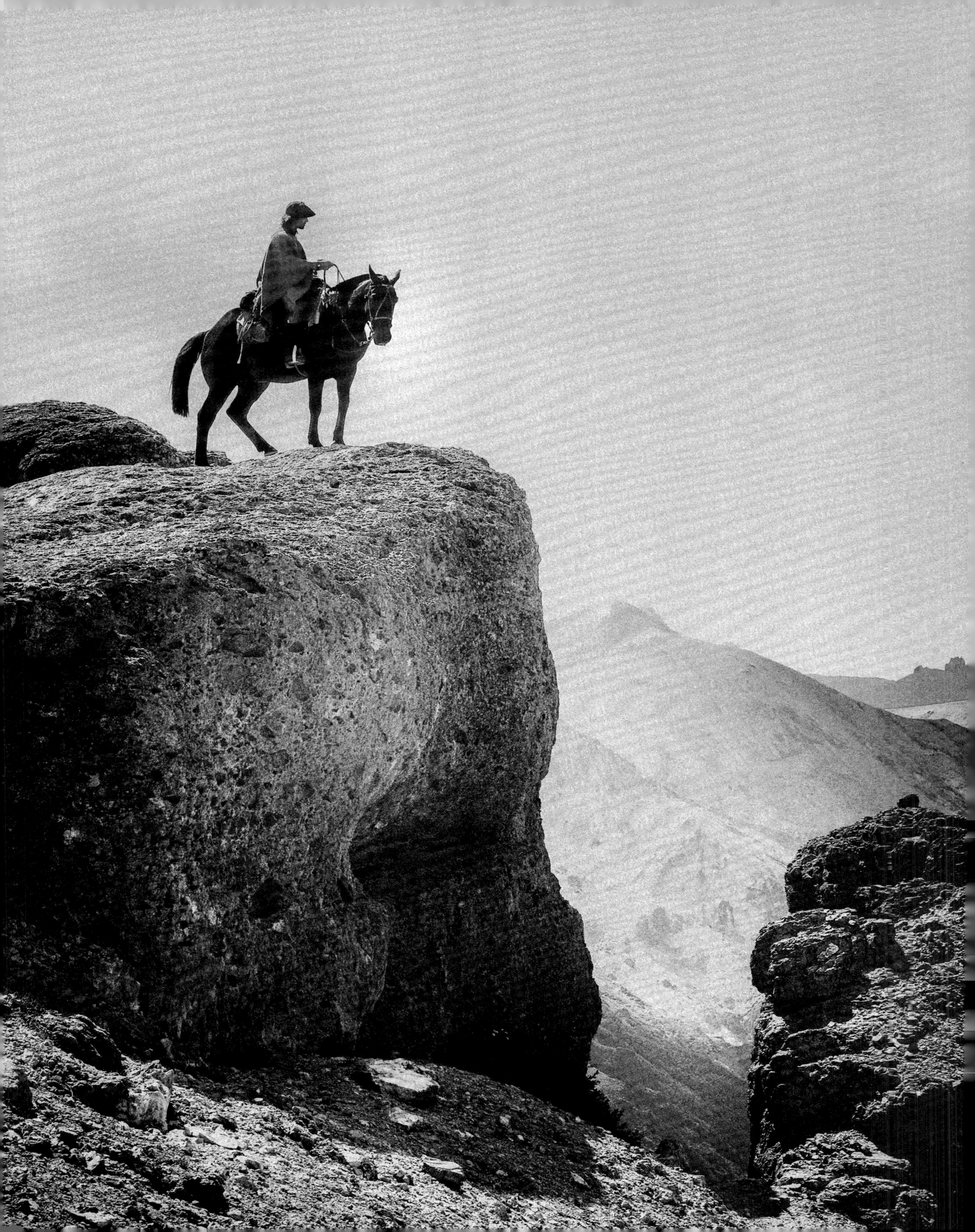

CHAPTER 11

INTO THE SADDLE AND OUT OF THIS WORLD

A GAUCHO ENCOUNTER

Argentina

40°29'17.78" S
71°22'43.90" W

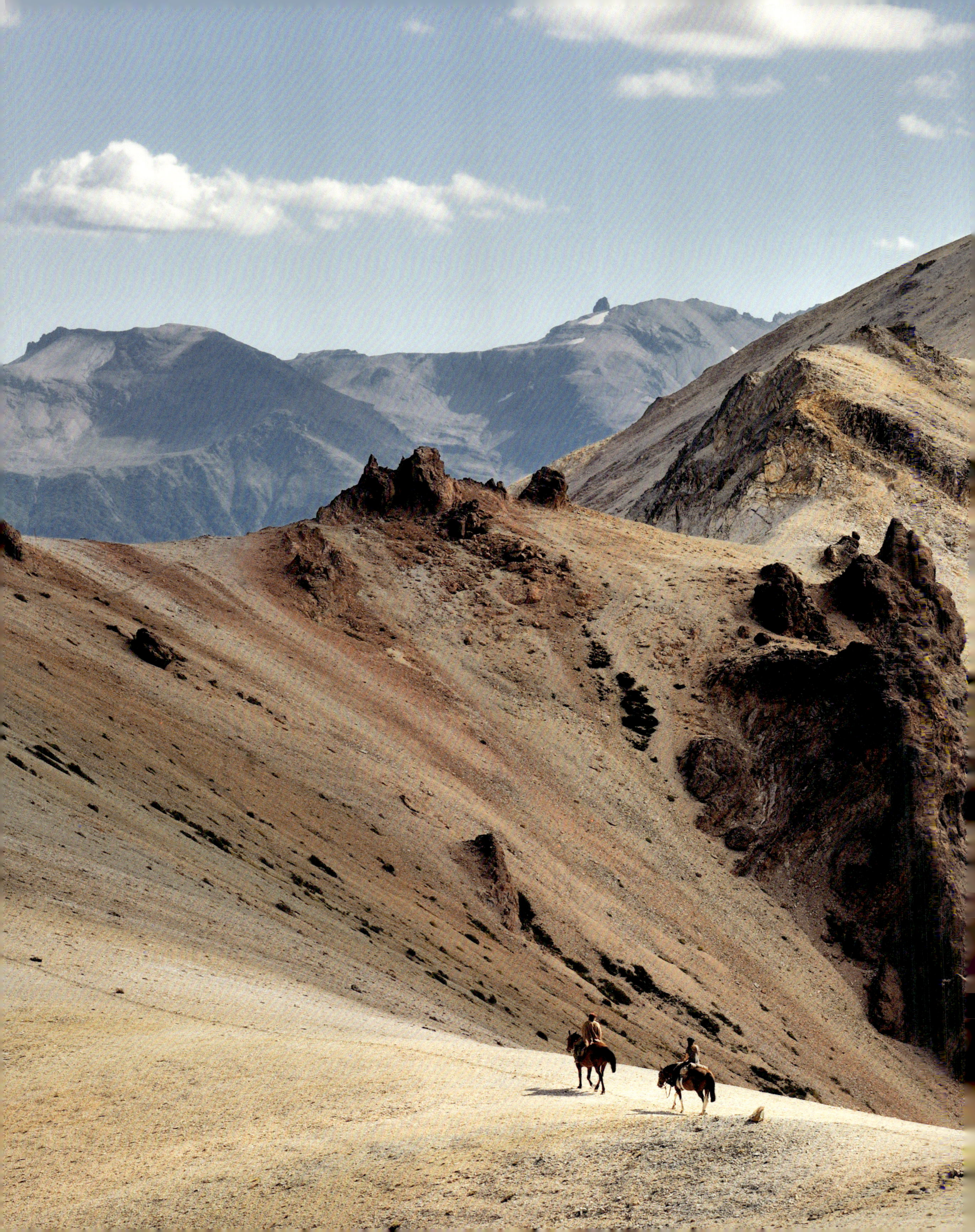

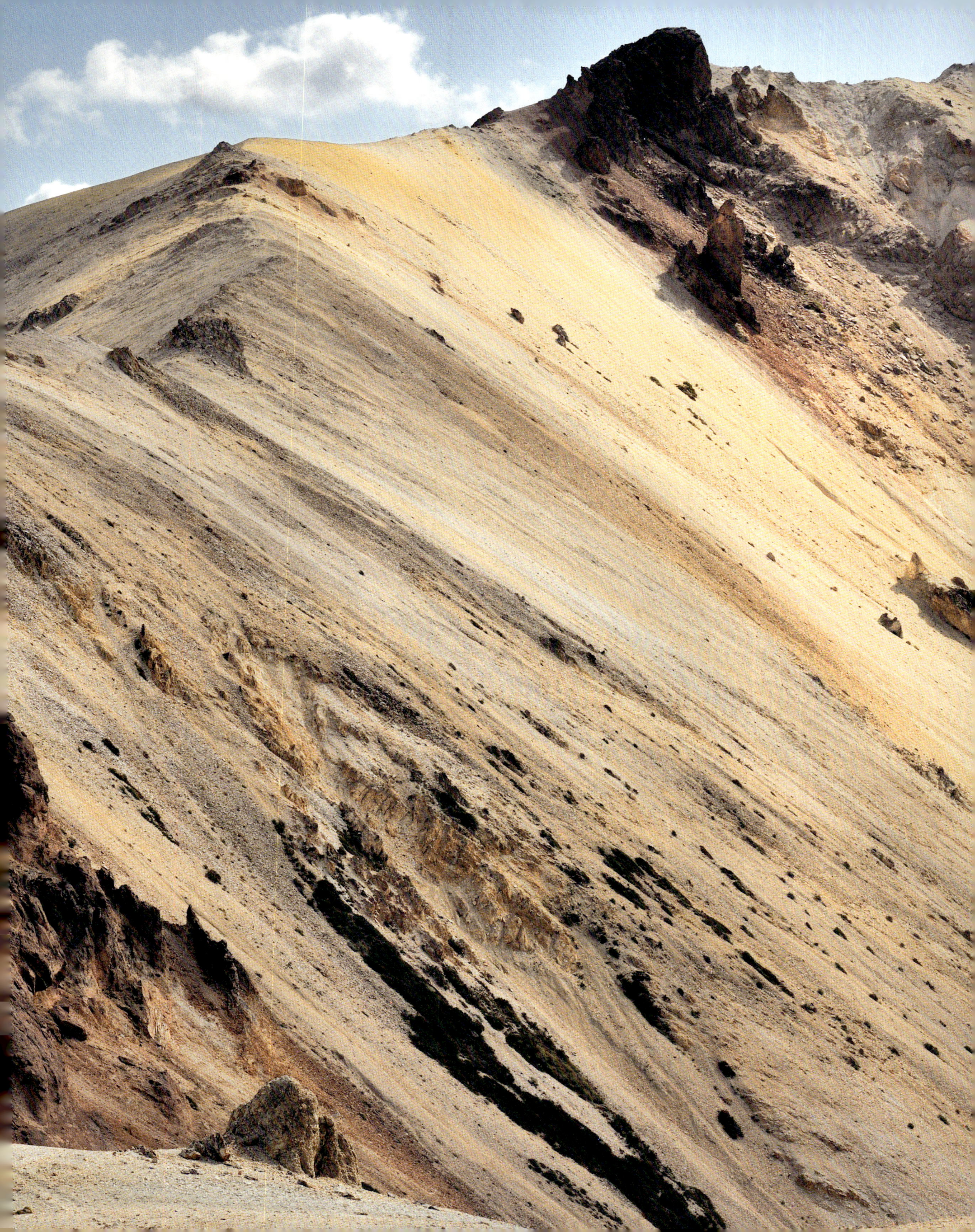

Into the Saddle and out of This World

A gaucho encounter

Country:	**Argentina**
Best time to travel:	**November to February**
Closest airport:	**San Martín de los Andes**
Price tag:	**€€€**
Style of travel:	**A smart and sensational horse-riding escapade**

Introduction

"No hour of life is wasted that is spent in the saddle." So said Winston Churchill. Certainly, no hour is wasted on this seven-day tour of Argentine Patagonia with one of South America's most sought-after guides. Jakob von Plessen, founder of Jakotango Riding Safaris, is the authority when it comes to leading travellers on horseback into the remotest parts of Patagonia. Born to Austrian immigrants von Plessen mixes rugged gaucho tradition with European finesse. A true cavalier in the Argentine pampas, rough and refined at the same time, he developed his guiding career during 12 years on the safari circuit in Kenya, before returning to Argentina to set up his own company. Riding with Jakotango is the ultimate Argentine adventure, with every journey stylishly outfitted by either von Plessen or one of his fellow guides, all of whom combine inherent knowledge of the unique landscape with superlative horsemanship, and offer a breathtaking perspective on emerald-green lakes, narrow knife-edge mountain passes, and endless pampas.

The Views

The adventure unfolds on a Manhattan-size private reserve bordering Nahuel Huapi, that bridges two national parks: Nahuel Huapi, the oldest national park in Argentina, and the 400,000-hectare- (1 million ac.) Lanín National Park, featuring the snow-covered Lanín volcano. From Buenos Aires, it is a two-hour flight south to the old colonial city of Bariloche, then a few hours' drive north in an off-road vehicle and a half-hour boat ride across a deep glacial lake where you are met by your horses and their gauchos to ride into the stylish Jakotango Base Camp, located on a 12,000-hectare (30,000 ac.) estancia (private cattle ranch). Here, guests can unwind on the veranda of a log-built cabin, next to a fire, while observing Criollos, the native horses of the pampas, grazing in the distance. A good night's sleep is advisable before mounting a Criollo and ascending the Pass of Tears, one of the most dramatic mountain crossings of this region. Riders bed down in comfortable tents, each with a proper bed and a wood-fired heater to chase away the chills of the Patagonian night. The terrain and landscape begs for traditional gaucho attire. Riders dress in loose trousers called bombachas, following Jakob, who also wears a beret stylishly tipped to the side and a facón (dagger) tucked into his belt. The Criollos are ridden in the South American tradition with the reins held in one hand. The horses' stamina, skill, and balance are legendary. The horses are experts in balancing on mountain ledges, preferring to decide for themselves which is the best way to climb up and climb (read: slide) down. As pebbles dislodged by hooves tumble into the abyss, the gauchos never flinch or misstep. Still, an experienced rider will find it impossible not to be nervous crossing the Pass of Tears. "Sit tight and lean back," Jakob will shout over the howling wind. "Let the horse do her thing!" Built by Jakob, the paths here — high, dramatic, narrow as the blade on a facón — lead riders between the peaks and through valleys and unspoiled forest to places they will feel they have discovered themselves. After passing above the tree line, the trek crosses a ridgeline of volcanic ash as soft and dense as sand — a stark and beautiful contrast with the blues of distant Andean lakes. Why do they ride the horses with one hand? So riders can use the other to drink from the Malbec-filled leather pouches they carry, tippling to toast the scenery as well as to summon courage. Jakob describes the Pass of Tears as 'vertigo unfriendly', but nothing can prepare you for the panorama across an alpine desert that glows orange, yellow, and pink and stretches to a green glacial

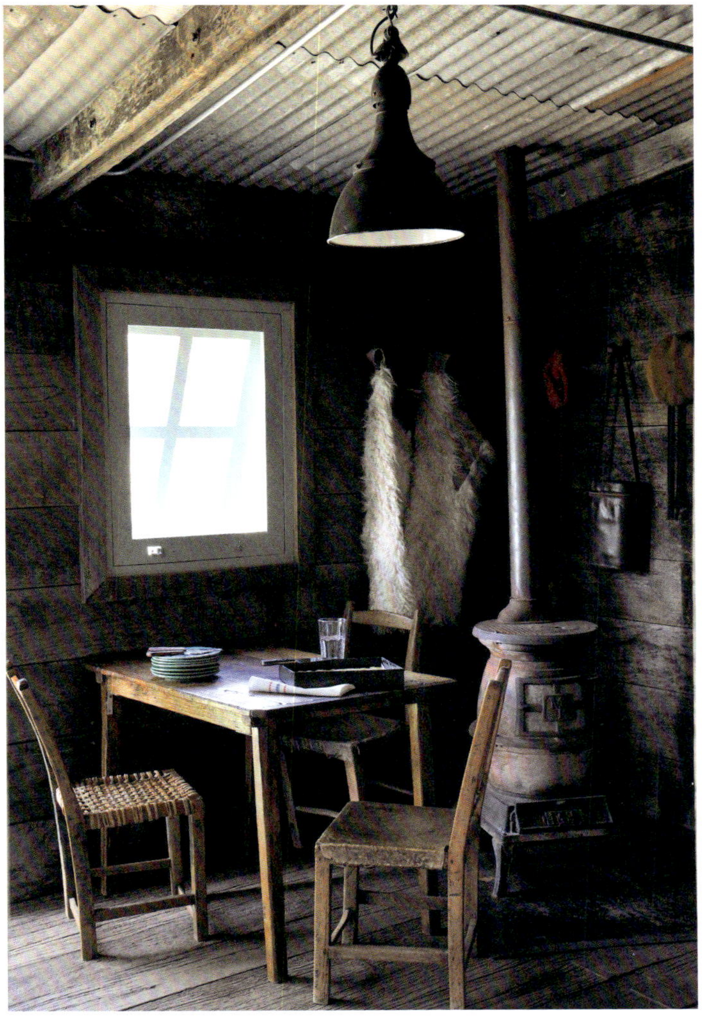

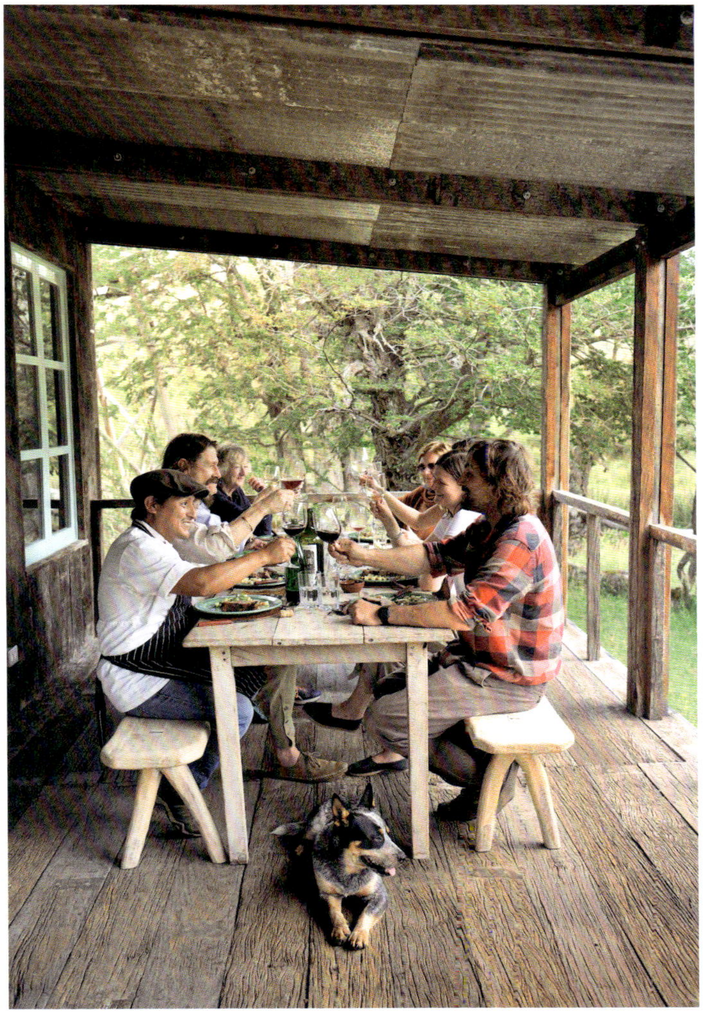

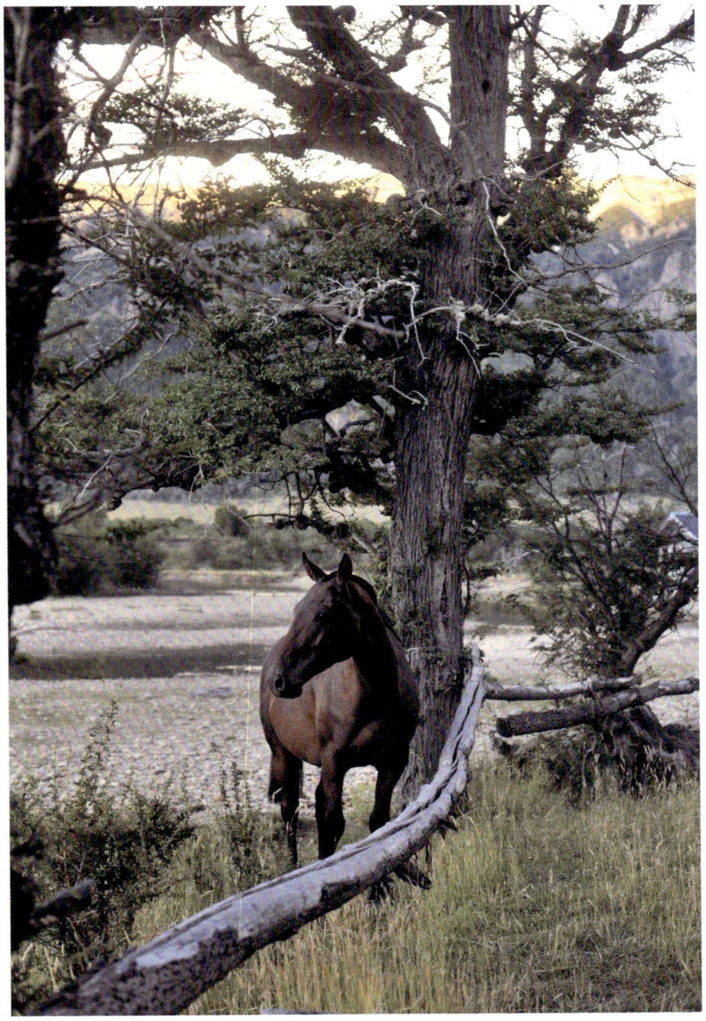

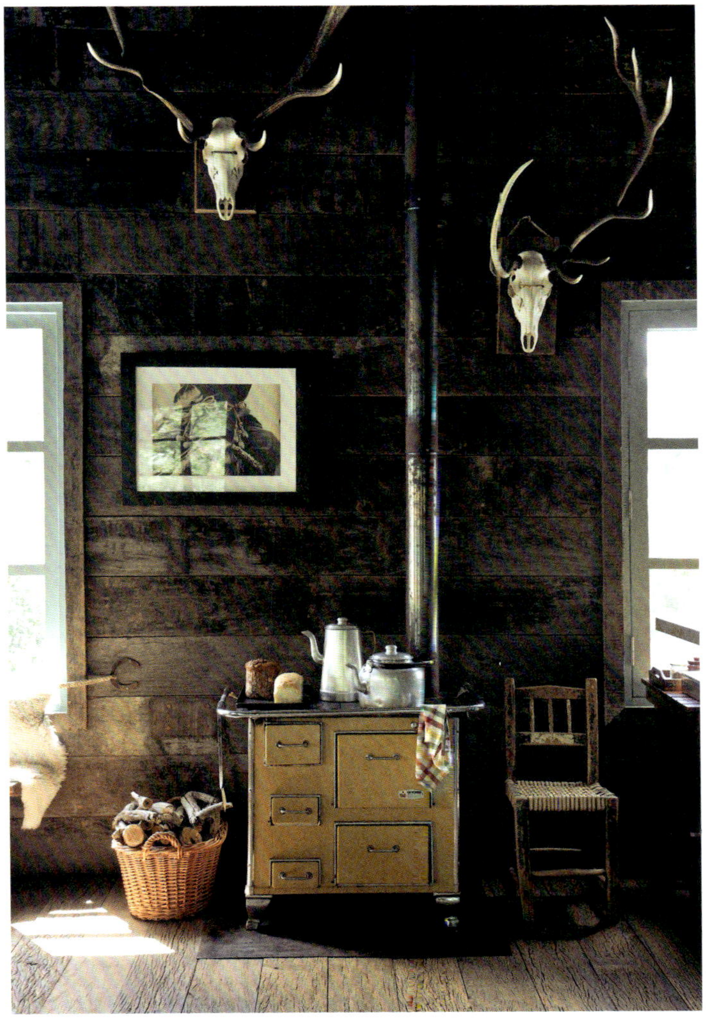

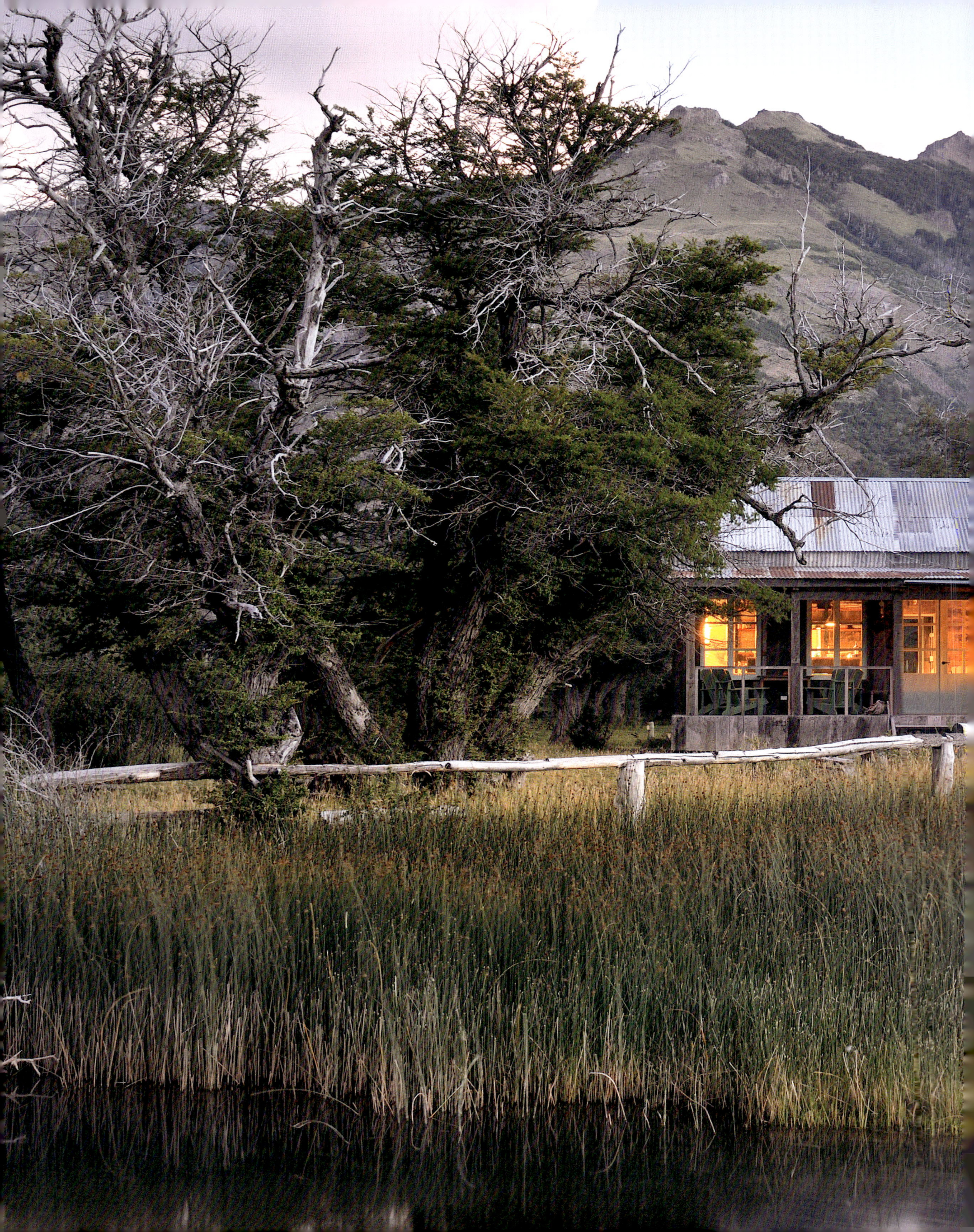

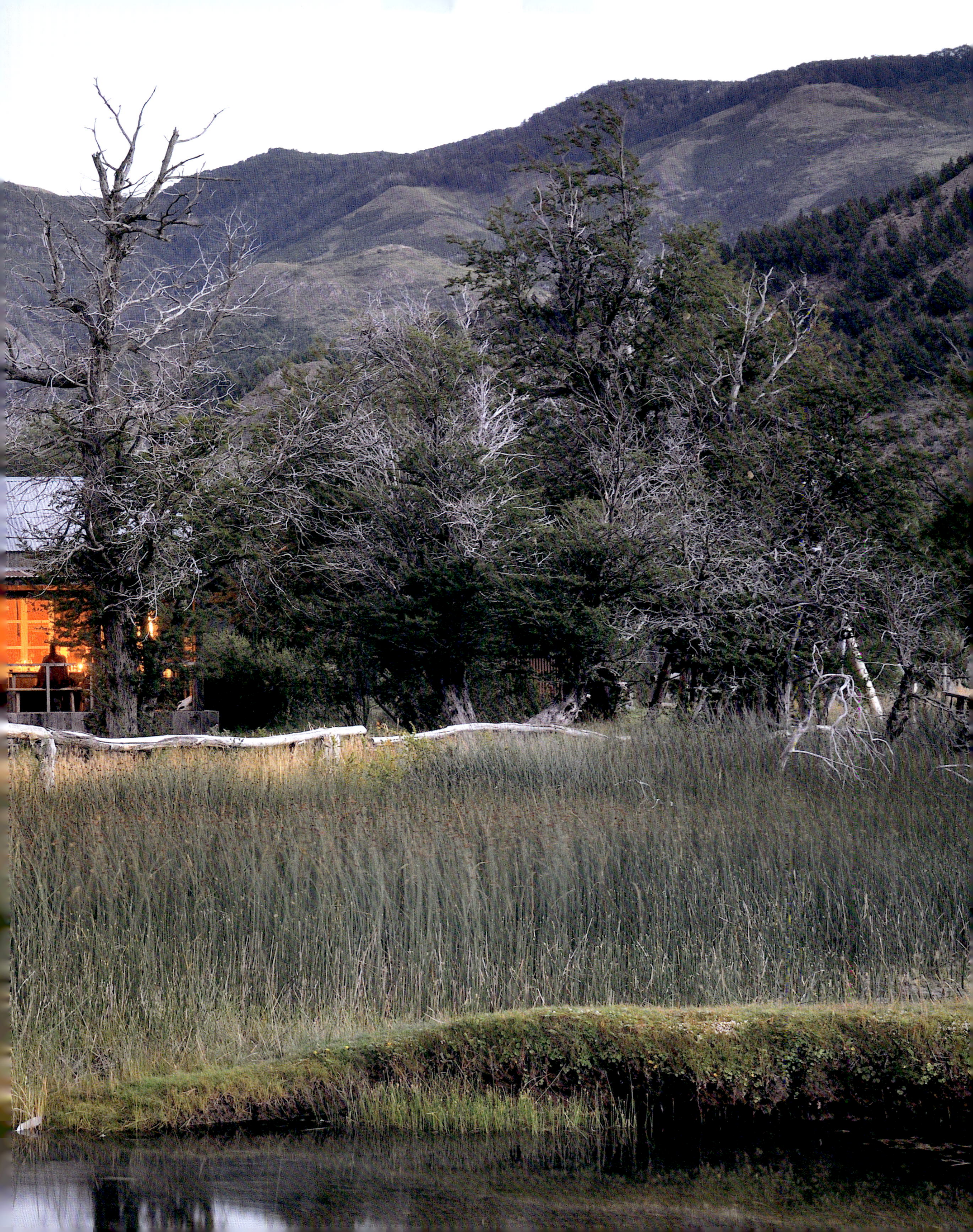

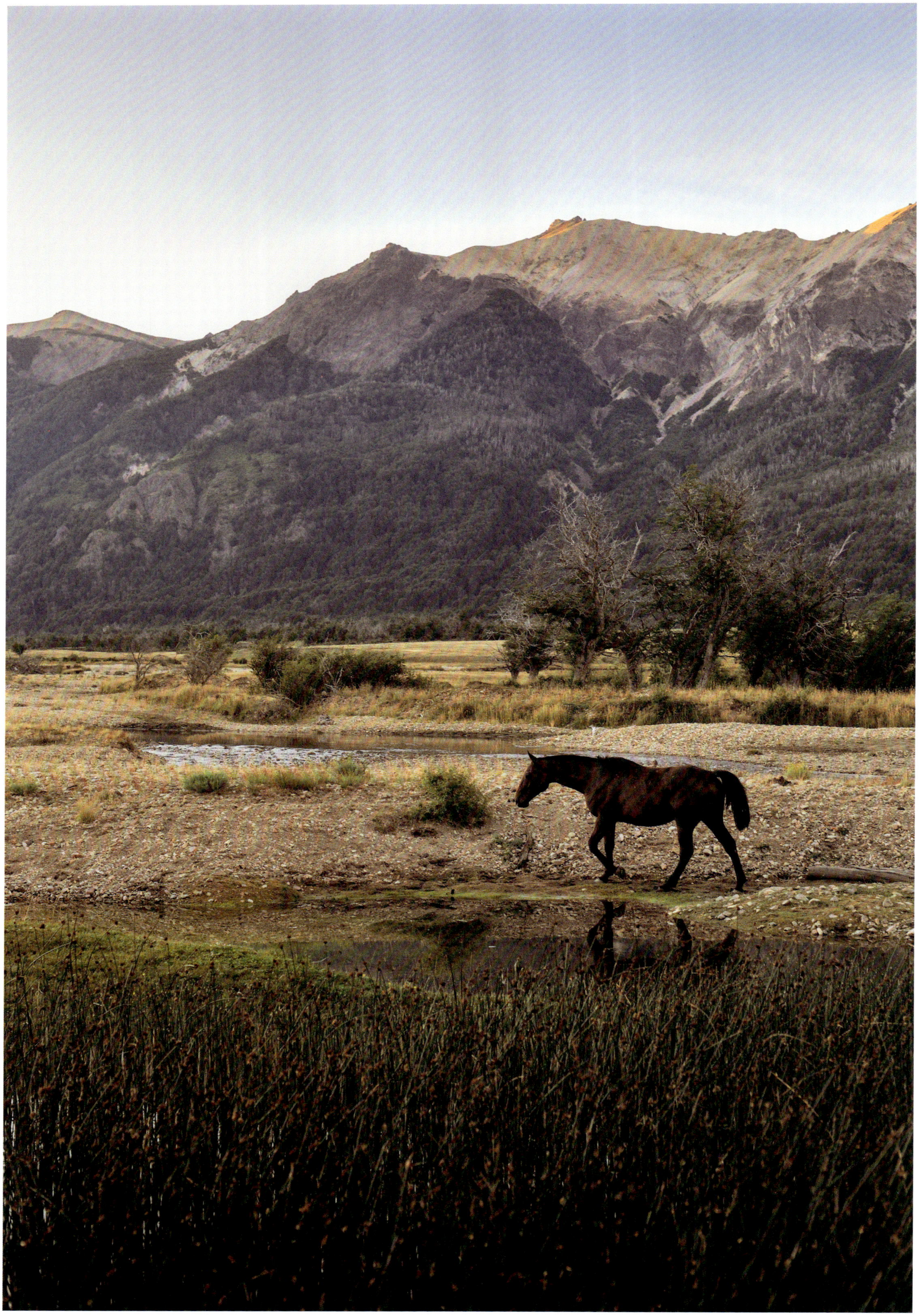

Previous spread In the old timber log cabin, wood-fired heaters and open fires chase away chilly nights. Think gaucho-chic, vintage curios and traditional patterned woollen blankets and long, lazy dinners eaten at wooden tables on rustic chairs. The veranda is the perfect place to relax after a long day in the saddle. In the middle distance, Criollos, the native horses of the pampas, graze and rest; soon they'll be ready to take guests back into the Patagonian wilderness.

valley. In Jakob's experienced hands, this condor's view over the heart of Patagonia soon replaces fear with awe. In the evening, base camp awaits with sizzling slabs of seasoned lamb grilled on the parrilla accompanied by flowing Malbec. Then it's back to the flickering candlelight of the log cabin and deep sleep. Every detail at Jakotango, from the enamel mugs and the iron kettle to the mounted deer skulls, seems to have jumped from the pages of an interiors magazine. Sluggish Wi-Fi is traded for siesta and conversation. Instead of 'status updates' one puzzles over the intricacies of the recado saddle, made of sheepskin and leather. Jakob's face has graced the occasional tabloid, and he was even cast in an advertisement for Dior Sauvage, which was set at the Jakotango base camp. With his movie-star charisma and art director's eye, he might not be the average gaucho, yet like Marito and Alberto, his fellow gauchos, both employed as guides — he has a deep understanding of horsemanship and of his own backyard. "When I was a kid there was a gaucho on this farm we used to go to," he says. "He lived alone in this tiny farmhouse with a dog and a couple of horses. He was a humble man, and what I most admired about him was that he took his job with so much pride and honour. You could see it in his posture when he was on the horse. In his face he had a Bible's worth of stories written in every wrinkle — burnt by the sun and the wind and the cold. And he just knew how to survive, and I thought to myself, 'I want to be like this.'" Morning comes fast, starting with a mug of black coffee boiled on a wood-burning stove. The horses emerge with impeccable manes groomed and saddled, as if preparing for a Buenos Aires polo match. Rough weather approaches, and soon the dust kicked up by their hooves is whipped away by a fierce wind. The temperature drops on another spectacular pass where there is nowhere to hide. "Even in this howling wind," Jakob says in the Dior Sauvage video, "I feel content. Not needing to be anywhere else gives you peace of mind." Soon, however, the Criollos are zigzagging through a mossy cathedral of native coihue trees. A hearty sandwich is served with wine, followed by a siesta for the horses, who are released from their saddles untacked, while the riders look for a spot to lay down on their recados and digest both lunch and the views they saw in the morning. To wake up, there is an icy bath in an emerald lake, and then the final ride begins. The road back to civilisation leads through a mossy forest to the shore of Lake Traful, where the barn of a third-generation settler family has been converted into comfortable sleeping quarters. Here, a gaucho named Felipe, his wife Marta, and their daughter live in isolation, reachable only by boat or on horseback, surrounded by the coihues, whose canopy shatters the dim light into pieces. Arriving here and savouring memories of the roasted lamb, the looming volcanoes, the companionship of the Criollos, and the meditative silence of base camp, the feeling of accomplishment and inner peace is overwhelming. Winston Churchill was right about spending time in the saddle. Indeed, four legs are better than two.

Creature Comforts
Jakotango's Base Camp is the dreamiest Patagonian refuge ever. It is a fine example of how 'rustic' doesn't have to be primitive or uncomfortable. Each private tent comes with a great bed, crisp linen, and an adjoining bathroom. Wood-fired stoves heat up the bedroom and are lit twice a day. Other camps are simpler, yet still comfortable, with modest tents or under-the-stars fly-camping also available.

Go Now
Patagonia is one of the world's last wild corners, still unmapped in places. The best way to explore it is on the back of a Criollo, riding with the gauchos, which is not just an activity but an authentic lifestyle revolving around the simple but intense joie de vivre of this remote region.

How to Go
Catch a flight to Bariloche from Buenos Aires. The Jakotango team will pick you up from the airport. Their horse safaris start from 700 Euro per person per day. Lodging, food, drinks, and guiding are included.

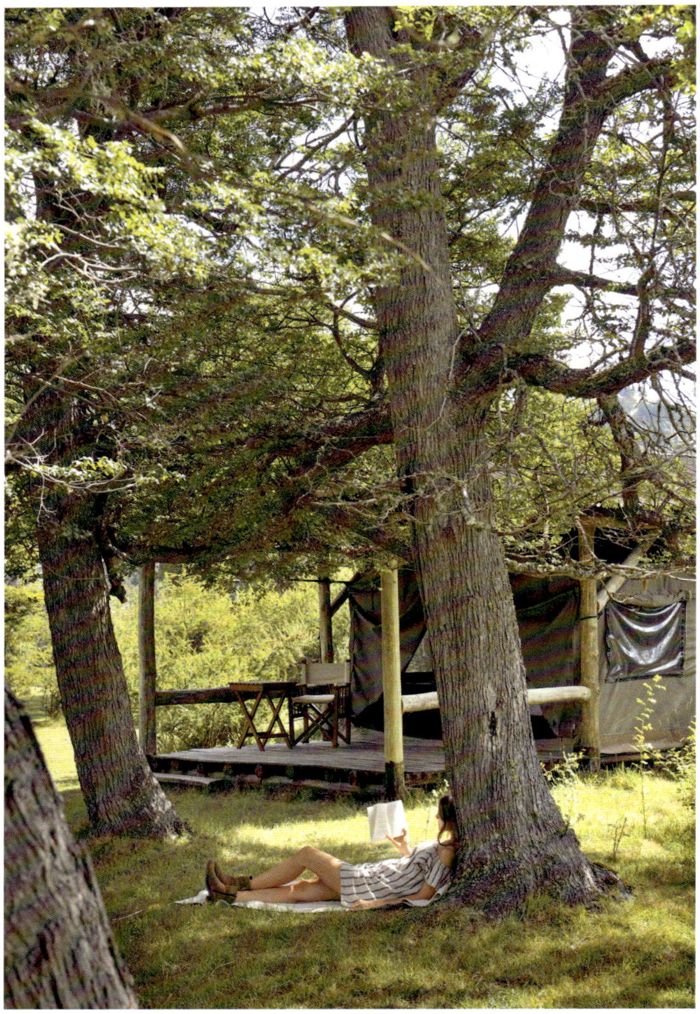

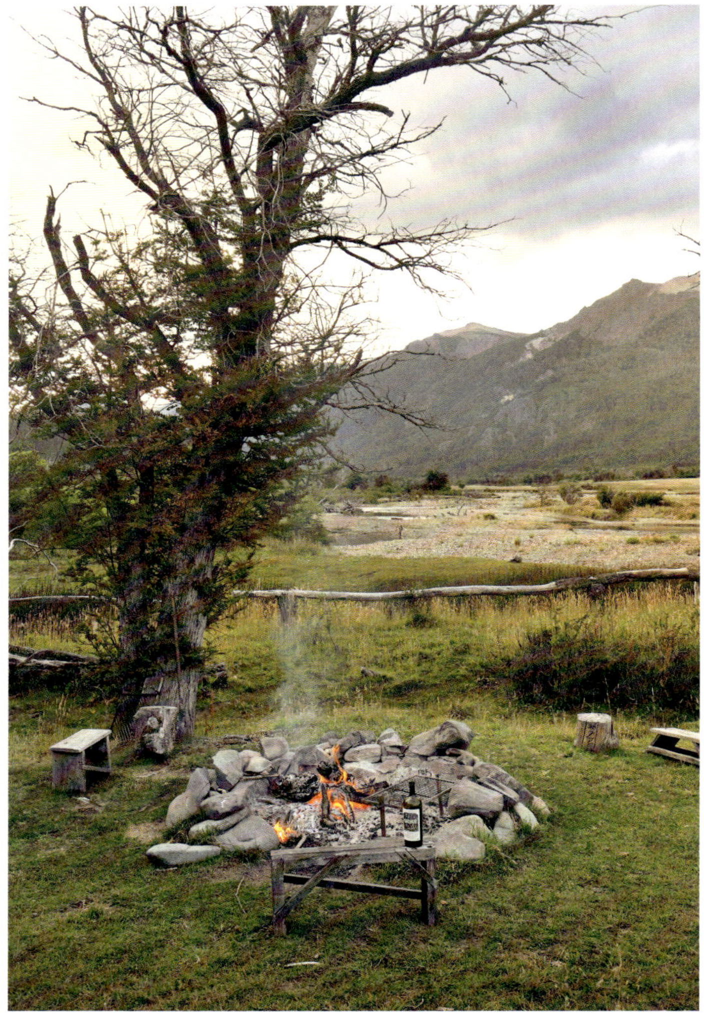

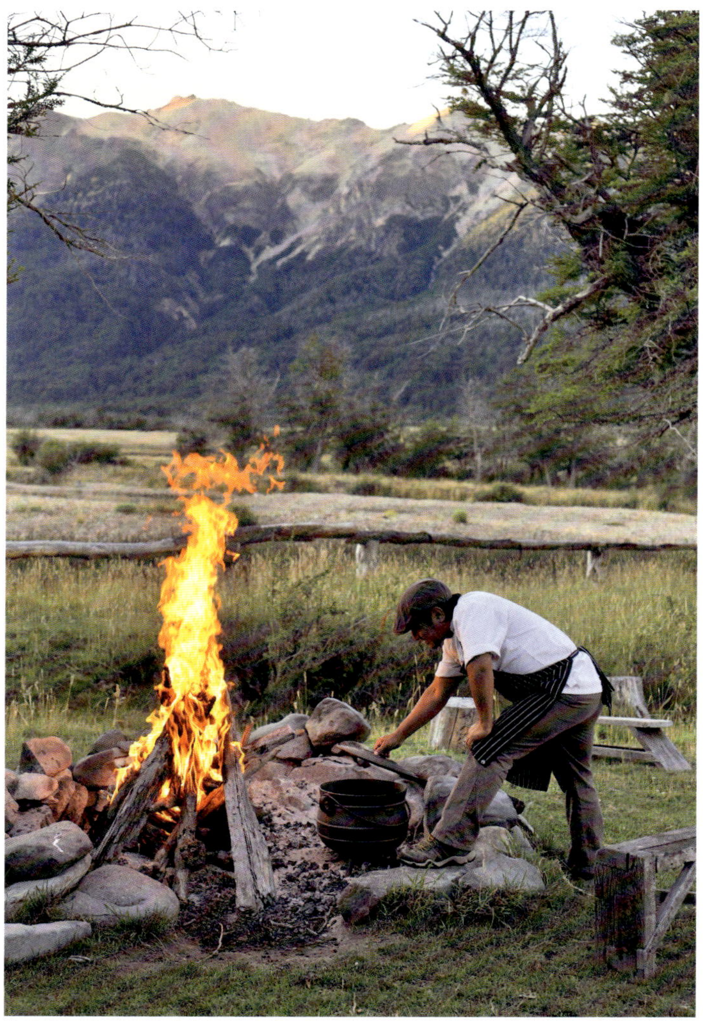

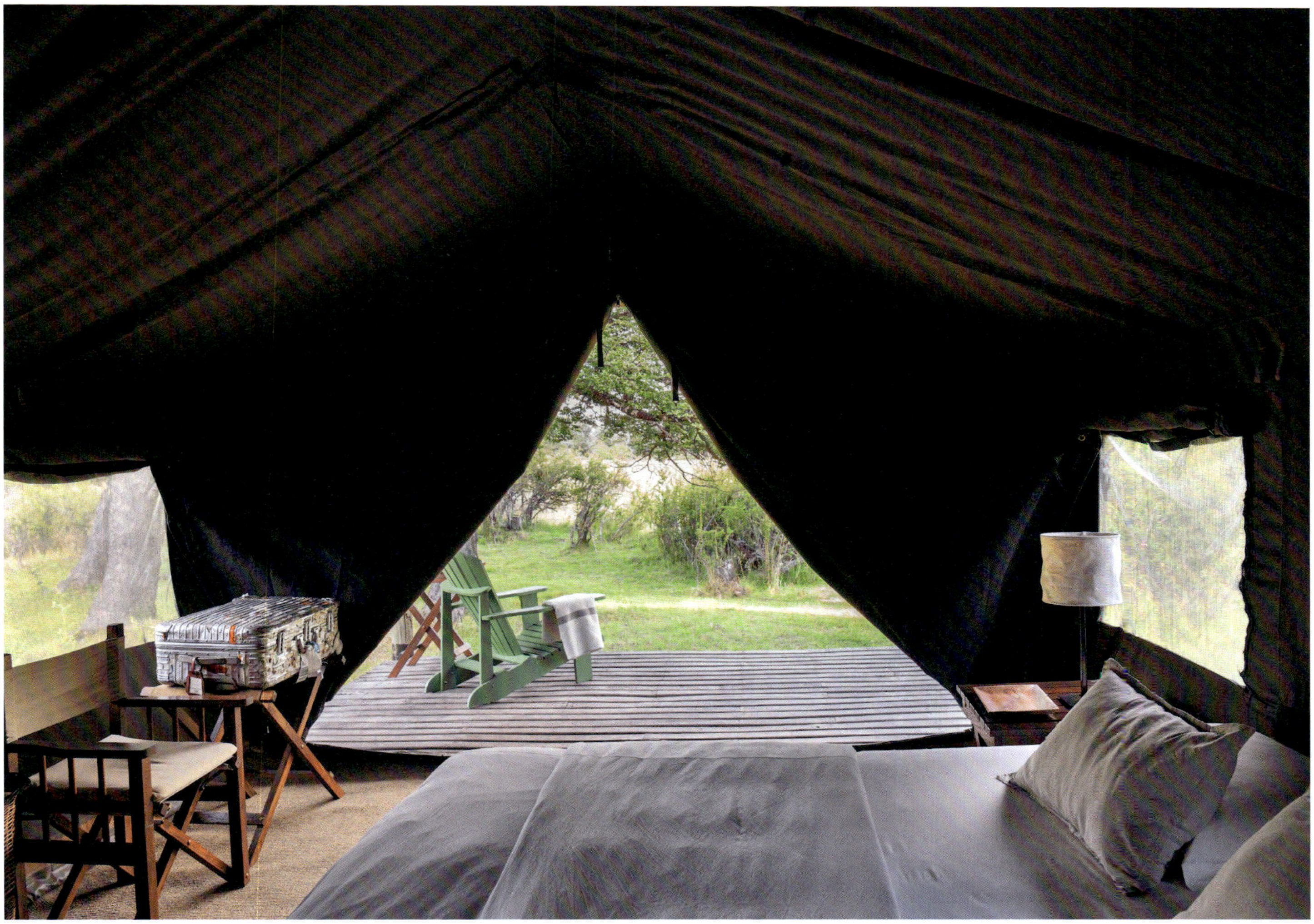

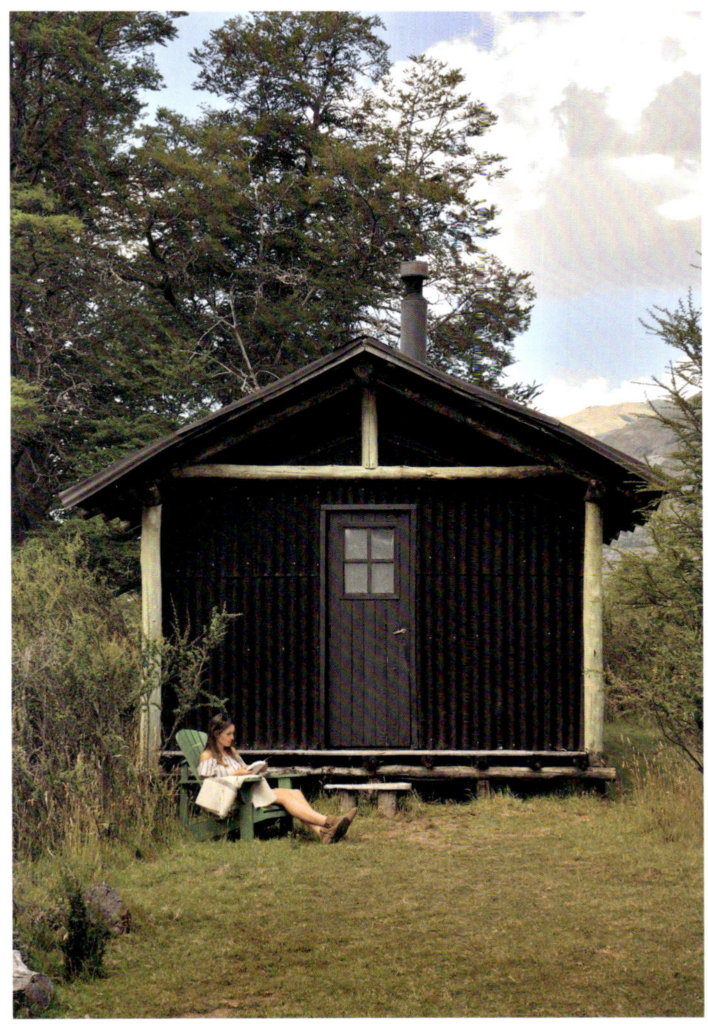

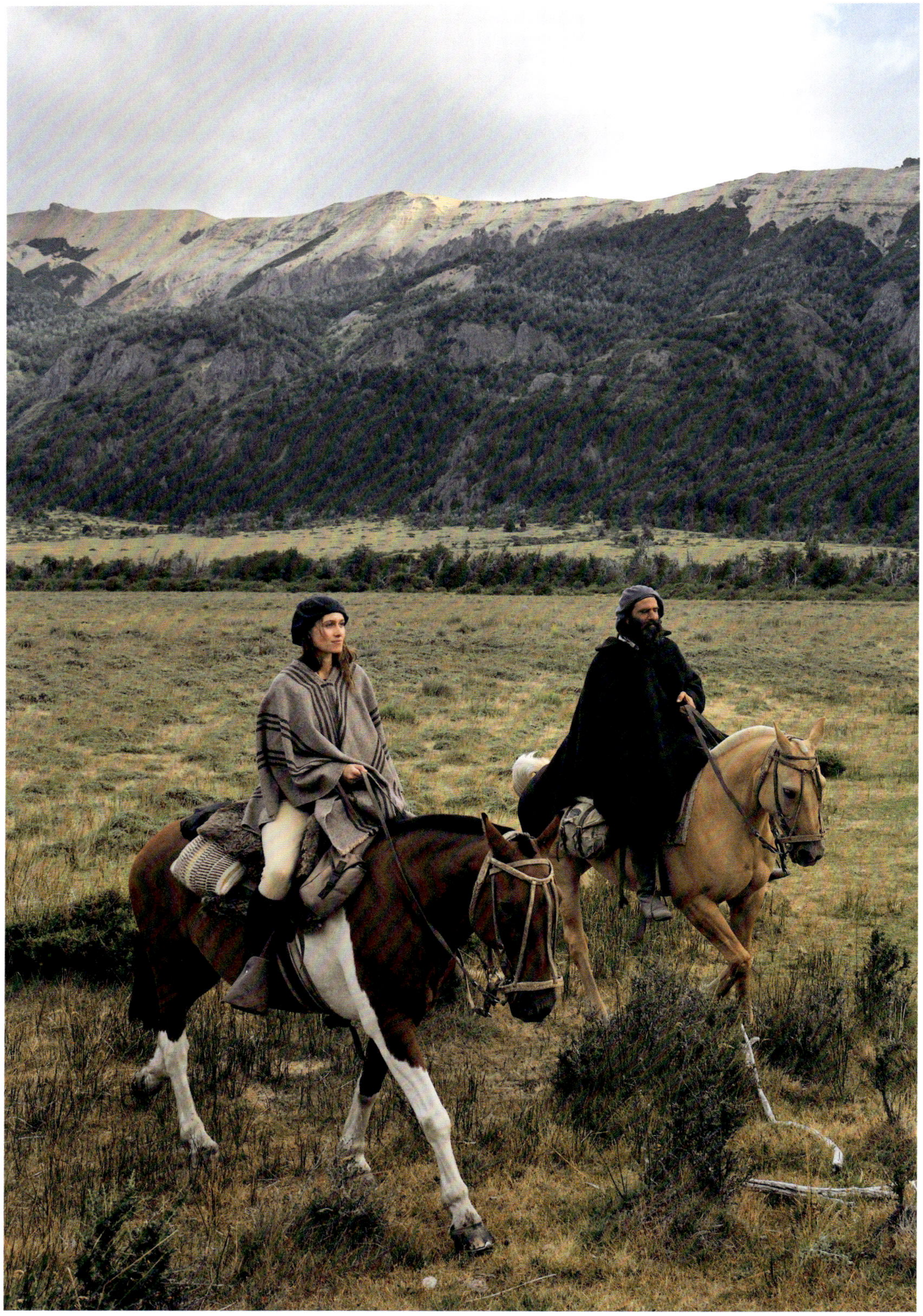

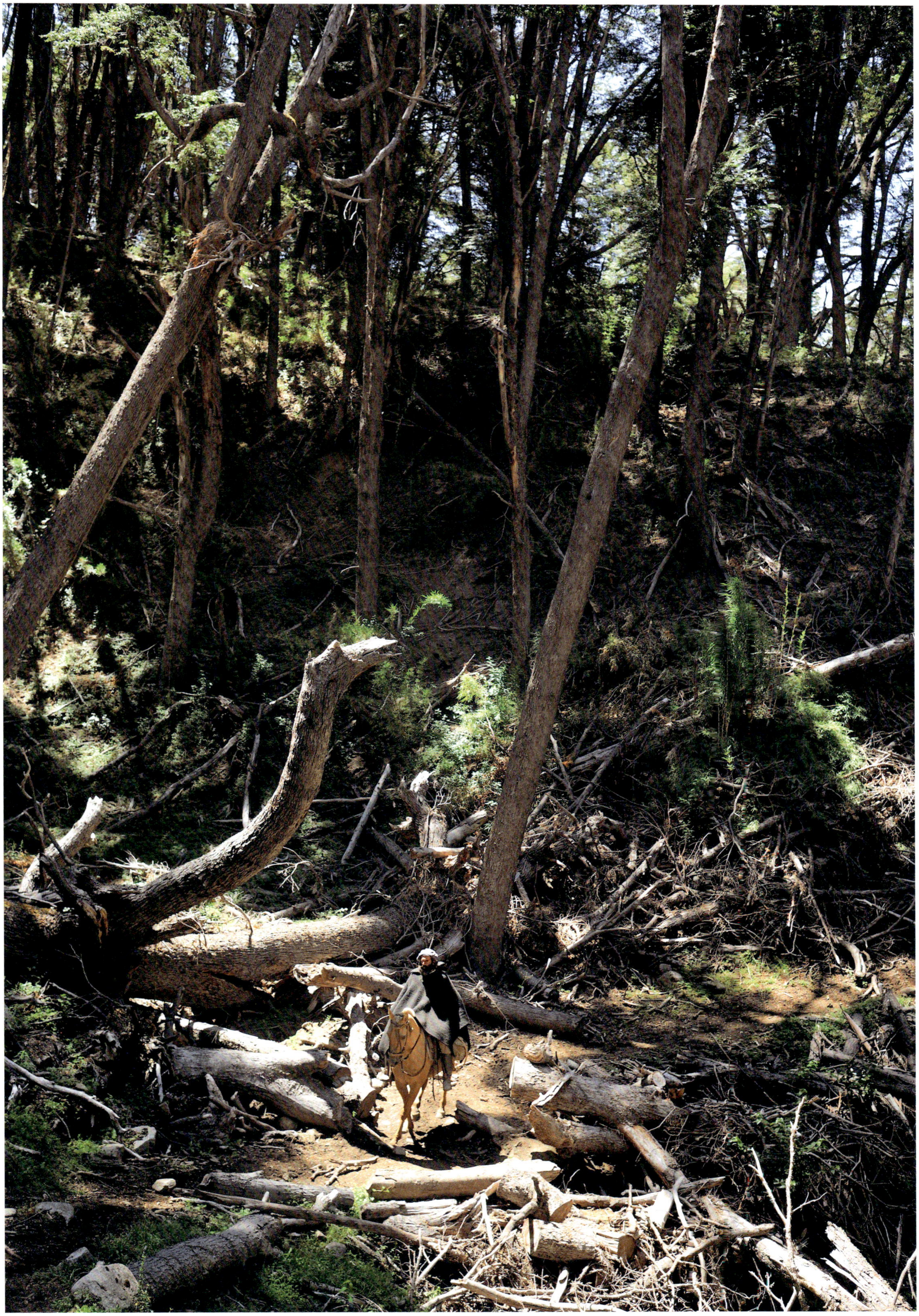

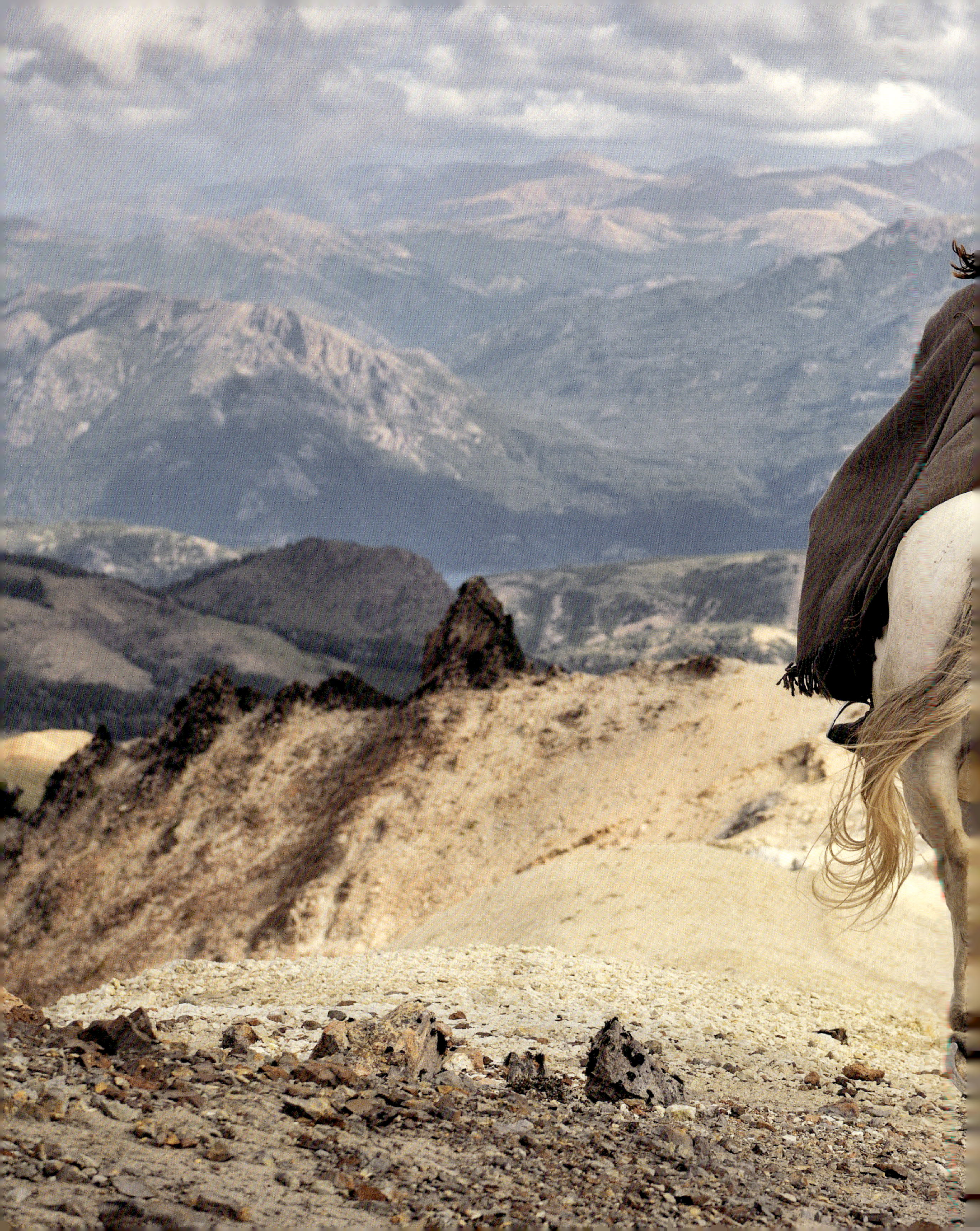

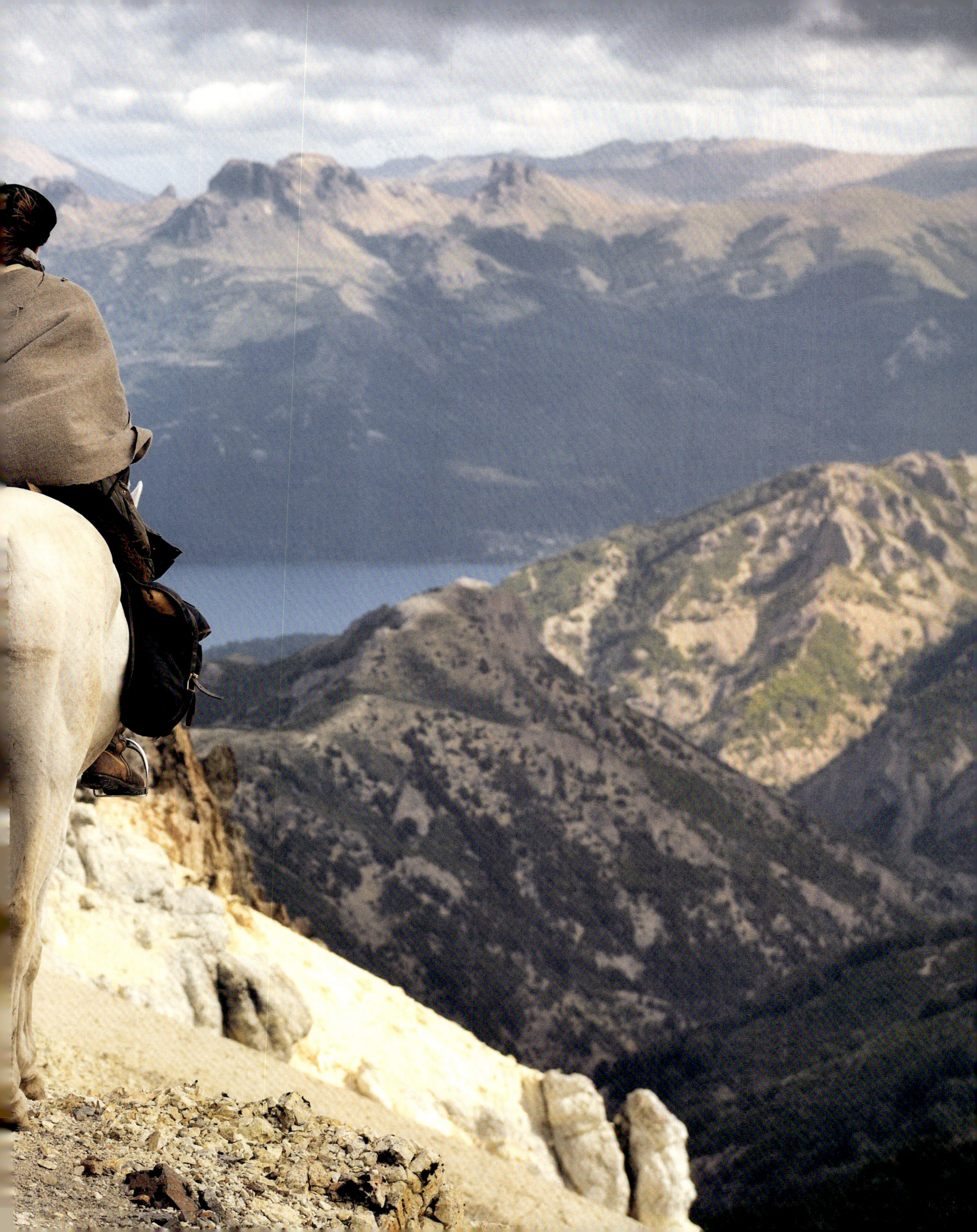

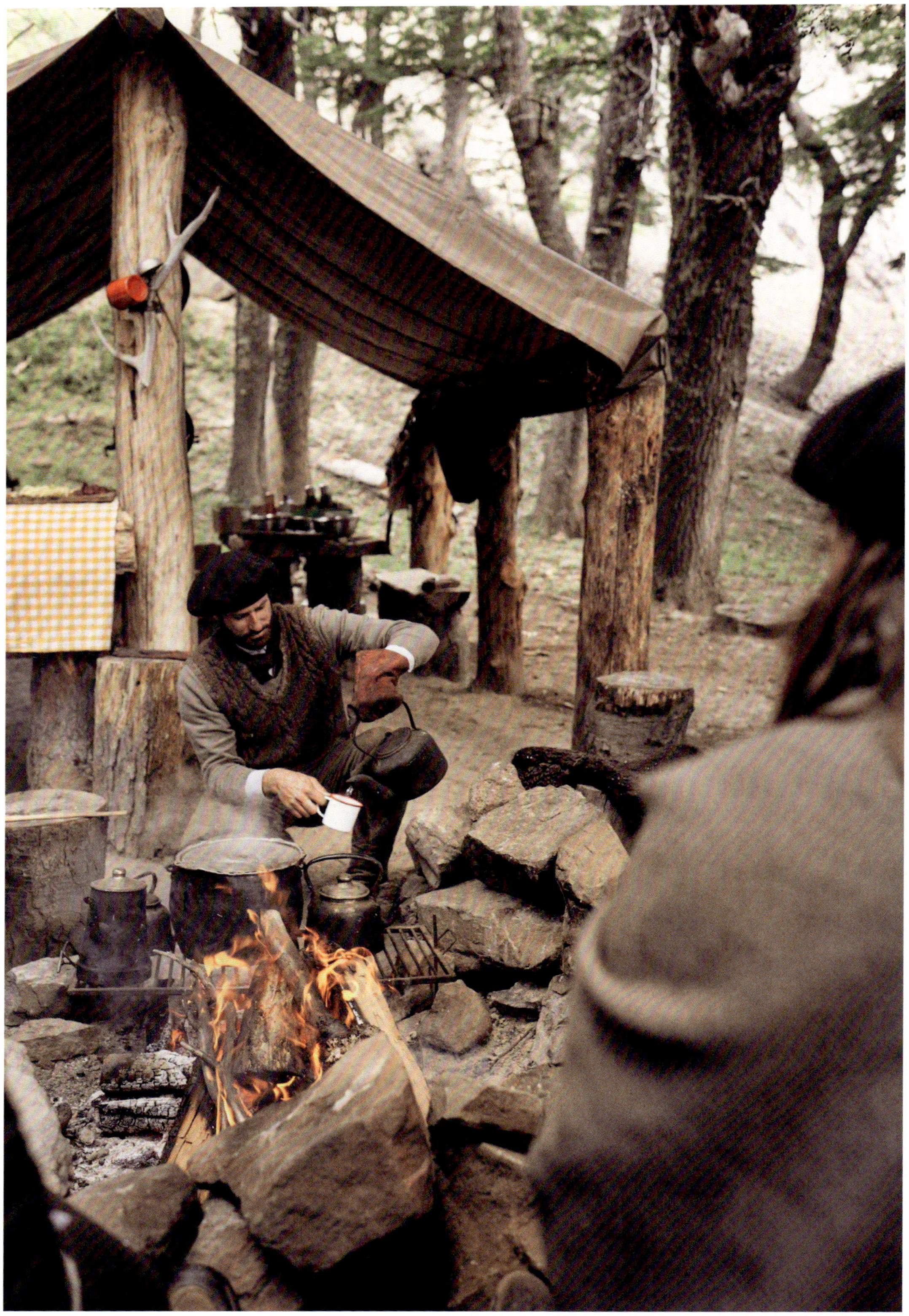

This spread The further south the journey goes, the more remote and disconnected from the modern world it feels. Here, gauchos live hermit lifestyles that are both simple and intense. There are no roads or trails here. Jakob built the paths himself. They pass the tree line and trek across a ridgeline of volcanic ash that is as flaky and shifty as sand. From the distance there are always the vast blue lakes and mountain passes which flow amidst the Andes.

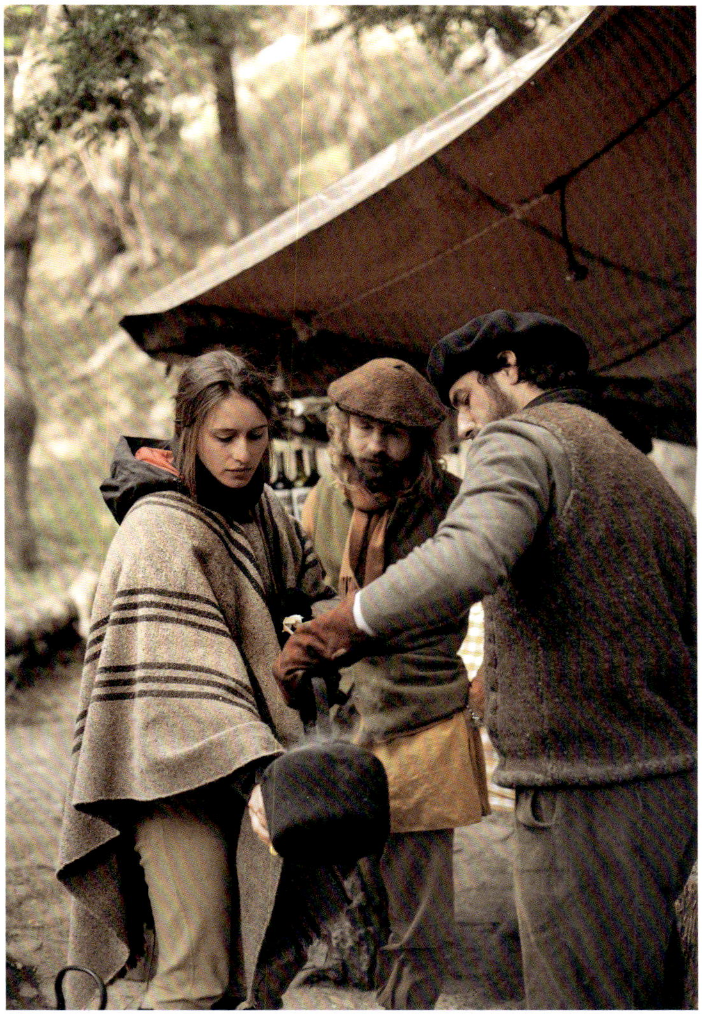
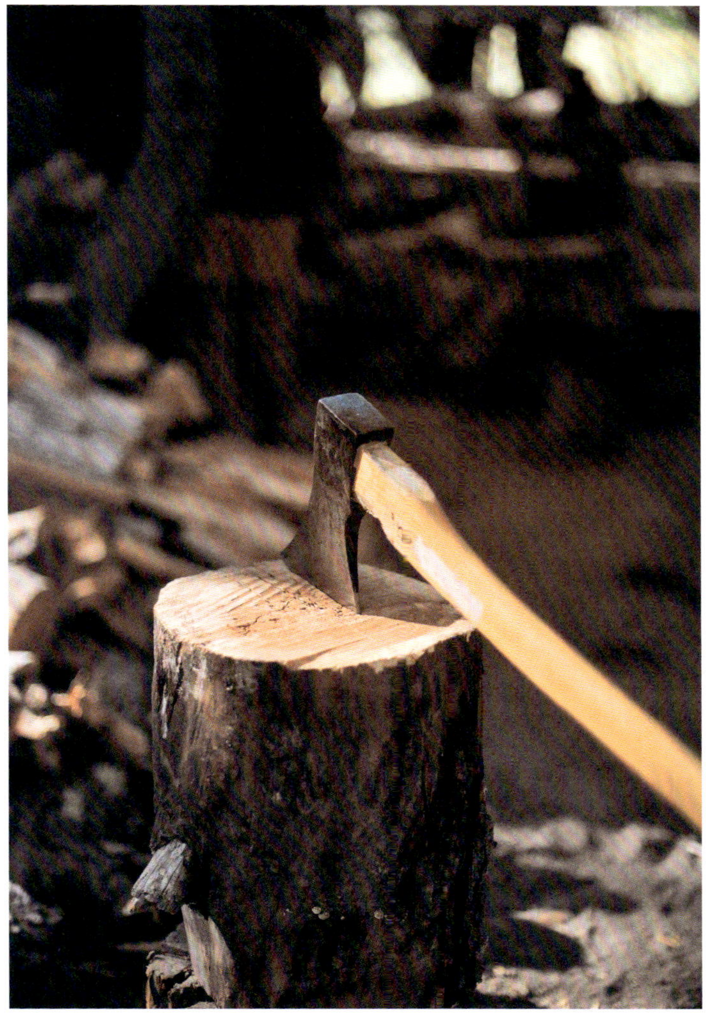
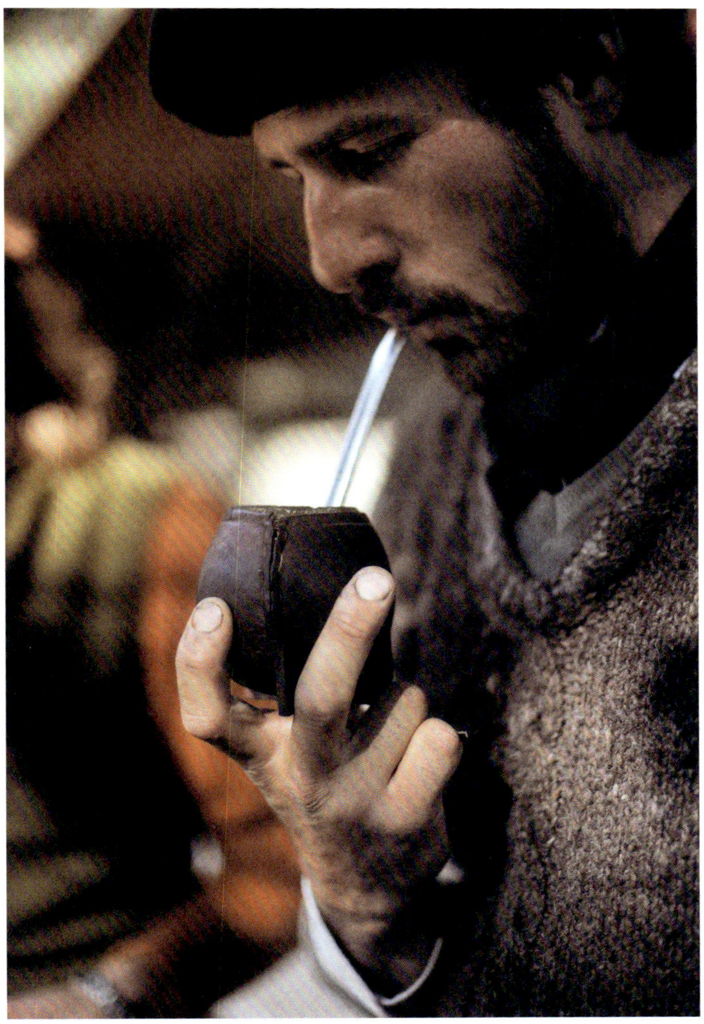
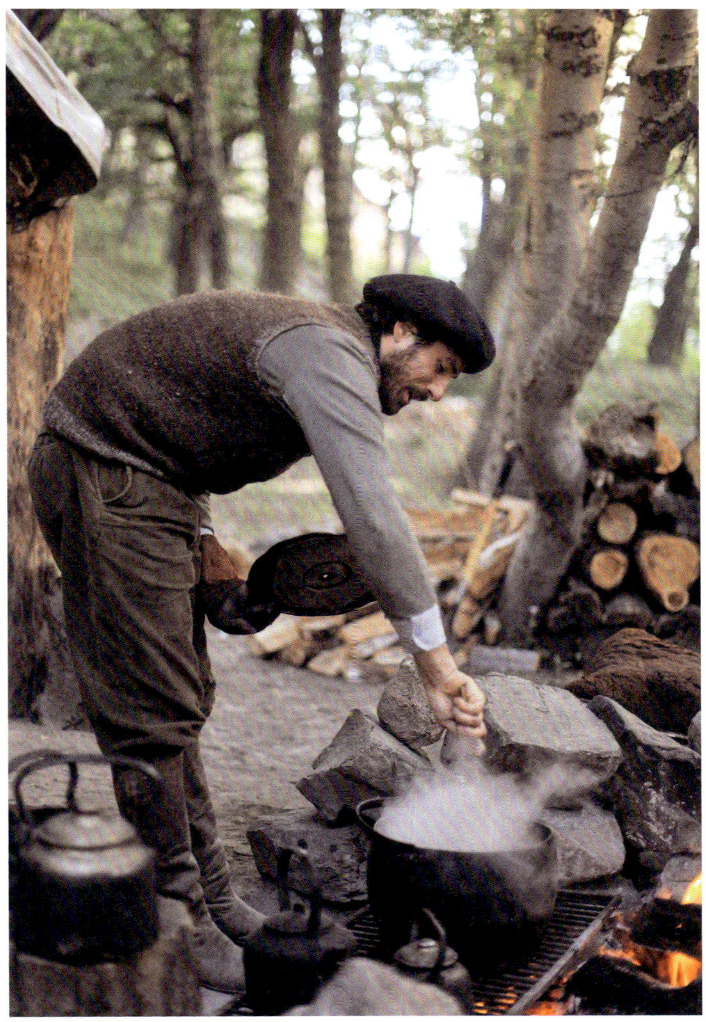

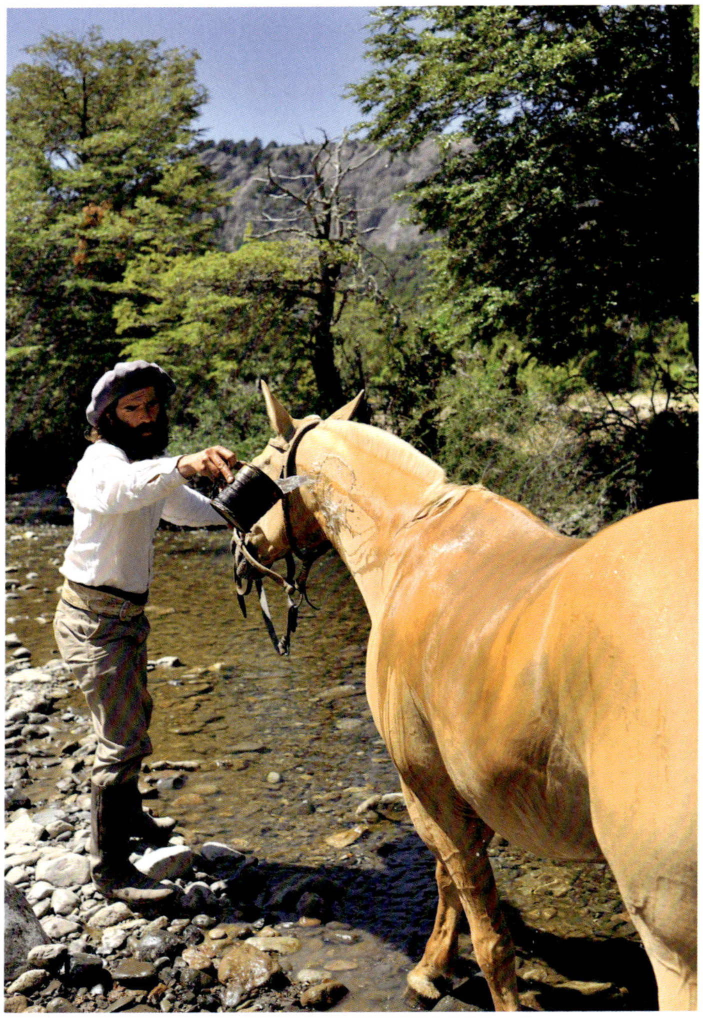

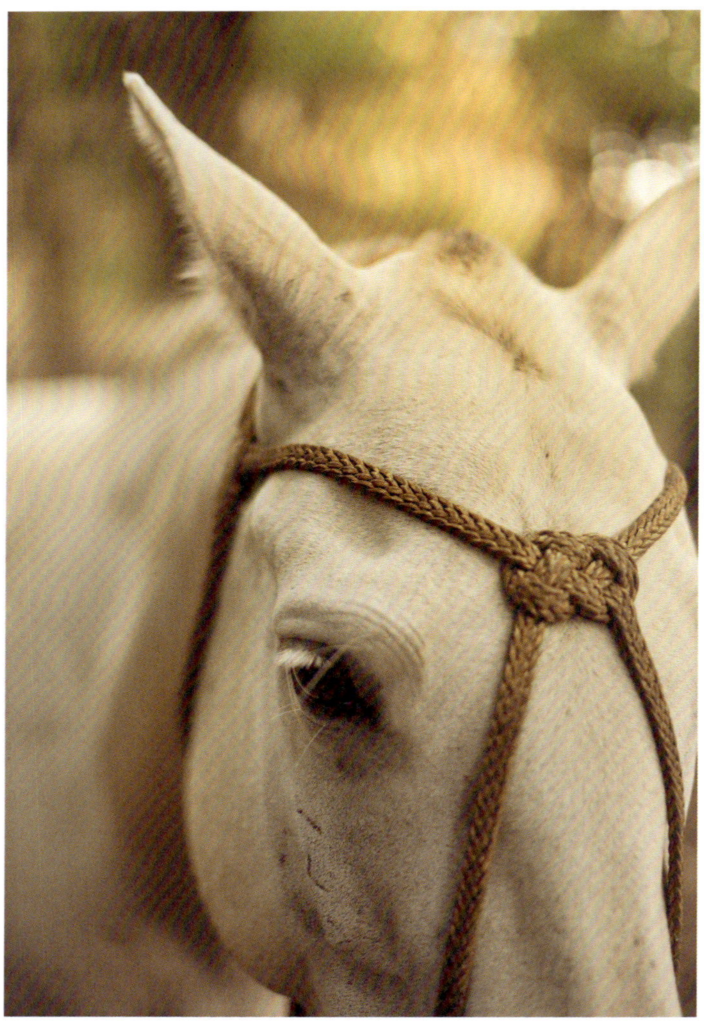

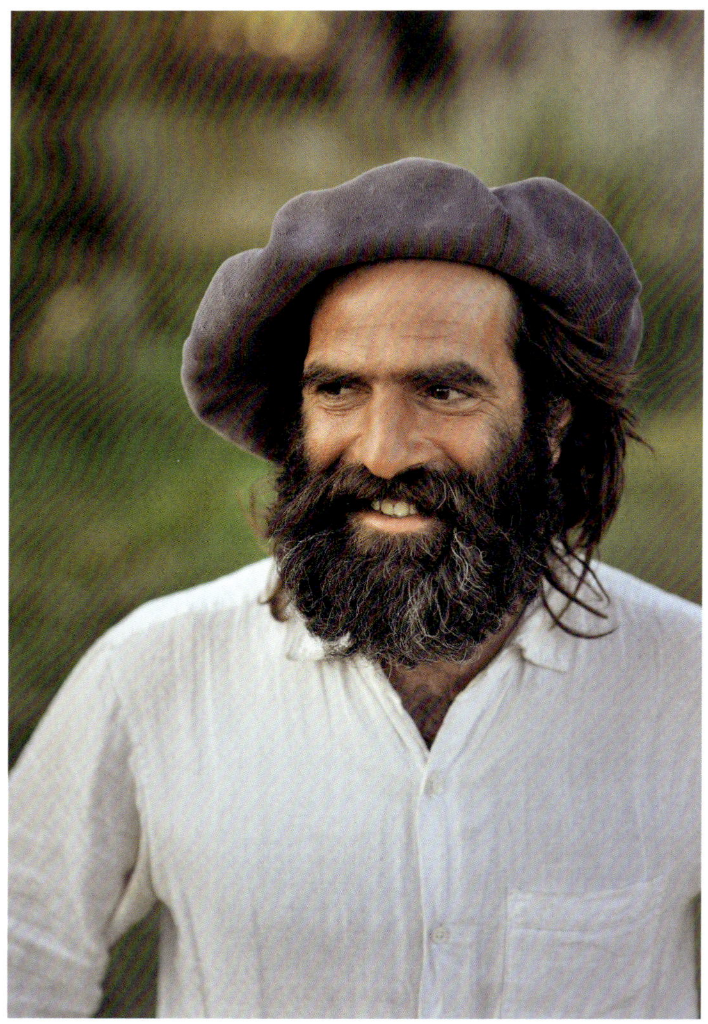

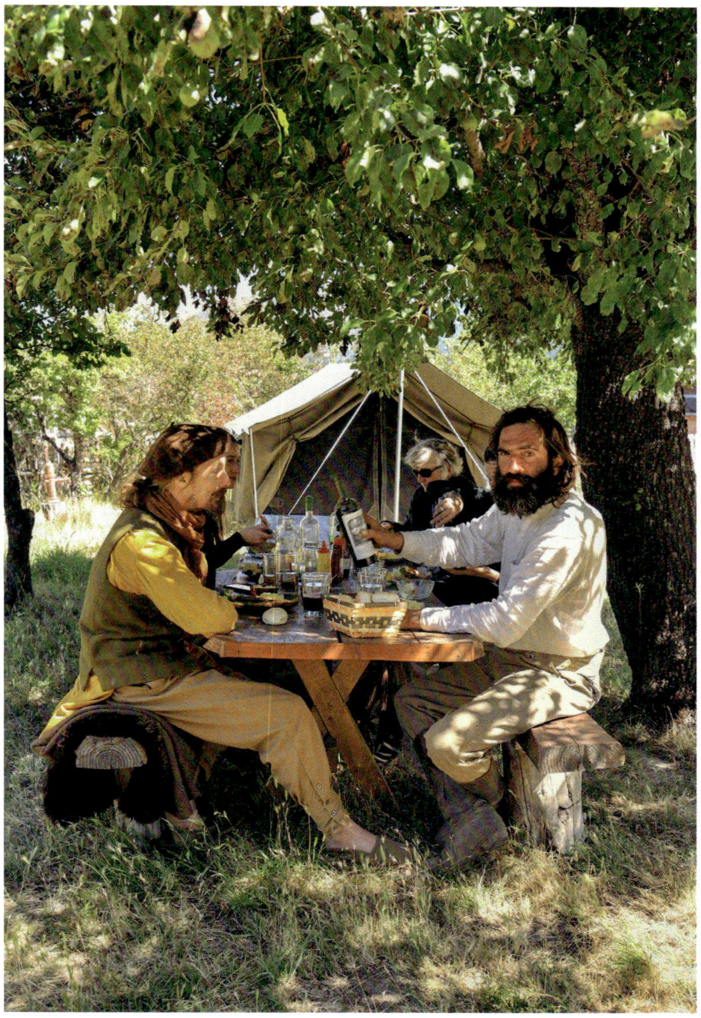

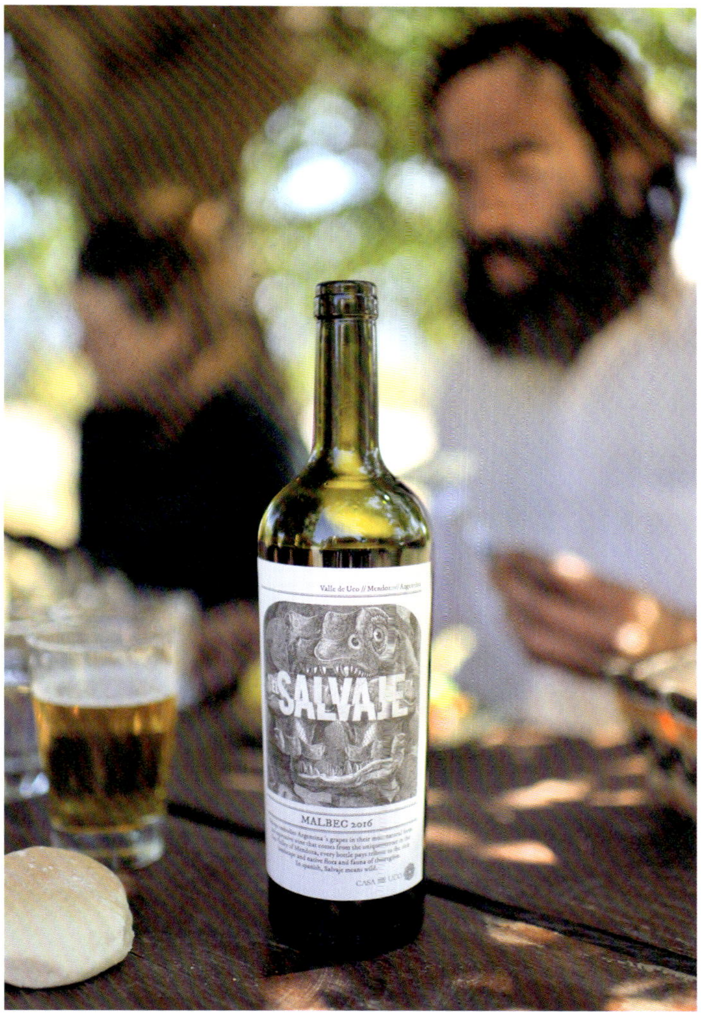
SALVAJE
MALBEC 2016

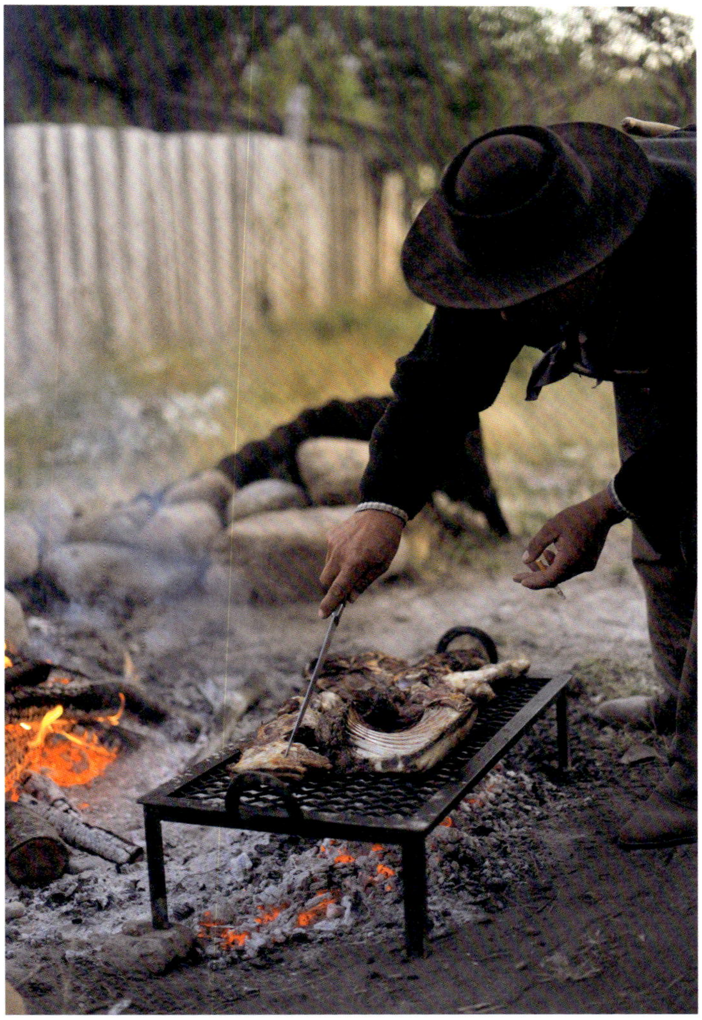

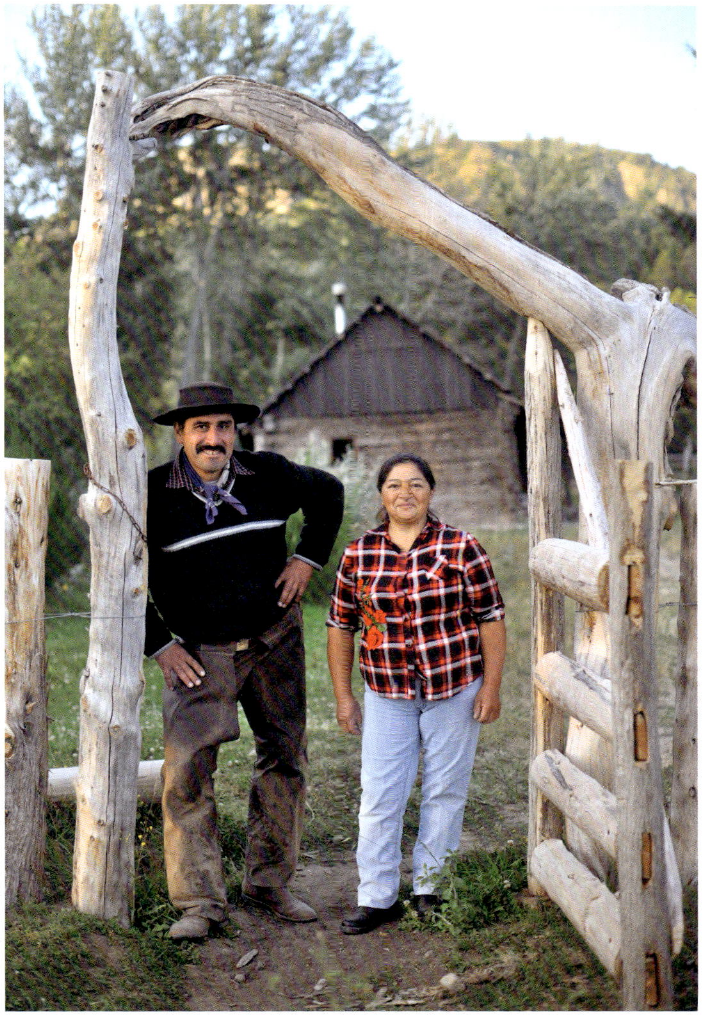

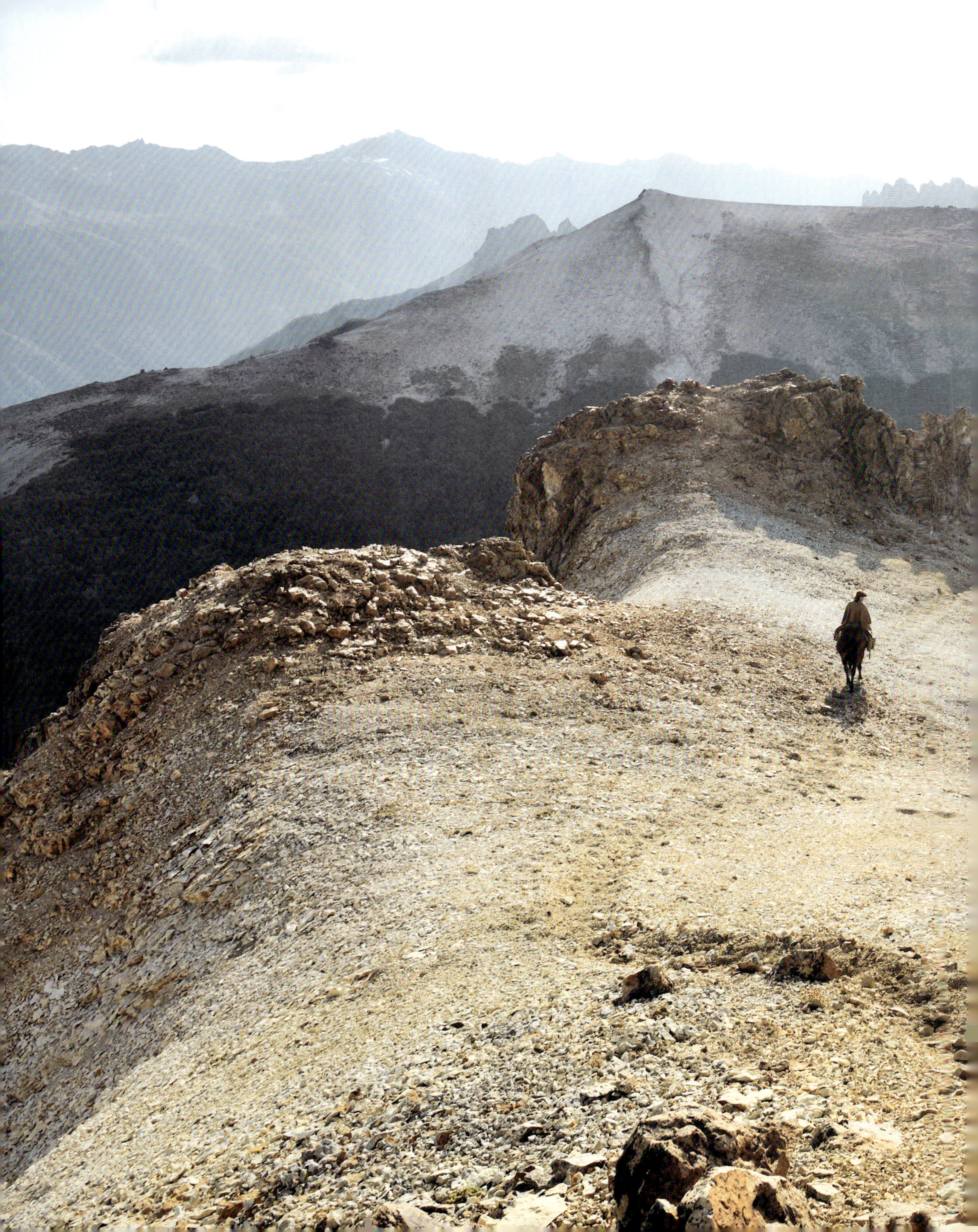

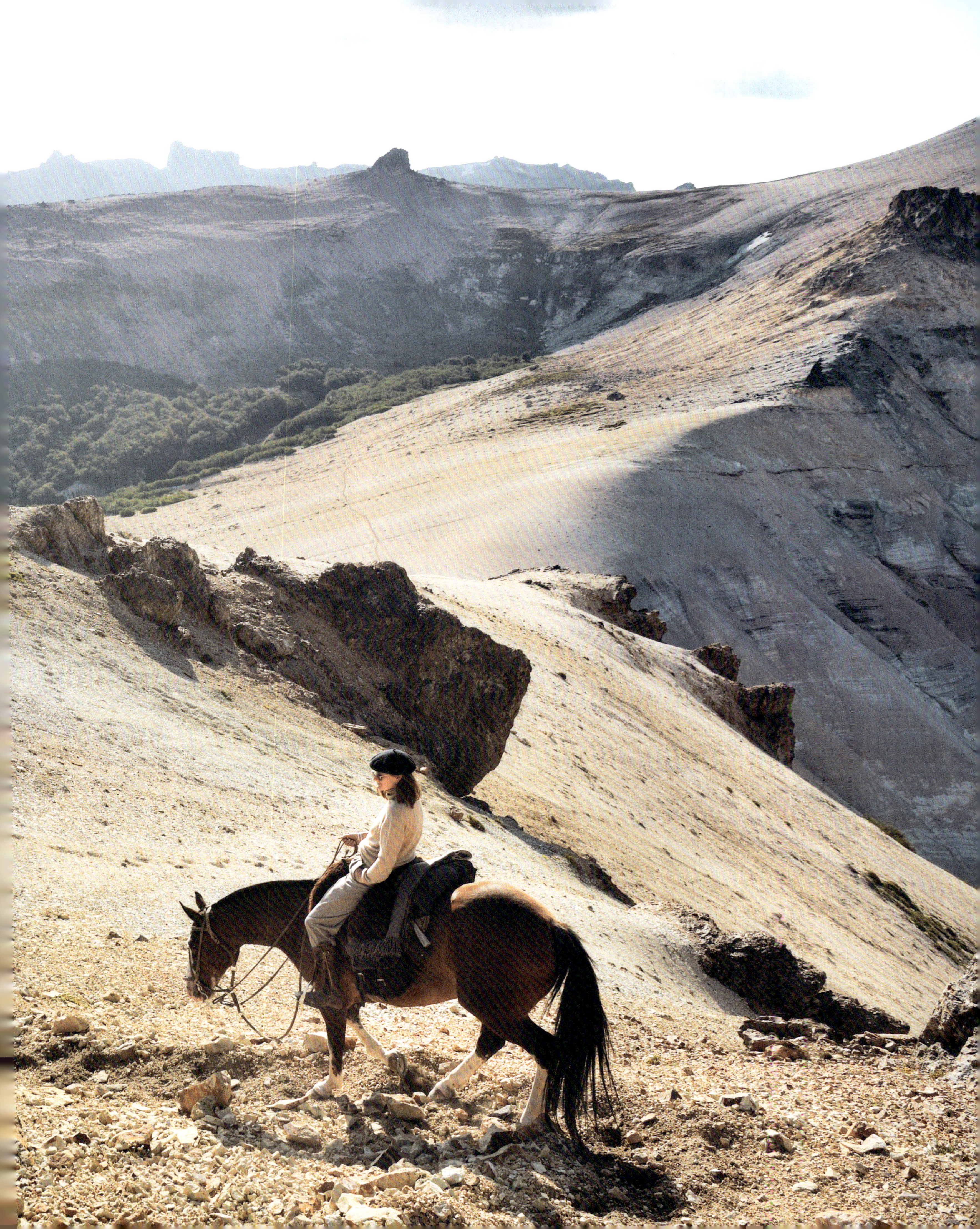

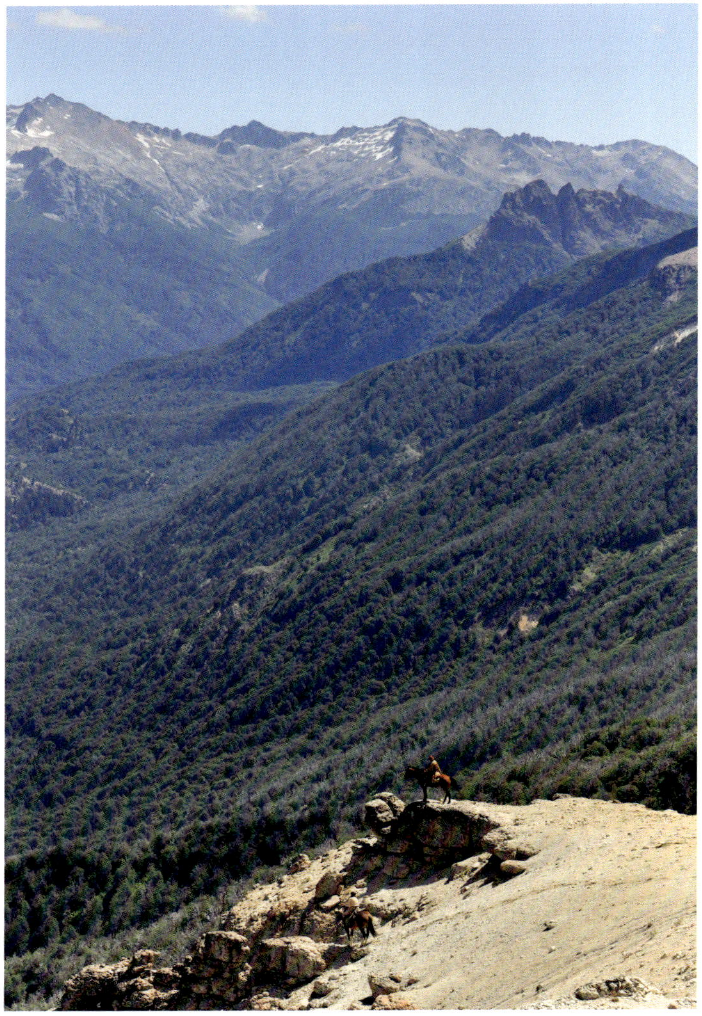

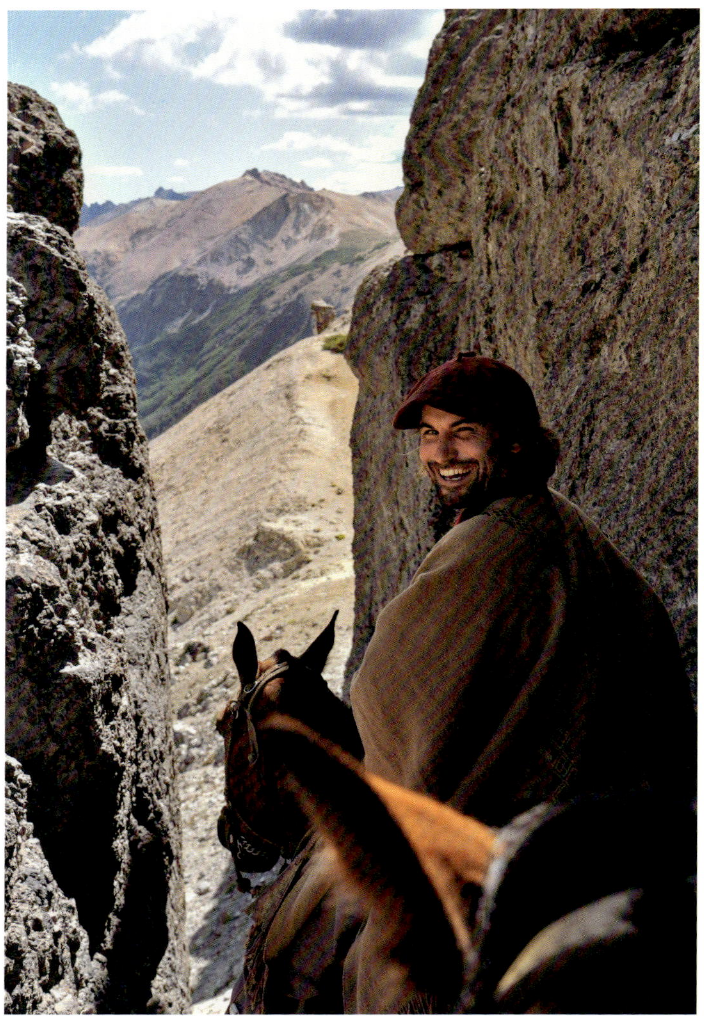

This spread On the dramatic Pass of Tears, a sharp mountain ridge covered in pebbles and loose sand, Criollos climb with surprising confidence and ease. The horses are experts in balancing on mountain tops and love to decide for themselves what the best way is to climb up and climb (read: slide) down. The advice is just to sit back and let the horse do its thing.

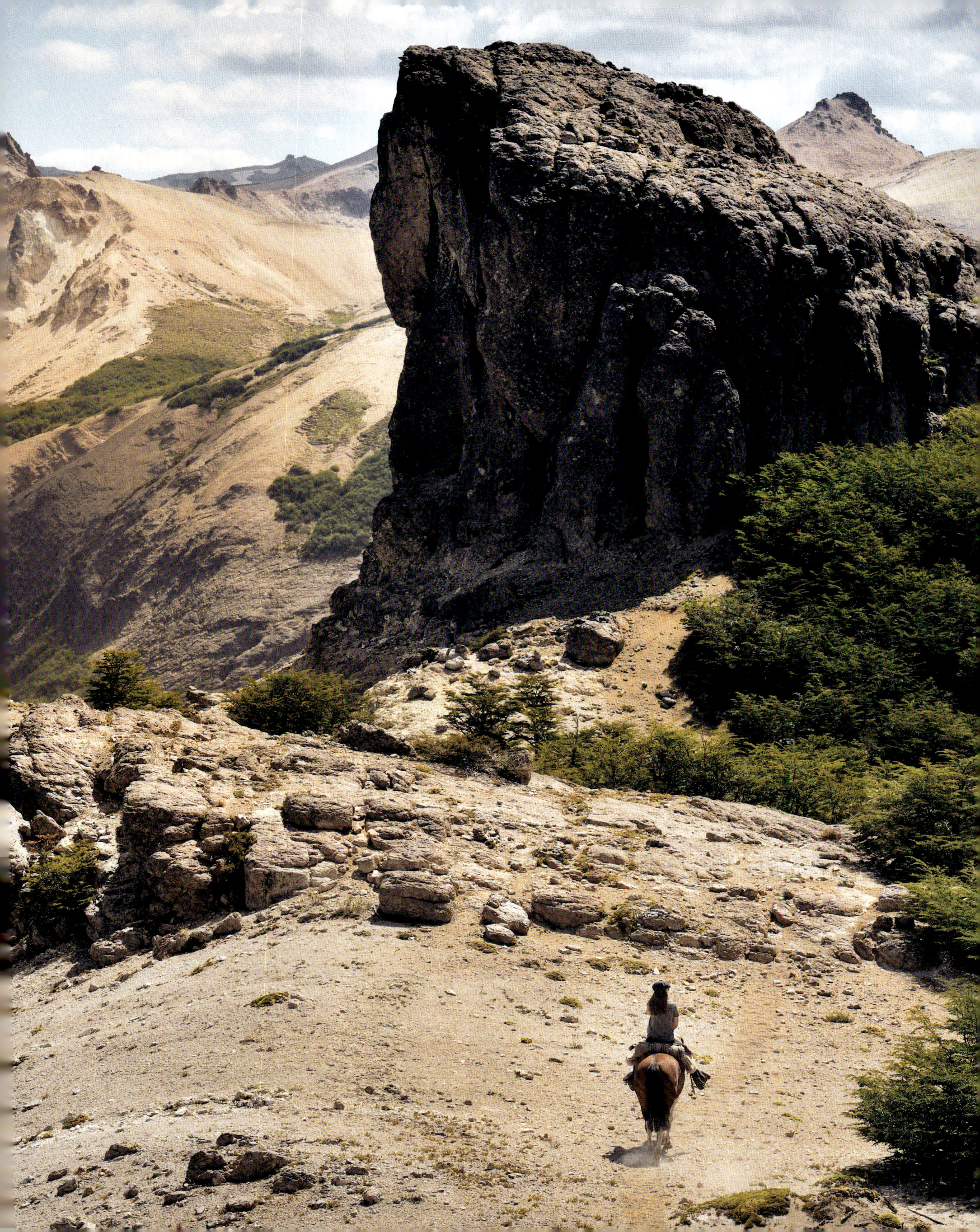

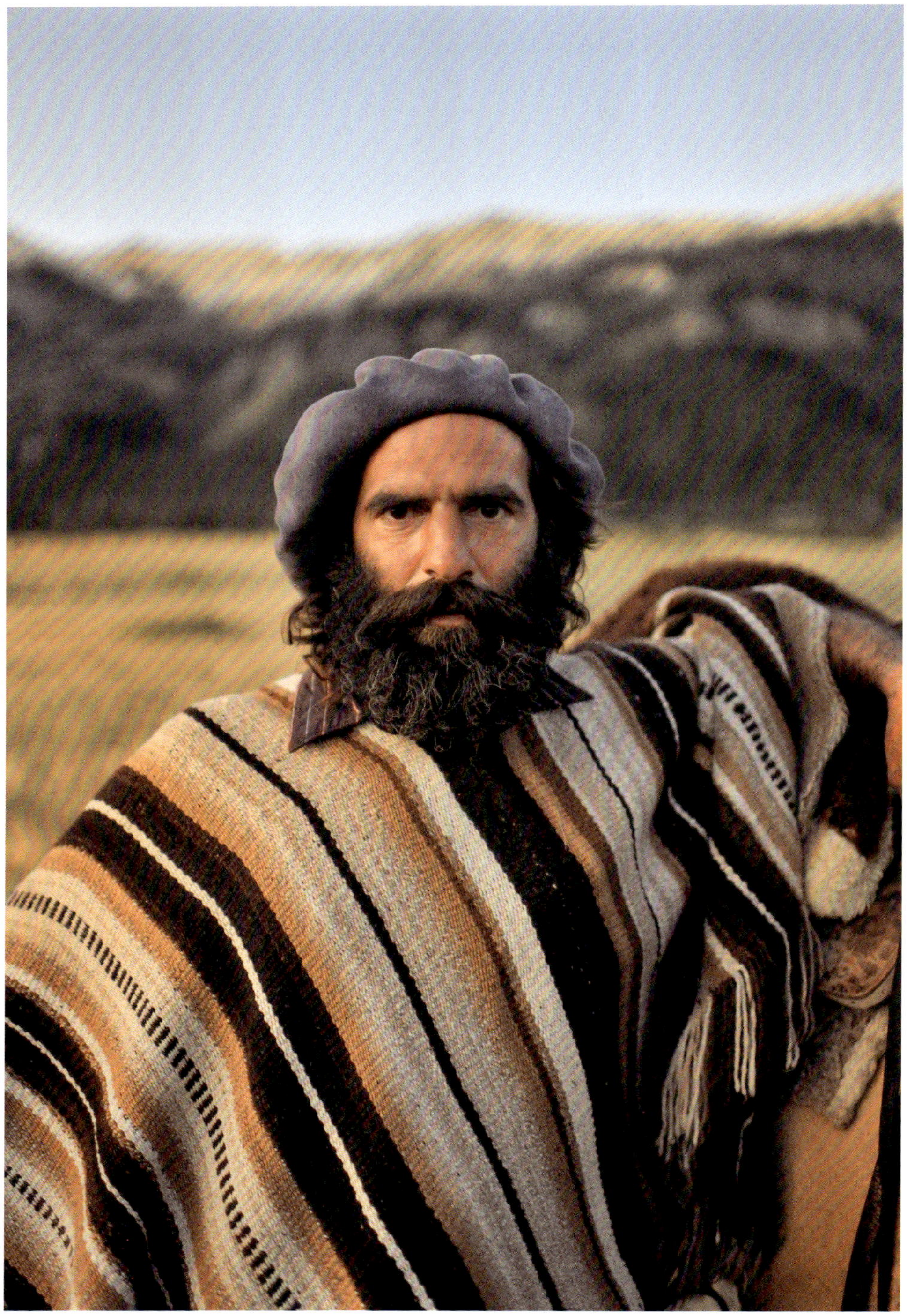

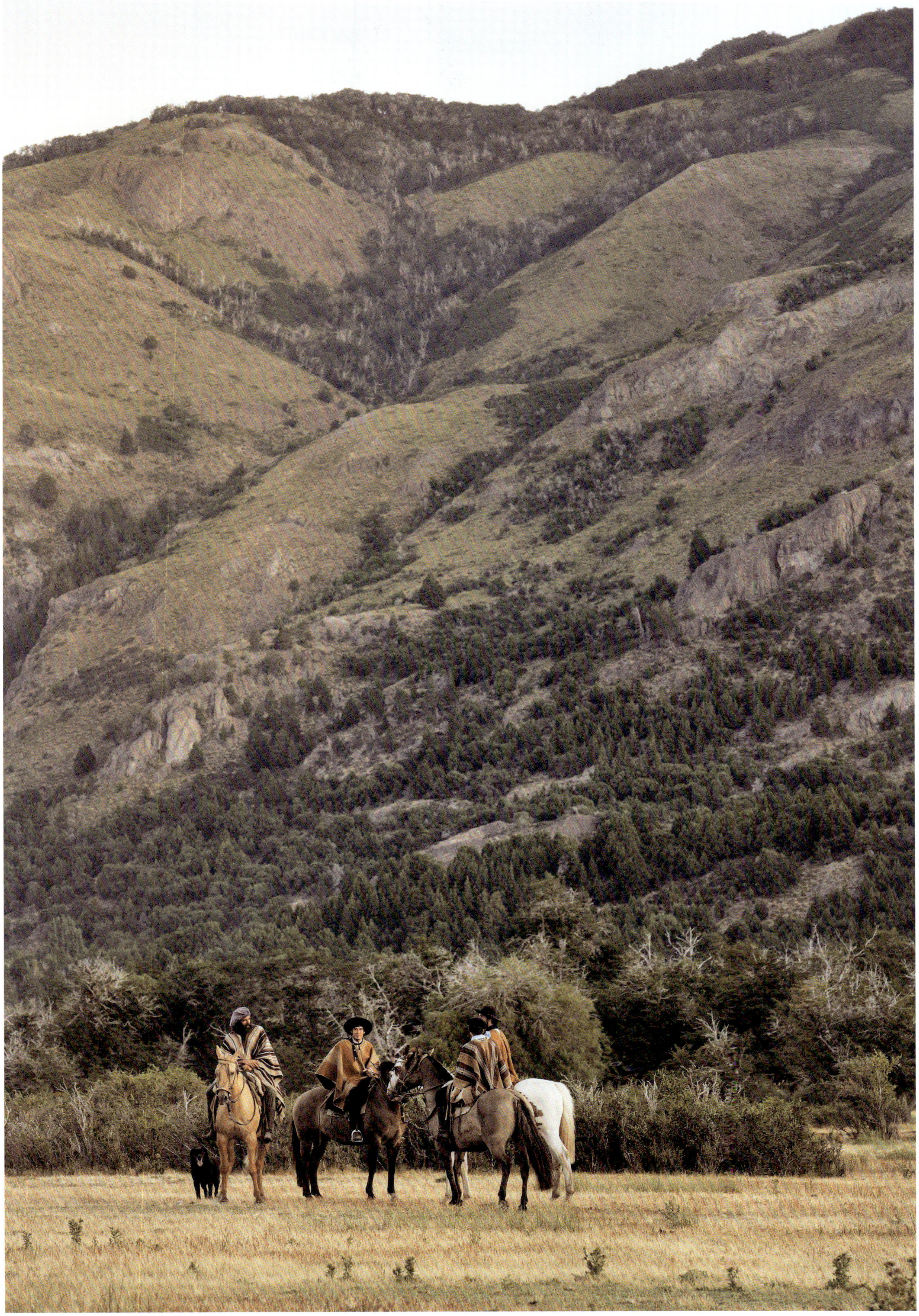

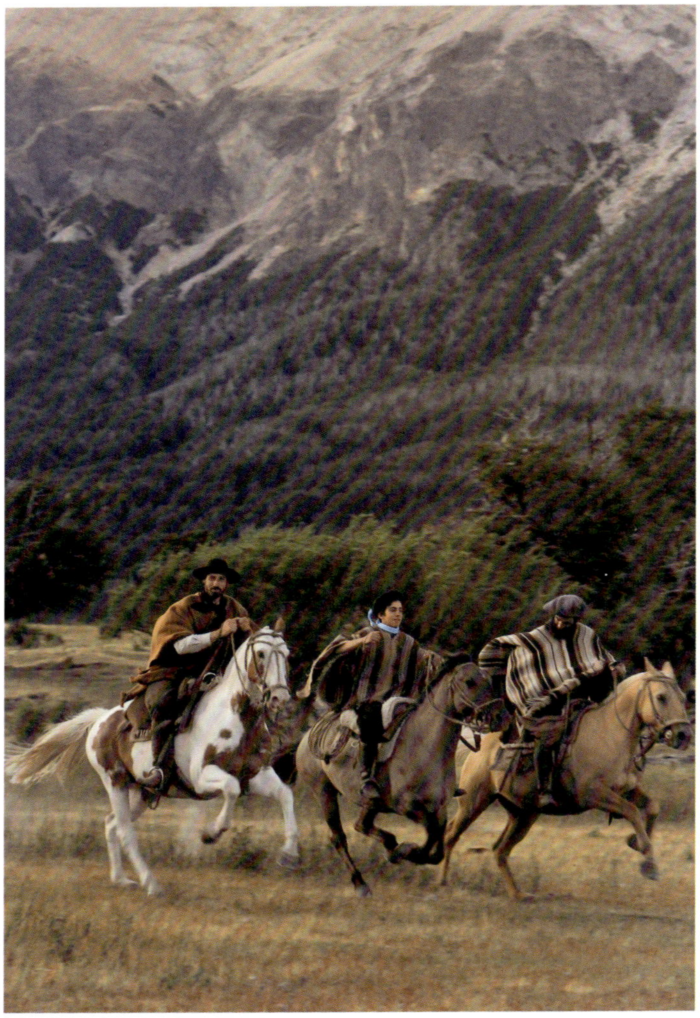

This spread Horses are indispensable for many aspects of traditional life in Patagonia, such as shepherding livestock to green pastures in remote valleys with no roads. Early in his life, Jakob von Plessen learned the arts of herding, branding, and lassoing from local gauchos. Perhaps most importantly, they taught him how to break a Criollo, a semi-wild horse breed that often doesn't encounter humans until a certain age.

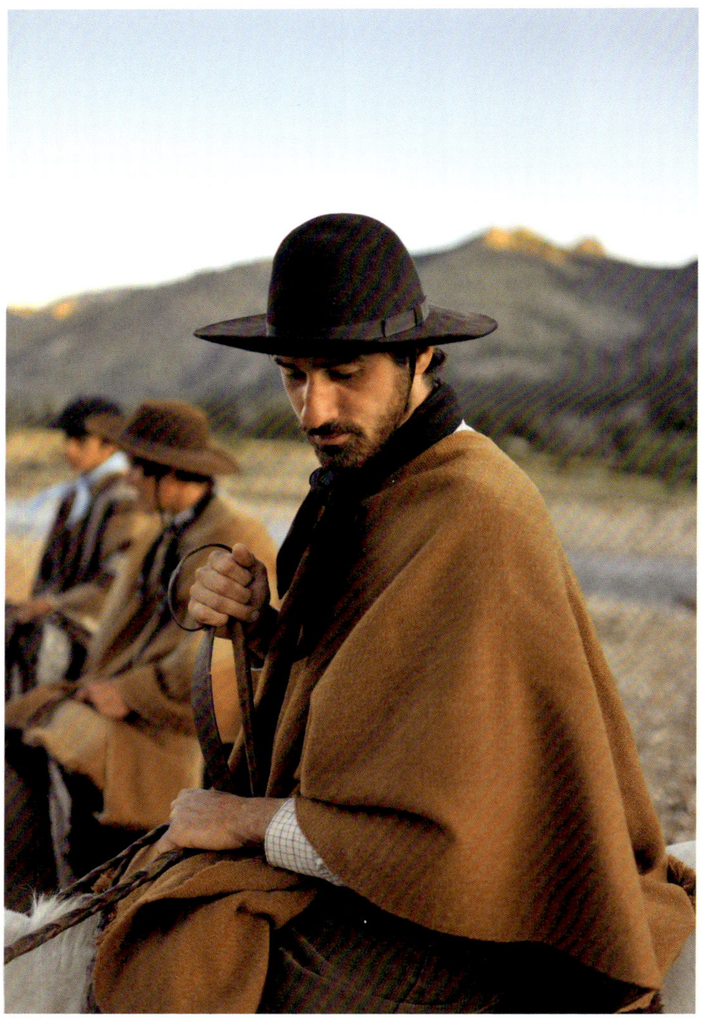

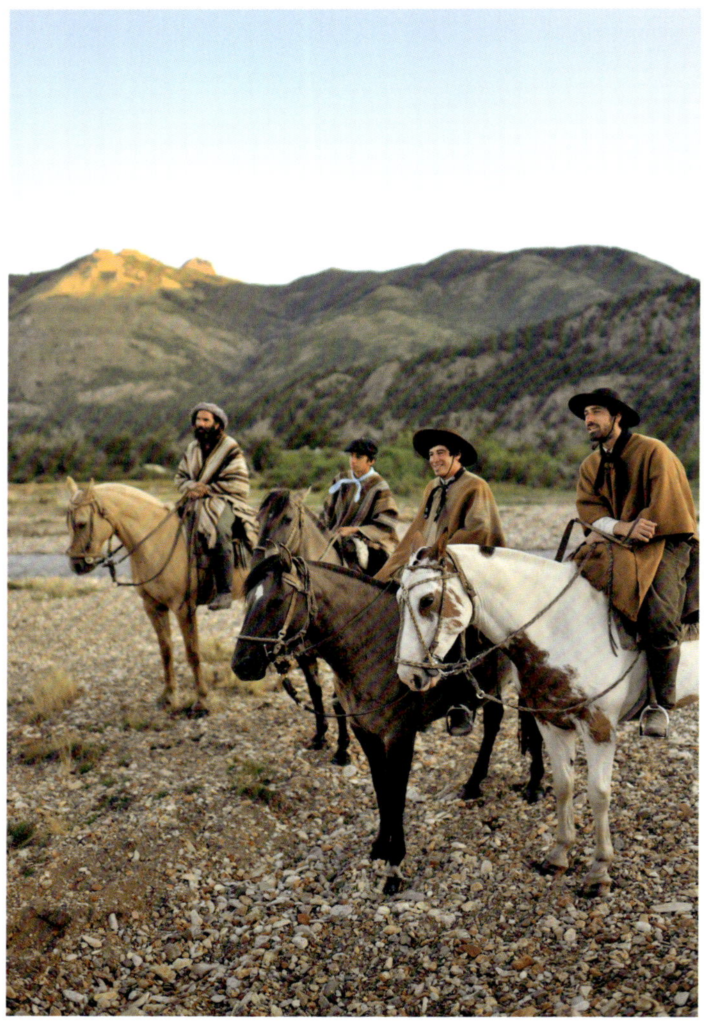

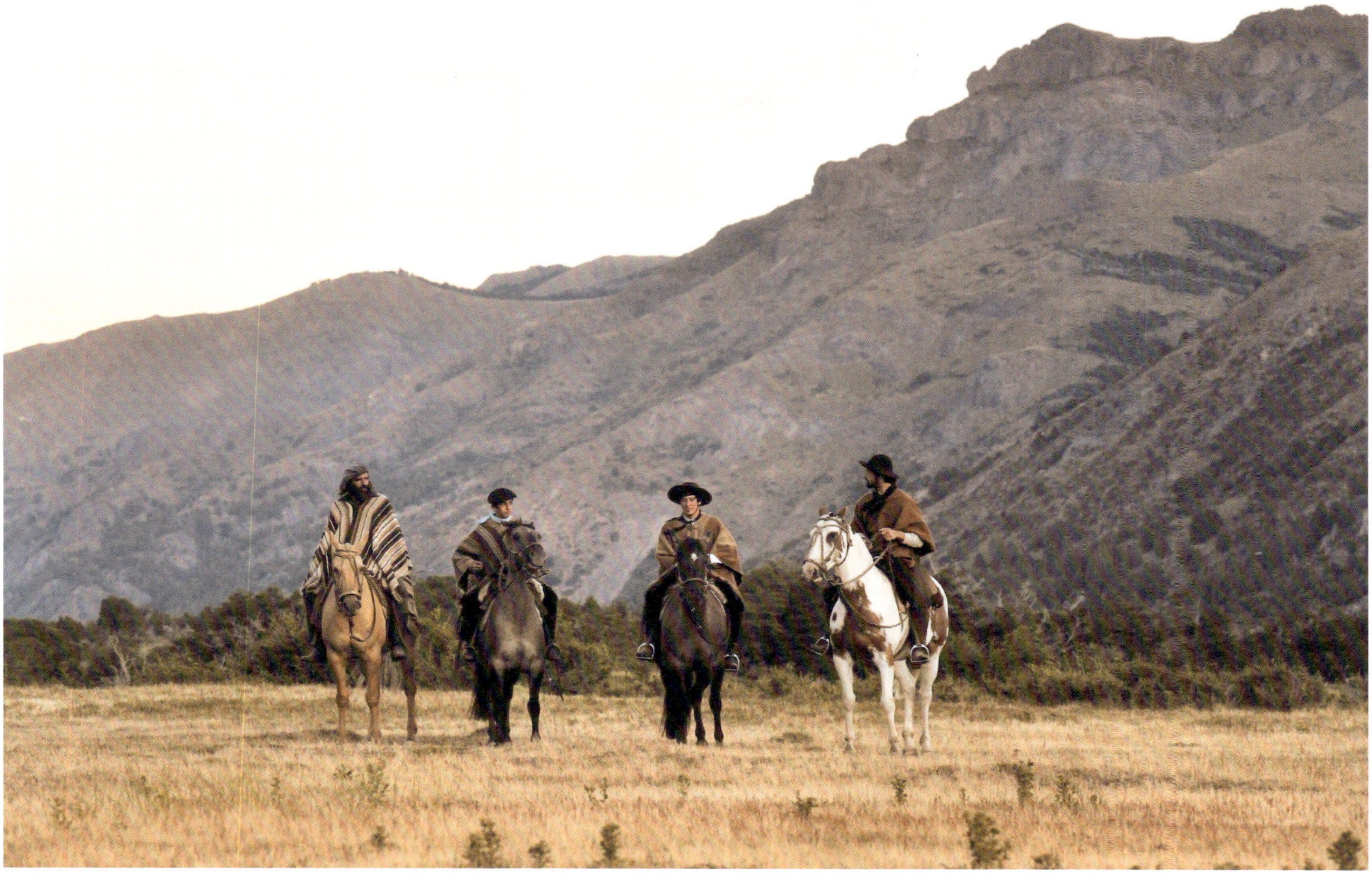

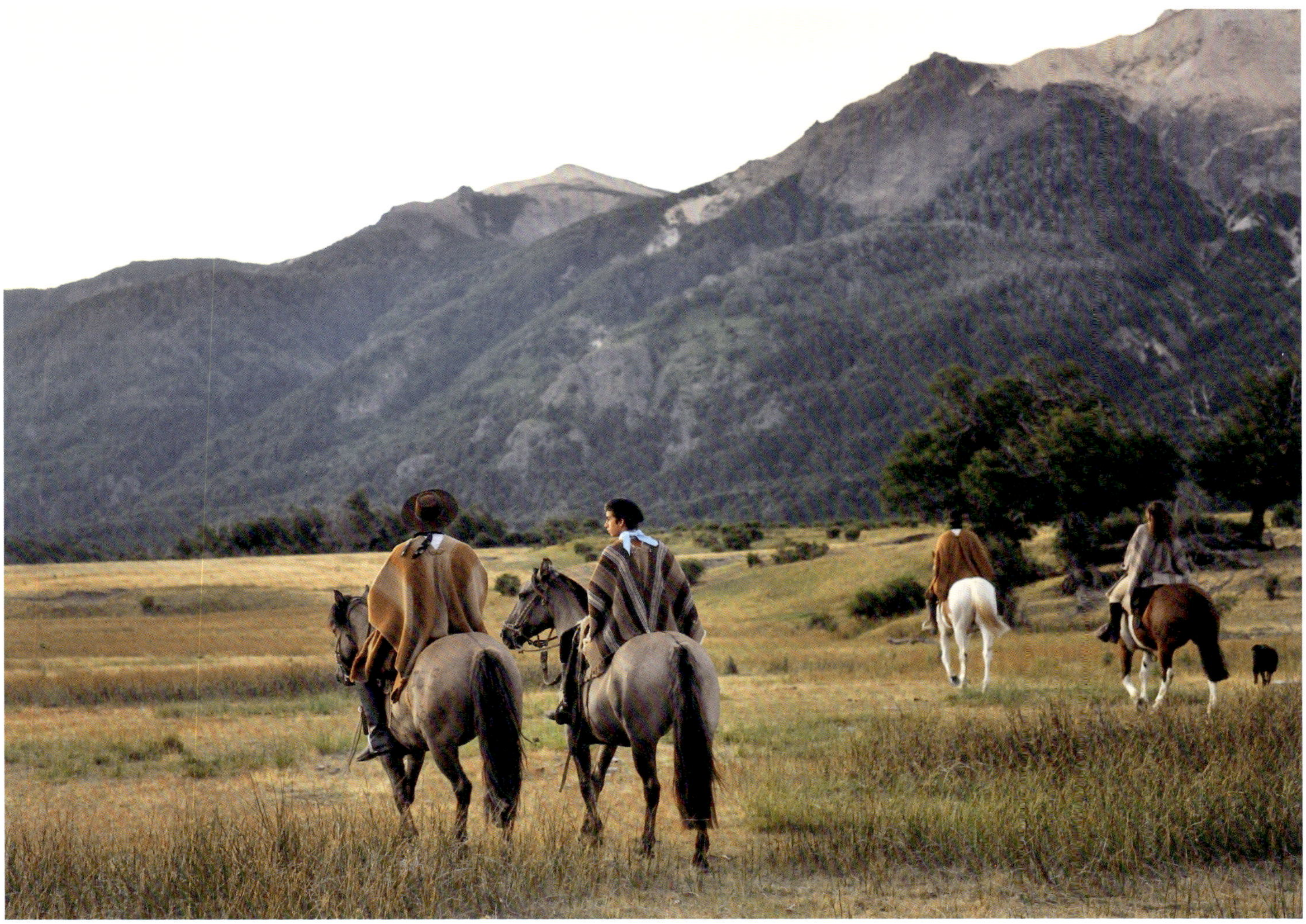

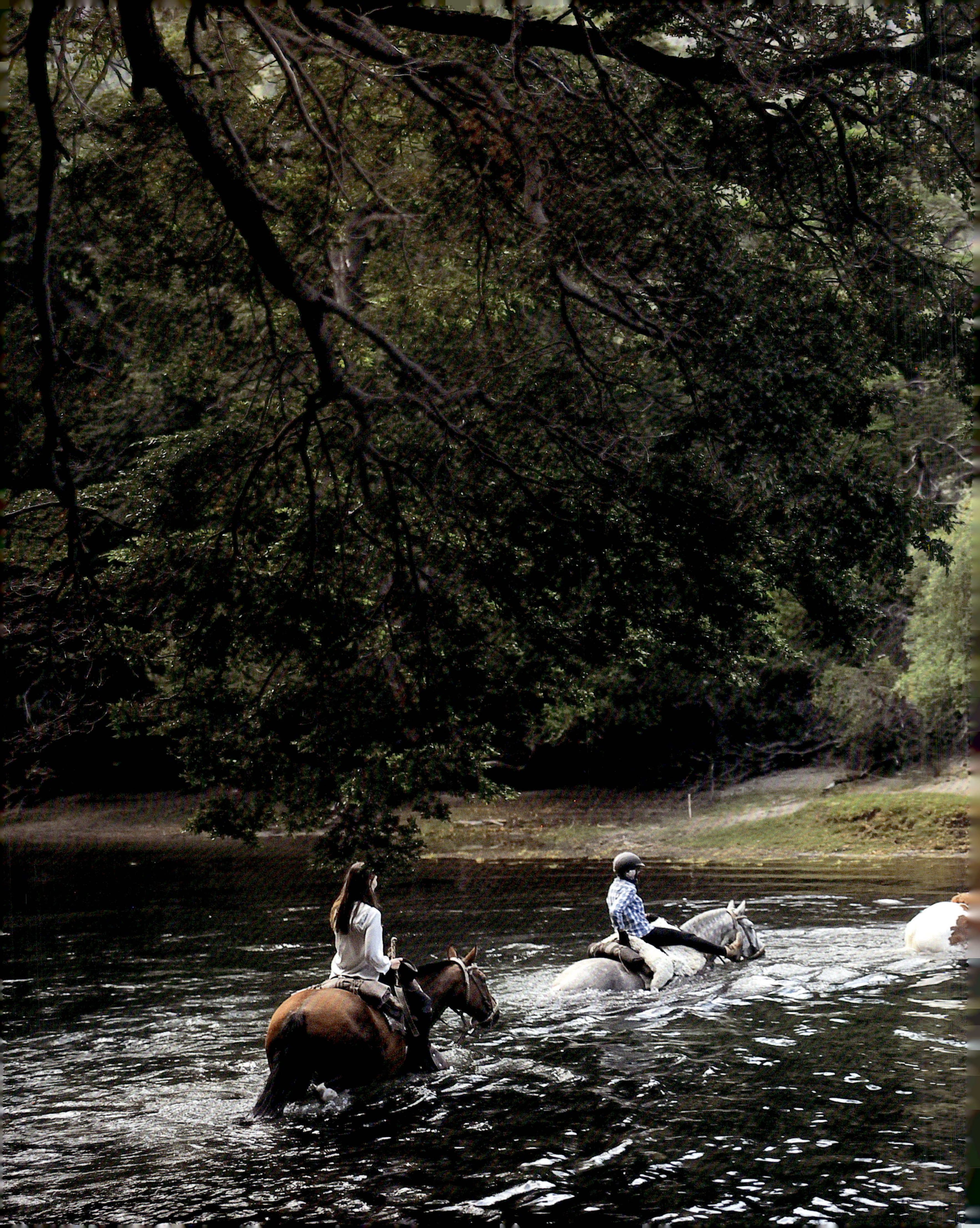

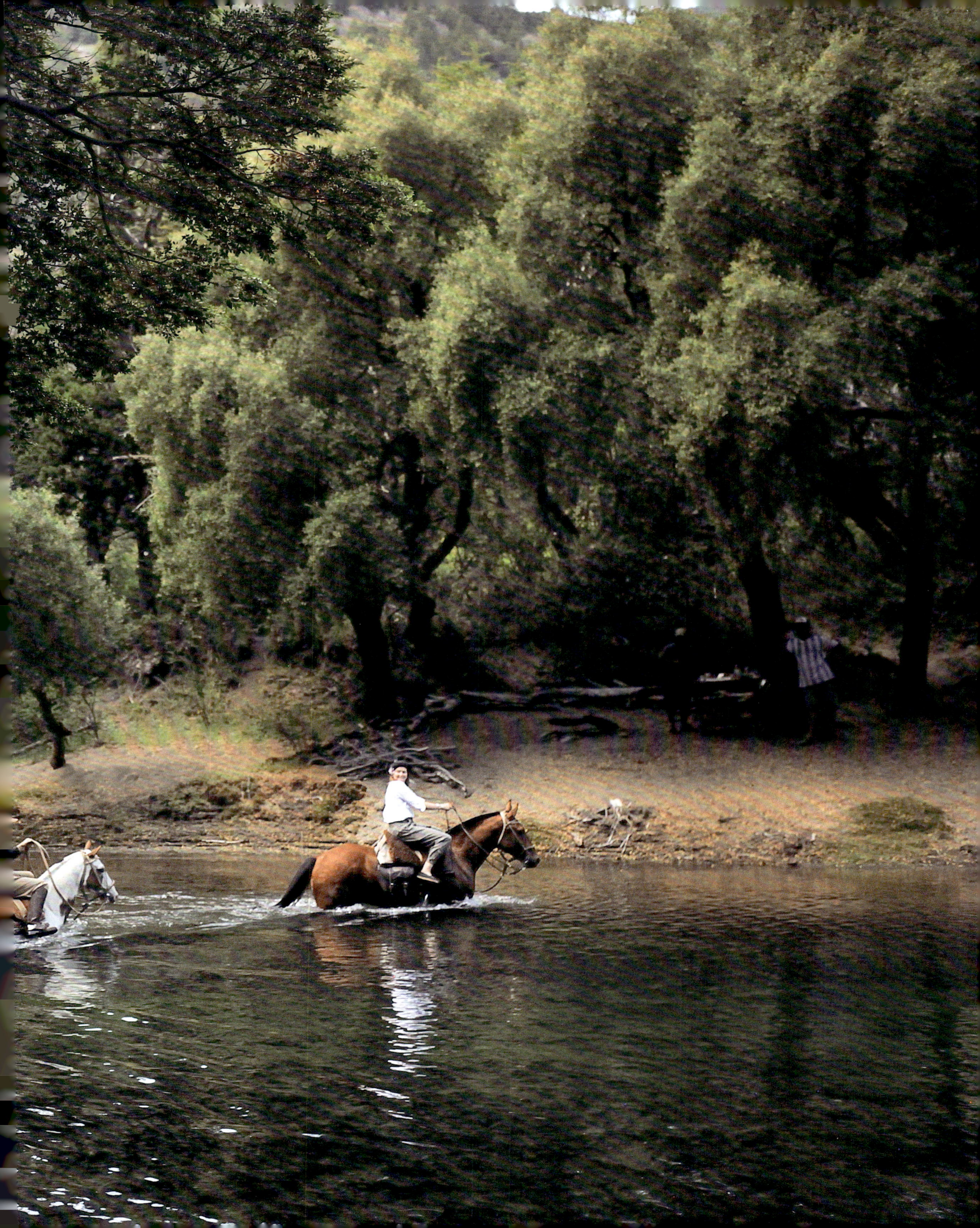

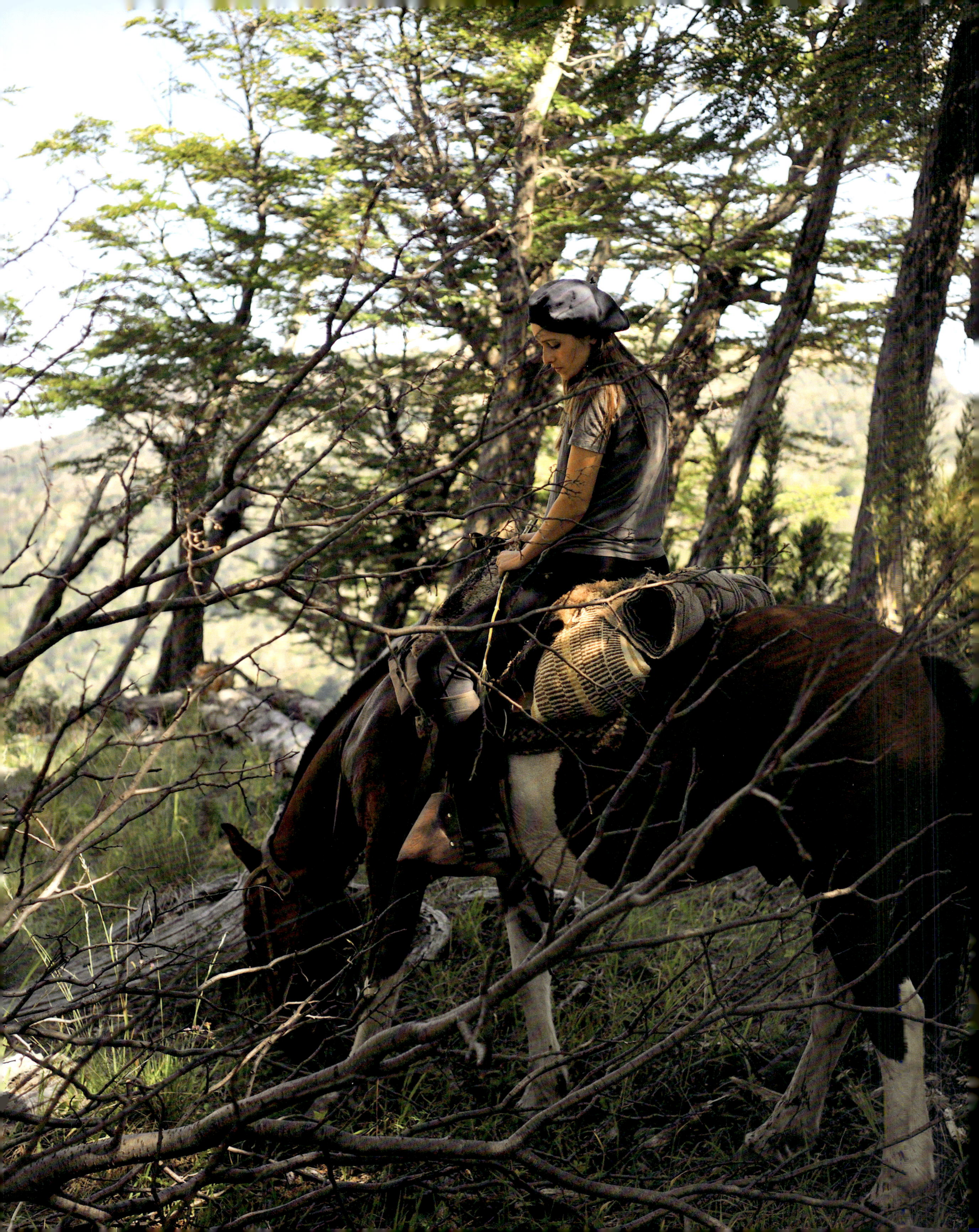

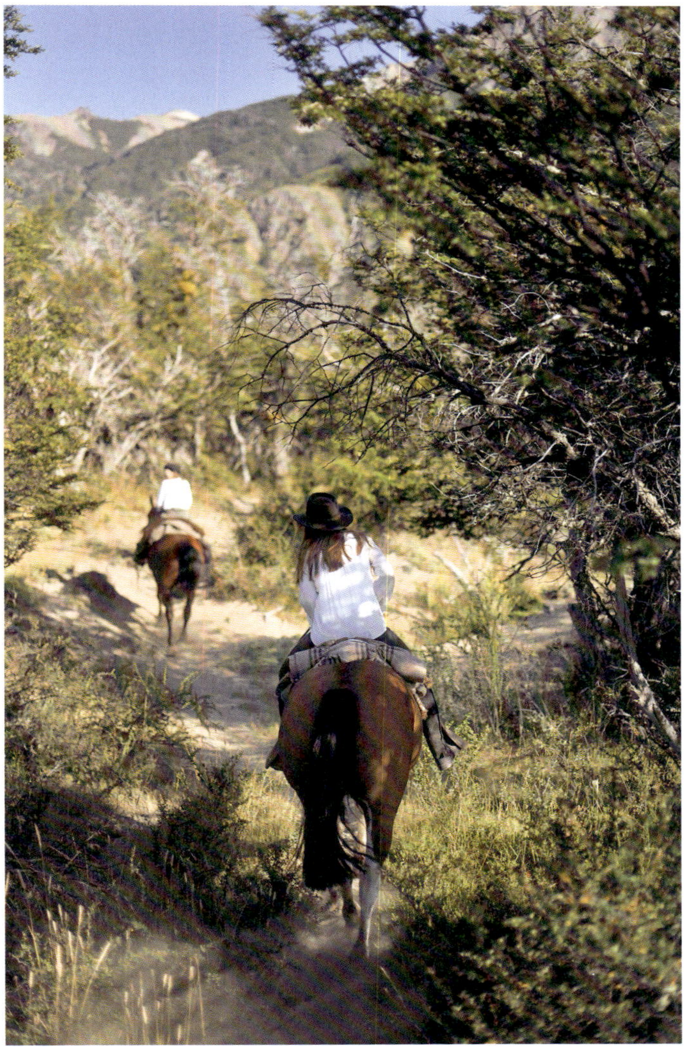 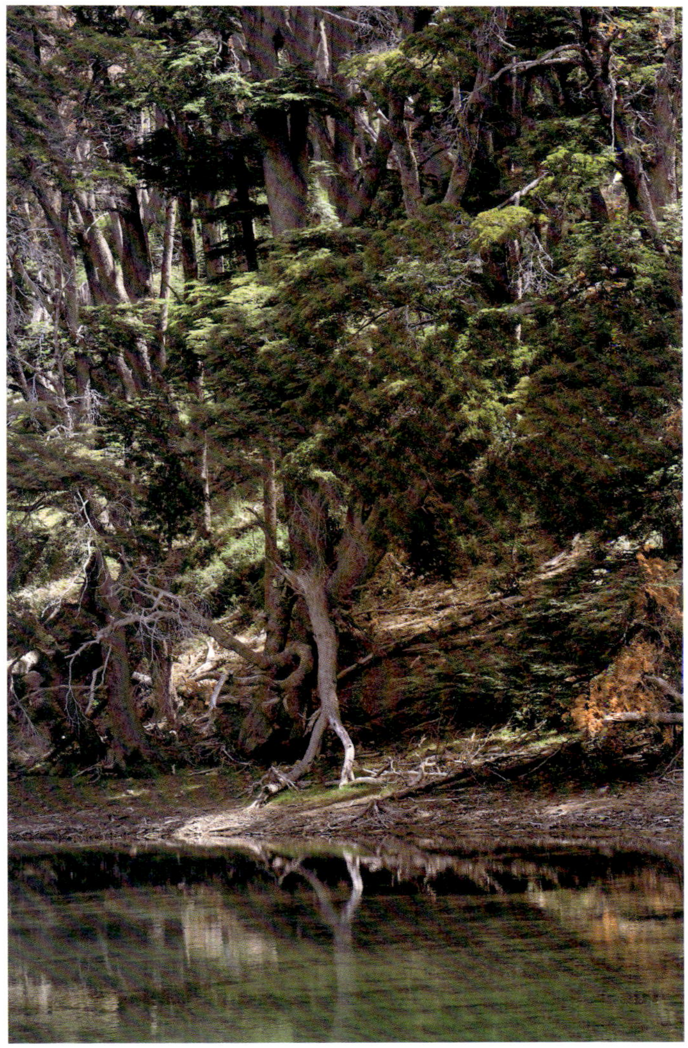

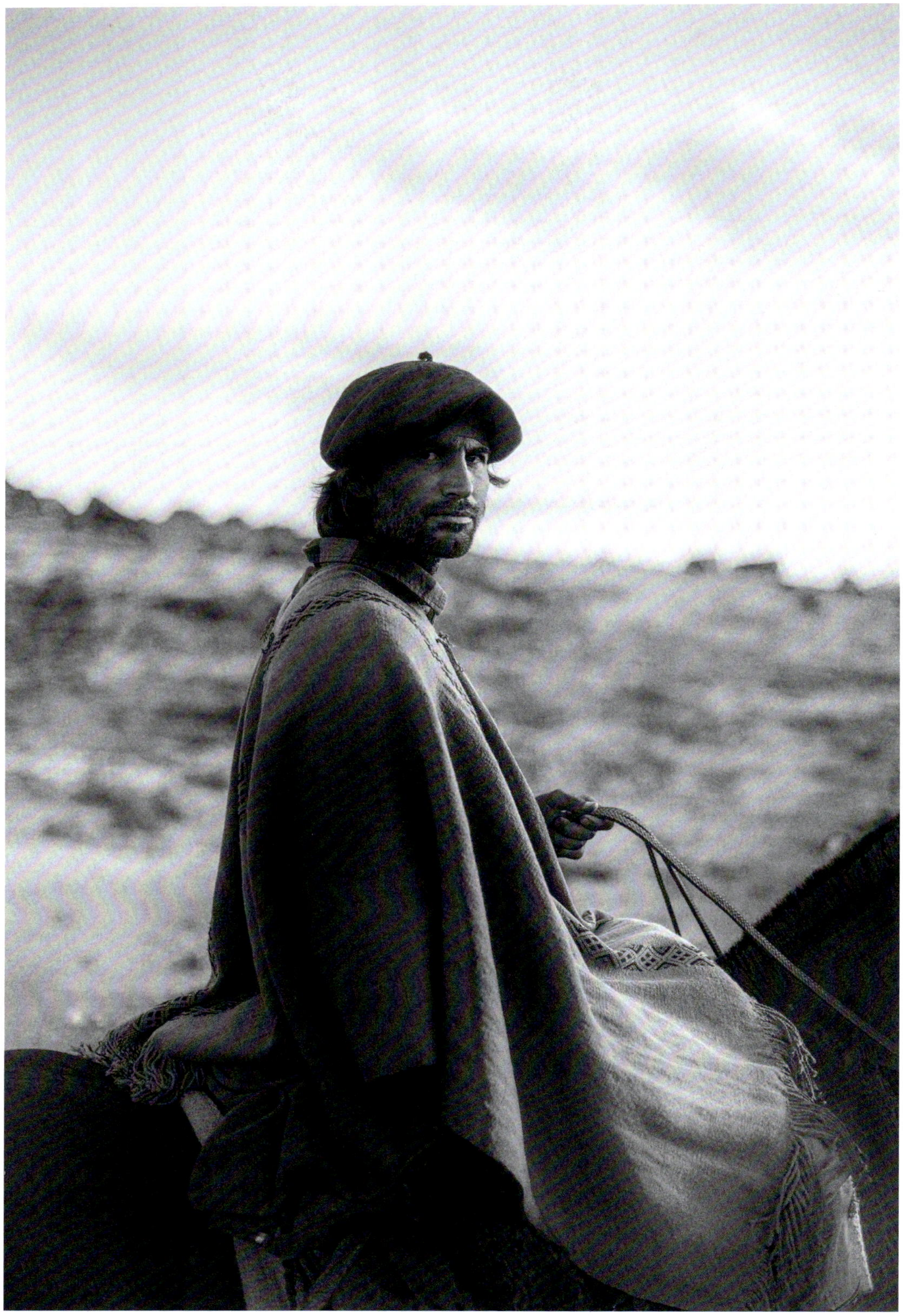

Above Jakob von Plessen still feels like a pioneer, even in modern times. The joy of riding in Patagonia is that everywhere little gems await discovery, from crystalline streams to cave paintings and fossils. Von Plessen got his outdoor skills while growing up, spending time with gauchos and learning the art of managing cattle, herding and using the lasso and, last but not least, training and learning to understand wild horses. After a gap year and 12 years of long horse-riding and guiding adventures in Kenya he returned home to set up Jakotango in one of the most remote parts of Argentine Patagonia. Von Plessen is one of the three founders of the Gaucho Derby, a 500-kilometre (310 mi.) race featuring dozens of riders and over 200 horses through high-altitude wilderness. His fellow founders are Rama Gregorio Arcal, a local estancia owner, and Tom Morgan, a British adventurer who organises adventures and races around the world.

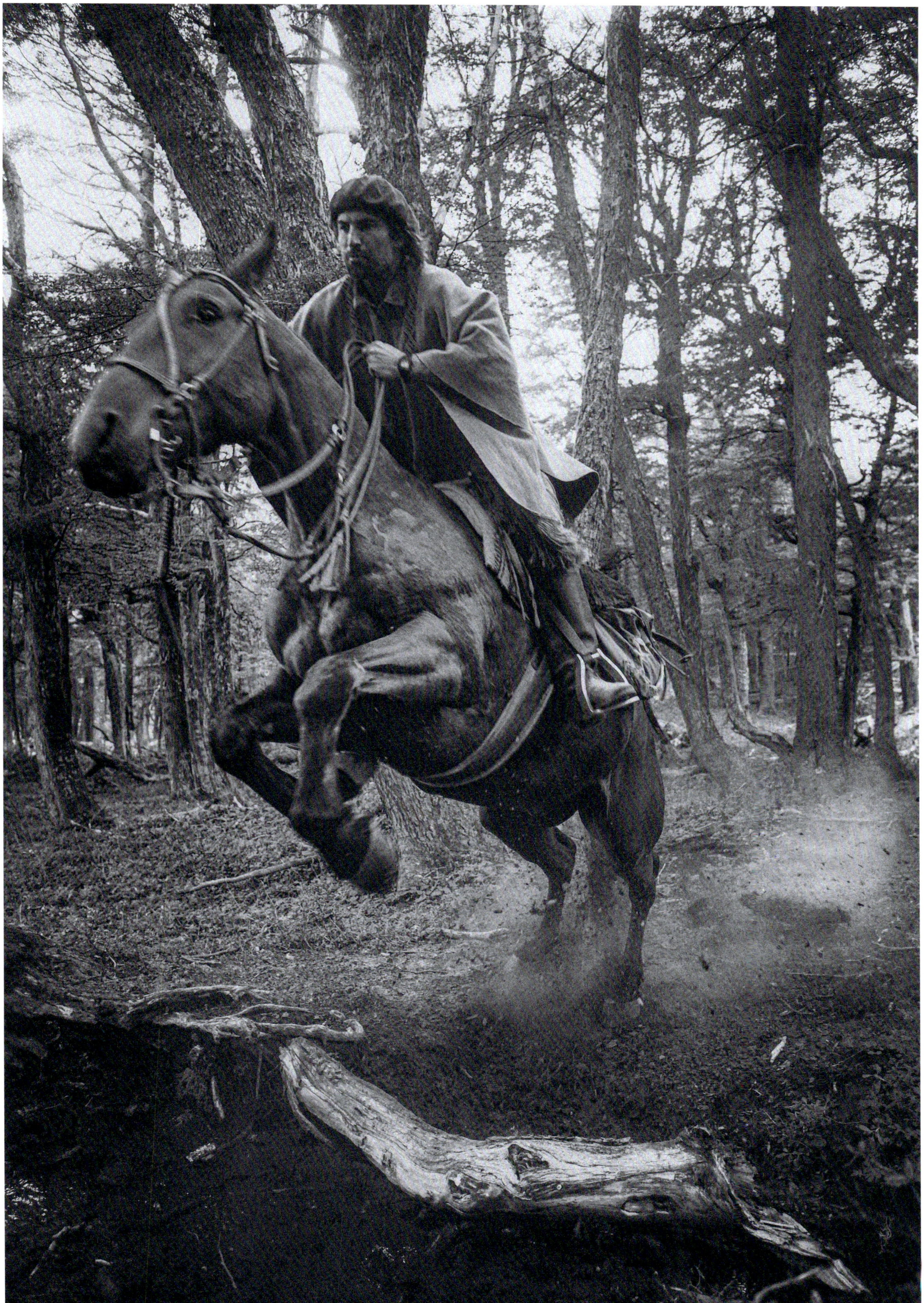

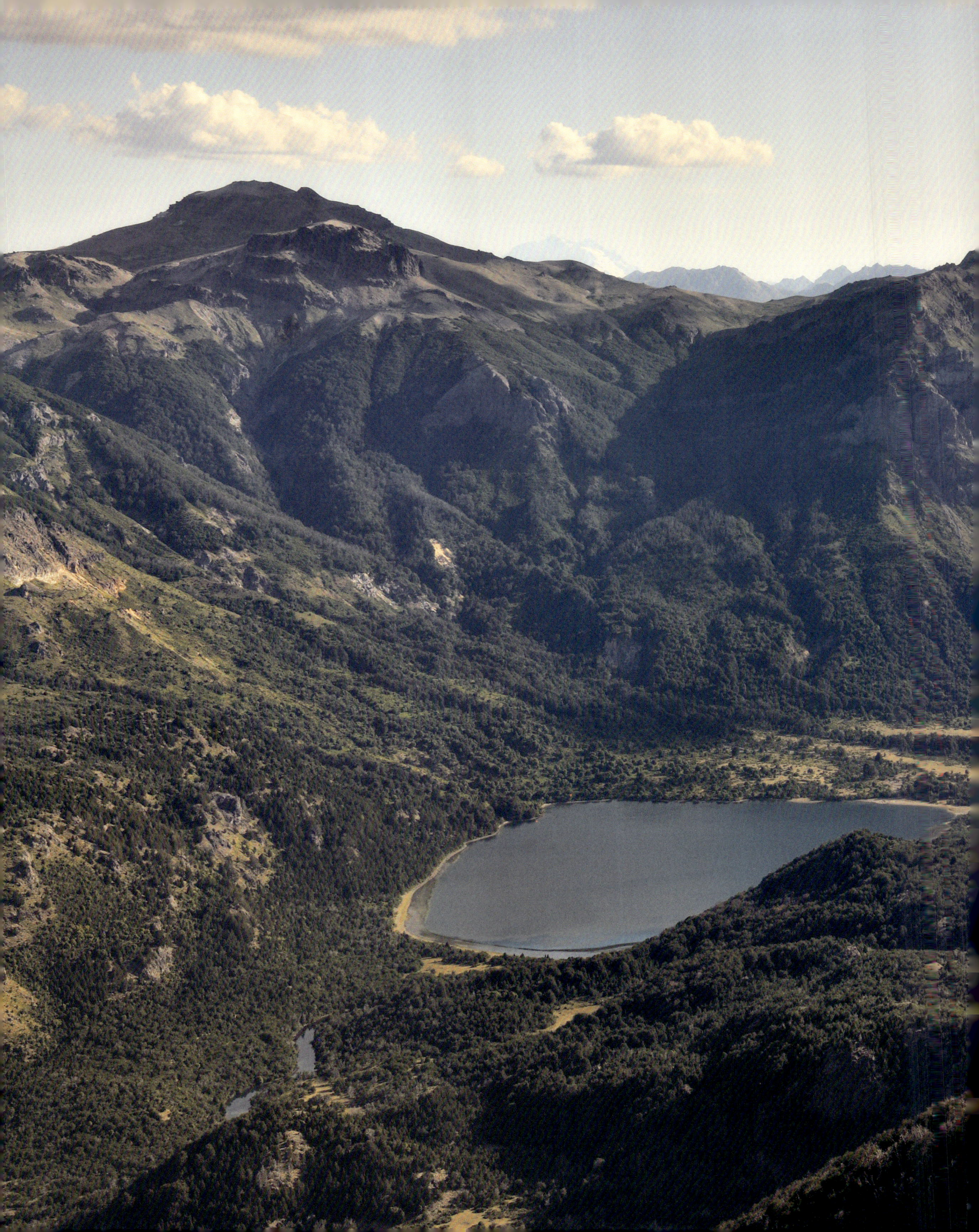

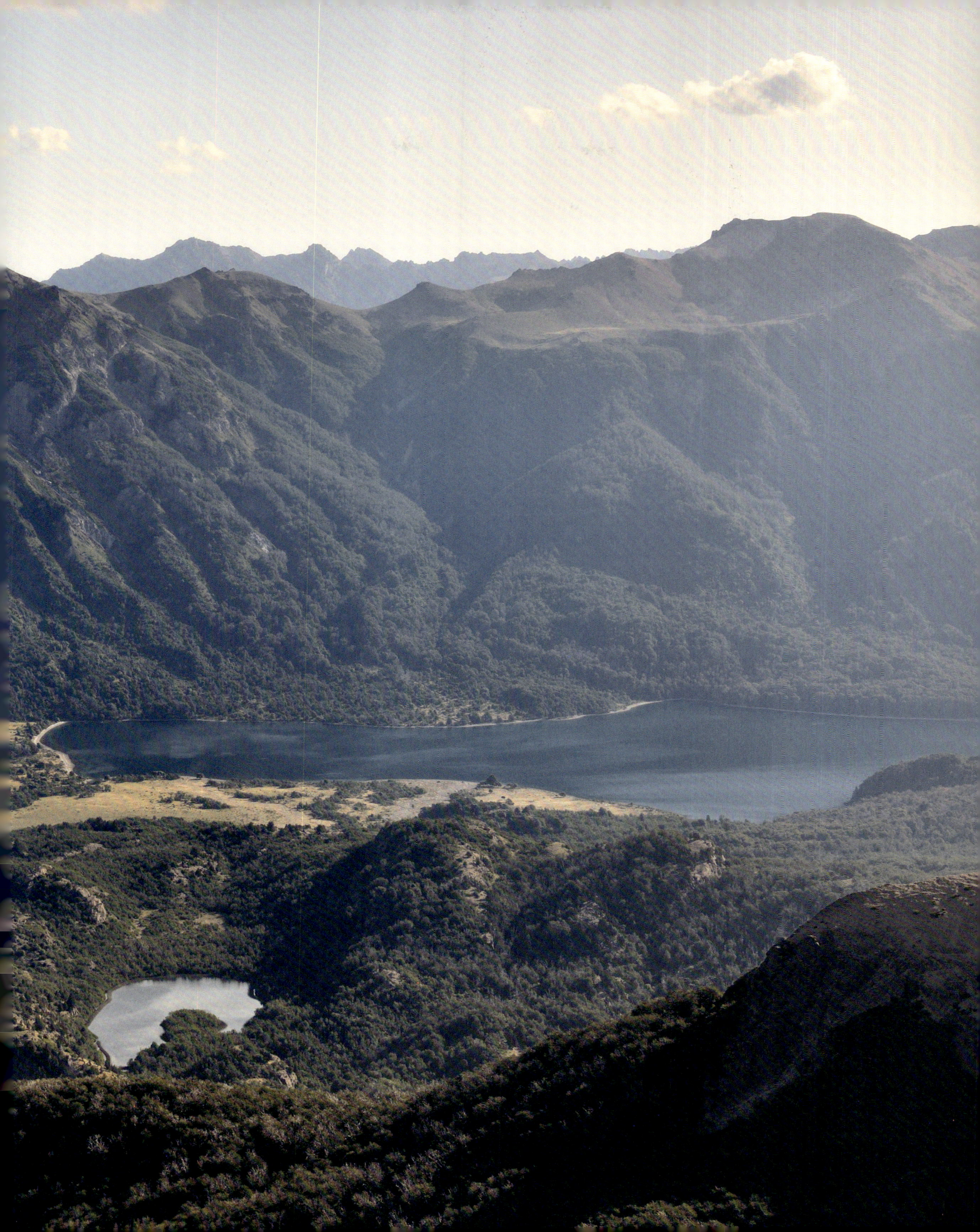

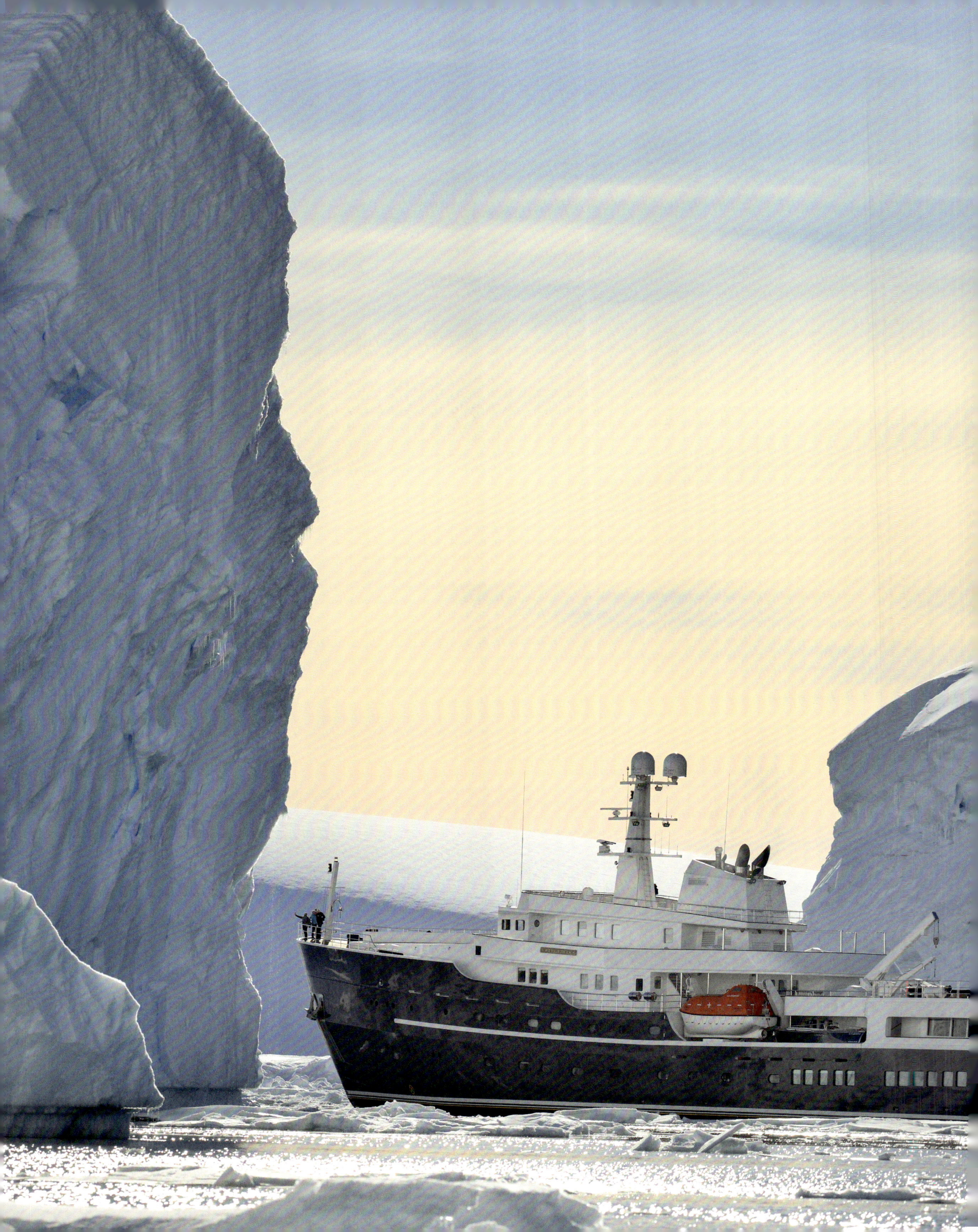

CHAPTER 12

A PRIVATE ESCAPE TO ANTARCTICA

FRONT-ROW-SEAT ANTARCTICA

Antarctica

62°11'31.94" S
58°58'56.13" W

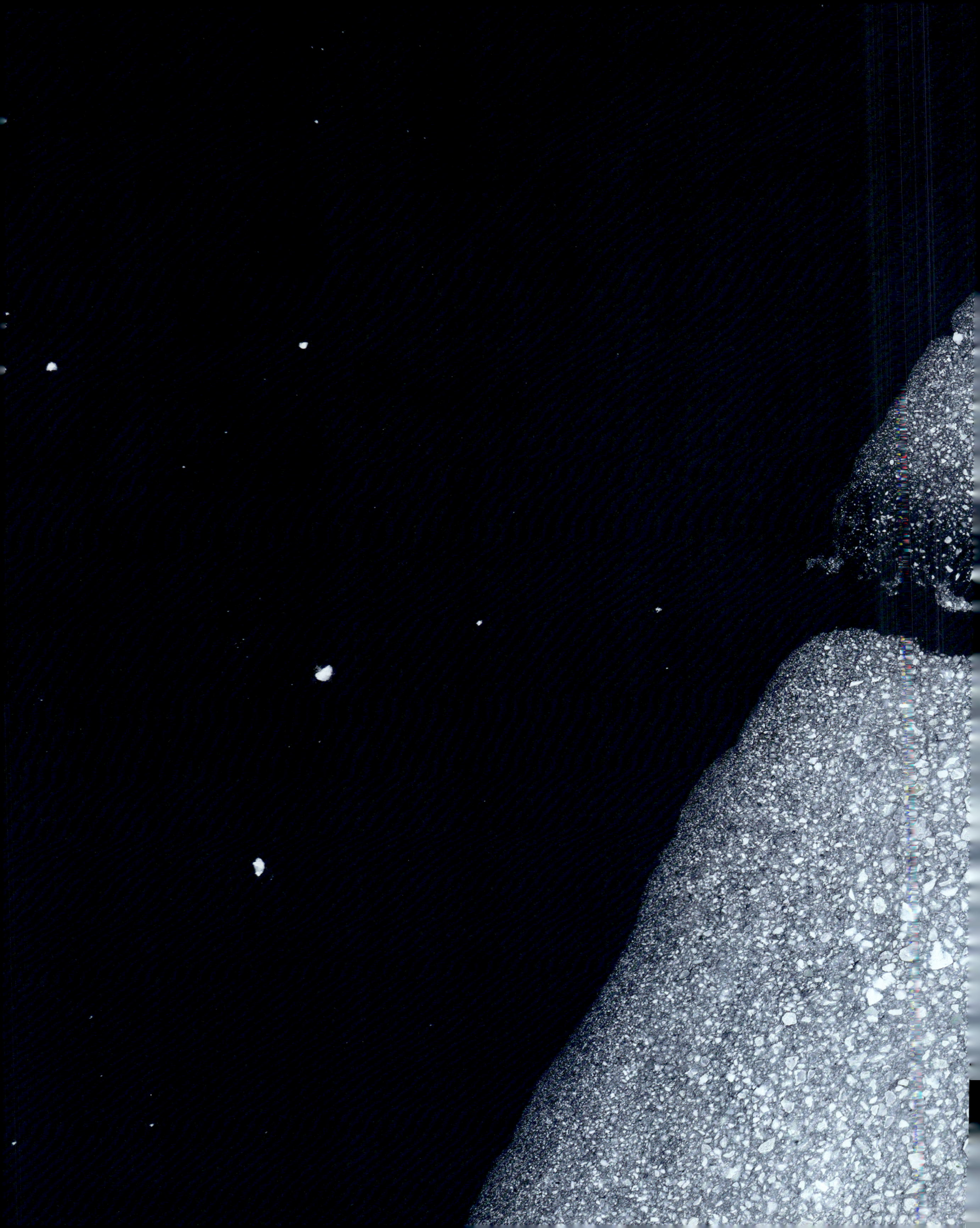

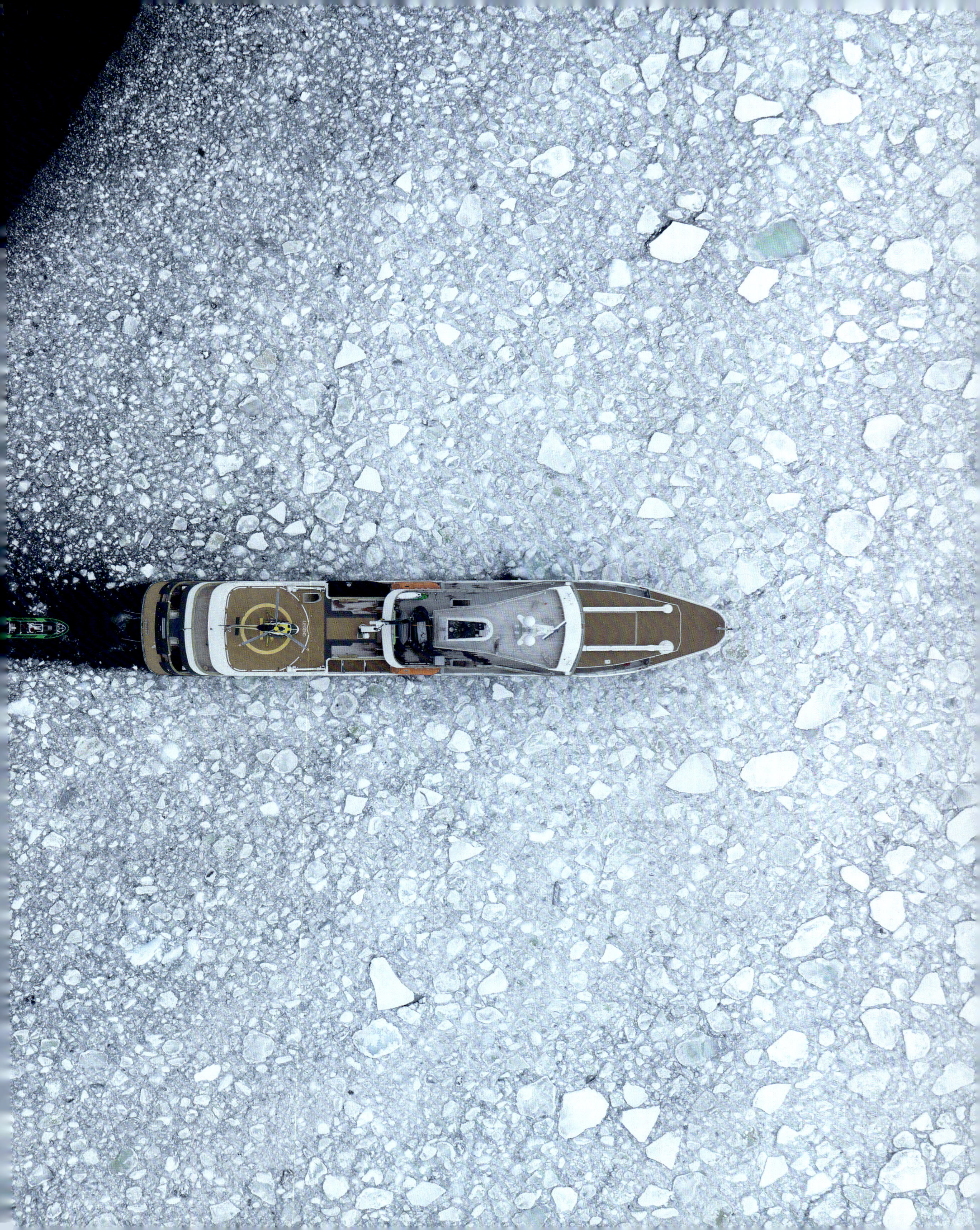

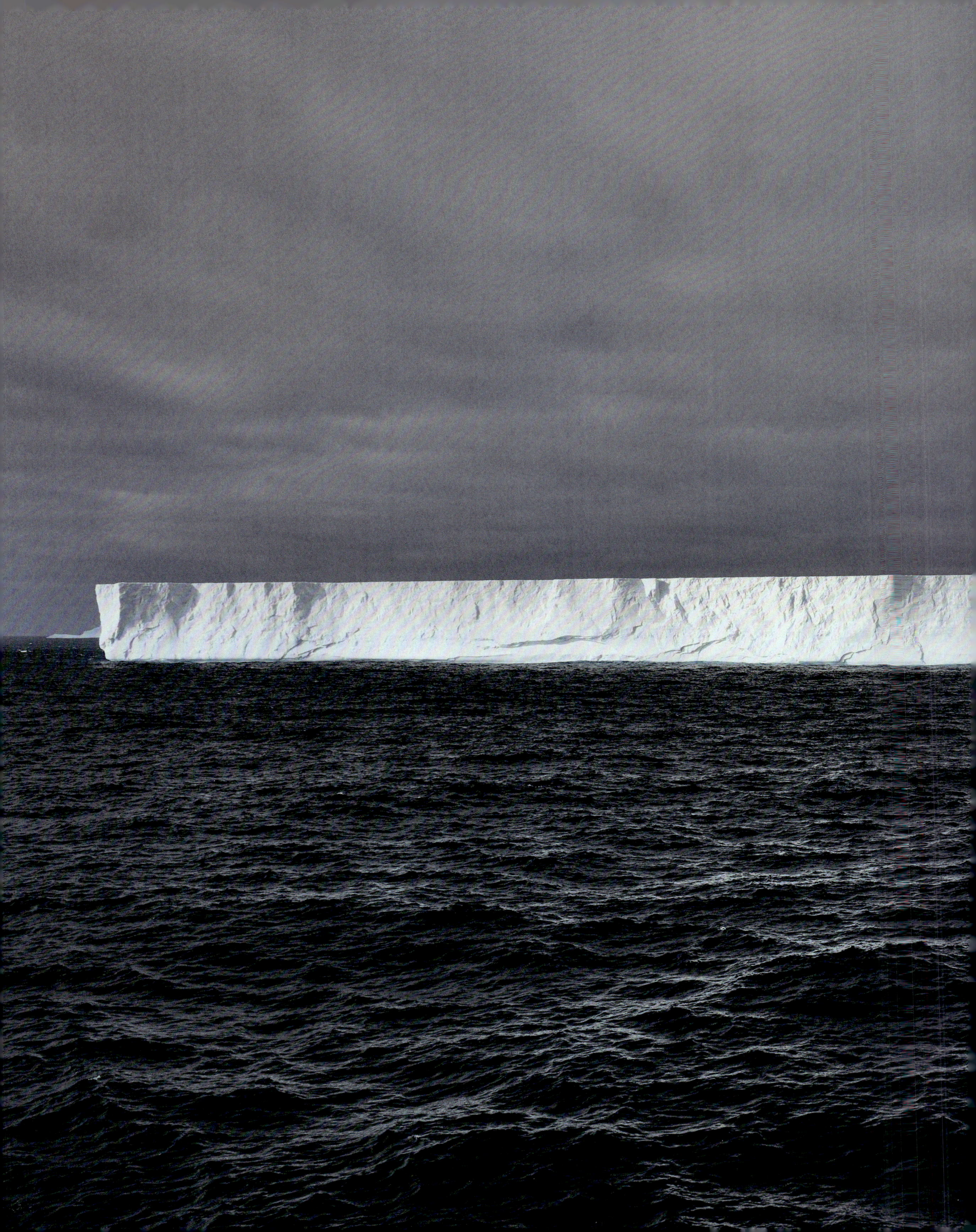

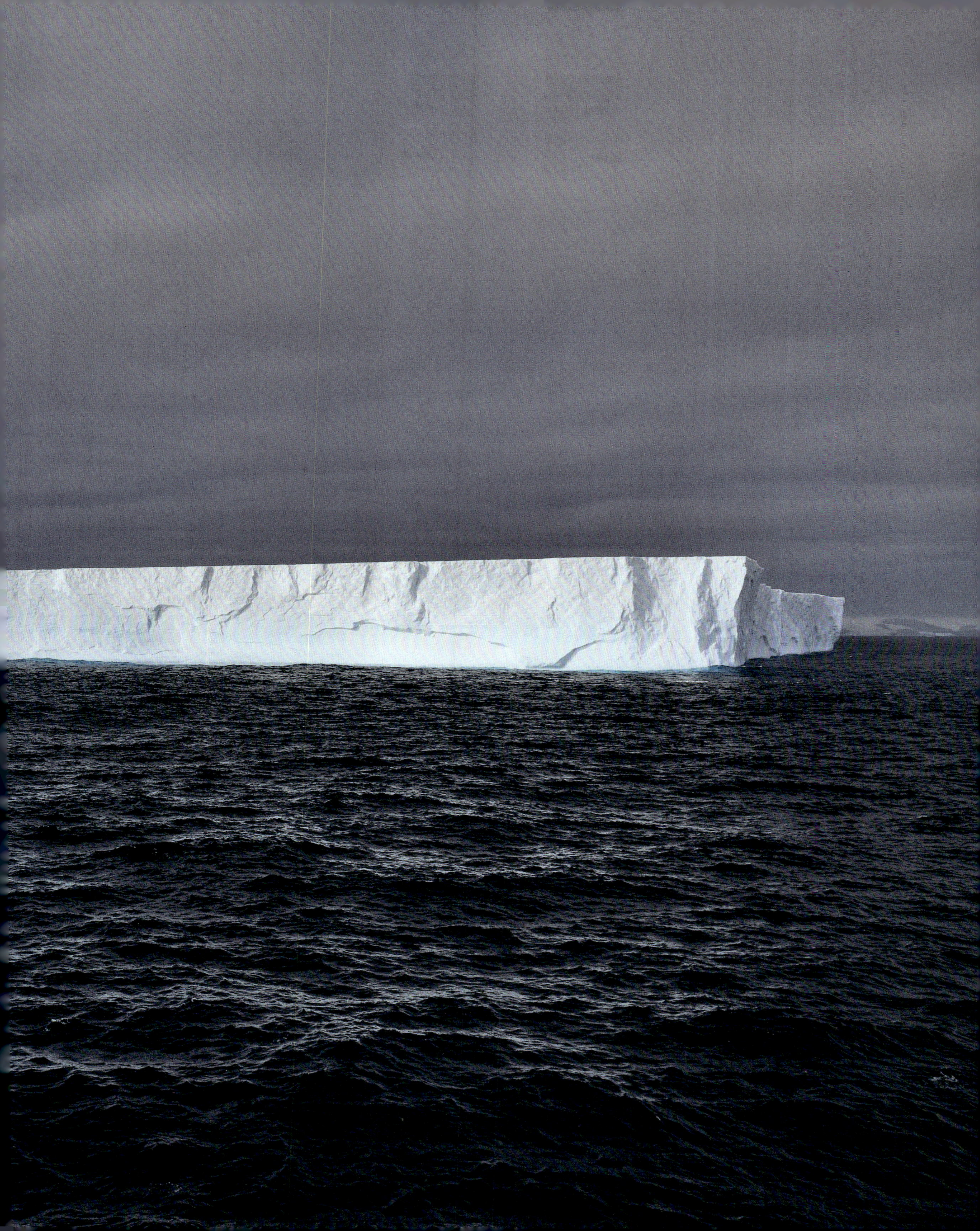

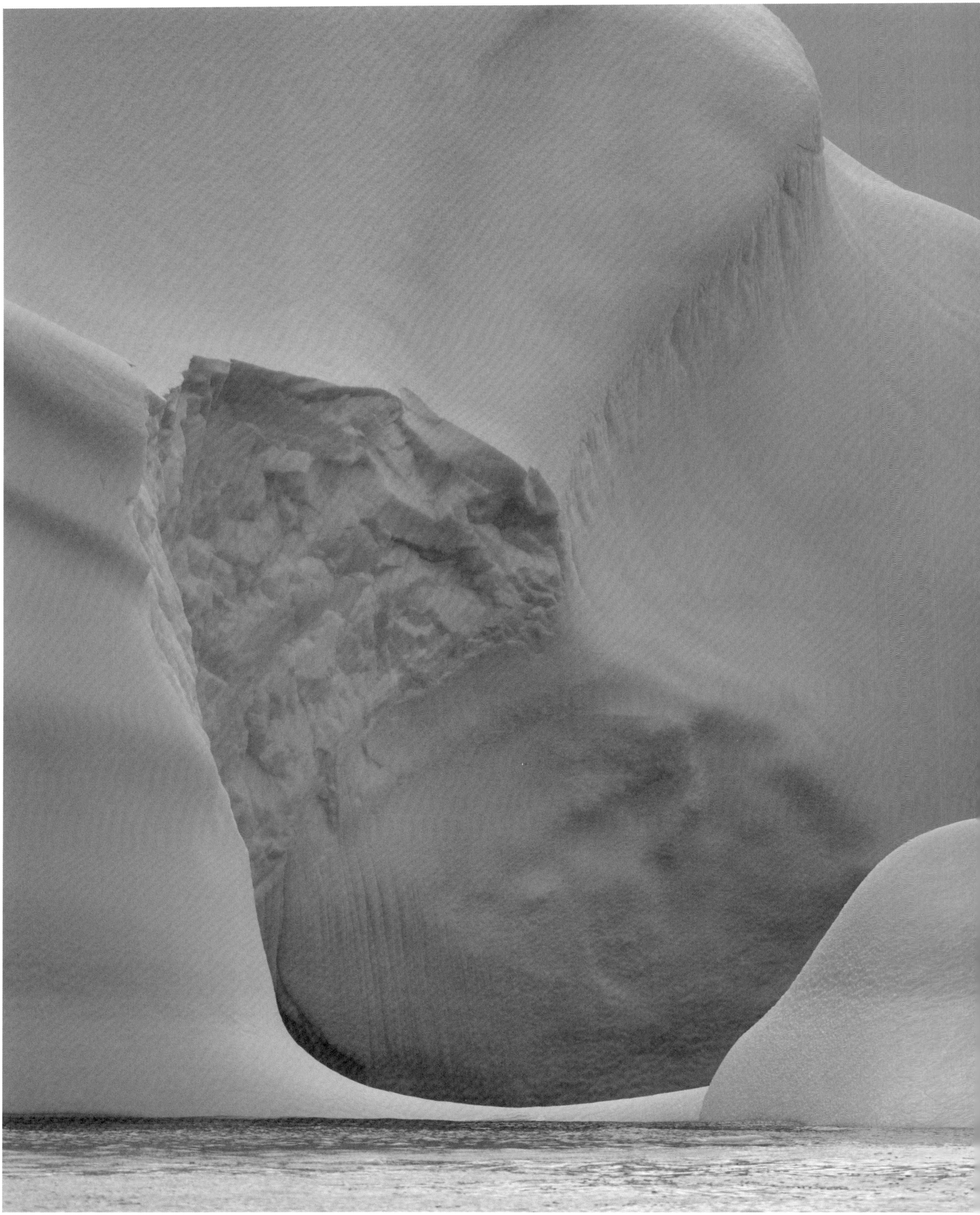

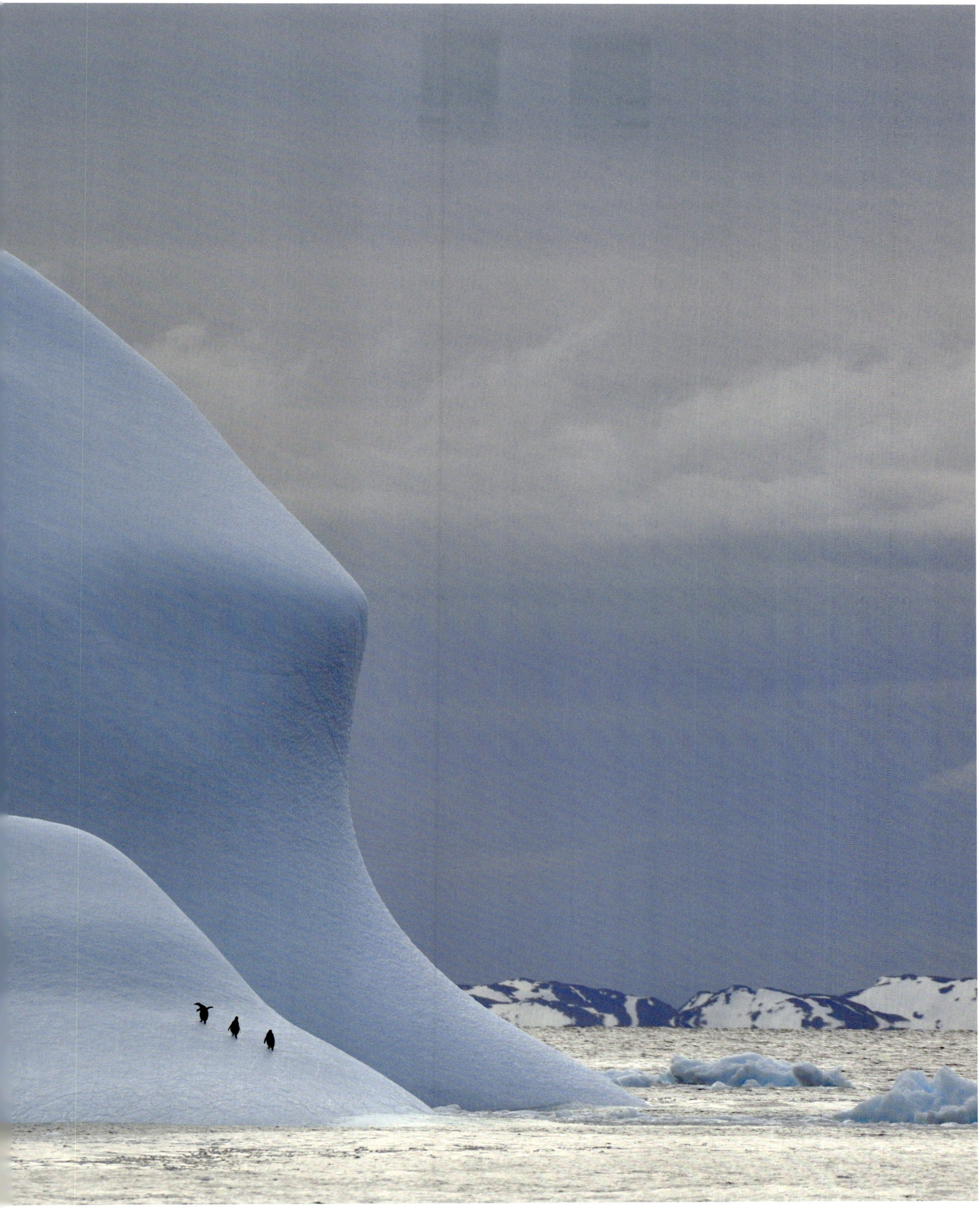

A Private Escape to Antarctica

Front-row-seat Antarctica

Region:	**Antarctica**
Best time to travel:	**November to March**
Closest airport:	**Punta Arenas, Chile**
Price tag:	**€€€€€**
Style of travel:	**Front-row-seat Antarctica**

Introduction

Legend, a unique former icebreaker converted into an elegant luxury yacht, takes small groups of guests to one of the most inaccessible places in the world, a place where, during the austral summer, the sun never sets and the sky is streaked pink. A place which is also the driest, highest, windiest and coldest of all the continents. In the distance, a pod of killer whales frolics in the dark, icy waters dotted with tiny fragments of ice, and on the horizon looms Mount Français, one of the highest peaks on the Antarctic Peninsula. *Legend* is sleek, dark, and elegant, a classic 77-metre- (253 ft.) long superyacht with teak decks, a helipad, and supercharged with a Targa 27 superyacht tender, two fast Icelandic Rafnar RIB high-tech boats, sturdy Zodiacs, and a five-seater submarine (maximum depth 200 metres/650 feet). The yacht also has room for a helicopter. Legend comes into her own in these icy waters. She accommodates 22 guests, more than the average chartered superyacht, but far fewer passengers than the expedition vessels that typically carry 100, sometimes even 200 passengers or more from the southernmost tip of Argentina to the Antarctic Peninsula. *Legend* has another unique benefit: the two-day crossing — out and back — of the tempestuous Drake Passage, and the inevitable boredom and seasickness associated with it, can be skipped in favour of a two-and-a-half-hour scenic flight that takes passengers from Punta Arenas in Chile to King George Island, in the South Shetland Islands, where the vessel awaits. It's hard to believe that swapping the modern world for a perfectly white, Antarctic wilderness can be this fast and easy. After less than three hours of travel, the first penguins have been spotted, pieces of iceberg can be seen bobbing in the bay, and all of this while champagne and canapés are served on the elegant teak deck of this private explorer yacht.

The Views

Fildes Bay, where *Legend* is anchored, is the starting point for the journey. Thanks to her ice-strengthened hull, otherwise unnavigable places like Iceberg Alley and Antarctic Sound, even the northern fringes of the Weddell Sea, aren't off limits. Less formidable craft often miss out on the emperor penguins who breed on Snow Hill Island, but not this yacht. *Legend* offers its passengers the freedom, time, and space to truly explore Antarctica. The freedom to make unscheduled stops, while respecting the strict rules that govern visits, is a privilege. And the difference is exceptional. Imagine sitting on a deck chair, enjoying the blissful solitude, and seeing a humpback whale surfacing from the icy waters. On *Legend*, the crew is able to seize such opportunities, halting the ship and lowering a tender with front-row seats to the action. While some penguin colonies and research stations along the iceberg-dotted Lemaire Channel are inaccessible because of passenger limits, *Legend* has a freer rein. Late one afternoon, in the spectacular, iceberg-filled Lemaire Channel, Legend nestles into a thick patch of ice in a remote bay. There's no wind. Minke whales feeding under the ice sheet ripple the mirror-like water beyond the floe edge. The sky is a spectacle of changing colours, and the otherworldly sound of distant avalanches mixes with the exhalations of breaching whales. At Trinity Island and Mikkelsen Harbour passengers board Zodiacs and fast Rafnar boats — made in Iceland and perfect for icy waters — for excursions to see Weddell seals sunbathing, and to watch colonies of squawking gentoo penguins in the shadow of an old whaling station. Nowadays, nature is left undisturbed in Antarctica. Tourism and research are the only reasons for human interaction. In the afternoon, after sailing several hours through a sensationally silent landscape, an alfresco lunch is served. The back of the

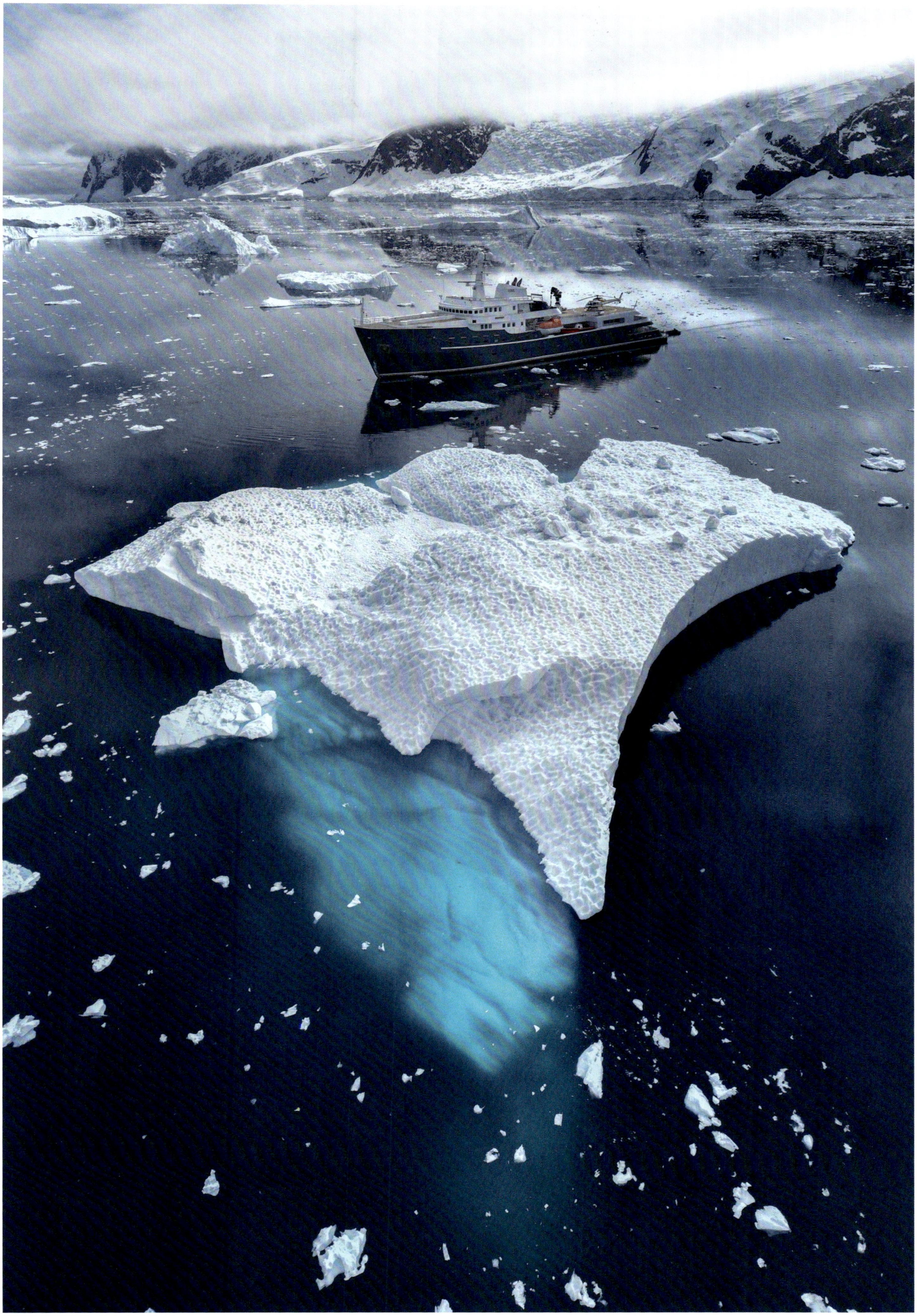

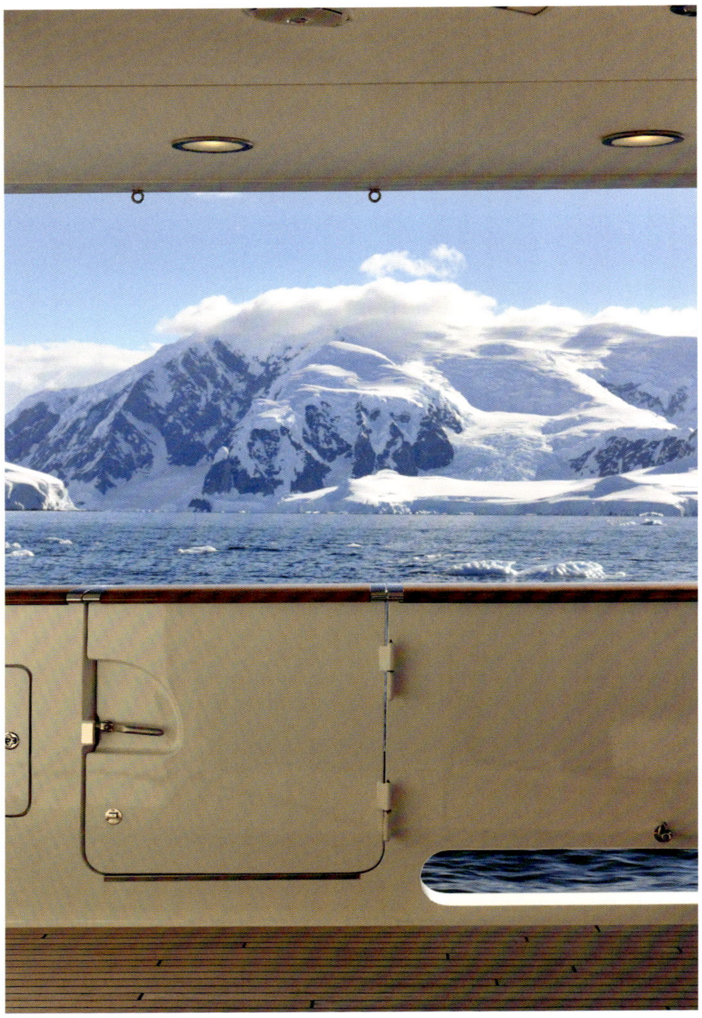
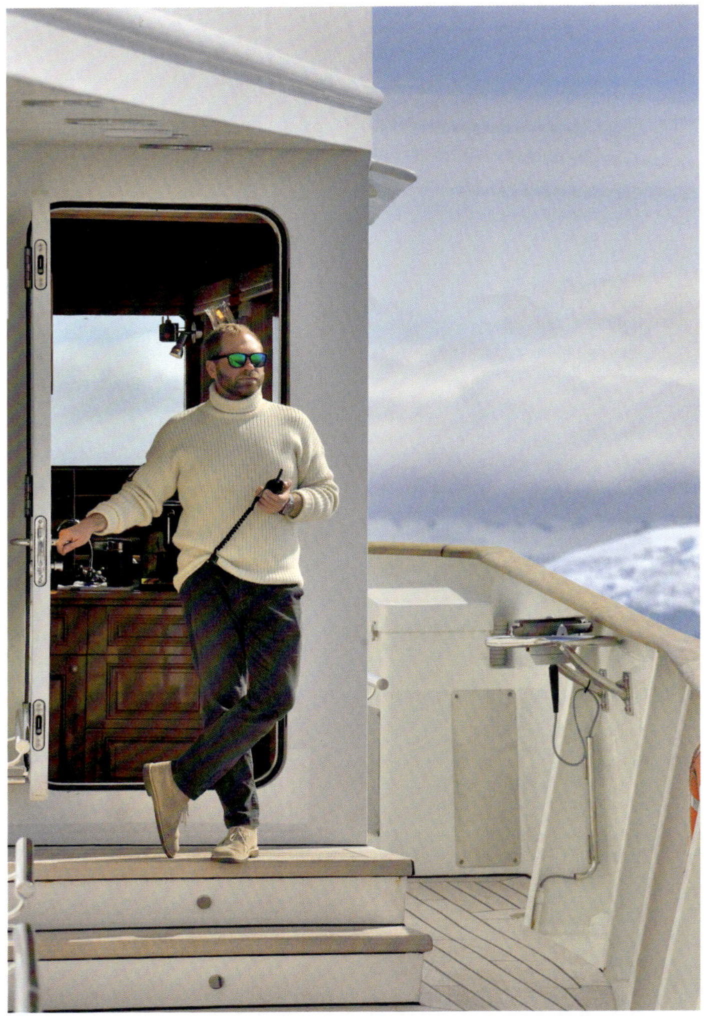
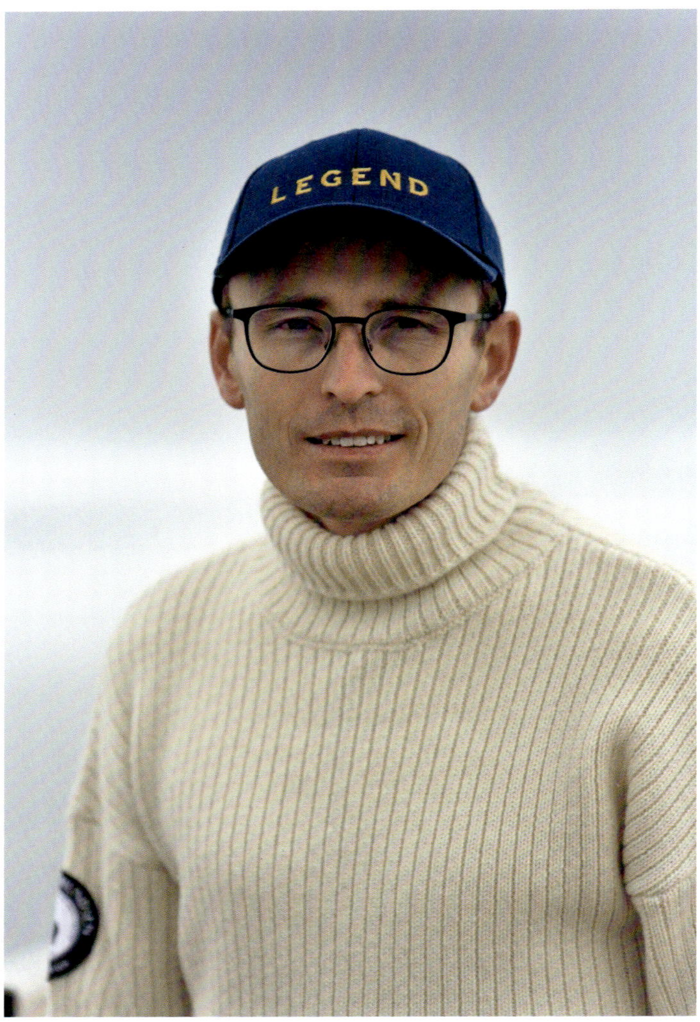
LEGEND
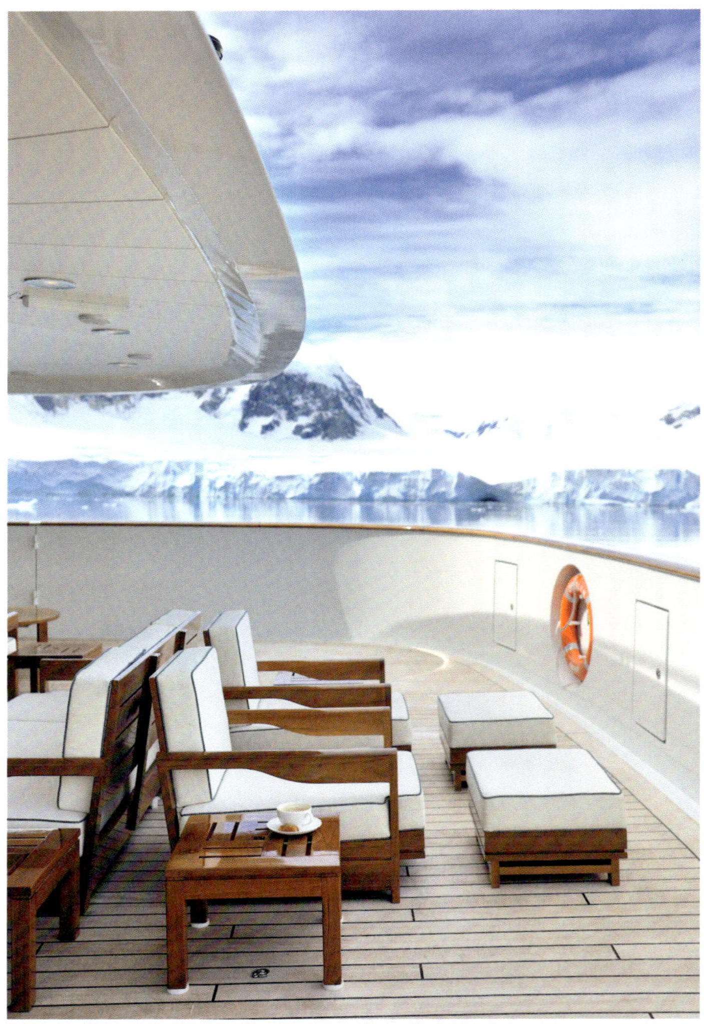

yacht is one of the best spots to pull up a deck chair, enjoy some sunshine and gaze over the still waters of this intoxicating landscape. Sometimes, leopard seals float by on giant shards of ice. As *Legend* passes by they may lift their heads for a second or two, before they doze off in the warm sun. After lunch, passengers depart on Zodiacs to slalom among the icebergs, listening to one of the most extraordinary sounds on the planet: the release of air pockets that have been trapped inside the melting ice for millions of years. In the distance, new icebergs are born as they break off ancient glacier walls. Another almost supernatural sound. Filled with Adélie and gentoo penguins, the Lemaire Channel is so photogenic that it is nicknamed Kodak Alley. Nearby is the Vernadsky Research Station, manned year-round by Ukrainian scientists who spend their free time playing snooker, listening to vinyl records, and drinking strong spirits to keep things warm and cheerful. Sometimes the scientists are invited to board passing ships, as is the case with *Legend*. While more seals and penguins float past, and the looking-glass waters reflect a slowly setting sun, the scientists mingle with passengers on *Legend*'s top deck, sharing an authentic South American parrilla (barbecue) washed down with lots of pisco sours. As the sun struggles to dip below the horizon, the ship sails into picture-perfect Paradise Bay. A cold night might deposit a layer of frazil, or super-thin ice, on the ocean surface. If you're lucky, the bay's glacial walls might rumble, sending ice crashing into the water. This is a hotspot for humpback and minke whales, occasionally joined by pods of orca (killer whales). Leaving the Antarctic Peninsula behind, *Legend* sails north towards Whaler's Bay on the famous Deception Island. The final highlight of the trip is Yankee Harbour, located between Glacier Bluff and Spit Point on the southwestern side of Greenwich Island. Here, 10 metres (33 ft.) above the beach, are 4,000 breeding pairs of nesting gentoo penguins. It's a nature-lover's paradise. As other ships prepare for the two-day return journey across the Drake Passage, *Legend* drops her guests back at King George Island for their short flight to the South American mainland. They bid farewell to the most surreal and remote polar landscape in the world.

Creature Comforts

M/Y Legend was originally built in 1974 as an Ice Class 1A FS icebreaker. In 2013, the ship was bought by a Dutchman seeking to offer chartered luxury cruises to Greenland and Antarctica. Now it is owned by an ex-CEO of Silicon Valley who charters her out for those wanting total privacy in remote corners of this planet. A two-year refit added a proper yacht interior including a Balinese spa, an outdoor hot tub with built-in rain curtain, a gymnasium, and a cinema for cosy evenings after long days in the cold. In the suites, the décor harks back to the time of the great explorers. Throughout the ship, there are plenty of little corners and spaces to enjoy polar vistas. The outdoor deck is a wind-free space with a teppanyaki grill where Arctic barbecues are served as you glide through the icy landscape.

Go Now

In Antarctica the planet's fragility is fully exposed. At the turn of the millennium this was still an exclusive destination with few visitors. Since then, its popularity has grown, and bigger ships have started to arrive. New regulations to keep Antarctica pristine and quiet are now set. One of the best ways to experience this remote white paradise — one of the world's most fragile regions — in style is on board the serene *M/Y Legend*.

How to Go

Legend can sail to Antarctica from anywhere, even from the Falkland Islands, while passing by the magnificent South Georgia Island, depending on the desires of the passengers. Those who prefer to skip the crossing of the infamous Drake Passage separating the tip of South America from the Antarctic Peninsula — which entails a long and rough two days at sea — can simply take the two-and-a-half-hour flight from Punta Arenas in Chilean Patagonia to King George Island, which is located just off the Antarctic Peninsula, and board the yacht there. Trips start from a 1.000.000 Euro for a week, all-inclusive.

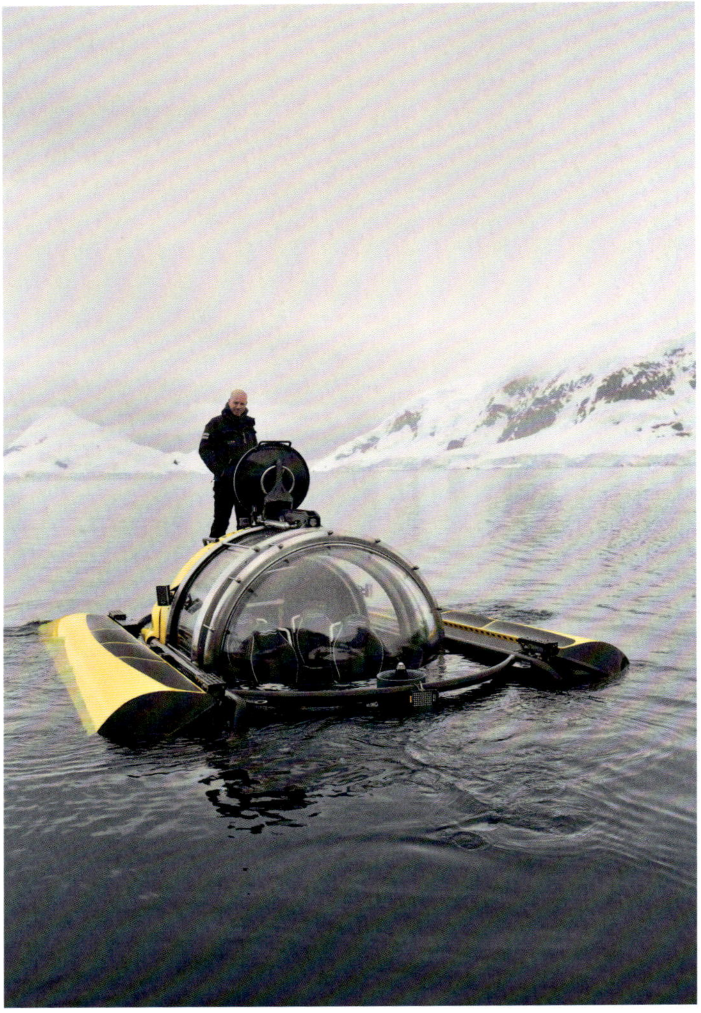

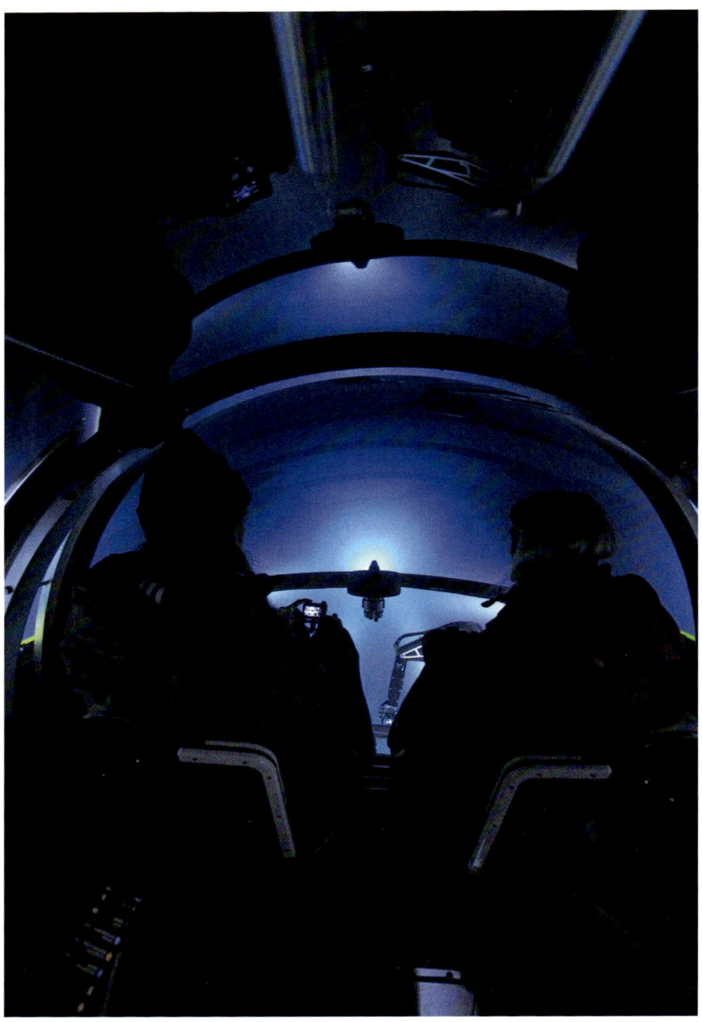

This spread *Legend* is one of the only private yachts built to explore Arctic waters, which are often frozen or full of icebergs. Passengers enjoy front-row seats to rarely seen polar landscapes on unique, small-scale journeys. The yacht is able to navigate these fragile places as they were meant to be — slowly, safely, and gently. *Legend* has been outfitted with a complete collection of extreme toys. One of them is the U-Boat Worx C Explorer 5 submersible. This small submersible can take up to four guests (plus a pilot) down to a depth of 200 metres (650 ft.) to see the mesmerizing underwater landscape of Antarctica. On board there are also four Zodiac MK5 Ribs for spectacular beach landings, and for some speed, two 'made in Iceland' Rafnar 850 RIB-boats.

Next spread A helicopter flight is the ultimate way to see the icebergs and glaciers along the coastline of the Antarctic peninsula. Watching icebergs floating in the icy ocean and witnessing the vast last landscape of the White Continent from a birds's-eye view is breathtaking.

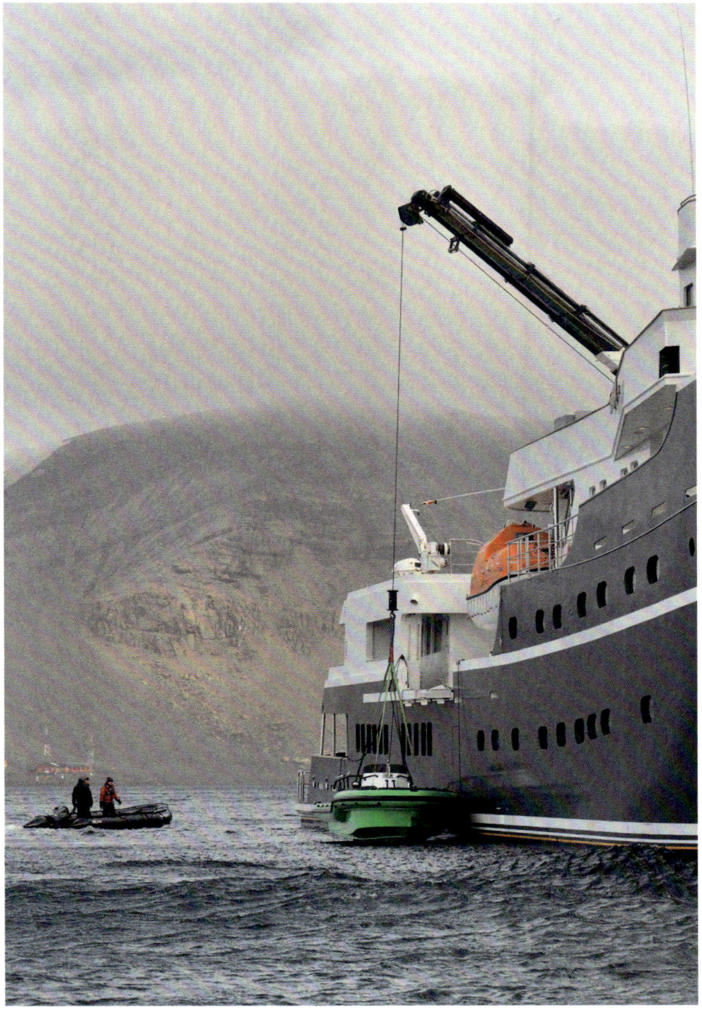

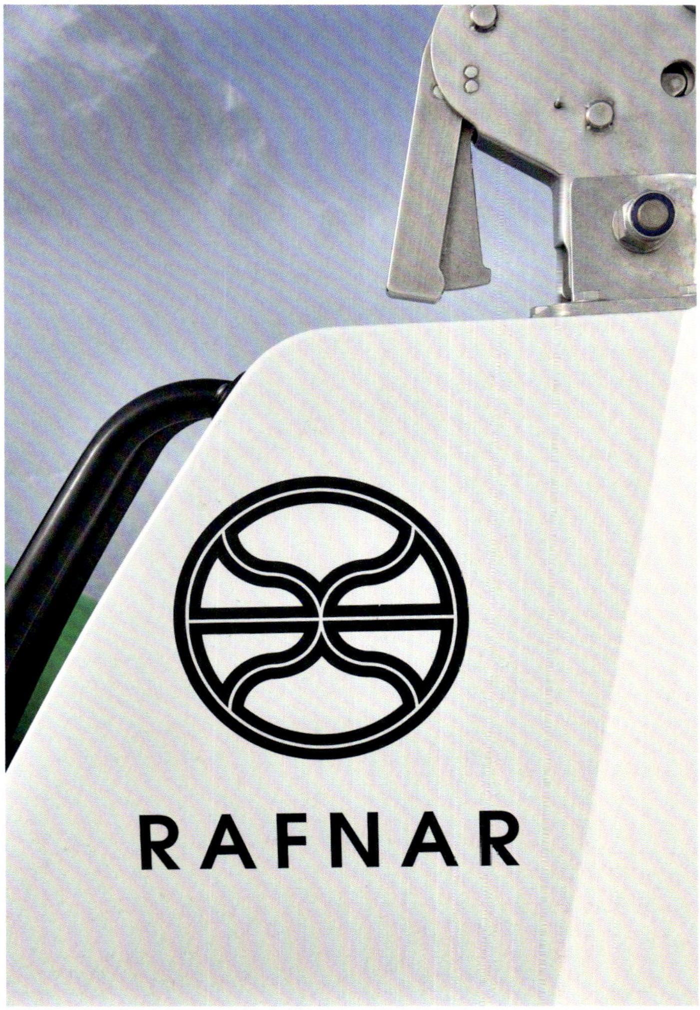
RAFNAR

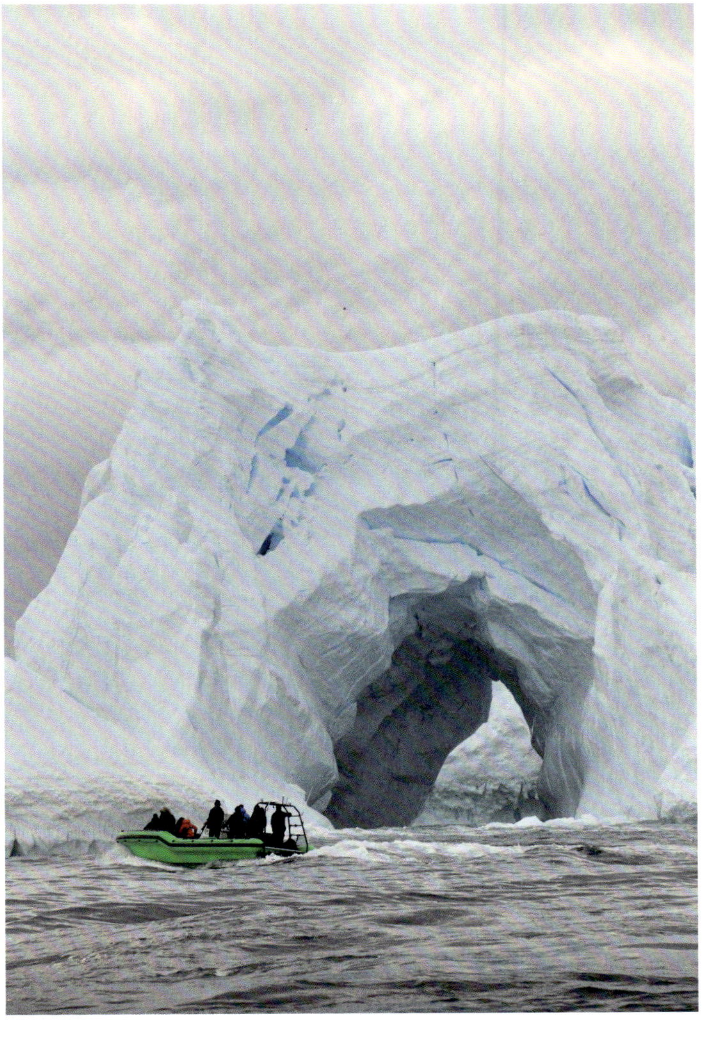

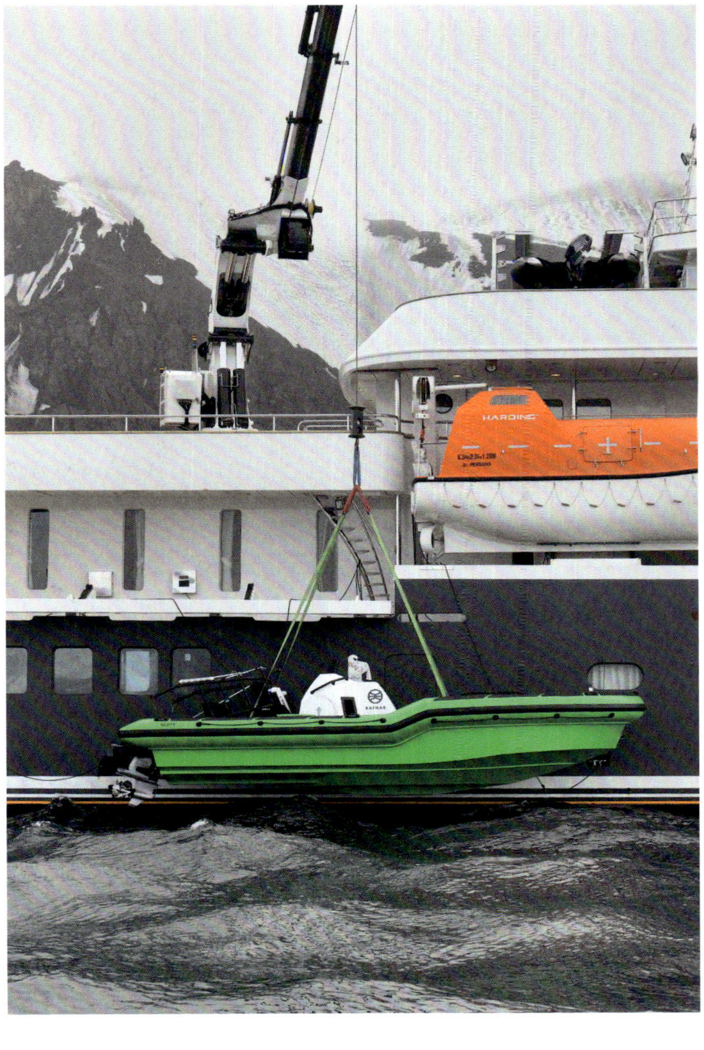
HARDING

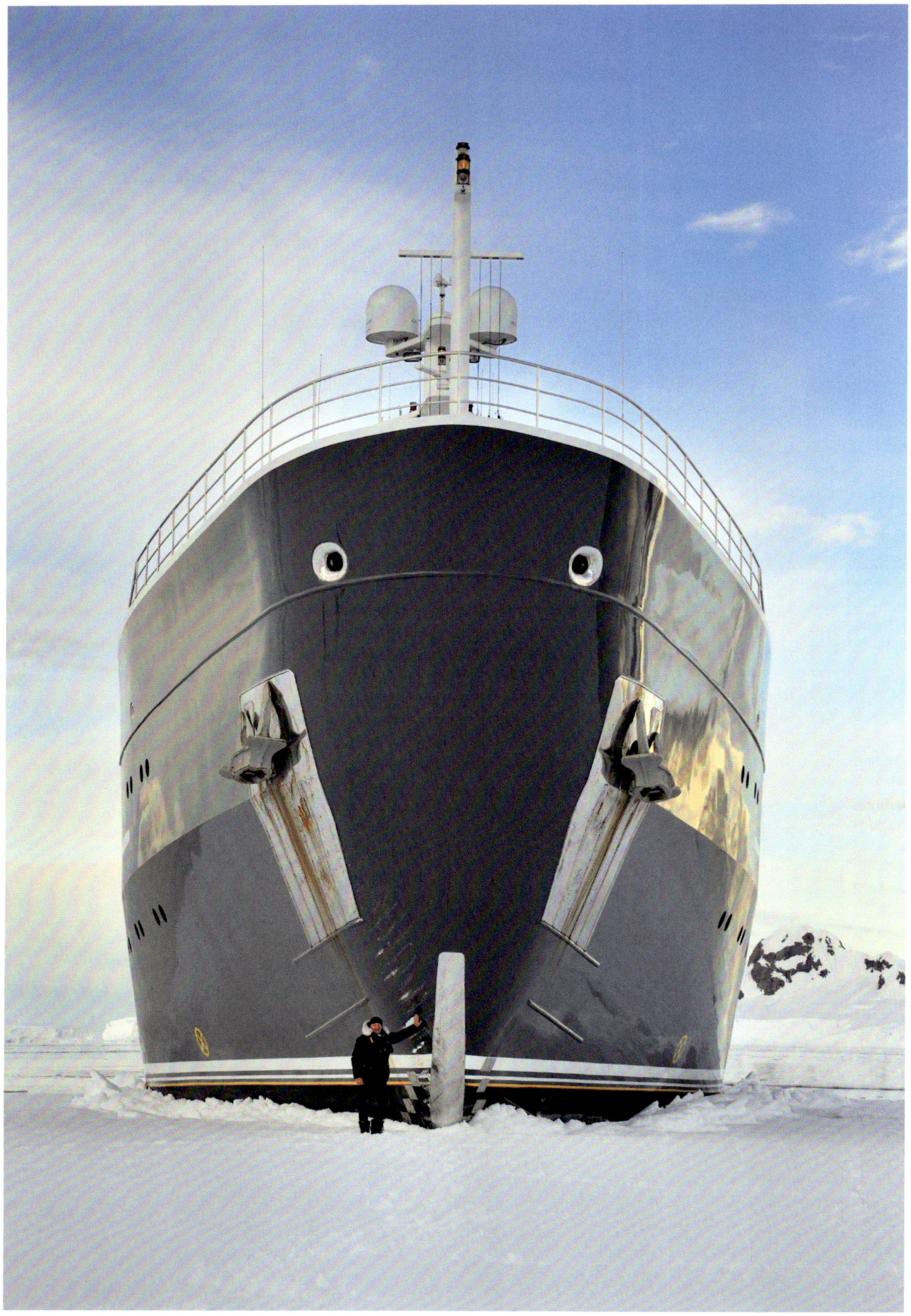

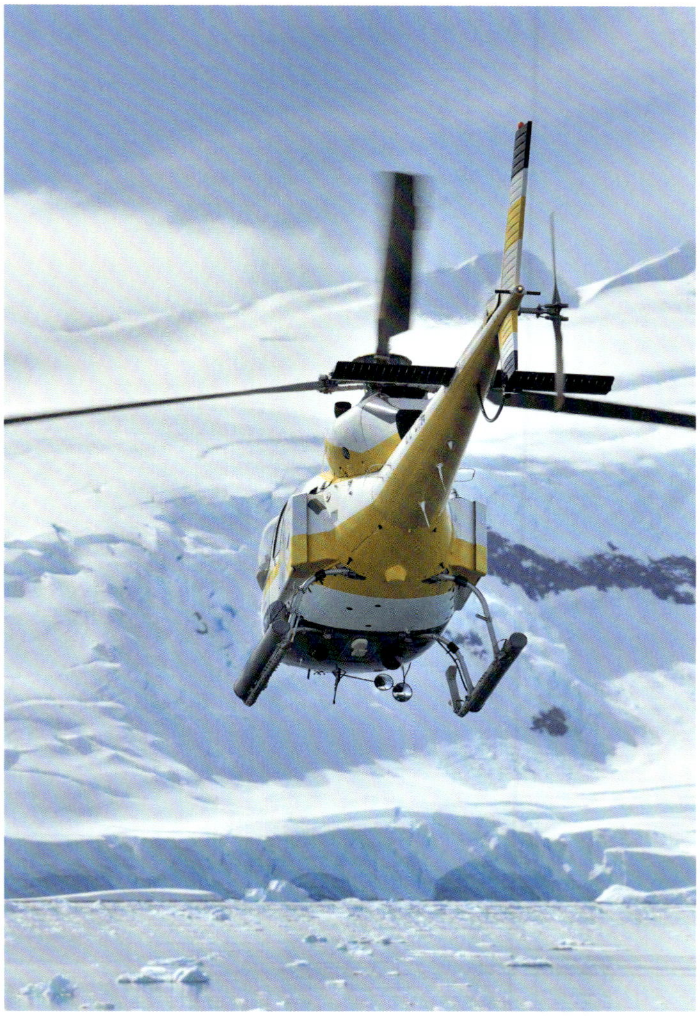

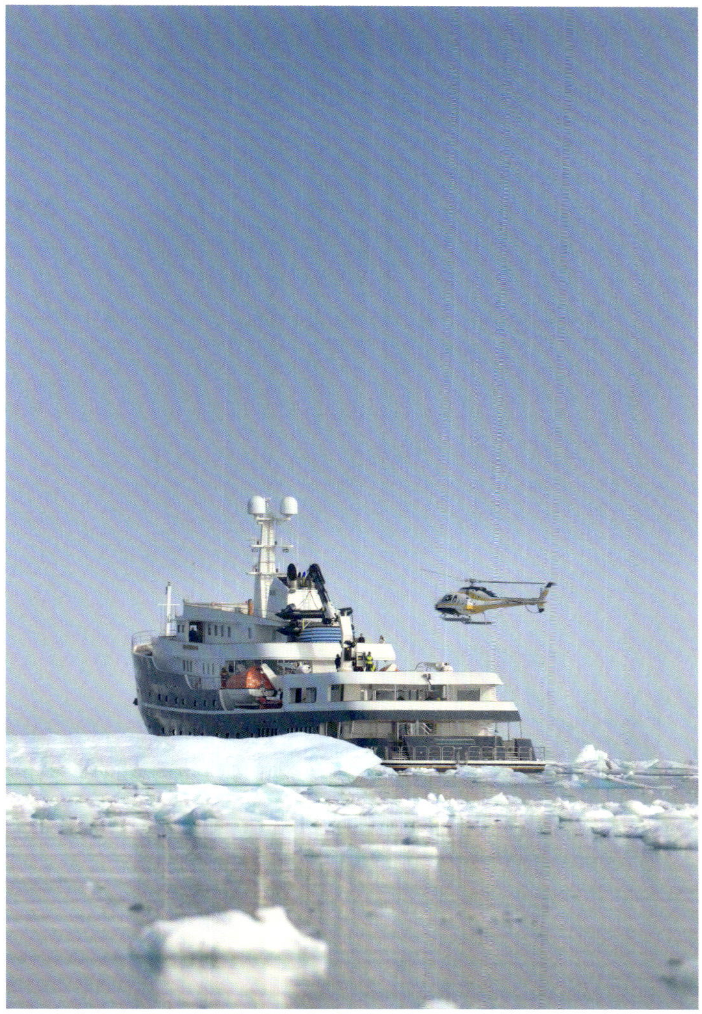

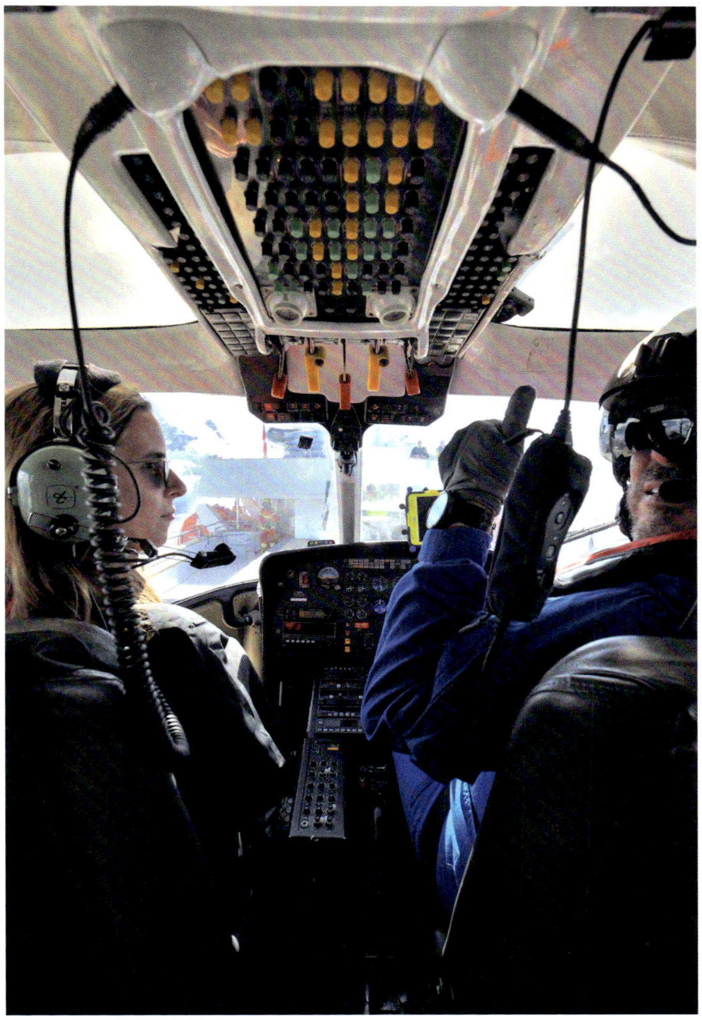

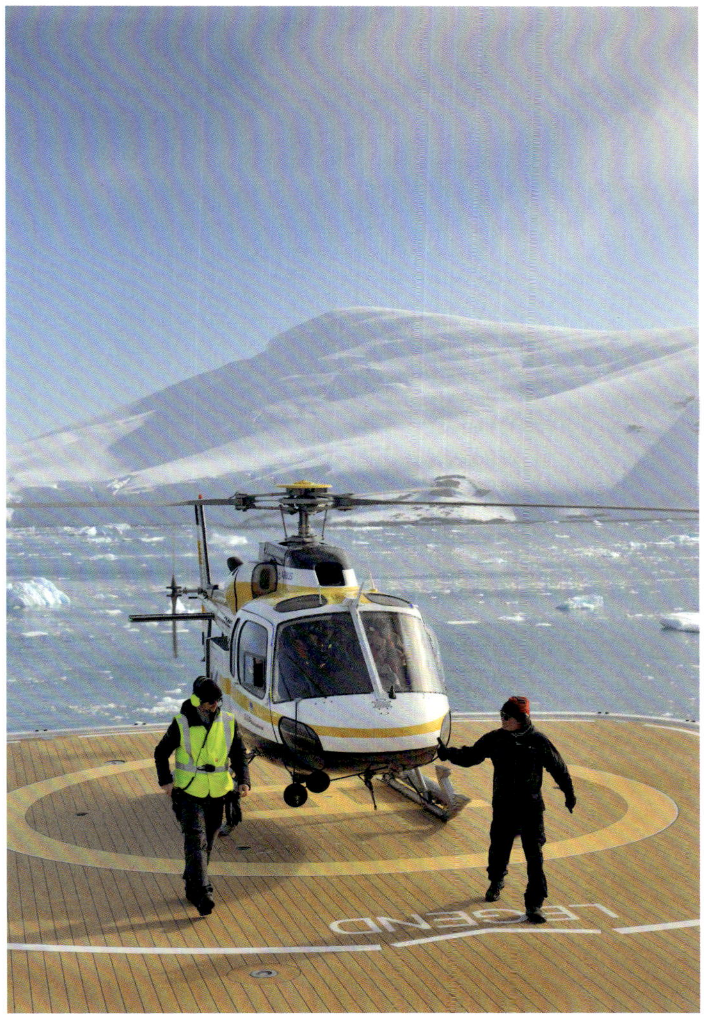
LEGEND

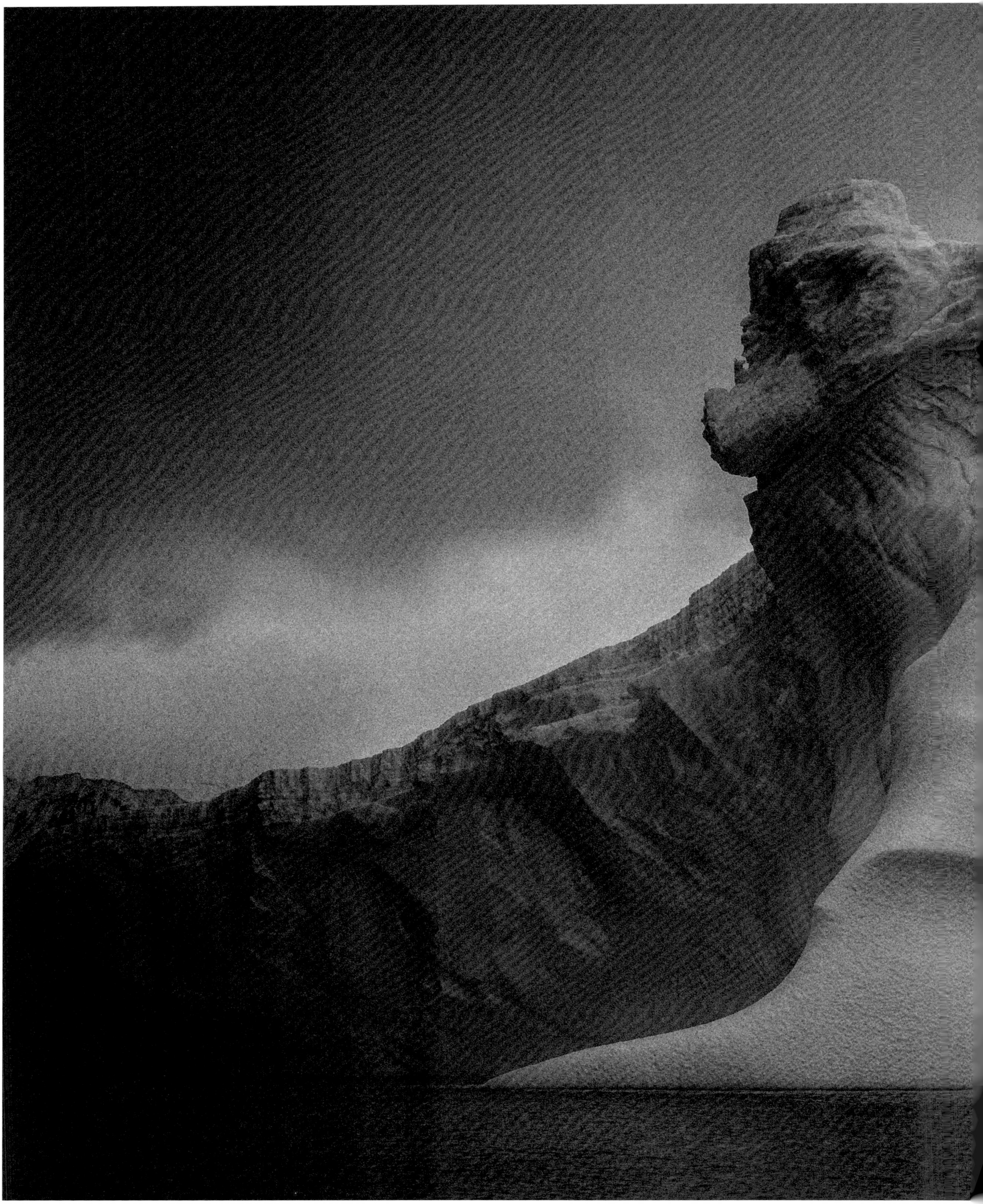

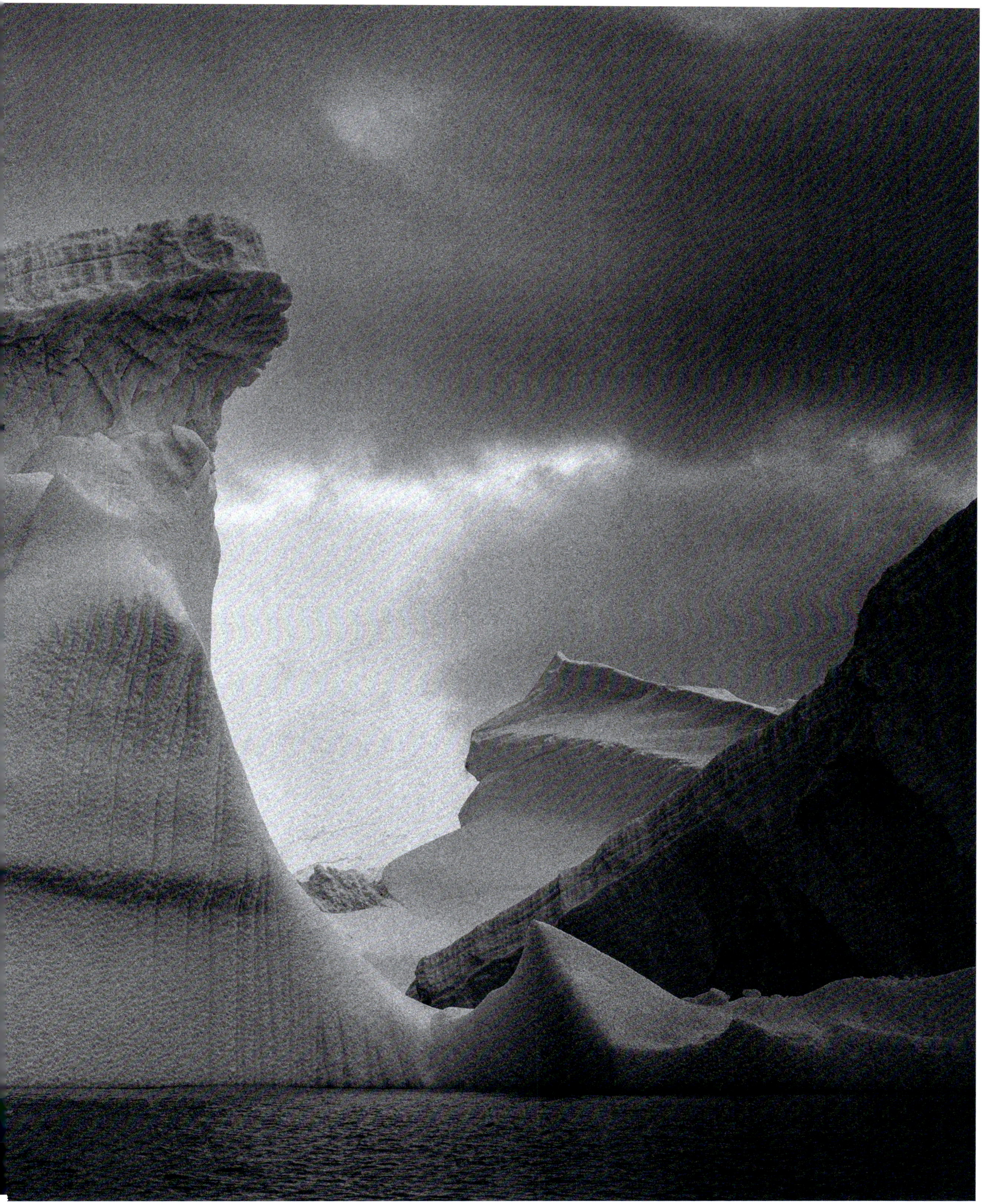

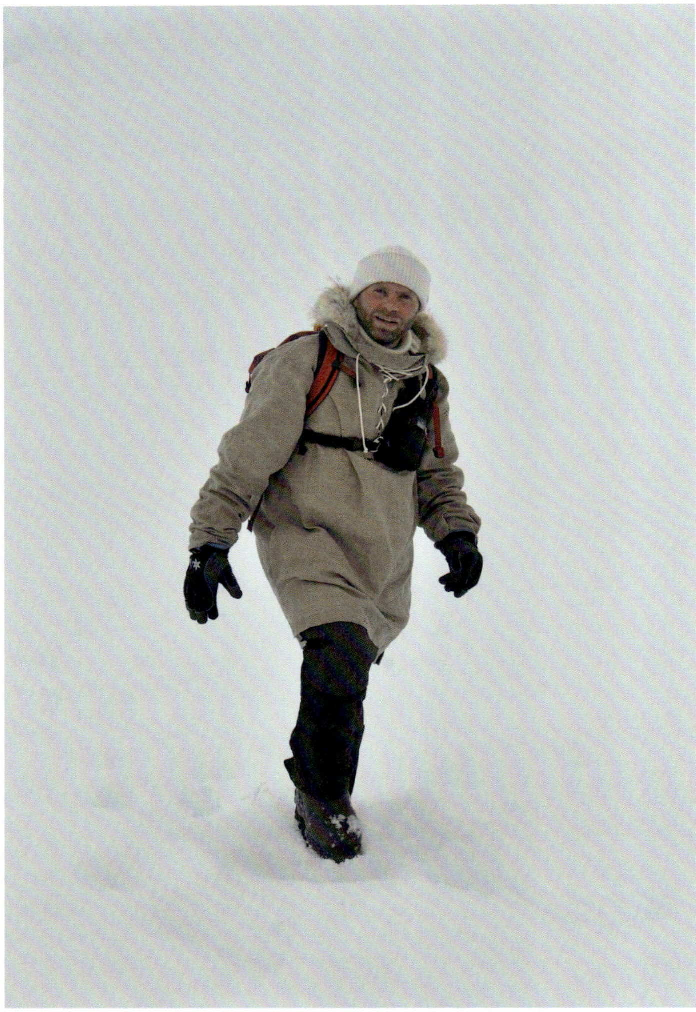

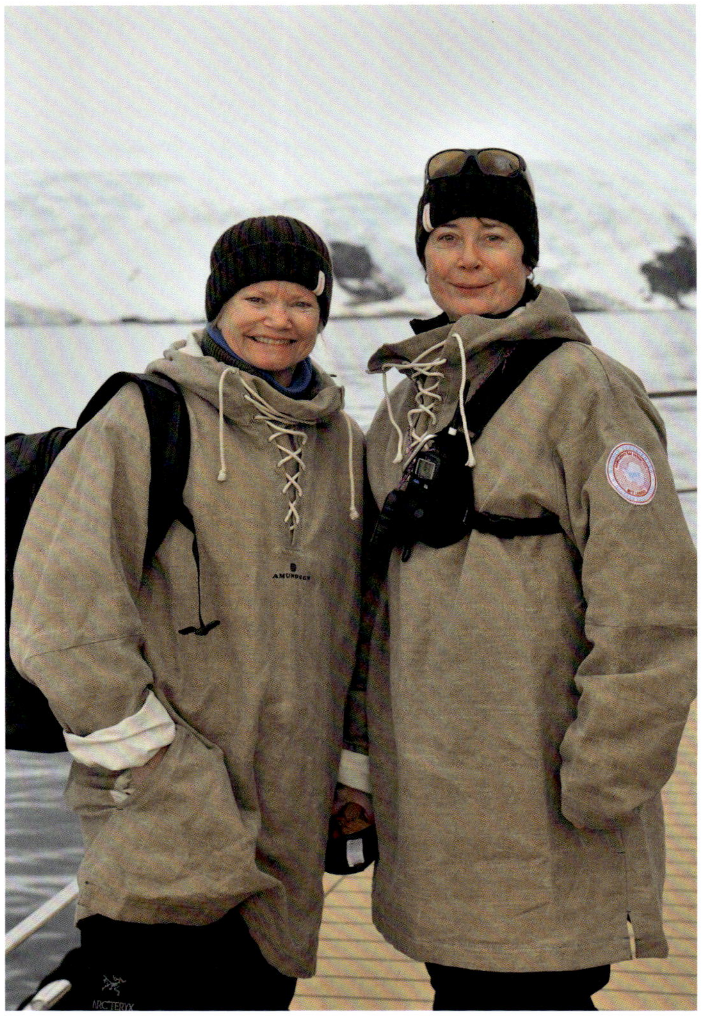

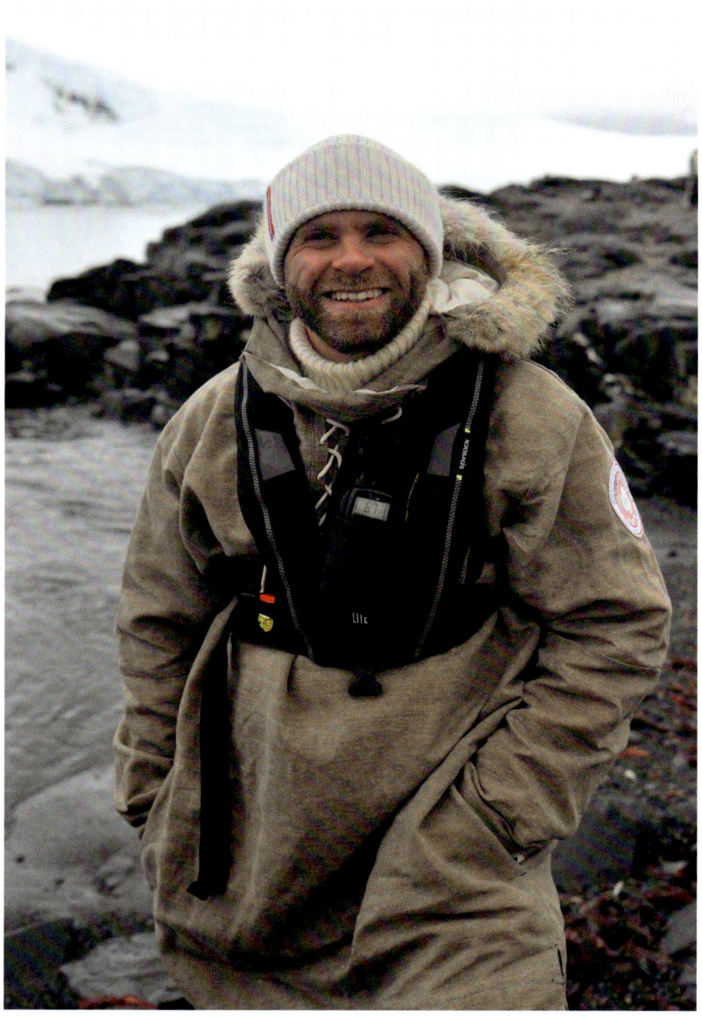

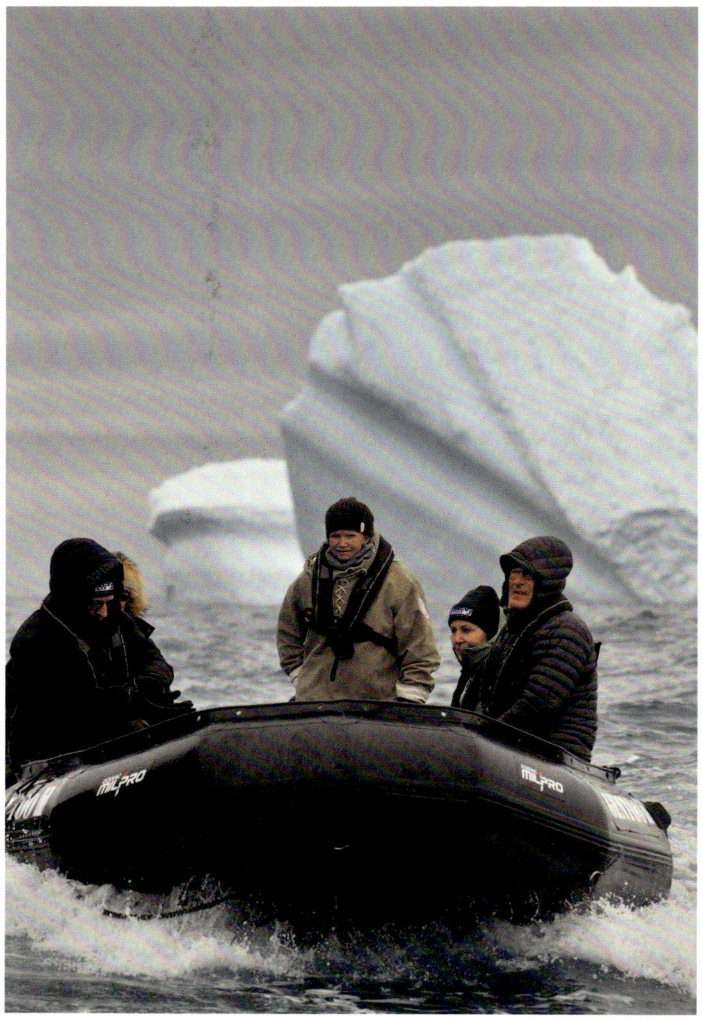

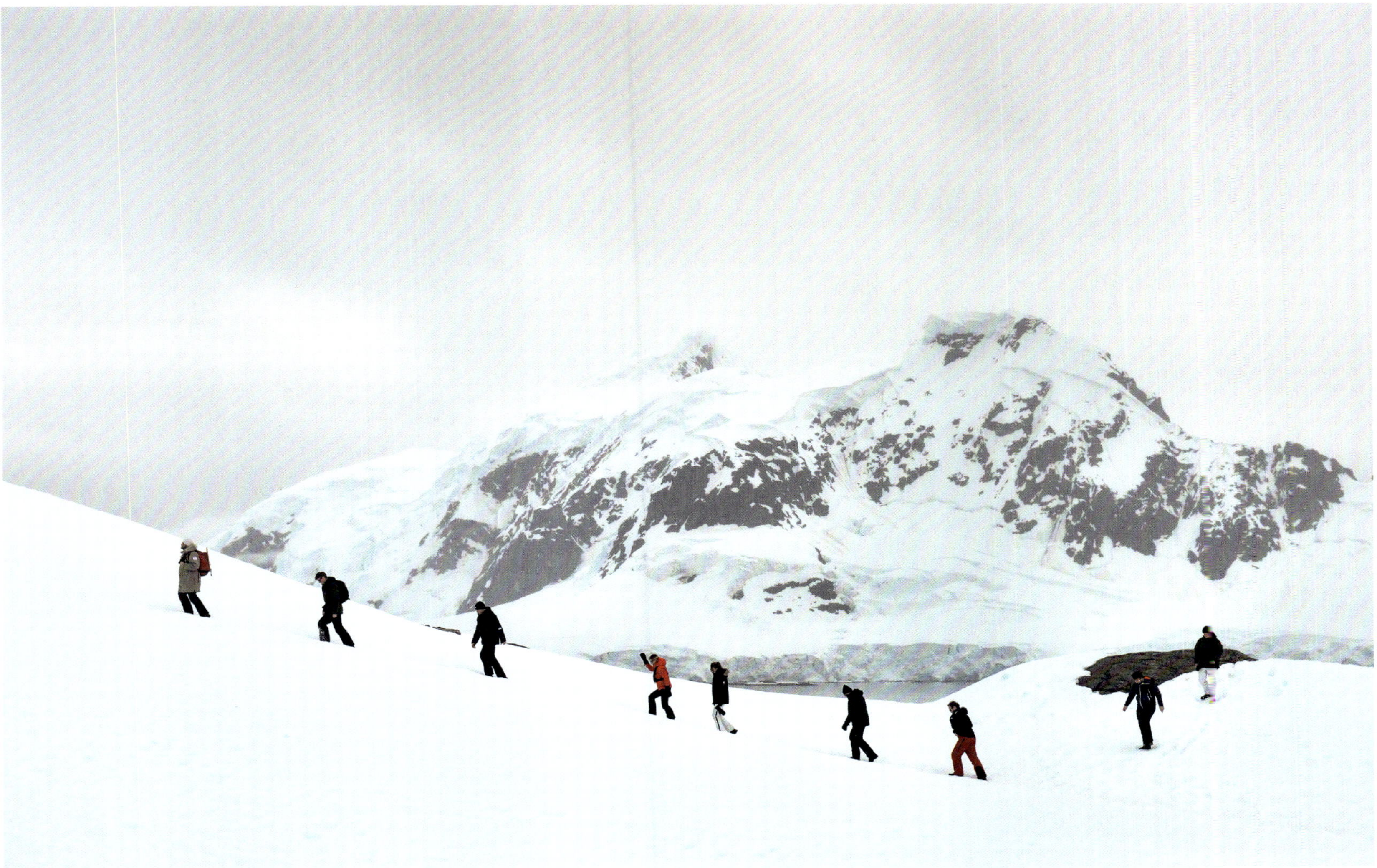

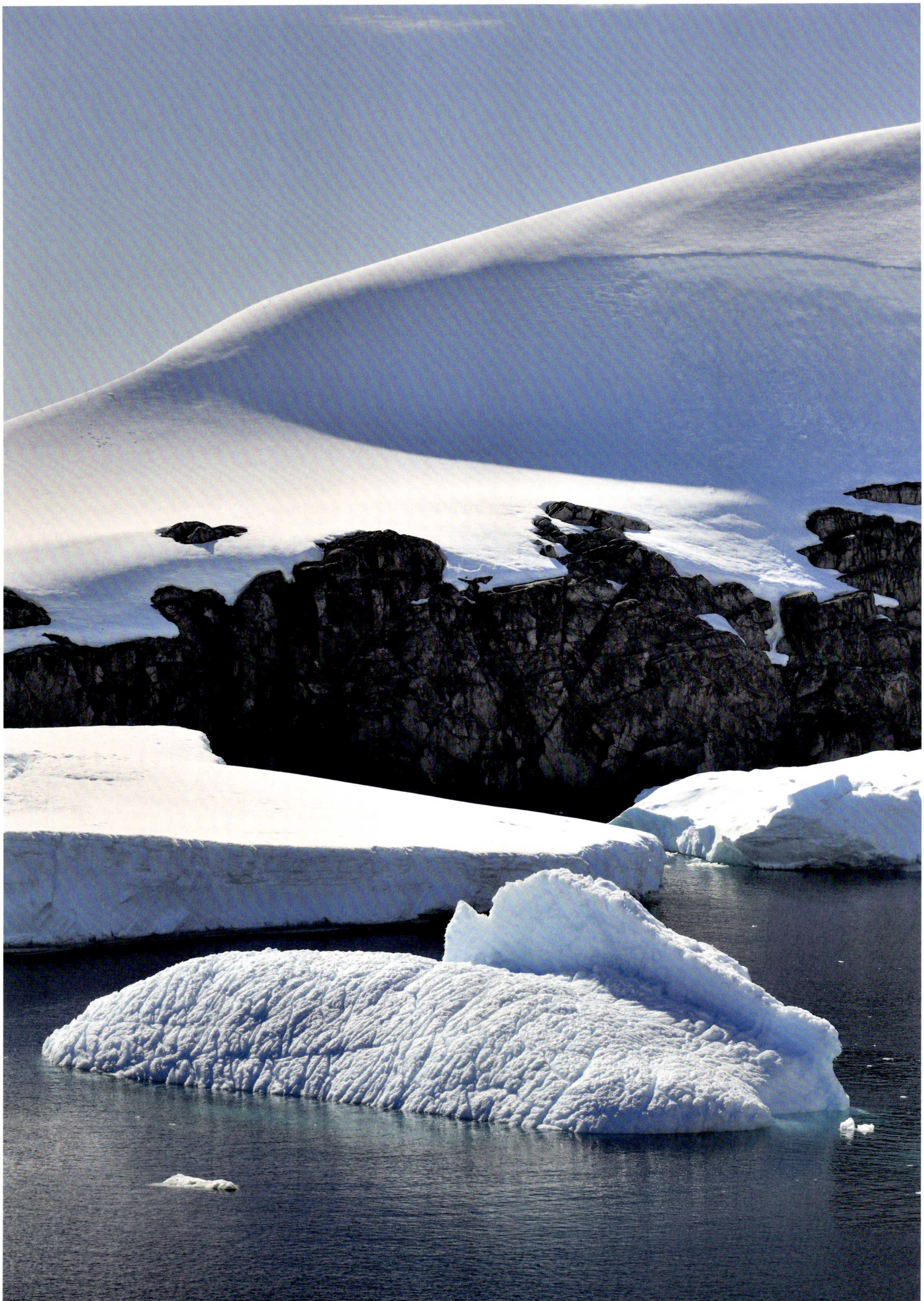

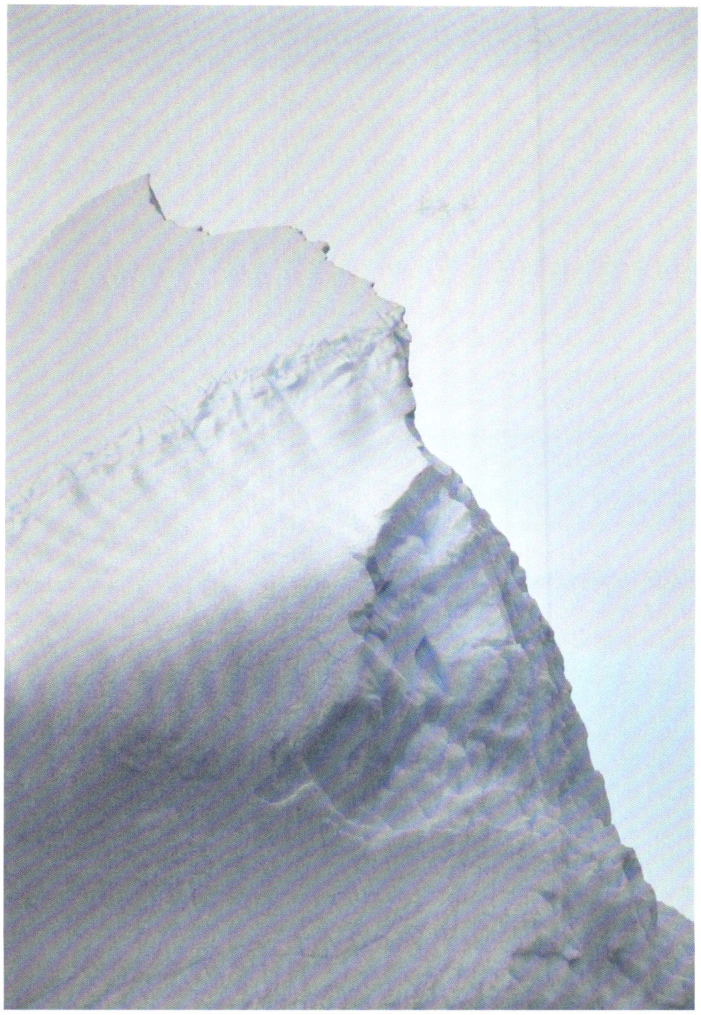
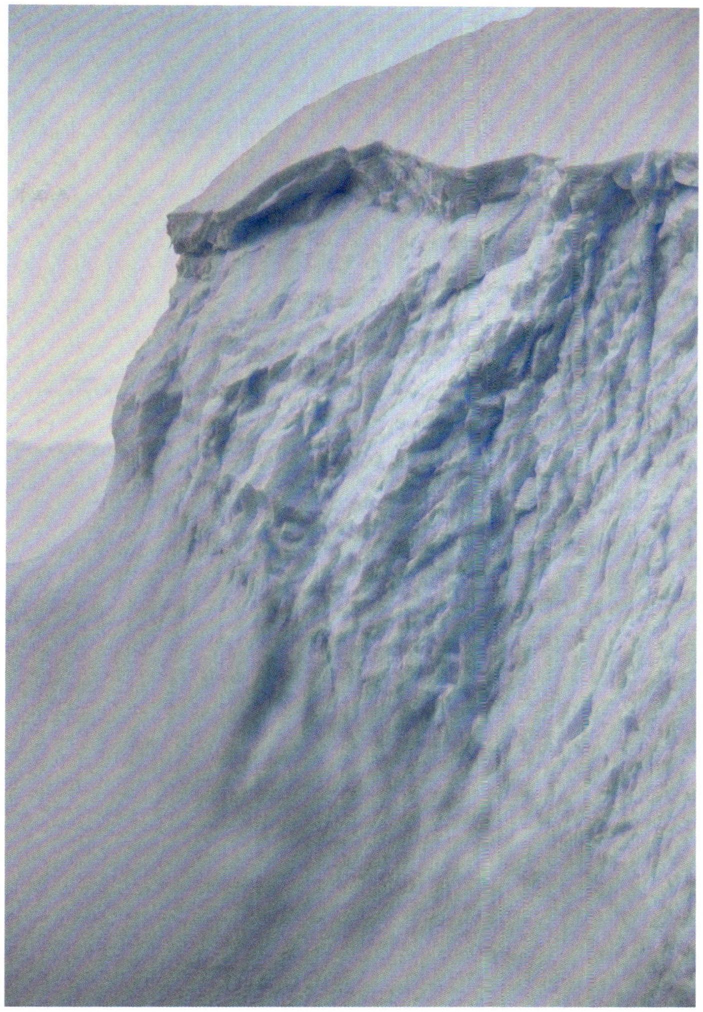
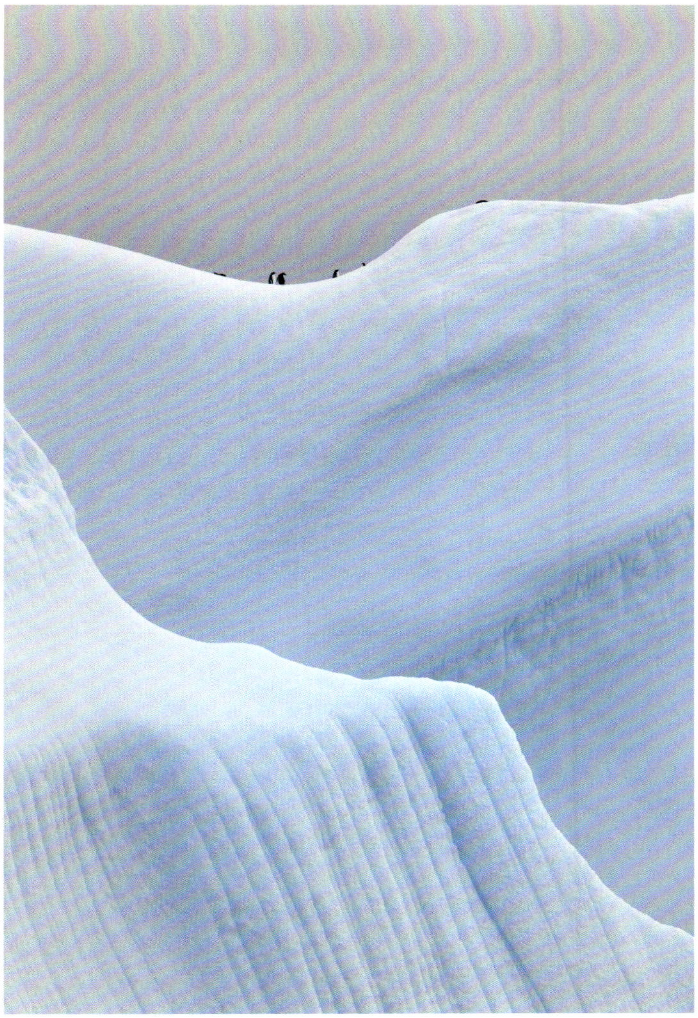
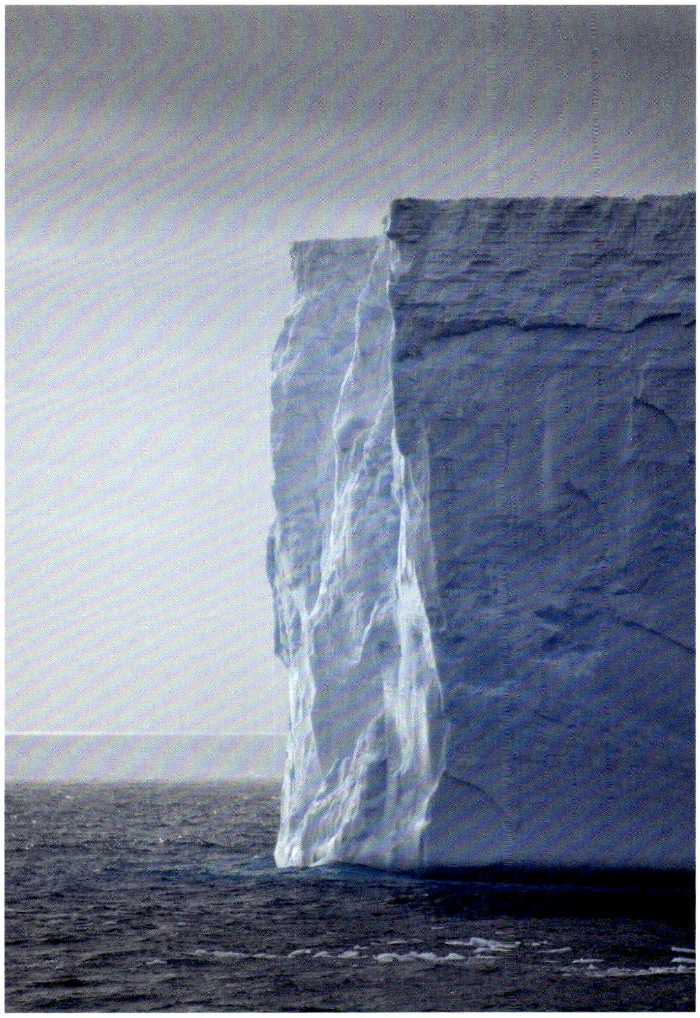

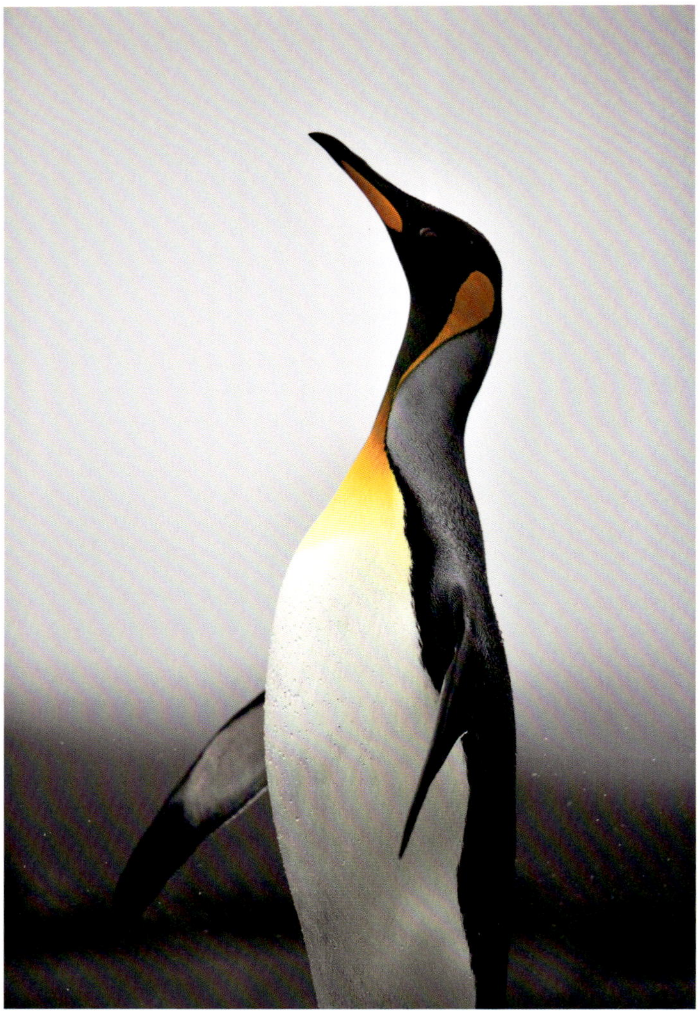

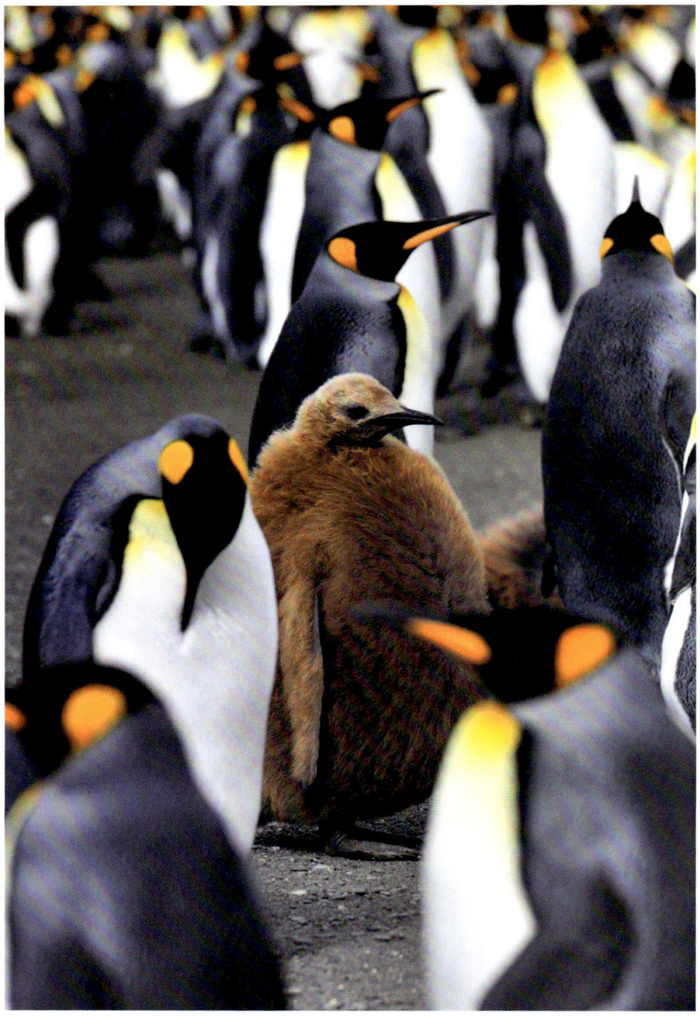

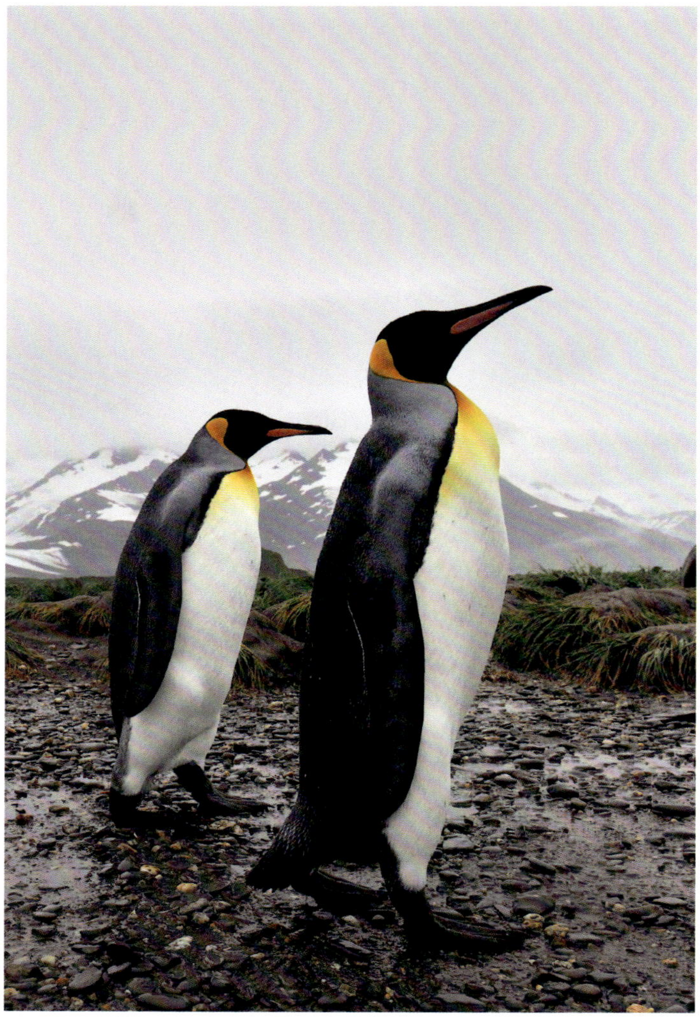

This spread Penguins are also found on the black beaches of South Georgia, where giant colonies of king penguins reside. Of the 18 species, five are found in Antarctica and another four on the sub-Antarctic islands. Emperor penguins, despite their majestic size (up to 40 kilos / 88 pounds) and longevity (they can live up to 40 years), are threatened with extinction. The Adélie penguin was first discovered in 1840 by the French explorer Jules Dumont d'Urville, who named the species after his wife Adèle. On ice, Adélie penguins waddle at 2.5 kilometres an hour (1.5 mph); in water, they are faster, reaching top speeds of 8 kilometres an hour (5 mph). Vast colonies of species such as gentoo, chinstrap, and macaroni penguins can also be spotted during this voyage around the Antarctic Peninsula. Witnessing and observing these waddling penguins is a delight; the waggling and squabbling, the diving into the ocean to feed, sliding from frozen hills or carrying little stones to make nests.

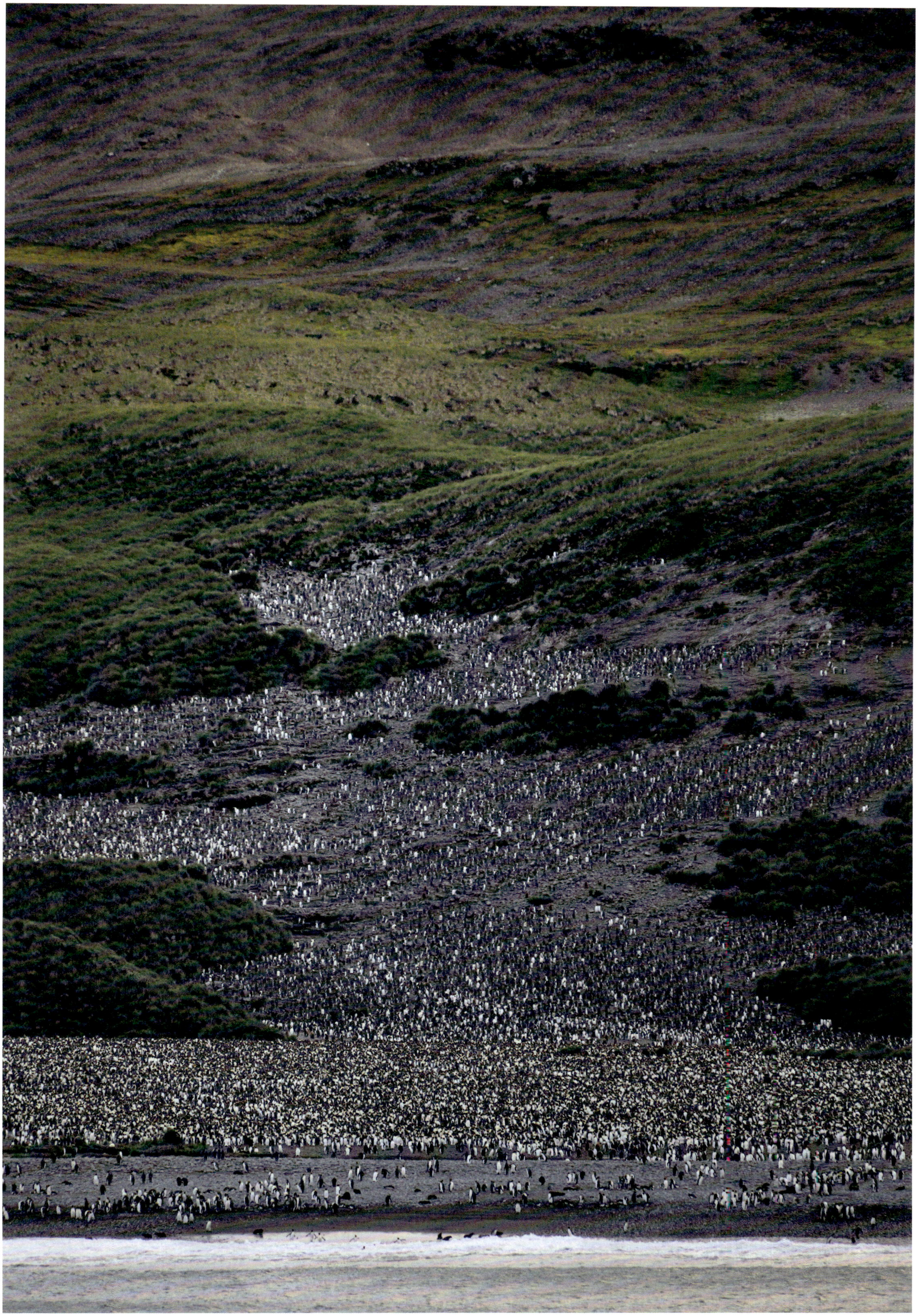

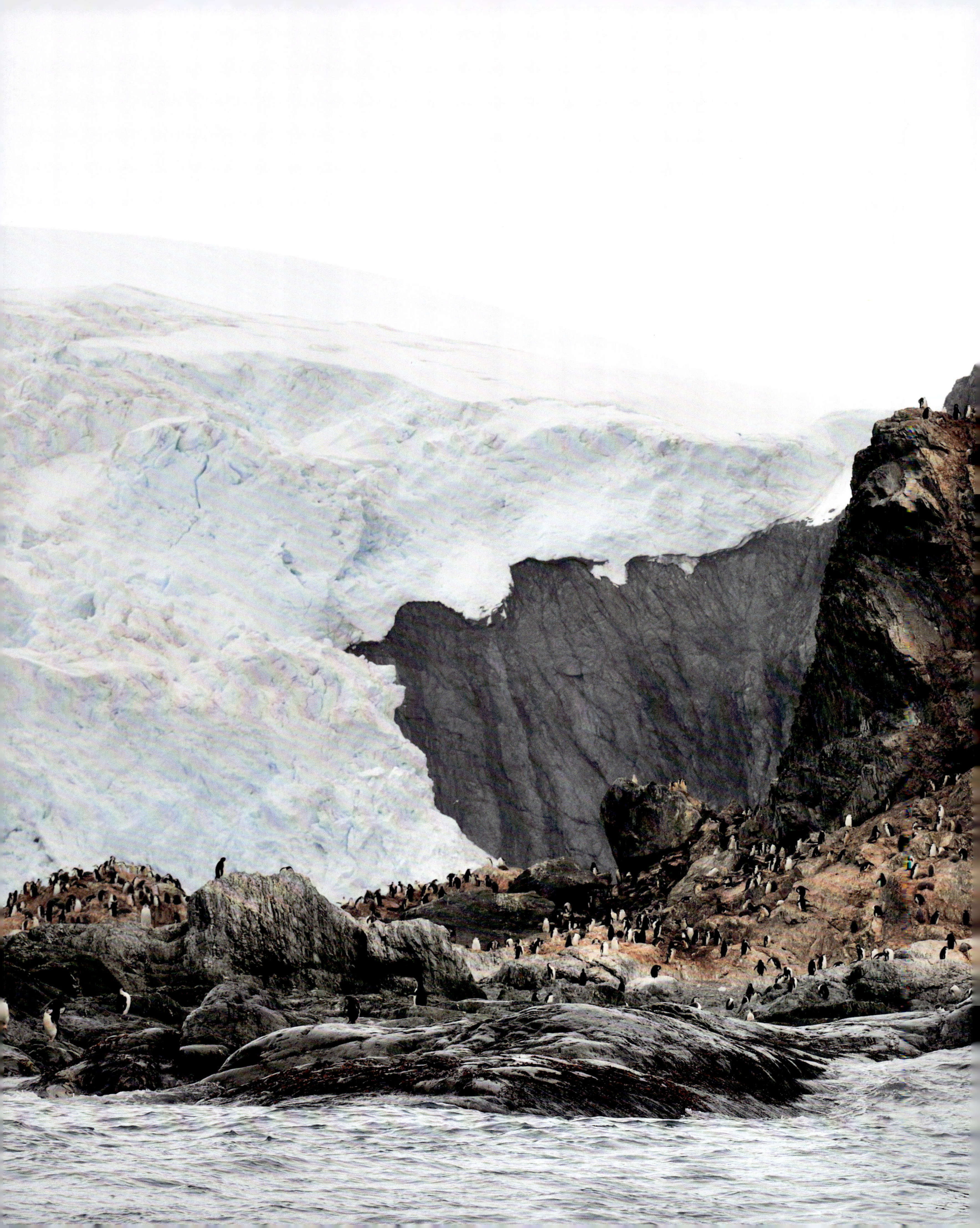

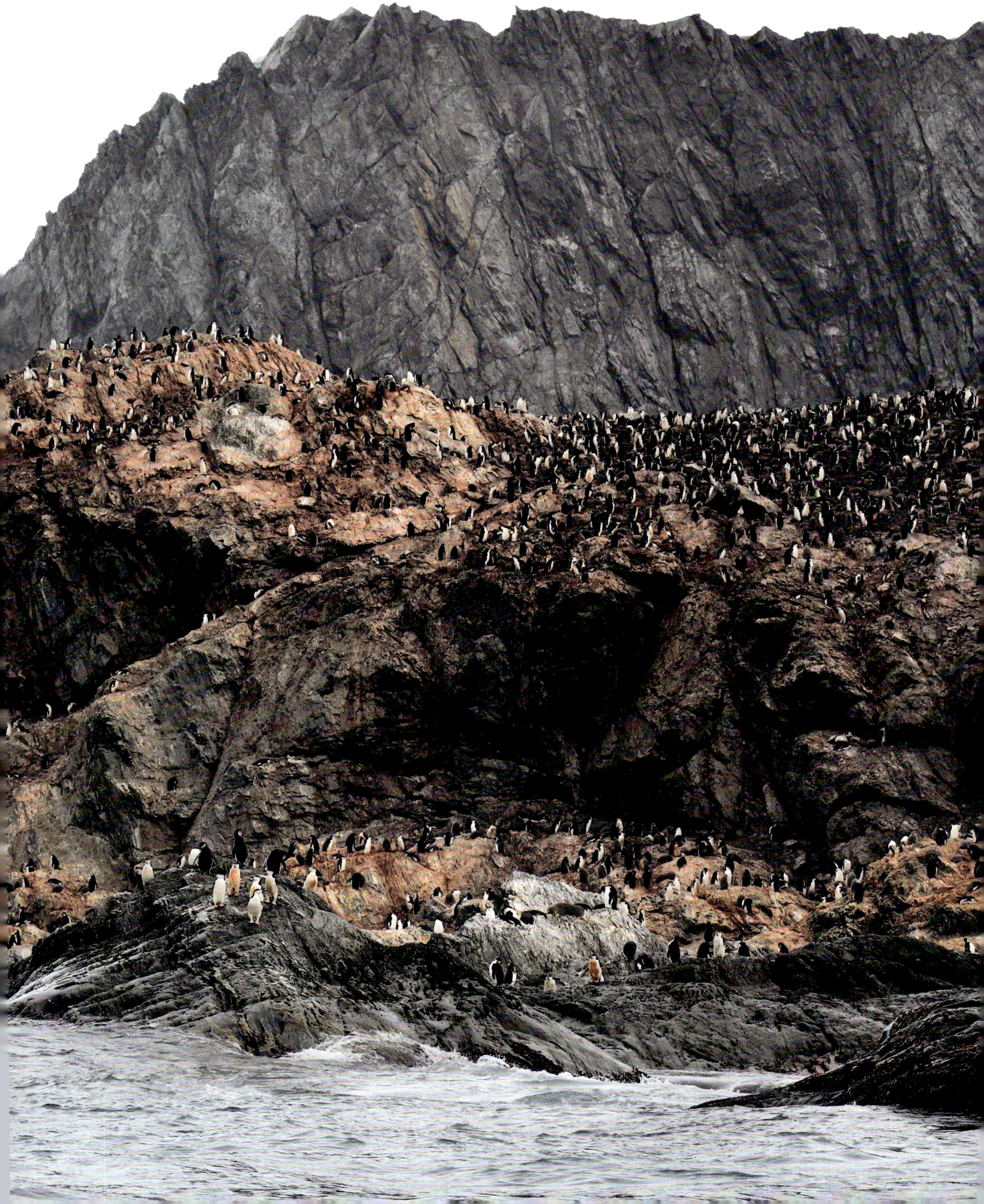

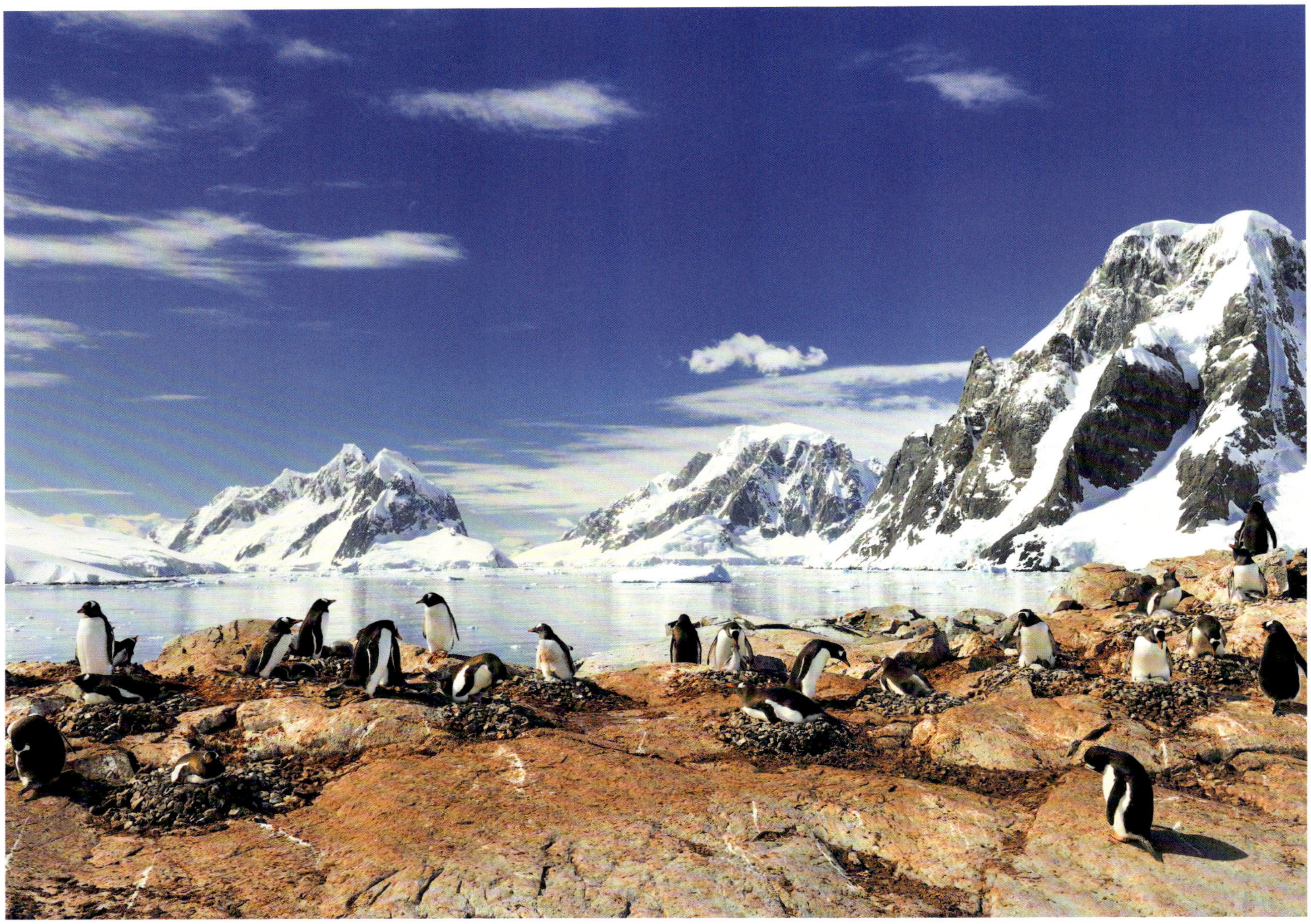

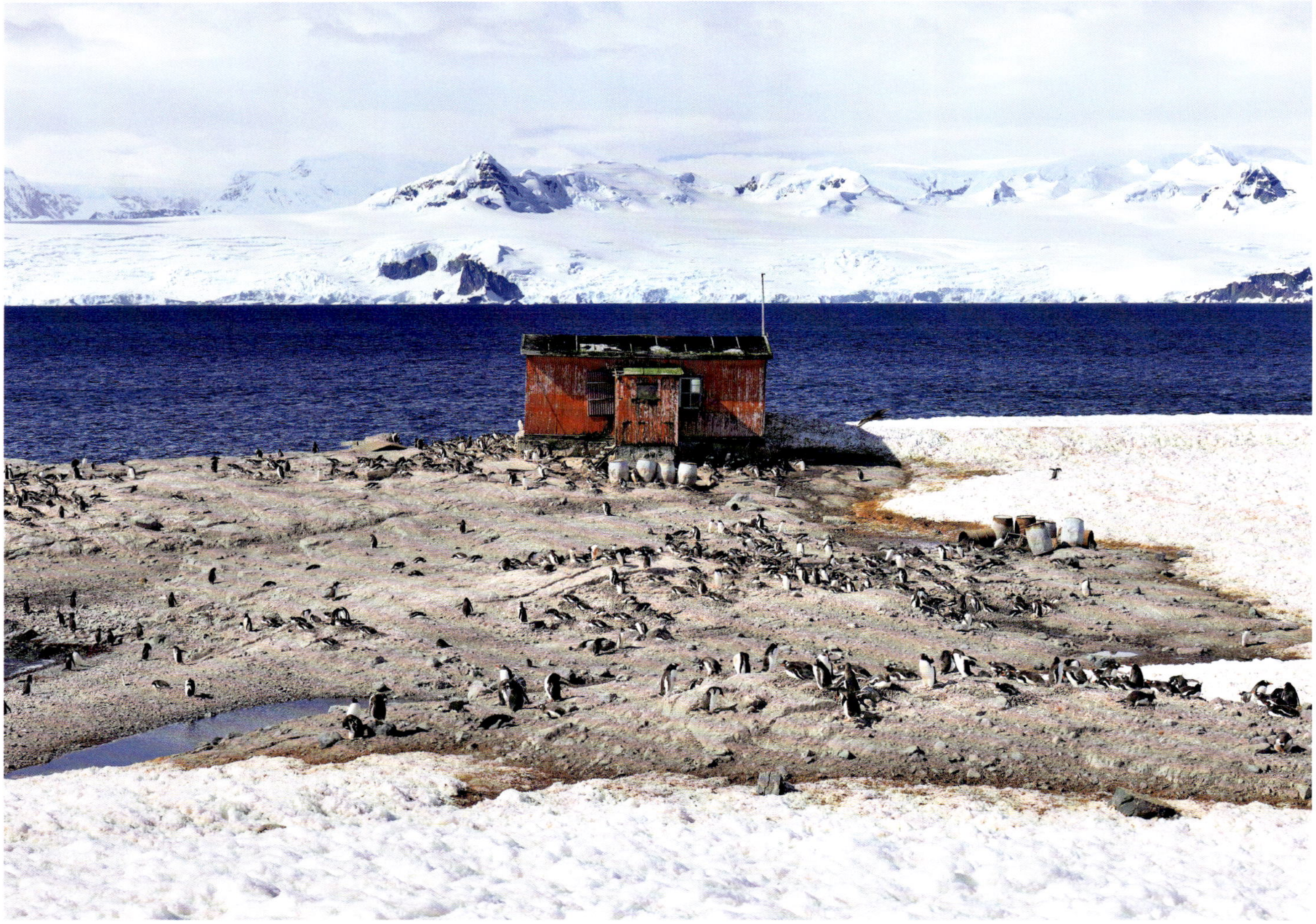

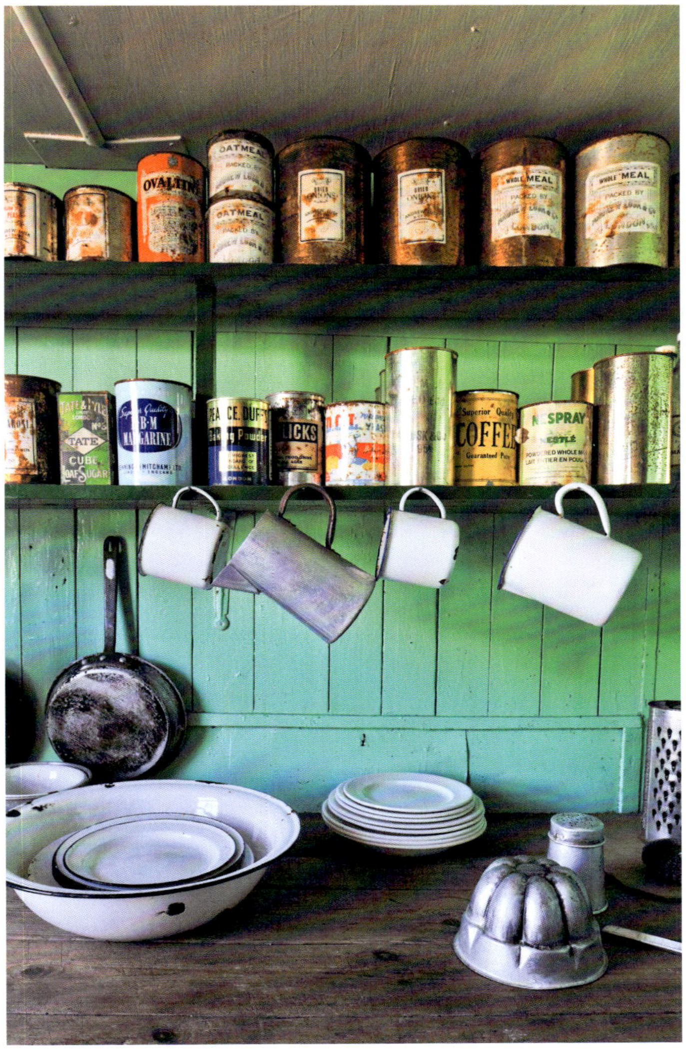

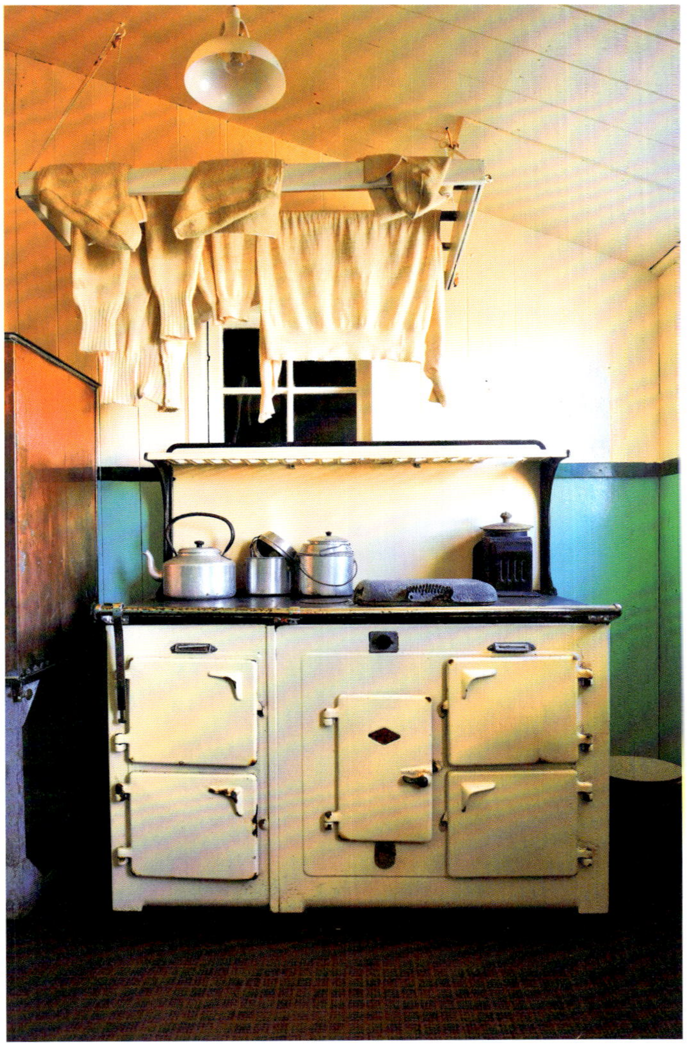

This spread Port Lockroy, a former British research station, was built in 1944 to facilitate geological, meteorological, and botanical surveys. From 1950 to 1962, when it closed, research of the ionosphere was conducted here. In 1996 it reopened as a small museum for visiting ships. The décor has been left untouched, offering a glimpse of life in Antarctica as it was before tourism. Deemed a historic site under the Antarctic Treaty, the station is managed by the UK's Antarctic Heritage Trust. This station is also the world's most southerly post office where roughly 90,000 postcards are processed in four months.

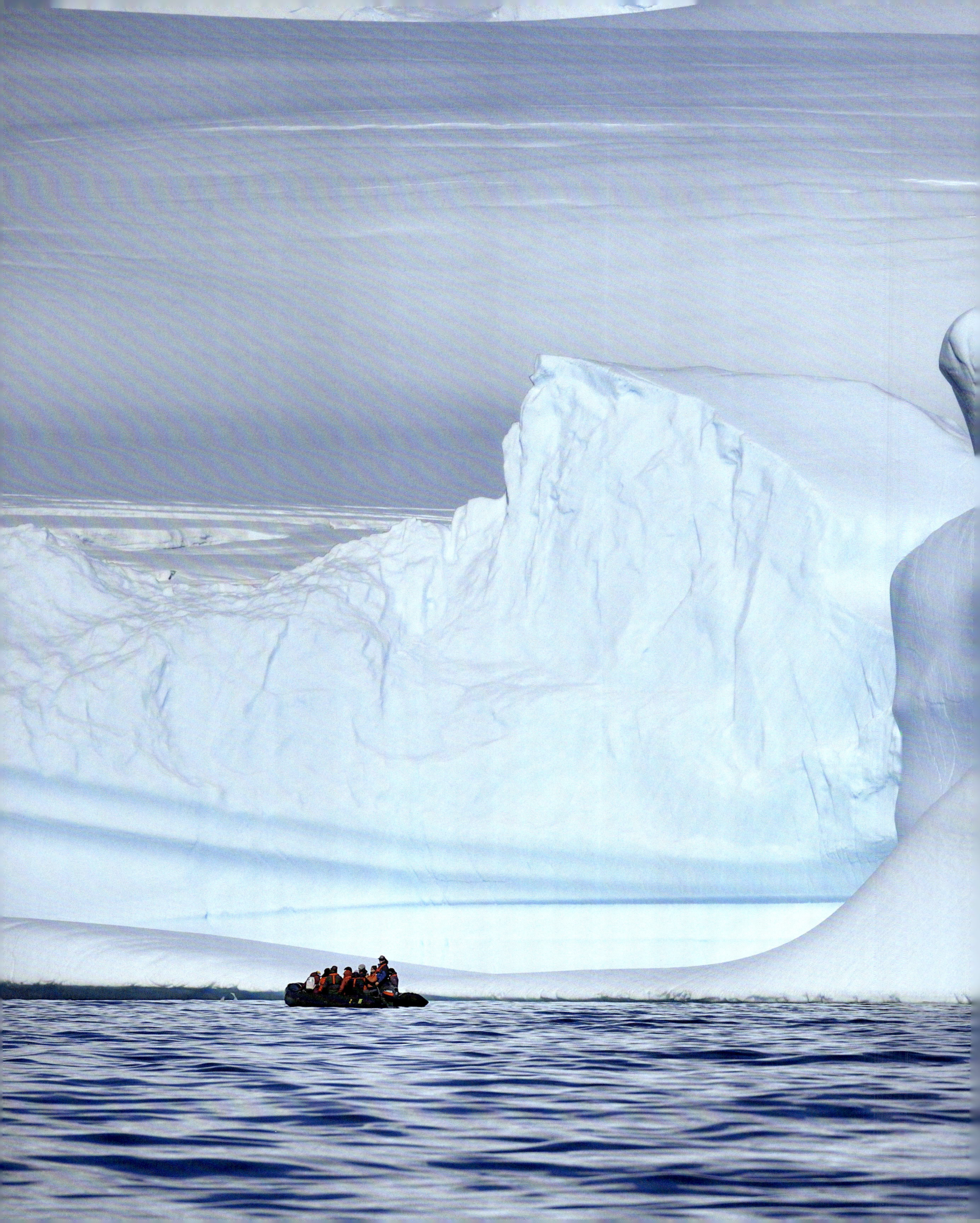

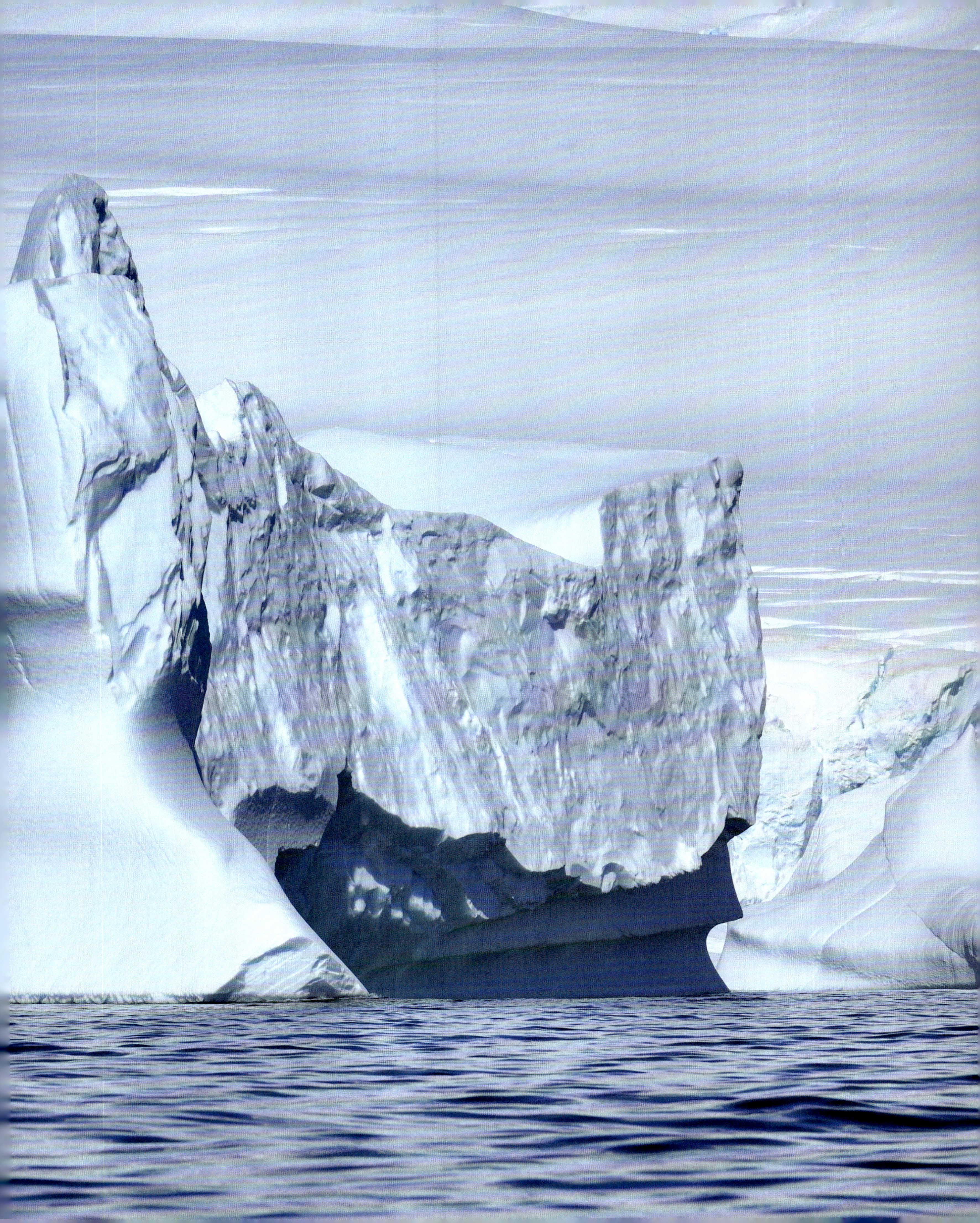

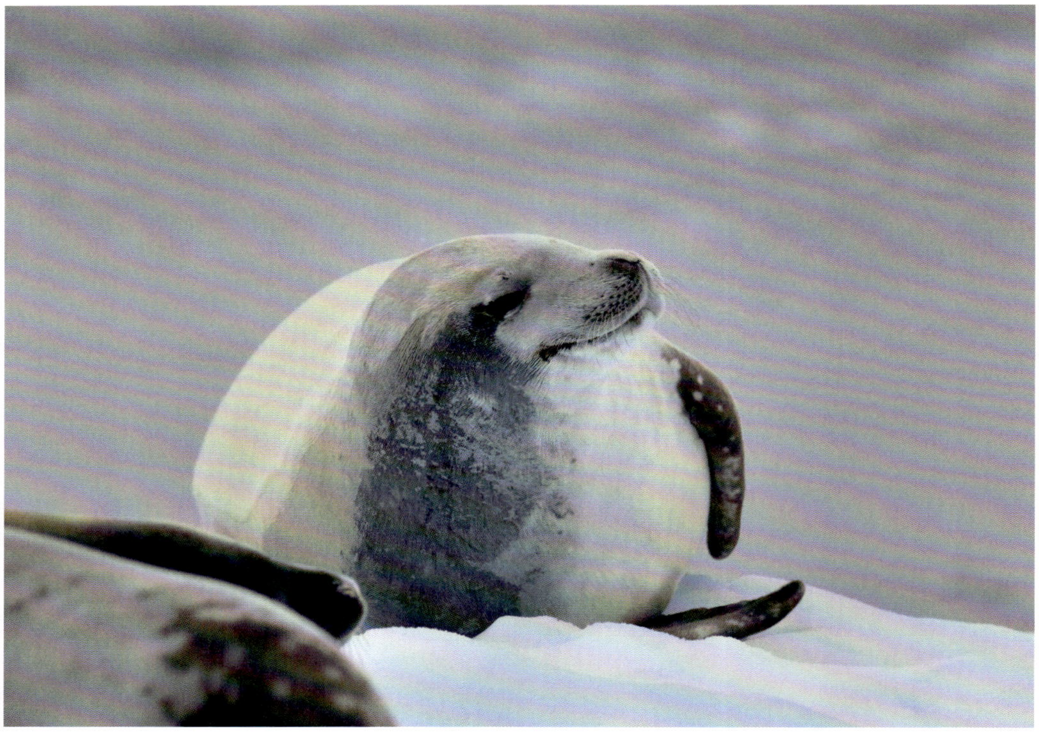

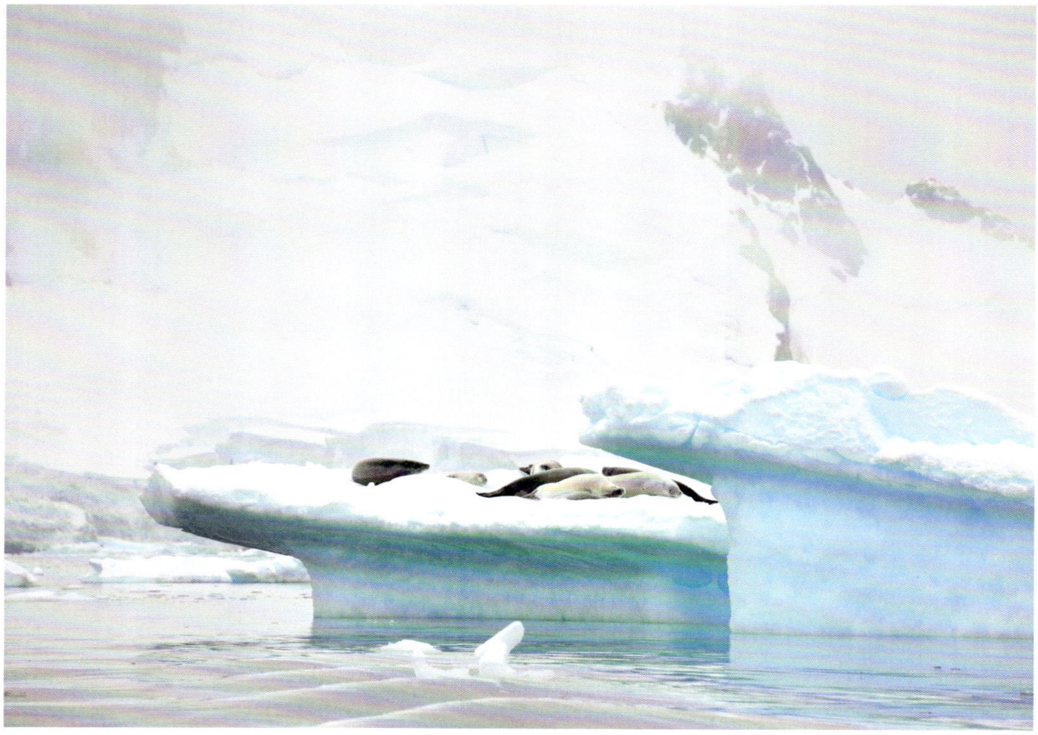

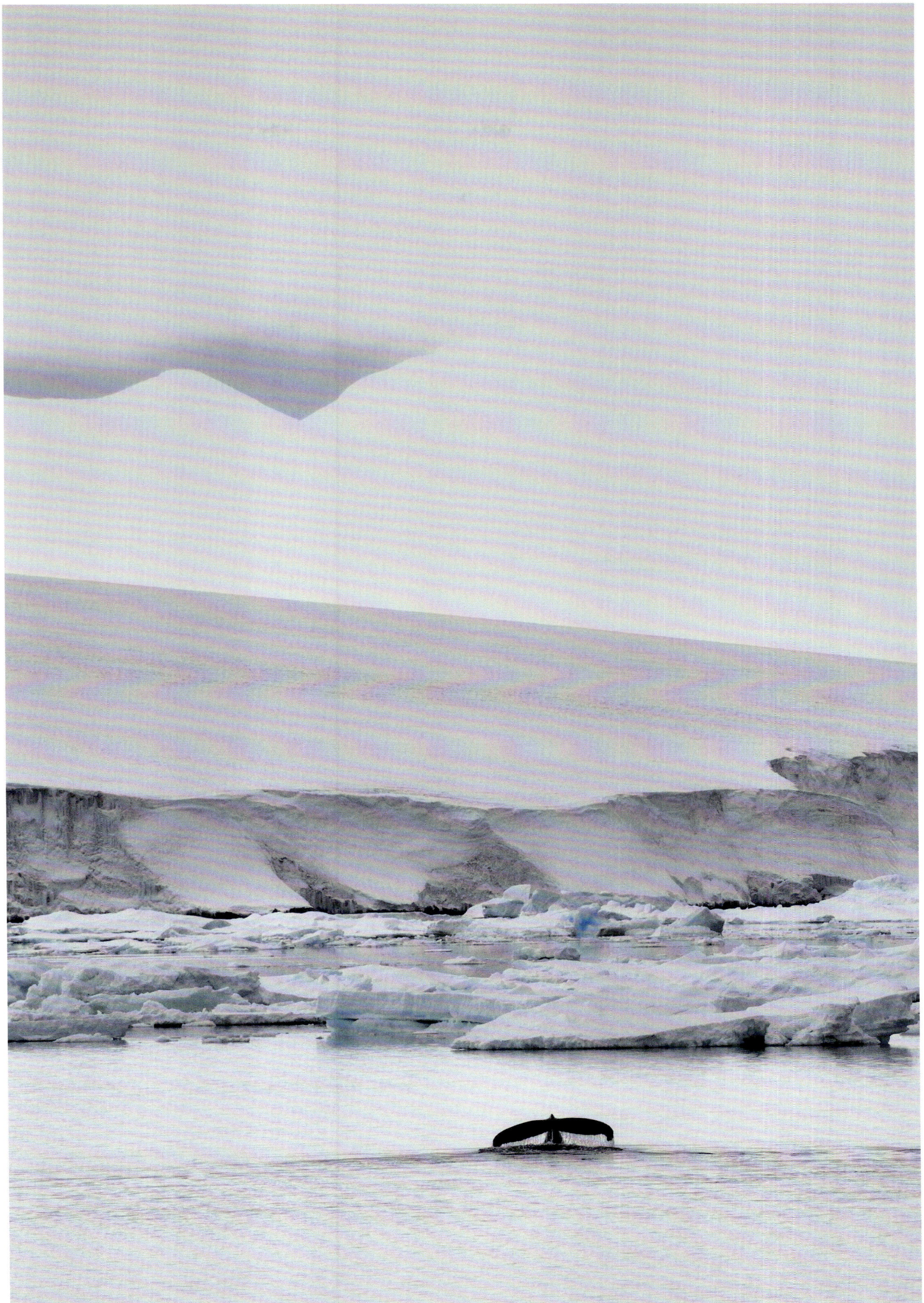

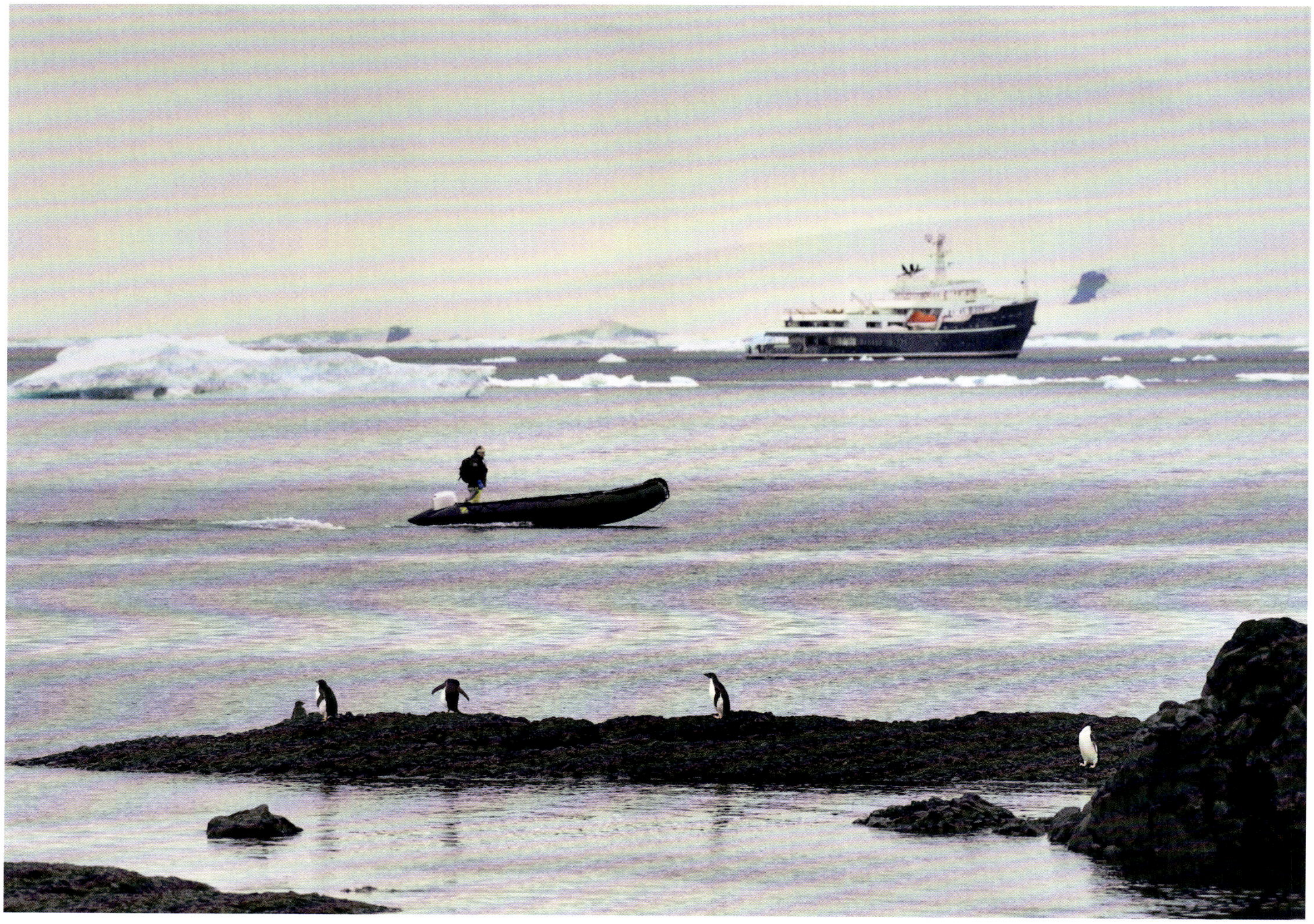

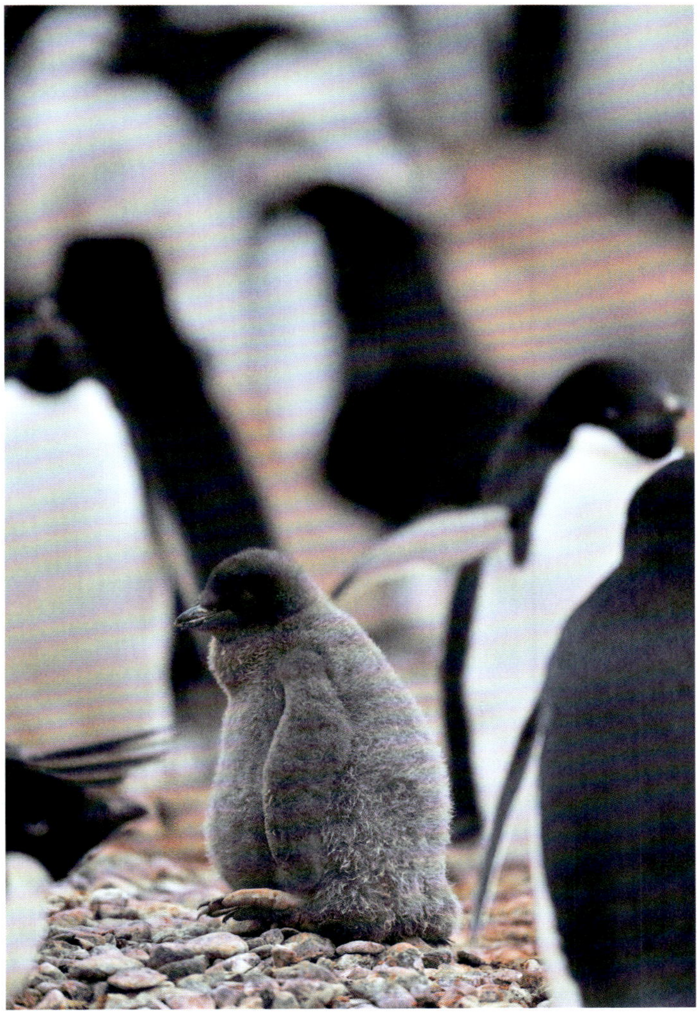
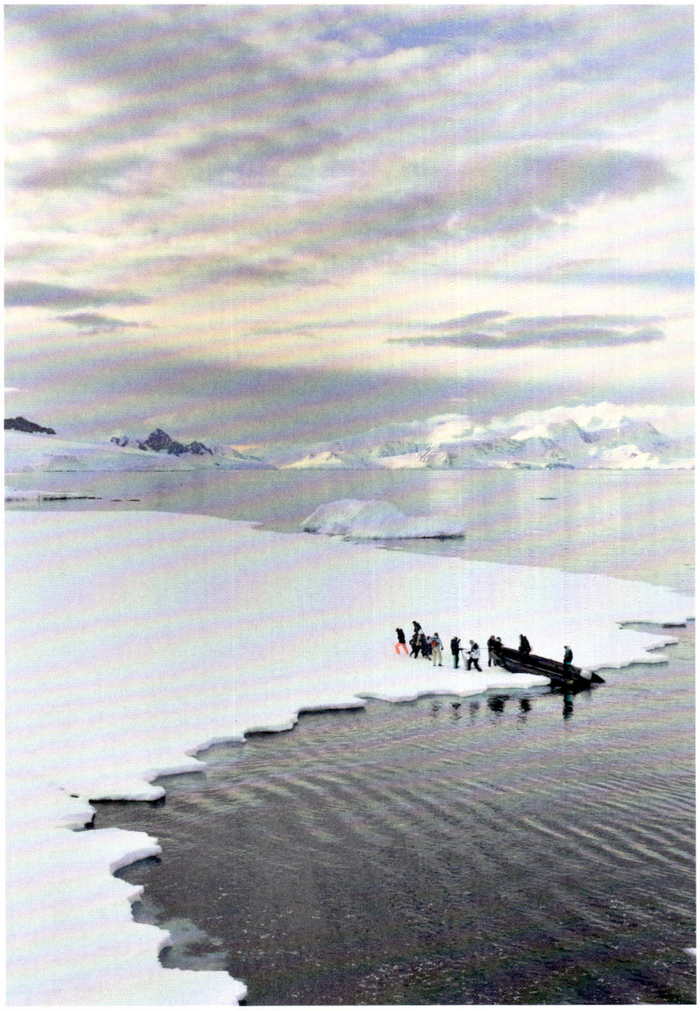
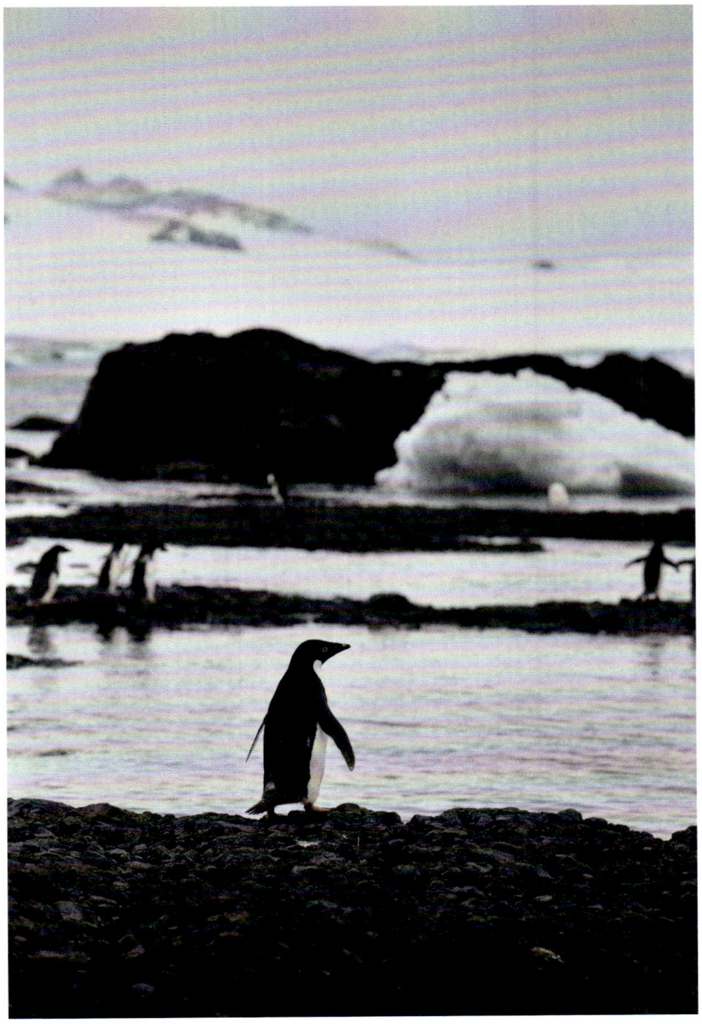
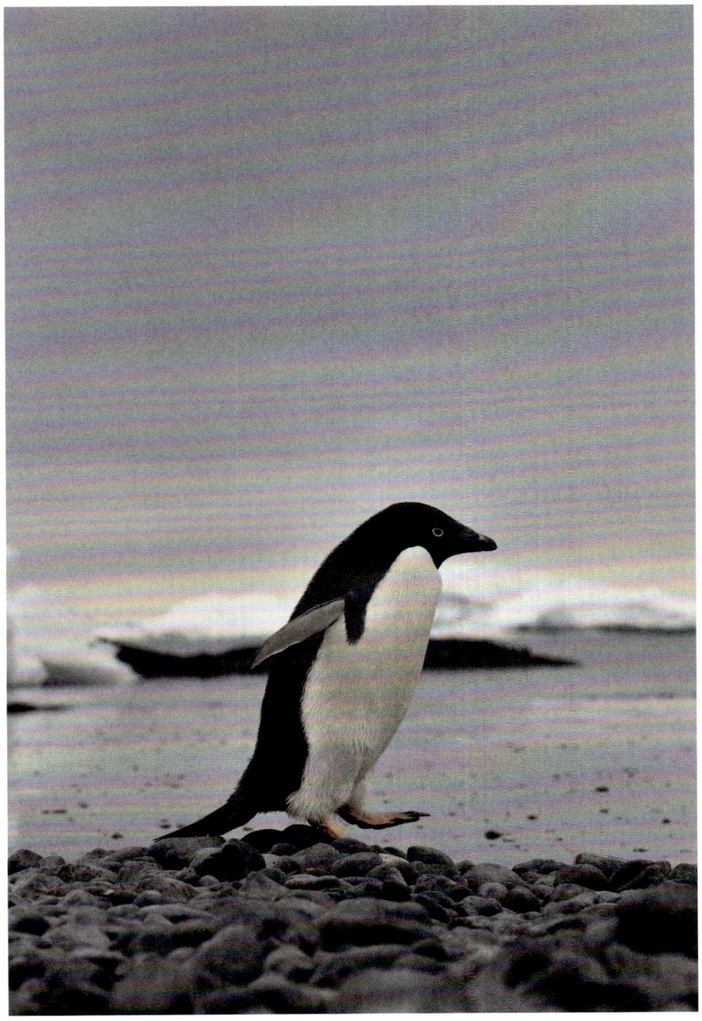

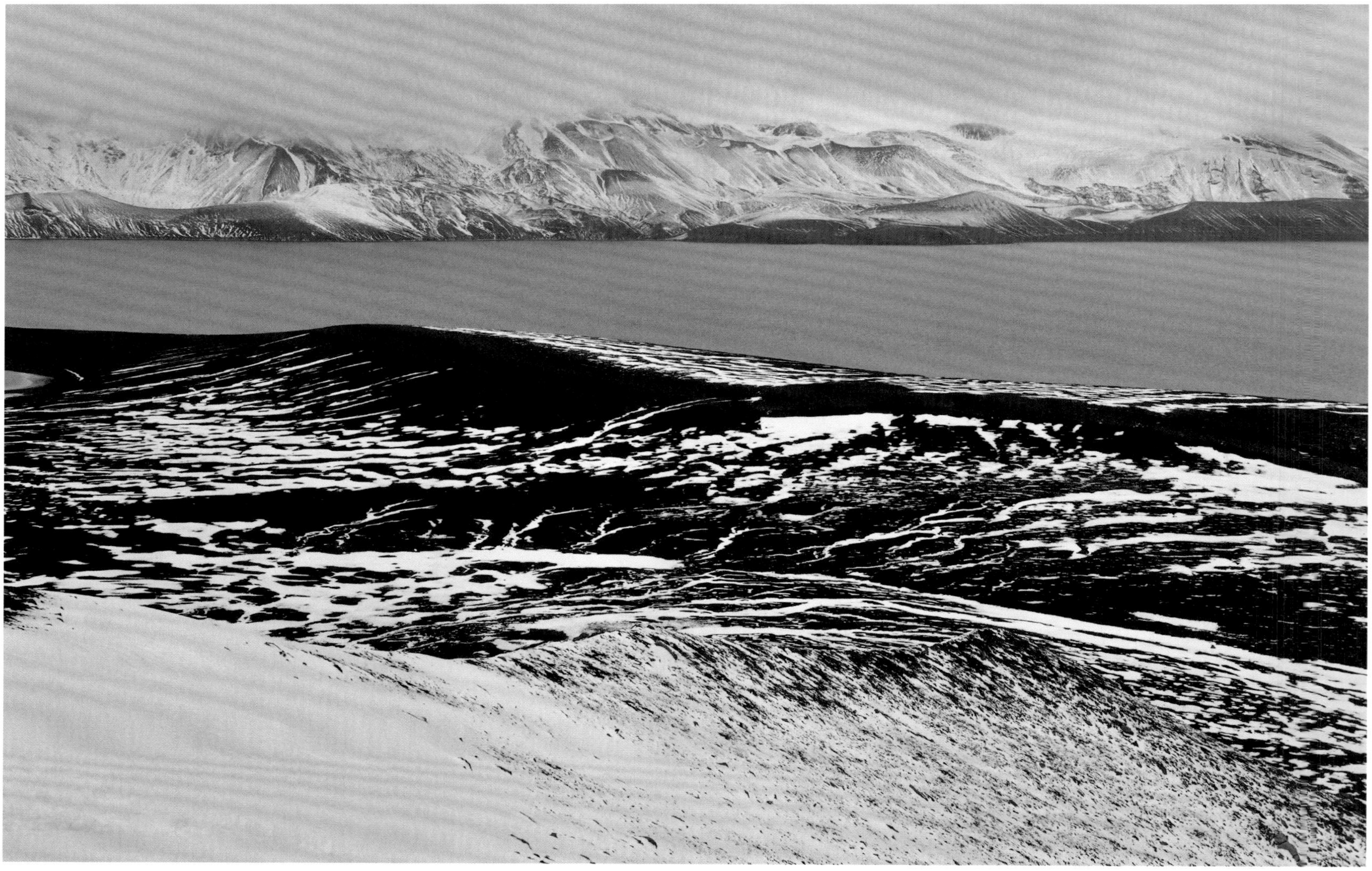

This spread Deception Island, located in the South Shetland Islands and northwest of the Antarctic Peninsula, is actually the top of an active volcano, forming a flooded caldera about 15 kilometres (9 mi.) wide and with a narrow entry called Neptune's Bellows. The last eruption was in 1970 and pockets of geothermal activity are made visible by clouds of steam. Whale and seal hunters, explorers and scientists have all called the island 'home'. Just off the volcanic beach in Whalers Bay are the remnants of the whaling industry — bleached whalebones and drums for storing oil and blubber — as well as gigantic steel barrels and the derelict dwellings from a British scientific station that was abandoned after the last eruption. This is the perfect spot to take a polar plunge from *Legend* and swim in the warm water of this flooded caldera, which is often surrounded by a spooky, low mist.

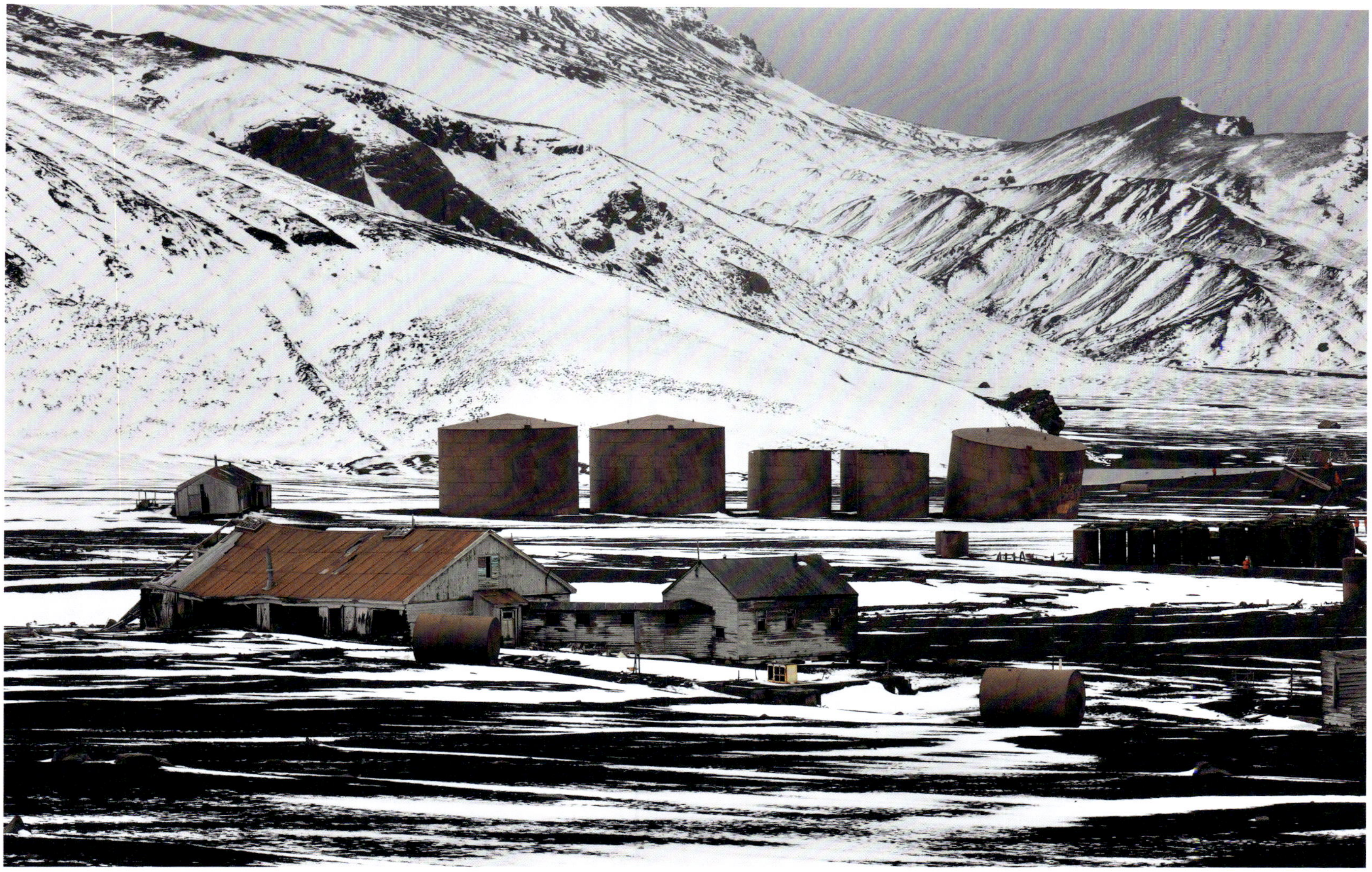

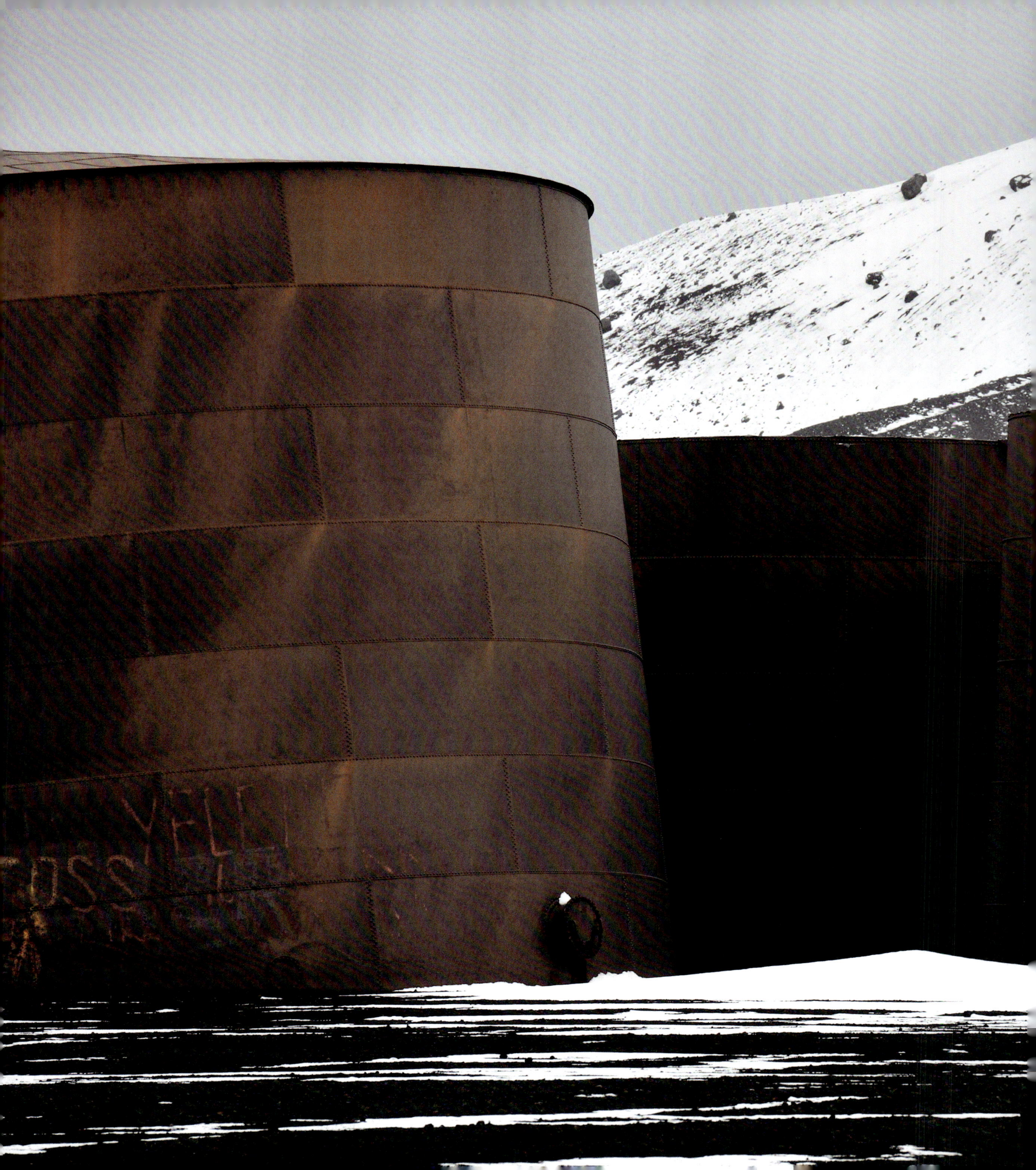

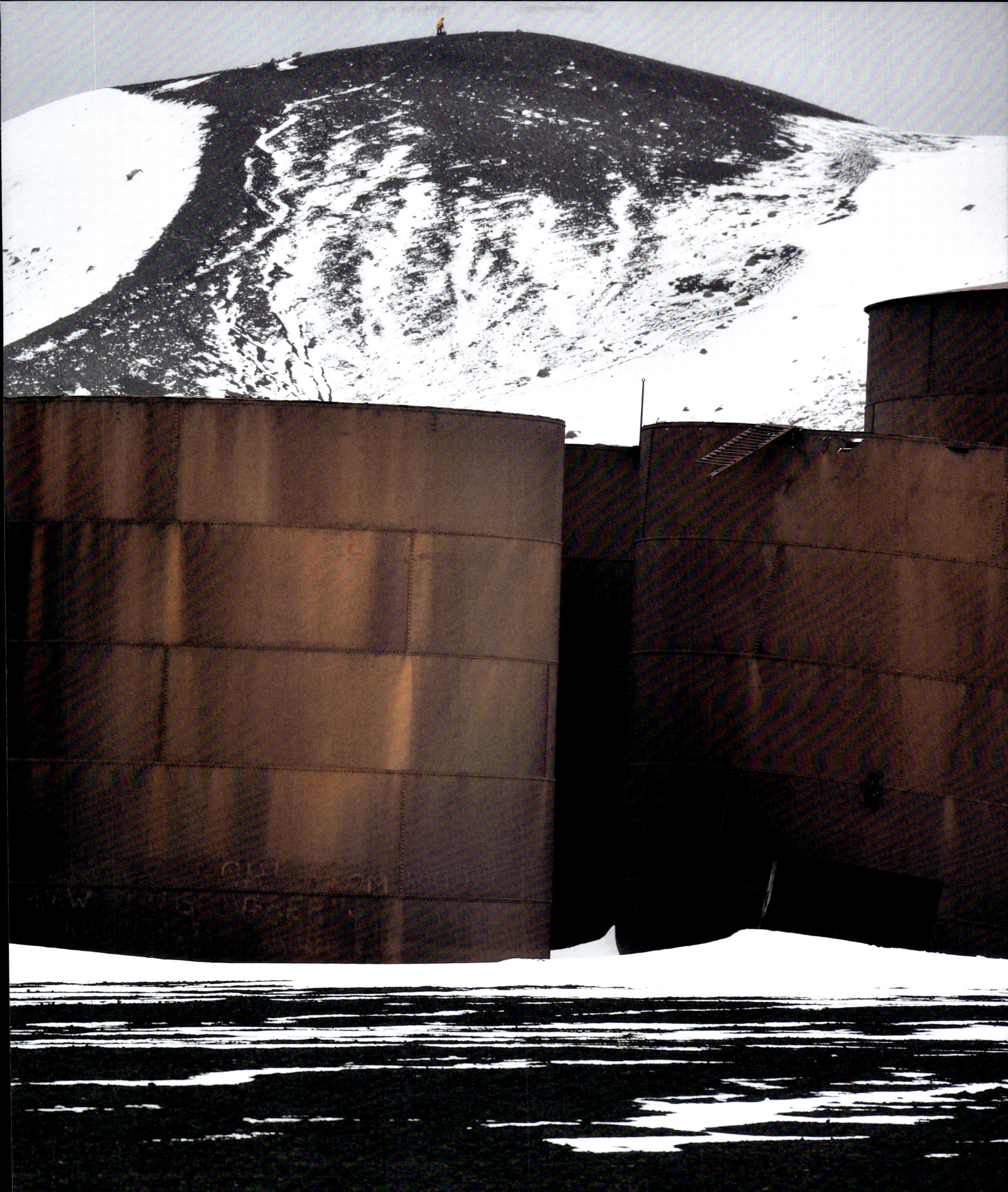

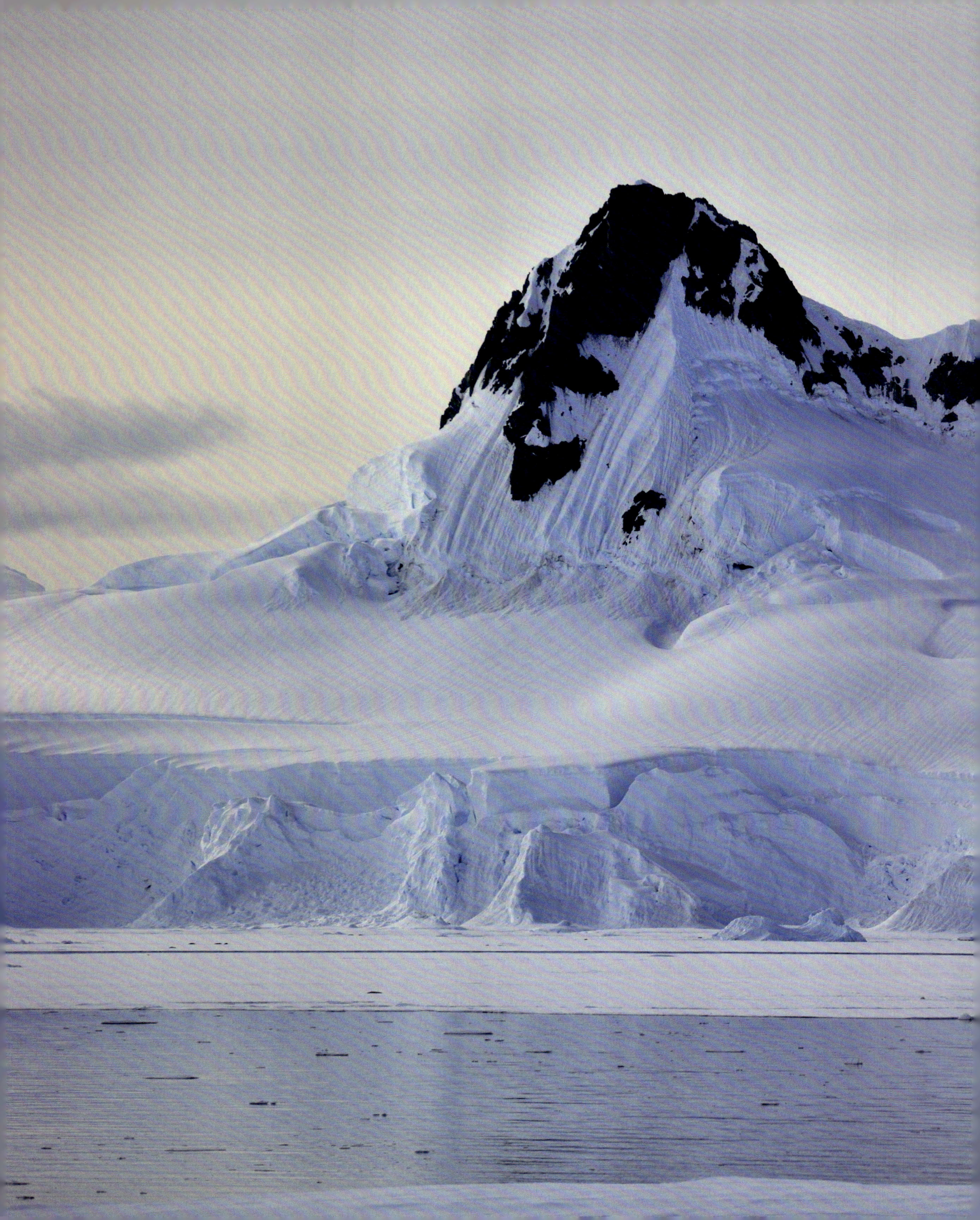

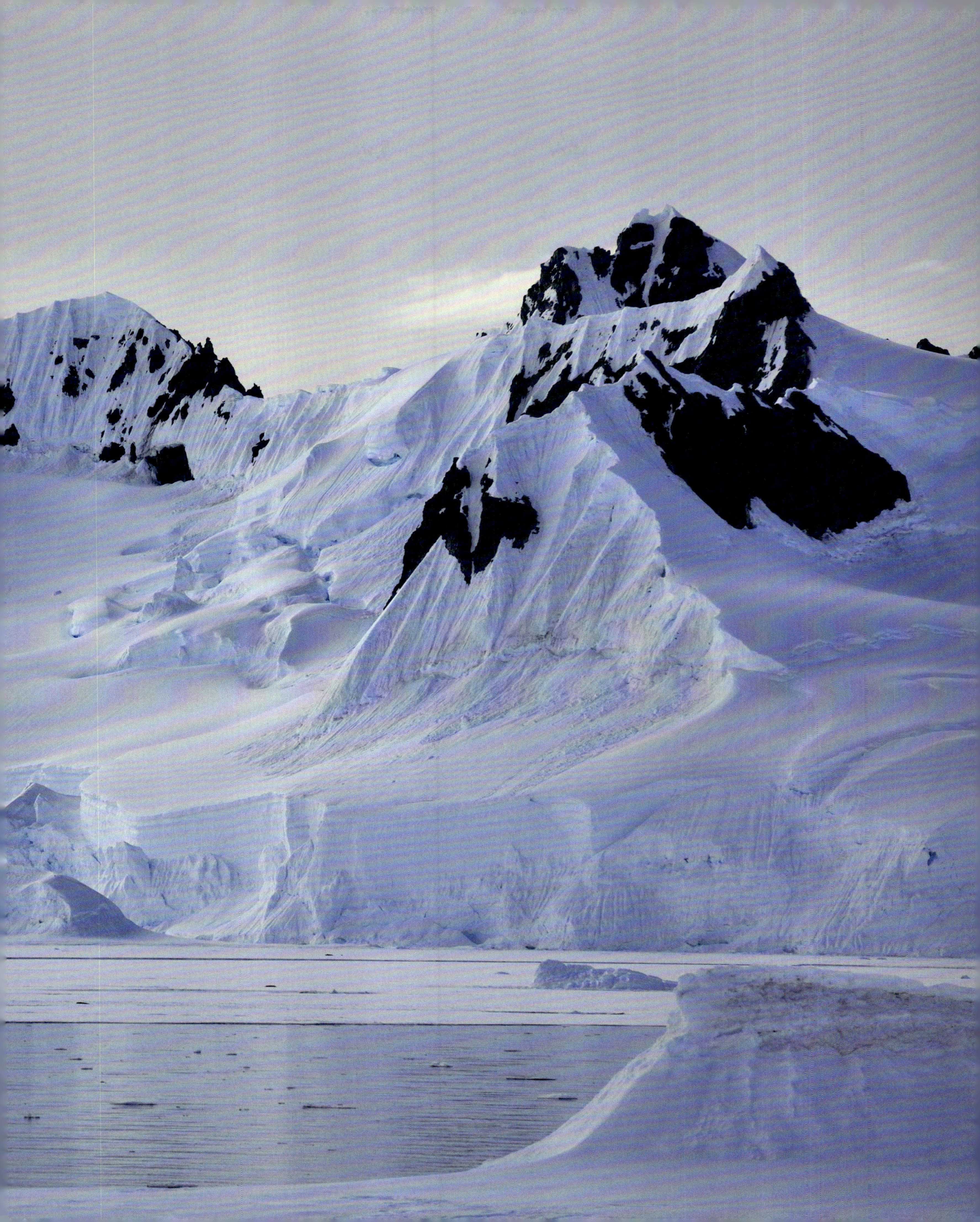

Biography

David De Vleeschauwer is a Belgian photographer with a career spanning two decades. He splits his time between the world's most remote places and a farm on the Belgium-France border. De Vleeschauwer's focus is on the unexpected, celebrating the diverse beauty of the world. He rarely treads in another's footsteps, choosing instead to travel to uncharted territories relatively untouched by modern tourism. Snorkelling with killer whales, travelling to the geographical poles or in the hostile reaches of North Korea, there is no environment too extreme for De Vleeschauwer's pursuit of the remote. He has travelled to more than 120 countries, working in temperatures ranging from -50°C to +50°C (-60°F to 120°F). His work has been featured in leading publications including *Monocle*, *The Wall Street Journal*, *Financial Times* and *National Geographic*. He shoots with Nikon and with 1850s collodion wet plates.

David De Vleeschauwer arranges one-off group trips each year to wherever his pioneering spirit is most drawn, following his natural instincts to stay ahead of the tourist curve. He is also available to organise and lead private trips anywhere in the world, from Antarctica with a superyacht to helicopter safaris in Botswana.

Want to travel along?
Drop him a line at david@remoteexperiences.com

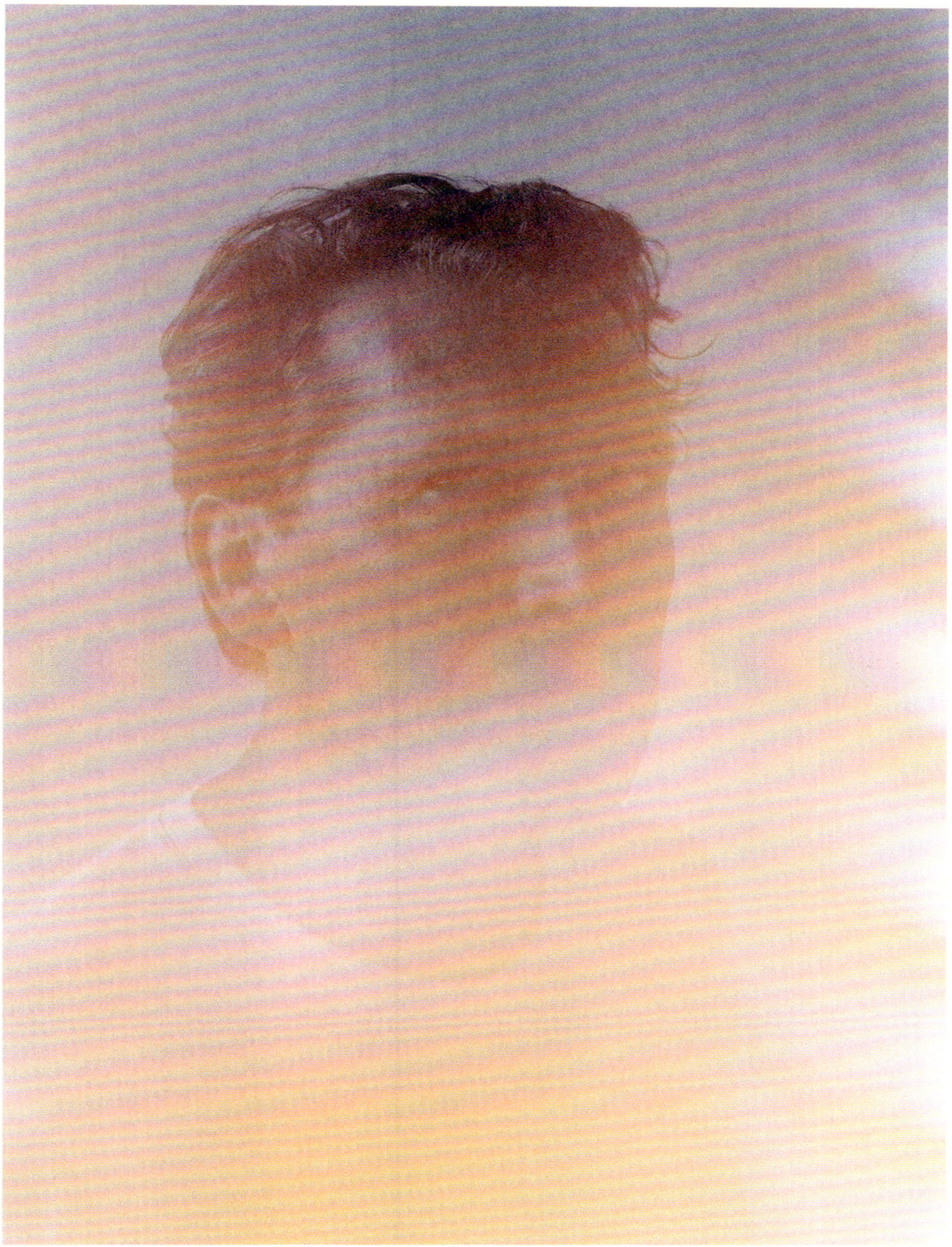

Portrait by Jonas Loellmann

Acknowledgements

Photography, art direction and layout
David De Vleeschauwer, Antwerp

Writers
Debbie Pappyn, Evora
Sophy Roberts, Dorset
Pico Iyer, Tokyo
Stanley Stewart, Rome
Jurriaan Teulings, Amsterdam

Editor
Ben Howe, New York

Production
Miguel De Groote, Kortrijk

Proofreading
Teresa Machan, London
Vicki Reeve, Cornwall

Drone pilot, Antarctica
Shelton Duprez, Toronto

Typeface
Plantin

If you have any questions or remarks please contact:

david@remoteexperiences.com

The Author
Debbie Pappyn is a journalist who contributes monthly to *Monocle* and other titles worldwide. Together with David she has been on the road for nearly two decades. As an in-the-know journalist, she has to keep her eye and pen sharp when travelling. It has become second nature to stay ahead of trends, and she has developed a natural instinct and curiosity for the next hot destination. Debbie's chief pursuit has always been to bring a crossover between experiential travel, adventurous explorations in style, and the quest for lesser-known places, secret boltholes, and under-the-radar destinations — from far-flung to cosmopolitan, but always in search for the less obvious or standard experiences.

Thank you to all the idealists and non-conformists who helped us travel to these amazing outposts.

And all those who still believe that travel is not only about changing sky, but about changing hearts and minds too.

A big thank you to our families and friends around the world, who we rarely see but never forget.

And finally our heartfelt thanks to Benedikt and Marlene Taschen for making this book happen.

– David De Vleeschauwer
www.remoteexperiences.com

EACH AND EVERY TASCHEN BOOK PLANTS A SEED!
TASCHEN is a carbon neutral publisher. Each year, we offset our annual carbon emissions with carbon credits at the Instituto Terra, a reforestation program in Minas Gerais, Brazil, founded by Lélia and Sebastião Salgado. To find out more about this ecological partnership, please check: www.taschen.com/zerocarbon
Inspiration: unlimited. Carbon footprint: zero.

To stay informed about TASCHEN and our upcoming titles, please subscribe to our free magazine at www.taschen.com/magazine, follow us on Instagram and Facebook, or e-mail your questions to contact@taschen.com

Hohenzollernring 53, D-50672 Köln
www.taschen.com

Printed in Slovakia
ISBN 978-3-8365-8602-3

Bon voyage